5th International Conference on Physics and Astrophysics of Quark Gluon Plasma 2005

(ICPAQGP 2005)

Journal of Physics: Conference Series Volume 50

Kolkata, India
8-12 February 2005

ISBN: 978-1-61738-296-3

5th International Conference on Physics and Astrophysics of Quark Gluon Plasma 2005

(ICPAQGP 2005)

Journal of Physics: Conference Series Volume 50

Kolkata, India
8-12 February 2005

ISBN: 978-1-61738-296-3

IOP Institute of Physics

Printed from e-media with permission by:

Curran Associates, Inc.
57 Morehouse Lane
Red Hook, NY 12571
www.proceedings.com

Some format issues inherent in the e-media version may also appear in this print version.

Printed by Curran Associates, Inc. (2010)

Institute of Physics
Dirac House, Temple Back
Bristol BS1 6BE UK

Phone: 44 1 17 929 7481
Fax: 44 1 17 920 0979

techtracking@iop.org

TABLE OF CONTENTS

EXPERIMENT I: RHIC

EXPERIMENT II: SPS

EXPERIMENT III: FUTURE FACILITIES

THEORY

The 5th International Conference on Physics and Astrophysics of Quark Gluon Plasma (ICPAQGP 2005) was held on 8–12 February 2005 at the Variable Energy Cyclotron Centre and Saha Institute of Nuclear Physics campus, Kolkata, India. The conference was enriched by the august presence of about 300 participants representing 18 countries across the globe. It had plenary talks and oral presentations, which form a part of these proceedings. Besides invited and contributed talks there were also a large number of poster presentations.

The conference was energized by discussions of fresh experimental data from RHIC on strong elliptic flow, jet quenching, single photon spectra etc. Moreover, new theoretical results were brought to the discussion forum during this conference. Colour glass condensates, hydrodynamical flow, jet quenching and sQGP were intensely debated by the participants. The highlight of ICPAQGP 2005 was the presentation of fresh experimental results from the RHIC-IV run.

The ICPAQGP series, since its inception in 1988, has placed emphasis on the role of quark matter in the fields of astrophysics and cosmology. The subsequent conferences held in 1993, 1997, 2001 and 2005 had also retained this focus. The conference was preceded by a Fest Colloquium in honour of Professor Bikash Sinha. Professor Sinha, regarded as the pioneer in establishing quark gluon plasma research in India, has successfully encouraged a group of young Indian researchers to devote themselves wholeheartedly to QGP research—both theoretical and experimental.

Members of the International Advisory Committee played a pivotal role mainly in the selection of speakers. The contributions of the Organizing Committee in all aspects, from selecting the contributory talks posters down to arranging local hospitality, were much appreciated. We thank the members of both committees for making ICPAQGP 2005 an interesting platform for scientific deliberation.

The ICPAQGP 2005 was supported financially by the Board of Research in Nuclear Science of the Department of Atomic Energy, Government of India.

ICPAQGP - 2005

Committees:-

Local Organising Committee:-

Madan M. Aggarwal, Chandigarh.
Jan-e Alam, Kolkata.
Debadesh Bandopadhyay, Kolkata.
Rakesh Bhandari, Kolkata.
Pratap Bhattacharyya, Kolkata.
Somenath Chakravarty, Kolkata.
Subhasis Chattopadhyay, Kolkata.
Sourendu Gupta, Mumbai.
Sanjay K. Ghosh, Kolkata.
Ashok Goyal, New Delhi.
Kamales Kar, Kolkata.
Bedangadas Mohanty, Kolkata.
Lalit K. Mangotra, Jammu.
Tapan K. Nayak, Kolkata.
Sashi C. Phatak, Bhubaneswar.
Sudhir Raniwala, Jaipur.
G. Relhan, Mumbai.
Pradip Roy, Kolkata.
Bikash Sinha, Kolkata.
Raghava Varma, Mumbai.

International Advisory Committee:-

P. Aurenche, Annecy.
J. P. Blaizot, Saclay.
P. Braun-Munzinger, Darmstadt.
W. Busza, Upton, NY.
J. Cleymans, Cape Town.
S. Dattagupta, Kolkata.
K. Eskola, Jyvaskyla.
C. Gale, Montreal.
R. Gavai, Mumbai.
H. A. Gustafsson, Lund.
H. Gutbrod, Darmstadt.
T. Hallman, Upton, NY.
J. W. Harris, Yale.
T. Hatsuda, Tokyo.
B. Jacak, Stony Brook.
B. K. Jain, Mumbai.
S. Kailas, Mumbai.
J. Kapusta, Minneapolis.
L. Kluberg, Palaiseau.
T. Kodama, Rio de Janiero.
J. Lattimer, Stony Brook.
P. Levai, Budapest.

T. Ludlam, Upton, NY.
L. McLerran, Upton, NY.
Y. Miake, Tsukuba.
B. Mueller, Durham, N.C.
S. Ozaki, Upton, NY.
T. Padmanabhan, Pune.
G. Paic, Mexico.
S. Raha, Kolkata.
V. S. Ramamurthy, New Delhi.
K. Redlich, Wroclaw.
L. Riccati, Torino.
D. H. Rischke, Frankfurt.
H. G. Ritter, Berkley.
P. V. Ruuskanen, Jyvaskyla.
H. Satz, Bielefeld/Lisbon.
Y. Schutz, Nantes.
J. Schukraft, Geneva.
E. Shuryak, Stony Brook.
B. Sinha, Kolkata.
D. K. Srivastava, Kolkata.
J. Stachel, Heidelberg.
R. Stock, Frankfurt.
H. Stoecker, Frankfurt.
I. Tserruya, Rehovot.
M. S. Turner, Chicago.
F. Videbaek, Upton, NY.
Y. P. Viyogi, Kolkata.
X. N. Wang, Berkley.
W. A. Zajc, Irvington.
G. Zinovjev, Kiev.

Institute of Physics Publishing Journal of Physics: Conference Series **50** (2006) 1–7
doi:10.1088/1742-6596/50/1/001 5th International Conference on Physics and Astrophysics of Quark Gluon Plasma

Ultrarelativistic heavy ion collisions Theoretical overview

Jean-Paul Blaizot[1]

ECT*, Villa Tambosi, strada delle Tabarelle 286, 38050 Villazzano (TN), Italy

E-mail: blaizot@ect.it

Abstract. This is a short review of some theoretical aspects of the physics of ultra-relativistic heavy ion collisions. I review the main properties of the QCD phase diagram and recent developments in the physics of high gluon densities in the hadronic wavefunctions at high energy. Then I comment salient results obtained at RHIC.

1. Introduction

Ultra-relativistic heavy ion collisions allow us to study the densest and hottest forms of matter that can be created in the laboratory, and to address fundamental questions related to the state of matter at temperatures and densities which may be considered as "extreme" on hadronic scales. On such scales, the dynamics is controlled by Quantum Chromodynamics (QCD) and the asymptotic freedom leads us to expect that, at high temperature and density, matter is a weakly interacting system of the QCD elementary degrees of freedom, the quarks and the gluons.

Much theoretical work has been devoted to understanding the properties of such a quark-gluon plasma in static, thermal equilibrium. For such conditions, lattice calculations provide reliable information "from first principles". Such calculations have long been limited to equilibrium properties at vanishing chemical potential. However, recently, various techniques have been proposed that allow calculations at finite, albeit small, chemical potential; also, algorithms based on the maximum entropy method are being developed to reconstruct real time information from Euclidean correlators, and hence access in particular the plasma transport properties.

Aside from the issues concerning the state of matter at large density, there is another fundamental question that is playing an increasingly important role in the study of heavy ion reactions: that of the structure of the wavefunction of a nucleus at high energy. It turns out that this question also leads us to the consideration of high density gluonic systems where non linear QCD effects can play an important role. The study of dense systems of partons offers exciting challenges, and indeed much theoretical developments have been taking place in this area during the last years.

The asymptotic form of the wavefunction is of course directly relevant to the study of heavy ion collisions, as it determines the "initial conditions" for the creation of matter and its subsequent evolution. But the knowledge of the wavefunction is not enough to explain how the partons initially present in the wavefunctions as virtual entities turn into active degrees

[1] Member of CNRS

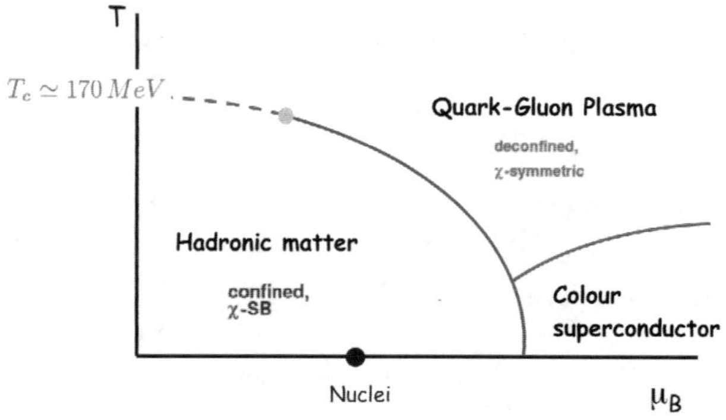

Figure 1. A schematic representation of the expected phase diagram of QCD matter in the plane temperature (T), baryon chemical potential (μ_B)

of freedom constituting matter with high energy density, leading eventually to a quark-gluon plasma. Much work is presently devoted to this question.

The two fundamental questions mentioned above, concerning the state of matter at high temperature or density and the form of the wavefunction at high energy, focus on extreme situations where simplicity can emerge, allowing for explicit calculations to be done from first principles. However the consideration of such idealized situations is clearly not enough to understand nuclear collisions. There, the field is largely driven by experiment, and theoretical efforts aim at getting a correct interpretation of the data. It is however encouraging to observe that as the collision energy increases, parton degrees of freedom play an increasingly important role in the collision dynamics, bringing the discussion closer to these fundamental issues alluded to earlier. This is particularly so for the observables that are directly sensitive to the early stages of the collisions. These are the observables on which I shall focus in the last part of this talk. In the next two sections, I say a few more words about the QCD phase diagram and on the hadronic wave functions with high gluon densities.

2. The QCD phase diagram and the Quark-Gluon Plasma

Much work on the phase diagram has been done since the original suggestions that, because of asymptotic freedom, hadronic matter should turn into free quark matter at sufficiently high density [1, 2].

A schematic picture of this phase diagram is presented in Fig. 1. Reliable information on the vertical axis (vanishing baryon chemical potential) is available from lattice gauge calculations (see for instance [3]). Recently various extrapolation techniques have been used to extend these calculations to finite baryon density, and yield evidence for a tri-critical point at some finite value of the baryon chemical potential [4].

At small temperature, but large baryon chemical potential, the phase diagram exhibits a rich structure dominated by the phenomena associated with color superconductivity [5]. It is not clear whether such phenomena have any chance to be observed in heavy ion collisions, but they could play an important role in the physics of compact stars, in particular they could affect the cooling of neutron stars.

Lattice calculations show that, as expected, the thermodynamical functions of QCD matter turn into those of free massless particles at asymptotically high temperatures. Analytical weak coupling techniques are capable of reproducing those lattice results down to temperatures of the order of $3T_c$ [6]. This gives support to a simple quasiparticle picture: at temperatures higher than $\sim 3T_c$, the dominant effects of the interactions can be absorbed in a renormalization of quasiparticle properties, leaving a weak residual interaction between the quasiparticles. In the vicinity of T_c, it is still unclear what the relevant degrees of freedom are. Effective theories in terms of Polyakov loops are being studied with encouraging results [7, 8, 9, 10]. There are also arguments, inspired by strong coupling calculations in string theories, according to which Coulomb bound states could play an important role [11]. It has also been suggested that the reduction of the pressure from the free gas limit above T_c is a strong coupling effect, analogous to that found in the strong coupling regime of supersymmetric Yang-Milss theories [12, 13]. Such theories have also been used to estimate the viscosity [14].

There is a word of caution that I would like to insert at this point. There is no doubt that interactions play an essential role near T_c. However, while supersymmetric Yang-Mills theories have nice properties, and offer in particular new tools to calculate in strong coupling, they differ from QCD in some essential aspects: the coupling constant does not run, and no phase transition occurs. Furthermore, we have indications from lattice calculations (from the measurement of $\epsilon - 3P$, [15]) that the physics near T_c is not conformal invariant. It is therefore unclear whether the strong coupling regime of supersymmetric Yang-Mills theories really describes the physics of QCD near its phase transition. In fact, recent QCD calculations using dimensional reduction techniques seem to indicate that the effective coupling constant may not be so large even close to T_c [16], at least not large enough to totally invalidate weak coupling calculations (with proper resummations, such as done in [6] for instance).

3. High gluon density in nuclear wavefunctions at high energy

When two nuclei collide at very high energy, they do not see each other as collections of nucleons bound together by nuclear forces. Rather, the relevant degrees of freedom involved in the early stages of a collision at sufficiently high energy are partons, mostly gluons, whose density grows as the energy increases (i.e., when x, their momentum fraction, decreases). This growth of the number of gluons in the hadronic wave functions is a phenomenon which has been well established at HERA [17]. One expects however that it should eventually "saturate" when non linear QCD effects start to play a role.

The existence of such a saturation regime has been predicted long ago[18, 19], together with estimates for the typical transverse momenta where it could occur in heavy ion collisions [20]. But it is only during the last decade that equations providing a dynamical description of the saturated regime have been obtained[21, 22] and [23, 24, 25, 26, 27]. A remarkable feature which emerges from the solution of these equations is that the dense, saturated system of partons to be found in hadronic wave functions at high energy has universal properties, the same for all hadrons or nuclei. This suggests that the early stages of hadronic collisions at sufficiently high energies are governed by universal wave functions whose properties could, in principle, be calculated from QCD.

The momentum scale Q_s that characterizes the onset of saturation is called the saturation momentum [28]. Partons with transverse momentum $Q > Q_s$ are in a dilute regime; those with $Q < Q_s$ are in the saturated regime. One may obtain an estimate of Q_s by looking at the scale at which QCD non linear effects become important. This occurs when field fluctuations are such that gluon kinetic energies become comparable to their interaction energies, that is when $\langle(\partial A)^2\rangle \sim \alpha_s\langle(A^2)^2\rangle(\approx \alpha_s\langle A^2\rangle^2)$, where $\langle A^2\rangle$ denotes the magnitude of the fluctuations of the gauge fields. The relevant dynamics is in the transverse plane, and the magnitude of the gradient is fixed by the transverse momentum Q; thus saturation sets in when $Q^2 \sim \alpha_s\langle A^2\rangle$. The magnitude of the gauge field fluctuations, $\langle A^2\rangle$, can be estimated from the particle number density in the transverse plane, i.e. $\langle A^2\rangle \sim \rho$, with ρ given by

$$\rho \sim \frac{xG(x,Q^2)}{\pi R^2}\ , \tag{1}$$

where R is the radius of the hadron, x the momentum fraction of the considered gluons and $G(x,Q^2)$ the gluon structure function. The condition $Q^2 \sim \alpha_s\langle A^2\rangle$ translates then into an equation for the saturation momentum

$$Q_s^2 \sim \alpha_s(Q_s^2)\frac{xG(x,Q_s^2)}{\pi R^2}\ . \tag{2}$$

Note that at saturation, naive perturbation theory breaks down, even though $\alpha_s(Q_s)$ may be small if Q_s is large: the saturation regime is a regime of weak coupling, but large density.

The saturation momentum increases as the gluon density increases. This may come from an increase of the gluon structure function as x decreases. The increase of the density may also come from the additive contributions of several nucleons in a nucleus. In large nuclei, one expects $xG_A(x,Q_s^2) \propto A$, and hence $Q_s^2 \propto \alpha_s A^{1/3}$, where A is here the number of nucleons in the nucleus. Thus, the saturation regime sets in earlier (i.e., at lower energy) in collisions involving large nuclei than in those involving protons. In fact, the parton densities in the central rapidity region of a Au-Au collision at RHIC are not too different from those measured in deep inelastic scattering at HERA. There is however one important difference: while at HERA these densities result from evolution with energy, at RHIC there is little evolution, at least in the central rapidity region, and the large densities result mostly from the additive contributions of the participant nucleons. Of course, one has the possibility at RHIC to explore various situations. In particular the study of dA collisions in the fragmentation region of the deuteron gives access to a regime where final state interactions play a minor role and where quantum evolution could be significant. Indeed, exciting results have been obtained in this regime, as we shall mention later.

There is another interesting feature of the saturation regime. Consider the number of partons occupying a small disk of radius $1/Q_s$ in the transverse plane. This is easily estimated by combining eqs. (1) and (2); one finds this number to be proportional to $1/\alpha_s$, a large number if α_s is small. In such conditions of large numbers of quanta, classical field approximations may become relevant to describe the nuclear wave-functions. This observation is at the basis of the McLerran-Venugopalan model [29]. The color glass formalism provides a more complete physical picture. It relies on the separation of the degrees of freedom into "hard", Lorentz contracted, and frozen color sources flying along the light-cone, and low x partons which are described by classical gauge fields $A^\mu(x)$ determined by solving the Yang-Mills equations with the source given by the frozen partonic configuration. An average over all acceptable configurations must be performed in order to calculate observables. In this average, the weight of a given configuration is a functional $W_{x_0}[\rho]$ of the density ρ of color sources which depends on the separation scale x_0 between the modes which are treated as frozen sources, and those which are treated as fields. As one lowers this separation scale, more and more modes are included among the frozen sources, and therefore the functional W_{x_0} evolves with x_0 according to a renormalization group equation, which leads to the non linear evolution equations alluded to earlier, namely the Balitsky-Kovchegov equation [21, 22] and the JIMLWK equation [23, 24, 25, 26, 27].

As a final remark, oriented towards the future, one may note that many of the phenomena uncovered at RHIC and related to the early stages of the collisions, should become more clearly visible at the LHC. There, with center of mass energies of 5.5 TeV per nucleon-nucleon collision collisions, the typical value of x at mid-rapidity will be about 5.10^{-4} and values as small as 10^{-5} could be reached at forward rapidities. The corresponding values of Q_s^2 range from 5 GeV^2 to 14 GeV^2. Such large values of the saturation momentum make the coupling constant α_s smaller than at RHIC, giving firmer grounds to the weak coupling expansion used in the color glass condensate framework. A corollary of this is that the gluon occupation number in the saturation region will be larger, making the classical description also more justified.

4. Some RHIC discoveries

Experiments done at RHIC complement those performed at the CERN SPS at lower energy. In the last few years, they have produced a lot of new results.

It is useful to keep in mind, when discussing the experimental data, that the "signatures" that have been proposed as diagnostics of quark-gluon plasma formation in the early days of the development of the field, are mostly based on the simple picture of non interacting particles. While this picture is expected to be correct at asymptotically high temperatures, as we have recalled earlier, it certainly does not reflect entirely the physics at temperatures around the transition, which are likely the temperatures achieved in present experiments. From that perspective, it is perhaps not too surprising to find sizeable interaction effects in the matter produced at RHIC, even if it is presumably dominated in the early stages by partonic degrees of freedom.

I shall focus mostly on the probes which are sensitive to the densest and hottest states of matter, as these are those which depend more directly on the partonic structure of the initial wave functions, and hence are more likely to be related to quantities that can be calculated directly from QCD with minimal phenomenological input.

- *Large energy densities* are achieved [30]. Of course, in an expanding system such as that produced in heavy ion collisions, the value of the energy density depends on the time at which it is measured. But even the most conservative estimates suggest that for the first couple of fm/c after the collision, the energy density exceeds that needed to turn hadronic matter into a quark-gluon plasma. It is also significant that the multiplicities of produced particles are smaller than had been anticipated by most models [31], suggesting that coherence effects in particle production are presumably important [32].

- *Collective behavior* is observed in a spectacular fashion through the so-called elliptic flow (see for instance [33] and [34]). This phenomenon occurs for non central collisions: the collision zone is then spatially asymmetric, with roughly the shape of an almond. If matter thermalizes and expands hydrodynamically, then the expansion will dominantly take place along the directions of the largest gradients, that is in the collision plane [35]. This is what has been observed. From a careful comparison of data with hydrodynamical calculations, it has been concluded that this phenomenon implies that the produced matter thermalizes at very early times [36]. Understanding how short thermalization times could be obtained from QCD remains a theoretical challenge[37, 38]. For a recent suggestion emphasizing the role of plasma instabilities, see [39, 40].

It is worth emphasizing at this point that hydrodynamics, while it can be considered as rather successful, in particular when it is combined with a cascade description of the late stages of the expansion [41], does not perfectly account for all the data. One can even argue that some data on the elliptic flow provide evidence for incomplete thermalization on the time scale needed to establish the elliptic flow, a time of order $\bar{R}$ with $\bar{R}$ a measure of the size of the interaction region in the transverse plane [42]. Without going into details, let us just mention that one may regard the quantity $(1/S)(dN/dy)$, where S is the area of overlap of the two nuclei in the transverse plane and dN/dy the multiplicity density, as a measure of the average number of collisions undergone by a particle at a time $t \sim \bar{R}$. If this is so, one would expect the elliptic flow, under appropriate conditions [42], to be independent of this parameter if the hydrodynamical regime is attained. Data clearly indicate the contrary [43]. A more systematic discussion of this issue is presented in [42].

- *High transverse momentum particles are suppressed.* A suppression by a factor 5, relative to the extrapolation based on independent nucleon-nucleon collisions, was observed in central Au-Au collisions for momenta up to $p_T \sim 10$ Gev [44, 45]. This, together with the disappearance of jet-like correlations between particles produced in opposite azimuthal directions [46], confirms the role of strong final state interactions of the produced hard particles with the surrounding matter, and favors the interpretation in terms of "jet quenching" [47, 48, 49, 50, 51]. This interpretation receives further support from the d-Au collisions: there final state interactions play a minor role and no jet quenching is observed in the central rapidity region [52, 53].

- *Hints of gluon saturation.* The last discovery that I shall mention concerns the production of particles in d-Au collisions at forward rapidity, that is in the fragmentation region of the deuteron [54]. There, one does not expect final state interactions to play a major role. The suppression of particle production observed by the BRAHMS collaboration [54] may constitute evidence for initial state effects, caused perhaps by the phenomenon of gluon saturation [55, 56] (a more completet discussion is presented in [57]).

A more comprehensive discussion of RHIC data and the status of their present theoretical interpretation can be found in [58].

5. Conclusions

We have now clear evidence that the matter produced in ultrarelativistic heavy ion collisions has an energy density which is larger, on the relevant time scales, by one or two orders of magnitude than the energy density of ordinary nuclear matter. The required conditions for the formation of the quark-gluon plasma are therefore clearly met.

Data on the elliptic flow, on "jet quenching", and many other, provide evidence for strong interaction effects in the produced matter, pointing out, although indirectly, for the increasingly important role of partonic degrees of freedom at RHIC.

It is sometimes considered as a "surprise" that the matter produced in nucleus-nucleus collisions does not behave as a gas of weakly interacting quarks and gluons, as would be the case for the quark-gluon plasma at high temperatures: but the temperatures probed at RHIC may simply not be high enough for that. Thus, in the same way as present lattice data indicate that the transition from hadronic matter to the quark-gluon plasma may be a smooth cross over rather than a sharp phase transition, it is not unconceivable that the "discovery" of the quark-gluon plasma may go through a smooth path, forcing us on the way to learn many of its intricate properties. But beautiful data are here, and more will come, to guide us in this endeavor.

Acknowledgments

Support from CEFIPRA, under the project No. 3104-3, is gratefully acknowledged.

References

[1] J. C. Collins and M. J. Perry, *Phys. Rev. Lett.* 34:1353, 1975.

[2] N. Cabibbo and G. Parisi, *Phys. Lett.* B59:67, 1975.

[3] F. Karsch, Lattice QCD at high temperature and density, *Lect. Notes Phys.* 583:209–249, 2002.

[4] Z. Fodor and S. D. Katz, *JHEP* 04:050, 2004.

[5] K. Rajagopal and F. Wilczek, The condensed matter physics of QCD, arXiv: hep-ph/0011333.

[6] J. P. Blaizot, E. Iancu, and A. Rebhan, *Phys. Rev.* D63:065003, 2001; *Phys. Rev.* D68:025011, 2003.

[7] A. Dumitru, Y. Hatta, J. Lenaghan, K. Orginos, and R. D. Pisarski, *Phys. Rev.* D70: 034511, 2004.

[8] P. N. Meisinger, M. C. Ogilvie and T. R. Miller, Phys. Lett. B **585**: 149, 2004.

[9] K. Fukushima, Phys. Lett. B **591**: 277, 2004.

[10] C. Ratti, M. A. Thaler and W. Weise, arXiv:hep-ph/0506234.
[11] Edward V. Shuryak, arXiv: hep-ph/0405066.
[12] G. T. Horowitz and A. Strominger, Nucl. Phys. B **360**: 197, 1991.
[13] S. S. Gubser, I. R. Klebanov and A. A. Tseytlin, Nucl. Phys. B **534**: 202, 1998.
[14] G. Policastro, D. T. Son and A. O. Starinets, Phys. Rev. Lett. **87**: 081601, 2001.
[15] G. Boyd, J. Engels, F. Karsch, E. Laermann, C. Legeland, M. Lutgemeier and B. Petersson, Nucl. Phys. B **469**:419,1996.
[16] M. Laine and Y. Schroder, JHEP **0503**: 067, 2005.
[17] S. Chekanov et al. (ZEUS), *Phys. Rev.* D67:012007, 2003.
[18] L. V. Gribov, E. M. Levin, and M. G. Ryskin, *Phys. Rept.* 100:1–150, 1983.
[19] A. H. Mueller and J.-W. Qiu, *Nucl. Phys.* B268:427, 1986.
[20] J. P. Blaizot and A. H. Mueller, *Nucl. Phys.* B289:847, 1987.
[21] I. Balitsky, *Nucl. Phys.* B463:99–160, 1996.
[22] Y. V. Kovchegov, *Phys. Rev.* D61:074018, 2000.
[23] J. Jalilian-Marian, A. Kovner, A. Leonidov, and H. Weigert, *Nucl. Phys.* B504:415–431, 1997; *Phys. Rev.*, D59:014014, 1999.
[24] Y. V. Kovchegov, *Phys. Rev.* D54:5463–5469, 1996.
[25] E. Iancu, A. Leonidov, and L. D. McLerran, *Nucl. Phys.* A692:583–645, 2001; *Phys. Lett.*, B510:133–144, 2001.
[26] H. Weigert, *Nucl. Phys.* A703:823–860, 2002.
[27] E. Ferreiro, E. Iancu, A. Leonidov, and L. McLerran, *Nucl. Phys.* A703:489–538, 2002.
[28] A. H. Mueller. *Nucl. Phys.* B558:285–303, 1999.
[29] L. D. McLerran and R. Venugopalan, *Phys. Rev.* D49:2233–2241, 1994; *Phys. Rev.* D49:3352–3355, 1994; *Phys. Rev.*, D50:2225–2233 1994.
[30] B. B. Back et al. (PHOBOS), arXiv: nucl-ex/0410022.
[31] K. J. Eskola, *Nucl. Phys.* A698:78–87, 2002.
[32] D.Kharzeev, E. Levin, and M. Nardi, arXiv: hep-ph/0111315.
[33] J. Adams et al. (STAR), *Phys. Rev.* C72:014904, 2005.
[34] S. S. Adler et al. (PHENIX), *Phys. Rev. Lett.* 91:182301, 2003.
[35] J.-Y. Ollitrault, *Phys. Rev.* D46:229–245, 1992.
[36] P. F. Kolb and U. Heinz, arXiv: nucl-th/0305084.
[37] A. H. Mueller, *Nucl. Phys.* B572:227–240, 2000.
[38] R. Baier, A. H. Mueller, D. Schiff, and D. T. Son, *Phys. Lett.* B502:51–58, 2001.
[39] P. Arnold, J. Lenaghan, G. D. Moore, and L. G. Yaffe, arXiv: nucl-th/0409068.
[40] A. Rebhan, P. Romatschke and M. Strickland, arXiv: hep-ph/0412016.
[41] D. Teaney, J. Lauret and E. V. Shuryak, Phys. Rev. Lett. **86**: 4783, 2001.
[42] R. S. Bhalerao, J. P. Blaizot, N. Borghini and J. Y. Ollitrault, *Phys. Lett.* B627:49, 2005.
[43] C. Alt *et al.* [NA49 Collaboration], Phys. Rev. C **68**, 034903 (2003).
[44] S. S. Adler et al. (PHENIX), *Phys. Rev. Lett.* 91:072301, 2003.
[45] J. Adams et al. (STAR) *Phys. Rev. Lett.* 91:172302, 2003.
[46] C. Adler et al. (STAR), *Phys. Rev. Lett.* 90:082302, 2003.
[47] X.-N. Wang and M. Gyulassy, *Phys. Rev. Lett.* 68:1480–1483, 1992.
[48] M. Gyulassy and X.-N. Wang, *Nucl. Phys.* B420:583–614, 1994.
[49] R. Baier, Y. L. Dokshitzer, S. Peigne, and D. Schiff, *Phys. Lett.* B345:277–286, 1995.
[50] R. Baier, Y. L. Dokshitzer, A. H. Mueller, S. Peigne, and D. Schiff, *Nucl. Phys.* B483:291–320, 1997.
[51] M. Gyulassy, I. Vitev, X.-N. Wang, and B.-W. Zhang, arXiv: nucl-th/0302077.
[52] S. S. Adler et al. (PHENIX), *Phys. Rev. Lett.* 91:072303, 2003.
[53] A. H. Tang. *J. Phys.* G30:S1235–S1238, 2004.
[54] I. Arsene et al. (BRAHMS), arXiv: nucl-ex/0410020.
[55] D. Kharzeev, E. Levin and L. McLerran, Phys. Lett. B **561**: 93, 2003.
[56] D. Kharzeev, Y. V. Kovchegov, and K. Tuchin, *Phys. Lett.* B599:23–31, 2004.
[57] J.-P. Blaizot and F. Gelis, arXiv: hep-ph/0405305.
[58] M. Gyulassy and L. McLerran, arXiv: nucl-th/0405013.

Institute of Physics Publishing Journal of Physics: Conference Series **50** (2006) 8–21
doi:10.1088/1742-6596/50/1/002 5th International Conference on Physics and Astrophysics of Quark Gluon Plasma

From Little Bangs to the Big Bang

John Ellis
Theory Division, Physics Department, CERN, CH-1211 Geneva 23, Switzerland
E-mail: john.ellis@cern.ch CERN-PH-TH/2005-070 astro-ph/yymmnnn

Abstract.
The 'Little Bangs' made in particle collider experiments reproduce the conditions in the Big Bang when the age of the Universe was a fraction of a second. It is thought that matter was generated, the structures in the Universe were formed and cold dark matter froze out during this very early epoch when the equation of state of the Universe was dominated by the quark-gluon plasma (QGP). Future Little Bangs may reveal the mechanism of matter generation and the nature of cold dark matter. Knowledge of the QGP will be an essential ingredient in quantitative understanding of the very early Universe.

1. The Universe is Expanding

The expansion of the Universe was first established by Hubble's discovery that distant galaxies are receding from us, with redshifts proportional to their relative distances from us. Extrapolating the present expansion backwards, there is good evidence that the Universe was once 3000 times smaller and hotter than today, provided by the cosmic microwave background (CMB) radiation. This has a thermal distribution and is very isotropic, and is thought to have been released when electrons combined with ions from the primordial electromagnetic plasma to form atoms. The observed small dipole anisotropy is due to the Earth's motion relative to this cosmic microwave background, and the very small anisotropies found by the COBE satellite are thought to have led to the formation of structures in the Universe, as discussed later [1].

Extrapolating further back in time, there is good evidence that the Universe was once a billion times smaller and hotter than today, provided by the abundances of light elements cooked in the Big Bang [2]. The Universe contains about 24 % by mass of ^{4}He, and somewhat less Deuterium, ^{3}He and ^{7}Li. These could only have been cooked by nuclear reactions in the very early Universe, when it was a billion times smaller and hotter than today. The detailed light-element abundances depend on the amount of matter in the Universe, and comparison between observations and calculations suggests that there is not enough matter to stop the present expansion, or even to explain the amount of matter in the galaxies and their clusters. The calculations of the light-element abundances also depend on the number of particle types, and in particular on the number of different neutrino types. This is now known from particle collider experiments to be three [3], with a corresponding number of charged leptons and quark pairs.

2. The Very Early Universe and the Quark-Gluon Plasma

When the Universe was very young: $t \to 0$, also the scale factor a characterizing its size would have been very small: $a \to 0$, and the temperature T would have been very large, with characteristic relativistic particle energies $E \sim T$. In normal adiabatic expansion, $T \sim 1/a$, and, while the energy density of the Universe was dominated by relativistic matter, $t \sim 1/T^2$. The following are some rough orders of magnitude: when the Universe had an age $t \sim 1$ second, the temperature was $T \sim 10,000,000,000$ degrees, and characteristic thermal energies were $E \sim 1$ MeV, comparable with the mass of the electron. It is clear that one needs particle physics to describe the earlier history of the Universe [1].

The very early Universe was presumably filled with primordial quark-gluon plasma (QGP). When the Universe was a few microseconds old, it is thought to have exited from this QGP phase, with the available quarks and gluons combining to make mesons and baryons. The primordial QGP would have had a very low baryon chemical potential μ. Experiments with RHIC reproduce cosmological conditions more closely than did previous SPS experiments, as seen in Fig. 1, and the LHC will provide [4] an even closer approximation to the primordial QGP. I shall not discuss here the prospects for discovering quark matter inside dense astrophysical objects such as neutron stars, which would have a much larger baryon chemical potential.

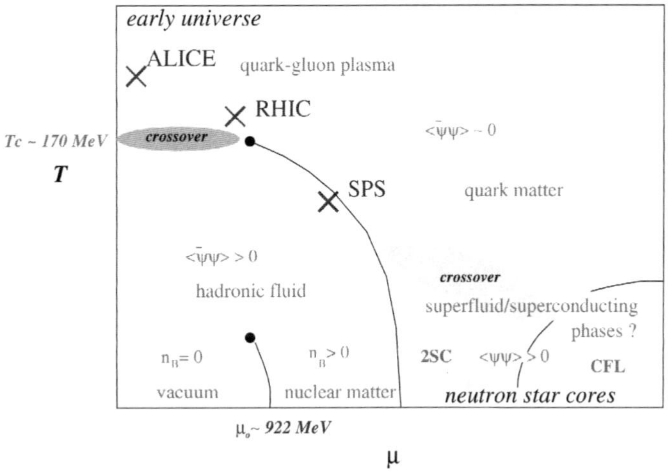

Figure 1. *The phase diagram of hot and dense QCD for different values of the baryon chemical potential μ and temperature T [5], illustrating the physics reaches of SPS, RHIC and the ALICE experiment at the LHC [4].*

To what extent can information about the early Universe cast light on the quark-hadron phase transition? The latest lattice simulations of QCD with two light flavours u, d and one moderately heavy flavour s suggest that there was no strong first-order transition. Instead, there was probably a cross-over between the quark and hadron phases, see, for example, Fig. 2 [5], during which the smooth expansion of the Universe is unlikely to have been modified substantially. Specifically, it is not thought that this transition would have induced inhomogeneities large enough to have detectable consequences today.

3. Open Cosmological Questions

The Standard Model of cosmology leaves many important questions unanswered. *Why is the Universe so big and old?* Measurements by the WMAP satellite, in particular, indicate that its age is about 14,000,000,000 years [6]. *Why is its geometry nearly Euclidean?* Recent data indicate that it is almost flat, close to the borderline for eternal expansion. *Where did the matter come from?* The cosmological nucleosynthesis scenario indicates that there is approximately one proton in the Universe today for every 1,000,000,000 photons, and no detectable amount

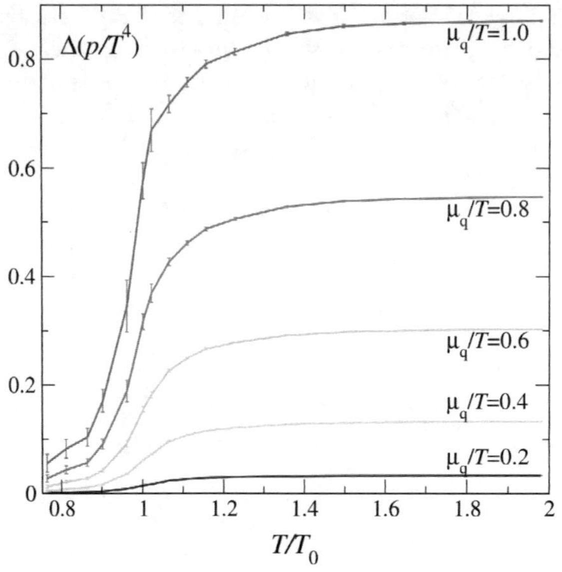

Figure 2. *The growth of the QCD pressure with temperature, for different values of the baryon chemical potential* μ *[5]. The rise is quite smooth, indication that there is not a strong first-order phase transition, and probably no dramatic consequences in the early Universe.*

of antimatter. *How did cosmological structures form?* If they did indeed form from the ripples observed in the CMB, how did these originate? *What is the nature of the invisible dark matter thought to fill the Universe?* Its presence is thought to have been essential for the amplification of the primordial perturbations in the CMB.

It is clear that one needs particle physics to answer these questions, and that their solutions would have operated in a Universe filled with QGP.

4. A Strange Recipe for a Universe

According to the 'Concordance Model' suggested by a multitude of astrophysical and cosmological observations, the total density of the Universe is very close to the critical value: $\Omega_{Tot} = 1.02 \pm 0.02$, as illustrated in Fig. 3 [6]. The theory of cosmological inflation suggests that the density should be indistinguishable from the critical value, and this is supported by measurements of the CMB. On the other hand, the baryon density is small, as inferred not only from Big-Bang nucleosynthesis but also and independently from the CMB: $\Omega_{Baryons} \sim$ few%. The CMB information on these two quantities comes from observations of peaks in the fluctuation spectrum in specific partial waves corresponding to certain angular scales: the position of the first peak is sensitive to Ω_{Tot}, and the relative heights of subsequent peaks are sensitive to $\Omega_{Baryons}$. The fraction Ω_m of the critical density provided by all forms of matter is not very well constrained by the CMB data alone, but is quite tightly constrained by combining them with observations of high-redshift supernovae [7] and/or large-scale structures [8], each of which favours $\Omega_{Matter} \sim 0.3$, as also seen in Fig. 3.

As seen in Fig. 4, there is good agreement between BBN calculations and astrophysical observations for the Deuterium and ^{4}He abundances [2]. The agreement for ^{7}Li is less striking, though not disastrously bad [1]. The good agreement between the corresponding

[1] It seems unlikely that the low abundance of ^{7}Li observed could have been modified significantly by the decays

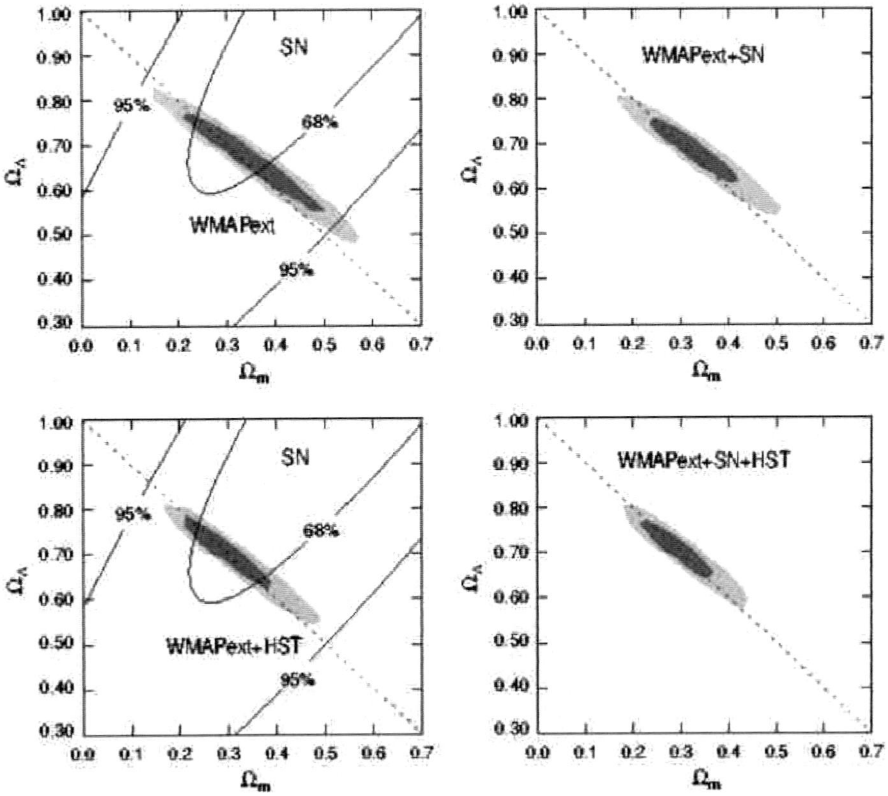

Figure 3. *The density of matter Ω_m and dark energy Ω_Λ inferred from WMAP and other CMB data (WMAPext), and from combining them with supernova and Hubble Space Telescope data [6].*

determinations of $\Omega_{Baryons}$ obtained from CMB and Big-Bang nucleosynthesis calculations in conventional homogeneous cosmology imposes important constraints on inhomogeneous models of nucleosynthesis. In particular, they exclude the possibility that $\Omega_{Baryons}$ might constitute a large fraction of Ω_{Tot}. Significant inhomogeneities might have been generated at the quark-hadron phase transition, if it was strongly first-order [10]. Although, as already discussed, lattice calculations suggest that this is rather unlikely, heavy-ion collision experiments must be the final arbiter on the nature of the quark-hadron phase transition.

5. Generating the Matter in the Universe

As was pointed out by Sakharov [11], there are three essential requirements for generating the matter in the Universe via microphysics. First, one needs a difference between matter and antimatter interactions, as has been observed in the laboratory in the forms of violations of C and CP in the weak interactions. Secondly, one needs interactions that violate the baryon and lepton numbers, which are present as non-perturbative electroweak interactions and in grand unified theories, but have not yet been seen. Finally, one needs a breakdown of thermal equilibrium, which is possible during a cosmological phase transition, for example at the GUT or electroweak scale, or in the decays of heavy particles, such as a heavy singlet neutrino ν_R [12]. The issue

of heavy particles [9]: it would be valuable to refine the astrophysical determinations.

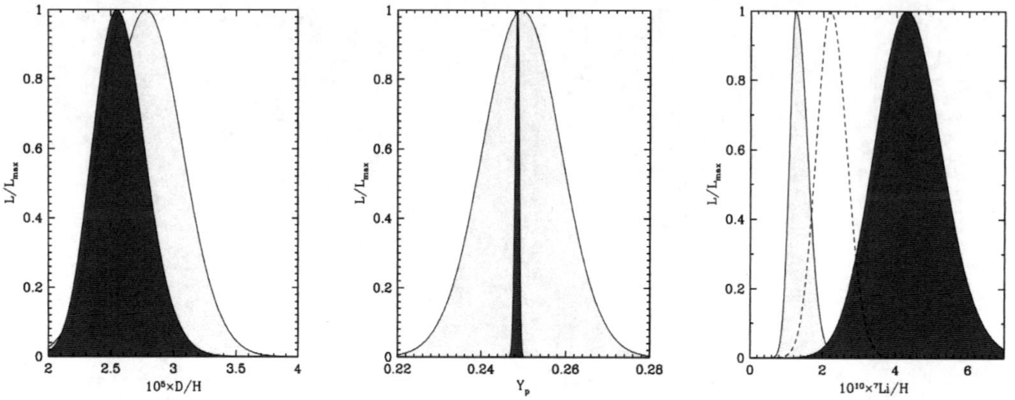

Figure 4. *Primordial light element abundances as predicted by BBN (light) and WMAP (dark shaded regions)[2], for (a) D/H, (b) the ^{4}He abundance Y_p and (c) ^{7}Li/H [2].*

then is whether we will be able to calculate the resulting matter density in terms of laboratory measurements. Unfortunately, the Standard Model C and CP violation measured in the quark sector seem unsuitable for baryogenesis, and the electroweak phase transition in the Standard Model would have been second order. However, additional CP violation and a first-order phase transition in an extended electroweak Higgs sector might have been able to generate the matter density [13], and could be testable at the LHC and/or ILC. An alternative is CP violation in the lepton sector, which could be probed in neutrino oscillation experiments, albeit indirectly, or possibly in the charged-lepton sector, which might be related more directly to the matter density [14].

In any case, detailed knowledge of the QGP equation of state would be necessary if one were ever to hope to be able to calculate the baryon-to-entropy ratio with an accuracy of a few percent.

6. The Formation of Structures in the Universe

The structures seen in the Universe - clusters, galaxies, stars and eventually ourselves - are all thought to have developed from primordial fluctuations in the CMB. This idea is supported visually by observations of galaxies, which look smooth at the largest scales at high redshifts, but cluster at smaller scales at low redshifts, as seen in Fig. 5 [15]. This scenario requires amplification of the small fluctuations observed in the CMB, which is possible with massive non-relativistic weakly-interacting particles. On the other hand, relativistic light neutrinos would have escaped from smaller structures, and so are disfavoured as amplifiers. Non-relativistic 'cold dark matter' is preferred, as seen in a comparison of the available data on structures in the Universe with the cosmological Concordance Model [8].

The hot news in the observational tests of this scenario has been the recent detection of baryonic ripples from the Big Bang [16], as seen in Fig. 6. These are caused by sound waves spreading out from irregularities in the CMB, which show up in the correlation function between structures in the (near-)contemporary Universe as features with a characteristic size. In addition to supporting the scenario of structure formation by amplification of CMB fluctuations, these observations provide measurements of the expansion history and equation of state of the Universe.

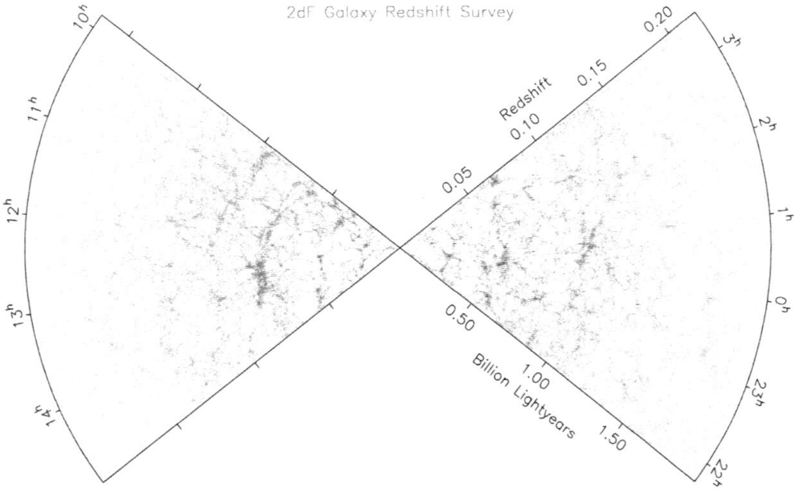

Figure 5. *The spatial distribution of galaxies observed in the 2dF galaxy redshift survey [15]. The distribution is quite smooth at large redshifts, and exhibits more structures at smaller redshifts.*

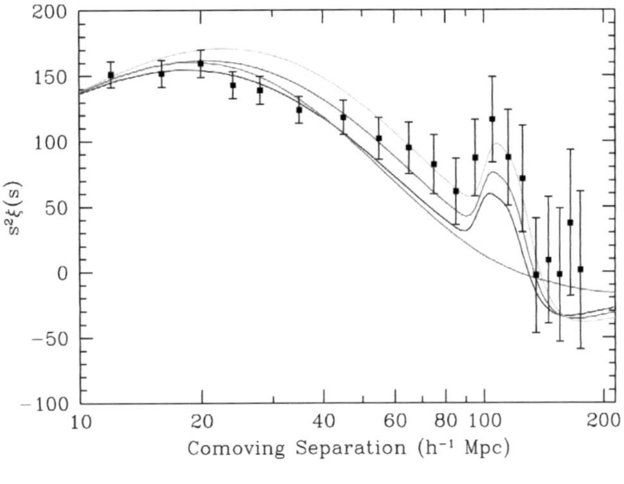

Figure 6. *The baryonic 'ripple' in the large-scale correlation function of luminous red galaxies observed in the Sloan Digital Sky Survey of galactic redshifts [16].*

7. Do Neutrinos Matter?

Oscillation experiments tell us that neutrinos have very small but non-zero masses [17, 18], and so must make up at least some of the dark matter. As already mentioned, since such light neutrinos move relativistically during the epoch of structure formation, they would have escaped from galaxies and not contributed to their formation, whereas they could have contributed to the formation of clusters. Conversely, the success of the cosmological Concordance Model enables one to set a cosmological upper limit on the sum of light neutrino masses, as seen in Fig. 7: $\Sigma_\nu m_\nu < 0.7$ eV [6], which is considerably more sensitive than direct laboratory searches. In the future, this cosmological sensitivity might attain the range indicated by atmospheric neutrino data [17].

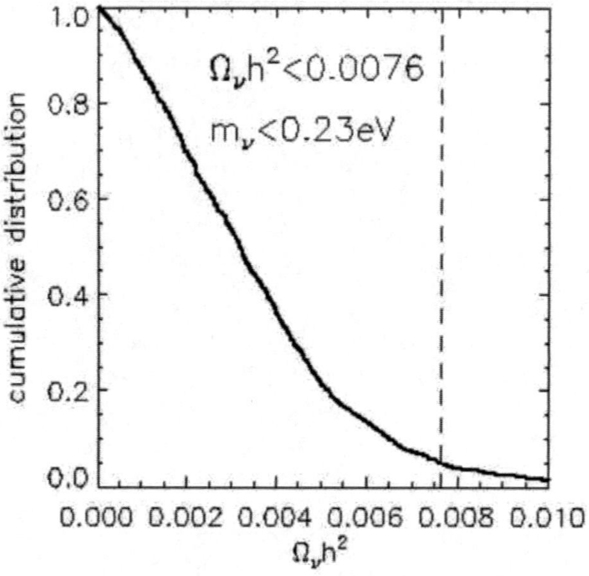

Figure 7. *The likelihood function for the total neutrino density $\Omega_\nu h^2$ derived by WMAP [6]. The upper limit $m_\nu < 0.23$ eV applies if there are three degenerate neutrinos.*

However, even if no dark matter effect of non-zero light neutrino masses is observed, this does not mean that neutrinos have no cosmological role, since unstable heavier neutrinos might have generated matter via the Sakharov mechanism [11].

8. Particle Dark Matter Candidates

Candidates for the non-relativistic cold dark matter required to amplify CMB fluctuations include the *axion* [19], TeV-scale *weakly-interacting massive particles* (WIMPs) produced thermally in the early Universe, such as the lightest supersymmetric partner of a Standard Model particle (probably the lightest neutralino χ), the *gravitino* (which is likely mainly to have been produced in the very early Universe, possibly thermally), and *superheavy relic particles* that might have been produced non-thermally in the very early Universe [20] (such as the 'cryptons' predicted in some string models [21]).

9. Supersymmetric Dark Matter

Supersymmetry is a very powerful symmetry relating fermionic 'matter' particles to bosonic 'force' particles [22]. Historically, the original motivations for supersymmetry were purely theoretical: its intrinsic beauty, its ability to tame infinities in perturbation theory, etc. The first phenomenological motivation for supersymmetry at some accessible energy was that it might also help explain the electroweak mass scale, by stabilizing the hierarchy of mass scales in physics [23]. It was later realized also that the lightest supersymmetric particle (LSP) would be stable in many models [24]. Moreover, it should weigh below about 1000 GeV, in order to stabilize the mass hierarchy, in which case its relic density would be similar to that required for cold dark matter [25]. As described below, considerable effort is now put into direct laboratory searches for supersymmetry, as well as both direct and indirect astrophysical searches.

Here I concentrate on the minimal supersymmetric extension of the Standard Model (MSSM), in which the Standard Model particles acquire superpartners and there are two doublets of Higgs fields. The interactions in the MSSM are completely determined by supersymmetry, but one must postulate a number of soft supersymmetry-breaking parameters, in order to accommodate the mass differences between conventional particles and their superpartners. These parameters include scalar masses m_0, gaugino masses $m_{1/2}$, and trilinear soft couplings A_0. It is often assumed that these parameters are universal, so that there is a single m_0, a single $m_{1/2}$, and a single A_0 parameter at the input GUT scale, a scenario called the constrained MSSM (CMSSM). However, there is no deep theoretical justification for this universality assumption, except in minimal supergravity models. These models also make a prediction for the gravitino mass: $m_{3/2} = m_0$, which is not necessarily the case in the general CMSSM.

As already mentioned, the lightest supersymmetric particle is stable in many models, this because of the multiplicative conservation of R parity, which is a combination of spin S, lepton number L and baryon number B: $R = (-1)^{2S-L+3B}$. It is easy to check that conventional particles have $R = +1$ and sparticles have $R = -1$. As a result, sparticles are always produced in pairs, heavier sparticles decay into lighter ones, and the lightest supersymmetric particle (LSP) is stable.

The LSP cannot have strong or electromagnetic interactions, because these would bind it to conventional matter, creating bound states that would be detectable as anomalous heavy nuclei. Among the possible weakly-interacting scandidates for the LSP, one finds the *sneutrino*, which has been excluded by a combination of LEP and direct searches for astrophysical dark matter, the lightest *neutralino* χ, and the *gravitino*. There are good prospects for detecting the neutralino or gravitino in collider experiments, and neutralino dark matter may also be detectable either directly or indirectly, but gravitino dark matter would be a nightmare for detection.

10. Constraints on Supersymmetry

Important constraints on supersymmetry are imposed by the absences of sparticles at LEP and the Tevatron collider, implying that sleptons and charginos should weigh > 100 GeV [26], and that squarks and gluinos should weigh > 250 GeV, respectively. Important indirect constraints are imposed by the LEP lower limit on the mass of the lightest Higgs boson, 114 GeV [27], and the experimental measurement of $b \to s\gamma$ decay [28], which agrees with the Standard Model. The measurement of the anomalous magnetic moment of the muon, $g_\mu - 2$, also has the potential to constrain supersymmetry, but the significance of this constraint is uncertain, in the absence of agreement between the e^+e^- annihilation and τ decay data used to estimate the Standard Model contribution to $g_\mu - 2$ [29]. Finally, one of the strongest constraints on the supersymmetric parameter space is that imposed by the density of dark matter inferred from astrophysical and cosmological observations. If this is composed of the lightest neutralino χ, one has $0.094 < \Omega_\chi h^2 < 0.129$ [6], and it cannot in any case be higher than this. For generic domains of the supersymmetric parameter space, this range constrains m_0 with an accuracy of a few per cent as a function of $m_{1/2}$, as seen in Fig. 8 [30].

11. The Relic Density and the Quark-Gluon Plasma

The accurate calculation of the relic density depends not only on the supersymmetric model parameters, but also on the effective Hubble expansion rate as the relic particles annihilate and

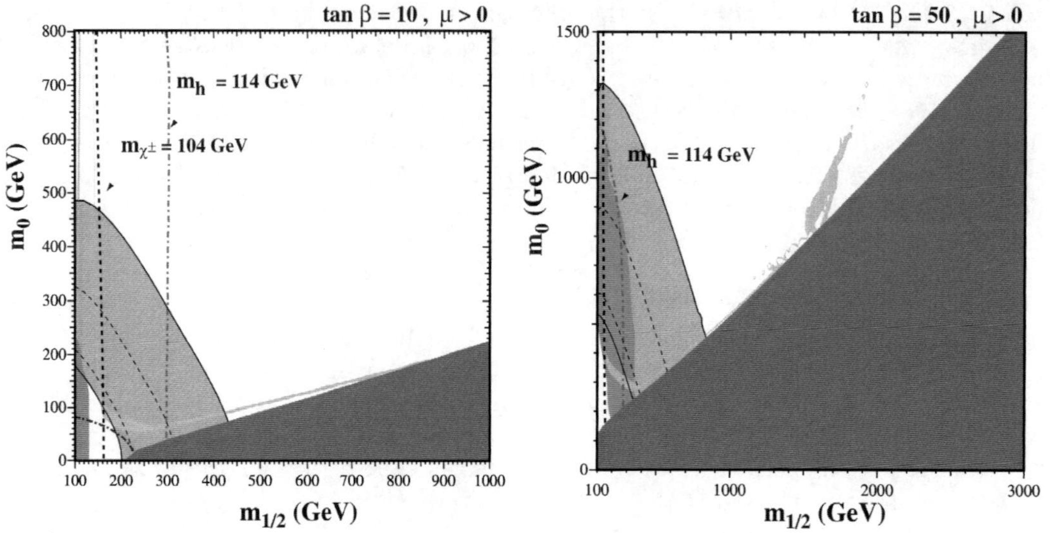

Figure 8. *The* $(m_{1/2}, m_0)$ *planes for (a)* $\tan\beta = 10$ *and* $\tan\beta = 50$ *with* $\mu > 0$ *and* $A_0 = 0, assuming m_t = 175$ *GeV and* $m_b(m_b)^{\overline{MS}}_{SM} = 4.25$ *GeV. The near-vertical (red) dot-dashed lines are the contours* $m_h = 114$ *GeV, and the near-vertical (black) dashed line is the contour* $m_{\chi^\pm} = 104$ *GeV. Also shown by the dot-dashed curve in the lower left is the corner excluded by the LEP bound of* $m_{\tilde{e}} > 99$ *GeV. The medium (dark green) shaded region is excluded by* $b \to s\gamma$*, and the light (turquoise) shaded area is the cosmologically preferred regions with* $0.094 \leq \Omega_\chi h^2 \leq 0.129$*. In the dark (brick red) shaded region, the LSP is the charged* $\tilde{\tau}_1$*. The region allowed by the E821 measurement of* a_μ *at the 2-*σ *level, is shaded (pink) and bounded by solid black lines, with dashed lines indicating the 1-*σ *ranges [30].*

freeze out of thermal equilibrium [25]:

$$\dot{n} + 3Hn = - < \sigma_{ann} v > (n^2 - n_{eq}^2).$$

This is, in turn, sensitive to the effective number of particle species:

$$Y_0 \simeq \left(\frac{45}{\pi}\right)^{\frac{1}{2}} \frac{1}{m M_P < \sigma_{ann} v >_{T_f} g_*^{1/2}(T_f)}$$

where

$$g_*^{1/2}(T) = \frac{H_{eff}}{g_{eff}}^{\frac{1}{2}} \left(1 + \frac{T}{3} \frac{\mathrm{d}\ln h_{eff}}{dT}\right).$$

To calculate this at the per-cent level, one needs to understand with comparable accuracy the equation of state of hot QCD around the freeze-out temperature [31], which is typically $\sim m_\chi/20 \sim$ a few GeV. Fig. 9 the comparison between the predictions of different equations of state for h_{eff}: it would be good to reach consensus on at the per-cent level.

12. Supersymmetry Searches at the LHC

A 'typical' CMSSM event should be relatively 'easy' to see at the LHC, as it should have a lot of 'missing' transverse energy carried away by invisible neutralinos, accompanied by several jets and/or leptons. Many studies have shown that the LHC should be able to cover most of the

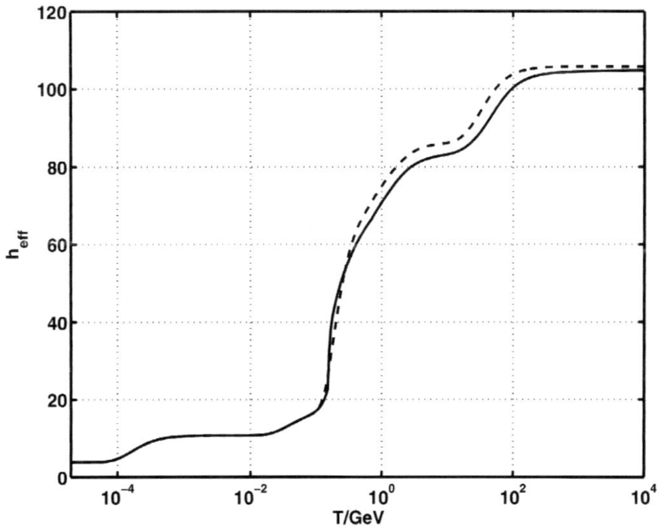

Figure 9. The factor $h_{\rm eff}(T)$ calculated using different equations of state [31].

region of parameter space allowed by cosmology [32], as seen in Fig. 10 [33]. Moreover, the LHC should be able to see several species of supersymmetric particles, which might make possible the *a priori* calculation of the relic density with an uncertainty comparable to that provided by astrophysical and cosmological data [34]. Minimal supergravity models might be even easier to discover, since in generic domains of parameter space the next-to-lightest supersymmetric particle would live a long time before it decayed into the gravitino [35], liberating a large amount of energy that could be measured in a dedicated detector [36].

13. Strategies for Detecting Supersymmetric Dark Matter

These include searches for the annihilations of relic particles in the galactic halo:$\chi\chi \rightarrow$ antiprotons or positrons, annihilations in the galactic centre: $\chi\chi \rightarrow \gamma + \ldots$, annihilations in the core of the Sun or the Earth: $\chi\chi \rightarrow \nu + \cdots \rightarrow \mu + \cdots$, and scattering on nuclei in the laboratory: $\chi A \rightarrow chiA$.

After some initial excitement, recent observations of cosmic-ray antiprotons are consistent with production by primary matter cosmic rays. Moreover, the spectra of annihilation positrons calculated in a number of CMSSM benchmark models [32] seem to fall considerably below the cosmic-ray background [37]. Some of the spectra of photons from annihilations in Galactic Centre, as calculated in the same set of CMSSM benchmark scenarios, may rise above the expected cosmic-ray background, albeit with considerable uncertainties due to the unknown enhancement of the cold dark matter density. In particular, the GLAST experiment may have the best chance of detecting energetic annihilation photons [37], as seen in the left panel of Fig. 11. Annihilations in the Solar System also offer detection prospects in some of the benchmark scenarios, particularly annihilations inside the Sun, which might be detectable in experiments such as AMANDA, NESTOR, ANTARES and particularly IceCUBE, as seen in the right panel of Fig. 11 [37].

The rates for elastic dark matter scattering cross sections calculated in the CMSSM are typically considerably below the present upper limit imposed by the CDMS II experiment, in both the benchmark scenarios and the global fit to CMSSM parameters based on present data [38]. However, if the next generation of direct searches for elastic scattering can reach

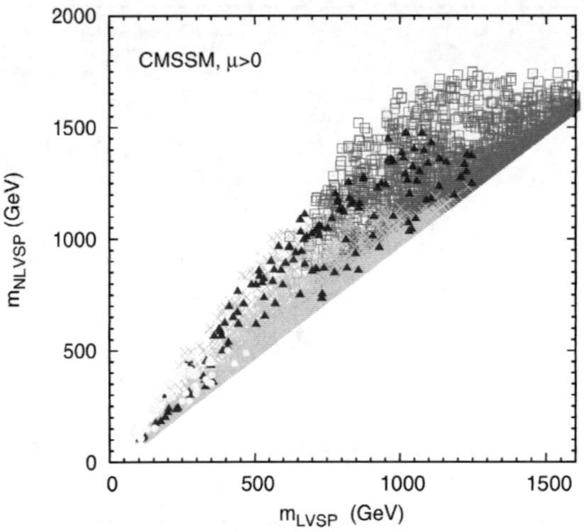

Figure 10. *Scatter plot of the masses of the lightest visible supersymmetric particle (LVSP) and the next-to-lightest visible supersymmetric particle (NLVSP) in the CMSSM. The darker (blue) triangles satisfy all the laboratory, astrophysical and cosmological constraints. For comparison, the dark (red) squares and medium-shaded (green) crosses respect the laboratory constraints, but not those imposed by astrophysics and cosmology. In addition, the (green) crosses represent models which are expected to be visible at the LHC. The very light (yellow) points are those for which direct detection of supersymmetric dark matter might be possible [33].*

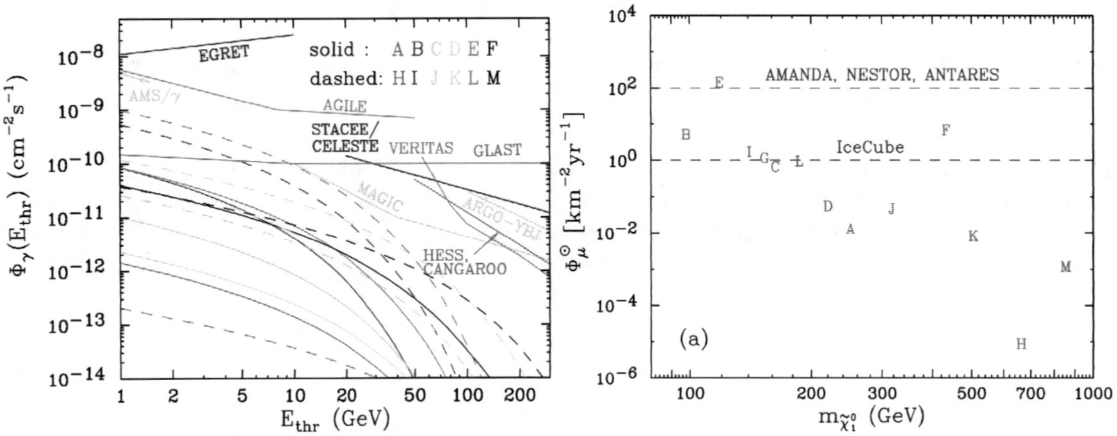

Figure 11. *Left panel: Spectra of photons from the annihilations of dark matter particles in the core of our galaxy, in different benchmark supersymmetric models [37]. Right panel: Signals for muons produced by energetic neutrinos originating from annihilations of dark matter particles in the core of the Sun, in the same benchmark supersymmetric models [37].*

a sensitivity of 10^{-10} pb, they should be able to detect supersymmetric dark matter in many supersymmetric scenarios. Fig. 12 compares the cross sections calculated under a relatively optimistic assumption for the relevant hadronic matrix element $\sigma_{\pi N} = 64$ MeV, for choices of CMSSM parameters favoured at the 68 % (90 %) confidence level in a recent analysis using the

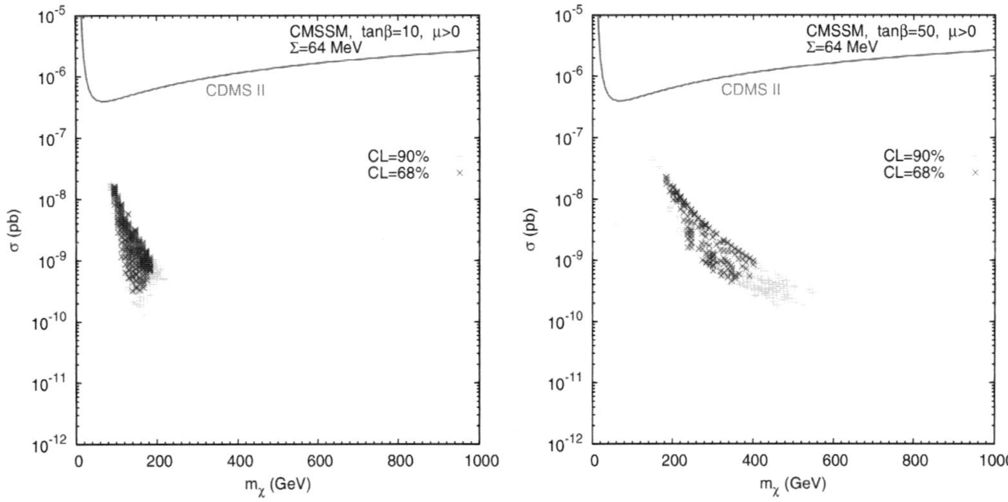

Figure 12. *Scatter plots of the spin-independent elastic-scattering cross section predicted in the CMSSM for (a)* $\tan\beta = 10, \mu > 0$ *and (b)* $\tan\beta = 50, \mu > 0$*, each with* $\sigma_{\pi N} = 64$ *MeV [38]. The predictions for models allowed at the 68% (90%) confidence levels [39] are shown by blue* $\times$ *signs (green + signs).*

observables $m_W, \sin^2\theta_W, b \to s\gamma$ and $g_\mu - 2$ [39].

14. Connections between the Big Bang and Little Bangs

Astrophysical and cosmological observations during the past few years have established a Concordance Model of cosmology, whose matter content is quite accurately determined. Most of the present energy density of the Universe is in the form of dark vacuum energy, with about 25 % in the form of dark matter, and only a few % in the form of conventional baryonic matter. Two of the most basic questions raised by this Concordance Model are the nature of the dark matter and the origin of matter.

Only experiments at particle colliders are likely to be able to answer these and other fundamental questions about the early Universe. In particular, experiments at the LHC will recreate quark-gluon plasma conditions similar to those when the Universe was less than a microsecond old [4], and will offer the best prospects for discovering whether the dark matter is composed of supersymmetric particles [40, 41]. LHC experiments will also cast new light on the cosmological matter-antimattter asymmetry [42]. Moreover, discovery of the Higgs boson will take us closer to the possibilities for inflation and dark energy.

There are many connections between the Big Bang and the little bangs we create with particle colliders. These connections enable us both to learn particle physics from the Universe, and to use particle physics to understand the Universe.

References

[1] J. R. Ellis, Lectures given at 16th Canberra International Physics Summer School on *The New Cosmology*, Canberra, Australia, 3-14 Feb 2003, arXiv:astro-ph/0305038; K. A. Olive, TASI lectures on *Astroparticle Physics*, arXiv:astro-ph/0503065.

[2] R. H. Cyburt, B. D. Fields, K. A. Olive and E. Skillman, Astropart. Phys. **23** (2005) 313 [arXiv:astro-ph/0408033].

[3] LEP Electroweak Working Group, `http://lepewwg.web.cern.ch/LEPEWWG/Welcome.html`.

[4] ALICE Collaboration, `http://pcaliweb02.cern.ch/NewAlicePortal/en/Collaboration/index.html`.

[5] C. R. Allton, S. Ejiri, S. J. Hands, O. Kaczmarek, F. Karsch, E. Laermann and C. Schmidt, Nucl. Phys. Proc. Suppl. **141** (2005) 186 [arXiv:hep-lat/0504011].
[6] D. N. Spergel *et al.* [WMAP Collaboration], Astrophys. J. Suppl. **148** (2003) 175 [arXiv:astro-ph/0302209].
[7] A. G. Riess *et al.* [Supernova Search Team Collaboration], Astron. J. **116** (1998) 1009 [arXiv:astro-ph/9805201]; S. Perlmutter *et al.* [Supernova Cosmology Project Collaboration], Astrophys. J. **517** (1999) 565 [arXiv:astro-ph/9812133].
[8] N. A. Bahcall, J. P. Ostriker, S. Perlmutter and P. J. Steinhardt, Science **284** (1999) 1481 [arXiv:astro-ph/9906463].
[9] J. R. Ellis, K. A. Olive and E. Vangioni, arXiv:astro-ph/0503023.
[10] K. Jedamzik and J. B. Rehm, Phys. Rev. D **64** (2001) 023510 [arXiv:astro-ph/0101292].
[11] A. D. Sakharov, Pisma Zh. Eksp. Teor. Fiz. **5** (1967) 32 [JETP Lett. **5** (1967 SOPUA,34,392-393.1991 UFNAA,161,61-64.1991) 24].
[12] M. Fukugita and T. Yanagida, Phys. Lett. B **174** (1986) 45.
[13] See, for example: M. Carena, M. Quiros, M. Seco and C. E. M. Wagner, Nucl. Phys. B **650** (2003) 24 [arXiv:hep-ph/0208043].
[14] See, for example: J. R. Ellis and M. Raidal, Nucl. Phys. B **643** (2002) 229 [arXiv:hep-ph/0206174].
[15] The 2dF Galaxy Redshift Survey, `http://www.mso.anu.edu.au/2dFGRS/`.
[16] D. J. Eisenstein *et al.*, arXiv:astro-ph/0501171; S. Cole *et al.* [The 2dFGRS Collaboration], arXiv:astro-ph/0501174.
[17] Y. Fukuda *et al.* [Super-Kamiokande Collaboration], Phys. Rev. Lett. **81** (1998) 1562 [arXiv:hep-ex/9807003].
[18] See, for example: Q. R. Ahmad *et al.* [SNO Collaboration], Phys. Rev. Lett. **87** (2001) 071301 [arXiv:nucl-ex/0106015].
[19] See, for example: S. Andriamonje *et al.* [CAST Collaboration], Phys. Rev. Lett. **94** (2005) 121301 [arXiv:hep-ex/0411033].
[20] D. J. H. Chung, E. W. Kolb and A. Riotto, Phys. Rev. D **59** (1999) 023501 [arXiv:hep-ph/9802238].
[21] K. Benakli, J. R. Ellis and D. V. Nanopoulos, Phys. Rev. D **59** (1999) 047301 [arXiv:hep-ph/9803333].
[22] J. Wess and B. Zumino, Nucl. Phys. B **70** (1974) 39.
[23] L. Maiani, *Proceedings of the 1979 Gif-sur-Yvette Summer School On Particle Physics*, 1; G. 't Hooft, in *Recent Developments in Gauge Theories, Proceedings of the Nato Advanced Study Institute, Cargese, 1979*, eds. G. 't Hooft *et al.*, (Plenum Press, NY, 1980); E. Witten, Phys. Lett. **B105** (1981) 267.
[24] P. Fayet, *Unification of the Fundamental Particle Interactions*, eds. S. Ferrara, J. Ellis and P. van Nieuwenhuizen (Plenum, New York, 1980), p.587.
[25] J. Ellis, J.S. Hagelin, D.V. Nanopoulos, K.A. Olive and M. Srednicki, Nucl. Phys. **B238** (1984) 453; see also H. Goldberg, Phys. Rev. Lett. **50** (1983) 1419.
[26] The Joint LEP 2 Supersymmetry Working Group, `http://lepsusy.web.cern.ch/lepsusy/`.
[27] LEP Higgs Working Group for Higgs boson searches, OPAL Collaboration, ALEPH Collaboration, DELPHI Collaboration and L3 Collaboration, Phys. Lett. B **565** (2003) 61 [arXiv:hep-ex/0306033]. *Search for neutral Higgs bosons at LEP*, paper submitted to ICHEP04, Beijing, LHWG-NOTE-2004-01, ALEPH-2004-008, DELPHI-2004-042, L3-NOTE-2820, OPAL-TN-744, `http://lephiggs.web.cern.ch/LEPHIGGS/papers/August2004_MSSM/index.html`.
[28] M.S. Alam et al., [CLEO Collaboration], Phys. Rev. Lett. **74** (1995) 2885 as updated in S. Ahmed et al., CLEO CONF 99-10; BELLE Collaboration, BELLE-CONF-0003, contribution to the 30th International conference on High-Energy Physics, Osaka, 2000. See also K. Abe *et al.*, [Belle Collaboration], [arXiv:hep-ex/0107065]; L. Lista [BaBar Collaboration], [arXiv:hep-ex/0110010]; C. Degrassi, P. Gambino and G. F. Giudice, JHEP **0012** (2000) 009 [arXiv:hep-ph/0009337]; M. Carena, D. Garcia, U. Nierste and C. E. Wagner, Phys. Lett. B **499** (2001) 141 [arXiv:hep-ph/0010003]; P. Gambino and M. Misiak, Nucl. Phys. B **611** (2001) 338; D. A. Demir and K. A. Olive, Phys. Rev. D **65** (2002) 034007 [arXiv:hep-ph/0107329]; T. Hurth, arXiv:hep-ph/0212304.
[29] G. W. Bennett *et al.* [Muon g-2 Collaboration], Phys. Rev. Lett. **92** (2004) 161802 [arXiv:hep-ex/0401008]; M. Davier, S. Eidelman, A. Hocker and Z. Zhang, Eur. Phys. J. C **31** (2003) 503 [arXiv:hep-ph/0308213]; K. Hagiwara, A. D. Martin, D. Nomura and T. Teubner, arXiv:hep-ph/0312250; J. F. de Trocóniz and F. J. Ynduráin, arXiv:hep-ph/0402285; K. Melnikov and A. Vainshtein, arXiv:hep-ph/0312226; M. Passera, arXiv:hep-ph/0411168.
[30] J. R. Ellis, K. A. Olive, Y. Santoso and V. C. Spanos, Phys. Lett. B **565** (2003) 176 [arXiv:hep-ph/0303043].
[31] M. Hindmarsh and O. Philipsen, arXiv:hep-ph/0501232.
[32] M. Battaglia *et al.*, Eur. Phys. J. C **22** (2001) 535 [arXiv:hep-ph/0106204].
[33] J. R. Ellis, K. A. Olive, Y. Santoso and V. C. Spanos, arXiv:hep-ph/0408118.
[34] M. Battaglia, A. De Roeck, J. R. Ellis, F. Gianotti, K. A. Olive and L. Pape, Eur. Phys. J. C **33** (2004) 273 [arXiv:hep-ph/0306219].

[35] J. R. Ellis, K. A. Olive, Y. Santoso and V. C. Spanos, Phys. Rev. D **70** (2004) 055005 [arXiv:hep-ph/0405110].
[36] J. L. Feng, S. Su and F. Takayama, Phys. Rev. D **70** (2004) 075019 [arXiv:hep-ph/0404231].
[37] J. R. Ellis, J. L. Feng, A. Ferstl, K. T. Matchev and K. A. Olive, Eur. Phys. J. C **24** (2002) 311 [arXiv:astro-ph/0110225].
[38] J. Ellis, K. A. Olive, Y. Santoso and V. C. Spanos, arXiv:hep-ph/0502001.
[39] J. R. Ellis, S. Heinemeyer, K. A. Olive and G. Weiglein, JHEP **0502** (2005) 013 [arXiv:hep-ph/0411216].
[40] ATLAS Collaboration, `http://atlas.web.cern.ch/Atlas/internal/Welcome.html`.
[41] CMS Collaboration, `http://cmsinfo.cern.ch/Welcome.html/CMScollaboration/CMScollaboration.html`.
[42] LHCb Collaboration, `http://lhcb.web.cern.ch/lhcb/`.

Institute of Physics Publishing Journal of Physics: Conference Series **50** (2006) 22–33
doi:10.1088/1742-6596/50/1/003 5th International Conference on Physics and Astrophysics of Quark Gluon Plasma

Probing Dense Partonic Matter at RHIC

Barbara Jacak for the PHENIX Collaboration
Department of Physics and Astronomy
Stony Brook University,
Stony Brook, NY 11794, USA
E-mail: `Barbara.Jacak@stonybrook.edu`

1. Introduction

Comprehensive analysis of the experimental data collected by the PHENIX experiment at the Relativistic Heavy Ion Collider (RHIC) during the first three years of operation has shown that a new state of matter is created in Au+Au collisions. This matter cannot be described in terms of ordinary color neutral hadrons [1]. Strongly interacting systems under extreme conditions of temperature and density have long been predicted to undergo a phase transition where their constituent quarks and gluons are no longer confined into hadrons.

To characterize the properties of this new matter, it is instructive to review approaches to characterize electromagnetic plasmas. Plasma physicists typically seek to determine the pressure, viscosity and equation of state, along with the thermalization time and extent of plasmas as they become experimentally accessible. It should be noted that not all electromagnetic plasmas are very long lived; the lifetime of plasmas created by high intensity laser light impinging upon a thin target is only a few nanoseconds.

These same quantities are of interest for the quark gluon plasma. They can be determined from measurements of collective behavior among particles emitted from the plasma or by probing with energetic particles transmitted through the plasma. Other plasma properties of interest for both electromagnetic and strong interaction plasmas are radiation rate, collision frequency, conductivity, opacity/transmision probability, and the Debye screening length. These are best determined using probe particles with De Broglie wavelengths short compared to the characteristic wavelength of the plasma. For electromagnetic plasmas, the probes of choice are energetic photons or electrons impinging upon the plasma from the outside. The analogous probes of the quark gluon plasma are high momentum quarks and gluons produced in the very first nucleon-nucleon collisions before the processes driving the system toward thermalization take place. In these proceedings, I will focus primarily on new measurements of such probes.

The PHENIX experiment is optimized to measure electromagnetic probes and high transverse momentum phenomena, along with global variables, multiparticle flows, and soft identified hadron spectra to understand the evolution of the produced matter over all relevant timescales. These diverse criteria required combining a large number of detection subsystems with a high bandwidth trigger and data acquisition system. A description of the PHENIX spectrometers can be found in reference [2].

2. Initial State

In order to search for effects of the matter upon the probes, it is essential to first benchmark the production of those probes in p+p collisions, and determine the effects of binding in the intial

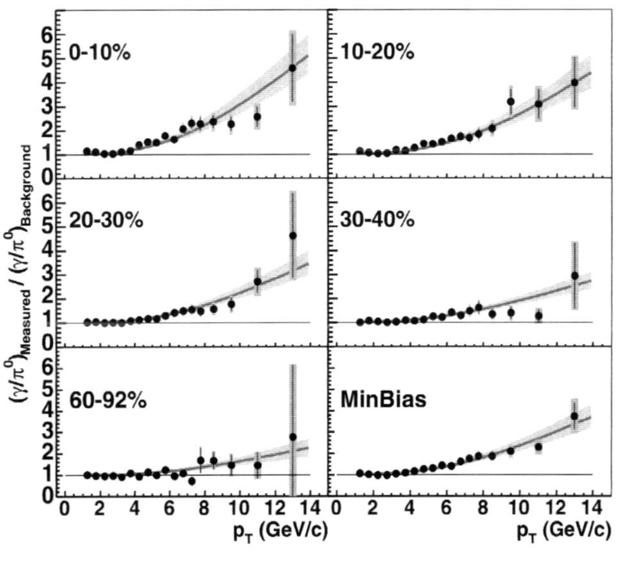

Figure 1. Double ratio of measured to background (γ/π^0) as a function of p_T for minimum bias and five centralities of Au+Au collisions at $\sqrt{s_{NN}} = 200$ GeV [6]. Solid curves are the ratio of pQCD predictions to the background photon yield based on the measured π^0 for each centrality class. The shaded regions around the curves indicate variation of the pQCD calculation for scale changes from $p_T/2$ to 2 p_T plus the $\langle N_{coll} \rangle$ uncertainty.

state nucleus upon the distribution of partons in the colliding nucleons. PHENIX measured the yields as a function of transverse momentum of π^0[3] and direct photons[4] in $\sqrt{s} = 200$ GeV p+p collisions. Both the yields and the p_T distributions are well reproduced by leading order pQCD[3, 4]..

As photons are not expected to interact with the color charges of quarks and gluons in the system, direct photon production in Au+Au collisions should also agree with calculations based upon pQCD. In PHENIX this expectation is tested in $\sqrt{s_{NN}} = 200$ GeV Au+Au collisions by measuring the ratio of all photons to those from π^0 decays, and plotting the double ratio of measured to simulated γ/π^0. As the simulation is performed starting with the measured π^0 spectrum, the π^0 contribution cancels in the double ratio, making it the ratio of inclusive to direct photons. This technique minimizes systematic uncertainties. Any significant deviation of the double ratio above unity indicates a direct photon excess [5]. Figure 1 shows this double ratio as a function of p_T in Au+Au collisions in different centrality bins [6]. An excess is observed at high p_T with a magnitude that increases with increasing centrality of the collision. The measured results are compared to same NLO pQCD predictions which agreed with the PHENIX p+p data [7], scaled by the number of binary nucleon collisions for each centrality class. The binary scaled predictions agree well with the measured direct photons[6]. The increasing ratio with centrality is attributed to the decreasing decay background due to π^0 suppression [8].

The agreement of the direct photon production rate in Au+Au collisions with NLO pQCD calculations indicates that the effects of nucleons being bound inside a nucleus do not preclude the use of perturbative QCD. This is remarkable, given the complicated nuclei involved. PHENIX has also measured the production of charmed mesons in Au+Au and p+p collisions. We find that the total cross section of charmed mesons also scales with the number binary nucleon collisions [9]. As the charm mass is large, $c - \bar{c}$ pairs can only be produced in high

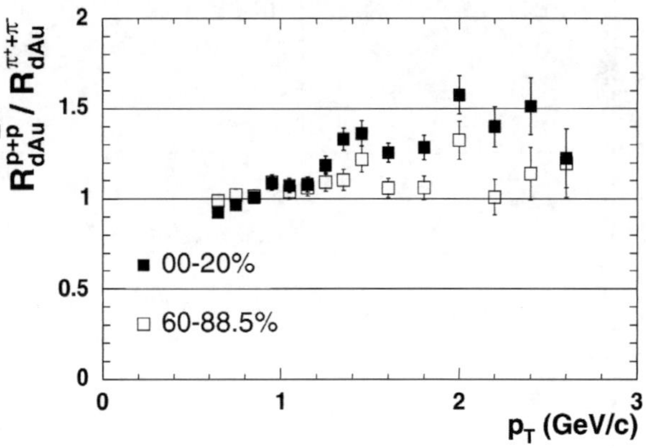

Figure 2. Ratio of the proton over the pion nuclear modification in d+Au collisions for central and peripheral events. Error bars indicate statistical errors only.

momentum transfer collisions, and should therefore be calculable with pQCD. The photon results suggest that binary collision scaling should hold for charm production as well, and PHENIX has established that it does.

Several initial state effects may be expected in collisions involving nuclei. These include nuclear shadowing, saturation of the gluon distribution, and multiple initial state scattering of the incoming partons. The collision scaling of high p_T photons and charm yields indicate that these effects mostly cancel for large q^2 processes at mid-rapidity. However, they should be more important for softer processes and at rapidities nearer those of the beams. A wide range of d+Au collision data was collected to quantify these effects. PHENIX can probe saturation effects via the rapidity dependence of hadron production, and initial state multiple scattering via the dependence of hadrons upon the number of collisions suffered by each deuteron participant. We find that calculations incorporating shadowing and initial state semi-hard scattering can reproduce the data on hadron production at midrapidity reasonably well [10].

However, no model of initial state multiple scatterings can predict the difference in baryon and meson yields in nucleon- or deuteron- nucleus collisions. Baryon yields at moderate p_T are considerably enhanced in central d+Au collisions compared to p+p. This contrasts with the very small enhancement observed for pions, and is illustrated in Figure 2. The figure shows the ratio of R_{dAu} for baryons to that of mesons in the most peripheral (60-88.5%) and most central (0-20%) d+Au collisions. The nuclear modification factor, R_{dAu} is defined as

$$R_{dAu} = \frac{(1/N_{dAu}^{evt}) d^2 N_{dAu}/dy dp_T}{T_{dAu} d^2 \sigma_{inel}^{pp}/dy dp_T},$$

where $T_{dAu} = \langle N_{coll} \rangle / \sigma_{inel}^{pp}$ describes the nuclear geometry and $d^2\sigma_{inel}^{pp}/dydp_T$ for p+p collisions is derived from the measured p+p cross section. $\langle N_{coll} \rangle$ is the average number of inelastic nucleon-nucleon collisions determined from a Glauber simulation. The figure shows that the difference between baryons and mesons reaches 50% in central collisions. More successful explanations of the baryon enhancement invoke hadronization by recombination of quarks from fragmenting jets with those drawn from the nearby nuclear medium [11].

The rapidity dependence of charged hadron production in d+Au collisions is measured by PHENIX using the muon arms, which cover $\eta = 1.4$ - 2.2. We detect stopped muons from hadron

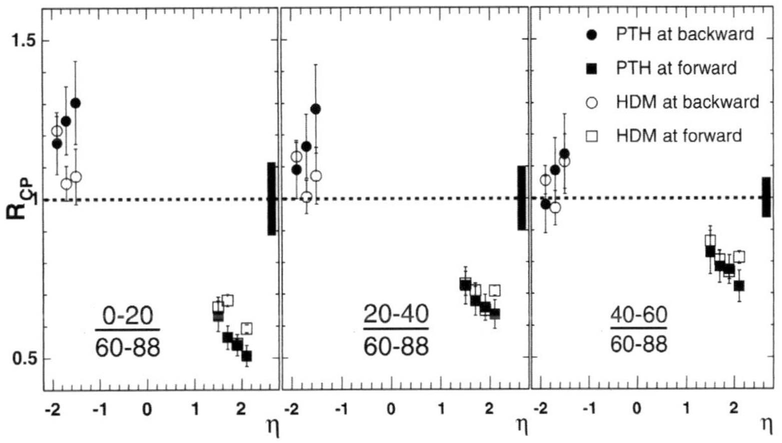

Figure 3. R_{cp} as a function of η for $1.5\langle p_T\rangle$ 4.0 GeV/c for different centrality classes.

decays and also hadrons which punch through the absorber and interact in the muon arm [12]. As comparable data from p+p collisions are as yet unavailable, we calculate a nuclear modification factor by comparing yields in central and peripheral collisions, scaled by the corresponding number of binary nucleon collisions in each centrality bin, $\langle N_{coll}\rangle$:

$$R_{cp} = \frac{\langle\left(\frac{dN}{d\eta dp_T}\right)^{Central}\rangle/\langle N_{coll}^{Central}\rangle}{\langle\left(\frac{dN}{d\eta dp_T}\right)^{Peripheral}\rangle/\langle N_{coll}^{Peripheral}\rangle}$$

R_{cp} integrated in the p_T range from 1.5 to 4.0 GeV/c is shown in Figure 3 as a function of η for the different centrality classes [13]. We observe that R_{cp} shows a suppression at forward rapidity (deuteron going direction) that is largest for the most central events. The opposite trend is observed at backward rapidity (Au going direction), where R_{cp} shows an enhancement that is also largest for the most central events. There is a weak p_T dependence, with slightly smaller R_{cp} at lower p_T. We observe a clear pseudorapidity dependence at forward rapidity with R_{cp} dropping further at larger η values; our results are consistent with the BRAHMS data [14]. Within the uncertainties, we are unable to discern any pseudorapidity dependence at the backward rapidity. Forward suppression is qualitatively consistent with several theories including shadowing or saturation effects in initial state multiple scattering, but also recombination [15, 16, 17]. The enhancement at backward rapidity has not yet been explained.

3. Thermalization

Turning now from the initial state, I discuss observables that tell about the approach to thermalization in Au+Au collisions. PHENIX measures many species of hadrons, including the ϕ meson via its decays to K^+K^- and e^+e^-. We find that the slope of the ϕ p_T distributions reconstructed in the two decay channels are consistent with each other, and are also consistent with a blast wave fit to π, K, and p spectra [18]. This observation supports the picture of a common expanding source for all low and intermediate momentum hadrons at midrapidity.

Good tests of the system's approach toward thermalization are the magnitude and species dependence of the elliptic flow. Elliptic flow is measured by Fourier analysis of the momentum distribution of emitted particles as a function of the azimuthal angle with respect to the plane of the Au+Au collision. The strength of the flow is quantified by the magnitude of the second

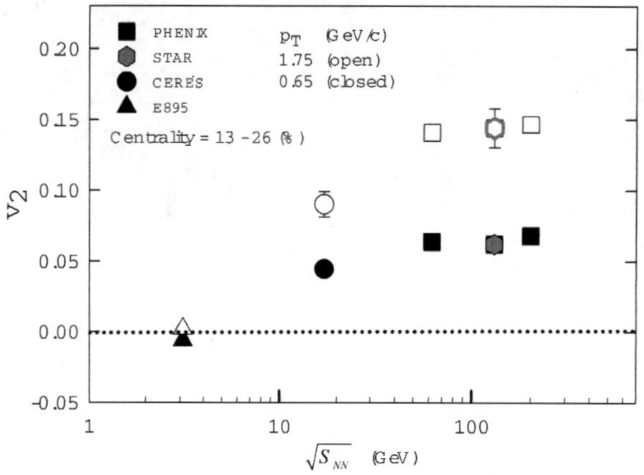

Figure 4. Differential v_2 vs. $\sqrt{s_{NN}}$ for charged hadrons in nucleus-nucleus collisions. Results are shows for a centrality class of 13-26% and p_T of 1.75 GeV/c (open symbols) and 0.65 GeV/c (closed symbols).

harmonic coefficient of the Fourier expansion, generally known as the v_2 parameter. PHENIX has measured v_2 for charged pions, kaons, protons, antiprotons, neutral pions, inclusive photons and electrons from the decay of charmed mesons. This is measured using several methods, including two particle correlations at midrapidity, two-particle cumulants, and the distribution of particles in the central arm with respect to the reaction plane determined by the beam-beam counters (BBC) at rapidity η = 3.0 - 3.9. Most of the measurements for identified particles use the PHENIX central arm spectrometers and the BBC reaction plane. We find that the different species exhibit approximately the same v_2 per constituent quark. v_2 rises as a function of p_T up to 1.5-2 GeV/c per constituent quark, then falls again at higher p_T. In $\sqrt{s_{NN}} = 62.4$ GeV Au+Au collisions, the observed v_2 per quark is quite similar to that in 200 GeV Au+Au collisions.

This is illustrated in Figure 4, which shows v_2 for inclusive charged hadrons at two p_T values (0.65 and 1.75 GeV/c), as a function of $\sqrt{s}$ for heavy ion collisions from AGS through RHIC energies [19]. The data are compared for relatively central collisions representing the upper 13-26% of the total cross section. For both p_T cuts, the magnitude of v_2 shows a significant increase with collision energy ($\approx$ 50 % increase from SPS to RHIC) up to $\sqrt{s_{NN}} = 62.4$ GeV. Thereafter, v_2 appears to saturate for larger beam energies. Given the fact that the energy density is estimated to increase by approximately 30% over the range $\sqrt{s_{NN}}$ = 62.4 - 200 GeV, this apparent saturation of v_2 may be indicative of the role of a rather soft equation of state. Such a softening could result from the production of mixed phase within this energy range.

Of course, for such a discussion to be applicable, the system must be in local thermal equilibrium. For thermalization to occur, the particles must interact and/or radiate. If such processes occur with sufficient frequency, the system should be describable by locally equilibrated fluid elements, with a velocity profile following $\beta_{\parallel}^{Fluid} = z/t$. This has been tested by calculations with hydrodynamical models of Au+Au collisions with initial conditions fixed by the collision geometry and energy density fixed by total particle and energy production; for more detail see discussion and references in [1]. In a hydrodynamic picture, the source of elliptic flow is the spatial anisotropy of the energy density in the transverse plane at the time hydrodynamics becomes valid (i.e. at the time that local thermal equilibrium is reached). If local equilibration

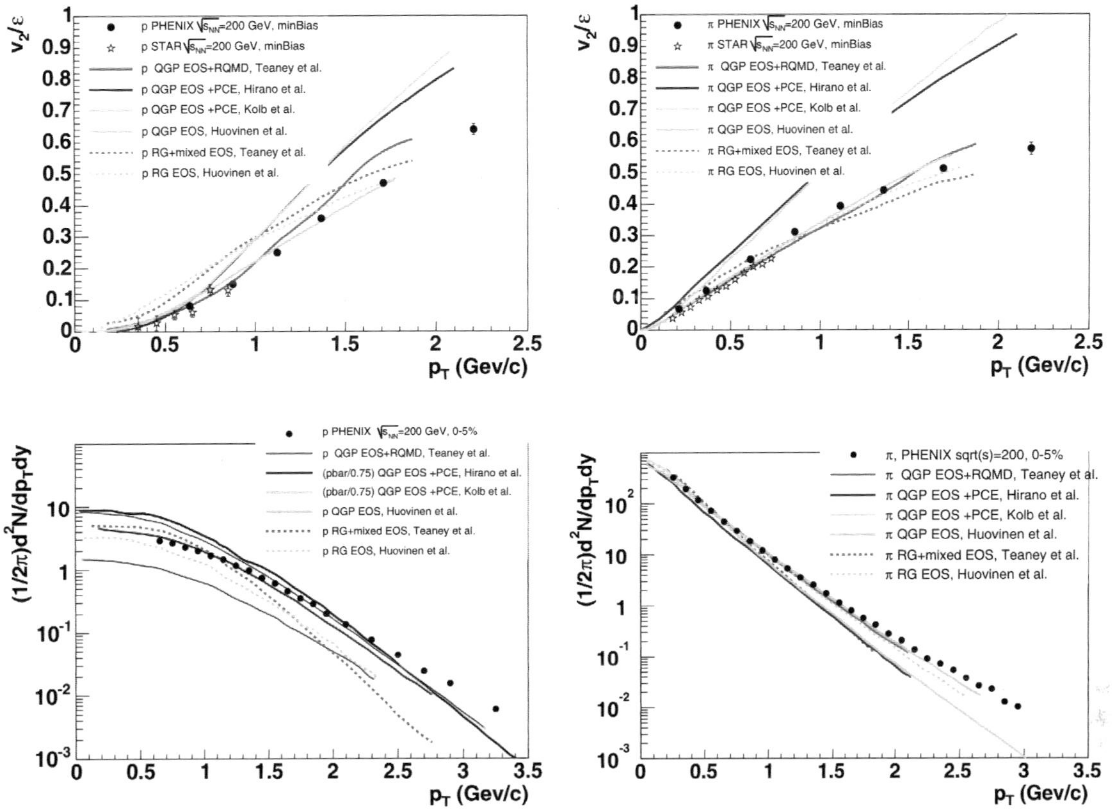

Figure 5. Top panels: proton and pion $1/\epsilon v_2(p_T)$ vs. p_T for minimum bias collisions at RHIC, compared with hydrodynamics calculations. Bottom panels: proton and pion p_T spectra for the most central (0-5%) collisions at RHIC compared with the same hydrodynamics calculations. See text and [1] for references to the data and hydrodynamics calculations.

and the onset of hydrodynamical behavior is delayed because interactions between the initially produced particles are weak at first, then the spatial anisotroy giving rise to elliptic flow is reduced. Consequently, the magnitude of the v_2 parameter is sensitive to the thermalization time. The thermalization time may be inferred from the hydrodynamics calculations constrained with data on the p_T and species dependence of elliptic flow and particle spectra.

Figure 5 shows the comparison of a number of hydrodynamics models [20, 21, 22, 23] with the data on pion and proton v_2 as a function of p_T for minimum bias Au+Au collisions, and pion and proton p_T spectra in central collisions [1]. The calculations including a phase transition from QGP to a hadronic phase are shown with solid lines, while calculations with no pure QGP phase are drawn with dashed lines. The four calculations that include a QGP phase all assume an ideal gas EOS for the QGP phase, a resonance gas for the hadronic phase, and connect the two using a first-order phase transition and a Maxwell construction. The calculations that do not include a QGP phase either have a hadronic plus mixed phase with the latent heat of the transition forced to infinity (Teaney)[20], or they use only a hadronic resonance gas (i.e. no mixed or QGP phases in one of Huovinen's calculations) [21]. The calculations also differ in their treatment of the final hadronic phase and freezeout. Hirano's [22] is the only 3D hydrodynamics calculation; this and

Kolb's [23] both allow for partial chemical equilibrium by chemically freezing out earlier than the kinetic freezeout. This has been done in order to reproduce the large proton yield measured at RHIC. In contrast, Huovinen[21] maintains full chemical equilibrium throughout the hadronic phase. Teaney[20] uses a hybrid model that couples the hadronic phase to RQMD to allow hadrons to freeze out according to their scattering cross sections; incorporating this step allows for chemical equilibrium to be broken in the hadronic phase. All four models have assumed ideal hydrodynamics, i.e. zero viscosity and zero mean free path.

From the top panels of Figure 5 it is clear that the four calculations that include a phase transition from the QGP phase to a hadronic phase (solid lines) reproduce the low p_T proton data better than the two hydro calculations without the QGP phase (dashed lines). The presence of a first order phase transition softens the equation of state, reducing the elliptic flow. At higher p_T there is quite some variation between the models. Part of this is due to the modeling of the final hadronic stage. It is notable that Kolb's and Hirano's calculations significantly overpredict v_2, and they both have similar partial chemical equilibrium assumptions in the late hadronic stage. Comparing to the transverse momentum spectra, we see that all the models reproduce the pion spectra below 1 GeV/c p_T. Calculations including a QGP phase do a better job reproducing the proton spectra, presumably because of increased transverse flow from the stronger early pressure gradients. The calculations with partial chemical equilibrium during the hadronic phase overpredict the proton spectra at low p_T.

All the models qualitatively reproduce the trends observed in the data, but significant sensitivity to the assumptions in the models is demonstrated. Thus is it important to reduce the model uncertainty in the final state to extract quantitative information on the equation of state during the reaction, including the possible softening of the equation of state due to the presence of a mixed phase. Nevertheless, we can use the calculations to extract limits for the thermalization time[1]. All the hydrodynamical models require quite short thermalization times, in the range of 0.6-1.0 fm/c, to reproduce the magnitude of the observed elliptic flow.

The general success of hydrodynamics models to explain the data provides strong evidence for local thermalization of the system, or at least for equipartition of the momenta, in the early stage of the collision. It is natural to ask whether the heavy quarks flow along with the light ones, or whether their large masses increase the time required for them to thermalize so much that they cannot receive the same velocity boost from the collective motion as the light quarks.

Electrons are a useful tool for the study of heavy quarks such as charm and bottom. PHENIX has measured single electron p_T spectra in Au+Au collisions at $\sqrt{s_{NN}} = 130$ GeV[24] and 200 GeV[9]. The results are consistent with semileptonic charm decays in addition to the decays of light mesons and photon conversions. The spectrum of electrons of non-photonic origin can be measured by adding a converter of well-calibrated thickness to determine the photonic electron rate and subtracting these from the inclusive electron spectrum. Alternatively, the measured π^0 and η spectra can be used to calculate the photonic sources of electrons, including conversions of photons from light meson decays. The calculated background spectrum is then subtracted from the measured electrons. The former technique was used in [9], and the latter to measure charm production in p+p collisions. We use the same approach to determine v_2 for open charm by measuring the distribution of electrons with respect to the reaction plane. Though electrons originating from decays of D mesons have a significant angular deviation from the original D meson directions, the extracted v_2 value remains well correlated with the v_2 of the D meson[25]. Consequently, PHENIX uses electrons and positrons to determine v_2 for heavy quarks. We subtract v_2 of electrons from photon conversions and Dalitz decays of light neutral mesons from the inclusive electron v_2[26].

Figure 6 shows the heavy flavor v_2 in minimum bias $\sqrt{s_{NN}} = 200$ GeV Au+Au collisions. The statistical error bars are propagated from the statistical uncertainties on the inclusive electron v_2, while the bands show the 1σ systematic uncertainty of the heavy flavor v_2. The result allows

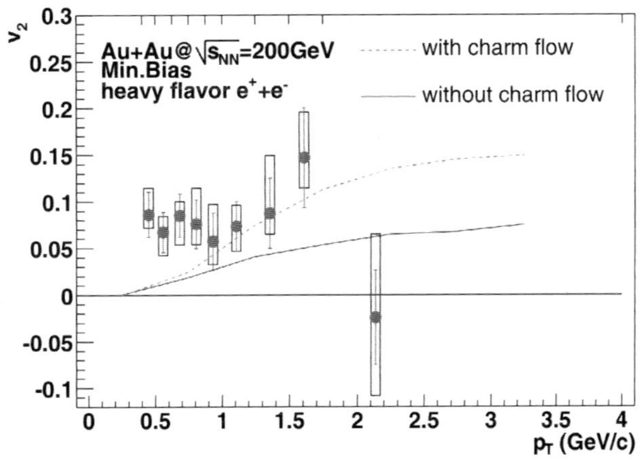

Figure 6. Elliptic flow (v_2) for electrons and positrons from heavy flavor decays, as a function of the electron or positron p_T, in minimum bias $sqrts_{NN}$ = 200 GeV Au+Au collisions. The lines correspond to calculations with two different charm flow scenarios[25]. The solid line has no rescattering of the initially produced charm quarks (i.e. no charm flow), while the dashed line reflects the effect of complete thermalization (maximum charm flow).

calculation of the confidence level for a non-zero v_2, yielding 90% confifdence that the heavy flavor electron v_2 is non-zero in the p_T range from 1 - 1.75 GeV/c. The v_2 of electrons from decays of D mesons have been predicted assuming that the D mesons are formed by charm quark coalescence with thermal light quarks. The solid line on Figure 6 assumes that the charm and anticharm quarks experience no reinteractions after they are formed in initial state hard processes. The second scenario, shown by the dashed line, assumes complete thermalization of the heavy quarks with the transverse flow of the bulk matter. Though the large statistical errors on this data sample do not allow exclusion of either scenario, the points do suggest that heavy quark flow may indeed be taking place.

4. Probes of the Partonic State

Given the various pieces of evidence pointing toward early local thermalization of the system, it is of great interest to use the short wavelength probes discussed in the Introduction to measure the properties of the hot, dense matter $\approx$ 1 fm/c after the Au nuclei traverse one another. Of paramount interest is how the the probes couple to the medium. It is already well established that hard scattered partons traversing the medium experience considerable energy loss[27]. But the exact nature of the interaction of the parton with the medium is not well studied experimentally. This question is particularly interesting for partons of moderate p_T, between 5 and 10 GeV/c, which may begin to hadronize in or near the medium.

Having seen the tantalizing result that heavy quarks may indeed flow along with the bulk of light quarks, it is natural to wonder whether heavy quarks lose energy in the medium as the light quarks do. It has been predicted that the energy loss of charm quarks should be less than that of light quarks, as the large charm quark mass mass decreases the phase space available for gluon radiation (i.e. "dead cone" effect)[28]. The dead cone should be even more significant for bottom quarks. More detailed calculations include also simulation of the effects of the charm p_T spectral shape and contributions from B meson decays; these predict that the nuclear modification factor for electrons from charm decays should be 0.6-0.8 for electrons with

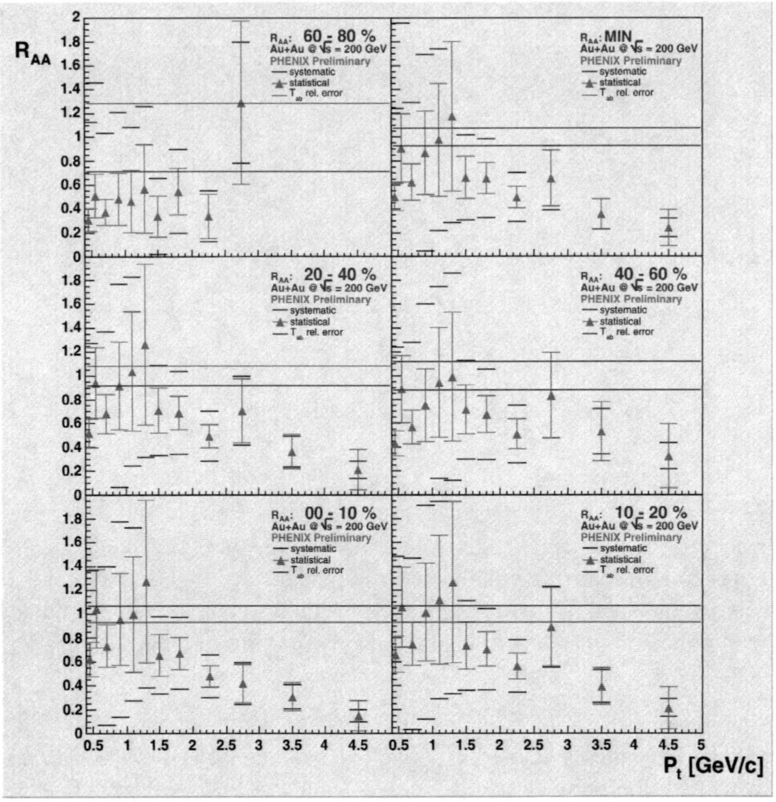

Figure 7. Nuclear modification factor, R_{AA}, for electrons and positrons from heavy quark decays in $/sqrts_{NN} = 200$ GeV Au+Au collisions. The error bars show statistical errors, while the bands around each point indicate systematic uncertainties. The bands centered about R_{AA} = 1.0 indicate the systematic uncertainty in the number of binary nucleon collisions in each centrality class.

$p_T \approx 2.5$ GeV/c [29]. This is a significant suppression, though less than that observed for light quarks[27, 8]. Armesto and co-workers take into account differences between the energy loss of quarks and gluons traversing the medium, and predict even larger suppression of electrons from heavy quarks[30]. Moore and Teaney point out that if the charm quarks demonstrate flow along with the bulk of the medium, this is evidence for thermalization of charm, and then the medium modifications of the charm spectrum should be substantial[31].

Figure 7 shows the nuclear modification factor, R_{AA} for electrons and positrons from heavy quark decays in $/sqrts_{NN} = 200$ GeV Au+Au collisions, as a function of the electron p_T. The figure shows minimum bias collisions in addition to 5 centrality classes. Electrons from heavy quark decays are measured by subtracting the p_T spectrum of electrons from photonic sources, calculated using the measured π^0 spectra, from the inclusive electron spectrum. The same subtraction technique is used for Au+Au and reference p+p spectra, but of course the input π^0 spectra are different. In the most peripheral collisions, statistical uncertainties limit the accessible p_T range, but where there are data, all the points are consistent with $R_{AA} = 1$, i.e. no suppression. In more central collisions, R_{AA} falls well below 1.0 for electron $p_T \geq 2$ GeV/c, providing clear evidence for charm energy loss. At the highest p_T of this measurement, the electron R_{AA} becomes nearly as small as that for π^0. This is remarkable, as electrons above

3.5 or 4.0 GeV/c p_T are expected to also include contributions from B meson decays, and B mesons should experience less energy loss than D mesons. Though the data are consistent with predictions from Armesto et al.[30] for medium densities at the extreme high end of those allowed by the observed light quark energy loss, the lack of observable dilution from B meson decays is intriguing.

Hard scattered partons are formed in the initial nucleon collisions, and traverse the hot, dense medium. These partons experience significant energy loss via medium-induced gluon radiation. The next question is - where does the radiated energy go? Though radiated gluons are nearly collinear with the parent parton, they are rather soft and may interact further with the gluon field in the medium. Though it is often assumed that fragmentation takes place outside the medium, the formation time for 2-4 GeV/c p_T hadrons is not very long, particularly for baryons. Thus fragmentation may begin in or near the medium and involve not only the quarks and antiquarks from gluons radiated as fragmentation begins, but also comoving gluons from medium-induced radiation, and indeed recombination of jet partons with partons from the medium itself[11].

PHENIX probes these processes using correlations of high transverse momentum particles, in the p_T range from 1 - 4 GeV/c. We select Au+Au collisions containing hard scattering events by requiring detection of a hadron with $2.5 \leq p_T \leq 4$ GeV/c. We then construct correlation functions between these trigger particles and associated particles with $1.0 \leq p_T \leq 2.5$ GeV/c, as a function of their azimuthal angle difference[32, 33]. We correct for the non-uniform PHENIX azimuthal acceptance by constructing an area normalized correlation function utilizing pairs from mixed events. To extract the yield of jet-induced pairs, we analyze the correlation function assuming that each hadron can be attributed either to a jet fragmentation source or to the underlying event. The underlying event has an azimuthal correlation arising from the fact that the single particle distributions respect the reaction plane of the event. This correlation is removed by modulating the background pair distribution with the inclusive v_2 value measured for the trigger and associated particle p_T range. Thus the correlation function is decomposed via

$$C(\Delta\phi) = b_0(1 + 2(v_2^T v_2^A)cos(2\Delta\phi)) + J(\Delta\phi)$$

The average level of the background, b_0 is fixed by assuming that the jet fragmentation yield of particle pairs is zero for at least one value of $\Delta\phi$. We refer to this as the ZYAM (zero yield at minimum) assumption. We have verified the validity of this assumption by independently estimating the b_0 values using the T*A combinatorial pair rate. Once the background level is determined, we extract conditional yields for particle pairs from jets.

The conditional yields of jet partners per trigger particle are shown in Figure 8[33]. For the most peripheral event sample, the jet associated yield distribution has the same shape as jet pairs in p+p collisions: a well-defined near side peak around $\Delta\phi = 0$ and a somewhat wider away side peak around $\Delta\phi = \pi$. For more central event samples, the shape of the near-side peak is essentially unchanged while the associated yield in the near-side peak increases, indicating some change in the fragmentation process. A much more dramatic change is visible in the away side peak, which is considerably broader in all the centrality classes more central than 60%. In mid-central and central collisons there is a local minimum at $\Delta\phi = \pi$. Though the existence of these local minima *perse* is not significant once we take the systematic errors on $\langle v_2^T v_2^A \rangle$ into account, it is clear that the away side peak in more central collisions has a very different shape than in peripheral collisions.

Convoluting the jet fragments' angles with respect to their parent partons and the acoplanarity between the two partons would yield a Gaussian-like shape in $\Delta\phi$, possibly broadened through jet quenching. The observed shapes in the away side peaks cannot result from such a convolution. The away side peaks are suggestive of recent theoretical predictions of dense medium effects on fragment distributions. These include combination of jet partons with medium partons accelerated by a density wave in the shocked medium[34, 35, 36], and jet

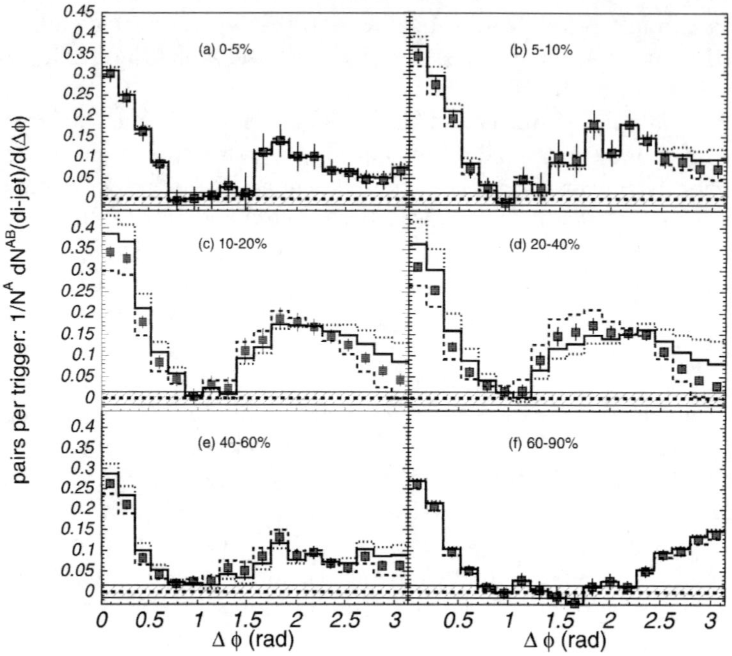

Figure 8. Jet fragment pair relative azimuthal angle distributions for different centralities in $\sqrt{s_{NN}} = 200$ GeV Au+Au collisions[33]. The yields are normalized per trigger particle. The shaded bands indicate the systematic error associated with the determination of the relative angle where the jet pair yield is zero. The dashed(solid) lines are the distributions that would result from increasing (decreasing) $\langle v_2^T v_2^A \rangle$ by one unit of the systematic error; the dotted curve would result from decreasing by two units.

asymmetries caused by the hydrodynamic flow of the underlying event[37]. The broadening of the away side jet implies that integration of the away side peak in a narrow angular range around $\Delta\phi = \pi$ yields fewer associated partners in central collisions than in peripheral Au+Au or in d+Au collisions. However, integrating over the entire broadened peak recovers the jet partners in the range 1.0 GeV/c $\leq p_T^A \leq$ 2.5 GeV/c used in this analysis.

5. Summary

The PHENIX collaboration has measured a large number of observables in Au+Au, d+Au and p+p collisions at RHIC. We have shown that direct photons and hard processes are calculable with perturbative QCD, yielding a calibrated source of short wavelength probes of the medium formed in nuclear collisions. Anomalous behavior of baryon yields is seen in d+Au collisions, suggesting that baryon formation is already subject to the influence of the nearby nucleus in that case.

The elliptic flow trends support the picture of rapid thermalization in Au+Au collisions. Furthermore, there are first indications that heavy quarks may participate in the collective flow with the light quarks. We have seen that heavy quarks lose energy in the hot dense medium, as may be expected if they approach local thermal equilibrium. Jet fragmentation is modified in central and semi-central Au+Au collisions. The data suggest that the lost energy may excite the medium and modify the formation of jet hadrons. The formation of moderate p_T baryons (from approximately 2-5 GeV/c) is expected to take place in, or near, the medium in Au+Au and d+Au collisions. Indeed the data show enhanced baryon production probability in both cases. This suggests that recombination of partons from the fragmenting jet and from the medium play an important role in hadronization.

Quantitative determination of the opacity of and collision frequency in the medium, along with its color Debye screening length awaits completion of the analysis of the high statistics run4 data. PHENIX has a billion events currently being analyzed.

6. References

[1] K. Adcox, *et al.* Nucl. Phys. **A757**, 184 (2005).
[2] K. Adcox *et al.*, Nucl. Instrum. Methods **A499**, 469 (2003) and references therein.
[3] S.S. Adler *et al.*, Phys. Rev. Lett. **91**, 241803 (2003).
[4] S.S. Adler *et al.*, Phys. Rev. D **71**, 071102R (2005).
[5] See talk of S. Bathe, these proceedings.
[6] S.S. Adler *et al.*, Phys. Rev. Lett. **94**, 232301 (2005).
[7] L.E. Gordon and W. Vogelsang, Phys. Rev. D **48**, 3136 (1993).
[8] S.S. Adler *et al.*, Phys. Rev. Lett. **91**, 072301 (2003).
[9] S.S. Adler *et al.*, Phys. Rev. Lett. **94**, 082301 (2005).
[10] A. Accardi and M. Gyulassy, Phys. Lett. **B586**, 244 (2004).
[11] R. Hwa and C.B. Yang, Phys. Rev. C **70**, 037901 (2004).
[12] See talk of A. Purwar, these proceedings.
[13] S. S. Adler *et al.*, Phys. Rev. Lett. **94**, 082302 (2005).
[14] I. Arsene *et al.*, Phys. Lett. **B595**, 209 (2004).
[15] D. Kharzeev, Y.V. Kovchegov and K. Tuchin, arXiv:hep-ph/0405045.
[16] R. Vogt, arXiv:hep-ph/0405060.
[17] R. Hwa, C.B. Yang and R.J. Fries, arXiv:nucl-th/0410111.
[18] S. S. Adler *et al.*. arXiv:nucl-ex/0410012 (accepted in Phys. Rev. C).
[19] S. S. Adler *et al.*. Phys. Rev. Lett. **94**, 232302 (2005).
[20] D. Teaney, J. Lauret, E.V. Shuryak, arXiv:nucl-th/0110037.
[21] P. Huovinen, P.F. Kolb, U.W. Heinz, P.V. Ruuskanen, and S.A. Voloshin, Phys. Lett. **B503**, 58 (2001).
[22] T. Hirano and K. Tsuda, Phys. Rev. C **66**, 054905 (2002).
[23] P. Kolb and U. Heinz, arXiv:nucl-th/0305084 and "Quark Gluon Plasma 3", Editors R.C. Hwa and X.N. Wang, World Scientific, Singapore.
[24] K. Adcox *et al.*. Phys. Rev. Lett. **88**, 082301 (2002).
[25] V. Greco, C.M. Ko and R. Rapp, Phys. Lett. **B595**, 202 (2004).
[26] S. S. Adler *et al.*. arXiv:nucl-ex/0503003.
[27] K. Adcox *et al.*. Phys. Rev. Lett. **88**, 022301 (2002).
[28] D. Kharzeev and Y. Dokshitzer, Phys. Lett. **B519**, 199 (2001).
[29] M. Djordjevic, M. Gyulassy and S. Wicks, arXiv:hep-ph/0410372.
[30] N. Armesto, S. Dainese, C. Salgado, and U. Wiedemann, arXiv: hep-ph/0501225.
[31] Guy Moore and Derek Teaney, arXiv:hep-ph/0412346.
[32] See talk of N. Ajitanand, these proceedings.
[33] S.S. Adler, et al. *et al.*. arXiv:nucl-ex/0507004.
[34] J. Casalderrey-Solana, E.V. Shuryak and D. Teaney, arXiv:hep-ph/0411315.
[35] H. Stoecker, Nucl. Phys. **A750**, 121 (2005).
[36] J. Ruppert and B. Muller, arXiv:hep-ph/0503158.
[37] N. Armesto, C. Salgado and U. Wiedemann, arXiv:hep-ph/0411341.

Institute of Physics Publishing Journal of Physics: Conference Series **50** (2006) 34–41
doi:10.1088/1742-6596/50/1/004 5th International Conference on Physics and Astrophysics of Quark Gluon Plasma

PHOBOS Overview

David J. Hofman[6] for the PHOBOS collaboration:
B.B.Back[1], M.D.Baker[2], M.Ballintijn[4], D.S.Barton[2], R.R.Betts[6], A.A.Bickley[7], R.Bindel[7], A.Budzanowski[3], W.Busza[4], A.Carroll[2], Z.Chai[2], M.P.Decowski[4], E.García[6], T.Gburek[3], N.George[2], K.Gulbrandsen[4], S.Gushue[2], C.Halliwell[6], J.Hamblen[8], M.Hauer[2], G.A.Heintzelman[2], C.Henderson[4], R.S.Hollis[6], R.Hołyński[3], B.Holzman[2], A.Iordanova[6], E.Johnson[8], J.L.Kane[4], N.Khan[8], P.Kulinich[4], C.M.Kuo[5], W.T.Lin[5], S.Manly[8], A.C.Mignerey[7], R.Nouicer[2,6], A.Olszewski[3], R.Pak[2], I.C.Park[8], C.Reed[4], C.Roland[4], G.Roland[4], J.Sagerer[6], H.Seals[2], I.Sedykh[2], C.E.Smith[6], M.A.Stankiewicz[2], P.Steinberg[2], G.S.F.Stephans[4], A.Sukhanov[2], M.B.Tonjes[7], A.Trzupek[3], C.Vale[4], G.J.van Nieuwenhuizen[4], S.S.Vaurynovich[4], R.Verdier[4], G.I.Veres[4], E.Wenger[4], F.L.H.Wolfs[8], B.Wosiek[3], K.Woźniak[3], B.Wysłouch[4]

[1] Argonne National Laboratory, Argonne, IL 60439-4843, USA
[2] Brookhaven National Laboratory, Upton, NY 11973-5000, USA
[3] Institute of Nuclear Physics PAN, Kraków, Poland
[4] Massachusetts Institute of Technology, Cambridge, MA 02139-4307, USA
[5] National Central University, Chung-Li, Taiwan
[6] University of Illinois at Chicago, Chicago, IL 60607-7059, USA
[7] University of Maryland, College Park, MD 20742, USA
[8] University of Rochester, Rochester, NY 14627, USA

E-mail: `hofman@uic.edu`

Abstract. A brief overview of the current results and conclusions from the PHOBOS experiment at the Relativistic Heavy Ion Collider (RHIC) is given. No evidence is found for non-monotonic behavior of observables measured by PHOBOS in the RHIC energy region. Convincing evidence is found that we have created a state of matter with high energy-density, that is nearly net-baryon free and is strongly interacting. The data are found to exhibit "simple" scaling behaviors, which include extended longitudinal scaling and scaling with the number of participating nucleons. The Au+Au collision charged particle data also exhibit a remarkable factorization of collision energy and geometry.

1. Introduction

This report focuses on results from the PHOBOS experiment at the Relativistic Heavy-Ion Collider (RHIC). A primary, and unique, strength of the PHOBOS detector is its ability to measure the total charged particle multiplicity due to the $\sim 4\pi$ acceptance of the detector. The detectors that allow the measurement of charged particle multiplicity ($dN_{ch}/d\eta$) over the broad pseudorapidity range ($|\eta| < 5.4$) are the midrapidity Vertex detector ($|\eta| < 1$), the centrally

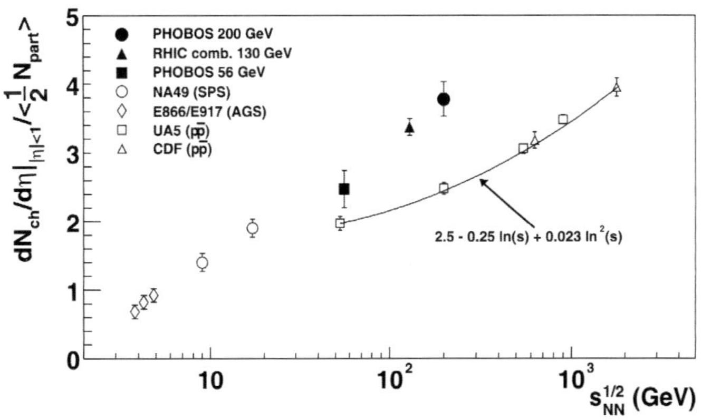

Figure 1. An early PHOBOS result showing the collision energy dependence of the midrapidity pseudorapidity density of charged particle production per participant pair in central heavy-ion collisions at RHIC (solid points) compared to both lower energy heavy-ion data from the SPS and AGS, as well as results from $\bar{\mathrm{p}}$+p collisions [2].

located Octagon barrel ($|\eta| < 3.2$), and a series of six Ring counters ($3.1 < |\eta| < 5.4$), all of which surround a thin beryllium beam-pipe. The physics capabilities of PHOBOS are expanded with a midrapidity 16-layer Spectrometer (located in an ≈ 2 Tesla double-dipole magnet), two TOF walls (for extended particle ID), along with a suite of additional detectors for triggering and centrality determination. The majority of the PHOBOS detector is based on the technology of silicon pad sensors, see Ref. [1] for details.

2. Evolution of Global Observables with Energy and Centrality

Already in 2002, with the availability of $\sqrt{s_{NN}}$ = 56, 130 and 200 GeV Au+Au collision data at RHIC, PHOBOS results demonstrated a lack of any dramatic non-logarithmic energy dependence in the midrapidity charged particle pseudorapidity density, as seen in Fig. 1. It is important to note that these measured total charged particle multiplicities are dominated by the emission of low p_T (≤ 1.5 GeV/c) particles. This early result for the bulk charged particle multiplicity has been strengthened by additional RHIC data at 19.6 GeV, which confirmed the measurement from the highest energy SPS heavy-ion data. The PHOBOS results for Au+Au collisions at 62.4 GeV will be forthcoming. The centrality dependence of these same midrapidity yields also exhibited no dramatic variations from the lowest to the highest RHIC collision energy, as seen in Fig. 10.

The energy and centrality dependence of elliptic flow (v_2) for bulk charged particle production is given in Figs. 2 and 3. The most striking feature of these data is the triangular shape for $v_2(\eta)$ with a relatively sharp peak at midrapidity, which is pointedly non-boost-invariant in character. In addition, a smooth evolution of the same basic $v_2(\eta)$ shape is seen both as a function of energy (Fig. 2) and centrality (Fig. 3). The lack of centrality dependence with shape in η is illustrated more clearly by the inset to Fig. 3, which shows that the ratio of peripheral data to central data is constant, within errors, in the pseudorapidity range of $-2 < \eta < 2$.

Expanding the study of charged particle production to include the transverse momentum dependence, PHOBOS finds a smooth evolution in the p_T dependence of charged particle production, as illustrated through measurement of the nuclear modification factor, R_{AA} (see Ref. [5] and Fig. 12). This lack of dramatic variation in the shape of R_{AA}, with both centrality

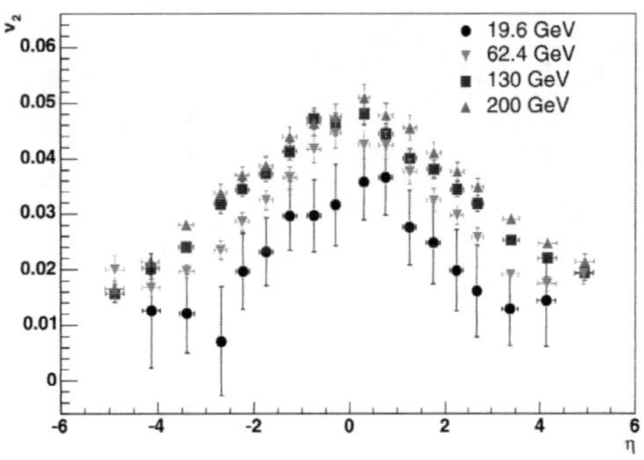

Figure 2. Pseudorapidity dependence of the elliptic flow parameter, v_2, for central Au+Au collisions at four different collision energies spanning the coverage of RHIC [3]. The data has been averaged over the most central 40% of the cross section.

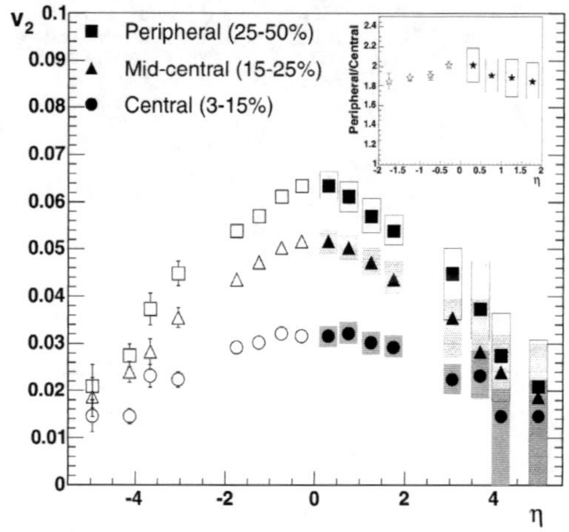

Figure 3. Centrality dependence of the elliptic flow parameter, v_2, for Au+Au collisions at $\sqrt{s_{NN}}$ = 200 GeV [4]. The inset shows the ratio of peripheral to central results.

and energy, persists past $p_T = 4$ GeV, currently the highest transverse momentum published by PHOBOS for Au+Au collisions.

We close this section with the mention of two additional PHOBOS results. First, a monotonic behavior of the HBT radii is seen in central Au+Au collisions at both 62.4 and 200 GeV for R_o, R_s and R_l, measured as a function of the pair transverse momentum, k_T [6]. In fact, the values of the HBT radii at equivalent k_T are identical at the two energies, within errors. Second, a smooth evolution of identified antiparticle to particle ratios near midrapidity is observed as a function of energy for central Au+Au collisions [7]. No evidence for any sudden change in baryon density is found. Instead there is a steady rise of the ratios towards unity with increasing collision energy.

Thus the picture that emerges from the global observables measured by PHOBOS in Au+Au collisions is one of remarkable regularity, even over a factor of ten change in collision energy.

3. Properties of the Created "State of Matter"

One of the first questions to be asked in studies of relativistic heavy-ions is; what is the energy-density of the created matter? A primary reason this value is of great interest stems from the remarkable theoretical study of high temperature QCD afforded by the techniques of lattice gauge theory. These theoretical calculations indicate that, for low baryon density, there is a phase transition in QCD matter at a critical temperature that corresponds to an energy density of ~ 1 GeV/fm^3 [9]. The antiparticle to particle ratios data tell us we have a low baryon density, with $\mu_B \approx 27$ MeV for central collisions at the highest RHIC energy [7]. The measurements of the charged particle production near midrapidity, coupled with knowledge of the average $\langle p_T \rangle$ of those particles, affords an estimate of the energy density created in the collision. The largest uncertainty is the equilibration time, $\tau_{\rm eq}$. Conservative estimates put $\tau_{\rm eq} \leq 2$ fm/c which leads to a lower limit for the energy density at full RHIC energy of ≥ 3 GeV/fm^3 [16]. This energy density is ~ 6 times the energy density inside nucleons, ~ 20 times that inside nuclei and quite comfortably above the QCD phase threshold of ~ 1 GeV/fm^3 predicted by the lattice calculations. It seems quite plausible that we have created a very interesting and new "state of

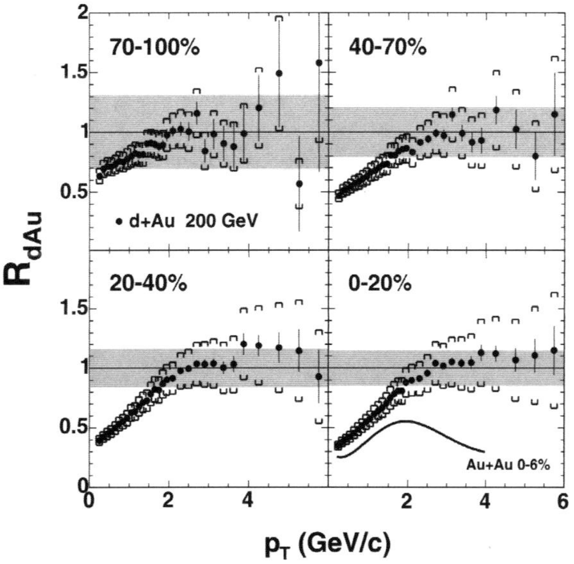

Figure 4. The centrality dependence of the nuclear modification factor, R_{dAu}, for d+Au collisions at $\sqrt{s_{NN}}$ = 200 GeV. For the most central 0-20% of the d+Au collisions it is also compared to R_{AA} for central (0-6%) Au+Au collisions, where it is clearly evident that at higher transverse momentum (p_T) the heavy-ion data is strongly suppressed relative to both the binary-scaling limit of 1, as well as to the d+Au data at the equivalent pseudorapidity range of $\eta = 0.2$ - 1.4 (in the direction of the deuteron). [8].

matter".

The properties of this matter appear to be rather strongly interacting in nature. In the PHOBOS data, this conclusion rests on the three pillars of; high p_T suppression (Fig. 4), the magnitude of elliptic flow (Fig. 5) and the measured low p_T yields of identified particles (Fig. 6). First, the suppression of high p_T particles in central Au+Au collisions, and the corresponding lack of suppression at midrapidity for d+Au, can be naturally understood as the result of a significant level of interaction between the produced high momentum particles and the created state of matter. Second, the magnitude of the elliptic flow parameter, v_2, is important because if there were no interactions, then there would be no mechanism for the assymmetric shape of the initial colliding nuclei overlap to impact on the final azimuthal distribution of the observed particles. Furthermore, at low p_T, the magnitude of v_2 reaches the hydrodynamic limit predicted for an ideal fluid, as shown by the solid line in Fig. 5. Third, the invariant yield of very low momentum particles does not exceed expectations based on simple extrapolation from higher p_T, as shown in Fig. 6. An enhancement of low p_T particles could be expected for a weakly-interacting system, but not for a strongly-interacting system, with the correspondingly large pressure gradients that act to accelerate particles with very low momenta.

Thus the defining properties of this produced state of matter appear to be a high energy-density, approaching a baryon-free environment, and strongly interacting; perhaps similar to an ideal fluid.

4. "Simple" Scaling Behaviors

The data measured by PHOBOS exhibit four distinct scaling behaviors. The first two, extended longitudinal scaling and scaling by the number of participants, are discussed below. Two

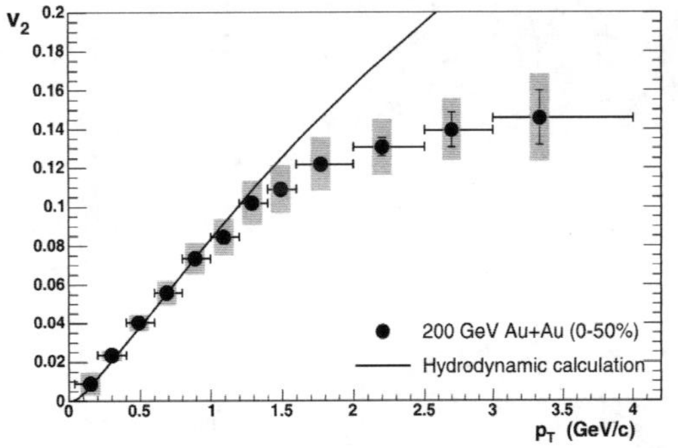

Figure 5. Transverse momentum (p_T) dependence of the elliptic flow parameter, v_2, for the most central 50% Au+Au collisions at $\sqrt{s_{NN}} = 200$ GeV [4].

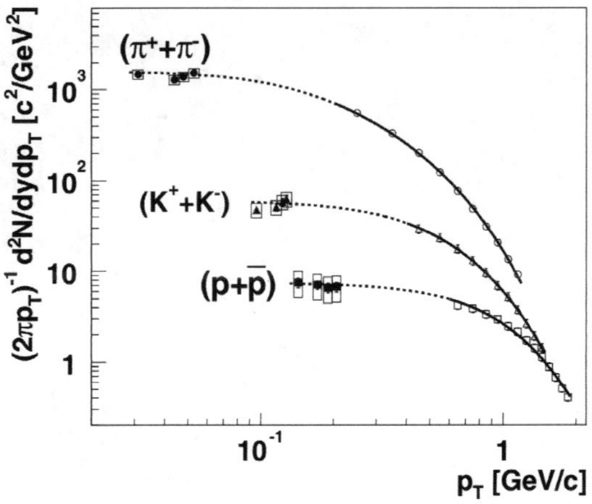

Figure 6. The invariant yield of very low momentum π, K and p (closed symbols) produced in central (0-15%) collisions as a function of p_T as measured by PHOBOS [10], and compared to extrapolations of fits to PHENIX measurements [11] at intermediate p_T for Au+Au collisions at $\sqrt{s_{NN}} = 200$ GeV.

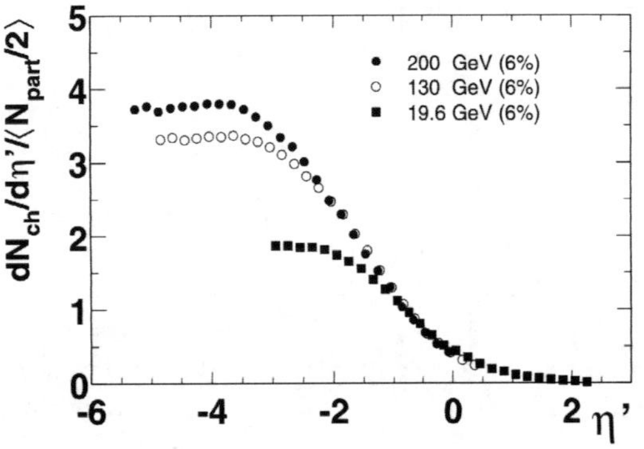

Figure 7. The pseudorapidity dependence of the charged particle density, $dN_{ch}/d\eta'$, for central (0-6%) Au+Au collisions plotted in the approximate rest frame of one of the colliding beams (i.e. $\eta' = |\eta| - y_{\text{beam}}$) [14].

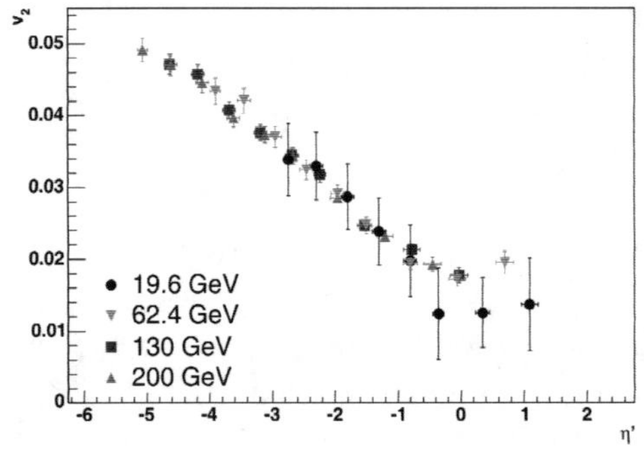

Figure 8. The pseudorapidity dependence of the elliptic flow parameter, v_2, for central (0-40%) Au+Au collisions plotted in the approximate rest frame of one of the colliding beams (i.e. $\eta' = |\eta| - y_{\text{beam}}$) [3].

additional behaviors, the apparant universality of total charged particle production [12] and the transverse mass scaling in particle spectra for d+Au collisions [13], will not be discussed here and the interested reader is encouraged to examine the cited references.

The feature of extended longitudinal scaling is apparent when either the charged particle pseudorapidity density, $dN_{ch}/d\eta$, or the elliptic flow parameter, v_2, is viewed in the approximate rest frame of one of the colliding nuclei, i.e. on an x-axis scale of $\eta' = |\eta| - y_{\text{beam}}$ (see Figs. 7 and 8). What is immediately observed is that the Au+Au data from different energies all converge near the beam rapidity, and furthermore, that this convergence not only extends significantly away from beam rapidity but also grows in extent with increasing collision energy. This feature

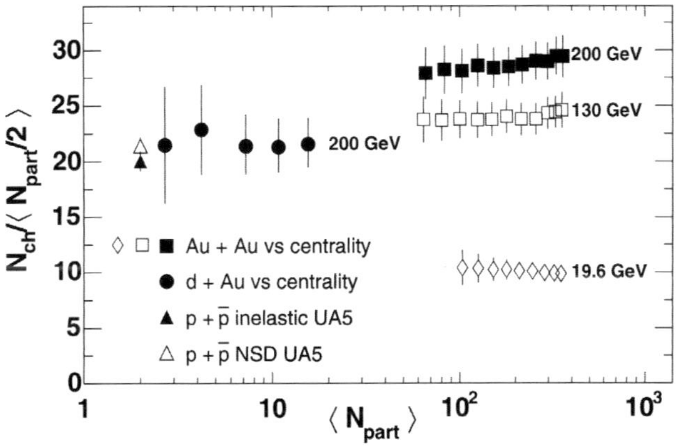

Figure 9. Centrality (N_{part}) dependence of the integrated charged particle production per participant pair for Au+Au collisions (at the three given collision energies) and for d+Au and $\overline{\text{p}}$+p at $\sqrt{s_{NN}} = 200$ GeV [16]. Note the logarithmic x-axis scale for $\langle N_{part} \rangle$.

is reminiscent of the more limited phenomena of "limiting fragmentation" first observed in $\overline{\text{p}}$+p collisions. In the PHOBOS data, the more general feature of extended longitudinal scaling is not only manifest in v_2 and $dN_{ch}/d\eta'$ for central Au+Au collisions, but also for non-central Au+Au $dN_{ch}/d\eta'$ data [14] as well as for $dN_{ch}/d\eta'$ in d+Au collisions [15].

As mentioned in the introduction, one of the strengths of the PHOBOS detector is the ability to measure the total charged particle production, N_{ch}. This result is given in Fig. 9 for Au+Au collisions at three different energies and for d+Au collisions at 200 GeV, all as a function of the centrality of the collision. The striking feature of this result is that N_{ch} scales amazingly well with the number of participant pairs to the extent that, within errors, N_{ch} per participant pair is flat for both Au+Au and d+Au collisions at all energies studied, as a function of the measured region of collision centrality. In addition, looking only at 200 GeV data, there is a difference in the overall scale of charged particle production between the $\overline{\text{p}}$+p and d+Au data, which themselves are in good agreement, and the Au+Au heavy-ion data, which has a larger N_{ch} per participant pair. This observation is discussed in more detail in Ref. [12].

5. Factorization of Energy and Centrality

The last feature of the PHOBOS data that we discuss is an apparent factorization of collision energy and centrality, first observed in the midrapidity charged particle pseudorapidity density [17]. The midrapidity $dN_{ch}/d\eta$ per participant pair for Au+Au collisions is given in Fig. 10 at the lowest (solid squares) and highest (solid circles) RHIC energies, along with lines from two calculations and one fit, see Ref. [18] for more details. The first indication of the factorization of energy and centrality can be seen from the simple two component fit (dotted lines)

$$\frac{dN_{ch}}{d\eta} = n_{pp}((1-x)\frac{\langle N_{part}\rangle}{2} + x\langle N_{coll}\rangle).$$

The free "fit" parameter is x; $N_{part}/2$ is the number of participant pairs, N_{coll} is the number of binary (nucleon-nucleon) collisions, and the overall scale is given by n_{pp} the yield measured in elementary p($\overline{\text{p}}$) + p collisions (also given by the open symbols). Although there are large (systematic) errors involved, we find the data from both energies to be consistent with the same "fit" parameter of $x \sim 0.13$, indicating a common factorization into a centrality dependent

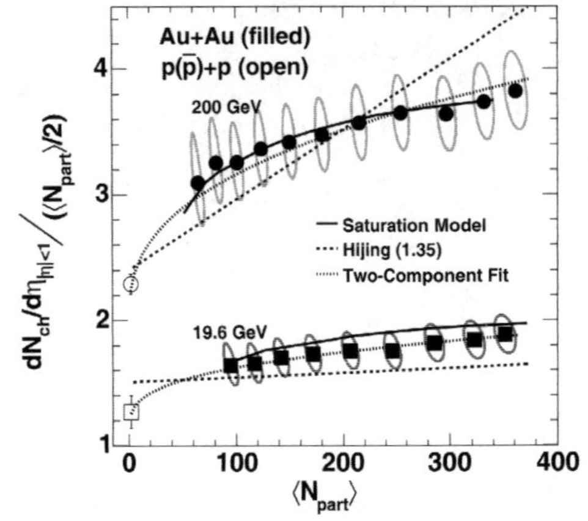

Figure 10. Centrality (N_{part}) dependence of the midrapidity pseudorapidity density per participant pair at $\sqrt{s_{NN}}$ = 19.6 and 200 GeV Au+Au collisions [18]. Error ellipses represent 90% C.L. limits.

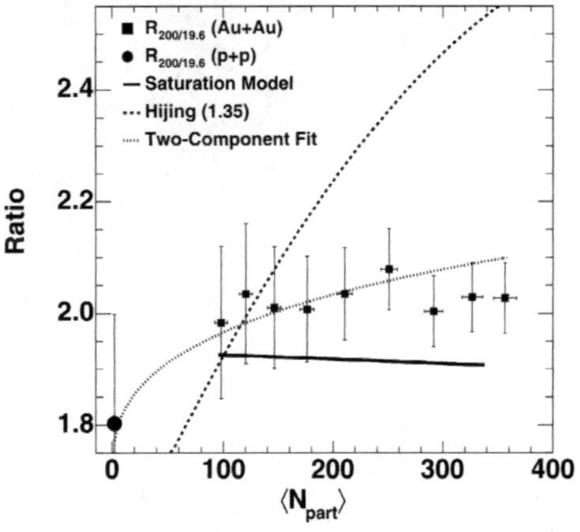

Figure 11. Centrality (N_{part}) dependence of the ratio, $\mathrm{R}_{200/19.6}$, of the midrapidity pseudorapidity density per participant pair for Au+Au collisions at $\sqrt{s_{NN}}$ = 19.6 and 200 GeV [18].

component (x) and a collision energy component (set by the overall scale of n_{pp}). This factorization is even more evident in Fig. 11, where the ratio of the data from both energies is given. In this case we see the remarkable feature that the ratio is flat with collision centrality.

The observation of N_{part} scaling in the total charged particle multiplicity and only a small fraction of binary collision (N_{coll}) scaling manifest in the midrapidity data motivates a closer examination of the binary collision scaling assumption that is inherent in the standard definition of the nuclear modification factor, R_{AA}. This is done in Fig. 12 through the use of two modified nuclear modification factors, $R_{AA}^{N_{part}}$ and $R_{PC}^{N_{part}}$. The factor $R_{AA}^{N_{part}}$, shown along the middle panels of Fig. 12, is the standard nuclear modification factor but with the Au+Au data normalized per participant pair, instead of per collision. The factor $R_{PC}^{N_{part}}$ (the bottom panels) is simply a ratio of peripheral Au+Au data to the most central (0-6%) one, with each normalized by their corresponding number of participants. We make two observations. First, comparing the top (R_{AA}) and middle ($R_{AA}^{N_{part}}$) panels, we see that yields normalized by N_{part} are simply less centrality dependent. We can start to see the same type of factorization evident in the "bulk" midrapidity yields, but now as a function of p_T. Second, comparing the middle ($R_{AA}^{N_{part}}$) and bottom ($R_{PC}^{N_{part}}$) panels, we see that by taking ratios of peripheral to central data independently for each energy gives a result that is very similar at both energies and only has a weak p_T dependence. In other words the data appears to factorize in both energy and centrality, now also as a function of p_T.

6. Conclusions

The data from RHIC thus far indicates that we have created a state of matter with high energy-density, that is nearly net-baryon free and that is strongly interacting. The transition to this new state of matter does not show abrupt changes in observables measured by PHOBOS. The data exhibit many "simple" scaling behaviors and factorizations that serve to unite the data. This suggests strong global constraints and illustrates the importance of the collision geometry in the initial state and the very early evolution of the colliding system.

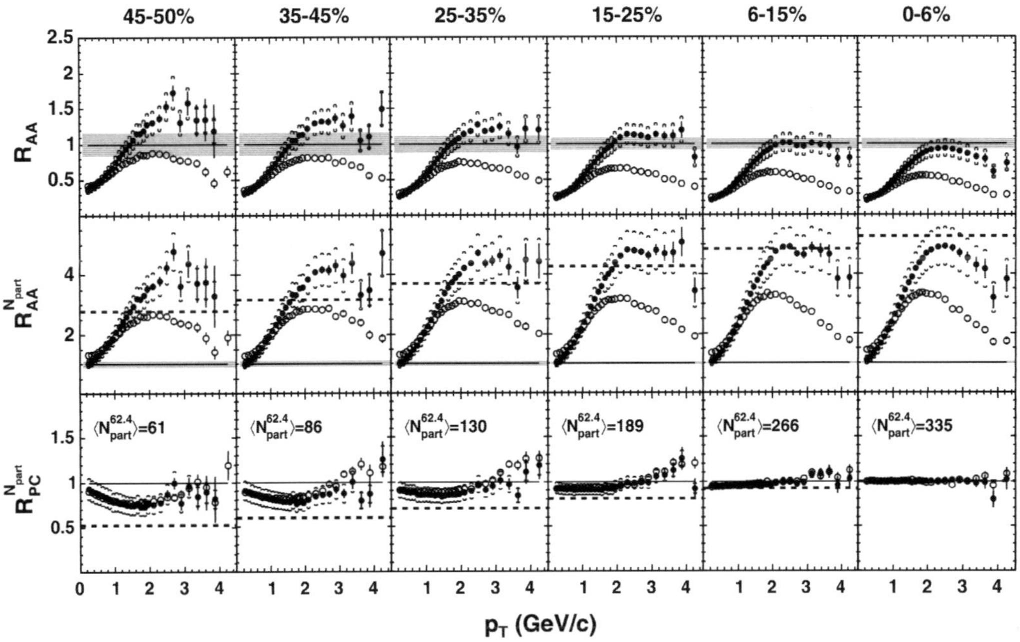

Figure 12. Centrality evolution of R_{AA} and two modified nuclear modification factors ($R_{AA}^{N_{part}}$ and $R_{PC}^{N_{part}}$) as a function of transverse momentum, p_T, for Au+Au collisions at $\sqrt{s_{NN}} = 62.4$ (closed circles) and 200 (open circles) GeV [5].

Acknowledgments

This work was partially supported by U.S. DOE grants DE-AC02-98CH10886, DE-FG02-93ER40802, DE-FC02-94ER40818, DE-FG02-94ER40865, DE-FG02-99ER41099, and W-31-109-ENG-38, by U.S. NSF grants 9603486, 0072204, and 0245011, by Polish KBN grant 1-P03B-062-27(2004-2007), and by NSC of Taiwan Contract NSC 89-2112-M-008-024.

References

[1] Back B B *et al.* 2003 *Nucl. Instrum. Methods* A **499** 603
[2] Back B B *et al.* 2002 *Phys. Rev. Lett.* **88** 022302
[3] Back B B *et al.* 2005 *Phys. Rev. Lett.* **94** 122303
[4] Back B B *et al.* 2004 *arXiv:nucl-ex/0407012*
[5] Back B B *et al.* 2005 *Phys. Rev. Lett.* **94** 082304
[6] Back B B *et al.* 2004 *arXiv:nucl-ex/0409001*
[7] Back B B *et al.* 2003 *Phys. Rev.* C **67** 021901(R)
[8] Back B B *et al.* 2003 *Phys. Rev. Lett.* **91** 072302
[9] Karsch F 2002 *Nucl. Phys.* A **698** 199
[10] Back B B *et al.* 2004 *Phys. Rev.* C **70** 051901(R)
[11] Adler S S *et al.* 2004 *Phys. Rev.* C **69** 034909(R)
[12] Back B B *et al.* 2003 *arXiv:nucl-ex/0301017*
[13] Veres G I *et al.* 2004 *J. Phys. (London)* G **30** S1143
[14] Back B B *et al.* 2003 *Phys. Rev. Lett.* **91** 052303
[15] Back B B *et al.* 2004 *arXiv:nucl-ex/0409021*
[16] Back B B *et al.* 2004 *arXiv:nucl-ex/0410022*
[17] Back B B *et al.* 2002 *Phys. Rev.* C **65** 061901(R)
[18] Back B B *et al.* 2004 *Phys. Rev.* C **70** 021902(R)

Institute of Physics Publishing Journal of Physics: Conference Series **50** (2006) 42–49
doi:10.1088/1742-6596/50/1/005 5th International Conference on Physics and Astrophysics of Quark Gluon Plasma

BRAHMS Overview

Ramiro Debbe† for the BRAHMS Collaboration

† Brookhaven National Laboratory, Upton NY, 11973

Abstract. A brief review of BRAHMS measurements of bulk particle production in RHIC Au+Au collisions at $\sqrt{s_{NN}} = 200GeV$ is presented, together with some discussion of baryon number transport. Intermediate p_T measurements in different collision systems (Au+Au, d+Au and p+p) are also discussed in the context of jet quenching and saturation of the gluon density in Au ions at RHIC energies. This report also includes preliminary results for identified particles at forward rapidities in d+Au and Au+Au collisions.

1. Introduction

The vast majority of particles produced in high energy hadron collisions (pp, AA) have transverse momenta with steeply falling distributions and consists mostly of pions. This bulk particle production is often refered to as "soft physics"; an up to now intractable non-perturbative domain of QCD that remains poorly understood. Even though these collisions can be modeled with varying degrees of accuracy, a solution from "first principles" is still being sought. The BRAHMS collaboration at RHIC has set out to measure the bulk particle production in as wide as possible range in rapidity with good particle identification and high momentum resolution. The wide rapidity range offers an almost complete coverage that allows for the inclusion of conservation laws whenever a description of the physics behind the collisions is proposed. This contribution starts with a brief summary of some of BRAHMS most relevant measurements of bulk particle production.

The production of jets at RHIC energies is well established and, one of the most dramatic discoveries in Au+Au collisions at full energy has been the suppression of those jets as they traverse a highly opaque medium formed by the collisions [1, 2]. Studies of jet production with the BRAHMS apparatus are done through the measurement of leading particles up to intermediate values of transverse momentum ($\sim 4GeV/c$). Our ability to identify the particles we measure, together with our high rapidity reach, make BRAHMS intermediate p_T measurements very relevant to study the longitudinal component of the new medium formed in Au+Au collisions. The high rapidity reach has also been instrumental in the study of charged particle production in d+Au collisions compared to incoherently added p+p collisions. The evolution of this comparison with rapidity and the centrality of the collision will be reviewed in this presentation. Finally, preliminary results from the analysis of Au+Au and d+Au collisions are presented in the last section. More details about BRAHMS results can be found in our "white paper" [3].

2. Experimental setup

The BRAHMS setup consists of two rotatable spectrometers, the mid-rapidity spectrometer (MRS) and the forward spectrometer (FS), complemented with an event characterization system used to determine the geometry of the collisions. The MRS spectrometer measures the momentum of charged particles with two tracking stations (time projection chambers) and a single dipole magnet. Particle identification in this spectrometer is done with time-of-flight hodoscopes and a threshold Cherenkov detector. Details about the performance of the MRS spectrometer can be found in Ref. [4]. The FS spectrometer measures the much higher momenta of charged particles produced at small angles with five tracking stations (two TPC and three drift chambers). Particle identification in the FS spectrometer is done with a complement of two time-of-flight hodoscopes, one threshold Cherenkov counter and a Ring Imaging Cherenkov detector. A detailed description of the BRAHMS experimental setup can be found in [5]. The geometry of the collisions is extracted from the multiplicity of charged particles measured in the $| \eta | \leq 2.2$ range [6]. The normalization of our measurements is obtained with minimum biased triggers designed to maximize the coverage of the inelastic cross section. In Au+Au collisions that trigger was defined with the Zero Degree Calorimeters, ZDC, and for the p+p and d+Au collisions, with a set of scintillators located around the beam pipe.

3. Rapidity densities

The rapidity densities for the most central data sample $(0 - 5\%)$ in Au+Au collisions at $\sqrt{s_{NN}} = 200 GeV$ have been obtained by integration of the invariant differential yields $\frac{1}{2\pi}\frac{d^2N}{p_T dp_T dy}$ at several intervals of rapidity. The measurement of very low values of transverse momentum p_T is limited by multiple scattering and particle decay, an interpolation is thus necessary in order to integrate the yields all the way down to zero p_T. Power law functions were used to integrate the pion distributions and single exponentials in $m_T - m_0$ for the kaon distributions [7]. Protons were fitted with single gaussians [8]. The result of these integrals is shown in Fig. 1. The shape of the produced particle densities (pions, kaons and anti-protons) is remarkably Gaussian. The width of the negative pion distribution is equal to: $\sigma_{\pi^-} = 2.29 \pm 0.02$. The widths of the rapidity densities of pions, kaons and anti-protons are very similar, and at the same time different from the ones associated with a single thermal source. A hydrodynamical description of the system is considered as the best approach to explain this wide distributions, even though hydrodynamical models [9], have not yet been able to reproduce the rapidity dependence of the so called elliptic flow [10].

Panel c of Fig. 1 shows the measured rapidity density for protons and anti-protons and panel d shows the net-proton density as function of rapidity after corrections from hyperon feed down. From this measurement an average rapidity loss of 2 units of rapidity has been deduced. The fact that the net proton yield around mid-rapidity is considerably different from zero, suggests that there are other mechanisms, besides valence quark transport that can move baryon number to mid-rapidity. Baryon junctions are considered as good candidates because their small-x component would be a natural way to place finite baryon number at $y \sim 0$ [11]. More details of this measurement can be found in ref. [8].

With the measurements described above, there is enough information to do an energy balance of the Au+Au collisions at $\sqrt{s_{NN}} = 200 GeV$. With reasonable assumptions about the particles that were not measured, and allowing for some $10 - 15\%$ error in the extrapolation up to beam rapidity, we find that 25 TeV out of the 33 TeV of total energy is found in produced particles. The integrated energy of the extrapolated net baryon distribution has been found to be $27 \pm 6\ GeV$ (see Ref. [8]), leaving $73 \pm 6\ GeV$, out of the 100 GeV of each incoming nucleon, available for particle production, well in agreement with the result obtained from the energy balance per particle species.

4. Intermediate p_T studies and the nuclear modification factors

As mentioned above, one of the most important results from RHIC is the suppression of intermediate p_T compared to appropriately scaled p+p collisions [1, 2]. Such reduction in

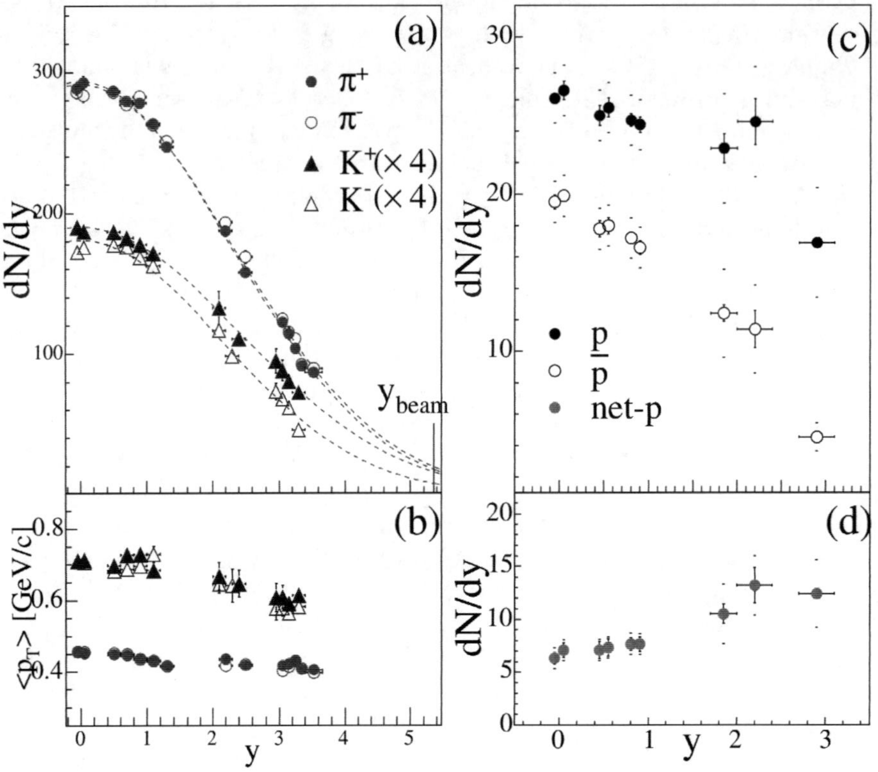

Figure 1. (a) Rapidity densities for pions and kaons. (b) Mean transverse momentum for pions, kaons and protons. (c) Rapidity density of protons and anti-protons. (d) Net proton rapidity densities corrected for hyperon feed-down. Small horizontal lines are used to display the systematic errors for the measurements displayed in panels c and d.

yield was soon identified as energy loss in a dense and opaque medium; partons with high fractional momentum traverse that medium while their energy is being degraded by multiple interactions (mainly gluon bremsstrahlung) and later hadronize into jets whose leading particles are then detected. Similar measurements performed in d+Au collisions at the same energy did not show the same strong suppression at mid-rapidity but rather an enhancement [12, 2]. This enhancement is understood as multiple elastic scatterings at the partonic level before the interactions that produce the jets whose leading particles are detected. The dominant source of the suppression measured in Au+Au collisions would then be the final state interactions with the opaque medium formed at the collision. BRAHMS extended a similar study to higher values of rapidity and found that the description of the enhancement measured at mid-rapidity in d+Au collisions was not applicable at forward rapidities, where in fact, we found again a suppression [13].

Figure 2 shows the nuclear modification factor defined as $R_{dAu} = \frac{1}{N_{coll}} \frac{\frac{dN^{dAu}}{dp_T d\eta}}{\frac{dN^{pp}}{dp_T d\eta}}$, where N_{coll} is the number of binary collisions estimated to be equal to 7.2 ± 0.6 for minimum biased d+Au collisions. Each panel shows the ratio calculated at a different η value. At mid-rapidity ($\eta = 0$), the nuclear modification factor exceeds 1 for transverse momenta greater than 2 GeV/c in a similar, although less pronounced way as Cronin's p+A measurements performed at lower

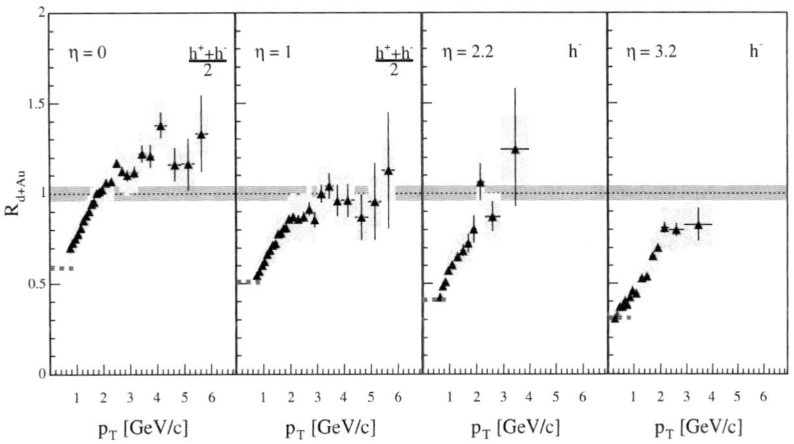

Figure 2. Nuclear modification factor for charged hadrons at pseudorapidities $\eta = 0, 1.0, 2.2, 3.2$. Statistical errors are shown with error bars. Systematic errors are shown with shaded boxes with widths set by the bin sizes. The shaded band around unity indicates the estimated error on the normalization to $\langle N_{coll} \rangle$. Dashed lines at $p_T < 1$ GeV/c show the normalized charged particle density ratio $\frac{1}{\langle N_{coll} \rangle} \frac{dN/d\eta(d+Au)}{dN/d\eta(pp)}$.

energies [14].

A shift of one unit of rapidity is enough to make the Cronin type enhancement disappear, and as the measurements are done at higher rapidities, the ratio becomes consistently smaller than 1 indicating a suppression in d+Au collisions compared to scaled p+p systems at the same energy.

In all four panels, the statistical errors, shown as error bars (vertical lines), are dominant specially in our most forward measurements. Systematic errors are shown as shaded rectangles. An additional systematic error is introduced in the calculation of the number of collisions N_{coll} that scales the d+Au yields to a nucleon-nucleon system. That error is shown as a 15% band at $R_{dAu} = 1$.

The four panels of Fig. 3 show the central $R_{CP}^{central}$ (filled symbols) and semi-central $R_{CP}^{semi-central}$ (open symbols) ratios for the four η settings. The evolution as function of rapidity seen in this figure is more obvious for samples of central events. Starting on the left panel corresponding to $\eta = 0$, the central events yields are systematically higher than those of the semi-central events, but at the highest pseudo-rapidity $\eta = 3.2$, the yields of central events are $\sim 60\%$ lower than the semi-central events for all values of transverse momenta. These results have generated much interest in the community because they appear as another indication of the onset of the so called Color Glass Condensate [15] at RHIC energies and high atomic numbers (A=179 for the gold ions) [16].

The analysis of the BRAHMS data collected from d+Au and p+p collisions is currently in progress, in particular, Fig. 4 presents the minimum bias nuclear modification R_{dAu} for anti-protons and negative pions at $\eta = 3.2$. The extraction of these ratios involved the following assumptions: at each p_T bin, one can extract the nuclear modification factor for identified particles, by multiplying the numerator and the denominator of this factor by the fractions of raw counts of identified particles to raw counts of negative particles for d+Au and p+p systems respectively:

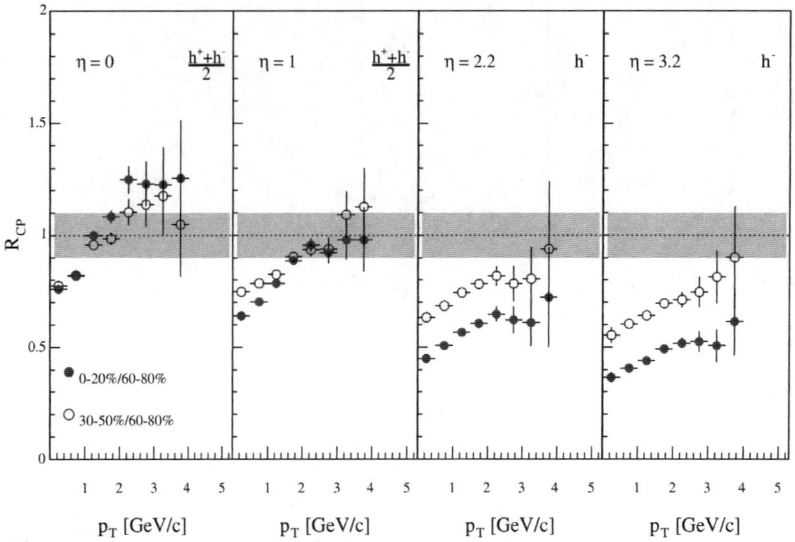

Figure 3. Central (full points) and semi-central (open points) R_{cp} ratios (see text for details) at pseudorapidities $\eta = 0, 1.0, 2.2, 3.2$. Systematic errors ($\sim 5\%$) are smaller than the symbols. The ratios at the highest pseudorapidities ($\eta = 2.2$ and 3.2) are calculated for negative hadrons. The uncertainty on the normalization of the ratios is displayed as a shaded band around unity. Its value has been set equal to the error in the calculation of N_{coll} in the most peripheral collisions (12%).

$$R^{\bar{p}}_{dAu} = R^{h^-}_{dAu}\frac{(\frac{\bar{p}}{h^-})^{dAu}}{(\frac{\bar{p}}{h^-})^{pp}} = \frac{1}{N_{coll}}\frac{\left(\frac{dn^{dAu}}{dp_T d\eta}\right)^{h^-}}{\left(\frac{dn^{pp}}{dp_T d\eta}\right)^{h^-}}\frac{\frac{\left(\frac{dn^{dAu}}{dp_T d\eta}\right)^{\bar{p}}}{\left(\frac{dn^{dAu}}{dp_T d\eta}\right)^{h^-}}}{\frac{\left(\frac{dn^{pp}}{dp_T d\eta}\right)^{\bar{p}}}{\left(\frac{dn^{pp}}{dp_T d\eta}\right)^{h^-}}} = \frac{1}{N_{coll}}\frac{\left(\frac{dn^{dAu}}{dp_T d\eta}\right)^{\bar{p}}}{\left(\frac{dn^{pp}}{dp_T d\eta}\right)^{\bar{p}}}$$

The ratio of raw counts is then equated with the ratio of differential yields in η and p_T, assuming that all corrections do cancel out. The ratio of raw counts was obtained with information from the Ring Imaging Cherenkov detector whose efficiency is high ($\sim 95\%$) but has not been included in this analysis. The errors shown for the nuclear modification factors of anti-protons and pions include the contributions from correlations between the parameters of the fits made to the ratios of raw counts. No attempt was made to estimate the contributions from anti-lambdas feed down to the anti-proton result. The remarkable difference between baryons and mesons has also been seen at RHIC energies around mid-rapidity, and has been related to parton recombination [17]. The same explanation could be offered for these high rapidity results, but it is difficult to imagine the existence of the thermal bath of partons at high rapidity in d+Au collisions.

At the time the invariant yields at $\eta = 3.2$ where shown for the first time [18], the clear difference between positive and negative charged particles, prompted some to associate the suppression seen in d+Au collisions at forward angles with the so called "beam fragmentation"

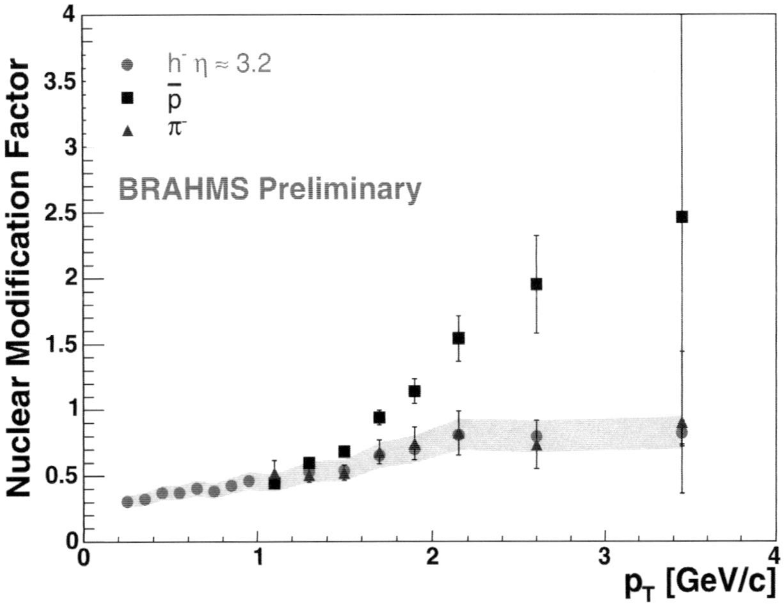

Figure 4. The nuclear modification factor R_{dAu} calculated for anti-protons (filled squares) and negative pions (filled triangles) at $\eta = 3.2$. The same ratio calcutated for negative particles at the same pseudo-rapidity [13] is shown with filled circles, and the systematic error for that measurement is shown as grey band.

[19]. More recent work based on NLO pQCD has indicated the difficulty in reconciling the data with the behavior of standard fragmentation functions [20]. Panel a of Fig. 5 shows that, there are as many protons as positive pions ($\sim$ 80% at $p_T \sim 2GeV/c$) in the positive particle distribution at $\eta = 3.2$. The abundance of baryons at this high energy and rapidity doesn't support the idea of baryon suppression in the fragmentation region [21, 22] where, because of their high energy, the quarks of the beam would fragment independently mostly into mesons. Panel b of the same figure shows that for all values of p_T, the fraction of negative pions is high and equal to 80%.

The suppression of particle production at mid-rapidity in Au+Au central collisions compared to incoherently added p+p interactions, or properly scaled disributions obtained from peripheral Au+Au collisions, is widely accepted as a necessary condition for momentum degradation by gluon radiation, as partons traverse an opaque medium before they hadronize. That momentum degradation is expected to be a function of the length of the parton's path inside the medium as well as the medium density [23]. In the above mentioned work, such partonic energy loss is encoded in a medium modified fragmentation function that has the standard (in vacuum) behavior, but only after the energy has been degraded by several interactions in the medium. These interactions are distributed in number according to a Poisson distribution with a mean value equal to $L/\sigma\rho$ where L is the parton's path length in the medium, σ is the interaction cross-section and ρ the medium density. The final fragmentation also includes the gluons radiated at each interaction. The pion rapidity density, which must be directly related to the medium density, changes by almost a factor of three between mid-rapidity and y = 3 (see Fig. 1). One would then expect less energy loss at y=3, by a comparable factor, but as can be seen in

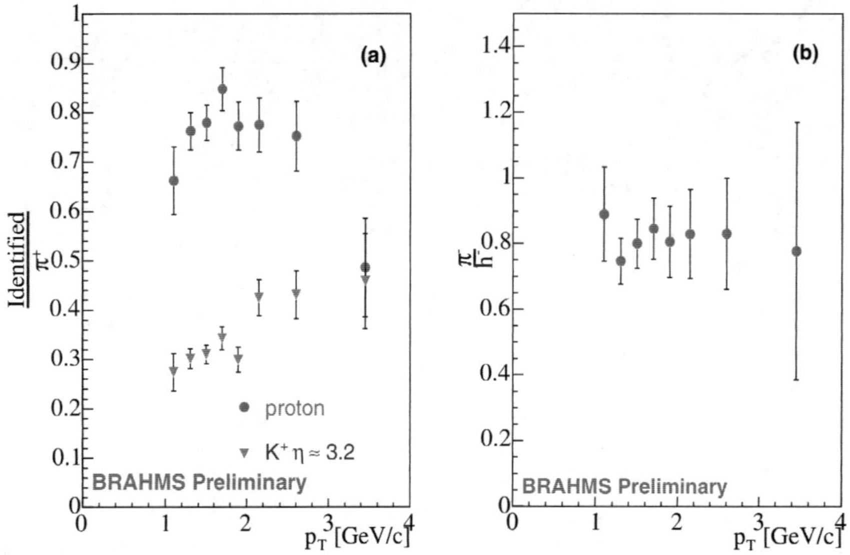

Figure 5. Some results with pid at $\eta = 3.2$: (a) Particle composition of positive charged hadrons at $\eta = 3.2$ (b) The fraction of negative hadrons identified as pions at the same pseudo-rapidity.

Fig. 6, there is no noticeable difference between the p_T suppressions at $\eta = 3.2$ and previous measurements at mid-rapidity and $\eta = 2.2$ [2]. This result may indicate the interplay between energy loss effects that get weaker at higher rapidities, and initial state related suppression that becomes stronger as the number of gluons in one of the Au ions is further reduced by gluon fusion. On the other hand, if the p_T distributions at forward rapidities are stepper than the ones measured at mid-rapidity, smaller energy losses would be magnified producing similar results as the measurements. Further analysis on this subject is currently in progress.

In summary, BRAHMS has studied the properties of bulk particle production as well as baryon number transport in Au+Au collisions at $\sqrt{s_{NN}} = 200GeV/c$. We have also compared charged particle production in Au+Au and d+Au collisions to similar production in p+p collisions at the same energy. Such comparisons, show an strong suppression at intermediate transverse momentum that is associated with the formation of a dense and opaque medium. In d+Au collisions at the same energy, such suppression does not appear at mid-rapidity, but is present at forward rapidities and is even more pronounced in central collisions.

5. Acknowledgments

This work was supported by the Office of Nuclear Physics of the U.S. Department of Energy, the Danish Natural Science Research Council, the Research Council of Norway, the Polish State Committee for Scientific Research (KBN) and the Romanian Ministry of Research.

References

[1] K. Adcox *et. al.*, PHENIX Collaboration, Phys. Rev. Lett. **88** 022301 (2002); S. S. Adler *et. al.*, STAR Collaboration, Phys. Rev. Lett. **89** 202301 (2002); B.B. Back, *et. al.*, PHOBOS Collaboration, Phys. Lett. B **578**, 297 (2004).
[2] I. Arsene *et al.*, BRAHMS Collaboration, Phys. Rev. Lett. **91**, 072305 (2003).
[3] I. Arsene *et al.*, BRAHMS Collaboration, Nucl. Phys. A **757** 1-27, nucl-ex/0410020 (2004).

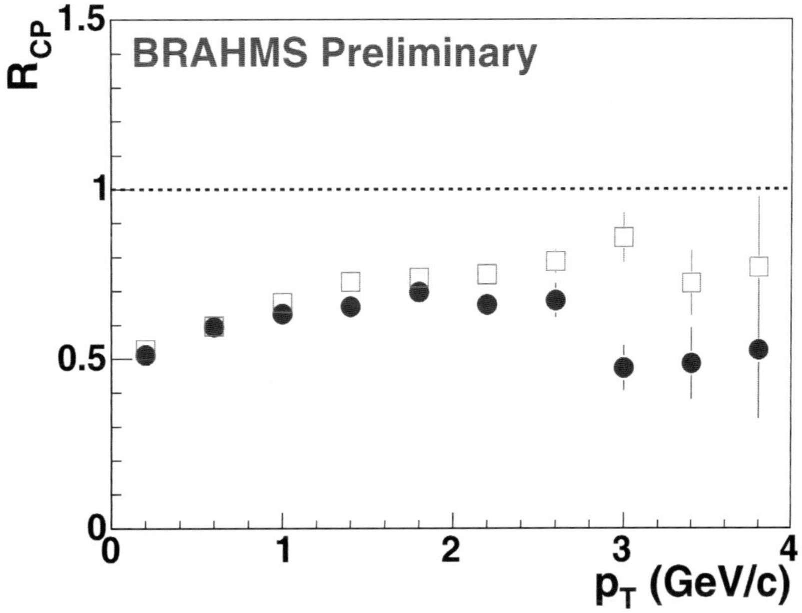

Figure 6. Rcp of charged particles from Au+Au at η = 3.2 Open squares: positives, filled circles: negatives.

[4] I. Arsene et al., submitted to Phys. Rev. C. nucl-ex/053010 (2005).
[5] M. Adamczyk *et al.*, BRAHMS Collaboration, Nuclear Instruments and Methods A **499** 437 (2003).
[6] Y.K. Lee *et al.*, Nuclear Instruments and Methods A **499** 437 (2003).
[7] I.G. Bearden *et. al.*, Accepted for publication Phys. Rev. Lett. (nucl-ex/0403050).
[8] I.G. Bearden *et al.* Phys. Rev. Lett. **93,** 102301 (2004).
[9] T. Hirano and Y. Nara, Nucl. Phys. A **743**, 305 (2004); T. Hirano and Y. Nara J. Phys. G**31**, S1, (2005).
[10] B.B. Back, *et. al.*, PHOBOS Collaboration, Phys. Rev. Lett. **89**, 222301, (2002).
[11] D. Kharzeev Phys. Lett. B **378**, 238-246, (1996).
[12] B.B. Back *et al.*, PHOBOS Collaboration, Phys. Rev. Lett. **91**, 72302 (2003); S.S. Adler *et al.*, STAR Collaboration, Phys. Rev. Lett. **91**, 72303 (2003); J. Adams *et al.*, PHENIX Collaboration, Phys. Rev. Lett. **91**, 72304 (2003).
[13] I. Arsene et al., Phys. Rev. Lett. **93**, 242303 (2004).
[14] D. Antreasyan *et al.*, Phys. Rev. D **19**, 764 (1979).
[15] L. McLerran and R. Venugopalan, Phys. Rev. D **49**, 2233 (1994), Phys. Rev. D **49**, 3352 (1994), Phys. Rev. D **50**, 2225 (1994), Phys. Rev. D **59**, 094002 (1999); Y. V. Kovchegov, Phys. Rev. D **54**, 5463 (1996), Phys. Rev. D **55**, 5445 (1997).
[16] D. Kharzeev, Y. V. Kovchegov and K. Tuchin Phys. Rev. D **68**, 094013, (2003), hep-ph/0307037; D. Kharzeev, E. Levin and L. McLerran, Phys. Lett. B **561**, 93 (2003); R. Baier, A. Kovner and U. A. Wiedemann Phys. Rev. D **68**, 054009, (2003); J. Albacete, *et al.* hep-ph/0307179.; A. Dumitru and J. Jalilian-Marian, Phys. Lett. B **547**, 15 (2002); J. Jalilian-Marian, A. Kovner, A. Leonidov, H. Weigert, Phys. Rev. D **59**, 014014 (1999). J. Jalilian-Marian *et al.* Phys. Lett. B **577**, 54-60 (2003), nucl-th/0307022;
[17] R. C. Hwa, C. B. Yang and R. J. Fries Phys. Rev C**71**, 024902, (2005).
[18] R. Debbe (for the BRAHMS Collaboration) J. Phys. G **30** S759-S765, (2004).
[19] M. Gyulassy Proceedings RIKEN Workshop Volume **57**, 141 Dec. 2-6 (2003).
[20] V. Guzey, M. Strikman, W. Vogelsang, Phys. Lett. D **603**, 173 (2004).
[21] A. Dumitru, L. Gerland, and M. Strikman hep-ph/0211324, (2003).
[22] A. Berera *et al.* Phys. Lett. B **403**, 1-7, (1997).
[23] X. N. Wang and Z. Huang, Phys. Rev. C **55**, 3047, (1997).

Institute of Physics Publishing Journal of Physics: Conference Series **50** (2006) 50–61
doi:10.1088/1742-6596/50/1/006 5th International Conference on Physics and Astrophysics of Quark Gluon Plasma

Photon measurements in forward rapidity in heavy-ion collisions

Yogendra Pathak Viyogi

Variable Energy Cyclotron Centre, 1/AF, Bidhan Nagar, Kolkata 700064

E-mail: viyogi@veccal.ernet.in

Abstract.
Contribution of a preshower photon multiplicity detector to the physics of ultra-relativistic nuclear collisions is reviewed and future possibilities at RHIC and LHC are discussed.

1. Introduction

One of the main goals of the high energy heavy-ion collision program is to look for the possible formation of Quark-Gluon Plasma (QGP). In the collision process the system may go through a phase transition from confined hadronic matter to a deconfined state of quarks and gluons which will be a transient state. The system then evolves from a very hot and dense QGP state to normal hadronic matter by undergoing cooling, expansion and hadronization. Information obtained from various global observables (e.g photon multiplicity (N_γ), charged particle multiplicity (N_{ch}) and transverse energy) is used to characterize the system formed in such collisions. Study of photon production has shown a great promise in studying the various aspects of the reaction mechanism in such collisions and dynamics of particle production [1, 2, 3, 4, 5, 6, 7, 8, 9, 10].

The measurement of photons in high energy physics experiments has been traditionally carried out using calorimeters. Due to large spatial density of produced particles in the forward rapidity region in ultra-relativistic nuclear collisions, and consequent overlap of showers, one cannot use calorimeters beyond a certain region. In such a situation, a limited goal of photon study can be achieved using a preshower detector having a relatively thinner converter and restricting the development of shower. The preshower detector can only measure the spatial distribution of photons but not their energies.

Such a preshower photon multiplicity detector (PMD) has been employed in SPS and RHIC experiments and is planned also for the LHC experiment. In addition to extending the phase space coverage for the measurement of pseudorapidity distributions, the PMD data on spatial distribution of photons, along with other measurement of charged particle multiplicity and event centrality, has been able to address the following important physics topics related to phase transition and chiral symmetry restoration:

- determination of the reaction plane and the probes of thermalization via studies of azimuthal anisotropy and flow;
- critical phenomena near the phase boundary leading to fluctuations in global observables like multiplicity and pseudorapidity distributions;

- signals of chiral-symmetry restoration (e.g. disoriented chiral condensates) through the measurement of charged-particle multiplicity ($N_{\rm ch}$) also in a common part of phase space and study of the observables $N_\gamma/N_{\rm ch}$ with full azimuthal coverage.

2. Photon Multiplicity Detector

The preshower detector for measuring photon multiplicity consists of a thin converter behind which is placed an array of sensitive detection elements (cells or pads). The thickness of the converter is chosen such that it provides a reasonably large signal for electromagnetic particles, restricts shower size to a few cells and the response to charged hadrons is limited to preferably one pad or cell. The thickness of the converter and the granularity (size, shape of one detection element) have to be optimized using Monte-Carlo simulations. A second plane of similar granularity may be placed in front of the converter for improving the discrimination between charged hadrons and photons.

2.1. Detector for the SPS experiments

The PMD for the fixed target SPS experiments was made using plastic scintillator pads, with scintillation light being transported to readout devices using wavelength shifting fibers. The light at the fibre ends were converted to electrical signals and readout using the image intensifier (II) and CCD camera systems obtained from the old UA2 experiment at CERN. The basic parameters of the two detectors, used in the WA93 and WA98 experiments, are summarized in Table 1. These had only one plane of detection elements behind a 3 X_0 thick lead converter.

2.1.1. PMD for the WA93 experiment : The first PMD, made for the WA93 experiment, consisted of 7500 plastic scintillator pads and had long WLS fibers for transporting scintillation light to the II+CCD camera readout system [11]. The pads were square in shape and having uniform size of 2 cm. These were placed in four quadrants of a light-tight box, each quadrant having 1875 pads arranged in a matrix of 50 $\times$ 38 with a space for 5 $\times$ 5 pads left for the hole of the beam pipe. The fiber bundle from each quadrant was read out using one camera system.

The detector, mounted at 10.5m from the target, covered the pseudorapidity region $2.8 \leq \eta \leq 5.2$ of which the region $3.3\leq \eta \leq 4.9$ had full azimuthal coverage. It was installed in 1991 and was used to study the collisions of 200 A.GeV sulphur ions with gold nuclei during 1991 and 1992 SPS run periods.

2.1.2. PMD for the WA98 Experiment : The PMD for the WA98 experiment was also based on the plastic scintillator pads, with light being transported using WLS fibres [12]. Based on the experience gained during the operation of the WA93 PMD, the fabrication technology was modified for improvements in light collection and reduction of the distortion of the image at the CCD. The WLS fibres of short length were coupled to long lengths of clear fibres to reduce light attenuation and a black paint was used in bundling the fibre ends to reduce spreading of the image on the CCD. To make the hit density nearly uniform over the detector, four different pad sizes, ranging from 15mm square to 25 mm square, were used, inner part having smaller pads and outer parts having larger pads. The detector consisted of 28 individual light-tight box modules, each having a matrix of 50$\times$38 scintillator pads read out using one image intensifier and CCD camera system.

The detector extended to an area of 21 sq.m. It was installed at 21.5 m from the target and covered the pseudo-rapidity region $2.5 \leq \eta \leq 4.2$, of which the region η= 3.2-4.0 had full azimuthal coverage. It is shown in Fig. 1. It was installed in the experiment in 1994 and took data for lead ion induced collisions at 158.A GeV energy during 1994 to 1996 SPS run periods.

Table 1. Parameters of WA93 and WA98 PMDs

Parameter	WA93	WA98
(η, ϕ) coverage	2.8-5.2 (2π in 3.3-4.9)	2.9-4.2 (2π in 3.2-4.0)
Number of readout cameras	4	28
Number of pads	7500	53200
Distance from target	10.5 m	21.5 m
Pad sizes (square shape)	20 mm	15 mm, 20 mm, 23 mm, 25 mm
Pad thickness	3 mm	3 mm
Light transport	WLS fibres	WLS + Clear fibres

2.2. Detector for colliders

For the next generation collider experiments at RHIC and LHC, it was soon realized that the PMD would have be made with a different technology. Scintillator pads with fibres were bulky and the readout systems were quite expensive. The collider environment required sleek detectors which could fit in a small space. It was also realized that with the increased multiplicity of produced particles, the detector design criteria had to be redefined, in particular charged particle hits on the preshower plane had to confined to a single pad or cell. Due to the presence of substantial number of electrons, the use of second plane in front of the converter also became necessary.

Using gas detectors, low cost readout could be obtained in the form of signal processing ASICs. However, existing gas detector technologies did not provide the solution to the containment of charged particle hits to a single detection unit.

An extensive R&D effort resulted in the design of a cellular honeycomb proportional counter, using Ar(70%)+CO_2(30%) as the sensitive medium and with novel ideas of extended cathode which provided almost uniform response to charged particles within the detector volume. This design offered all the advantages of lower cost and ease of construction for large area and highly granular detector system. The results of such an R&D are summarized in Ref. [13, 14].

The basic parameters of the PMD for the RHIC and LHC experiments are given in Table 2. In these detectors also we used 3 X_0 thick lead converter.

2.2.1. PMD for the STAR experiment at RHIC : The STAR PMD [15] consists of a set of super-modules arranged in two vertical halves so that each half moves independently around the beam pipe. The super-modules are of various shapes but all are made up of rhombus unit modules having 24×24 cells. The number of unit modules in the supermodules vary from 4 to 9. There are two planes of the sensitive detectors, one in front of the converter (acting as charged particle veto) and one behind the converter which registers preshower signals. There are 12 supermodules in each plane of the detector.

The signal processing for STAR PMD is done using GASSIPLEX chips, which has 16 channels of preamplifier and shaper and provides analog multiplexed output. One front-end electronics (FEE)-Board has 4 chips and is connected to a group of 64 cells in a 8×8 matrix on the detector. The digitization and readout of the analog multiplexed signals is done using C-RAMS modules. The track and hold flag is generated using a pre-trigger signal based on beam-beam counter in STAR because the actual Level 0 signal arrives rather late. Each block of C-RAMS handles signals from a chain of 27 FEE boards, i.e., for 1728 cells. There are in all 48 chains read out using 24 C-RAMS modules.

The detector is mounted in the STAR experiment at 540cm from the center of the TPC and nominally covers the pseudorapidity region 2.3-3.8 with full azimuthal acceptance.

Fully instrumented preshower plane of the STAR PMD is shown in Fig. 2 in the data taking mode when the two halves are touching each other.

2.2.2. PMD for the ALICE experiment at the LHC : The PMD for the ALICE experiment consists of hexagonal cells having 0.23 cm^2 cross-section and 5 mm depth. Although originally proposed to use the cells identical to those used in the STAR PMD, the detector geometry was modified later due to constraints of available space within the ALICE experiment [16, 17].

The detector consists of rectangular shaped unit modules having 4608 cells. These unit modules are placed in supermdules which are separate gas-tight enclosures. There are two types of unit modules and super-modules, the difference being in the number of cells in the row and column, in one there are 48 rows and 96 columns while in the other there are 96 rows and 48 columns. There are 4 supermodules in each plane, arranged in such a way that each forms one quadrant of the detector. The detector can be opened as two halves around the beam pipe.

The design of signal processing electronics for the PMD is based on that of the tracking chambers of dimuon spectrometer in ALICE. The signals from the cells are processed using MANAS chips which provide analog multiplxed output for 16 channels. The readout of the analog multiplexed signals is performed, after digitization, using another custom built chip MARC and the CROCUS system.

The PMD will be mounted at 360cm from the center of the TPC in ALICE and will nominally cover the pseudorapidity region 2.3-3.5 with full azimuthal acceptance.

Figure 1. View of the WA98 PMD from the converter side. The sections on four sides were suitably inclined to provide near-normal incidence to incoming particles.

Figure 2. View of the preshower plane of STAR PMD from the RHIC tunnel side.

2.3. Photon reconstruction

The photon reconstruction is achieved by clustering the hits and then applying suitable algorithm to reject charged hadron hits. For the WA93/WA98 PMDs, where only the preshower plane of the detector was used, a simple algorithm based on a threshold on the cluster signal was used for the discrimination [5]. A more detailed algorithm based on neural network technique has

Table 2. Parameters of STAR and ALICE PMDs

Parameter	STAR@RHIC	ALICE@LHC
(η, ϕ) coverage	2.3-3.8, 2π	2.3-3.5, 2π
Number of planes	Preshower, veto	preshower, veto
Number of super-modules	24	8
Number of unit modules	144	48
Number of cells	83K	220K
Distance from vertex	540 cm	360 cm
Cell cross-section	1 cm^2	0.23 cm^2
Cell depth	8 mm	5 mm

also been investigated for PMD with charged particle veto [18] and used in the study of physics performance of ALICE PMD [13].

Due to interactions of charged hadrons in lead converter and the nature of the discrimination algorithm, the set of clusters finally labeled as 'photons' have some contaminants. The efficiency of photon reconstruction is usually around 70% and the purity of the photons in the accepted sample is also around 70%. These numbers vary with the system, centrality and pseudorapidity as studied using event generators and Monte-Carlo methods.

3. Physics results

The measurement of photon distributions in the forward rapidity region has made significant contributions to the physics at the SPS energy. The lower p_T cutoff for the PMD is only about 30 MeV/c [3, 13], thus enabling it to detect a large fraction of the photon spectrum. The presence of a charged particle multiplicity detector (SPMD) overlapping in phase space with the PMD has helped to greatly enhance the scope of physics studies using the PMD.

3.1. Pseudorapidity and multiplicity distributions

Pseudorapidity distribution of photons has been measured at various centralities using the PMD in WA93 experiment for S+Au collisions at 200.A GeV [3]. In general the shape of the distribution is well reproduced by the event generators like VENUS [19] but the absolute magnitude is somewhat underpredicted. Using the measurement of photon multiplicity by the PMD and the transverse e.m. energy by the MIRAC calorimeter, it was possible to study the centrality dependence of $< p_T >$ of photons [2].

The WA98 PMD measured photon production for Pb+Pb, Pb+Nb and Pb+Ni systems at 158.A GeV projectile energy. A systematic study of these data showed that VENUS again underpredicts the magnitude of pseudorapidity distributions [5]. Fig. 3 shows the minimum bias inclusive photon cross-section for the three systems along with the predictions of VENUS event generator with default parameters. The shape of these distributions is governed by the collision geometry. For asymmetric collisions of Nb and Ni targets small shoulders are present around N_γ of 300 and 200 respectively. This shoulder is produced when a decrease in the impact parameter leads to little increase in particle production and the cross-sections for these small impact parameter collisions pile up at a fixed N_γ.

The pseudorapidity distributions for the three systems studied are shown in Fig. 4 for various centralities. For the symmetric Pb+Pb collisions filled symbols represent the measured data and open symbols are reflections around η_{cm} (=2.92). The histograms show VENUS predictions. The discrepancy between data and VENUS is ~10% for central collisions around mid-rapidity and

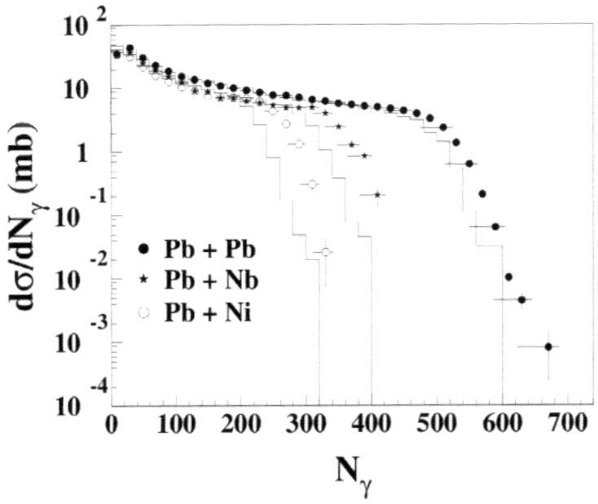

Figure 3. Minimum bias inclusive photon cross sections for Pb+Ni, Pb+Nb, and Pb+Pb reactions at 158·AGeV. Solid histograms are the corresponding distributions obtained from the VENUS event generator.

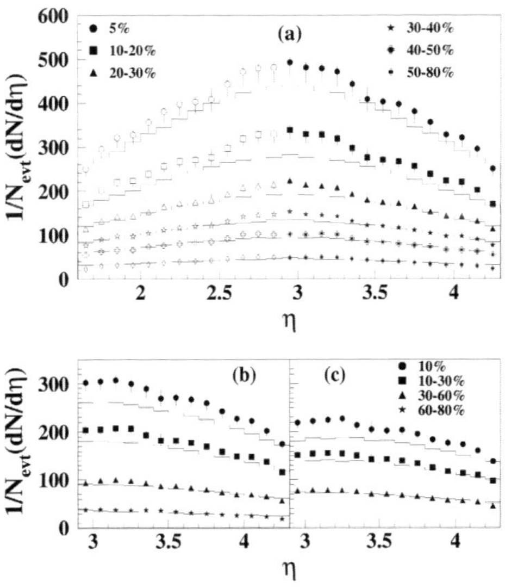

Figure 4. Pseudo-rapidity distributions of photons at various centralities in Pb induced reactions at 158·AGeV on (a) Pb, (b) Nb and (c) Ni targets. The solid histograms are the corresponding distributions obtained from the VENUS event generator.

decreases at larger impact parameters. For asymmetric systems the discrepancy between data and VENUS is comparatively larger.

An important observation of these systematic studies of photon production over a wide range of collision systems and centralities has been the scaling relation for total number of photons. It is found that this scales as a power law dependence on the number of participants, N_{part}, the exponent being 1.12±0.03 [5].

First results on photon distribution in the forward region at 62.4 GeV energy recently became available with the PMD in the STAR experiment [10]. These are described in detail in the article by B. Mohanty et al. in these proceedings [20]. At RHIC the HIJING model underpredicts photon production as measured by the PMD. However the AMPT model compares favorably well.

In Fig. 5 we have plotted the photon pseudorapidity distribution per participant nucleon in Au+Au collisions at 62.4A GeV as a function of η-y_{beam} for central collisions. Also superposed are the data from the WA98 experiment for the Pb+Pb collisions at 17.A GeV c.m. energy, from the WA93 experiment for S+Au collisions at 20.A GeV c.m. energy and from the UA5 experiment for *pp* collisions at 546 GeV. The data at SPS and at RHIC are found to be consistent with each other, suggesting that photon production follows limiting fragmentation behaviour. It is further found that the data are consistent with centrality- and energy- independent beahaviour similar to the results for identified charged particles [20].

3.2. Azimuthal anisotropy and flow

In the search for evidence of the phase transition from hadronic matter to quark gluon plasma in relativistic heavy-ion collisions, it is important to establish that thermalization occurs in the

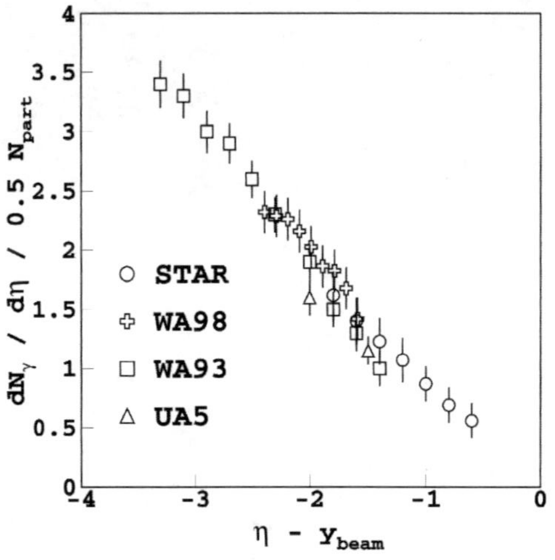

Figure 5. Limiting fragmentation as seen in the photon production.

reacting system. In non-central collisions the overlap volume, when projected onto the transverse plane, is not azimuthally symmetric. It has a smaller size along the direction of impact parameter than in the perpendicular direction. If thermal equilibrium is reached, collective flow develops with a velocity proportional to the pressure gradient, which is larger along the direction of impact parameter than along the perpendicular direction. Matter is thus expected to flow preferentially in the reaction plane which should result in an azimuthal anisotropy of the distribution of particles. In head-on collisions the anisotropy should disappear, even if thermalization occurs, due to the azimuthal symmetry of the overlap volume. Thus the study of anisotropy as a function of centrality assumes significance in the study of phase transitions.

The first observation of collective flow at the SPS energy came from the PMD data [1]. The anisotropy parameter[1] $\overline{\alpha}$ is plotted in Fig. 6 for the four centrality bins for the data (filled circles) and for the simulated data (open circles). The vertical bars indicate statistical errors only. The horizontal bars indicate the rms multiplicity of the bin. The estimated systematic errors are indicated by the bracket on each data point. The simulated events from VENUS event generator show a small anisotropy similar in magnitude to the estimated systematic error. The non-zero values are essentially due to π^0 decay correlations and finite multiplicity effects.

The extracted true anisotropies in the data are small but significantly greater than the simulation which does not include flow effects. The observed anisotropy decreases with decreasing impact parameter from $\overline{\alpha} \approx 0.07$ for peripheral collisions to $\overline{\alpha} \approx 0.045$ for central collisions. This dependence is similar, but the values are roughly half as large, as has been predicted by hydrodynamical model calculations [21].

The effect of photon correlation due to π^0 decay and other detector related effects on the deduced anisotropy values from photon measurements have been investigated in detail [22]. It has been shown that the measured anisotropy in photons can be used to deduce the anisotropy in parent neutral pions and a scaling relation between these two has been extracted in terms of an experimentally measurable quantity.

This concept has been tested and established in the analysis of data from the WA98 experiment [9]. For the study of anisotropy using data from the WA98 PMD, all the decay and detector effects were considered within a Monte-Carlo framework using the measured flow

[1] At that time the nomenclature of various flow coefficients were not standardized. In Ref. [1] the word 'directed flow' has been used for what is now commonly understood as elliptic flow and the notation $\overline{\alpha}$ is actually v_2.

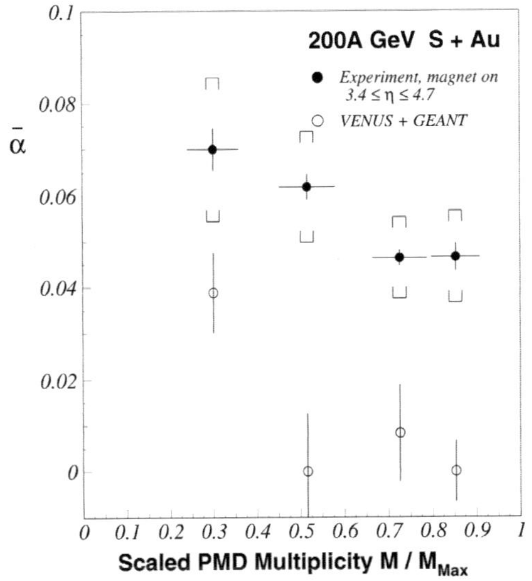

Figure 6. The true anisotropy $\bar{\alpha}$ as a function of the centrality (scaled multiplicity) for the experimental data (filled circles) and VENUS + GEANT simulations (open circles).

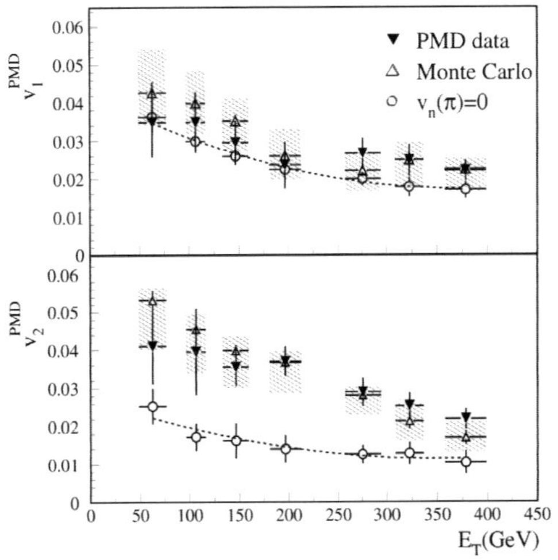

Figure 7. First order, v_1, and second order, v_2, anisotropy coefficients of the azimuthal distributions of photon hits for different centralities characterized by the measured transverse energy. The boxes indicate the extent of the estimated systematic errors.

values of charged particles in the same experiment within an overlapping part of phase space and assuming that π^0 flow value is the same as that for charged particles (which are predominately charged pions). The results are shown in Fig. 7 for both the PMD data and the simulation using input π^0 anisotropy. The photon anisotropy coefficients extracted from the simulated PMD data are consistent within errors with the measured PMD results.

The use of PMD data to determine the event plane gives a strong π^0 decay effect. This is shown by open circles in Fig. 7 which show simulation results of anisotropy in PMD for isotropic π^0 emission. It is seen that the first order anisotropy coefficients are dominated by the decay effect. Although second order coefficients also have significant decay contribution, the measured values are much higher.

This is the first time that a fully consistent result is obtained both from charged particle anisotropy studies and photon anisotropy studies which are in agreement with each other.

3.3. Chiral symmetry restoration and disoriented chiral condensates

The restoration of chiral symmetry and its subsequent breaking through a phase transition has been predicted to create Disoriented Chiral Condensates (DCC) [23]. This phenomenon has been predicted to cause anomalous fluctuations in the relative production of charged and neutral pions in high energy hadronic and nuclear collisions in certain phase space regions (domains). It is predicted that the neutral pion fraction $f = \frac{N_{\pi^o}}{N_{\pi^o}+N_{\pi^\pm}}$ should follow a $1/\sqrt{f}$ distribution in the case of DCC formation in contrast to the binomial distribution for generic particle production. A number of methods has been proposed to study the formation of DCC (see references in [23]).

The formation of DCC has been studied extensively using the data from the two detectors, SPMD and PMD, measuring respectively the multiplicity of charged particles and photons in overlapping part of phase space in the WA98 experiment.

3.3.1. Correlation and wavelet analysis : DCC has been studied in the WA98 experiment by various methods, both in the full overlapping phase space of the two detectors and also within smaller azimuthal domains [4, 6, 8]. Approaches based on direct correlation between charged particle and photon multiplicities event-by-event and discrete wavelet transform method have been used to study the distributions and their widths (rms values) looking for evidence of non-statistical fluctuation when compared with various mixed event formulations.

By constructing a set of mixed events to study various detector effects, the analysis showed that there exist non-statistical fluctuation individually in photon and charged particle production. However a correlated fluctuation has not been observed by this method of analysis. These results have been further compared with the predictions of a simple DCC model where DCC effect is introduced in the particle distribution of an event from a VENUS event generator at freezout by pairwise flipping of charged and neutral pions according to the $1/\sqrt{f}$ probability in a selected $(\eta - \phi)$ domain. The resulting distribution, after allowing π^0 to decay, is then analysed using the same correlation and wavelet analysis methods. An upper limit on the production of DCC-like fluctuation is estimated to be 10^{-2} for $\Delta\phi$ between 45-90° and 3×10^{-3} for $\Delta\phi$ between 90-135° for the top 5% centrality.

Further studies have also been carried out for the effect of centrality [8]. The upper limit curves are shown in Fig. 8 as a function of the azimuthal window for two centralities, top 5% (solid lines) and top 5-10% (dashed lines). It is observed that the fluctuations may be more at lower centralities.

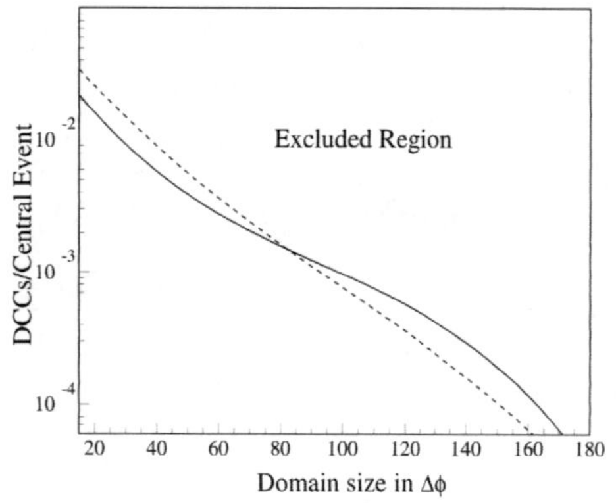

Figure 8. The 90% confidence level upper limit on DCC formation for central Pb+Pb collisions at 158A GeV, as a function of domain size in azimuthal angle within the context of a simple DCC model and the measured photon and charged particle multiplicities in the interval $2.9 \leq \eta \leq 3.75$.

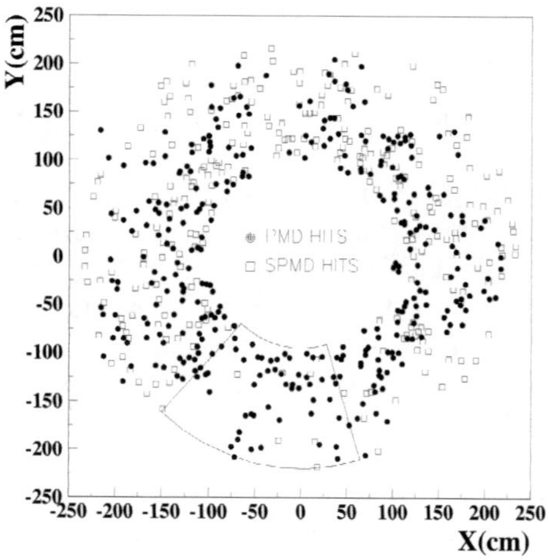

Figure 9. Plot showing photon hits (PMD) and charged particle hits (SPMD) in an azimuthal plane. The marked patch shows a region where number of charged particles are very less compared to photons.

3.3.2. Sliding window method : Several methods have been developed [24, 25] for the study of unusual structures in events arising due to non-statistical fluctuations resulting from phase

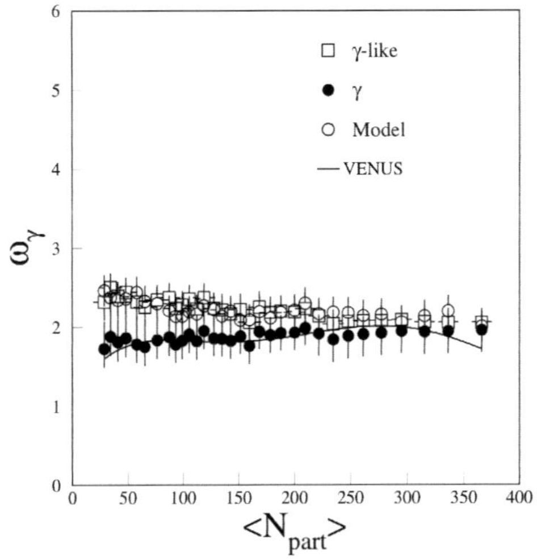

Figure 10. The relative fluctuations, ω_γ of photons as a function of number of participants. These are compared to calculations from a participant model and those from VENUS event generator.

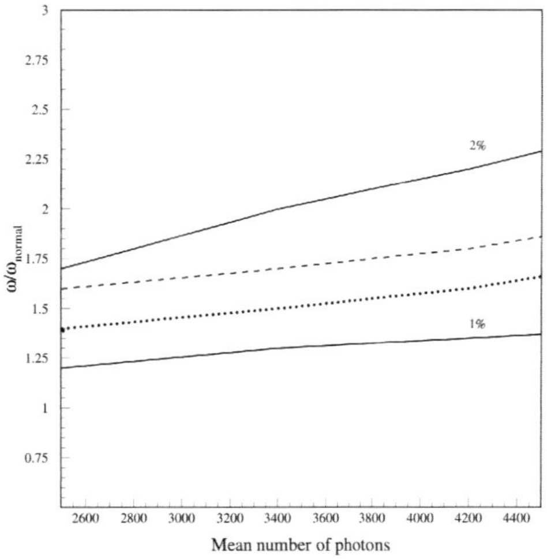

Figure 11. Fluctuation ω/ω_{normal} as a function of centrality. The dotted and dashed lines show the limits due to detector effects for a standalone PMD and for PMD with upstream material in the ALICE environment.

transitions. The sliding window method (SWM) is particularly appealing for the study of charge-neutral fluctuations and search for Centauro and anti-Centauro like events through model-independent direct observation. Some details of the method and preliminary results for the WA98 experiment are given in Ref. [25]. A typical event with an anti-Centauro like patch identified by the SWM is shown in Fig. 9, displaying PMD hits (filled circles) and SPMD hits (open boxes), projected onto the PMD plane, within the overlapping (η, ϕ) zone. The marked $60°$ patch corresponds to $f = 0.8$, this has only 7 charged particle hits as compared to 55 photon hits.

The percentage of events having exotic patches in data, with f-values beyond 4.5σ of the generic f- distribution, is found to be 0.39 ± 0.016, as compared to 0.081 ± 0.007 in mixed events and 0.013 ± 0.008 in GEANT simulated VENUS events. These patches are almost uniformly distributed in azimuth. Although a direct comparison with the results of upper limits, described in the previous section is not possible, the present value of 0.39% seems within the expected range. The present results are based on direct observation of event structure and are model-independent. The results, however, are preliminary and further checks are in progress to study the detector related effects as this is an important observation.

3.4. Multiplicity fluctuation

With the production of large number of particles on an event-by-event basis in high energy heavy ion collisions, it is very important to study anomalous fluctuations of multiplicity which is a major signal of QGP phase transition [26]. It has been proposed that event-by-event multiplicity fluctuations can arise from density fluctuations or droplet formation due to first order phase transition [27]. Large fluctuations in multiplicity are also expected at the tricritical point of QCD phase diagram [28]. Event-by-event fluctuations in the multiplicities of photons and charged particles have been studied in the WA98 experiment at CERN-SPS [7]. Fig. 10

shows the relative fluctuations, ω_γ $(= \sigma^2/N_\gamma)$ of photons as a function of mean number of participants. The data presented show the fluctuations in $\gamma - like$ clusters (open squares) and photons after correction for efficiency and purity (filled circles).

These are compared to calculations from a participant model [29] and those from VENUS event generator. The centrality dependence of charged particle multiplicity fluctuation in the measured data is found to agree reasonably well with those obtained from the participant model. However, for photons the multiplicity fluctuation has been found to be lower compared to those obtained from the participant model.

4. Future possibilities

The study of anisotropy of produced particles will remain a major goal of the PMD. One can deduce the flow values for neutral pions and compare with those for charged pions measured using other subsystems. The PMD will also provide the event plane information for the study of anisotropic emission of particles in any other rapidity window, both in STAR and in ALICE experiments. The usefulness of PMD has already been studied in the context of anisotropic emission of J/Ψ particles in the ALICE dimuon spcetroemter [30].

The STAR experiment has a forward time projection chamber (FTPC) in a pseudorapidity region which is overlapping with the coverage of the PMD. The FTPC provides a measure of charged particle multiplicity as well as their momenta. Using the momentum selection of charged particles, the study of DCC-like charged neutral fluctuation should be possible with improved results as has been shown by simulation studies [31]. It is shown that the p_T selection of charged particles enhances the DCC signal considerably more than that of photons.

In the ALICE experiment at the LHC, the much higher multiplicity of produced particles should result in highly reduced statistical fluctuation and consequently enhanced sensitivity to dynamical fluctuation. Possible physics studies with the PMD in ALICE have been presented in Refs. [13, 16]. The forward multiplicity detector (FMD) in ALICE has the (η, ϕ) coverage overlapping with those of the PMD. The FMD-PMD pair is expected to be used for the study of charged-neutral fluctuation at the LHC energy.

The sensitivity of the PMD to small fluctuations has been studied using the fluctuation measure $\omega = \sigma_\gamma^2/N_\gamma$ in a fast simulation technique where dynamical fluctuation has been introduced in the particle distribution to make it wider. The resulting distribution is again analyzed for the same fluctuation measure. Fig. 11 shows the ratio of fluctuation measure after and before introducing the dynamical component plotted as a function of the mean multiplicity of photons in PMD. It is found that dynamical fluctuation as low as 2% can be observed with the PMD even at lower multiplicities. As the statistical fluctuation decreases with increasing multiplicity, the difference between the limits set by detector effects and the dynamical fluctuation increases, providing more sensitivity at higher multiplicity.

5. Summary

Using a fine granularity preshower photon multiplicity detector (PMD), important contribution has ben made to the study of ultra-relativistic heavy ion collisions. Beginning with a detector made of plastic scintillator pads and wavelength shifting optical fibres for the SPS experiments, the PMD for the RHIC and LHC colliders has been made in the form of honeycomb gas proportional counters using a novel concept of extended cathode. Among the important studies at the SPS are first observation of collective flow and a detailed study of localized charged-neutral fluctuation as a probe for DCC. In addition the PMD's contribution include study of pseudo-rapidity distributions of photons, scaling of photon multiplicity with number of participating nucleons, centrality dependence of $< p_T >$ of photons and event-by-event fluctuations in photon multiplicity. The PMD is taking data at the RHIC collider and is expected to be installed in ALICE experiment at the LHC in 2007. The RHIC program has already paid dividends in the study of limiting fragmentation of identified particles. The study of DCC will be continued at the RHIC using the PMD and the FTPC in STAR experiment and at the LHC using the PMD and the FMD in the ALICE experiment. At the LHC, larger multiplicity will be helpful in studying the event-by-event fluctuation in global observables with better precision.

Acknowledgments
The experimental program for the study of quark-gluon plasma at CERN and BNL is funded by the Department of Atomic Energy and the Department of Science and Technology of the Government of India. In addition the program has also derived financial support from the University Grants Commission and the Council of Scientific and Industrial Research of the Government of India, the STAR Collaboration, the CERN NMS grants, and the Indo-German Exchange Program in various stages of the work. We acknowledge the collaborative work of all the team members, past and present, of the WA93 collaboration, WA98 collaboration, STAR collaboration and ALICE Collaboration. We acknowledge the extensive support received from the engineering and technical manpower and computer facility personnel at BNL and CERN and also at all the collaborating institutions, namely VECC Kolkata, IOP Bhubaneswar, Panjab University Chandigarh, University of Rajasthan Jaipur, University of Jammu, Jammu and GSI Darmstadt.

References

[1] Aggarwal M.M. et al, WA93 Collaboration, 1997, *Phys. Lett.* **B403** 207.
[2] Aggarwal M.M. et al, WA93 Collaboration, 1997, *Phys. Lett.* **B404** 207.
[3] Aggarwal M.M. et al, WA93 Collaboration, 1998, *Phys. Rev.* **C58** 1146.
[4] Aggarwal M.M. et al, WA98 Collaboration, 1998, *Phys. Lett.* **B420** 169.
[5] Aggarwal M.M. et al, WA98 Collaboration, 1999, *Phys. Lett.* **B458** 422.
[6] Aggarwal M.M. et al, WA98 Collaboration, 2001, *Phys. Rev.* **C64** 011901.
[7] Aggarwal M.M. *et al.*, WA98 Collaboration, 2002, *Phys. Rev.* **C 65** 054912.
[8] Aggarwal M.M. et al, WA98 Collaboration, 2003, *Phys. Rev.* **C67** 044901.
[9] Aggarwal M.M. *et al.*, WA98 Collaboration, 2005, *Eur. Phys. J* **C41** 287.
[10] Adams J. *et al.*, STAR Collaboration, 2005, *Phys. Rev. Lett.* **95** 062301.
[11] Aggarwal M.M. et al., 1996, *Nucl. Instr. Meth.* **A372** 143.
[12] Aggarwal M.M. et al., 1999, *Nucl. Instr. Meth.* **A424** 395.
[13] ALICE Technical Design Report on Photon Multiplicity Detector ,1999, CERN/LHCC 99-32.
[14] Aggarwal M.M. et al., 2002, *Nucl. Instr. Meth.* **A488** 131.
[15] Aggarwal M.M. et al., 2003, *Nucl. Instr. Meth.* **A499** 751.
[16] ALICE Technical Design Report on Photon Multiplicity Detector, Addendum-1, CERN/LHCC 2003-038 (2003).
[17] Carminati F. et al., Physics Performance Report of ALICE Vol. I, 2004, *Jour. Phys. G, Nucl. Part. Phys.* **30** 1654.
[18] Chattopadhyay S., Ahammed Z. and Viyogi Y.P., 1999, *Nucl. Instrum. Meth.* **A421** 558.
[19] Werner K., 1993 *Phys. Rep.* **C232** 87.
[20] Mohanty B. et al, STAR Collaboration, contribution in these proceedings.
[21] Ollitrault J.-Y., 1992, *Phys. Rev.* **D46** 229.
[22] Raniwala R., Raniwala S., Viyogi Y.P., 2000, *Phys. Lett.* **B489** 9.
[23] Mohanty B. and Serreau B., 2005, *Phys. Rept.* **414** 263 and references therein.
[24] Das A.C. and Viyogi Y.P., 1996, *Phys.Lett.* **B380** 437; Aggarwal M.M., Bhatia V.S., Das A.C. and Viyogi Y.P., 1998, *Phys.Lett.* **B438** 357.
[25] Aggarwal M.M.,WA98 Collaboration, 2003, *Pramana* **60** 987.
[26] Heiselberg H., 2001, *Phys. Rep.* **351** 161.
[27] Van Hove L., 1984, *Z. Phys.* **C 21** 93; Kapusta J.I., Vischer A.P., 1995, *Phys. Rev.* **C 52** 2725.
[28] Stephanov M., Rajagopal K. and Shuryak E., 1998, *Phys. Rev. Lett.* **81** 4816.
[29] Baym G. and Heiselberg H., 1999, *Phys. Lett.* **B469** 7; Baym G. et al., *Phys. Lett.* **B219** 205.
[30] Morsch A., Raniwala R., Raniwala S., Viyogi Y.P., 2001, ALICE Internal Note ALICE-INT-2001-22.
[31] Mohanty B., Nayak T.K., Mahapatra D.P. and Viyogi Y.P., 2004, *Int. J. Mod. Phys.* **A19** 1453.

Institute of Physics Publishing Journal of Physics: Conference Series **50** (2006) 62–69
doi:10.1088/1742-6596/50/1/007 5th International Conference on Physics and Astrophysics of Quark Gluon Plasma

Signatures for strongly coupled Quark-Gluon Plasma

Edward Shuryak

Department of Physics and Astronomy,
Stony Brook NY 11794 USA

E-mail: `shuryak@tonic.physics.sunysb.edu`

Abstract. Dramatic changes had occurred with our understanding of Quark-Gluon Plasma, which is now believed to be rather strongly coupled, sQGP for short. Hydrodynamical behavior is seen experimentally, even for rather small systems (rather peripheral collisions). From elliptic flow the interest is shifting to even more sophysticated observable, the *conical flow*, created by quenched jets. The exact structure of sQGP remains unknown, at the moment the best picture seem to be a liquid made partly of binary bound states. As we discuss at the end, those can be possibly seen in the dilepton spectra, as "new vector mesons" above T_c.

1. Introduction: a strongly coupled QGP

Although some speculations about further running of the QCD coupling constant at $T > T_c$ to higher values were around for a while (see e.g. Fig.5 in [1]), the main arguments for a new QGP regime at $T = (1-3)T_c$ are: (i) a triumph of hydrodynamical description of RHIC collisions, explaining strong radial and elliptic flows [2, 3]; (ii) discovery of multiple bound states of heavy and light quarks at $T > T_c$, from the lattice correlators [4] and directly from Schreodinger/Klein-Gordon/Dirac equations [5, 6] with properly defined (lattice-based) potentials [7]. These general arguments are by now well accepted in the community, see e.g. [8] and the "experimental white papers", and so I would not discuss them here in any detail, just make a couple of introductory remarkes.

One is a reply to the question raized in the opening talk by J.P.Blaizot, who wandered why do we know that equilibration happens as early as at time $\tau \sim 1/2fm$. The most direct asnwer to that is that hydro works well till very peripheral collisions, abruptly failing only when the initial "almond" is less than 1 fm wide.

Another comment is on an inevitable question: *How strong is strong enough?* The answer, suggested in a "strong coupling" paper [5], is that it should be sufficiently strong to *bind* quasiparticles together. If so, lines on the phase diagram were bound states gets *marginal* (have zero binding) separate the sQGP region. Presumably, the most stable is the most attractive gg singlet channel. These lines are *not* singularities of the thermodynamical quantities: as marginal states become virtual the scattering amplitude remains continuous.

One particular bound state, J/ψ, got special attention. According to lattice results [4] using MEM method and our calculations based on lattice potentials [6] it does not melt for $T < 2.5 - 3T_c$.

An important new feature sQGP scenario [6] is and possible existence of hundreds of *colored* bound states, such as qg_3 and $(gg)_8$, which outnumber the colorless ones by about 10:1. This prediction is not yet tested on the lattice[1], but it will be. Calculations presented in that work solved the "pressure puzzle", explaining where additional contribution to that of (not-so-light) quasiparticles comes from. New binary bound states help to get the EoS right, see [6].

This is of course only a beginning of the road toward a full understanding of sQGP. Studies related to 3-body physics (known as Efimov effect) have revealed that if there are marginal 2-body states, there must also be 3-body ones, some of them even deeply bound. A really manybody studies are not yet done, and it may well be that sQGP is polymerized into some chains, or crystallize into a (locally ordered) liquid. Color degrees of freedom are numerous and provide lots of possibilities, so a lot of hard work is ahead to understand the sQGP structure.

As experience of other fields shows, global EoS is probably too insensitive a measure to tell the difference between all those models, at least with the level of accuracy we now have. *Transport properties* like *viscosity* is a different matter, they are known to be quite sensitive to coupling and can vary by several orders of magnitude for known liquid and plasmas. Its theory is still missing, and a lot of efforts would be needed to have it.

One early idea suggested in [5] is based on the*marginal states* with small binding, which may be related to large scattering length and hydro behavior. This idea is further confirmed by experiments with cold atoms, where precisely this mechanism – known as the Feshbach resonance – is used. Not only the elliptic flow in agreement with hydrodynamics was observed [9], but more recent experiments such as [10] have revealed that in fact frequency of the two lowest (z and r-modes) of oscillations in elliptic trap both agree with hydro prediction better than a percent. Furthermore, the damping-to-frequency ratio dives from $\sim .1$ to about 10^{-3} near the Feshbach resonance. It means that there hydro oscillations may be repeated about a thousand times, before dissipative viscose effects take over and kill it. And all of it happens, let me repeat, where the interaction is the *strongest*, corresponding to viscosity as strikingly small [11], as it is for QGP at RHIC.

Another connection on which I would like to provide an update is a connection to $\mathcal{N}$=4 SUSY YM theory and AdS/CFT correspondence. Let me just remind that this theory has zero beta function and the coupling does not run, and for all values from 0 to infinity the finite-T version of it is in the same Coulomb (QGP-like) phase. In [12] we argued that matter is made not from quasiparticles, which have large masses $M_{q,g} \sim \sqrt{\lambda}T$ (where $\lambda = g^2 N_c >> 1$ is 't Hooft gauge coupling) but of deeply bound binary states of those with much smaller masses $M_{mesons} \sim T$ which we found for relativistically rotating states with large angular momentum. Recently there was a progress in introducing fundamental massive quarks into AdS/CFT theory, living on a separate D7 brane, with results for "charmonium" spectroscopy in strong coupling [13]. It was found that the "s-wave" states with zero orbital momentum do not "fall on the center" but survive and do have masses $M_{mesons} \sim M_q/\lambda^{1/2}$, the same scale as needed to populate the heatbath (as advocated in our paper). Thus a splitting between $J/\psi, \eta_c$ and χ states is large and parametric in strong coupling. Some dreams about using the AdS/CFT correspondence for dynamical description of high energy collisions start to be discussed.

2. Where does the energy of the quenched jets go?

Quenching of high-p_t QCD jets leads to deposition of large energy into matter. If it is a near-perfect non-dissipative liquid, this energy should propagate outward as a "conical flow" similar to well known sonic booms from supersonic planes. This idea is developed in a recent paper[2] [14].

In what follows, we would like to treat separately two different types of energy losses: (i) the *radiative* losses, producing mostly relativistic gluons and (ii) the *scattering/ionization* losses,

[1] It is based on expected Casimir scaling of forces in all channels.

[2] As I learned in Kolkata, it was also suggested earlier by H.Stocker, see his talk for details.

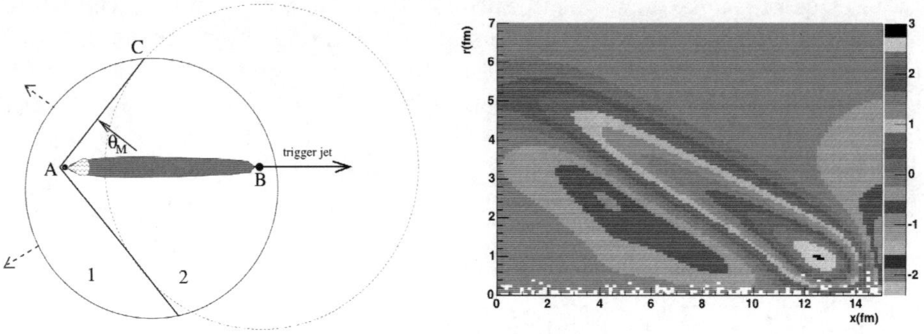

Figure 1. (a) A schematic picture of flow created by a jet going through the fireball. The trigger jet is going to the right from the origination point (the black circle at point B) from which sound waves start propagating as spherical waves (the dashed circle). The companion quenched jet is moving to the left, heating the matter and thus creating a cylinder of additional matter (shaded area). The head of the jet is a "non-hydrodynamical core" of the QCD gluonic shower, formed by the original hard parton (black dot A). The solid arrow shows a direction of flow normal to shock cone at the angle θ_M, the dashed arrows show the direction of the flow after shocks hit the edge of the fireball. (b) Distribution of the radial flow velocity v_r in the $x - r$ plane (units used are fm) for a jet propagating through the diameter of Au. We used $dE/dx = 1\,GeV/fm$ and the width of the Gaussian $\sigma = 1\,fm$.

which deposit energy in the medium. The radiative losses [15, 16] is the dominant mechanism. However both the primary parton and radiated gluons move with about the same speed (of light) so from the viewpoint of matter those can be treated as one external object. What is now more important are elastic energy losses (first studied by Bjorken[17]) and also those due to "ionization" of new bound states in sQGP. The latter contribution were recently calculated by Zahed and myself [18]. These mechanisms deposit additional energy, momentum and entropy into the matter, and their combined magnitude is $dE/dx = 2 - 3\,GeV/fm$. Thus a jet up to 20-30 GeV would be quenched, if it passesg through the diameter of the fireball.

Let us start a discussion of associated collective effects with *the energy scales involved.* While the total CM energy in a AuAu at RHIC is very large, about 40 TeV compared to the energy of a jet (typically 5-20 GeV), the jet energy is transverse. The total transverse energy of all secondaries per one unit of rapidity $dE_\perp/dy \sim 600\,GeV$. Most of it is thermal, with only about 100 GeV being related to collective motion. Furthermore, the so called elliptic flow is a $\sim 1/10$ asymmetry and therefore it carries energy $\sim 10\,GeV$ which is quite comparable to that lost by jets in question. Since elliptic flow was observed and studied in detail, we conclude that conical flow should be observable as well.

Fig.1(a) explains a view of the process, in a plane transverse to the beam. Two oppositely moving jets originate from the hard collision point B. Due to strong quenching, the survival of the trigger jet biases it to be produced close to the surface and to move outward. This in turn forces its companion to move inward through matter and to be maximally quenched.

The energy deposition starts at point B, thus a spherical sound wave appears (the dashed circle in Fig.1(a)). Further energy deposition is along the jet line, and is propagating with a speed of light, till the leading parton is found at point A at the moment of the snapshot. As is well known, the interference of perturbations from a supersonically moving body (such as a supersonic jet plane or meteorite) create conical flow behind the shock waves. The angle θ_M defined in the figure is given by a simple geometrical condition: distance $AB = ct$ while CB is the distance travelled by a shock. Since the velocity of the shock depends on its intensity, the cone should in fact be somewhat rounded near its top: that is ignored in the figure. The region near the head of the jet, which we would refer to as a "non-hydrodynamical core", which is constantly producing gluons, which emit new ones etc: the whole shower is a complicated nonlinear phenomenon which should obviously be treated via tools of quantum field theory. The energy/momentum flow through the core boundary we will treat phenomenologically, identifying it with dE/dx of the second type, which we will approximate by a time-independent constant. This should thus lead to stationary-state conical flow, depending on $x - t$ only.

Hydrodynamical equations we use are linear since the whole energy deposited by the jets can be treated as small compared to the total energy of the medium. We will use cylindrical coordinates with x along the jet axis and r orthogonal to it. For simplicity we will assume that the perturbed medium is homogeneous and at rest. In this linearized approximation the nonzero components of the stress-energy tensor are

$$\delta T^{00} = \delta\epsilon \qquad \delta T^{0i} = (\epsilon + p)v^i \qquad \delta T^{ij} = \delta p \delta^{ij} \tag{1}$$

where ϵ and p are the internal energy and pressure and $\vec{v}$ is the velocity field of the perturbation. Thus, by recalling that $\frac{\partial p}{\partial e} = c_s^2$ (with c_s velocity of sound) the energy and momentum conservation equation $\partial_\mu \delta T^{\mu\nu}$ can be written as

$$\begin{aligned} \partial_0 \delta T^{00} + \partial_i \delta T^{0i} &= 0 \\ \partial_0 \delta T^{0i} + c_s^2 \partial_i \delta T^{00} &= 0 \end{aligned} \tag{2}$$

The initial conditions are set by the the process of thermalization of the energy lost by the jet. As already mentioned, this thermalization process is very complicated and should take place at distances of the order Γ_s from the production point. As Γ_s is also the typical size of the liquid cells, we will simply consider that there is a variation of the energy and momentum at the position of the particle.

This initial condition can be easily expressed if we first concentrate in the perturbation due to the propagation of the particle in an interval dt around its position. In this case the previous conditions lead to

$$\begin{aligned} \delta T^{00}(t_0, x) = \Delta E \delta^3(\vec{x}(t_0) - \vec{x}) \\ \delta T^{0x}(t_0, x) = \Delta P \delta^3(\vec{x}(t_0) - \vec{x}) \end{aligned} \tag{3}$$

where we have assumed that the particle moves in the x direction at the speed of light ($c = 1$) and $\vec{x}(t)$ is the trajectory of the particle ($x = t$). We assume also that the energy and momentum loss are equal and, as we look for a constant drag, we set $\Delta E = \frac{dE}{dx}dt$. Assuming that c_s is constant, it can be shown that the solution for these conditions is given by:

$$\begin{aligned} \delta T^{00} = \frac{1}{c_s}\frac{dE}{dx} dt_0 \left(\partial_t - \partial_x\right) \Theta(t - t_0) \\ \frac{\left[\delta\left(R - c_s(t - t_0)\right) - \delta\left(R + c_s(t - t_0)\right)\right]}{4\pi R} \end{aligned} \tag{4}$$

Where $R = |\vec{x} - \vec{x}(t_0)|$, the distance from the observation point to the emission point at the intersection of the past "sound cone" and a jet world line. The reader can readily identify the argument of the derivatives as the retarded Green function of the wave equation, and be warned that R depends on observation time and space point. The calculation is close to that done for electromagnetic waves (Cerenkov radiation), except that sound is a longitudinal excitation.

In general, let a jet be born at the initial point t_i and dies the final point t_f. A point-like source creates two spheres of perturbation, with the Mach cone tangent to both:

$$\begin{aligned}\delta T^{00} &= \qquad (5)\\ &\frac{1}{c_s^2}\frac{dE}{dx}\left(\frac{\delta(t-t_i-R_{t_i}/c_s)}{4\pi R_{t_i}} - \frac{\delta(t-t_f-R_{t_f}/c_s)}{4\pi R_{t_f}}\right)\\ &- \frac{2}{c_s^2}\frac{dE}{dx}\partial_x \frac{\Theta(t-t_i-R(t)/c_s)\Theta(t_f-t+R(t)/c_s)}{4\pi\sqrt{(x-t)^2-(\frac{1}{c_s^2}-1)r^2}}\end{aligned}$$

when integrating over the whole space, the two spherical terms give the total deposited energy $\frac{dE}{dx}(t_f - t_i)$. The conical term apparently has no energy. In order to explain why, we prefer to regularize the problem, introducing a distributed source, e.g. a Gaussian one, see the paper for details.

How can the effect be observed? The most important feature of conical flow is its direction, at the so called Mach angle (6). Let us start with the simplest case of central collisions, in which the cross section of the fireball is a circle and the elliptic flow is absent. Most quenched jets going through it will pass a length comparable to its diameter, and thus the high-energy jets which are not quenched completely and reach through the fireball will produce conical flow during time $\tau \sim 10 - 15 fm/c$. The angle θ_M defined in the figure is given by a simple geometrical condition: the distance travelled by jet during the proper time interval.

$$cos\theta_M = \frac{1}{c\tau}\int c_s dt \qquad (6)$$

The appropriate value for the speed of sound is a weighted average of that for QGP ($c_s \approx 1/\sqrt{3}$), the mixed phase $c_s \approx 0$ and a "resonance gas", $c_s \approx \sqrt{.2}$, $\bar{c}_s \approx .33$. We thus conclude that in this case the emission angle of the conical flow should be at the angles about 70 degrees relative to the jet, or in radians

$$\theta_{emission} = \pi \pm arccos(\bar{c}_s/c) \approx 1.9, 4.3 \qquad (7)$$

In Fig.2(left) we show preliminary data from STAR experiment. The peak near angle zero consists of particles from a triggered outward moving jet, while particles from the companion jet should result in a peak near $\Delta\phi = \pi$. Such peak is clearly seen in pp collisions (a), in which no matter is present: but in AuAu collision (b) one finds instead a *double-peaked* distribution with a *minimum* at the original jet direction $\Delta\phi = \pi$. Remarkably, as Fig.2(b) shows, this value $\pi \pm 1.2$ nicely agrees with the observed positions of the two maxima of the distribution of the secondaries, relative to the direction of the triggered hard charged hadron.

At this conference B.Jacak have for the first time reported analogous data from PHENIX experiment (see her talk). It selects companions with $p_t > 1\,GeV$ and different centrality classes. Five out of six of them (except the most peritheral bin) show even sharper peaks at the same angle as we predict. Furthermore, after the conference STAR reported spectra of associated partiles, measuring in particular their mean $< p_t > (\phi)$. It also has maxima in the direction away from $\phi = \pi$, where one finds a clear a minimum, consistent in magnitude with the mean transverse momentum in average background events. Thus, all known facts strongly support our explanation[3] of the peaks.

[3] It has been argued at this conference by I.Vitev (see his talk) that radiation effects calculated with Landau-Pomeranchuck-Migdal effect can also produce a maximum at large angles, similar to Mach angle we discuss. Those calculations however used static (infinitely massive) scatterers. If this unrealsitic approximation is lifted, a maximum presumably corresponds to an angle 90 degrees in CM. To make it as large as 70 degrees observed, one would need a combined mass of all target partons participating in the collision to be $M \sim 10\,GeV$ or so, which is hard to imagine.

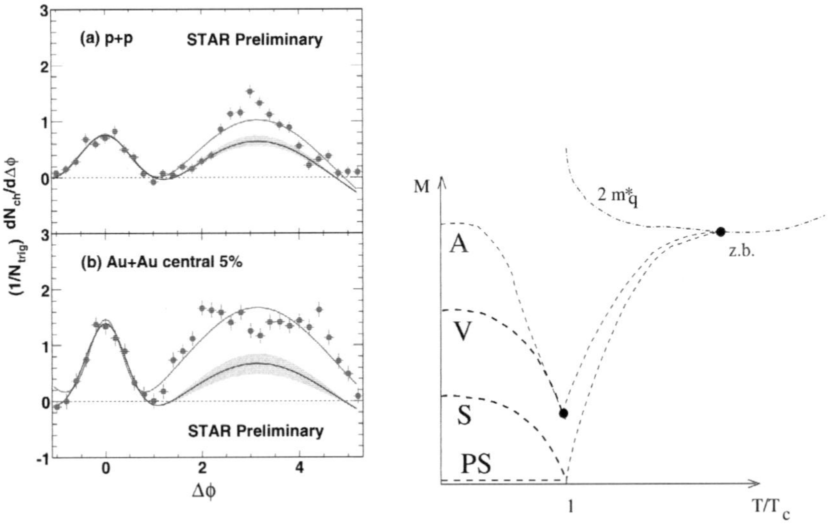

Figure 2. (left) The background subtracted distribution of secondaries in azimuthal angle $\Delta\phi$ which are associated with triggered hard charged hadron $p_t > 4\,GeV$, by STAR collaboration [20]. The figure (a) is for p+p and (b) for 5% most central Au+Au collisions. (right) Schematic T-dependence of the masses of $\bar{q}q$ states. A, V, S and PS stand for axial, vector, scalar and pseudoscalar states. The dash-dotted line shows a behavior of twice the quasiparticle mass. Two black dots indicate places where we hope the dilepton signal may be observable.

3. Direct observation of binary bound states in sQGP via dileptons?

Photons and dileptons are "penetrating probes", providing information about all stages of heavy ion collisions [21]. It was then assumed that the invariant mass ($M^2 = (p_{l+} + p_{l-})^2$) spectra of penetrating probes produced in QGP should be monotonously decreasing with the dilepton mass, contrasting with peaks from hadronic stage due to familiar vector mesons $\rho, \omega, \phi, J/\psi$.... Now the logic of it is changed [19]: if meson-like bound states continue to exist at $T > T_c$ one may see them in dilepton spectra as well. Furthermore, even at high T when there are no bound states, there still exists a near-threshold enhancement, which may still be detected and used to infer the value of the quasiparticle masses and the strength of their interaction.

The observability of new states is obscured by the fact that their mass is "sliting" with T, from very large masses $M \sim 1.5 - 2\,GeV$ where the states gets unbound to normal values at freezeout. One may thus wander if the very existence of the peaks can be observed. We hope it can be done in two *endpoints* mass regions:

(i) $M \sim 1.5 - 2\,GeV$ corresponding to $T \approx T$(zero binding) and the edge of the quasiparticle continuum, at the initial QGP at RHIC.

(ii) $M \sim 0.5\,GeV$, where we expect to see the contributions of the modified vector mesons at $T \approx T_c$.

The basic idea is very simple: the probability of $\bar{q}q \to l^+l^-$ process is enhanced by the initial state attractive interaction. Attractive interaction obviously correlate $\bar{q}$ and q in space and increases the chances to find $\bar{q}q$ close to each other and annihilate. In such general form, the enhancement persists whether the potential is deep enough to make bound levels or not, and whether quark quasiparticles have small or large width.

In the case of pure Coulomb interaction the relevant parameter $z = \pi\alpha/v$ contains the ratio of the coupling strength to velocity. The enhancement is given by the well known Gamow factor[4]

[4] The sign in exponent is the opposite to that on the original Gamow factor, for alpha-particle interaction with

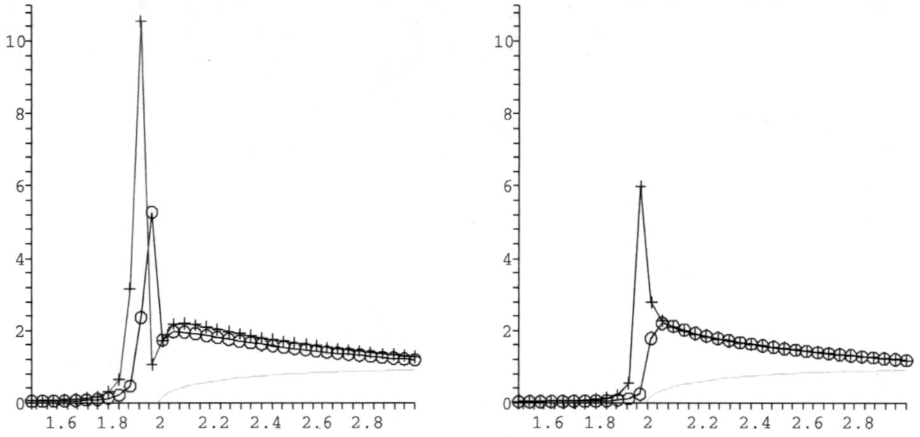

Figure 3. Modification of the spectral density versus the invariant mass in M_q units for different temperatures: (a) 1.2 Tc (cross) and 1.4 Tc (circle), (b) 1.7 Tc (cross) and 3 Tc (circle) and the correction due to quark mass (line).

$F_{Gamow} = \frac{z}{1-exp(-z)}$. Note that the result is $\sim 1/v$ at small v (large z) and 1 at small z. It cancels the velocity in the phase space and makes the production rate to jump to a $finite$ value at the threshold.

New (sQGP) part of this picture is at $T > T_c$, which starts and ends at points marked by small black circles. Two of the black dots, at the mixed phase $T = T_c$, have a benefit of long time spent there during expansion. The same is true for the third black point, due to near-constancy of the mass of weakly bound states at $T = 1.5 - 2T_c$. The near-threshold bump is at about the same location at higher T as well. Thus one may hope that the structures corresponding to these endpoints (black circles at Fig.2(right) may be detected.

In it we have shown only states made of u, d quarks, ignoring the strange one. Needless to say, those exist and can be also produced. The peaks of $\bar{s}s$ ϕ-like states should be shifted in mass upward by $2m_s \approx 250 - 300\,MeV$, but their contribution to dilepton spectra is proportional to square of the electric charge $q_s^2 = 1/9$, which is 3 times smaller than the average of $q_u^2 = 4/9, q_d^2 = 1/9$. Strange states are more promising for pseudoscalar/scalar signals to which we turn below.

Omitting technical part, let me just say that non-relativistic Green´s function $G_{E+i\Gamma_t}(r, \bar{r})$ obeys the usual Schrodinger equation:

$$[-\frac{1}{m}\vec{\nabla}^2 + V(\vec{r}) - (E + i\Gamma)]G_{E+i\Gamma}(r, \bar{r}) = \delta^3(\vec{r} - \vec{\bar{r}}) \tag{8}$$

with inter-particle potential at relevant T $V(\vec{r})$ extracted from the lattice. The benefit of using this method is that one avoid summations over levels, and bound states are automatically included together with scattering ones.

Let us now explain our units. Since the effective mass of quark quasiparticle in the temperature interval considered is not known accurately, we use it as our basic energy unit. In plots twice this value appear as a threshold, to which we tentatively ascribe to it a particularly simple value $2M_q = 2\,GeV$: the reader should however be warned that this is a "GeV" in

nuclei or that used for HBT correlations, corresponded to a *repulsive* Coulomb force.

quotation marks, to be rescaled appropriately later when the value of quark effective masses in QGP will be better known.

What we learned from these calculations is how the spectral density changes as the bound states become less and less bound until the system arrives to zero binding (zero binding point). In figure 3 we show the shape of the dilepton spectral density for different temperature (1.2 T_c, 1.4 T_c, 1.7 T_c, 3 T_c). One can observe how exactly enhancement in the bound state and threshold region changes. The hight of this enhancement depends on the proximity of the level; at 1.7 T_c, where the bound level is close to threshold, we observe a big modification of the spectral function at $2GeV$. Note that proximity of the level to threshold happens in rather wide T interval, roughly between 1.5 T_c to 2 T_c (the zero binding point [5]. Note also that the enhancement is still seen at temperatures as high as 3 T_c, when all bound states have already been melted.

The issue of the observability of these peaks depend on their width: its magnitude is being addressed right now and I am not yet ready to report those. The pictures above only have "numerical width" for plotting, and thus are not realistic. However it is very likely that the width is not overwhelming to preclude their observation.

Acknowledgments.
I am thankful to organizers for making this important conference such a success. Wish to express once again my gratitude to Bikash Sinha for all his efforts over the years, and wish him to go on with the same energy for many years to come. I am greatful to my collaborators G.Brown,J.Casalderrey-Solana, D.Teaney and I.Zahed. This work was partially supported by the US-DOE grants DE-FG02-88ER40388 and DE-FG03-97ER4014.

[1] E.Shuryak, Nucl.Phys.A715:289-298,2003 hep-ph/0205031

[2] D. Teaney, J. Lauret and E. V. Shuryak, Phys. Rev. Lett. **86**, 4783 (2001) [arXiv:nucl-th/0011058]. "A hydrodynamic description of heavy ion collisions at the SPS and RHIC," arXiv:nucl-th/0110037.

[3] P.F. Kolb, P.Huovinen, U. Heinz, H. Heiselberg, Phys. Lett. **B500** (2001) 232. More details in P. F. Kolb and U. Heinz, nucl-th/0305084.

[4] M. Asakawa and T. Hatsuda, Nucl. Phys. **A715** (2003) 863c;S. Datta, F. Karsch, P. Petreczky and I. Wetzorke, hep-lat/0208012.

[5] E. V. Shuryak and I. Zahed, Phys. Rev. C **70**, 021901 (2004) [arXiv:hep-ph/0307267].

[6] E. V. Shuryak and I. Zahed, Phys. Rev. D **70**, 054507 (2004) [arXiv:hep-ph/0403127].

[7] O. Kaczmarek, S. Ejiri, F. Karsch, E. Laermann and F. Zantow, hep-lat/0312015.

[8] M. Gyulassy and L. McLerran, arXiv:nucl-th/0405013.

[9] K.M. O'Hara *et al.*, Science **298** (2002) 2179; T. Bourdel et al, Phys. Rev. Lett. **91** (2003) 020402.

[10] M.Bartenstein *et al.*, cond-mat/0403716.

[11] B. A. Gelman, E. V. Shuryak and I. Zahed, arXiv:nucl-th/0410067.

[12] E.V. Shuryak and I. Zahed, Phys. Rev. **D69** (2004) 014011. [arXiv:hep-th/0308073].

[13] M. Kruczenski, D. Mateos, R. C. Myers and D. J. Winters, JHEP **0307**, 049 (2003) [arXiv:hep-th/0304032]. S. Hong, S. Yoon and M. J. Strassler, JHEP **0404**, 046 (2004) [arXiv:hep-th/0312071].

[14] J. Casalderrey-Solana, E. V. Shuryak and D. Teaney, arXiv:hep-ph/0411315.

[15] M. Gyulassy and M. Plumer, Phys. Lett. **B243** (1990) 432; X. N. Wang, M. Gyulassy and M. Plumer, Phys. Rev. **D51** (1995) 3436; G. Fai, G. G. Barnafoldi, M. Gyulassy, P. Levai, G. Papp, I. Vitev and Y. Zhang, hep-ph/0111211.

[16] R. Baier, Y. L. Dokshitzer, S. Peigne and D. Schiff, Phys. Lett. **B345** (1995); R. Baier, Y. L. Dokshitzer, A. H. Mueller and D. Schiff, JHEP **0109** (2001) 033.

[17] J. D. Bjorken, FERMILAB-PUB-82-059-THY; D. A. Appel, Phys. Rev. **D33** (1986) 717; J. P. Blaizot and L. D. McLerran, Phys. Rev. **D34** (1986) 2739.

[18] E. V. Shuryak and I. Zahed, "Ionization of binary bound states in a strongly coupled quark-gluon plasma," arXiv:hep-ph/0406100.

[19] J. Casalderrey-Solana and E. V. Shuryak, arXiv:hep-ph/0408178.

[20] Fuqiang Wang for STAR collaboration, (Quark Matter 2004), J.Phys.G30:S1299-S1304,2004; nucl-ex/0404010

[21] E. V. Shuryak, Phys. Lett. **B78** (1978) 150, Yadernaya Fizika (Sov.J. of Nucl.Physics) **28** (1978) 796. Phys.Rep. **61** (1980) 71.

Institute of Physics Publishing Journal of Physics: Conference Series **50** (2006) 70–78
doi:10.1088/1742-6596/50/1/008 5th International Conference on Physics and Astrophysics of Quark Gluon Plasma

The Color Glass Condensate: A classical effective theory of high energy QCD

Raju Venugopalan

Physics Department, Brookhaven National Laboratory, Upton, NY 11973, USA

E-mail: raju@bnl.gov

Abstract. The Color Glass Condensate is a theory of the dynamical properties of partons in the Regge limit of QCD: $x_{\rm Bj} \rightarrow 0$, $Q^2 >> \Lambda^2_{\rm QCD}$ = fixed and the center of mass energy squared $s \rightarrow \infty$. We provide a brief introduction to the theoretical ideas underlying the Color Glass Condensate. We also discuss how these ideas provide a unified framework to discuss both Deeply Inelastic Scattering (DIS), and hadronic collisions (from pp to AA) in QCD.

1. Introduction

The study of the properties of the strong interactions in the asymptotic Bjorken limit of momentum transfer squared $Q^2 \rightarrow \infty$, the center of mass energy squared $s \rightarrow \infty$, and the Bjorken variable $x_{\rm Bj} \approx Q^2/s$ = fixed has proved to be one of the most creative ideas in theoretical physics [1]. Relatively little work has been done in the other high energy limit, namely, $x_{\rm Bj} \rightarrow 0$, $s \rightarrow \infty$ and Q^2 = fixed. This limit of the strong interactions, which we shall call the Regge limit, was studied intensively in the 60's and indeed led eventually to string theory. The reason these studies fell into disfavor in the strong interactions was that there was no small parameter in these studies (in modern parlance, $Q^2 \leq \Lambda^2_{\rm QCD}$).

With the advent of the collider era, we can now probe a wide window of physics where $s >> Q^2 >> \Lambda^2_{\rm QCD}$. In fact, this "window" describes the bulk of the high energy cross-section. One therefore has finally the possibility of studying the properties of the Regge limit of the theory using *weak coupling methods*. In this limit, the hadron behaves like matter that's dense but weakly coupled-not dissimilar to much of condensed matter physics [2].

In Regge asymptotics, the number of partons increases rapidly due to QCD bremsstrahlung. This growth is described, in the leading logarithmic approximation in x, by the BFKL equation [3]. Since the typical size of the partons in this limit is of order $1/Q^2$, the hadron becomes closely packed when the number of partons is of order R^2Q^2. In fact, this corresponds to an occupation number $f \sim 1/\alpha_S$. When the density of partons is of this order, repulsive many body "recombination" and screening effects compete with QCD Bremsstrahlung leading to a saturation of the number of partons in the hadron's wavefunction [4, 5, 6]. The saturation of partons of different sizes happens at different values of x. The scale at which this occurs is the saturation scale $Q_s(x)$–a dynamically generated semi-hard scale that controls the dynamics of physics in this regime of QCD.

In the language of the Operator Product Regime (OPE), the line $Q \equiv Q_s(x)$ in the x-Q^2 plane denotes the regime beyond which (when approached from high Q^2) higher twist effects become important. Recall that the OPE is best formulated in the Bj-limit where higher twists are power suppressed and can be forgotten. The opposite is true in the Regge limit. Since the number of twist operators grows (nearly) exponentially with the twist, the OPE quickly becomes unwieldy. Thus to describe physics in this regime we need a new organizing principle in QCD beyond the OPE.

In the following section, we shall describe an effective field theory formalism which provides such an organizing principle. We shall discuss key features of this theory which are best summarized by the label Color Glass Condensate (CGC). The quantum evolution of the theory at high energies is described by renormalization group equations. These have remarkable features which we shall outline. We next discuss the applications of this theory to both DIS and hadronic scattering. The theory predicts novel universal observables which can be measured in both. Finally, we discuss the consequence of this approach for a theoretical description of the quark gluon plasma.

2. A classical effective theory for high energy QCD

There is a formal Born-Oppenheimer separation between large x and small x modes [8] for a quantum field theory on the light cone. These are respectively the slow and fast modes in the effective theory. Thus on the time scale of the "wee" parton small x fields, the large x partons can be viewed as static charges. Since these are color charges, they cannot be integrated out of the theory but must be viewed as sources of color charge for the dynamical wee fields. With this dynamical principle in mind, one can write down an effective action for wee partons in QCD at high energies [7]. The generating functional of wee partons has the form

$$\mathcal{Z}[j] = \int [d\rho]\, W_{\Lambda^+}[\rho] \left\{ \frac{\int^{\Lambda^+} [dA]\delta(A^+) e^{iS[A,\rho] - j\cdot A}}{\int^{\Lambda^+} [dA]\delta(A^+) e^{iS[A,\rho]}} \right\} \tag{1}$$

where the wee parton action has the form

$$S[A,\rho] = \frac{-1}{4} \int d^4x\, F_{\mu\nu}^2 + \frac{i}{N_c} \int d^2x_\perp dx^- \delta(x^-) \times \mathrm{Tr}\left(\rho(x_\perp) U_{-\infty,\infty}[A^-]\right) . \tag{2}$$

In Eq. 1, ρ is a classical color charge density of the static sources and $W[\rho]$ is a weight functional of sources (which sit at momenta $k^+ > \Lambda^+$: note, $x = k^+/P^+_{\text{hadron}}$). The sources are coupled to the dynamical wee gluon fields (which in turn sit at $k^+ < \Lambda^+$) via the gauge invariant term [1] which is the first term on the RHS of Eq. 2. The second term in Eq. 2 is the QCD field strength tensor squared-thus the wee gluons are treated in full generality in this effective theory, formulated in the light cone gauge $A^+ = 0$. The source j is an external source-derivatives taken with respect to this source (with the source then put to zero) generate correlation functions in the effective theory.

The argument for why the sources are classical is subtle and follows from a coarse graining of the effective action to only include modes of interest. For large nuclei, or at small x, the wee partons couple to a large number of sources. For a large nucleus, it can be shown explicitly that this source density is classical [10]. Further, it was conjectured that the weight functional for a large nucleus was a Gaussian in the source density (corresponding to the quadratic Casimir operator) [7, 11]. This was shown explicitly recently to be the correct-albeit with a small correction proportional to the cubic Casimir operator which generates Odderon excitations [10]. For a large nucleus, the variance of the Gaussian distribution, the color charge squared per unit area μ_A^2, proportional to $A^{1/3}$, is a large scale-and is the only scale in the effective action [2]. Thus

[1] An alternative form is obtained in Ref. [9]-it reproduces the BFKL equation more efficiently.

[2] μ_A^2 is simply related in the classical theory to the saturation scale Q_s^2 via the relation $Q_s^2 = \alpha_S N_c \mu_A^2 \ln(Q_s^2/\Lambda^2_{\text{QCD}})$

for $\mu_A^2 >> \Lambda_{\rm QCD}^2$, $\alpha_S(\mu_A^2) << 1$, and one can compute the properties of the theory in Eq. 1 in weak coupling.

The Yang-Mills equations can be solved analytically to obtain the classical field of the nucleus as a function of ρ: $A_{\rm cl.}(\rho)$ [7, 11, 12]. From the generating functional in Eq. 1, one obtains for the two point correlator,

$$< AA >= \int [d\rho]\, W_{\Lambda^+}[\rho]\, A_{\rm cl.}(\rho)\, A_{\rm cl.}(\rho)\,. \tag{3}$$

From this expression, one can determine (for Gaussian sources) the occupation number $\phi = dN/\pi R^2/dk_\perp^2 dy$ of wee partons in the classical field of the nucleus. For $k_\perp >> Q_s^2$, one has the Weizsäcker-Williams spectrum $\phi \sim Q_s^2/k_\perp^2$, while for $k_\perp \leq Q_s$, one has a resummation to all orders in $k_\perp$, which gives $\phi \sim \frac{1}{\alpha_S}\ln(Q_s/k_\perp)$. (The behavior at low $k_\perp$ can, more accurately, be represented as $\frac{1}{\alpha_S}\Gamma(0,z)$ where Γ is the incomplete Gamma function and $z = k_\perp^2/Q_s^2$.) A nice expression for the classical field of the nucleus containing these two limits is given in Ref. [13].

We are now in a position to discuss why a high energy hadron behaves like a Color Glass Condensate [2]. The "color" is obvious since the degrees of freedom, the partons, are colored. It is a glass because the stochastic sources (frozen on time scales much larger than the wee parton time scales) induce a stochastic (space-time dependent) coupling between the partons under quantum evolution (to be discussed in the next section)-this is analogous to a spin glass. Finally, the matter is a condensate since the wee partons have large occupation numbers (of order $1/\alpha_S$) and have momenta peaked about Q_s. As we will discuss, these properties are enhanced by quantum evolution in x. The classical field retains its structure-while the saturation scale grows: $Q_s(x') > Q_s(x)$ for $x' < x$.

Small fluctuations about the effective action in Eq. 2 were computed in Ref. [14]. These gave large corrections of order $\alpha_S \ln(1/x)$ which suggested that the Gaussian weight functional was fragile under quantum evolution of the sources. A Wilsonian renormalization group (RG) approach systematically treats these corrections [15]. The basic recipe is as follows. Begin with the generating functional in Eq. 1 at some Λ^+, with an initial source distribution $W[\rho]$. Perform small fluctuations about the classical saddle point of the effective action, integrating out momentum modes in the region $\Lambda'^+ < k^+ < \Lambda^+$, ensuring that Λ'^+ is such that $\alpha_S \ln(\Lambda^+/\Lambda'^+) << 1$. The action reproduces itself at the new scale Λ'^+, albeit with a charge density $\rho' = \rho + \delta\rho$, and $W_{\Lambda^+}[\rho] \longrightarrow W_{\Lambda'^+}[\rho']$. The change of the weight functional $W[\rho]$ with x is described by the JIMWLK- non-linear RG equation [15] which we shall not write explicitly here.

The JIMWLK equations form an infinite hierarchy (analogous to the BBGKY hierarchy in statistical mechanics) of ordinary differential equations for the gluon correlators $< A_1 A_2 \cdots A_n >_Y$, where $Y = \ln(1/x)$ is the rapidity. The expectation value of an operator $\mathcal{O}$ is defined to be

$$< \mathcal{O} >_Y = \int [d\alpha] \mathcal{O}[\alpha] W_Y[\alpha]\,, \tag{4}$$

where $\alpha = \frac{1}{\nabla_\perp^2}\rho$. The corresponding JIMWLK equation for this operator is

$$\frac{\partial < \mathcal{O}[\alpha] >_Y}{\partial Y} = < \frac{1}{2} \int_{x_\perp, y_\perp} \frac{\delta}{\delta \alpha_Y^a(x_\perp)} \chi^{ab}_{x_\perp, y_\perp}[\alpha] \frac{\delta}{\delta \alpha_Y^b(y_\perp)} \mathcal{O}[\alpha] >_Y \,. \tag{5}$$

χ here is a non-local object expressed in terms of path ordered (in rapidity) Wilson lines of α [2]. This equation is analogous to a (generalized) functional Fokker-Planck equation, where Y is the "time" and χ is a generalized diffusion coefficient. This equation illustrates the stochastic properties of operators in the space of gauge fields at high energies. For the gluon density, which is proportional to a two-point function $< \alpha^a(x_\perp)\alpha^b(y_\perp) >$, one recovers the BFKL equation in the limit of low parton densities. The theory is conformal so it is not inconceivable that the full hierarchy is exactly solvable. Preliminary numerical solutions exist [16] but much work remains in that direction. A mean field solution deep in the saturation regime [17] shows that the weight functional there is a non local Gaussian with a variance proportional to $k_\perp^2$ for $k_\perp^2 < Q_s^2$.

For large N_c and large A ($\alpha_S^2 A^{1/3} >> 1$), the expectation value of the product of traces of Wilson lines factorizes into the product of the expectation values of the traces:

$$< \mathrm{Tr}(V_x V_z^\dagger)\mathrm{Tr}(V_z V_y^\dagger) > \longrightarrow < \mathrm{Tr}(V_x V_z^\dagger) > < \mathrm{Tr}(V_z V_y^\dagger) > , \tag{6}$$

where $V_x = \mathcal{P}\exp\left(\int dz^- \alpha^a(z^-, x_\perp)T^a\right)$. Here $\mathcal{P}$ denotes path ordering in x^- and T^a is the SU(3) generator in the adjoint representation. In Mueller's dipole picture [3], the cross-section for a dipole scattering off a target can be expressed in terms of these 2-point dipole operators as [18]

$$\sigma_{q\bar{q}N}(x, r_\perp) = 2 \int d^2b\, \mathcal{N}_Y(x, r_\perp, b)\,, \tag{7}$$

where $\mathcal{N}_Y$, the imaginary part of the forward scattering amplitude, is defined to be $\mathcal{N}_Y = 1 - \frac{1}{N_c} < \mathrm{Tr}(V_x V_y^\dagger) >_Y$. Note that the size of the dipole, $\vec{r}_\perp = \vec{x}_\perp - \vec{y}_\perp$ and $\vec{b} = (\vec{x}_\perp + \vec{y}_\perp)/2$. The JIMWLK equation for the two point Wilson correlator is identical in this large A, large N_c mean field limit to an equation derived independently by Balitsky and Kovchegov-the Balitsky-Kovchegov equation [20], which has the operator form

$$\frac{\partial \mathcal{N}_Y}{\partial Y} = \bar{\alpha}_S\, \mathcal{K}_{\mathrm{BFKL}} \otimes \left\{\mathcal{N}_Y - \mathcal{N}_Y^2\right\}\,. \tag{8}$$

Here $\mathcal{K}_{\mathrm{BFKL}}$ is the well known BFKL kernel. When $\mathcal{N} << 1$, the quadratic term is negligble and one has BFKL growth of the number of dipoles; when $\mathcal{N}$ is close to unity, the growth saturates. The approach to unity can be computed analytically [21]. The B-K equation is the simplest equation including both the Bremsstrahlung responsible for the rapid growth of amplitudes at small x as well as the repulsive many body effects that lead to a saturation of this growth.

A saturation condition, say $\mathcal{N} = 1/2$, determines the saturation scale. One obtains $Q_s^2 = Q_0^2 \exp(\lambda Y)$, where $\lambda = c\alpha_S$ with $c \approx 4.8$. The saturation condition affects the overall normalization of this scale but does not affect the power λ. In fixed coupling, the power λ is large and there are large pre-asymptotic corrections to this relation-which die off only slowly as a function of Y. BFKL running coupling effects change the behavior of the saturation scale completely-one goes smoothly at large Y to $Q_s^2 = Q_0^2 \exp(\sqrt{2b_0 c(Y + Y_0)})$ where b_0 is the coefficient of the one-loop QCD β-function. The state of the art computation of Q_s is the work of Triantafyllopoulos, who obtained Q_s by solving NLO-resummed BFKL in the presence of an absorptive boundary (which corresponds to the CGC) [22]. The pre-asymptotic effects are much smaller in this case and the coefficient $\lambda \approx 0.25$ is very close to the value extracted from saturation model fits to the HERA data [23]. No analytical solution of the BK equation exists in the entire kinematic region but there have been several numerical studies at both fixed and running coupling [25, 24, 26]. These studies suggest that the solutions have a soliton like structure and that the saturation scale has the behavior discussed here. Geometrical scaling of solutions is seen for a wide window in rapidities. Running coupling effects, as suggested, are important and make the results of the computations more physically plausible.

The soliton like structure is no accident, as was discovered by Munier and Peschanski [27]. They noticed that the BK-equation, in a diffusion approximation, bore a formal analogy to

[3] See also Ref. [19].

the FKPP equation describing the propagation of unstable non-linear wavefronts in statistical mechanics [29]. In addition, the full BK-equation lies in the universality class of the FKPP equation. This enables one to extract the universal properties such as the leading pre-asymptotic terms in the expression for the saturation scale. It was realized [28] that a stochastic generalization of the FKPP equation-the sFKPP equation-could provide insights into impact parameter dependent fluctuations [30] in high energy QCD beyond the BK-equation. This is a very active area of research now, with several groups hunting for the Pomeron loops responsible for these fluctuations.

To summarize, the Color Glass Condensate is a weak coupling effective theory describing the properties of hadron wavefunctions in QCD at high energies. Renormalization group equations-the JIMWLK equations-describe the behavior of multi-parton correlations in the hadron wavefunction as a function of rapidity. The theory has stochastic features closely analogous to the propagation of unstable non-linear wave fronts in statistical mechanics. Recent work is focused on trying to understand possible corrections beyond JIMWLK at low parton densities-which may be responsible for Pomeron loops. We now turn to the applications of this theory to hadronic scattering.

3. Hadronic scattering and $k_\perp$ factorization in the Color Glass Condensate

At collider energies, a new window opens up where $\Lambda_{\rm QCD}^2 << M^2 << s$. In principle, cross-sections in this window can be computed in the usual collinear factorization language-however, one needs to sum up large logarithmic corrections in s/M^2. An alternative formalism is that of $k_\perp$-factorization [31, 32], where one has a convolution of $k_\perp$ dependent "un-integrated" gluon distributions from the two hadrons with the hard scattering matrix. In this case, the in-coming partons from the wavefunctions have non-zero $k_\perp$. The rapidity dependence of the unintegrated distributions is given by the BFKL equation. However, unlike the structure functions, it has not been proven that these unintegrated distributions are universal functions.

At small x, both the collinear factorization and $k_\perp$ factorization limits can be understood in a systematic way in the framework of the Color Glass Condensate. Rather than a convolution of probabilities, one has instead a collision of classical gauge fields. The expectation value of an operator $\mathcal{O}$ can be computed as

$$< \mathcal{O} >_Y = \int [d\rho_1]\,[d\rho_2]\,W_{x_1}[\rho_1]\,W_{x_2}[\rho_2]\,\mathcal{O}(\rho_1,\rho_2)\,, \tag{9}$$

where $Y = \ln(1/x_F)$ and $x_F = x_1 - x_2$. All operators at small x can be computed in the background classical field of the nucleus at small x. Quantum information, to leading logarithms in x, is contained in the source functionals $W_{x_1(x_2)}[\rho_1(\rho_2)]$. The operator $\mathcal{O}$ can be expressed in terms of gauge fields $A^\mu[\rho_1,\rho_2](x)$.

Inclusive gluon production in the CGC is computed by solving the Yang-Mills equations $[D_\mu, F^{\mu\nu}]^a = J^{\nu,a}$, where $J^\nu = \rho_1\,\delta(x^-)\delta^{\nu+} + \rho_2\,\delta(x^+)\delta^{\nu-}$, with initial conditions given by the Yang-Mills fields of the two nuclei before the collision. These are obtained self-consistently by matching the solutions of the Yang-Mills equations on the light cone [33]. The initial conditions are determined by requiring that singular terms in the matching vanish. Since we have argued in Section 2 that we can *compute* the Yang-Mills fields in the nuclei before the collision, the classical problem is in principle completely solvable. Quantum corrections not enhanced by powers of $\alpha_S \ln(1/x)$ can be included systematically. The terms so enhanced are absorbed into the weight functionals $W[\rho_{1,2}]$. As we will now discuss, hadronic scattering in the CGC can be studied through a systematic power counting in the density of sources in powers of $\rho_{1,2}/k_{\perp;1,2}^2$.

3.1. Gluon and quark production in the dilute/pp regime: ($\rho_{p1}/k_\perp^2\ \rho_{p2}/k_\perp^2 << 1$)

The power counting here is applicable either to a proton at small x, or to a nucleus (whose parton density at high energies is enhanced by $A^{1/3}$) at large transverse momenta. The relevant quantity here is Q_s, which, as one may recall, is enhanced both for large A and small x. As long as $k_\perp >> Q_s >> \Lambda_{\mathrm{QCD}}$, one can consider the proton or nucleus as being dilute.

To lowest order in $\rho_{p1}/k_\perp^2$ and $\rho_{p2}/k_\perp^2$, inclusive gluon production was computed in the CGC framework in $A^\tau = 0$ gauge [33] and subsequently in the Lorentz gauge $\partial_\mu A^\mu = 0$ [34]. For $Q_s << k_\perp$, the inclusive cross-section is expressed as the product of two unintegrated ($k_\perp$ dependent) distributions times the matrix element for the scattering. This well known result for gluon production is substantially modified, as we shall discuss shortly, by high parton density effects in the nuclei.

$k_\perp$ factorization is a good assumption at large momenta for quark pair-production. This was worked out in the CGC approach by François Gelis and myself [35]. The result for inclusive quark pair production can be expressed in $k_\perp$ factorized form. The matrix element can be shown to be identical to the result derived in the $k_\perp$-factorization approach [31, 32]-thereby establishing that well known results can be recovered in the limit of low parton densities.

3.2. Gluon and quark production in the semi-dense/pA region ($\rho_p/k_\perp^2 << 1\ \rho_A/k_\perp^2 \sim 1$).

The power counting here is best applicable to asymmetric systems such as proton-nucleus collisions, which naturally satisfies the power counting for a wide range of energies. Of course, as one goes to extremely high energies, it is conceivable that the parton density locally in the proton can become comparable to that in the nucleus. In the semi-dense/pA case, one solves the Yang–Mills equations $[D_\mu, F^{\mu\nu}] = J^\nu$ with the light cone sources $J^{\nu,a} = \delta^{\nu+}\,\delta(x^-)\,\rho_p^a(x_\perp) + \delta^{\nu-}\,\delta(x^+)\,\rho_A^a(x_\perp)$, to determine the gluon field produced-to lowest order in the proton source density and to all orders in the nuclear source density.

Inclusive gluon production in this framework was first computed by Kovchegov and Mueller [36] and shown to be $k_\perp$ factorizable in Ref. [37]. In Ref. [38], the gluon field produced in pA collisions was computed explicitly in Lorentz gauge $\partial_\mu A^\mu = 0$. . Our result is exactly equivalent to that of Dumitru & McLerran in $A^\tau = 0$ gauge [43]. The well known "Cronin" effect is obtained in our formalism and can be simply understood in terms of the multiple scattering of a parton from the projectile with those in the target. The Cronin effect, its quantum evolution and comparison with experiment was discussed at this meeting in Yuri Kovchegov's talk and will therefore not be discussed here.

Quark production in p/D-A collisions can be computed with the gauge field in Lorentz gauge [39]. The field is decomposed into the sum of 'regular' terms and 'singular' terms; the latter contain $\delta(x^+)$. The regular terms are the cases where a) a gluon from the proton interacts with the nucleus and produces a $q\bar{q}$-pair outside, b) the gluon produces the pair which then scatters off the nucleus. Naively, these would appear to be the only possibilities in the high energy limit where the nucleus is a Lorentz contracted pancake. However, in the Lorentz gauge, one has terms identified with the singular terms in the gauge field which correspond to the case where the quark pair is both produced and re-scatters in the nucleus! [4]. Unlike gluon production, neither quark pair-production nor single quark production is strictly $k_\perp$ factorizable. The pair production cross-section can however still be written in $k_\perp$ factorized form as a product of the unintegrated gluon distribution in the proton times a sum of terms with three unintegrated distributions, $\phi_{g,g}$, $\phi_{q\bar{q},g}$ and $\phi_{q\bar{q},q\bar{q}}$. These are respectively proportional to 2-point, 3-point and 4-point correlators of the Wilson lines we discussed previously. For instance, the distribution $\phi_{q\bar{q},g}$ is the product of fundamental Wilson lines coupled to a $q\bar{q}$ pair in the amplitude and adjoint Wilson lines coupled to a gluon in the complex conjugate amplitude. For large transverse momenta or large mass

[4] Related work for single quark production has been discussed in Refs. [41, 42].

pairs, the 3-point and 4-point distributions collapse to the unintegrated gluon distribution, and we recover the previously discussed $k_\perp$-factorized result for pair production in the dilute/pp-limit. Single quark distributions are straightforwardly obtained and depend only on the 2-point quark and gluon correlators and the 3-point correlators. For Gaussian sources, as in the MV-model, these 2-,3- and 4-point functions can be computed exactly as discussed in Ref. [39]. In Ref. [40], we used these results to explicitly study the magnitude of violation of $k_\perp$-factorization.

Our results, coupled with the previous results for inclusive and diffractive distributions in DIS suggest that at small x dipole and multipole correlators of Wilson lines can be extracted in both DIS and hadronic collisions. These operators are gauge invariant and process independent. The renormalization group running of these operators will be a powerful and sensitive harbinger of new physics.

3.3. Gluon and quark production in the dense/AA region ($\rho_{A1}/k_\perp^2 = \rho_{A2}/k_\perp^2 \sim 1$).
In nucleus-nucleus collisions, $\rho_{1,2}/k_\perp^2 \sim 1$. Unlike gluon production in the pp and pA cases, $k_\perp$-factorization breaks down in the AA-case [46, 45]. This is because the classical field comes in with a factor $1/g$-thus each insertion on the gluon is of order O(1). One cannot therefore factor the quantum evolution of the initial wavefunctions into unintegrated gluon distributions unlike the pA case.

Nevertheless, the problem of nuclear collisions is well defined in weak coupling and can be solved numerically [46, 47]. The numerical simulations thus far assume Gaussian initial conditions as in the MV model. These are good initial conditions for central Gold-Gold collisions at RHIC where the typical x is of order 10^{-2}. They are not good initial conditions at the LHC where the typical x at central rapidities will be at least an order of magnitude lower. In that case, one has to use solutions of JIMWLK RG equations [16]. The numerical lattice formalism of Ref. [46] is ideal for computing particle production in the forward light cone by matching the Wilson lines from each of the nuclei on the light cone.

We restrict ourselves to discussing numerical solutions with Gaussian initial conditions. The saturation scale Q_s (which is an input in the numerical solutions in this approximation) and the nuclear radius R are the only parameters in the problem. The energy and number respectively of gluons released in a heavy ion collision of identical nuclei can therefore be simply expressed as

$$\frac{1}{\pi R^2}\frac{dE}{d\eta} = \frac{c_E}{g^2} Q_s^3\,;\quad \frac{1}{\pi R^2}\frac{dN}{d\eta} = \frac{c_N}{g^2} Q_s^2\,, \tag{10}$$

where (up to 10% statistical uncertainity) we compute numerically $c_E = 0.25$ and $c_N = 0.3$. Here η is the space-time rapidity. The number distributions of gluons can also be computed in this approach. Remarkably, one finds that a) the number distribution is infrared finite, and b) the distribution is well fit by a massive Bose-Einstein distribution for $k_\perp/Q_s < 1.5$ GeV with a "temperature" of $\sim 0.47 Q_s$ and by the perturbative distribution $Q_s^4/k_\perp^4$ for $k_\perp/Q_s > 1.5$.

The MV model when applied to heavy ion collisions correctly predicted the initial multiplicity at RHIC [46] and was remarkably successful in explaining rapidity distributions and the centrality dependence of multiplicities [48]. However, it soon became clear that the CGC alone was not sufficient to explain the RHIC data since a) it could not explain the RHIC v_2 data and b) it predicted a suppression in D-A collisions at RHIC which disagreed with the RHIC data [44]. This failure of the CGC (here meaning quantum evolution as opposed to the MV model-which has no evolution) strongly suggested that final state interactions are important at RHIC. This corroborates the remarkable success of hydrodynamic models. Why do predictions of bulk features-the multiplicity [46] and rapidity and centrality dependence [48] do so well then? If hydrodynamic behavior sets in early, and viscous effects are small, the bulk features from the initial conditions will be preserved by hydrodynamic flow [49]. Many puzzles remain. We don't understand early thermalization or why viscous corrections are small.

The RHIC data on the multiplicity and transverse energy of produced hadrons combined with Eq. 10 place strong constraints on what Q_s can be. If Q_s is too small, we find, absurdly, that the initial transverse energy is less than the final measured transverse energy. If Q_s is too large, we find that the initial multiplicity of gluons is greater than the final multiplicity of hadrons. These constraints therefore allow us to place the bound that $1.3 < Q_s < 2$ GeV. This bound is consistent with an $A^{1/3}$ extrapolation of the Golec-Biernat–Wusthoff fit of Q_s^2 to the HERA data [23]. A simple extrapolation gives $Q_s \approx 1.4$ GeV.

The transition to the QGP from the CGC remains as an outstanding theoretical problem. Due to the rapid expansion of the system, the occupation number of modes falls well below one on time scales of order $1/Q_s$. A necessary condition for thermalization is that momentum distributions should be isotropic. The CGC initial conditions are very anisotropic with $< p_\perp >\sim Q_s$ and $< p_z >\sim 0$. How does this isotropization take place? All estimates of final state re-scattering of partons formed from the melting CGC give longer times than what the RHIC collisions seem to suggest [50].

Collective instabilities [51], analogous to the well known Weibel instabilities in plasma physics, can speed up themalization [52, 53]. Starting from very anisotropic (CGC-like) initial conditions, these instabilities drive the system to isotropy on very short time scales, of order $1/Q_s$ in some estimates. An equally interesting problem is that of chemical equilibration. One would expect, in weak coupling, the production of quarks to be suppressed. However, since the CGC produces strong fields of order $1/g$, it could drive the system to chemical equilibrium. First steps have been taken to study this problem [54, 55] by numerically solving the Dirac equation in the background field of the two nuclei.

4. Open Issues in the CGC

The CGC is a framework to think about problems in high energy QCD. A topic of much excitement among theorists recently is whether there are contributions beyond the JIMWLK equations-in particular those that generate "Pomeron loops". We addressed the issue of $k_\perp$-factorization and why "dipole" and "multipole" operators are relevant variables at high energies. Can one derive factorization theorems in this framework analogous to those derived previously for Collinear Factorization? Turning to phenomenology, we have the beginnings of a consistent phenomenological picture of the CGC and the QGP in D-A and A-A collisions. For this to become a quantitative science, we need to understand the problem of thermalization from first principles in QCD. It is a difficult task but by no means an impossible one.

Acknowledgements

This work was supported in part by DOE Contract No. DE-AC02-98CH10886 and by a research grant from the Alexander Von Humboldt Foundation. I would like to further acknowledge the kind hospitality of the Institute for theoretical physics-II at Hamburg University and the European Center for Theoretical Nuclear Physics and related areas (ECT⋆).

5. References

[1] F. Wilczek, hep-ph/0502113.

[2] E. Iancu and R. Venugopalan, hep-ph/0303204, in QGP3, Ed. R. Hwa and X.-N. Wang, World Scientific (2004).

[3] E. A. Kuraev, L. N. Lipatov and V. S. Fadin, Sov. Phys. JETP **45**, 199 (1977) [Zh. Eksp. Teor. Fiz. **72**, 377 (1977)]; I. I. Balitsky and L. N. Lipatov, Sov. J. Nucl. Phys. **28**, 822 (1978) [Yad. Fiz. **28**, 1597 (1978)].

[4] L.V. Gribov, E.M. Levin, M.G. Ryskin, Phys. Rept. **100**, 1 (1983).

[5] A.H. Mueller, J-W. Qiu, Nucl. Phys. **B 268**, 427 (1986).

[6] J. P. Blaizot and A. H. Mueller, Nucl. Phys. B **289**, 847 (1987).

[7] L. D. McLerran and R. Venugopalan, Phys. Rev. D **49**, 2233 (1994); *ibid.*, 3352, (1994); *ibid.*, **50**, 2225 (1994).
[8] L. Susskind, Phys. Rev. **165**, 1535, (1968).
[9] J. Jalilian-Marian, S. Jeon and R. Venugopalan, Phys. Rev. D **63**, 036004 (2001).
[10] S. Jeon and R. Venugopalan, Phys. Rev. D **70**, 105012 (2004); hep-ph/0503219.
[11] Yu.V. Kovchegov, Phys. Rev. **D 54**, 5463 (1996); *ibid.*, **55**, 5445 (1997).
[12] J. Jalilian-Marian, A. Kovner, L.D. McLerran, H. Weigert, Phys. Rev. **D 55**, 5414 (1997).
[13] D. Triantafyllopoulos, hep-ph/0502114.
[14] A. Ayala, J. Jalilian-Marian, L.D. McLerran, R. Venugopalan, Phys. Rev. **D 52**, 2935 (1995); *ibid.*, **53**, 458 (1996).
[15] J. Jalilian-Marian, A. Kovner, A. Leonidov, H. Weigert, Nucl. Phys. **B 504**, 415 (1997); J. Jalilian-Marian, A. Kovner, A. Leonidov, H. Weigert, Phys. Rev. **D 59**, 014014 (1999); E. Iancu, A. Leonidov, L.D. McLerran, Nucl. Phys. **A 692**, 583 (2001); E. Ferreiro, E. Iancu, A. Leonidov, L.D. McLerran, Nucl. Phys. **A 703**, 489 (2002).
[16] K. Rummukainen and H. Weigert, Nucl. Phys. A **739**, 183 (2004).
[17] E. Iancu, K. Itakura and L. McLerran, Nucl. Phys. A **724**, 181 (2003).
[18] A. H. Mueller, Nucl. Phys. B **415**, 373 (1994).
[19] N. N. Nikolaev and B. G. Zakharov, Z. Phys. C**49**, (1991) 607.
[20] I. Balitsky, Nucl. Phys. **B 463**, 99 (1996); Yu.V. Kovchegov, Phys. Rev. **D 61**, 074018 (2000).
[21] E.M. Levin, K. Tuchin, Nucl. Phys. **B 573**, 833 (2000).
[22] D. N. Triantafyllopoulos, Nucl. Phys. B **648**, 293 (2003).
[23] K. Golec-Biernat, M. Wüsthoff, Phys. Rev. **D 59**, 014017 (1999).
[24] N. Armesto, M.A. Braun, Eur. Phys. J. **C 20**, 517 (2001).
[25] K. Golec-Biernat, L. Motyka and A. M. Stasto, Phys. Rev. D **65**, 074037 (2002).
[26] J. L. Albacete, N. Armesto, A. Kovner, C. A. Salgado and U. A. Wiedemann, Phys. Rev. Lett. **92**, 082001 (2004); arXiv:hep-ph/0408216.
[27] S. Munier and R. Peschanski, Phys. Rev. Lett. **91**, 232001 (2003); Phys. Rev. D **69**, 034008 (2004); *ibid.*, D **70**, 077503 (2004).
[28] E. Iancu, A. H. Mueller and S. Munier, arXiv:hep-ph/0410018.
[29] R. A. Fischer, Ann. Eugenics, **7**, (1937) 355; A. Kolmogorov, I. Petrovsky, and N. Piscounov, *Moscow Univ. Bull. Math.*, **A1**, (1937) 1.
[30] A. H. Mueller and A. I. Shoshi, Nucl. Phys. B **692**, 175 (2004).
[31] S. Catani, M. Ciafaloni, F. Hautmann, Nucl. Phys. B **366**, 135 (1991).
[32] J. C. Collins and R. K. Ellis, Nucl. Phys. B **360**, 3 (1991).
[33] A. Kovner, L. D. McLerran and H. Weigert, Phys. Rev. D **52**, 3809 (1995).
[34] Y. V. Kovchegov and D. H. Rischke, Phys. Rev. C **56**, 1084 (1997).
[35] F. Gelis and R. Venugopalan, Phys. Rev. D **69**, 014019 (2004).
[36] Y. V. Kovchegov and A. H. Mueller, Nucl. Phys. B **529**, 451 (1998).
[37] D. Kharzeev, Yu. Kovchegov, K. Tuchin, Phys. Rev. D **68**, 094013 (2003).
[38] J. P. Blaizot, F. Gelis and R. Venugopalan, Nucl. Phys. A **743**, 13 (2004).
[39] J. P. Blaizot, F. Gelis and R. Venugopalan, Nucl. Phys. A **743**, 57 (2004).
[40] H. Fujii, F. Gelis and R. Venugopalan, hep-ph/0502204; hep-ph/0504047.
[41] K. Tuchin, Phys. Lett. B **593**, 66 (2004); hep-ph/0504133.
[42] B. Z. Kopeliovich, J. Raufeisen and A. V. Tarasov, Phys. Rev. C **62** 035204 (2000); N. N. Nikolaev and W. Schafer, arXiv:hep-ph/0411365.
[43] A. Dumitru and L. D. McLerran, Nucl. Phys. A **700**, 492 (2002).
[44] D. Kharzeev, E. Levin and L. McLerran, Phys. Lett. B **561**, 93 (2003).
[45] I. Balitsky, arXiv:hep-ph/0409314.
[46] A. Krasnitz and R. Venugopalan, Nucl. Phys. B **557**, 237 (1999); Phys. Rev. Lett. **86**, 1717 (2001); A. Krasnitz, Y. Nara and R. Venugopalan, Phys. Rev. Lett. **87**, 192302 (2001).
[47] T. Lappi, Phys. Rev. C **67**, 054903 (2003).
[48] D. Kharzeev and M. Nardi, Phys. Lett. B **507**, 121 (2001); D. Kharzeev and E. Levin, Phys. Lett. B **523**, 79 (2001).
[49] T. Hirano and Y. Nara, Nucl. Phys. A **743**, 305 (2004).
[50] R. Baier, A. H. Mueller, D. Schiff and D. T. Son, Phys. Lett. B **502**, 51 (2001); Phys. Lett. B **539**, 46 (2002).
[51] S. Mrowczynski, Phys. Lett. B **314**, 118 (1993).
[52] P. Arnold, J. Lenaghan and G. D. Moore, JHEP **0308**, 002 (2003).
[53] P. Romatschke and M. Strickland, Phys. Rev. D **68**, 036004 (2003).
[54] D. D. Dietrich, Phys. Rev. D **70**, 105009 (2004); *ibid.*, **68**, 105005 (2003).
[55] F. Gelis, K. Kajantie and T. Lappi, arXiv:hep-ph/0409058.

Institute of Physics Publishing
doi:10.1088/1742-6596/50/1/009

Journal of Physics: Conference Series **50** (2006) 79–86
5th International Conference on Physics and Astrophysics of Quark Gluon Plasma

QCD Phenomenology and Heavy Ion Physics

Yuri V. Kovchegov

Department of Physics, The Ohio State University, Columbus OH, 43210, USA

E-mail: yuri@mps.ohio-state.edu

Abstract. We describe application of saturation/Color Glass Condensate physics to heavy ion collisions, concentrating on several important observables. We show how simple saturation-inspired assumptions about particle production in heavy ion collisions allow one to accurately describe total charged particle multiplicity as a function of the collision's centrality, of the center of mass energy, and of the particle rapidity. We will then describe how predictions of saturation physics can be tested by studying energy/rapidity dependence of particle spectra in p(d)A collisions. We concentrate on the nuclear modification factor R^{pA} for gluon production. We show that at moderately high energy/rapidity the nuclear modification factor R^{pA} exhibits Cronin enhancement. As the energy/rapidity increases, R^{pA} decreases. At sufficiently high energy/rapidity R^{pA} becomes less than 1 for all values of p_T indicating the onset of suppression of gluon production due to quantum evolution effects. These predictions were confirmed by RHIC data.

1. Introduction

In this talk we will discuss saturation/Color Glass Condensate physics [1, 2, 3, 4] in the context of heavy ion collisions. Saturation physics is based on the fact that small-x wave functions of ultrarelativistic hadrons and nuclei are characterized by a hard scale Q_s, known as the saturation scale [1, 2, 3, 4]. The scale Q_s arises due to *saturation* of high partonic densities at small-x and is an increasing function of energy and atomic number of the nucleus [5, 6, 7, 8, 4]. This large scale makes the strong coupling constant small $\alpha_s(Q_s) \ll 1$ leading to dominance of the classical gluonic fields in high energy processes [4, 9]. Therefore, the dominant gluon field in high energy collisions is given by the classical field of the scattering color charges [10]. Corresponding gluon production cross section was found for pA collisions in [11] and the effects of quantum small-x evolution [5] were included in it in [12]. The gluon production cross section for heavy ion collisions (AA) at the classical level has been studied both numerically [13] and analytically [14]. It is not quite clear at present how to include the effects of nonlinear quantum evolution [5] in the results of [13, 14]. Below we will show how one can understand many properties of the RHIC data on particle multiplicities in heavy ion collisions using general saturation principles without an exact knowledge of gluon production cross section in AA [15, 16, 17].

We will proceed by discussing particle production in $p(d)A$ collisions, where, due to absence of final state interactions, predictions of saturation physics can be tested directly. Concentrating on the nuclear modification factor R^{pA} we argue that the quasi-classical gluon production cross section calculated in [11] leads to Cronin enhancement [18, 19, 20, 21, 22]. This regime is probably relevant for mid-rapidity particle production in $\sqrt{s} = 200$ GeV dAu collisions at RHIC. At higher (forward) rapidities the effects of CGC small-x evolution would lead to suppression of R^{pA} and disappearance of Cronin effect [23, 18, 24]. The suppression predicted in

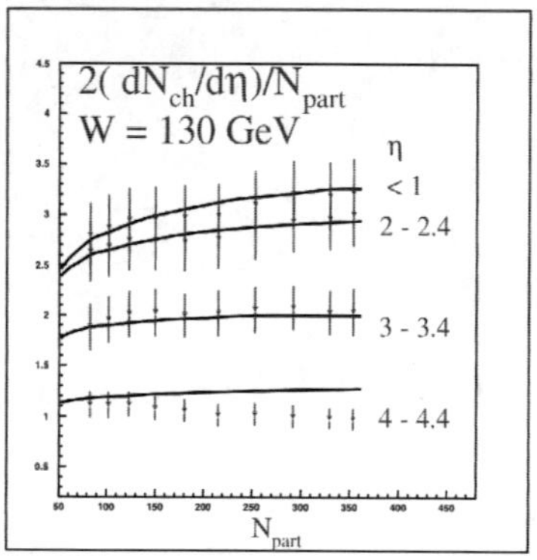

Figure 1. Saturation model fit of the PHOBOS data on the total charged particle multiplicity at mid-rapidity as a function of centrality at $\sqrt{s} = 130$ GeV taken from [16].

[23, 18, 24] has been observed experimentally in [25, 26, 27, 28, 29], confirming the expectation of saturation/Color Glass physics. We conclude by presenting a quantitative saturation-inspired model [30] describing the data of [26].

2. Particle Multiplicity from Saturation Approach

2.1. Multiplicity at Mid-Rapidity Versus Centrality

As was argued in [4, 10], the gluon production in AA collisions is described by the classical fields. Classical field $A_\mu \sim 1/g$ leads to produced gluon multiplicity

$$\frac{dN}{d^2k\, d^2b\, dy} \sim \langle A_\mu A_\mu \rangle \sim \frac{1}{\alpha_s}. \tag{1}$$

Thus, gluon transverse momentum spectrum described by a single scale Q_s can be written as

$$\frac{dN}{d^2k\, d^2b\, dy} = \frac{1}{\alpha_s} f(k_\perp/Q_s) \tag{2}$$

with some unknown function $f(k_\perp/Q_s)$ to be determined by actual calculations. Integrating over k and b yields

$$\frac{dN}{dy} = \text{const}\, \frac{1}{\alpha_s}\, \pi\, R^2\, Q_s^2. \tag{3}$$

where the value of the constant is determined from $f(k_\perp/Q_s)$ and is not important here. Following [15, 16] we assume that the scale for the coupling constant in Eq. (3) is set by Q_s. (This step, of course, goes beyond the classical limit and assumes that Eq. (3) is valid even when running coupling corrections are included.) Then, as $R^2 \sim A^{2/3}$ and if $Q_s^2 \sim A^{1/3}$ [4, 9], together with running coupling

$$\alpha_s(Q_s) = \frac{1}{b\, \ln Q_s^2/\Lambda^2} \sim \frac{1}{\ln A} \tag{4}$$

we conclude from Eq. (3) that

$$\frac{1}{A}\frac{dN}{d\eta} \sim \ln A. \tag{5}$$

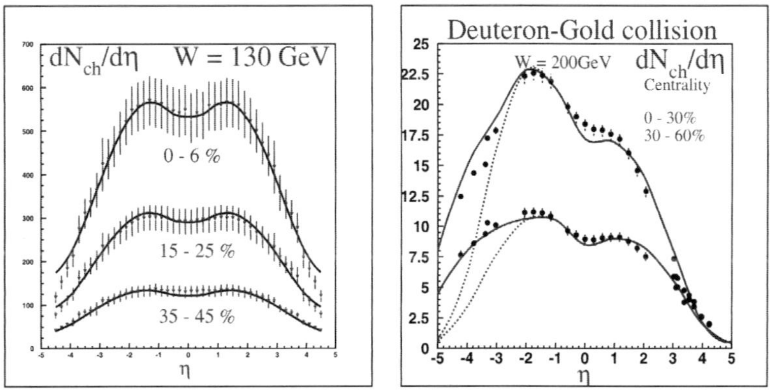

A **B**

Figure 2. (A) Saturation model fit of the PHOBOS data on total charged particle multiplicity in $AuAu$ collisions as a function of rapidity and centrality at $\sqrt{s} = 130$ GeV taken from [16]. (B) Prediction of the same model compared to BRAHMS dAu data.

For heavy ion experiments at different collision centralities we substitute A by the number of participants N_{part} so that Eq. (5) becomes

$$\frac{1}{N_{part}} \frac{dN}{d\eta} \sim \ln N_{part}. \tag{6}$$

Eq. (6) allowed the authors of [15] to correctly *predict* the behavior of particle multiplicity at mid-rapidity at RHIC as a function of centrality at $\sqrt{s} = 130$ GeV. A fit of particle multiplicity at various values of rapidity at $\sqrt{s} = 130$ GeV taken from [16] is shown along with PHOBOS data in Fig. 1.

2.2. Multiplicity as a Function of Energy

Eq. (3) also allows one to test whether the scaling of total particle multiplicity with energy is consistent with saturation/Color Glass predictions. Using the fact that $Q_s^2 \sim 1/x_{Bj}^{\lambda}$ [5, 7] in Eq. (3) and, for the moment, dropping the slower Q_s-dependence in α_s leads to

$$\frac{dN/d\eta(\sqrt{s_1})}{dN/d\eta(\sqrt{s_2})} = \left(\frac{\sqrt{s_1}}{\sqrt{s_2}}\right)^{\lambda}. \tag{7}$$

Using PHOBOS data for total charge multiplicity at $\sqrt{s} = 130$ GeV for most central collisions

$$\frac{dN}{d\eta}(\sqrt{s} = 130\,\text{GeV}) = 555 \pm 12(stat) \pm 35(syst) \tag{8}$$

together with $\lambda = 0.25 \div 0.3$ obtained in [31] by analyzing HERA data, Kharzeev and Levin [16] *predicted* the total charged particle multiplicity at $\sqrt{s} = 200$ GeV to be

$$\frac{dN}{d\eta}(\sqrt{s} = 200\,\text{GeV}) = 616 \div 634 \tag{9}$$

in agreement with the later measured PHOBOS result

$$\frac{dN}{d\eta}(\sqrt{s} = 200\,\text{GeV}) = 650 \pm 35(stat). \tag{10}$$

2.3. $\frac{dN}{d\eta}$ *Versus* η *and* N_{part}

Describing rapidity distribution of the produced particles requires a little modeling of the spectra. Assuming k_T-factorization for particle production cross section

$$\frac{d\sigma^{AA}}{d^2k\, dy} = \frac{2\,\alpha_s}{C_F}\frac{1}{\underline{k}^2}\int d^2q\,\phi_A(\underline{q})\,\phi_A(\underline{k}-\underline{q}) \tag{11}$$

along with saturation-inspired unintegrated gluon distribution functions ($\phi_A(\underline{k}) \sim \alpha_s/\underline{k}^2$ if $k_\perp > Q_s$ and $\phi_A(\underline{k}) \sim S_\perp/\alpha_s$ if $k_\perp < Q_s$) the authors of [16] produced an impressive fit of the PHOBOS data on charged particle multiplicities as functions of rapidity and centrality at $\sqrt{s} = 130$ GeV shown in Fig. 2A. The *predictions* made in [16] for similar multiplicity data at $\sqrt{s} = 200$ GeV were also in good agreement with the later published BRAHMS data. Recently, the same model was successfully used to predict rapidity distribution of charged particle multiplicity in dAu collisions, as shown in Fig. 2B.

Phenomenological success of the saturation models presented above does not contradict the possibility of strong final state interactions leading to formation of quark-gluon plasma. As was argued in [32], thermalization in the saturation framework would not introduce any fundamentally new scales leaving Eqs. (3) and (6) practically unchanged. (In fact, thermalization of [32] appears only to introduce a slowly-varying extra factor of $\alpha_s^{-2/5}(Q_s)$ in Eq. (3).) Late stage interactions are also not very likely to significantly modify the rapidity distribution of Fig. 2 due to causality constraints.

3. Particle Production in p(d)A Collisions

Here we will discuss particle production and transverse momentum spectra in dAu collisions. The first step in understanding particle (gluon) production in pA collisions in the saturation/Color Glass Condensate framework is finding the classical gluon field of the colliding proton and nucleus in McLerran-Venugopalan model. The corresponding gluon production cross section in pA was constructed in [11] yielding

$$\frac{d\sigma^{pA}}{d^2k\, dy} = \int d^2b\, d^2x\, d^2y\, \frac{1}{(2\pi)^2}\frac{\alpha_s C_F}{\pi^2}\frac{\underline{x}\cdot\underline{y}}{\underline{x}^2\underline{y}^2}\, e^{-i\underline{k}\cdot(\underline{x}-\underline{y})}$$

$$\times\left[1 - e^{-\underline{x}^2 Q_{s0}^2 \ln(1/x_T\Lambda)/4} - e^{-\underline{y}^2 Q_{s0}^2 \ln(1/y_T\Lambda)/4} + e^{-(\underline{x}-\underline{y})^2 Q_{s0}^2 \ln(1/(\underline{x}-\underline{y})_T\Lambda)/4}\right], \tag{12}$$

where $\underline{k}$ and y are the produced gluon's transverse momentum and rapidity, $\underline{b}$ is the proton's impact factor, Q_{s0} is the saturation scale in McLerran-Venugopalan model and $\underline{x}$, $\underline{y}$ are two-dimensional gluon transverse position vectors which are integrated over.

To find the nuclear modification factor

$$R^{pA}(\underline{k}, y) = \frac{\frac{d\sigma^{pA}}{d^2k\, dy}}{A\,\frac{d\sigma^{pp}}{d^2k\, dy}}. \tag{13}$$

resulting from Eq. (12) one can integrate it explicitly over $\underline{x}$ and $\underline{y}$ neglecting the logarithms in the exponents obtaining

$$R^{pA}(k_T) = \frac{\underline{k}^4}{Q_{s0}^2}\left\{-\frac{1}{\underline{k}^2} + \frac{2}{\underline{k}^2}\,e^{-\underline{k}^2/Q_{s0}^2} + \frac{1}{Q_{s0}^2}\,e^{-\underline{k}^2/Q_{s0}^2}\left[\ln\frac{Q_{s0}^4}{4\,\Lambda^2\underline{k}^2} + \mathrm{Ei}\left(\frac{\underline{k}^2}{Q_{s0}^2}\right)\right]\right\}. \tag{14}$$

The ratio $R^{pA}(k_T)$ is plotted in Fig. 3 for $\Lambda = 0.2\, Q_{s0}$. It clearly exhibits an enhancement at high-k_T characteristic of Cronin effect. Similar conclusions have been reached by other authors [19, 20, 21, 22]. To prove that Eq. (12) leads to Cronin effect one can expand it at high k_T:

$$R^{pA}(k_T) = 1 + \frac{3}{2}\frac{Q_{s0}^2}{\underline{k}^2}\ln\frac{\underline{k}^2}{\Lambda^2} + \ldots, \qquad k_T \to \infty. \tag{15}$$

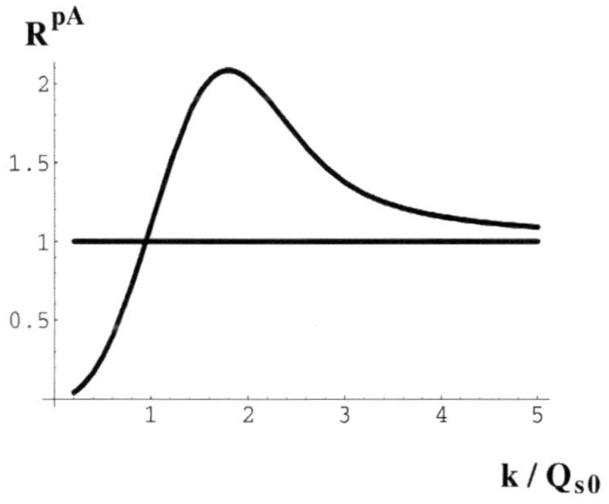

Figure 3. Nuclear modification factor R^{pA} plotted as a function of k_T/Q_{s0} for gluon production in the quasi-classical McLerran-Venugopalan model as found in [11]. The cutoff is $\Lambda = 0.2\, Q_s$.

The plus sign in front of the first correction to 1 on the right hand side of Eq. (15) implies that R^{pA} approaches 1 from above at high k_T indicating Cronin enhancement. Eq. (15), together with Fig. 3, can also be used to illustrate the dependence of the Cronin maximum on centrality. As one can easily see from Eqs. (15) and (14) the position of the Cronin maximum is given by $k_T \sim Q_{s0}$ and the height of the maximum scales as $R^{pA}(k_T \sim Q_{s0}) \sim \ln Q_{s0}$. Since $Q_{s0}^2 \sim A^{1/3}$ we conclude that the position and height of the Cronin peak are increasing functions of the centrality of pA collisions.

Before including quantum evolution into Eq. (12), let us first note that Eq. (12), even though it includes multiple rescatterings, can still be written in k_T-factorized form (!) [12, 18]

$$\frac{d\sigma^{pA}}{d^2k\, dy} = \frac{2\,\alpha_s}{C_F}\frac{1}{\underline{k}^2}\int d^2q\, \phi_p(\underline{q}, 0)\, \phi_A(\underline{k} - \underline{q}, 0), \tag{16}$$

with the unintegrated "gluon distributions" given by

$$\phi_A(x, \underline{k}^2) = \frac{C_F}{\alpha_s\,(2\pi)^3}\int d^2b\, d^2r\, e^{-i\underline{k}\cdot\underline{r}}\, \nabla_r^2\, N_G(\underline{r}, \underline{b}, y = \ln 1/x), \tag{17}$$

and

$$\phi_p(x, \underline{k}^2) = \frac{C_F}{\alpha_s\,(2\pi)^3}\int d^2b\, d^2r\, e^{-i\underline{k}\cdot\underline{r}}\, \nabla_r^2\, n_G(\underline{r}, \underline{b}, y = \ln 1/x). \tag{18}$$

Here $N_G(\underline{r}, \underline{b}, y = \ln 1/x)$ and $n_G(\underline{r}, \underline{b}, y = \ln 1/x)$ are forward scattering amplitudes of a gluon dipole of size r located at impact parameter $\underline{b}$ on a nucleus and a proton correspondingly. Eq. (12) is reproduced by using $N_G(\underline{r}, \underline{b}, 0) = 1 - e^{-\underline{r}^2 Q_{s0}^2 \ln(1/r_T\Lambda)/4}$ and $n_G(\underline{r}, \underline{b}, 0) = \underline{r}^2 \Lambda^2 \ln(1/r_T\Lambda)$ [18].

As was shown in [12], Eq. (16) makes inclusion of quantum small-x evolution straightforward. A tedious analysis shows that the inclusion of evolution preserves k_T-factorization of Eq. (16) yielding the full answer for inclusive cross section [12]

$$\frac{d\sigma^{pA}}{d^2k\, dy} = \frac{2\,\alpha_s}{C_F}\frac{1}{\underline{k}^2}\int d^2q\, \phi_p(\underline{q}, Y - y)\, \phi_A(\underline{k} - \underline{q}, y), \tag{19}$$

where Y is the full rapidity interval between the proton and the nucleus. The gluon distribution functions in Eq. (19) are still given by Eqs. (17) and (18), but now with n_G given by the solution

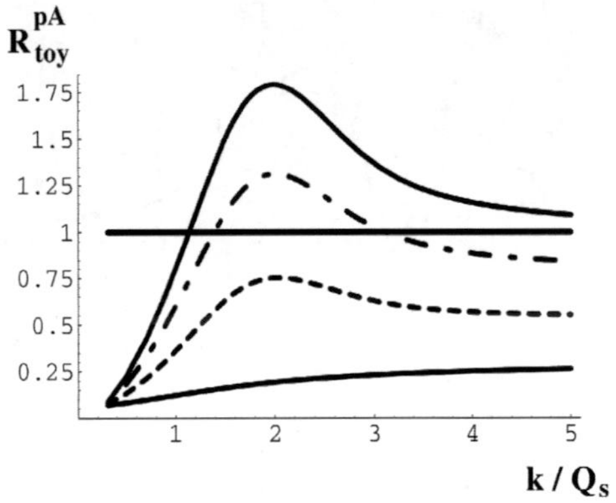

Figure 4. R^{pA} plotted as a function of k_T/Q_s for (i) McLerran-Venugopalan model, which is valid for moderate energies/rapidities (upper solid line); (ii) our toy model for very high energies/rapidities from [18] (lower solid line); (iii) an interpolation to intermediate energies/rapidities (dash-dotted and dashed lines).

of the BFKL equation and with N_G given by the solution of the non-linear evolution equation [5].

To analyze the nuclear modification factor given by Eq. (19) one has to separately study three regions of transverse momentum: (i) $k_T > k_{geom}$, (ii) $Q_s(y) < k_T \leq k_{geom}$ and (iii) $k_T \sim Q_s(y)$. (For $k_T \ll Q_s(y)$ one always has suppression due to parton saturation [18], as one can see already in Fig. 3.) $Q_s(y)$ is the saturation scale (such that $Q_{s0} = Q_s(y=0)$) and k_{geom} is the scale describing the border of the geometric scaling region [7] $k_{geom} \approx Q_s^2(y)/Q_{s0}$. As was shown in [18], in region (i) the double logarithmic approximation applies giving

$$R^{pA}(k_T, y)\Big|_{k_T > k_{geom}} \sim \exp\left[2\sqrt{2\,\overline{\alpha}_s\, y}\left(\sqrt{\ln\frac{k_T}{Q_{s0}}} - \sqrt{\ln\frac{k_T}{\Lambda}}\right)\right] < 1 \tag{20}$$

since $Q_{s0} \gg \Lambda$. Inside the geometric scaling region (ii) nuclear modification factor becomes [18]

$$R^{pA}(k_T, y)\Big|_{Q_s(y) < k_T \leq k_{geom}} \sim A^{-1/6} \exp\left[\frac{\ln^2\frac{k_T}{\Lambda} - \ln^2\frac{k_T}{Q_{s0}}}{14\,\zeta(3)\,\overline{\alpha}_s\, y}\right] \ll 1, \tag{21}$$

so that at very high y one gets $R^{pA} \sim A^{-1/6} \ll 1$ for $A \gg 1$, reproducing the original prediction of the initial state suppression made in [23]. Finally, to understand what happens to Cronin maximum one has to study R^{pA} at $k_T = Q_s(y)$. The analysis done in [18] gives the nuclear modification factor in the region (iii)

$$R^{pA}(k_T = Q_s(y), y) \sim \exp\left\{4\,\overline{\alpha}_s\, y\left(1 - \sqrt{1 + \frac{\ln A^{1/6}}{2\,\overline{\alpha}_s\, y}}\right)\right\} \ll 1. \tag{22}$$

Eq. (22) at very high y also gives $R^{pA} \sim A^{-1/6}$, similar to Eq. (21), indicating that as energy increases Cronin maximum should gradually decrease and eventually disappear completely. Eqs. (20), (21) and (22) demonstrate that at very high energy/rapidity the nuclear modification factor R^{pA} should become less than 1 for all k_T. Since $R^{pA} \sim A^{-1/6}$ at very large y, one should also expect that R^{pA} will become a decreasing function of centrality at high energies/rapidities.

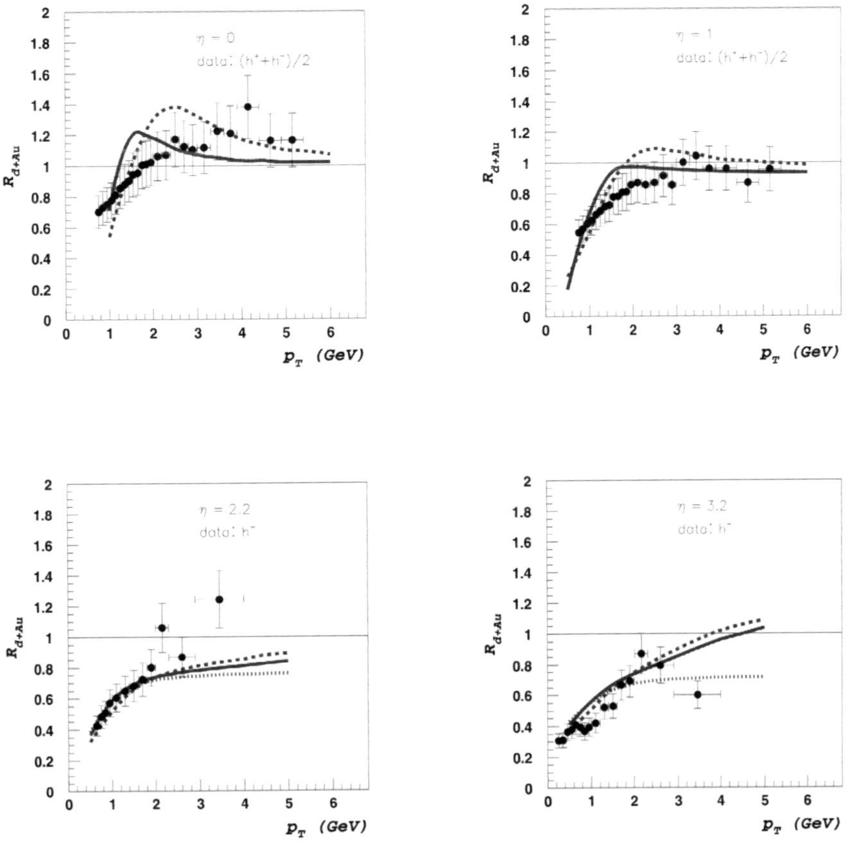

Figure 5. Nuclear modification factor R_{dAu} of charged particles for different rapidities. The fit uses the model described in [30]. Data is from [26].

Our conclusions regarding the variation of R^{pA} with energy/centrality are summarized in Fig. 4. The top solid curve in Fig. 4 is the same as in the quasi-classical approximation shown in Fig. 3. It corresponds to moderately high energy/rapidity. As energy/rapidity increases R^{pA} decreases (dash-dotted and dashed lines) and the Cronin peak flattens, eventually approaching a flat curve (lower solid line) which has suppression at all p_T. Similar conclusions have been reached in [24].

Using a simple model for dipole amplitudes N_G and n_G in Eqs. (17) and (18) and adding a valence quark contribution to Eq. (19) one can fit the R_{dAu} data at various rapidities reported by the BRAHMS collaboration [25, 26], as shown in Fig. 5 [30]. At mid-rapidity ($\eta = 0$) the nuclear modification factor R^{dAu} in Fig. 5 exhibits Cronin enhancement. This indicates that particle production at mid-rapidity at RHIC is best described by the multiple rescatterings in McLerran-Venugopalan model. At higher rapidities BRAHMS data exhibits strong suppression with R^{dAu} becoming less than 1 for all p_T at pseudorapidity $\eta = 3.2$ in qualitative agreement with our predictions shown in Fig. 4 [18] and in quantitative agreement with our fit from [30] shown in Fig. 5. This indicates the onset of quantum evolution characteristic of saturation/Color Glass Condensate at forward rapidities.

[1] L. V. Gribov, E. M. Levin and M. G. Ryskin, Phys. Rept. **100**, 1 (1983).

[2] A. H. Mueller and J. w. Qiu, Nucl. Phys. B **268**, 427 (1986).

[3] J. P. Blaizot and A. H. Mueller, Nucl. Phys. B **289**, 847 (1987).

[4] L. D. McLerran and R. Venugopalan, Phys. Rev. D **49**, 2233 (1994) [arXiv:hep-ph/9309289]; Phys. Rev. D **49**, 3352 (1994) [arXiv:hep-ph/9311205]; Phys. Rev. D **50**, 2225 (1994) [arXiv:hep-ph/9402335].

[5] I. Balitsky, Nucl. Phys. B **463**, 99 (1996) [arXiv:hep-ph/9509348]; Y. V. Kovchegov, Phys. Rev. D **60**, 034008 (1999) [arXiv:hep-ph/9901281].

[6] J. Jalilian-Marian, A. Kovner, A. Leonidov and H. Weigert, Nucl. Phys. B **504**, 415 (1997) [arXiv:hep-ph/9701284]; Phys. Rev. D **59**, 014014 (1999) [arXiv:hep-ph/9706377]; Phys. Rev. D **59**, 034007 (1999) [Erratum-ibid. D **59**, 099903 (1999)] [arXiv:hep-ph/9807462]; J. Jalilian-Marian, A. Kovner and H. Weigert, Phys. Rev. D **59**, 014015 (1999) [arXiv:hep-ph/9709432]; E. Iancu, A. Leonidov and L. D. McLerran, Phys. Lett. B **510**, 133 (2001) [arXiv:hep-ph/0102009]; E. Iancu and L. D. McLerran, Phys. Lett. B **510**, 145 (2001) [arXiv:hep-ph/0103032]; E. Ferreiro, E. Iancu, A. Leonidov and L. McLerran, Nucl. Phys. A **703**, 489 (2002) [arXiv:hep-ph/0109115]; E. Iancu, K. Itakura and L. McLerran, Nucl. Phys. A **708**, 327 (2002) [arXiv:hep-ph/0203137]; for a review see E. Iancu and R. Venugopalan, arXiv:hep-ph/0303204.

[7] E. Levin and K. Tuchin, Nucl. Phys. B **573**, 833 (2000) [arXiv:hep-ph/9908317]; E. Iancu, K. Itakura and L. McLerran, Nucl. Phys. A **708**, 327 (2002) [arXiv:hep-ph/0203137].

[8] A. H. Mueller and D. N. Triantafyllopoulos, Nucl. Phys. B **640**, 331 (2002) [arXiv:hep-ph/0205167]; D. N. Triantafyllopoulos, Nucl. Phys. B **648**, 293 (2003) [arXiv:hep-ph/0209121]; A. H. Mueller, Nucl. Phys. A **724**, 223 (2003) [arXiv:hep-ph/0301109].

[9] Y. V. Kovchegov, Phys. Rev. D **54**, 5463 (1996) [arXiv:hep-ph/9605446]; Phys. Rev. D **55**, 5445 (1997) [arXiv:hep-ph/9701229]; J. Jalilian-Marian, A. Kovner, L. D. McLerran and H. Weigert, Phys. Rev. D **55**, 5414 (1997) [arXiv:hep-ph/9606337].

[10] A. Kovner, L. D. McLerran and H. Weigert, Phys. Rev. D **52**, 6231 (1995) [arXiv:hep-ph/9502289]; Y. V. Kovchegov and D. H. Rischke, Phys. Rev. C **56**, 1084 (1997) [arXiv:hep-ph/9704201].

[11] Yu. V. Kovchegov and A. H. Mueller, Nucl. Phys. B **529**, 451 (1998) [arXiv:hep-ph/9802440].

[12] Yu. V. Kovchegov and K. Tuchin, Phys. Rev. D **65**, 074026 (2002) [arXiv:hep-ph/0111362].

[13] A. Krasnitz and R. Venugopalan, Phys. Rev. Lett. **84**, 4309 (2000) [arXiv:hep-ph/9909203].

[14] Y. V. Kovchegov, Nucl. Phys. A **692**, 557 (2001) [arXiv:hep-ph/0011252].

[15] D. Kharzeev and M. Nardi, Phys. Lett. B **507**, 121 (2001) [arXiv:nucl-th/0012025]; D. Kharzeev, E. Levin and M. Nardi, arXiv:hep-ph/0111315.

[16] D. Kharzeev and E. Levin, Phys. Lett. B **523**, 79 (2001) [arXiv:nucl-th/0108006].

[17] D. Kharzeev, E. Levin and M. Nardi, Nucl. Phys. A **730**, 448 (2004) [Erratum-ibid. A **743**, 329 (2004)] [arXiv:hep-ph/0212316].

[18] D. Kharzeev, Y. V. Kovchegov and K. Tuchin, Phys. Rev. D **68**, 094013 (2003) [arXiv:hep-ph/0307037].

[19] R. Baier, A. Kovner and U. A. Wiedemann, Phys. Rev. D **68**, 054009 (2003) [arXiv:hep-ph/0305265].

[20] J. Jalilian-Marian, Y. Nara and R. Venugopalan, Phys. Lett. B **577**, 54 (2003) [arXiv:nucl-th/0307022].

[21] B. Z. Kopeliovich, J. Nemchik, A. Schafer and A. V. Tarasov, Phys. Rev. Lett. **88**, 232303 (2002) [arXiv:hep-ph/0201010] and references therein.

[22] A. Accardi and M. Gyulassy, arXiv:nucl-th/0308029; X. N. Wang, Phys. Rev. C **61** (2000) 064910; Phys. Lett. B **565** (2003) 116; G. G. Barnafoldi, G. Papp, P. Levai and G. Fai, arXiv:nucl-th/0307062.

[23] D. Kharzeev, E. Levin and L. McLerran, Phys. Lett. B **561**, 93 (2003) [arXiv:hep-ph/0210332].

[24] J. L. Albacete, N. Armesto, A. Kovner, C. A. Salgado and U. A. Wiedemann, Phys. Rev. Lett. **92**, 082001 (2004) [arXiv:hep-ph/0307179].

[25] R. Debbe, [BRAHMS Collaboration], talk given at the APS DNP Meeting at Tucson, AZ, October, 2003; R. Debbe [BRAHMS Collaboration], arXiv:nucl-ex/0403052.

[26] I. Arsene *et al.* [BRAHMS Collaboration], arXiv:nucl-ex/0403005.

[27] M. Liu [PHENIX Collaboration], Talk at Quark Matter 2004 Conference, Oakland, California, January 2004; arXiv:nucl-ex/0403047.

[28] P. Steinberg [PHOBOS Collaboration], Talk at Quark Matter 2004 Conference, Oakland, California, January 2004.

[29] L. Barnby [STAR Collaboration], Talk at Quark Matter 2004 Conference, Oakland, California, January 2004; arXiv:nucl-ex/0404027.

[30] D. Kharzeev, Y. V. Kovchegov and K. Tuchin, Phys. Lett. B **599**, 23 (2004) [arXiv:hep-ph/0405045].

[31] K. Golec-Biernat and M. Wusthoff, Phys. Rev. D **59**, 014017 (1999) [arXiv:hep-ph/9807513].

[32] R. Baier, A. H. Mueller, D. Schiff and D. T. Son, Phys. Lett. B **539**, 46 (2002) [arXiv:hep-ph/0204211].

Institute of Physics Publishing Journal of Physics: Conference Series **50** (2006) 87–94
doi:10.1088/1742-6596/50/1/010 5th International Conference on Physics and Astrophysics of Quark Gluon Plasma

Results from Lattice QCD

Rajiv V. Gavai[3]

Department of Theoretical Physics, Tata Institute of Fundamental Research,
Homi Bhabha Road, Mumbai 400 005, India

Abstract. I present our recent results on the critical end point in the μ_B-T phase diagram of QCD with two flavours of light dynamical quarks and compare them with similar results from other groups. Implications for a possible energy scan at the RHIC are discussed. I also comment briefly on the new results of great relevance to heavy ion collisions from finite temperature lattice QCD simulations on speed of sound, specific heat and on the fate of J/ψ.

1. Introduction

Lattice Quantum Chromo Dynamics (QCD), defined on a space-time lattice, has been our best and most reliable tool to extract non-perturbative physics of the strongly interacting theory for over a decade now. In many cases, such as the hadron spectra or the weak decay constants, the focus of lattice QCD has now shifted to obtaining more precise results. In the fields of strong interest to this conference, namely Relativistic Heavy Ion Collisions and the possible transition to the new Quark-Gluon Plasma (QGP) phase, also there are similar examples : the quark-hadron transition temperature [1], T_c, the Wróblewski Parameter [2], λ_s and the the equation of state [1]. Already a lot of reliable theoretical information on these physical observables of interest has been predicted by lattice QCD and again the focus is now for better precision. Due to the strong influence of dynamical fermions on thermodynamics and the nature of the phase transition, one still has some way to go before precise prediction for realistic world of two light (u and d) and one heavy (s) flavour become available. Nevertheless, the continuing progress in lattice QCD makes it perhaps a reachable goal. I shall be unable to review all the important recent lattice results in the time available. I have therefore chosen to focus on those with thrust on something new, either conceptually or quantitatively. In particular, I shall cover the new results on the T-μ_B phase diagram and the physically interesting question of J/ψ-dissolution/persistence. An interesting theoretical issue these days has been the comparison of predictions of conformally invariant theories with lattice QCD results. I shall show one such comparison along with new results for speed of sound.

2. QCD Phase Diagram

The first principles exploration of the QCD phase diagram has the lattice QCD partition function as its focal point. Assuming three flavours of quarks, u, d, and s and denoting by μ_f the corresponding chemical potentials, it is defined by,

$$\mathcal{Z} = \int DU \quad \exp(-S_G) \prod_{f=u,d,s} \text{Det } M(m_f, \mu_f) \quad . \tag{1}$$

[3] E-mail: gavai@tifr.res.in

Note that the quark mass m_f and the corresponding chemical potential μ_f enter only through the Dirac matrix M for each flavour. The baryon and isospin chemical potentials are related to those for quarks by $\mu_B = \mu_u + \mu_d + \mu_s$ and $\mu_3 = \mu_u - \mu_d$.

Progress in unravelling the phase diagram using lattice QCD has been slow due to so-called "fermion problem": for nonzero μ, the determinant in eq. (1) is complex. This renders both analytical and numerical investigations very difficult. Recently, progress was made by a key observation. Following the phase transition from the $\mu_B = 0$ axis into the μ_B-T plane is relatively easier. Various methods have been employed to achieve this. In addition to the original re-weighting method [3] in two couplings, imaginary chemical potential [4] and Taylor series expansion [5] have been used to find and locate the critical end point, expected for a world with two light flavours on the basis of a variety of model considerations [6]. I shall describe below our results obtained by using the Taylor series method. One of its primary advantages is the ease of taking the continuum and thermodynamic limit which is mandatory for all such lattice computations. This is especially difficult for the re-weighting method [3] due to the $\exp(-\Delta S)$ factor it has which becomes exponentially small for bigger lattices. Furthermore, the discretization errors propagate in an unknown manner in a re-weighting computation, making reliable estimates in continuum limit very difficult. A much better control of the systematic errors is, on the other hand, achieved in the method we employ. We study volume dependence of the Taylor coefficients at several temperature T to i) bracket the critical region and then to ii) track its change as a function of volume. As shown in Figure 1, one expects to pin down the critical region much more precisely this way in going from a smaller volume V_1 to a bigger one V_2.

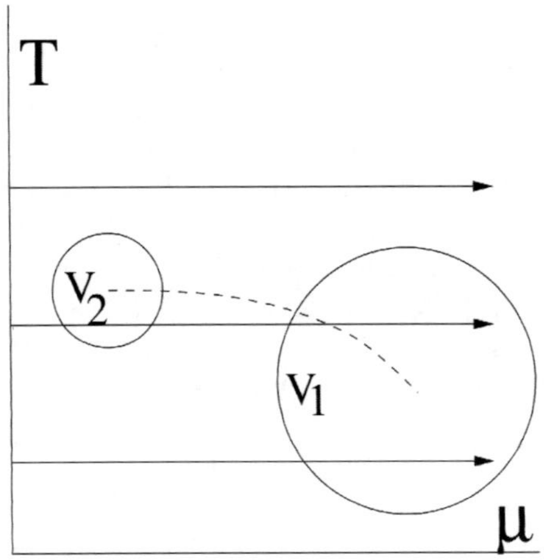

Figure 1. Volume dependence of the critical region.

Various quark, baryon and isospin densities and the corresponding susceptibilities can be obtained from eq. (1) as:

$$n_i = \frac{T}{V}\frac{\partial \ln \mathcal{Z}}{\partial \mu_i}, \qquad \chi_{ij} = \frac{T}{V}\frac{\partial^2 \ln \mathcal{Z}}{\partial \mu_i \partial \mu_j} \quad . \tag{2}$$

Higher order susceptibilities, χ_{n_u,n_d}, can be similarly defined as $n_u(n_d)$ partial derivatives of $\ln \mathcal{Z}$ with respect to $\mu_u(\mu_d)$. Using them, the pressure P has the following expansion in μ:

$$\frac{\Delta P}{T^4} \equiv \frac{P(\mu,T)}{T^4} - \frac{P(0,T)}{T^4} = \sum_{n_u,n_d} \chi_{n_u,n_d} \frac{1}{n_u!}\left(\frac{\mu_u}{T}\right)^{n_u} \frac{1}{n_d!}\left(\frac{\mu_d}{T}\right)^{n_d} \tag{3}$$

Quark number susceptibilities in (2) are crucial for many quark-gluon plasma signatures which are based on fluctuations in globally conserved quantities such as baryon number or electric charge. Theoretically, they serve as an important independent check on the methods and/or models which aim to explain the large deviations of the lattice results for pressure $P(\mu{=}0)$ from the corresponding perturbative expansion. This has been discussed elsewhere in details. Here we will be concerned with using them with the higher order terms to learn about the critical end point. Indeed, a series for baryonic susceptibility χ_B can be constructed using them, whose radius of convergence gives the nearest critical point [7]. Limited by the finite (and even small) number of terms available, one uses standard tests for series convergence on them to do the best one can. Successive estimates for the radius of convergence can be obtained from these using $r_{n+2} = \sqrt{\chi_B^n/\chi_B^{n+2}}$. We use terms up to 8th order in μ, i.e., the estimates from 2/4, 4/6 and 6/8 terms in eq.(3). It may be noted here that a similar construction of a series goes through for the off-diagonal susceptibility, χ_{11} as well. The ratio χ_{11}/χ_{20} can be shown to yield the ratio of widths of the measure in the imaginary and real directions at $\mu = 0$, i.e., a measure of the difficulty of the 'fermion sign' problem mentioned above. With some care, this argument can be generalized to nonzero μ.

2.1. *Our Results*

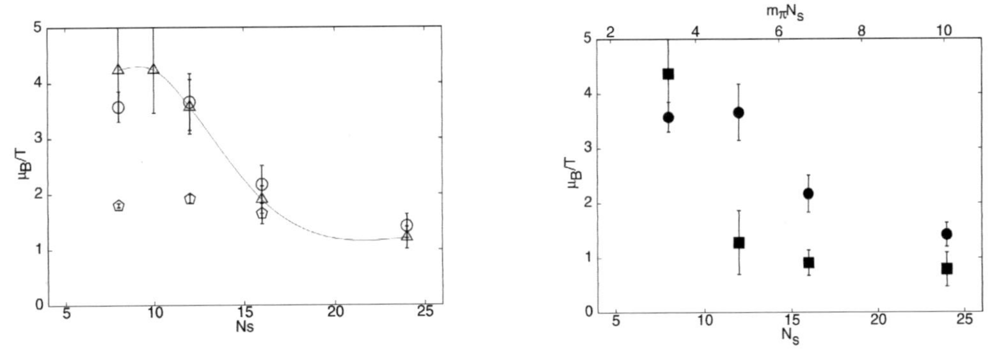

Figure 2. The radius of convergence as a function of N_s ($m_\pi N_s$). The left panel shows results obtained by comparing the 4th and 6th order at $T/T_c = 0.9$ (pentagons), 0.95 (circles) and 1 (triangles). The right panel shows results obtained at $T/T_c = 0.95$ using the ratio of the 4th and 6th order NLS (boxes) and the 6th and 8th order NLS (circles).

Our results [7] were obtained by simulating full QCD with two flavours of dynamical staggered quarks of mass $m/T_c = 0.1$, where T_c is the transition temperature for zero chemical potential, on lattices of size $4 \times N_s^3$, with N_s = 8, 10, 12, 16 and 24. R-algorithm [8] with a trajectory length of one unit of molecular dynamics time on N_s = 8 was used for simulations of full QCD. The trajectory length was scaled with N_s on larger spatial lattices. Earlier investigations have pinned down the physical parameters for our simulations at zero chemical potential [9] : $m_\rho/T_c = 5.4 \pm 0.2$ and $m_\pi/m_\rho = 0.31 \pm 0.01$. We used the canonical methods to fix the lattice spacing a in physical units for which we made runs on symmetric 16^4 lattices. Their details as well as other details about evaluation of the higher order susceptibilities can be found in [7]. Here I shall be brief and state only that our simulations were made at T/T_c = 0.75(2), 0.80(2), 0.85(1), 0.90(1), 0.95(1), 0.975(10), 1.00(1), 1.05(1), 1.25(1), 1.65(6) and 2.15 (10) with a typical statistics of 50-100 in the units of (maximum) autocorrelation length.

Using terms of up to 8th order in μ in the expansion in eq. (3), we obtained estimates for the radius of convergence shown in Figure 2. The left panel shows r_6 as a function of N_s at three temperatures whereas the right panel exhibits r_6 and r_8 at $T/T_c = 0.95$. The latter also shows on top the spatial lattice size in units of inverse pion length. As seen from the left panel of the figure, our estimate of the radius of convergence for the smaller lattices with $N_s \sim 10$ is consistent with the critical end point estimate of the re-weighting method [3] with appropriate size, i.e, $N_s m_\pi$ = 3-4. Note that $m_\pi/m_\rho = 0.31$ in that case as well. Figure 2 shows, however, large finite size effects for these small lattices and a strong change in the estimate around $N_s m_\pi \sim 6$. Such finite size effects are expected theoretically [10] and have been seen in other cases such as chiral condensates[11] and lattice QCD evaluations of structure functions [12]. A key consequence of our use of larger volumes than before is that we find a shift in the critical end point to $\mu_B/T \sim 1-2$, as is evident from Figure 2. Recently, similar computations but up to 6th order have been reported on a large lattice with $N_s m_\pi \sim 15$, and were also discussed in this conference [13]. These use, however, large quark masses, and consequently a a large $m_\pi/m_\rho \sim 0.7$. It will be interesting to see if lowering their quark mass leads to results similar to ours.

2.2. Measure of Sign Problem

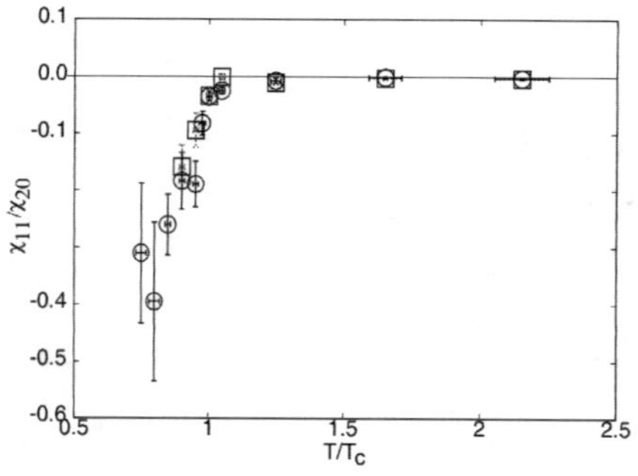

Figure 3. The ratio of quark number susceptibilities χ_2 and χ_{11} on 4×16^3 (circles) and 4×24^3 (boxes) lattices.

Figure 3 displays the ratio of the two susceptibilities χ_{11} and χ_{20} as a function of temperature on two different lattice sizes. This ratio, which is a measure of the seriousness of sign problem, shows how the increase in both its value and fluctuations make the simulations progressively more difficult as one lowers the temperature to T_c and below. The success of the current round of finite density lattice simulations can also be traced from this figure to the fact that all current efforts, regardless of the details of their methods, attempt to explore the phase diagram starting from T_c, the transition point at zero chemical potential. An important check one can make in the Taylor series expansion technique is that each coefficient in the Taylor expansion must be volume independent. Since the actual evaluation of these terms entails cancellation of divergences, this is indeed a nontrivial check which has been made successfully [7]. In fact, care is needed in extracting any equation of state at finite chemical potential from these coefficients : a systematic study on different volumes (spatial lattices) with a subsequent extrapolation to infinite volume is necessary to ensure that some peak-like behaviour in, e.g., χ_{40}, is real and does not give spurious results for thermodynamic quantities.

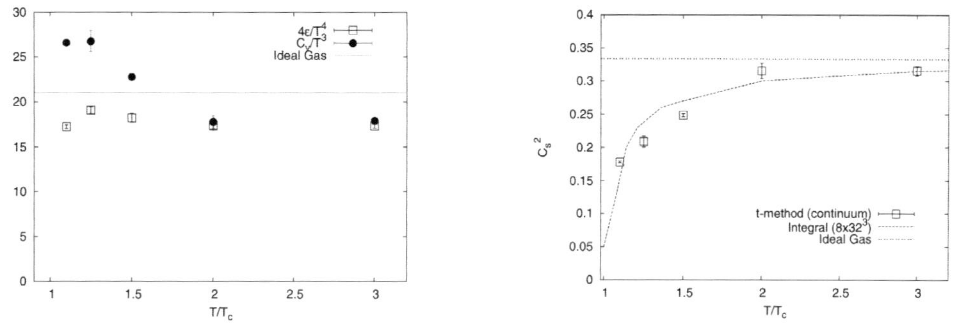

Figure 4. Specific heat C_v and 4ϵ (left) and the speed of sound c_s as a function of temperature for quenched QCD. See text for more details.

3. Speed of Sound

An important physical property of the strongly interacting matter at finite temperature, which has been computed [14] since long using lattice techniques, is the the speed of sound. Its relevance to many aspects of heavy ion physics, such as the elliptic flow or the hydrodynamical studies can hardly be overemphasized. A related quantity of interest is the specific heat which has been related to event-by-event fluctuations of the transverse momentum (p_T). There are various measures of these fluctuations currently in use in the literature. It may actually be better to relate them to the specific heat in order to make them both more physical as well as uniform in convention.

A recent development in this direction on the lattice is a new way to obtain these by relating them to the temperature derivative of an anomaly measure $\Delta/\epsilon \equiv (\epsilon - 3P)/\epsilon$ [15]. Using lattices with 8, 10, and 12 temporal sites ($38^3 \times 12$ and 38^4 lattices) and with statistics of 0.5-1 million iterations, ϵ, P, s, C_s^2 and C_v were obtained in continuum using this differential method at $2T_c$ and $3T_c$ for quenched QCD. Sizable deviations from the corresponding ideal gas results were found although C_v was consistent with 4ϵ, as expected in a weak coupling scenario. On the other hand the result for entropy density at $3T_c$ also seemed to agree with a strong coupling (super Yang-Mills) prediction [16] : $s/s_0 = f(g^2N_c)$, where $f(x) = \frac{3}{4} + \frac{45}{32}\zeta(3)(2x^{-3/2}) + \cdots$ and $s_0 = \frac{2}{3}\pi^2 N_c^2 T^3$. However, curiously it too failed at $2T_c$.

New results for the specific heat and the speed of sound, obtained by using a new method, called the t-method, were presented at this conference in a poster [17]. While the differential method [18] for obtaining thermodynamic quantities has lesser systematic errors and is convenient to use, it was abandoned in favour of the integral method [19] due to the former's problem of yielding negative Pressure. It was shown that this problem of negative pressure can be completely avoided even in the differential method by writing the derivatives with respect to temperature and volume using a parameter t : the old (Bielefeld) method [18] corresponds to $t = 1$ and leads to negative pressure but $t = 0$ results in positive pressure always. Figure 4 shows these new results of [17] for the specific heat (the left panel) and the speed of sound (right panel). While the corresponding pressure is not shown here, it is positive even around T_c, an input which went in getting the speed of sound (squares), shown in Figure 4. Note that these continuum results have been obtained by simulating lattices with many lattice spacings and then extrapolating to zero unlike the results for the integral method (dotted line) in Figure 4 which are for $N_t = 8$. The interesting and noteworthy features are : i) the specific heat does tend to peak near T_c and is widely different there from the weak coupling expectation of 4ϵ, ii) for $T \geq 2T_c$, this weak coupling expectation is fulfilled but the resulting value is very different from that for an ideal gas, iii) speed of sound is about 10 % smaller than the ideal gas for higher temperatures but drops close to T_c, and iv) the results are in good agreement with the integral method but could differ significantly in the neighbourhood of T_c. Further investigations are needed to see if c_s^2 dips close to zero near T_c.

4. Persistence of J/ψ

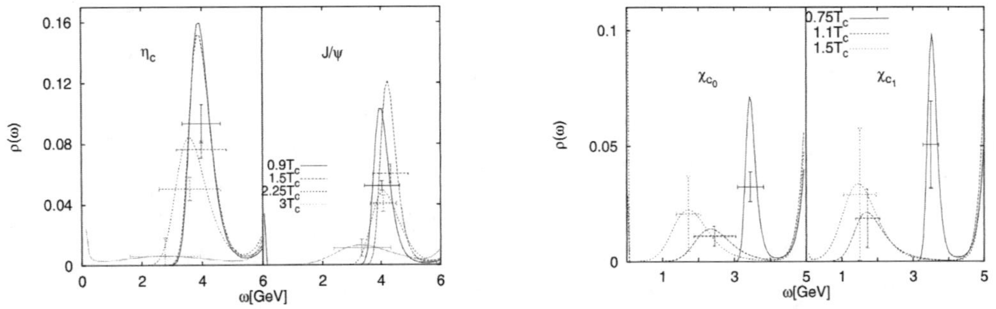

Figure 5. Spectral functions [20] of the vector J/ψ and scalar χ mesons as a function of energy at various temperatures. See text for more details.

An intriguing result on the J/ψ and similar heavy quarkonia has recently emerged [20, 21] from the lattice QCD investigations. As is well known, Matsui and Satz [22] proposed J/ψ suppression as a signal of quark-gluon plasma. Quarkonium potential model calculations with an ansatz for temperature dependence of the heavy quark-antiquark potential lead to the result that J/ψ and χ_c states dissolve in QGP by $1.1T_c$. The impressive results from the CERN NA50 [23] did suggest an anomalous suppression of J/ψ in Pb-Pb collisions, bringing forth detailed theoretical questions on this suppression mechanism and the above mentioned ansatz. In particular, one approach taken by the lattice experts [20, 21] has been to study the spectral function of these states in QGP to look for their dissolution from first principles. The recognition of the maximum entropy method (MEM) as a reliable tool to extract the spectral functions from temporal lattice correlators at zero temperature lead to its application at finite temperature. Caution is, however, called for in doing so since nonzero and high temperatures are obtained by making the temporal lattice smaller whereas MEM works better for longer temporal lattices.

Figure 4 shows the results of the Bielefeld group [20], obtained by simulating lattices from $48^3 \times 12$ to $64^3 \times 24$ in sizes in the quenched approximation of no dynamical quarks. The vertical bars are drawn to indicate the size of the error on the integrated spectral function in the area shown by the horizontal bars. One observes that the χ_c does seem to indeed dissolve by $1.1T_c$, as expected from naive potential model considerations. However, the J/ψ and η_c persist up to 2.25 T_c and are gone only at $3T_c$. Similar results have been obtained by other groups [21] as well, although there seems to be some disagreement on the precise value of the temperature at which the J/ψ peak vanishes.

While these results have to be still confirmed in a full QCD simulation with light dynamical quarks, they already can have strong consequences for the J/ψ signal. Recall that only about 30-40 % of the J/ψ come through χ decays (with a substantial fraction from χ_2 which has not yet been investigated), thus the predicted suppression will be rather small until temperatures close to $3T_c$ are reached (or energy densities 81 times that of critical energy density is reached), which seems to be a tall order for the current round of CERN and BNL experiments. An interesting observable consequence could, however, be a change of the suppression pattern as a function of $\sqrt{s}$ or the energy density reached.

5. Summary

The QCD phase diagram in $T - \mu_B$ plane has begun to emerge. Since all results so far have been obtained on $N_t = 4$, i.e, very coarse, lattices, several quantitative issues related to continuum limit still have to wait. It may be noted that the existence of equally legitimate [24] different prescriptions of introduction of the chemical potential on the lattice can, and do, lead to even larger lattice artifacts on such coarse lattices. Nevertheless, it is very heartening to find that different methods lead to a similar qualitative picture.

Our results, using Taylor expansion, permit us to take the thermodynamic limit, unlike other methods. Our investigation of volume dependence suggests $N_s m_\pi > 6$ approximates the thermodynamic volume limit reasonably. The μ_B/T of the critical end point, identified from the radius of convergence using up to, and including the 8th order terms, drops strongly around that volume. We find that $\mu_B/T \sim 1 - 2$ is indicated for the critical point with a corresponding temperature of $\simeq 0.95 T_c$, where T_c is the transition temperature for the baryonfree case.

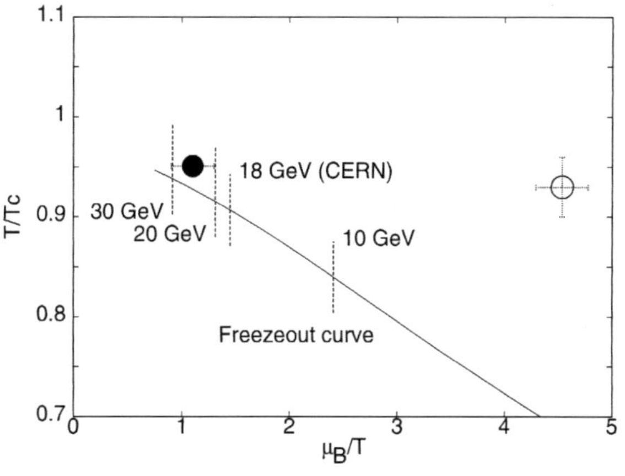

Figure 6. The QCD phase diagram with a freeze-out curve superimposed. The filled circle denotes the estimate of the critical point which has been obtained in [7]. The open circle is an earlier estimate from [3] using smaller lattices and nearly the same quark mass.

Figure 6 shows our results, together with the earlier determination [3] on small volumes but similar quark mass, i.e, the pion mass. A freeze-out curve [25], obtained by treating a resonance gas as ideal, has been superimposed on the Figure. From the marked values of the CM energy, $\sqrt{s}$, per nucleon needed to reach that point on the curve, one notes that the critical end-point may be in the reach of future RHIC energy scans.

I also presented new continuum results on the speed of sound and the specific heat for quenched QCD. A new method, which revives the earlier differential method but without its drawback of negative Pressure was used to obtain these results. Interesting quantitative agreement was found with strong coupling predictions from super Yang-Mills theory at $3T_c$ but not at $2T_c$.

Finally, the intriguing persistence of J/ψ in QGP was discussed. The MEM technique has been used to obtain the spectral function at finite temperature. While it showed the peak in the χ spectral function to go away by $1.1 T_c$, the peak in the vector spectral function persists up to 2.25-3T_c. Extension of these simulations to incorporate dynamical quarks would be crucial to be sure of the phenomenological consequences of these results.

[1] E. Laermann and and O. Philipsen, Ann. Rev. Nucl. Sci. **53** (2003) 163.

[2] R. V. Gavai and S. Gupta, hep-ph/0502198.

[3] Z. Fodor and S. Katz, JHEP **0203** (2002) 014.

[4] Ph. de Forcrand and O. Philipsen, *Nucl. Phys.* **B 642** (2002) 290; M.-P. Lombardo and M. D'Elia, *Phys. Rev.* **D 67** (2003) 014505.

[5] C. R. Allton et al., *Phys. Rev.* **D 68** (2003) 014507; R. V. Gavai and S. Gupta, *Phys. Rev.* **D 68** (2003) 034506.

[6] A. Barducci et al., *Phys. Lett.* **B 231** (1989) 463; M. A. Halasz et al., *Phys. Rev.* , **D 58** (1998) 096007; J. Berges and K. Rajagopal, *Nucl. Phys.* **B 538** (1999) 215.

[7] R. V. Gavai and S. Gupta, hep-lat/0412035.

[8] S. Gottlieb et al., *Phys. Rev.* **D 35** (1987) 2531.

[9] S. Gottlieb et al., *Phys. Rev. Lett.* **59** (1987) 1513; S. Gottlieb et al, *Phys. Rev.* **D 38** (1988) 2245.

[10] H. Leutwyler and A. Smilga, *Phys. Rev.* **D 46** (1992) 5607.

[11] S. Gupta and R. Ray, *Phys. Rev.* **D70** (2004) 114015.

[12] I. Wetzorke et al., *Nucl. Phys.* Proc. Suppl. **140** (2005) 393 (hep-lat/0409142).

[13] C. R. Allton et al., hep-lat/0501030

[14] R. V. Gavai and A. Gocksch, *Phys. Rev.* **D 33** (1986) 614; K. Redlich and H. Satz, *Phys. Rev.* **D 33** (1986) 3747.

[15] R. V. Gavai, S. Gupta and S. Mukherjee, hep-lat/0412036.

[16] S. S. Gubser, I. R. Klebanov and A. A. Tseytlin, *Nucl. Phys.* **B 534** (1998) 202.

[17] R. V. Gavai, S. Gupta and S. Mukherjee, Poster at this conference.

[18] J. Engels et al., *Nucl. Phys.* **B 205 [FS5]** (1982) 545.

[19] J. Engels et al., *Phys. Lett.* **B 252** (1990) 625.

[20] S. Datta et al., *Phys. Rev.* **D 69** (2004) 094507.

[21] M. Asakawa and and T. Hatsuda, *Phys. Rev. Lett.* **92** (2004) 012001; T. Umeda et al. Eur. Phys. J. **C 39** (2005) 9 (hep-lat/0211003).

[22] T. Matsui and H. Satz, *Phys. Lett.* **B 178** (1986) 416.

[23] M.C. Abreu et al.,The NA50 collaboration, *Phys. Lett.* **B 477** (2000) 28.

[24] R. V. Gavai *Phys. Rev.* **D 32** (1985) 519.

[25] J. Cleymans, private communication.

Institute of Physics Publishing
doi:10.1088/1742-6596/50/1/011

Journal of Physics: Conference Series **50** (2006) 95–102
5th International Conference on Physics and Astrophysics of Quark Gluon Plasma

Fluctuations and Correlations in Heavy Ion Collisions

V. Koch

Lawrence Berkeley National Laboratory
Berkeley, CA 94720, U.S.A.

Abstract. An overview of the physics of event by event fluctuations in heavy ion collisions is provided. We will discuss what the measurement of fluctuations and correlation can tell us about the system created in these collisions.

1. Introduction

The study and analysis of fluctuations and correlations are an essential method to characterize a physical system. In general, one can distinguish between several classes of fluctuations. On the most fundamental level there are quantum fluctuations, which arise if the specific observable does not commute with the Hamiltonian of the system under consideration. These fluctuations probably play less a role for the physics of heavy ion collisions. Second, there are "dynamical" fluctuations reflecting the dynamics and responses of the system. They help to characterize the properties of the bulk (semi-classical) description of the system. Examples are density fluctuations, which are controlled by the compressibility of the system. Finally, there are "trivial" fluctuations induced by the measurement process itself, such as finite number statistics etc. These need to be understood, controlled and subtracted in order to access the dynamical fluctuations which tell as about the properties of the system.

While fluctuations are related to the variance of a given observable[1], correlations are accessible via the co-variances. The measurement of these co-variances will provide information about the independence of the degrees of freedom considered.

Fluctuations are also closely related to phase transitions. The well known phenomenon of critical opalescence is a result of fluctuations at all length scales due to a second order phase transition. First order transitions, on the other hand, give rise to bubble formation, i.e. density fluctuations at the extreme.

The most efficient way to address fluctuations of a system created in a heavy ion collision is the study of event-by-event (E-by-E) fluctuations, where a given observable is measured on an event-by-event basis and the fluctuations are studied over the ensemble of the events. In most cases (namely when the fluctuations are Gaussian) this analysis is equivalent to the measurement of two particle correlations over the same region of acceptance [1]. Consequently, fluctuations tell us about the 2-point functions of the system, which in turn determine the response of the system to external perturbations.

In the framework of statistical physics, which appears to describe the bulk properties of heavy ion collisions up to RHIC energies, fluctuations measure the susceptibilities of the system. These susceptibilities also determine the response of the system to external forces. For example, by measuring fluctuations of the net electric charge in a given rapidity interval, one obtains information on how this (sub)system would respond to applying an external (static) electric field. In other words, by measuring fluctuations one gains access to the same fundamental properties of the system as "table top" experiments dealing with macroscopic probes. In the later case, of course, fluctuation measurements would be impossible.

2. Correlations and Fluctuations in a thermal system

Given the free energy $F = -T \log Z$ of a system, its fluctuations are controlled by the susceptibilities, which are the second derivatives of the free energy with respect to the appropriate conjugate variables,

$$\chi_{i,j} = -\frac{1}{V} \frac{\partial^2}{\partial \mu_i \mu_j} F. \tag{1}$$

As an example, the fluctuations of the net charge are given by

$$\left\langle \delta Q^2 \right\rangle = -T \frac{\partial^2}{\partial \mu_Q^2} = TV \chi_{Q,Q}. \tag{2}$$

Covariances are related to the off diagonal susceptibilities, $i \neq j$, for example

$$\langle (\delta B)(\delta S) \rangle = TV \chi_{B,S} = -T \frac{\partial^2 F}{\partial \mu_B \mu_S}. \tag{3}$$

Note that for vanishing baryon number chemical potential, these susceptibilities can be calculated in Lattice QCD. In this case it is more convenient to study quark number susceptibilities $\chi_{f,f'}$ where f and f' represent the different quark flavors u,d and s. These have been calculated in lattice QCD [2, 3] and provide interesting insight into the structure of the matter above the phase transition.

Recently the correlation between baryon number and strangeness has been proposed as a useful diagnostic for the structure of high temperature matter [4]. The authors observed that in case of uncorrelated quarks strangeness is carried exclusively by quarks, and thus is tightly correlated to the baryon number. In a hadron gas or in a bound-state picture of the QGP [5] strangeness can be carried by mesons or $s\bar{q}$ states, which do not carry baryon number. This will reduce the correlations. In ref. [4] the following ratio is considered

$$C_{BS} = \equiv -3 \frac{\sigma_{BS}}{\sigma_S^2} = -3 \frac{\langle BS \rangle - \langle B \rangle \langle S \rangle}{\langle S^2 \rangle - \langle S \rangle^2} = -3 \frac{\langle BS \rangle}{\langle S^2 \rangle}, \tag{4}$$

which gives $C_{BS} = 1$ for uncorrelated quarks, and $C_{BS} \simeq 0.66$ and $C_{BS} \simeq 0.62$ for a hadron gas and a bound-state QGP, respectively. In terms of the flavor susceptibilities

$$C_{BS} = 1 + \frac{\chi_{us} + \chi_{ds}}{\chi_{ss}}. \tag{5}$$

Using the results obtain in quenched Lattice QCD [2] one obtains $C_{BS} = 1$, consistent with uncorrelated quarks. At the same time the equation of state (EOS) differs from that of free quarks an gluons. However, the EOS can be reproduced in a quasi-particle picture [6, 7] suggesting an effective mean field description for the matter at high temperature. Incidentally the quasiparticle picture and the EOS indicate a predominantly repulsive interaction, which might partially account for the observed flow at RHIC.

Let us point out that C_{BS} is accessible in experiment via event-by-event fluctuations [4], although a precise determination would require neutron detection capabilities.

3. Transverse momentum and charge fluctuations

The field of event-by-event fluctuations is relatively new to heavy ion physics and ideas and approaches are just being developed. So far, most of the analysis has concentrated on transverse momentum and the net charge fluctuations.

The pioneering event-by-event studies have been carried out by the NA49 collaboration. They have analyzed the fluctuations of the mean transverse momentum [8] and the kaon to pion ratio [9]. Both measurements have been carried out at at the CERN SPS at slightly forward rapidities. In both cases mixed events can essentially account for the observed signal, leaving little room for genuine dynamical fluctuations.

Transverse momentum fluctuations should be sensitive to temperature/energy fluctuations [10, 11]. These in turn provide a measure of the heat capacity of the system [12] since

$$\left\langle \delta E^2 \right\rangle = \frac{\partial^2}{\partial \beta^2} \log Z = -T^3 \frac{\partial^2}{\partial T^2} F = T^2 C_V. \tag{6}$$

As the QCD phase transition is associated with a maximum of the specific heat, the temperature fluctuations should exhibit a minimum in the excitation function. It has also been argued [13, 14] that these fluctuations may provide a signal for the long range fluctuations associated with the tri-critical point of the QCD phase diagram. In the vicinity of the critical point the transverse momentum fluctuations should increase, leading to a maximum of the fluctuations in the excitation function.

Transverse momentum fluctuations have been analyzed by several experiments at different bombarding energies. At SPS energies the NA49 collaboration measured transverse momentum fluctuations in the forward rapidity region and found no significant deviation from pure statistics [8]. Similarly, at RHIC energies, the PHENIX collaboration also reports no significant non-statistical transverse momentum fluctuations [15]. In contrast to that the CERES collaboration finds fluctuations larger than those from mixed events [16] at SPS energies and at RHIC the STAR collaboration reports significant deviations from mixed events [17]. To which extent this can be attributed to the different acceptance regions covered by these experiments remains to be investigated.

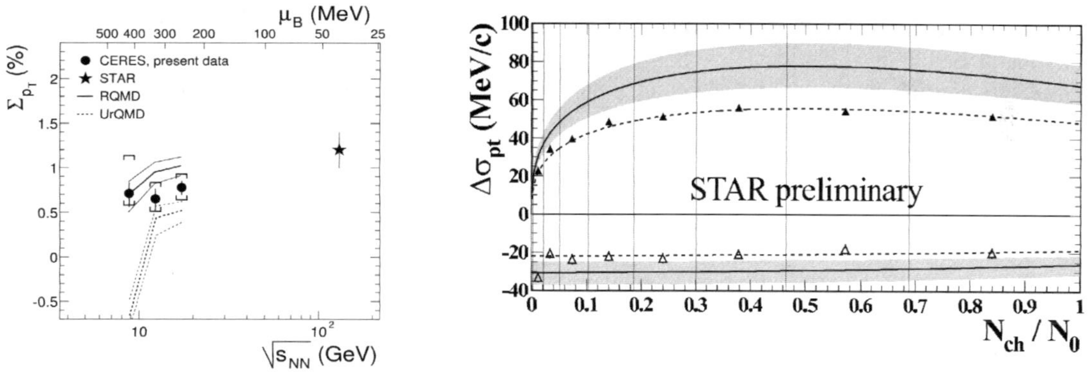

Figure 1. Left:Excitation function for transverse momentum fluctuations from [16]. Right: Centrality dependence of p_t fluctuations (from [17]).

The obvious question is why are the p_t fluctuations small and similar for all bombarding energies. This could have a simple explanation. Note, that experiments always subtract mixed events. If the system is similar to an ideal gas, i.e. does not carry significant correlations, the fluctuations are just those of the mixed events [24] and thus not "net" p_t fluctuations remain. Now a hadron gas as well as a uncorrelated QGP are rather similar to an ideal gas. Only if the system is prepared close to the phase transition, where the heat capacity C_V has a maximum, we expect the p_t fluctuations to deviate significantly from that of an ideal gas. So it could very well be the case that the phase transition is either between SPS and RHIC energies, or, as suggested by the rather interesting preliminary data on K/π fluctuations [18], it is located at the lower SPS energies (20-30 AGeV beam energy).

Another observable of interest are so-called charge fluctuations, since they provide a signature for the existence of a de-confined Quark Gluon Plasma phase [19, 20]. The essential idea is that in a QGP the charge carriers are the quarks, which posses fractional charge. Since charge fluctuations are proportional to the square of the charge

$$\left\langle \delta Q^2 \right\rangle = q^2 \left\langle (\delta N)^2 \right\rangle, \tag{7}$$

the ratio of charge fluctuation over entropy

$$\frac{\langle \delta Q^2 \rangle}{S} \sim \frac{\langle \delta Q^2 \rangle}{\langle N_{charge} \rangle} \tag{8}$$

is sensitive to the fractional charges in a QGP. In ref. [20] the observable

$$D \equiv 4 \frac{\langle \delta Q^2 \rangle}{\langle N_{charge} \rangle} \tag{9}$$

has been proposed and is has been shown that $D = 4$ for an uncorrelated pion gas, $D \simeq 3$ for a resonance gas [21] and $D \simeq 1 - 1.5$ for a Quark Gluon Plasma, respectively. Since the electric charge is conserved globally, and thus does not fluctuate, experimental measurements need to be corrected for charge conservation effects. These become significant once a sizeable fraction of the final state particles are taken into account. Several prescriptions for these corrections have been proposed [22, 23] which all agree in the limit of small acceptance. A detailed discussion can be found in [24].

In the mean time charge fluctuation have been analyzed by several experiments. PHENIX [25] at RHIC which measures with a small rapidity acceptance, finds charge fluctuations consistent with a resonance gas, if extrapolated to larger acceptance. STAR, which has a large acceptance also finds charge fluctuations consistent with a resonance gas [26]. CERES [27] and NA49 [28], which both measure at SPS energies, report preliminary results on charge fluctuations, which are consistent with a pure pion gas. However, at the SPS the overall rapidity distribution is rather narrow, so that the correlation effect of the resonances gets lost when correcting for charge conservation [29]. But certainly, none of the measurements is even close to the prediction for the QGP.

These findings have prompted ideas, that possibly a constituent quark plasma, without gluons, has been produced [30]. However, the measurement of additional observables would be needed in order to distinguish this from a hadronic gas.

But maybe the present range of Δy is so small, that the charge fluctuations have time to assume the value of the resonance gas. As shown in [20] and [31], the larger the rapidity interval considered, the longer the relaxation time for the charge fluctuations. Thus, maybe even larger acceptance is needed to recover the QGP value. This is also suggested by a model calculation using several event generators. As shown in Fig. 2, the results for the parton cascade arrive at the predicted value for the QGP only for $\Delta y \geq 3$. None of the present experiments has such a coverage yet and thus a detailed analysis of D as a function of Δy is needed, before any firm conclusions can be drawn.

We note that a recombination picture [33] for the hadronization should not affect the charge fluctuations, as recombination is local in momentum space and thus introduces negligible rapidity shifts of the charges.

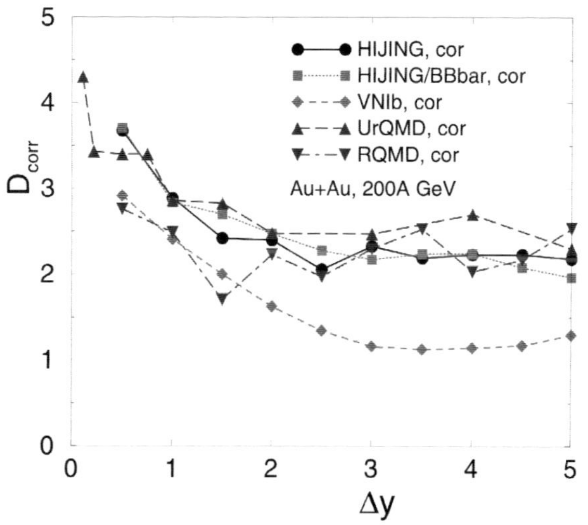

Figure 2. Charge fluctuation from several event generators [32]

Another way to access the non-trivial correlations of the system is the so called balance function [34, 35]. The balance function for charged particle for instance is defined as

$$B(\eta|\Delta\eta) = \frac{1}{2}\left[\frac{\langle N_{+-}(\eta|\Delta\eta)\rangle}{\langle N_{-}(\Delta\eta)\rangle} + \frac{\langle N_{-+}(\eta|\Delta\eta)\rangle}{\langle N_{+}(\Delta\eta)\rangle} - \frac{\langle N_{++}(\eta|\Delta\eta)\rangle}{\langle N_{+}(\Delta\eta)\rangle} - \frac{\langle N_{--}(\eta|\Delta\eta)\rangle}{\langle N_{-}(\Delta\eta)\rangle}\right], \tag{10}$$

where $N_{+-}(\eta|\Delta\eta)$ is the number of unlike-sign pairs which are η apart from each other within the rapidity window $\Delta\eta$. It essentially measures the average distance in rapidity over which a given charge is neutralized (balanced). It is related to the above charge fluctuations in that the latter can be expressed as an integral over the charge balance function [35]. The balance function measurement at $\sqrt{s} = 130$ GeV has been reported by the STAR collaboration[36]. Going from peripheral to central collisions, the width of balance function steadily decreases. The trend is what one would expect if more of the system is filled with a QGP as the collision becomes more central. However, since the reduction is only about 20 % going from most peripheral to most central, it is not yet clear whether this signals the presence of a QGP, constituent quark clusters [37] or more mundane effect such as the strong flow. For instance in [38] the measured balance functions, along with particle ratios and spectra, could be explained in an expanding hadron gas model.

As detailed in [24] all these event-by-event fluctuation observables can be derived from underlying basic correlator

$$\Delta_{\alpha,\beta}(p,q) = \langle n_\alpha(p) n_\beta(q)\rangle\,, \tag{11}$$

which gives the correlation between particles with quantum numbers α and β and momenta p and q respectively. The difference between momentum fluctuation charge fluctuations etc. is then simply the choice of α and β and the weighting functions this correlator is folded with. Also, in order to remove effects from finite number statistics, so called dynamical fluctuations are extracted by either subtracting [39] or dividing [14] the result obtained with and uncorrelated basic "correlator",

$$\Delta^0_{\alpha,\beta}(p,q) = \delta_{\alpha,\beta}\delta_{p,q} n_\alpha(p). \tag{12}$$

Recently is has been pointed out [40, 41] that this can also be achieved by generalized factorial moments.

4. Kaon to pion ratio and its fluctuation

Recently NA49 has reported interesting results on the excitation function of the inclusive kaon to pion ratio as well as its fluctuations [18, 42]. Around 30 AGeV beam energy the inclusive K/π ratio shows a maximum and at the same energy the preliminary data on the fluctuations of the K/π-ratio increase to about 8%, which is a factor of 4 larger then at 160 AGeV [18]. While the fluctuations at 160 AGeV can be accounted for by resonance decays [21] the value of 8% are impossible to reproduce in a hadron gas. Actually, as at lower bombarding energies global strangeness conservation plays a significant role, one would expect some small negative value for the dynamical fluctuations. Instead the fluctuations are large and positive. One possibility to account for these large fluctuation would be domain formation due to spinodal instabilities [43] with subsequent locking of strangeness [44]. In this scenario the large fluctuations of strangeness in the QGP are frozen in during the sudden breakup of the system due to spinodal instabilities. This scenario also provides for a small increase of the inclusive K/π-ratio.

5. Equilibrium

Another important question, which might be addressed by the study of fluctuation is equilibration. While measured particle abundances are well described by a hadron gas in chemical equilibrium [45] this is also the case for collision of elementary particle such as proton-proton or $e^+ - e^-$. Thus simple phase-space dominance a la Fermi [46] needs to be ruled out [47]. In other words, how can we distinguish between a superposition of essentially independent nucleon-nucleon collisions as depicted in Fig. 3(a) and a system which equilibrated over the entire volume (Fig. 3(b))? In the absence of any correlations, the partition functions factorizes and thus the two system are indistinguishable.

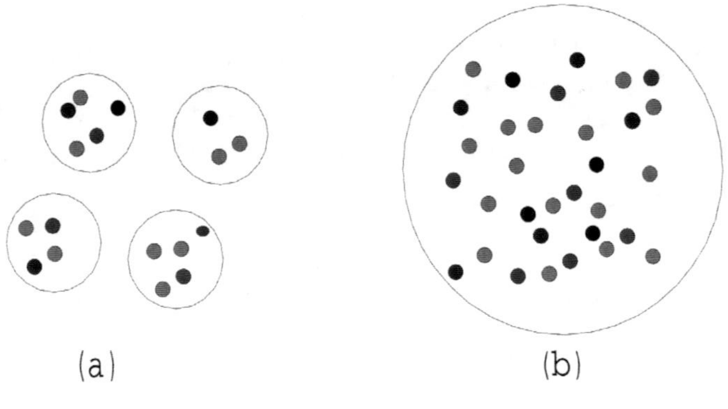

Figure 3. Individual nucleon-nucleon collisions (a) and "true" matter generated in a nucleus-nucleus collision (b).

At low energies (~ 1 AGeV) strangeness conservation introduces such correlations and leads to unique predictions for the second factorial moment of the kaon abundance [48]. At higher energies, however, explicit strangeness conservation becomes less relevant and has only a small effect on the single particle yield. Only if, for some reason, the domain over which strangeness (or any other conserved quantum number) is conserved is so small, that it contains of the order of one conserved quantum, conservation laws still affect particle abundances. In case of strangeness this has been demonstrated in [49]. However, once the strangeness correlation volume, i.e the volume over which strangeness is conserved is larger than twenty times that of a nucleon, the particle abundances are simply a superposition of the sub-domains and sensitivity to the size of the correlation volume is lost. In order to establish strangeness correlation volumina comparable with the system size, many-particle correlations need to be measured. As demonstrated in [50] a definitive measurement of equilibration at RHIC energies would require the measurement of five Omega-baryon coincidences. To which extend this is feasible remains to be seen. Two-particle correlations, which are often discussed as a possible means to establish the degree of equilibrium, are dominated by Poisson statistics and thus are misleading.

6. Conclusions

In this contribution we have discussed the physics of fluctuations in the context of heavy ion collisions. As this is a developing field, this should be considered as a snapshot of our present understanding. We have argued that fluctuations are indeed a new tool to investigate the properties of the strongly interacting matter at high temperature. As an example we have shown that baryon-number strangeness correlations can be utilized to address the correlations among quark flavors, and that charge fluctuations are sensitive to the fractional charges of the quarks, and thus server as a signature for the QGP. The measurement of momentum fluctuations, on the other hand should give us an idea about the heat capacity of the system. Furthermore, if the system is created close to a second order phase transition point, the associated long range fluctuations should be observable in event-by-event observables. Spinodal instabilities associated with a first order phase transition are also detectable in fluctuation observables. Also the question of equilibration can be addressed via fluctuations. At RHIC energies, however, this requires rather difficult measurements of many (> 5) particle correlations.

Acknowledgments

The author thanks A. Majumder and J. Randrup for a fruitful collaboration on some of the topics discussed here. This work was supported by the Director, Office of Science, Office of High Energy and Nuclear Physics, Division of Nuclear Physics, and by the Office of Basic Energy Sciences, Division of Nuclear Sciences, of the U.S. Department of Energy under Contract No. DE-AC03-76SF00098.

References

[1] A. Bialas and V. Koch, Phys. Lett. **B456**, 1 (1999), [arXiv:nucl-th/9902063].
[2] R.V. Gavai and S. Gupta, Phys. Rev. D **67**, 034501 (2003) [arXiv:hep-lat/0211015].
[3] C.R. Allton *et al.*, Phys. Rev. D **71**, 054508 (2005) [arXiv:hep-lat/0501030].
[4] V. Koch, A. Majumder, and J. Randrup [arXiv:nucl-th/0505052].
[5] E.V. Shuryak and I. Zahed, Phys. Rev. D **70**, 054507 (2004) [arXiv:hep-ph/0403127].
[6] J.P. Blaizot, E. Iancu and A. Rebhan, Quark Gluon Plasma 3, 60-122, Hwa, R.C. (ed.), World Scientfic, Singapore 2004. [arXiv:hep-ph/0303185].
[7] A. Peshier, B. Kämpfer and G. Soff, Phys. Rev. D **66**, 094003 (2002) [arXiv:hep-ph/0206229].
[8] NA49, H. Appelshauser *et al.*, Phys. Lett. **B459**, 679 (1999), [arXiv:hep-ex/9904014].
[9] NA49, S. V. Afanasev *et al.*, Phys. Rev. Lett. **86**, 1965 (2001), [arXiv:hep-ex/0009053].
[10] L. Stodolsky, Phys. Rev. Lett. **75**, 1044 (1995).
[11] E. V. Shuryak, Phys. Lett. **B423**, 9 (1998), [arXiv:hep-ph/9704456].
[12] L. Landau and L. Lifshitz, *Statistical Physics* (Pergamon Press, New York, 1980).
[13] M. A. Stephanov, K. Rajagopal, and E. V. Shuryak, Phys. Rev. Lett. **81**, 4816 (1998), [arXiv:hep-ph/9806219].
[14] M. A. Stephanov, K. Rajagopal, and E. V. Shuryak, Phys. Rev. **D60**, 114028 (1999), [arXiv:hep-ph/9903292].
[15] PHENIX, K. Adcox *et al.*, Phys. Rev. **C66**, 024901 (2002), [arXiv:nucl-ex/0203015].
[16] CERES, D. Adamova *et al.*, Nucl. Phys. **A727**, 97 (2003), [arXiv:nucl-ex/0305002].
[17] STAR, J. Adams *et al.*, Adams *et al.* [STAR Collaboration], arXiv:nucl-ex/0308033.
[18] C. Roland *et al.* [NA49 Collaboration], J. Phys. G **30**, S1381 (2004) [arXiv:nucl-ex/0403035].

[19] M. Asakawa, U. W. Heinz, and B. Muller, Phys. Rev. Lett. **85**, 2072 (2000), [arXiv:hep-ph/0003169].
[20] S. Jeon and V. Koch, Phys. Rev. Lett. **85**, 2076 (2000), [arXiv:hep-ph/0003168].
[21] S. Jeon and V. Koch, Phys. Rev. Lett. **83**, 5435 (1999), [arXiv:nucl-th/9906074].
[22] M. Bleicher, S. Jeon, and V. Koch, Phys. Rev. **C62**, 061902 (2000), [arXiv:hep-ph/0006201].
[23] C. Pruneau, S. Gavin, and S. Voloshin, (2002), [arXiv:nucl-ex/0204011].
[24] S. Jeon and V. Koch, Quark-Gluon Plasma 3, 430-490, Hwa, R.C. (ed.), World Scientfic, Singapore 2004 [arXiv:hep-ph/0304012].
[25] PHENIX, K. Adcox *et al.*, Phys. Rev. Lett. **89**, 082301 (2002), [arXiv:nucl-ex/0203014].
[26] J. Adams *et al.*, Phys. Rev. **C68**, 044905 (2003).
[27] H. Appelshauser *et al.* [CERES Collaboration], Nucl. Phys. A **752**, 394 (2005) [arXiv:nucl-ex/0409022].
[28] C. Blume *et al.*, (2002), [arXiv:nucl-ex/0208020].
[29] J. Zaranek, Phys. Rev. **C66**, 024905 (2002), [arXiv:hep-ph/0111228].
[30] A. Bialas, Phys. Lett. **B532**, 249 (2002), [arXiv:hep-ph/0203047].
[31] E. V. Shuryak and M. A. Stephanov, Phys. Rev. **C63**, 064903 (2001), [arXiv:hep-ph/0010100].
[32] Q. H. Zhang, V. Topor Pop, S. Jeon, and C. Gale, Phys. Rev. **C66**, 014909 (2002), [arXiv:hep-ph/0202057].
[33] R.J. Fries, B. Müller, C. Nonaka, and S.A. Bass Phys. Rev. Lett. **90**, 202303 (2003).
[34] S. A. Bass, P. Danielewicz, and S. Pratt, Phys. Rev. Lett. **85**, 2689 (2000), [arXiv:nucl-th/0005044].
[35] S. Jeon and S. Pratt, Phys. Rev. **C65**, 044902 (2002), [arXiv:hep-ph/0110043].
[36] J. Adams *et al.* [STAR Collaboration], Phys. Rev. Lett. **90**, 172301 (2003) [arXiv:nucl-ex/0301014].
[37] A. Bialas, (2003), [arXiv:hep-ph/0308245].
[38] P. Bozek, W. Broniowski, and W. Florkowski, (2003),[arXiv:nucl-th/0310062].
[39] M. Gazdzicki and S. Mrowczynski, Z. Phys. **C54**, 127 (1992).
[40] F. Jinghua and L. Lianshou, (2003), [arXiv:hep-ph/031030].
[41] A. Bialas, (2003), [arXiv:hep-ph/0310348].
[42] M. Gazdzicki *et al.* [NA49 Collaboration], J. Phys. G **30**, S701 (2004) [arXiv:nucl-ex/0403023].
[43] J. Randrup, Phys. Rev. Lett. **92** 122301 (2004).
[44] V. Koch, A. Majumder, and J. Randrup in preparation for Phys. Rev. C
[45] P. Braun-Munzinger, K. Redlich and J. Stachel, Quark Gluon Plasma 3, 491-599, Hwa, R.C. (ed.), World Scientfic, Singapore 2004. [arXiv:nucl-th/0304013].
[46] E. Fermi, Progr. Theoret. Phys. **5**, 570 (1950).
[47] V. Koch, Nucl. Phys. **A715**, 108 (2003),[arXiv:nucl-th/0210070].
[48] S. Jeon, V. Koch, K. Redlich, and X. N. Wang, Nucl. Phys. **A697**, 546 (2002), [arXiv:nucl-th/0105035].
[49] S. Hamieh, K. Redlich, and A. Tounsi, Phys. Lett. **B486**, 61 (2000), [arXiv:hep-ph/0006024].
[50] A. Majumder and V. Koch, Phys. Rev. C **68**, 044903 (2003) [arXiv:nucl-th/0305047].

Institute of Physics Publishing
doi:10.1088/1742-6596/50/1/012
Journal of Physics: Conference Series **50** (2006) 103–110
5th International Conference on Physics and Astrophysics of Quark Gluon Plasma

Particle Multiplicities and Correlations

G Roland†[1]

† Massachusetts Institute of Technology, Cambridge, MA, USA

Abstract. Studies of charged hadron multiplicities have provided important information about the physics of ultra-relativistic heavy ion collisions. The systematics of charged particle multiplicities in high energy collisions, i.e. nucleon-nucleon, e^+e^- and nucleus-nucleus reactions, as a function of collision energy and centrality, suggest that in heavy ion collisions the multiplicity distributions are determined in the initial collision stage and that the overall produced multiplicity is limited by coherence or destructive interference in this early stage. In this paper, I will present some of the experimental results supporting these conclusions and point out features of the data that still await an explanation in dynamical models of the collision process. I will discuss future possibilities to help answer these questions, including measurements at the Large Hadron Collider and studies of multiplicity correlations.

1. Introduction

Heavy-ion collisions provide the best opportunity to test the predictions of QCD for matter under conditions close to the expected QCD phase-transition or resembling the conditions in the early universe. However, the system created in a heavy-ion collision is different from that simulated in lattice calculations or the early universe in several important aspects. The small size of the available nuclei sets the scale of the temporal and spatial extent of the high-density system created in the collision. The asymmetry of the incoming system, with its momentum in the direction of the incoming beams, is reflected throughout the evolution of the collision process. Most estimates suggest a lifetime of the collision system of 10-20 fm/c, from the initial high density state until freeze-out occurs, i.e. the collision density becomes so low that the produced particles essentially enter a free-streaming motion and are finally detected by the experiments.

The main experimental and theoretical challenge is then to reconstruct the properties of the earlier stages of the system from experimental observables based on the final-state particles. It is therefore essential to study collision systems that specifically probe the different stages of this evolution by varying the system size and collision energy.

[1] rolandg@mit.edu

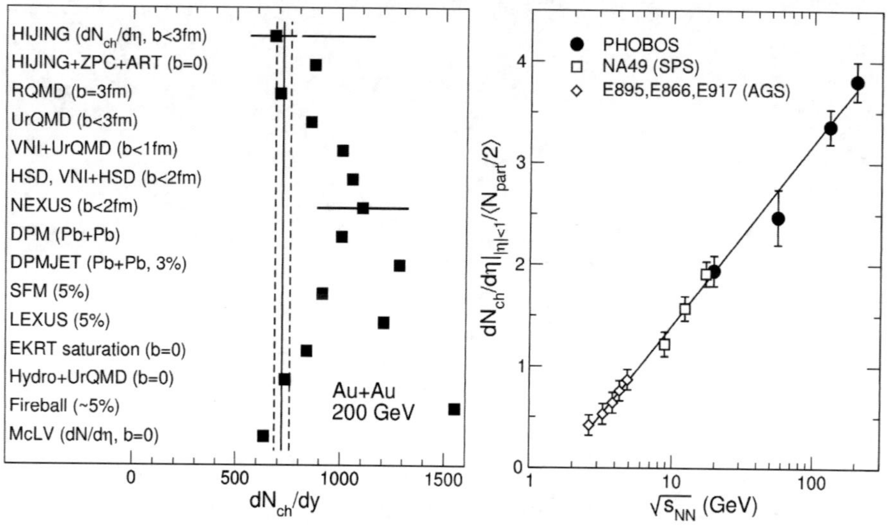

Figure 1. Charged particle multiplicity density per participant near mid-rapidity for central nucleus-nucleus collisions. Left: Comparison to predictions in various theoretical models. Right: Energy dependence.

In the past three years, RHIC has delivered Au+Au collisions at $\sqrt{s_{NN}}$ = 19.6, 56, 130 and 200 GeV and d+Au and proton-proton collisions at 200 GeV. As the results shown below will illustrate, the information from these systematic surveys, in combination with data from the AGS and CERN fixed target programs, is essential in forming a consistent dynamical picture of the evolution of a heavy ion collision.

The emphasis in this paper is on the systematics of particle multiplicities in these collisions. A priori, a mapping of the observed multiplicity to a specific stage of the collisions seems difficult, as in principle all stages of the collision could contribute to particle production. A close examination of the available data however, suggests that the multiplicity of charged particles and the distribution in pseudo-rapidity are determined very early on in the collision and provide a window into the dynamics of the initial state.

2. Multiplicity Systematics

2.1. Particle Density near Mid-Rapidity

The first physics results from RHIC were measurements of pseudo-rapidity densities of charged hadrons, $dN_{ch}/d\eta$, near mid-rapidity in central Au+Au collisions at collision energies of $\sqrt(s_{NN})$ = 56 and 130 GeV [1]. Like many of the RHIC data published since then, these first results proved surprising, showing a much smaller increase over multiplicities from lower energies than expected in the vast majority of theoretical approaches. Further measurements at the full RHIC energy of 200 GeV showed that in fact the mid-rapidity particle density appears to rise logarithmically from the AGS to the highest RHIC energies [2]. A summary of these measurements for central Au+Au (Pb+Pb) collisions as a function of collision energy is shown in the left hand plot of Fig. 1, while the left plot shows a compilation of predictions in comparison to data at 200 GeV [3, 4].

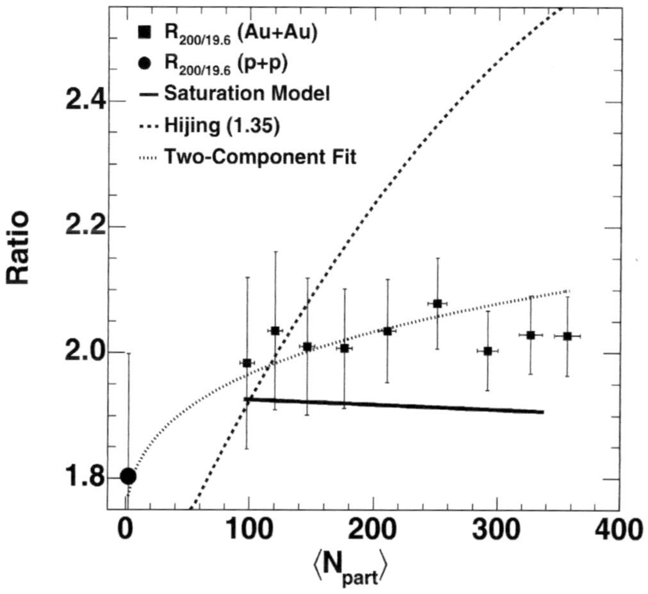

Figure 2. Ratio of the charged hadron mid-rapidity multiplicities at 200 and 19.6 GeV as a function of collision centrality.

Clearly, the charged hadron pseudo-rapidity density is related to the initial energy density of the system, as the initial volume of the system can be determined from the known size of the nuclear overlap region, in combination with an estimate of the longitudinal size at the time of equilibration, which is typically taken to be ≈ 1 fm. The corresponding energy density can be estimated to be on the order of 3-5 GeV/fm^3 or higher, if one assumes faster equilibration. While this estimate is still comfortably higher than the predictions for the critical energy density from lattice QCD of 1 GeV/fm^3, it was a surprise that the data fell at the lowest end of the range of predictions.

It has been suggested that the relatively low multiplicity seen at RHIC is a consequence of parton saturation, based on the idea that at high energies the density of low-x gluons in the transverse plane of the colliding nuclei will no longer allow them to interact independently. Rather, they will form a "Color Glass Condensate" [5], with the resultant coherent interaction of the constituents of the nuclei limiting the growth of particle multiplicity as a function of collision energy. Model calculations based on the idea of parton saturation, in combination with local parton-hadron duality, have had impressive success in describing the energy, centrality and rapidity dependence of charged hadron production in nucleus-nucleus collisions.

The importance of these ideas is illustrated in Fig. 2, which shows the ratios of the charged hadron mid-rapidity multiplicities at 200 and 19.6 GeV as a function of centrality[6]. This ratio is found to be flat, in sharp contrast to predictions of models like HIJING that describe charged particle production as a superposition of an empirical soft component and energy- and centrality dependent contributions from hard scattering processes. In this model, as shown in Fig. 2, the experimentally seen factorization of energy and centrality dependence is strongly violated. In contrast, parton saturation model calculations include this factorization approximately [5] or explicitly [7].

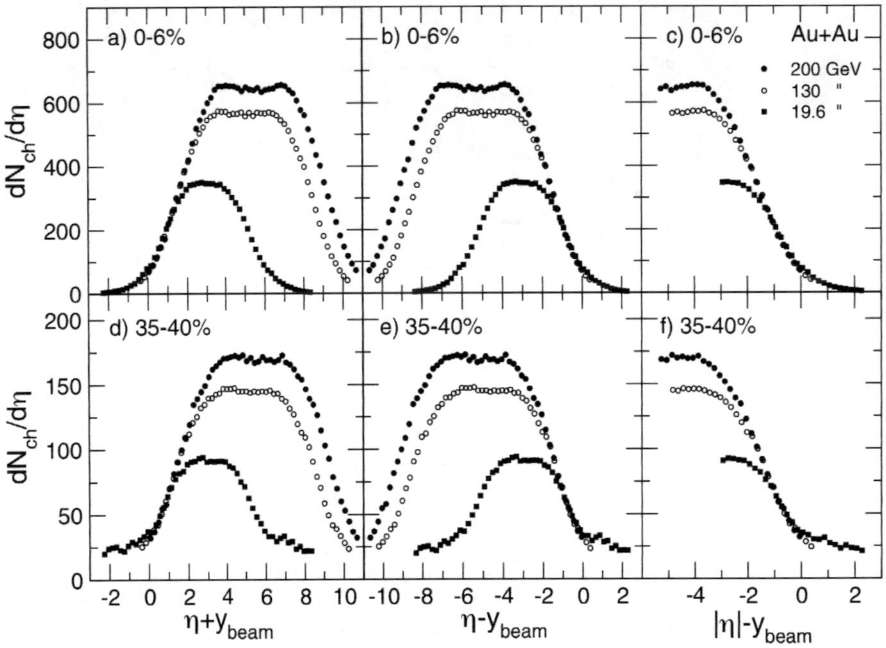

Figure 3. Pseudo-rapidity distributions of charged hadrons for central (top) and peripheral (bottom) events, viewed in the restframes of the colliding nuclei for collision energies of 19.6, 130 and 200 GeV. The distributions exhibit an extended longitudinal scaling, independent of collision energy ("limiting fragmentation").

2.2. Limiting Fragmentation

One of the interesting observations in multiplicity distributions at RHIC was first pointed out by the BRAHMS collaboration. They found [8] that the pseudo-rapidity distributions for charged particles over a large range of pseudo-rapidity were independent of the collision energy, when viewed in the restframe of one of the colliding nuclei. This observation was confirmed and extended by PHOBOS [9] and recently by STAR [10]. Measurements from PHOBOS are shown in Fig. 3. The observed extended longitudinal scaling of particle production is known from p+p collisions over a large range of collision energies, as well as from p+A collisions. It is usually termed "limiting fragmentation". What is surprising in nucleus-nucleus collisions is that the scaling behavior seen in these elementary collisions is preserved in the complicated dynamical environment of A+A collisions. It in fact appears to control particle production over a large part of longitudinal phase space, several units away from what is normally considered the fragmentation region. Furthermore, the longitudinal scaling seen in pseudo-rapidity distributions is also observed in the anisotropy of azimuthal particle distributions relative to the reaction plane of peripheral Au+Au collisions. Data from PHOBOS have shown that v_2 drops quickly for particles away from mid-rapidity. Although qualitative arguments can be made that this is related to the dropping particle density and resulting incomplete thermalization, no dynamical 3-dimensional calculations exist that successful describe these data. Comparing data from 4 different energies at RHIC, shows that the average v_2 parameter follows a very simple scaling behavior, as shown in Fig. 4. This observation suggests that at LHC even higher values of v_2 will be observed. Eventually, a complete understanding of the dynamical evolution of the system will need to explain this remarkably simple scaling behavior.

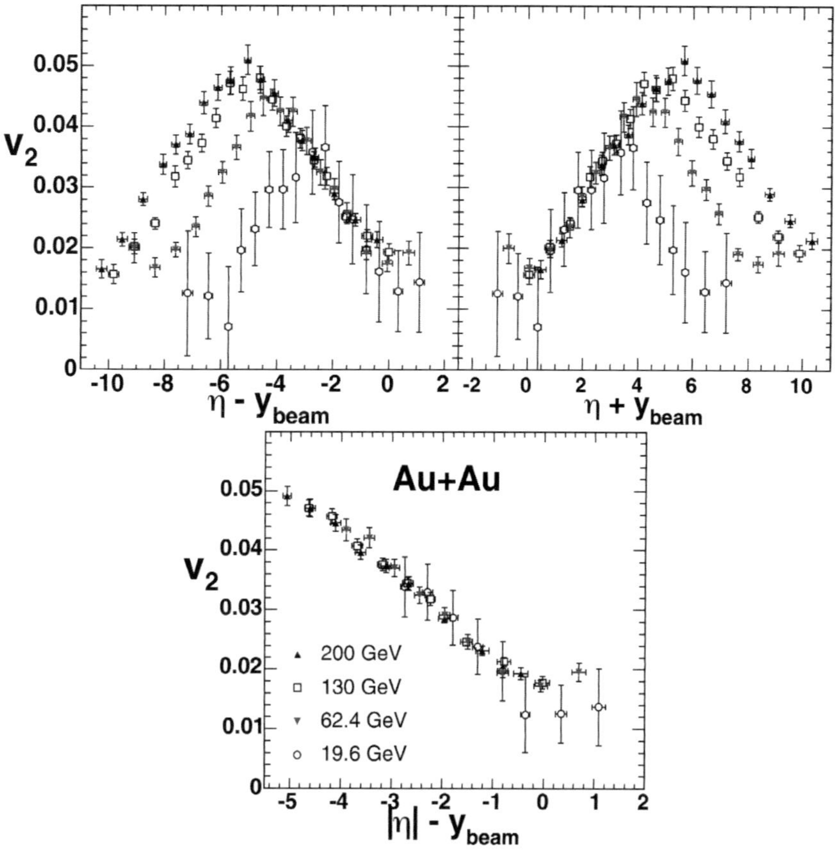

Figure 4. Elliptic flow parameter v_2 as a function of η' for Au+Au collisions from 19.6 to 200 GeV.

Based on these scaling observations, the data taken at RHIC allow an empirical prediction of the expected $dN/d\eta$ distribution at LHC. This prediction, shown in Fig. 5, is based on the logarithmic energy dependence of the mid-rapidity pseudo-rapidity (Fig. 1), in combination with the observed "limiting fragmentation" scaling of charged particle $dN/d\eta$ distributions. The resulting mid-rapidity charged hadron $dN/d\eta$ for central Pb+Pb collisions at the LHC is only ≈ 1200, far below most model expectations (and the design specifications of the experiments). It is important to note that a higher mid-rapidity multiplicity, e.g. following a power-law energy dependence, would either imply a violation of limiting fragmentation, which has been found to hold up to collision energies of 900 GeV in $p+p$ collisions, or a radically different shape of the $dN/d\eta$ distributions at the LHC. If however this empirical prediction holds, it will imply an even stronger "destructive interference" in the initial state particle production than suggested in parton saturation models. Clearly, the contribution to particle production from hard processes is expected to increase rapidly when going to LHC energies. A test of the factorization observed at low energies will be one of the fascinating day-1 measurements once the LHC comes online.

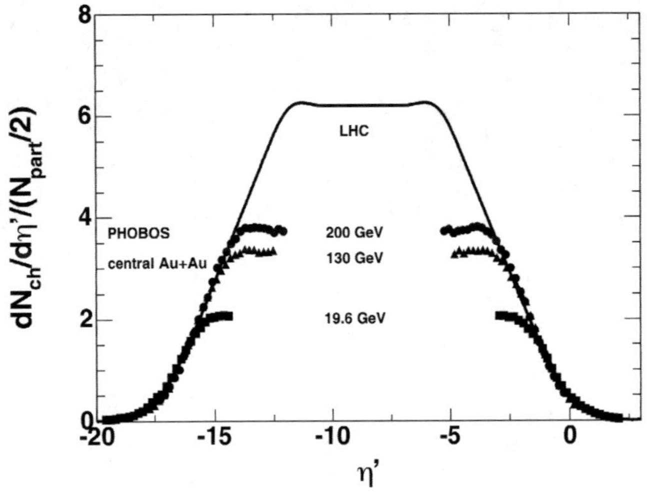

Figure 5. Empirical prediction for the pseudo-rapidity distribution in Pb+Pb collisions at the LHC ($N_{part} \approx 340$), based on the observed scaling features in Au+Au collisions at RHIC.

2.3. Multiplicity and the Initial State

As mentioned before, the multiplicity of charged hadrons could be expected to be a convolution of contributions from all different stages of the collision process. In this case, few conclusions could be drawn from the measurements directly, and the data could at best be used to test or calibrate models of the collision. However, several observations suggest that in fact the particle multiplicities are defined in the very early stage of the collision process and are little changed in the further evolution of the collision, in accordance with the idea of an isentropic expansion of the initial system.

Support to this claim is given by the success of pure initial state models based on the idea of parton saturation in describing the overall energy dependence of particle densities, the evolution of multiplicity distributions with centrality and the factorization of energy and centrality dependence. As such models can account (after normalization at one energy) for the relatively low overall multiplicities, it is reasonable to conclude that the contribution from further stages of the collision to the overall particle multiplicity is small. Further evidence comes from the survival of scaling behavior like e.g. limiting fragmentation seen in p+p and p+A collisions in the A+A environment, which would require a fortuitous cancellation of dynamical effects, if particle production was dominated by the later collision stages.

Finally, it was observed that the energy-dependence and absolute magnitude of the mid-rapidity particle density per participant pair in central Au+Au collisions coincide, within the experimental uncertainties, with those in e^+e^- collisions at the same collision energies [11]. Similarly, the total charged particle multiplicity per pair of participating nucleons for A+A collisions above $\sqrt{s_{NN}} \approx 20$ GeV is within 10% of the total charged particle multiplicity in e^+e^- collisions at the same collision energy and that of p+p collisions, if those are corrected for the leading particle effect.

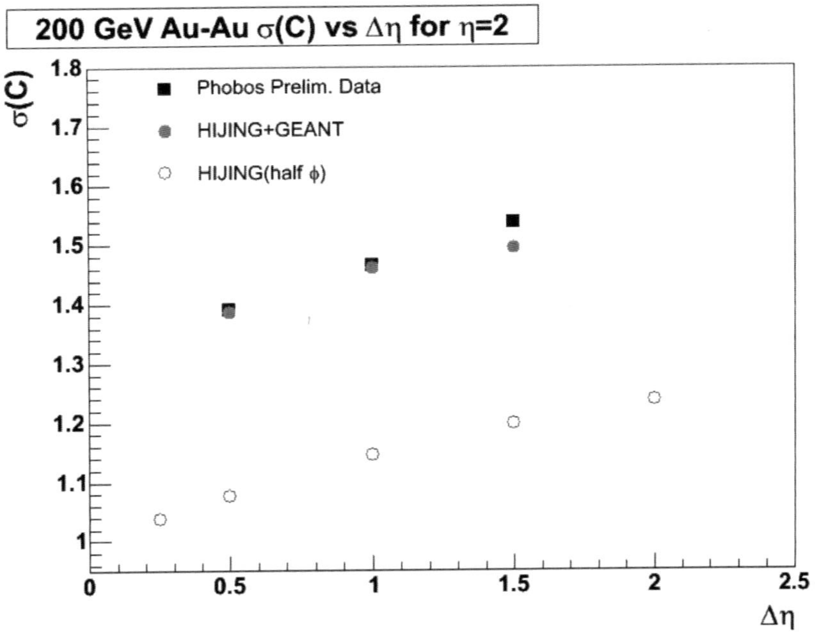

Figure 6. Forward-backward multiplicity fluctuations in central Au+Au collisions as a function of pseudo-rapidity bin-size, in comparison to HIJING model calculations.

In summary, these observations naturally hint at an universal mechanism of particle production in the strong interaction, defining particle multiplicities early on in the collision process. Measurements at the LHC, with a factor 30 higher collision energies, might yield the final clues for understanding these connections.

3. Multiplicity Correlations

The study of high p_T particle production has played a central role in developing our understanding of heavy-ion collisions. One of the most promising directions is the development of correlation studies that elucidate the how the fast partons interact with the medium [12]. Obviously, the energy of the fast partons is not lost, but redistributed inside the collision system. Differential studies of this transport process are beginning to emerge [13]. Studies of multiplicity fluctuations, as seen in Fig. 6 [14], show that the features of these fluctuations as a function of centrality, rapidity and size of the rapidity bins can be approximately reproduced in models like HIJING, provided that the correlations due to initial hard scattering processes are included. The studies suggest that the event-wise fluctuations in the multiplicity distributions are dominated by particle correlations introduced through initial hard scattering processes. Similarly, fluctuations in event-by-event average transverse momentum can be explained as results of the correlations at intermediate p_T from the same processes. In essence, most of the dynamical fluctuation signals observed at RHIC can be traced to the same physical origin.

One of the major tasks for the next years will therefore be to develop a phenomenology that will allow us to relate the observed final state correlations to the known seeds of fluctuations provided by the initial hard scattering processes and to use this to quantitatively extract properties of the dense medium. A good analogy for this experimental and theoretical program is given by recent progress in cosmology, where the final state correlation structure of e.g. the cosmic microwave background is used to quantitatively constrain parameters of the cosmic evolution. This analogy also suggests that this will be a long-term program, requiring input from a systematic survey of collision energies and collision systems, as well as a close connection between theory and experiment to find a common language in which to express the observed and calculated correlation structure.

4. Conclusions

The results from nearly two decades of experimental high-energy heavy-ion physics demonstrate convincingly that a unique, high-density medium is formed in these collisions. Many of the properties of the medium, such as the universality of entropy production, the presence of coherent effects limiting particle production, the large initial energy density, the isentropic nature of the expansion and the dominance of correlations originating from initial hard scattering processes can be inferred from studies of multiplicity distributions and multiplicity fluctuations. Still, many experimental and theoretical questions remain. Of highest importance are measurements related to the nature of the medium before hadronization and a theoretical understanding of how the initial equilibrated medium is formed and how the eventual decay into hadrons proceeds. Many of the answers will come from tests of our present models at the much higher collision energies accessible at the LHC. In addition, a comprehensive study of particle correlations over the full accessible phase space and using tagged initial hard processes, will help us to develop a more complete picture of the collision dynamics. At the same time, we can probably expect that the rich harvest of surprising experimental results will continue in these future studies.

[1] B B Back et al. Phys. Rev. Lett. **85**, 3100 (2000)
[2] B B Back et al. Phys. Rev. Lett. **88**, 22302 (2002)
[3] K J Eskola, Nucl. Phys. **A698** (2002) 78, arXiv:hep-ph/0104058.
[4] B B Back et al. arXiv:nucl-ex/0410022.
[5] L D McLerran, R Venugopalan, Phys. Rev. **D49** 2233 (1994),
D Kharzeev and B Levin, Phys. Lett. **B523** 79 (2001),
A Krasnitz, R Venugopalan, Phys. Rev. Lett 86 1717 (2001),
D Kharzeev, E Levin, L. McLerran, arXiv:hep-ph/0210332.
[6] B B Back, et al., Phys. Rev. C 70 (2004) 021902R.
[7] N Armesto, C A Salgado and U A Wiedemann, arXiv:hep-ph/0407018.
[8] I G Bearden *et al.* Phys. Rev. Lett. **88** 202301 (2002).
[9] B B Back *et al.*, Phys. Rev. Lett. **91**, 052303 (2003).
[10] J Adams *et al.* arXiv:nucl-ex/0502008, submitted
[11] B B Back *et al.*, arXiv:nucl-ex/0301017, submitted to Phys. Rev. C.
[12] C Adler *et al.*, Phys. Rev. Lett. **90** 082302 (2003).
[13] see M Kopytine, these proceedings.
[14] K Wozniak *et al.* J Phys G30 1377 (2004)

Institute of Physics Publishing Journal of Physics: Conference Series **50** (2006) 111–118
doi:10.1088/1742-6596/50/1/013 5th International Conference on Physics and Astrophysics of Quark Gluon Plasma

Heavy ion collisions: Correlations and Fluctuations in particle production.

Sergei A. Voloshin

Department of Physics and Astronomy, Wayne State University,
666 W. Hancock, Detroit, Michigan 48201

Abstract. Correlations and fluctuations (the latter are directly related to the 2-particle correlations) is one of the important directions in analysis of heavy ion collisions. At the current stage of RHIC exploration, when the details matter, basically any physics question is addressed with help of correlation techniques. In this talk I start with a general introduction to the correlation and fluctuation formalism and discuss weak and strong sides of different type of observables. In more detail, I discuss the two-particle p_t correlations/$\langle p_t \rangle$ fluctuations. In spite of not observing any dramatic changes in the event-by-event fluctuations with energy, which would indicate a possible phase transition, such correlations measurements remain an interesting and important subject, bringing valuable information. Lastly, I show how radial flow can generate characteristic azimuthal, transverse momentum and rapidity correlations, which could qualitatively explain many of recently observed phenomena in nuclear collisions.

1. Introduction

The physics of high energy heavy ion collisions attracts strong attention of the physics community as creation of a new type of matter, the Quark-Gluon Plasma, is expected in such collisions. During just a few years of the BNL RHIC operation many new phenomena has been observed, such as strong elliptic flow [1], and suppression of the high transverse momentum two particle back-to-back correlations [2]. Note that both of the above mentioned discoveries at RHIC have been made with help of correlations techniques: in this case analyzing azimuthal correlations. Further understanding of the physics processes responsible for both phenomena, such as what happens with the energy of the away-side jet or whether the observed constituent quark scaling [3, 4, 5] in elliptic flow reflects the hadronization via quark coalescence, require further dedicated correlation studies. In general, as we try to understand the details of system evolution we will rely more and more on correlation measurements.

Many collaborations recently have performed measurements of either multiplicity (such as net charge, K/π ratio) or mean transverse momentum fluctuations [6, 7]. The interest was driven by a possibility to observe a dramatic change in fluctuation pattern with collision energy and/or centrality indicating the phase transition. There exist theoretical prediction both for an increase of fluctuations [8] as well as for decrease in fluctuations [9, 10] (the latter is due to an increase in number of effective degrees of freedom in the QGP). At this time many new observables were introduced in the analysis. Unfortunately, the various observables used by experiments almost render a valid comparison of the results difficult or even impossible. The reason for that is a persistent confusion about properties of different observables used to measure fluctuations. Due to the importance of inter experiment comparison of the results, in the next section I try to give a short summary of the question along with my own recommendations for future analyzes. This part is somewhat technical, and a reader more interested in physical results may skip it and start with section II.

2. Correlation functions and fluctuations

In a multi-particle process the correlations in particle production are described via two and many particle densities [11, 12]. In this talk I use only single and two-particle densities; those are normalized respectively by the mean multiplicity and the mean number of pairs in a given region of momentum space:

$$\int_{\Delta X} dx \rho^{(1)}(x) = \langle n \rangle \, ; \quad \int_{\Delta X} dx_1 \int_{\Delta X} dx_2 \, \rho^{(2)}(x_1, x_2) = \langle n(n-1) \rangle \, . \tag{1}$$

Particle densities in general depend on a particle three momentum, e.g. rapidity, transverse momentum, and azimuthal angle, some of the arguments can be integrated out; for brevity I use just x. From $\rho^{(1)}(x)$ and $\rho^{(2)}(x_1, x_2)$ one constructs different types of correlation functions:

$$C(x_1, x_2) \equiv \rho^{(2)}(x_1, x_2) - \rho^{(1)}(x_1)\, \rho^{(1)}(x_2); \; B(x_1, x_2) \equiv \frac{C(x_1, x_2)}{\rho^{(1)}(x_1)}; \; R(x_1, x_2) \equiv \frac{C(x_1, x_2)}{\rho^{(1)}(x_1)\rho^{(1)}(x_2)} \tag{2}$$

Roughly speaking $C(x_1, x_2)$ has a meaning of a distribution of correlated pairs. The correlation function $B(x_a, x_b)$ corresponds to the distribution of particles b under condition that particle a is found at x_a. Often, a particle of type a is called a "trigger" particle and particle of type b is called an "associated" particle. When particles a and b carry opposite charges, such as electric, baryon, or strangeness, $\int_{\infty}^{\infty} B(x_a, x_b) dx_b = 1$, as there must be one associated particle somewhere in the momentum space that "balances" the charge of the trigger particle. Then $B(x_a, x_b)$ is often referred to as *balance function*. The correlation function $R(x_1, x_2)$ may be interpreted as the "probability" that a given pair is correlated. This type of the correlation function is used most often, as it is easiest to measure. Defined as a ratio, it is one of the so-called *robust* quantities, which do not depend on single particle reconstruction efficiencies.

Studying the correlations in nuclear collisions one usually is interested if those are different from the correlations in elementary collision. If one superimposes a few, N_{coll}, nucleon-nucleon collisions together:

$$\rho^{(1),AA}(x) = N_{coll}\, \rho^{(1),NN}(x) \tag{3}$$

$$\rho^{(2),AA}(x_1, x_2) = N_{coll}\, \rho^{(2),NN}(x_1, x_2) + N_{coll}(N_{coll} - 1)\, \rho^{(1),NN}(x_1)\rho^{(1),NN}(x_2). \tag{4}$$

Then, the balance function, $B(x_1, x_2)$, does not change at all, and the signal in $R(x_1, x_2)$ gets diluted: $R^{AA}(x_1, x_2) = R^{NN}(x_1, x_2)/N_{coll}$.

All of the above correlation functions can be calculated at fixed multiplicities. In this case they usually carry subscript n, e.g. $C_n(x_1, x_2)$, and called *semi-inclusive*. In nuclear collision analyses it is not possible to fix the multiplicity of individual nucleon-nucleon collisions, and one should compare the results with *inclusive* NN correlation function, although one should be careful comparing very peripheral collisions selected on the basis of total multiplicity. In this case the approximation of correlations in individual NN collision by inclusive correlation function can break down.

2.1. Fluctuations

Many different quantities have been suggested for fluctuation analyses.Before further discussion, let us formulate in general terms what constitutes a good observable. It should be (i) sensitive to the physics under study, (ii) it should be defined at the "theoretical" level, be apparatus/experiment independent, (iii) have clear physical meaning, and (iv) it should not be limited in scope, provide new venues for further study. If most of the proposed observables do satisfy the first requirement, very few satisfy (ii)–(iv). It made basically impossible to compare many of the results, even published, one to another. Comparing different observables I argue that the "old fashion" correlation functions, and, similar the two particle transverse momentum correlations, $\langle \Delta p_{t,1} \Delta p_{t,2} \rangle$, are the best in many respects.

2.1.1. Multiplicity fluctuation measures How the two, correlations and fluctuations, are related to each other? Let us look at the so-called reduced variance:

$$\omega_n \equiv \frac{\sigma_n^2}{\langle n \rangle} = 1 + \langle n \rangle \tilde{R}_{\Delta X}; \quad \tilde{R}_{\Delta X} = \frac{\int_{\Delta X} dx_1 \int_{\Delta X} dx_2 \rho_2(x_1, x_2)}{\int_{\Delta X} dx_1 \int_{\Delta X} dx_2 \rho_1(x_1) \rho_1(x_2)}, \tag{5}$$

where the last equation shows the relation to two particle density. If $\rho_1(x_1)$ does not vary much over the region ΔX, $\tilde{R}_{\Delta X}$ gives an *average* of the correlation function $R(x_1, x_2)$ over the momentum region used in the analysis. It follows that the reduced variance deviates from unity (which is interpreted as the value for *statistical* fluctuations) if and only if the correlation term $\tilde{R}$ equals zero.

Note that ω depends on the product of $\langle n \rangle$ and $\tilde{R}$. Given $\tilde{R}$ is a robust quantity, the mean event multiplicity is not, it depends on detector efficiency, track quality cuts, etc. Another feature of ω is that it can be different even if the underlying correlations are the same. Let us compare fluctuations of charged particles with fluctuations of only negative (or only positive) particles. Even if the correlations between particles have no any dependence on the particle charge, the reduced variance would show different values simply because only "half" of the particles are being considered. Lastly note that the reduced variance can not be independent of the size of the rapidity region ΔX as it sometimes assumed (it is possible only if $R(x_1, x_2) \propto \delta(x_1 - x_2)$ which is not physical) as often people compare its values at different ΔX.

Why one want to study ($\langle n \rangle R$) instead of just R? The reason is that this product should stay constant as function of centrality in a case when AA collision is considered as a superposition of independent nucleon-nucleon collisions. In this case the correlation functions scales as $1/N_{coll}$, $\langle n \rangle \propto N_{coll}$, and the product remains constant. The deviation from constant is easy to observe. Experimentally, the centrality is often obtained by measuring multiplicity, and it was wrongly assumed that such quantities as reduced variance do not depend on multiplicity in general. It would be desirable to report both quantities separately. Instead of $\langle n \rangle$ it would be better to use dn/dy, a quantity totally corrected for efficiency and only weakly dependent on ΔX. Besides, if $\tilde{R}$ is known, one can check how it scales also other quantities, such as number of participating nucleons, N_{part}, or the number of binary collisions, or something else that might be related to the number of sources of particle production.

To avoid the so called "volume fluctuations" (the effect of mixture of different impact parameter collisions in one event sample) the fluctuations in *particle ratios* are often used, e.g. K/π, where K and π stand for the particle respective multiplicities in a given event. Again, many "details" often forgotten. First, even in the original paper [9], the fluctuations of particle ratios are "reduced" (approximated) to the following

$$\frac{\sigma_{K/\pi}^2}{\langle K/\pi \rangle^2} \approx \frac{\sigma_K^2}{\langle K \rangle^2} + \frac{\sigma_\pi^2}{\langle \pi \rangle^2} - 2\frac{\langle K\pi \rangle - \langle K \rangle \langle \pi \rangle}{\langle K \rangle \langle \pi \rangle} = \frac{1}{\langle K \rangle} + \frac{1}{\langle \pi \rangle} + (\tilde{R}_{KK} + \tilde{R}_{\pi\pi} - 2\tilde{R}_{K\pi}). \tag{6}$$

Note, that we started with statistically rather bad defined quantity as any ratio of two random numbers is (what do you do if denominator is zero?) and ended up with a much better behaved quantity defined via the correlation function. The question is why one would want to deal with ratios when even theoretical predictions are based on a really different quantity, which only approximately corresponds to fluctuations in ratios, namely only in the case when multiplicities

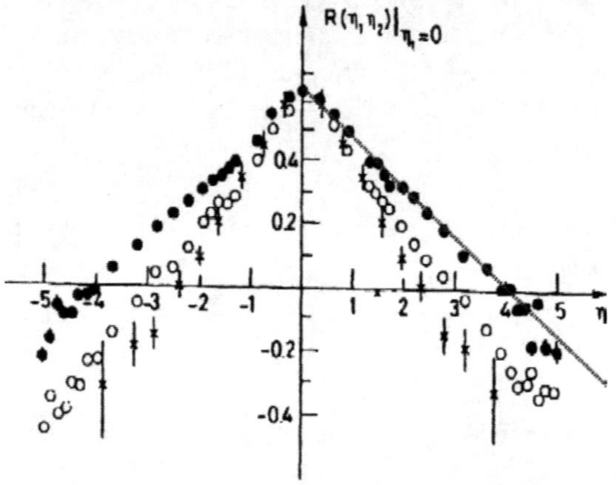

Figure 1. The charged particle correlation function $R(\eta_1, \eta_2)$ for $\eta_1 = 0$ at various ISR energies [12]. $\sqrt{s}$= 14, 23, and 63 GeV.

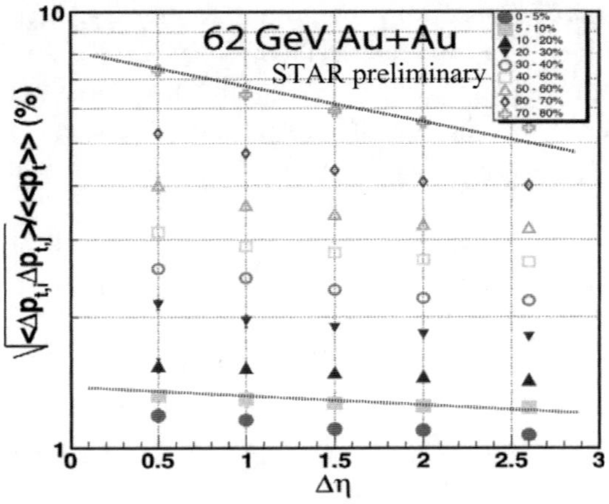

Figure 2. $\langle \Delta p_{t,1} \Delta p_{t,2} \rangle$ for different centralities as function of the size of the pseudorapidity region [18].

of particles used in the denominator is large? It is much better to report directly the correlation of the type $(\tilde{R}_{KK} + \tilde{R}_{\pi\pi} - 2\tilde{R}_{K\pi})$ [15]. The latter is easier to measure (robust quantity) and can be analyzed for particles with very low mean multiplicities (for example, p and $\bar{p}$).

2.2. *Mean p_t fluctuation measures*

Mean p_t event by event fluctuation analyses are often performed by calculating the event-wise mean transverse momentum, an overall event average, and the variance:

$$\langle p_t \rangle \equiv \frac{1}{n} \sum_i p_{t,i}; \ \langle\langle p_t \rangle\rangle \equiv \frac{1}{N_{ev}} \sum_j \langle p_t \rangle ; \ \sigma^2_{\langle p_t \rangle} = \left\langle \langle p_t \rangle^2 \right\rangle - \langle\langle p_t \rangle\rangle^2 = \frac{\sigma^2_{p_t,incl}}{n} + \frac{n-1}{n} \langle \Delta p_{t,1} \Delta p_{t,2} \rangle , \tag{7}$$

where the first term in $\sigma^2_{\langle p_t \rangle}$ is defined by a single particle p_t spectrum and is attributed to *statistical* fluctuations (the fluctuations in a case of independent particle production with the same single-particle distributions). The non-trivial fluctuations (non-statistical, sometimes called *dynamical*) are included in the second term, $\langle \Delta p_{t,1} \Delta p_{t,2} \rangle$ [14]. All the measures of fluctuations used by different experiments are related to this second term (two particle transverse momentum correlation (covariance)), although sometimes in a rather complicated way. In many cases such a relation involves tracking and analysis efficiencies, and if those are not reported separately, the final results become not comparable to other experiments/measurements.

If all events in the event sample have the same multiplicity, then the equations above provide *semi-inclusive* mean transverse momentum and correspondingly $\langle \Delta p_{t,1} \Delta p_{t,2} \rangle$. If the multiplicity vary within the event sample, there are two possibilities: one can average the semi-inclusive quantities over all events (what is used most often, I would call it event-by-event (EbyE) average) or calculate true *inclusive* quantities

$$\overline{p_t} \equiv \frac{\sum_{events} \sum_i^n p_{t,i}}{N_{events} \langle n \rangle}; \ \ \overline{\Delta p_{t,1} \Delta p_{t,2}} \equiv \frac{\sum_{events} \sum_{i<j} (p_{t,i} - \overline{p_t})(p_{t,j} - \overline{p_t})}{N_{events} \langle n(n-1) \rangle}. \tag{8}$$

The latter cam be written directly using two particle density

$$\overline{\Delta p_{t,1} \Delta p_{t,2}} = \frac{\int_{\Delta X} dx_1 \int_{\Delta X} dx_2 \rho^{(2)}(x_1, x_2) \Delta p_{t,1} \Delta p_{t,2}}{\langle n(n-1) \rangle} \tag{9}$$

Although quantitatively both definitions (EbyE and inclusive) would give almost indistinguishable values (unless the multiplicity is too low and event sample include events with very different multiplicities), theoretically, it is easier to analyze inclusive correlations. Therefore my preference would be to "migrate" from the EbyE to inclusive observables.

Note also that $\langle \Delta p_{t,1} \Delta p_{t,2} \rangle$ can be easily generalized for study of correlation between particles of different type and for differential measurements such as $\Delta p_{t,1}(\eta_1, \phi_1) \Delta p_{t,2}(\eta_2, \phi_2)$. Similar to the multiplicity correlation, its centrality dependence can be tested against different hypothesis such scaling with rapidity density, number of participants, etc.

3. Mean p_t fluctuations/correlations at RHIC

.

Recently STAR Collaboration has reported results on integrated two particle p_t correlations for Au+Au collisions at different collisions energies [18]. Fig. 2 shows the dependence of the two particle p_t correlations on $\Delta\eta = 2|\eta_{max}|$ the size of the pseudorapidity window taken around midrapidity. The results for different centralities are shown starting from the most peripheral (on the top, the strongest correlations) to the most central. The dependence on the size of the pseudorapidity region is rather weak. At higher collision energy ($\sqrt{s_{NN}} = 200$ GeV) the dependence is even weaker. Under assumption that spectra dependence on transverse momentum and pseudorapidity factorize, according to Eq. 9, the pseudorapidity dependence of $\langle \Delta p_{t,1} \Delta p_{t,2} \rangle$ should follow the one of correlation function $R(\eta_1, eta_2)$. Then $\langle \Delta p_{t,1} \Delta p_{t,2} \rangle \propto \tilde{R}$. For a simple estimate one can fit the correlation function by the form $R(\Delta\eta) \propto 1 - \alpha|\Delta\eta|$, which translates into $\tilde{R} \propto 1 - 4/3\alpha Y$, where $Y = (\Delta\eta)_{max}/2$. The blue lines shown in Figures 1 and 2 correspond to the same $\alpha = 0.16$ and describe the data well.

Note that the slope of $\langle \Delta p_{t,1} \Delta p_{t,2} \rangle$ dependence on Y changes with centrality. It is noticeably smaller for central collisions. Taking into account the relation to the correlation function, it means that the correlations in more central collision widens in η. The same conclusion can be derived directly from the differential measurements of the two particle p_t correlation on ($\Delta\eta$, $\Delta\phi$) presented at this conference (see Fig. 6 in [16]). I argue below that such correlations can be due to radial transverse flow.

Fig. 3 shows STAR results on the centrality and incident energy dependence of $\sqrt{\langle \Delta p_{t,1} \Delta p_{t,2} \rangle} / \langle p_t \rangle$. Remarkably the results in this form (two particle relative transverse momentum correlations) exhibit almost no dependence on the collision energy. The insert shows the results for the most central collision where $\sqrt{\langle \Delta p_{t,1} \Delta p_{t,2} \rangle} / \langle p_t \rangle \approx 1.2\%$ from lower SPS (Pb+Pb collisions, CERES Collaboration [7]) to the top RHIC energy. The centrality dependence of $\langle \Delta p_{t,1} \Delta p_{t,2} \rangle$ roughly follows the expectations for a dilution of the correlations as more and more independent nucleon-nucleon collisions mixed up. For more detailed analysis of the centrality dependence note that in Eq. 9 both numerator and denominator depend on centrality. Taking this into account one finds

$$\langle \Delta p_{t,1} \Delta p_{t,2} \rangle_{AA} \approx D_{N_{coll}} \langle \Delta p_{t,1} \Delta p_{t,2} \rangle_{NN}\,;\; D_{N_{coll}} = \frac{\langle n(n-1) \rangle_{NN}}{(N_{coll}-1)\langle n \rangle_{NN}^2 + \langle n(n-1) \rangle_{NN}}. \quad (10)$$

The factor D takes into account the dilution of the correlations due to a mixture of particles from N_{coll} uncorrelated NN collisions, and that in an individual NN collision the mean number of particle pairs, $\langle n(n-1) \rangle_{NN}$, on average is larger than $\langle n \rangle_{NN}^2$. At ISR energies in central rapidity region $\langle n(n-1) \rangle_{NN} \approx 1.66 \langle n \rangle_{NN}^2$ [12] (see Fig. 1). Taking into account the dilution factor $D_{N_{coll}}$ one find that the correlations increase for about 50% with centrality relative to the expectations based on the superposition of independent NN collisions (see Fig. 6 below).

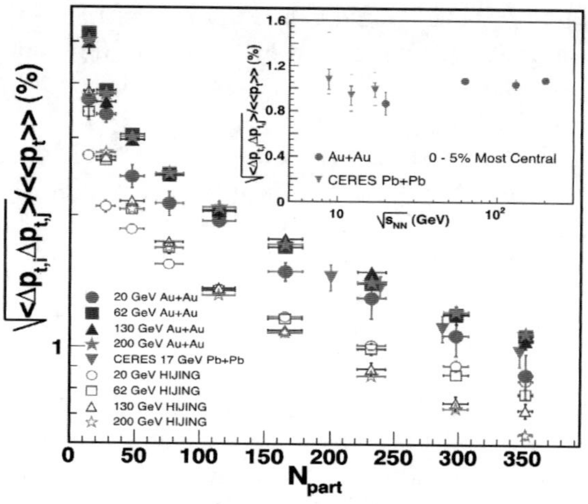

Figure 3. $\sqrt{\langle \Delta p_{t,1}\Delta p_{t,2}\rangle}/\langle p_t\rangle$ as a function of centrality for Au+Au collisions at different collision energies [18].

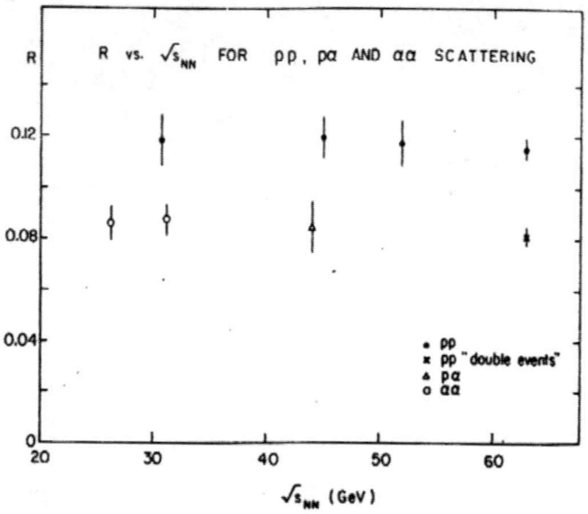

Figure 4. $\sqrt{\langle \Delta p_{t,1}\Delta p_{t,2}\rangle}/\langle p_t\rangle$ (calculated as non-statistical part in $\langle p_t\rangle$ event-by-event fluctuations) for p+p and $\alpha+\alpha$ collisions [17].

4. Transverse radial flow and two-particle correlations

At the first stage of a AA collision many individual nucleon-nucleon collision happen. Parton re-interactions lead to pressure build-up and the system undergoes longitudinal and transverse expansion. Transverse flow in the system creates strong position-momentum correlations in the transverse plane: further from the center axis of the system a particle is produced initially, on average the larger push it gets from other particles during the system evolution. As all particles produced in the same NN collision have initially the *same spatial* position in the transverse plane, they get *on average* the same push and thus become correlated. This picture leads to many distinctive phenomena, most of which can be studied by means of two (and many-) particle correlations [25].

The single particle spectra are affected by radial flow such that the mean transverse momentum is mostly sensitive to the *average* expansion velocity squared $\langle p_t\rangle_{AA} \approx \langle p_t\rangle_{NN} + \alpha\langle v^2\rangle$, (see Fig. 5 below) and to much lesser extend to the actual velocity profile (dependence of the expansion velocity on the radial distance from the center axis of the system). The two-particle transverse momentum correlations [14], $\langle \Delta p_{t,1}\Delta p_{t,2}\rangle$, measure the *variance* in collective transverse expansion velocity, and thus are more sensitive to the actual velocity profile. We employ a thermal model [13] for further calculations. In this model particles are produced by freeze-out of the thermalized matter at temperature T, approximated by a boosted Boltzmann distribution. Assuming boost-invariant longitudinal expansion and freeze-out at constant proper time, one finds

$$\frac{dn}{d\mathbf{p}_t} \sim \int d\rho_t d\phi_b \rho_t^{2/n-1} J(\mathbf{p}_t; T, \rho_t, \phi_b); \quad J(\mathbf{p}_t; T, \rho_t, \phi_b) \equiv m_t K_1(\beta_t) e^{\alpha_t \cos(\phi_b-\phi)}, \tag{11}$$

where ρ_t is the transverse flow rapidity, ϕ_b is the boost direction, $\alpha_t = (p_t/T)\sinh(\rho_t)$, and $\beta_t = (m_t/T)\cosh(\rho_t)$. It also assumes a uniform matter density within a cylinder, $r < R$, and a power law transverse rapidity flow profile $\rho_t \propto r^n$. Two particle spectrum for particles originating from the same NN collision in this picture can be written as

$$\frac{dn_{pair}}{d\mathbf{p}_{t,1} d\mathbf{p}_{t,2}} \sim \int d\rho_t d\phi_b \rho_t^{2/n-1} J(\mathbf{p}_{t,1}; T, \rho_t, \phi_b) J(\mathbf{p}_{t,2}; T, \rho_t, \phi_b). \tag{12}$$

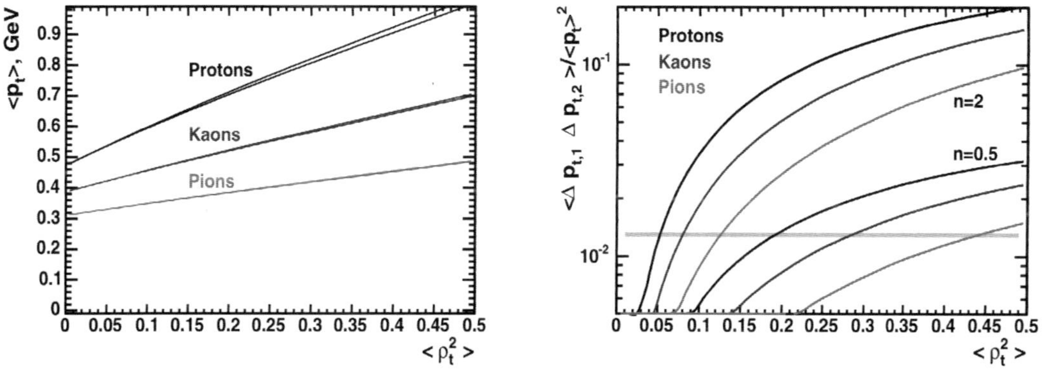

Figure 5. (color online) Mean transverse momentum and two particle p_t correlations according to the blast wave calculations. $T = 110$ MeV.

It additionally assumes that during the expansion time (before the freeze-out) the particles produced originally at the same spatial position do not diffuse far one from another compared to the system size. The results of the numerical calculations based on the above equations are presented in Fig. 5 as function of $\langle \rho_t^2 \rangle = \langle \rho_t \rangle^2 (4n+4)/(2+n)^2$. The results are shown for two different velocity (transverse rapidity) profiles, $n = 2$, and $n = 0.5$. One observes that indeed for all the particle types presented, $\langle p_t \rangle$ depends very weakly on the actual profile. On opposite, the correlations are drastically different for two cases studied.

In Fig. 6 we compare our estimates with STAR preliminary data [18] on two particle p_t correlations (taking into account the dilution factor, Eq. 10). We use $\langle \rho_t \rangle$ and T parameters from [23] and assume $N_{coll} = N_{part}/2$ and $\langle \Delta p_{t,1} \Delta p_{t,2} \rangle / \langle p_t \rangle^2 = 0.011$ (about 10% smaller than measured at ISR [17] (Fig. 4). It is observed that the transverse flow with $n = 1$ produces too strong correlations.

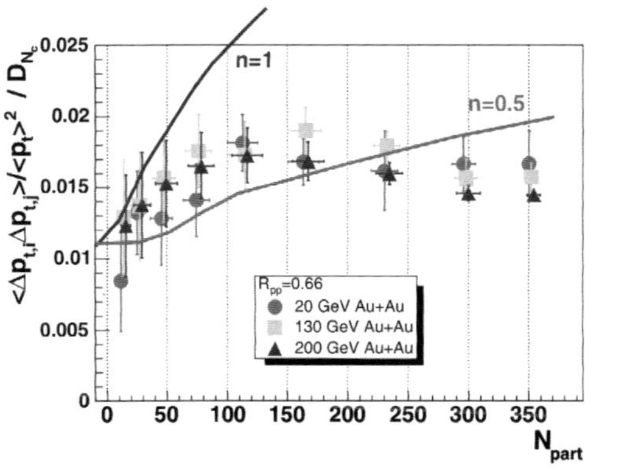

Figure 6. Comparison of the Blast Wave calculations for two different velocity profiles with preliminary STAR data [18]. Relation $\langle n(n-1) \rangle_{NN} = 1.66 \langle n \rangle_{NN}^2$ has been used.

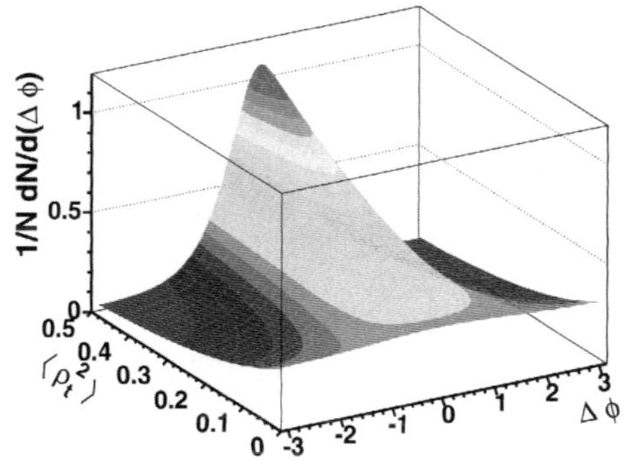

Figure 7. (color online) Two pion $\Delta\phi$ distribution as function of $\langle \rho_t^2 \rangle$ in the Blast Wave model. Linear velocity profile and $T = 110$ MeV have been assumed.

While transverse flow generates in general an elongation of the pseudorapidity correlations (along with narrowing in azimuthal angle) it should lead to narrowing of the *charge balance function* [19] due to the increase in mean p_t as [20]: $\Delta p_z = m_t \sinh(\Delta y) \approx m_t \Delta y \approx const.$

Quantitatively, the effect is consistent with experimentally observed narrowing for about 15 – 20% of the balance function width with centrality [21] and with centrality dependence of the net charge fluctuations [22]. As all particles from the same NN collisions are pushed in the same direction they become correlated in azimuthal space. The correlations can become really strong for large transverse flow as shown in Fig. 7 (for particles originated from the *same* NN collision). Our estimates show that the azimuthal correlations generated by transverse expansion could be a major contributor to the non-flow azimuthal correlations [24].

The above described picture of AA collisions has many interesting observable effects, only a few mentioned here. The picture become even richer if one looks at the identified particle correlations. Many questions require a detailed model study, but the approach opens a potentially very interesting possibility to address the initial conditions and the subsequent evolution of the system created in an AA collision.

5. Conclusion

The correlation techniques are a very powerful tool in our quest to understand multiparticle production in nuclear collision. They have been proven to lead to many discoveries in the past and promise even more to come.

Acknowledgments. Discussions with R. Bellwied, S. Gavin, C. Pruneau, S. Pratt, and U. Heinz are gratefully acknowledged. This work was supported in part by the U.S. Department of Energy Grant No. DE-FG02-92ER40713.

[1] STAR Collaboration, Ackermann K H et al. 2001 Phys. Rev. Lett. **86** 402
[2] STAR Collaboration, Adler C et al. 2003 Phys. Rev. Lett. **90** 082302
[3] Voloshin S A 2003 Nucl. Phys. A **715**, 379.
[4] STAR Collaboration, Adams J et al. 2004 Phys. Rev. Lett. **92**, 052302
[5] Molnar D and Voloshin S A 2003 Phys. Rev. Lett. **91**, 092301
[6] Mitchell J T 2004 J. Phys. G **30**, S819
[7] Appelshauser H 2004 J. Phys. G **30**, S935
[8] Stephanov M A, Rajagopal K and Shuryak E V 1999 Phys. Rev. D **60** 114028
[9] Baym G and Heiselberg H 1999 Phys. Lett. B **469** 7; Jeon S and Koch V 2000 Phys. Rev. Lett. **85** 2076
[10] Asakawa M, Heinz U W and Muller B 2000 Phys. Rev. Lett. **85** 2072
[11] Whitmore J 1976 Phys. Rept. **27** 187
[12] Foa L 1975 Phys. Rept. **22** 1
[13] Schnedermann E, Sollfrank J and Heinz U 1993 Phys. Rev. C **48**, 2462
[14] Voloshin S A, Koch V and Ritter H G 1999 Phys. Rev. C **60**, 024901
[15] Pruneau C, Gavin S and Voloshin S 2002 Phys. Rev. C **66**, 044904
[16] Kopytine M [STAR Collaboration] this proceedings, arXive:nucl-ex/0504006
[17] Braune K et al. 1983 Phys. Lett. **123B** 467
[18] Westfall G D [STAR collaboration] 2004 J. Phys. G **30**, S1389; Adams J et al. arXive:nucl-ex/0504031
[19] Bass S A, Danielewicz P and Pratt S 2000 Phys. Rev. Lett. **85**, 2689
[20] Scott Pratt and Sen Cheng 2003 Phys. Rev. C **68**, 014907
[21] STAR Collaboration, Adams J et al. 2003 Phys. Rev. Lett. **90**, 172301
[22] STAR Collaboration, Adams J et al. 2003 Phys. Rev. C **68**, 044905
[23] STAR Collaboration, Adams J et al. 2004 Phys. Rev. C **92**, 112301
[24] STAR Collaboration, Adler C et al. 2002 Phys. Rev. C **66**, 034904
[25] Voloshin S A 2003 arXiv:nucl-th/0312065

Institute of Physics Publishing | Journal of Physics: Conference Series **50** (2006) 119–126
doi:10.1088/1742-6596/50/1/014 | 5th International Conference on Physics and Astrophysics of Quark Gluon Plasma

Jet quenching in relativistic heavy ion collisions

Ivan Vitev

Los Alamos National Laboratory, Theory Division and Physics Division, Mail Stop H846
Los Alamos, NM 87545, USA

E-mail: ivitev@lanl.gov

Abstract.
Parton propagation in dense nuclear matter results in elastic, inelastic and coherent multiple soft scattering with the in-medium color charges. Such scattering leads to calculable modifications of the hadron production cross section that is evaluated in the framework of the perturbative QCD factorization approach. Final state medium-induced gluon bremsstrahlung is arguably the most efficient way of suppressing large transverse momentum particle production in nucleus-nucleus collisions. The observed hadronic attenuation, known as jet quenching, can be related to the properties of the medium, such as density and temperature, and carries valuable information about the early stages of heavy ion reactions. Non-Abelian energy loss in the quark-gluon plasma can be studied in much greater detail through the modification of the two particle back-to-back correlations. Perturbative calculations give good description of the redistribution of the lost energy in lower transverse momentum particles and predict significant increase of the correlation width of away-side di-hadrons. In contrast, energy loss in cold nuclear matter was found to be small but for large values of Feynman-x is expected to complement the dynamical higher twist shadowing in experimentally observable forward rapidity hadron suppression.

1. Energy loss in hot and dense nuclear matter

Jet quenching, the suppression of high transverse momentum hadron production relative to the expectation from p+p collisions scaled by the number of elementary nucleon-nucleon interactions, has been predicted [1] to be one of the key signatures for the creation of a hot and dense quark-gluon plasma (QGP) in ultra-relativistic nuclear collisions. Since, the non-Abelian energy loss of fast quarks and gluons in dense nuclear matter has been studied in great detail in a variety of theoretical approaches [2]. Its key features can be illustrated on the example of the reaction operator formalism [3], which systematically expands energy loss in terms of the correlations between multiple scattering centers.

For A+A collisions, the dominant final state energy loss follows the hard partonic scattering, which takes place on a time scale $t_0 \sim 1/Q$. Even in the absence of a medium, the large virtuality leads to non-Abelian bremsstrahlung. For real gluons of small and moderate frequencies ω [3]

$$|\mathcal{M}_c|^2 \propto \frac{dN^g_{\rm vac}}{d\omega d\sin\theta^* d\delta} \approx \frac{C_R\alpha_s}{\pi^2}\frac{1}{\omega\sin\theta^*} \,. \tag{1}$$

In Eq. (1) $\theta^* = \arcsin(k_\perp/\omega)$ is the angle relative to the jet axis, δ is the azimuthal cone angle (both illustrated in right panel of Fig. 1), α_s is the strong coupling constant and $C_R = 3\,(4/3)$ for gluon (quark) jets, respectively. Virtual gluon corrections remove the $\omega \to 0$ infrared singularity in the cross sections in accord with the Kinoshita-Lee-Nauenberg theorem [4] but the collinear

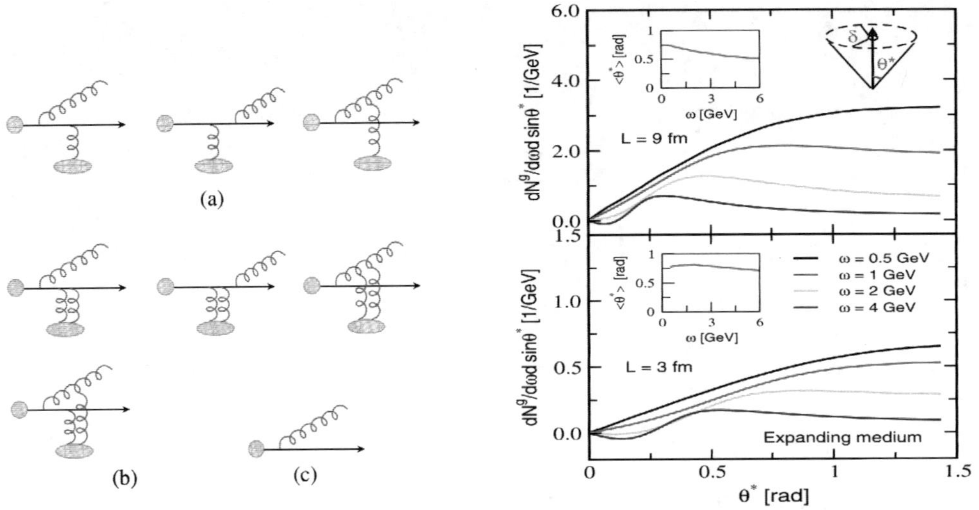

Figure 1. Left panel: diagrams contributing to the lowest order in opacity energy loss expansion. Right panel: the angular distribution of medium-induced bremsstrahlung of $E = 6$ GeV gluon jet for fixed values of the radiative gluon energy $\omega = 0.5$, 1, 2, 4 GeV. Top and bottom panels represent (1+1)D Bjorken expanding medium of transverse size $L = 9$ fm and $L = 3$ fm, respectively. Inserts show $\langle\theta^*\rangle$ versus ω.

$\theta^* \to 0$ divergence has to be regulated or subtracted in the parton distribution functions (PDFs) and the fragmentation functions (FFs).

In contrast, the final state medium-induced bremsstrahlung spectrum is both collinear and infrared safe. To first order in the mean number of soft interactions in the plasma the distribution of gluons in angle and frequency can be written as [5]

$$|\mathcal{M}_a|^2 + 2\mathcal{R}e\mathcal{M}_b^\dagger\mathcal{M}_c \propto \frac{dN^g_{\rm med}}{d\omega d\sin\theta^* d\delta} \approx \frac{2C_R\alpha_s}{\pi^2}\int_{z_0}^{L}\frac{d\Delta z}{\lambda_g(z)}\int_0^\infty dq_\perp\, q_\perp^2\frac{1}{\sigma_{el}}\frac{d\sigma_{el}}{d^2q_\perp}(z)\int_0^{2\pi}d\alpha$$
$$\frac{\cos\alpha}{q_\perp^2 - 2q_\perp\omega\sin\theta^*\cos\alpha + \omega^2\sin^2\theta^*}\left[1-\cos\left(\frac{(q_\perp^2 - 2q_\perp\omega\sin\theta^*\cos\alpha + \omega^2\sin^2\theta^*)\Delta z}{2\omega}\right)\right]. \tag{2}$$

In Eq. (2) $\alpha = \angle(\vec{k}_\perp, \vec{q}_\perp)$, $\lambda_g(z)$ is the position-dependent gluon mean free path and L is the transverse size of the medium. The momentum transfers $\vec{q}_\perp$ are distributed according to the normalized elastic scattering cross section $\sigma_{el}^{-1}d\sigma_{el}/d^2q_\perp(z) = \mu_D^2(z)/\pi(q_\perp^2+\mu_D^2(z))^2$. In this model $\langle q_\perp^2\rangle \propto \mu_D^2(z)$ and for a quark-gluon plasma in local thermal equilibrium $\mu_D^2(z) \sim 4\pi\alpha_s T^2$. For the case of (1+1)D dynamical Bjorken expansion of the QGP $\mu_D(z) = \mu_D(z_0)\,(z_0/z)^{1/3}$, $\lambda_g(z) = \lambda_g(z_0)\,(z/z_0)^{1/3}$. At small angles θ^* both the direct and double Born terms [3] in Eq. (2) are divergent. However, they come with different signs $\propto (-i)^2 = -1$, $\propto i(-i) = 1$, equivalent to a phase difference π. The soft and collinear divergences then exactly cancel [5].

From Eq. (2) the gluon distribution is not only finite when $\theta^* \to 0$ but vanishes on average due to the uniform angular distribution of momentum transfers from the medium, $\int_0^{2\pi} d\alpha\,\cos\alpha = 0$. We have checked that for physical gluons of $k_\perp \leq \omega$ the cancellations discussed here persist to all orders in the mean number of scatterings [3]. The small frequency and small angle spectral behavior of $dN^g_{\rm med}/d\omega d\sin\theta^* d\delta$ remains under perturbative control. Given the vastly different angular behavior of the vacuum and the medium-induced gluon bremsstrahlung, Eqs. (1) and (2), it is critical to identify the phase space where cancellation of the color currents induced by the hard and soft scattering occurs. We fix the parameters of the medium to $\mu_D(z_0) = 1.5$ GeV and $\lambda_g(z_0) = 0.75$ fm at initial formation time $z_0 = 0.25$ fm. Since small frequency emission is suppressed, we use only a moderate $\alpha_s = 0.25$. Triggering on high p_{T_1} hadron directs its parent parton "c" away from the medium and places the collision point of the lowest order (LO) ab $\to$ cd underlying perturbative process [6] close to the periphery of the nuclear overlap region. Then, it is the back-scattered jet "d" that traverses the QGP. For large nuclei, such as Au and Pb, path lengths $L = 9$ (3) fm are used to illustrate central (peripheral) collisions, respectively. We limit gluon emission to the forward jet hemisphere, $0 \leq \theta^* \leq \frac{\pi}{2}$.

The angular distribution of medium-induced radiation for $E = 6$ GeV gluon jet for select values of ω is shown in Fig. 1. We find that gluon emission is strongly suppressed within a cone of opening angle $\theta^* \simeq 0.25$ rad. The broad gluon distribution can be characterized by the mean emission angle

$$\langle \theta^* \rangle = \int_0^1 \theta^* \frac{dN^g_{\rm med}}{d\omega d\sin\theta^*} d\sin\theta^* \left[\int_0^1 \frac{dN^g_{\rm med}}{d\omega d\sin\theta^*} d\sin\theta^* \right]^{-1},$$

given in the insert of Fig. 1. We emphasize that the angular spectrum is distinctly non-Gaussian and the mechanism that determines its form is based on the destructive interference of color currents from hard and soft parton scattering rather than a slow random walk of the gluon in θ^*. The large angle distribution or radiative quanta is evident for both large and small frequencies ω.

In the case of heavy quark creation and propagation the vacuum [7] and the Bertsch-Gunion radiation [8] are modified by the large mass M_q. The mass enters in the propagators as follows, $k_\perp^2 \to k_\perp^2 + \omega^2 M_q^2/E^2$, and suppresses the gluon bremsstrahlung for $\sin\theta^* \leq M_q/E$. Such depletion can only be effective if gluon radiation peaks at $\theta^* \to 0$. Since for light quarks the final state medium-induced non-Abelian bremsstrahlung is *not* dominated by small angle emission, see left panel of Fig. 1, effects other than the "dead cone" control the energy loss pattern of heavy versus light quarks.

Qualitatively, for light quarks the behavior of the energy loss as a function of the density and the size of the system can be summarized to first order in opacity as follows:

$$\begin{aligned} \Delta E \;\; &\approx \int dz \frac{C_R \alpha_s}{2} \frac{\mu^2}{\lambda_g} z \ln \frac{2E}{\mu^2 L} = \int dz \frac{9 C_R \pi \alpha_s^3}{4} \rho^g(z) \ln \frac{2E}{\mu^2 L} \\ &= \left\{ \begin{array}{ll} \frac{9 C_R \pi \alpha_s^3}{8} \rho^g L^2 \ln \frac{2E}{\mu^2 L}, & \text{static} \\ \frac{9 C_R \pi \alpha_s^3}{4} \frac{1}{A_\perp} \frac{dN^g}{dy} L \ln \frac{2E}{\mu^2 L}, & (1+1)D \end{array} \right. . \end{aligned} \tag{3}$$

Eq. (3) assumes energetic jets and neglects kinematic bounds. Here, $A_\perp$ is the transverse size of nuclear matter. For static systems ΔE depends quadratically on the nuclear size. For the case of longitudinal Bjorken expansion this dependence is reduced to linear [3] but the energy loss is sensitive to the initial parton rapidity density dN^g/dy. Understanding the effective color charge density dependence of ΔE the is the key to jet tomography [9].

The calculated energy loss can be incorporated in the perturbative QCD factorization approach as a modification of the single and double inclusive hadron production cross sections. To lowest order in the double collinear limit these are given by:

$$\begin{aligned} \frac{d\sigma^{h_1}_{NN}}{dy_1 dp_{T_1}} &= K \sum_{abcd} p_{T_1} \int \frac{dz_1}{z_1^2} D_{h_1/c}(z_1) \int dx_a \frac{\phi_{a/N}(x_a)\phi_{b/N}(x_b)}{x_a x_b S} \left[\frac{1}{x_a S + U/z_1} \right] \\ &\quad \times 2\pi\alpha_s^2 |\overline{M}_{ab\to cd}|^2 , \end{aligned} \tag{4}$$

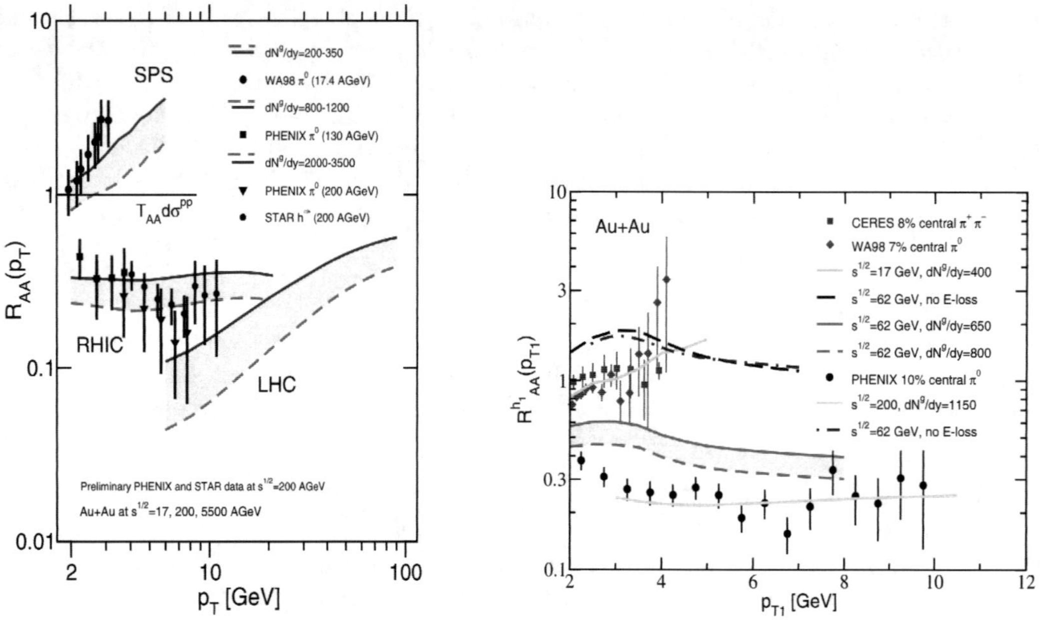

Figure 2. Left panel: nuclear modification factor $R^{h_1}_{AA}(p_T)$ for single inclusive pion production at SPS, RHIC and the LHC. Right panel: the same nuclear modification for the RHIC Au+Au run at c.m. energy of 62 AGeV.

$$
\begin{aligned}
\frac{d\sigma^{h_1h_2}_{NN}}{dy_1dy_2dp_{T_1}dp_{T_2}d\Delta\varphi} &= K\sum_{abcd}\int\frac{dz_1}{z_1}D_{h_1/c}(z_1)\left[D_{h_2/d}(z_2)\delta(\Delta\varphi-\pi)\right]\frac{\phi_{a/N}(x_a)\phi_{b/N}(x_b)}{x_ax_b\,S^2} \\
&\quad\times 2\pi\alpha_s^2|\overline{M}_{ab\to cd}|^2 \,. \qquad (5)
\end{aligned}
$$

In Eq. (5) $K=2$ is a next-to-leading order K-factor, $x_{a,b}=p_{a,b}/p_{N_a,N_b}$ are the momentum fractions of the incoming partons and $z_{1,2}=p_{h_1,h_2}/p_{c,d}$ are the momentum fractions of the hadronic fragments. We use standard lowest order Gluck-Reya-Vogt PDFs [10] and Binnewies-Kniehl-Kramer FFs [11]. Renormalization, factorization and fragmentation scales are suppressed everywhere for clarity. The spin (polarization) and color averaged matrix elements $|\overline{M}_{ab\to cd}|^2$ are given in [6].

Dynamical nuclear effects in multi-particle production can be studied through the ratio [12]

$$
R^{(n)}_{AB}=\frac{d\sigma^{h_1\cdots h_n}_{AB}/dy_1\cdots dy_nd^2p_{T_1}\cdots d^2p_{T_n}}{\langle N^{\rm coll}_{AB}\rangle\, d\sigma^{h_1\cdots h_n}_{NN}/dy_1\cdots dy_nd^2p_{T_1}\cdots d^2p_{T_n}}\,. \qquad (6)
$$

Centrality dependence is implicit in Eq. (6) and the modified cross section per average collision $d\sigma^{h_1\cdots h_n}_{AB}/\langle N^{\rm coll}_{AB}\rangle$ can be calculated from Eqs. (4) and (5) including the numerically evaluated energy loss. Calculation of the nuclear modification in central Au+Au collisions [9] at $\sqrt{s_{NN}}=17,200,5500$ GeV are shown in the left panel of Fig. (1). The right panel of Fig. (1) extends the GLV jet quenching predictions [13] to the intermediate RHIC energy of $\sqrt{s_{NN}}=62$ GeV. In this case parton energy loss, driven by the medium density dN^g/dy, is smaller but its effect at high p_T is amplified by the steepness of the underlying partonic spectra. Similar results for the light pion suppression were found in [14]. Baryons, however, are expected to be significantly less suppressed. In fact, the p/π^+ ratio was predicted to be larger than at the

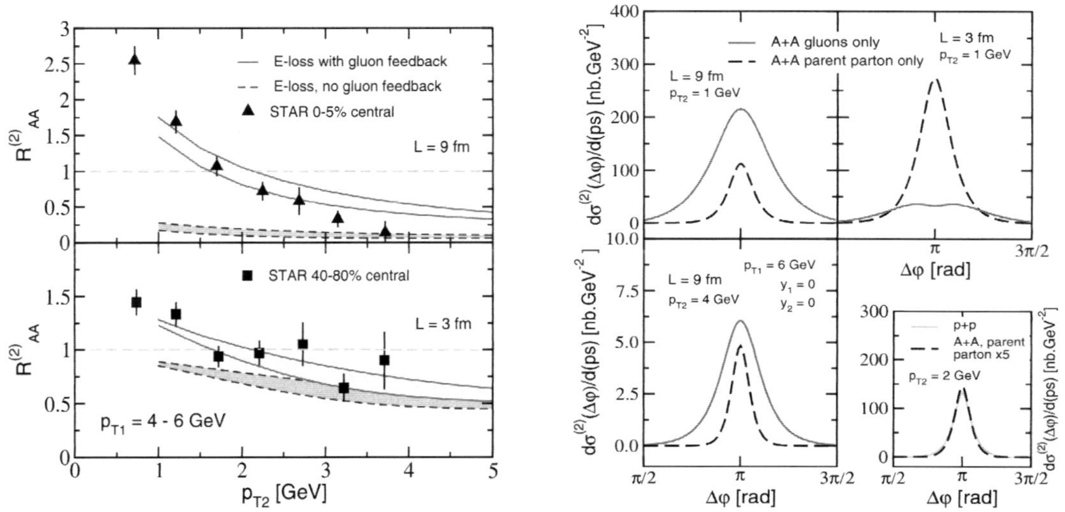

Figure 3. Left panel: nuclear modification $R^{(2)}_{AA}(p_{T_1}, p_{T_2})$ in central and peripheral Au+Au collisions at RHIC of the back-to-back di-hadron correlations with and without the contribution of medium-induced bremsstrahlung. Right panel: angular correlations of hadrons dominated by large angle gluon emission. Solid and dashed lines give *separately* the contribution of the radiative gluons and the attenuated parent parton.

top RHIC energy [15]. Relating the observed high p_T hadron attenuation [16, 17, 18, 19] to the energy density in central Au+Au collisions through the assumption of local thermal equilibrium we find that at times $\tau_0 \simeq 0.6$ fm $\epsilon_0 \simeq 15 - 20$ GeV/fm^3 at $\sqrt{s_{NN}} = 200$ GeV, well in excess of the critical energy for deconfinement ($\epsilon_c \simeq 1$ GeV/fm^3).

More detailed information about the properties of the quark-gluon plasma can be extracted via two particle correlation analysis. At present, a key question for perturbative QCD phenomenology is whether the medium induced gluon bremsstrahlung can significantly alter the di-hadron correlations measured at RHIC [20, 21]. We naturally focus on the away-side $|\Delta\varphi| \geq \frac{\pi}{2}$ case, where medium effects are the largest. A modification that does not change the $\Delta\varphi$-integrated cross section is vacuum- and medium-induced acoplanarity. The deviation of jets from being back-to-back in a plane perpendicular to the collision axis arises from the soft gluon radiation and transverse momentum diffusion in dense nuclear matter [22]. In the approximation of collinear fragmentation, the width of the away-side hadron-hadron correlation function can be related to the accumulated di-jet transverse momentum squared in the φ-plane, $\sin\sqrt{\frac{2}{\pi}}\sigma_{\rm Far} = \sqrt{\frac{2}{\pi}\langle k_T^2\rangle_\varphi}/p_{\perp d}$. Assuming a Gaussian form,

$$f_{\text{vac. or med.}}(\Delta\varphi) = \frac{1}{\sqrt{2\pi}\sigma_{\rm Far}} e^{-\frac{(\Delta\varphi-\pi)^2}{2\sigma^2_{\rm Far}}} , \qquad (7)$$

a good description of $|\Delta\varphi| \geq \frac{\pi}{2}$ correlations measured in elementary p+p collisions at $\sqrt{S} = 200$ GeV [18] requires a large $\langle k^2_{T\,\rm vac}\rangle_\varphi = 5$ GeV^2 for the di-jet pair with away-side scattered quark (and a 2.25 larger value for a scattered gluon). Additional broadening arises from the interactions of the jet in the QGP that ultimately lead to the reported energy loss, $\langle k^2_{T\,\rm hot}\rangle = 2\mu_D^2(z_0)/\lambda_{q,g}(z_0)\ln\frac{L}{z_0}$, although only half is projected on the φ-plane, $\langle k_T^2\rangle_\varphi = \langle k^2_{T\,\rm vac}\rangle_\varphi + \frac{1}{2}\langle k^2_{T\,\rm hot}\rangle$.

Two competing mechanisms do, however, change the p_{T_2} dependence of the perturbative cross section, Eq. (5). First is the the parent jet "d" fractional energy loss $\epsilon = \Delta E_d/E_d$, which we here for simplicity consider on average and evaluate by integrating Eq. (2). It leads to a rescaling of the hadronic fragmentation momentum fraction $z_2 \to z_2/(1-\epsilon)$ [2]. If the energy loss is large, a second mechanism is invoked as a consequence. Hadronic fragments of the radiative gluons will increase the probability of finding low and moderate p_{T_2} particles associated with the interacting jet [23].

To calculate di-hadron correlations, we first map the jet structure of a hard 90^0- scattered parton on rapidity $y \approx \eta = -\ln\tan(\theta/2)$ (θ being the angle relative to the beam axis) and azimuth ϕ,

$$\tan^2\theta^* = \cot^2\theta + \tan^2\phi\,,\ \tan\delta = -\frac{\cot\theta}{\tan\phi}\,,\ \left|\frac{\partial(\sin\theta^*,\delta)}{\partial(\theta,\phi)}\right| = \frac{1}{\sin^2\theta\cos^2\phi}\frac{(\tan^2\phi+\cot^2\theta)^{-1/2}}{(1+\tan^2\phi+\cot^2\theta)^{3/2}}\,.$$

The approximately flat rapidity distribution of the away-side jet near $y_2 = 0$ can be used to sum over all emission angles $\theta \in (\theta_{\min}, \theta_{\max}) \subset (0,\pi)$ yielding

$$\frac{dN^g_{\text{med}}}{d\omega d\phi} = \int_{\theta_{\min}}^{\theta_{\max}} d\theta \left[\frac{dN^g_{\text{med}}}{d\omega d\sin\theta^* d\delta} \left|\frac{\partial(\sin\theta^*,\delta)}{\partial(\theta,\phi)}\right| \right]\,. \tag{8}$$

It is critical to note that projection on a plane coincident with the jet cone axis (the ϕ-plane in Eq. (8) is one example) efficiently masks the $\theta^* \to 0$ void in the angular distribution of medium-induced gluons reported in Fig. 1. Our conclusion is independent of the physical mechanism that depletes the parton (or particle) multiplicity in a cone around the jet axis.

The end analytic result for the modification to Eq. (5) per average nucleon-nucleon collision in the heavy ion environment can be derived from the energy sum rule for all hadronic fragments from the jet,

$$\begin{aligned} D_{h_2/d}(z_2)\delta(\Delta\varphi-\pi) \quad \Rightarrow \quad & \frac{1}{1-\epsilon} D_{h_2/d}\left(\frac{z_2}{1-\epsilon}\right) f_{\text{med.}}(\Delta\varphi) + \frac{p_{T_1}}{z_1}\int_0^1 \frac{dz_g}{z_g} D_{h_2/g}(z_g) \\ & \times \int_{-\pi/2}^{\pi/2} d\phi\, \frac{dN^g_{\text{med}}(\phi)}{d\omega d\phi} f_{\text{vac.}}(\Delta\varphi-\phi)\,. \end{aligned} \tag{9}$$

Here, $z_g = p_{T_2}/\omega$.

Numerical results, shown in the left panel of Fig. 3, correspond to triggering on a high $p_{T_1} = 4$ - 6 GeV pion and measuring all associated $\pi^+ + \pi^0 + \pi^-$. Depletion of hadrons from the quenched parent parton alone leads to a large suppression of the double inclusive cross section – a factor of 5 - 10 in central and a factor of 1.5 - 2 in peripheral reactions with weak p_{T_2} dependence. Hadronic feedback from the medium-induced gluon radiation, however, completely changes the nuclear modification factor $R^{(2)}_{AA}$. It now shows a clear transition from a quenching of the away-side jet at high transverse momenta to enhancement at $p_{T_2} \leq 2$ GeV, a scale significantly larger than the one found in [23]. STAR data in central and peripheral Au+Au collisions [20] is shown for comparison.

Two-particle distributions in A+A reactions, calculated from Eqs. (5) and (9) for a $p_{T_1} =$ 6 GeV trigger pion and two different $p_{T_2} = 1,\ 4$ GeV associated pions, are shown in the Fig. (3). Qualitatively, the medium-induced gluon component to the cross section controls the growth of the correlation width in central and semi-central nuclear collisions. Quantitatively, the effect should be even larger than the one estimated here, which is limited by the imposed $0 < \theta^* < \frac{\pi}{2}$ constraint. Experimental measurements of significantly enhanced widths for $|\Delta\varphi| \geq \frac{\pi}{2}$ two-particle correlation in A+A collisions should thus point to copious hadron production from medium-induced large angle gluon emission.

2. Dynamical shadowing and energy loss in cold nuclear matter

Single and double inclusive hadron production in reactions involving nuclei is also modified by dynamical shadowing effects. Suppression of the particle production rates arises from the coherent final state scattering of the struck small-x parton with several nucleons and the generation of dynamical parton mass $m^2_{dyn} = \xi^2 A^{1/3}$. Here, $\xi^2 = 0.09 - 0.12$ GeV^2 for quarks is the scale of higher twist extracted from deeply inelastic scattering in heavy targets [24]. For electroweak vector meson exchange processes in DIS resumming the power corrections leads to an effective rescaling in the values of Bjorken-x. For p+A and A+A reactions the dynamical mass is amplified for the case of final state gluon scattering by the Casimir ratio $C_A/C_F = 2.25$. It leads to significantly larger power correction effects in hadronic reactions. Effectively,

$$x_B \to x_B \left(1 + \frac{\xi^2 (A^{1/3} - 1)}{Q^2} \right) , \qquad x \to x \left(1 + \frac{C_d \xi^2 (A^{1/3} - 1)}{-\hat{t}} \right) , \tag{10}$$

where the normalization relative to the proton is already included in Eq. (10) and $C_d = 1$ (2.25) for quarks (gluons) respectively.

The left panel of Fig. 4 shows the nuclear modification $R_{dA}(p_T)$ for hadron production in d+Au reactions [12]. We emphasize that the effect is large at small p_T and relatively small at high p_T. Note that at forward rapidity in Eq. (10) $-\hat{t} = p_T^2$. Data on forward y hadron production from BRAHMS is also shown [25]. The right panel of Fig. 4 presents the broadening and attenuation of the two particle back-to-back correlations. Such effects are found to be large at forward rapidity and at small p_{T_1}, p_{T_2} [12]. The coherent multiple scattering discussed here take place (in the c.m. of the collision) over a short interval $t \sim R_A/\gamma \ll R_A$ and for the case of A+A reactions precedes and complements the energy loss in the QGP which develops at time scales $t \sim R_A$.

In cold nuclear matter energy loss is anticipated to be significantly smaller when compared to the one in the quark-gluon plasma. It has been experimentally verified through measurements in $d + Au$ reactions at RHIC at $\sqrt{s_{NN}} = 200$ GeV where a small enhancement consistent with Cronin multiple scattering [26] has been observed [27]. It as been argued, however, that the radiative energy loss induced by the scattering of fast on-shell partons in nuclear matter evaluated by Bertsch and Gunion [28] can be implemented as a "Sudakov" suppression factor $S(x_F) \approx 1 - x_F$ [29] and can give a good description of the A^α, $\alpha < 1$ suppression in a large class of observed cross sections in hadronic reactions at forward x_F.

3. Conclusions

In summary, we calculated the nuclear modification of inclusive moderate and high p_T particle production and the back-to-back particle correlations in the framework of the perturbative QCD factorization approach, augmented by inelastic jet interactions in the quark-gluon plasma. Jet tomographic analysis of hadron attenuation points to initial energy density of $15 - 20$ GeV/fm^3 in $\sqrt{s_{NN}} = 200$ GeV central Au+Au collisions. At RHIC energies we found that the medium-induced gluon radiation determines the $|\Delta\phi| \geq \frac{\pi}{2}$ two-particle yields and the width of their correlation function to surprisingly high transverse momentum $p_{T_2} \sim 10$ GeV. The predicted transition from back-to-back jet enhancement to back-to-back jet quenching is quantitatively compatible with the recent STAR data. In cold nuclear matter at RHIC coherent final state multiple scattering gives the dominant contribution to the attenuation of single and double inclusive hadron production. Energy loss, however, may also play an important role at forward x_F.

Acknowledgments: This work is supported by the J. R. Oppenheimer Fellowship of the Los Alamos National Laboratory and by the US Department of Energy.

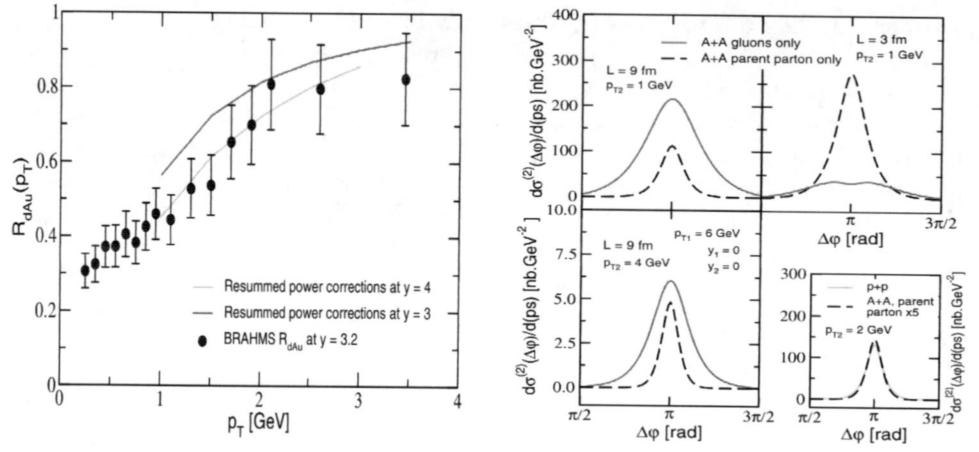

Figure 4. Left panel: modification of the single inclusive hadron production from resummed nuclear enhanced power corrections in forward rapidity d+Au collisions at RHIC. Right panel: modification of the back-to-back di-hadron correlations versus rapidity, centrality and p_T in $\sqrt{s_{NN}} = 200$ GeV d+Au reactions.

References:

[1] X. N. Wang and M. Gyulassy, Phys. Rev. Lett. **68**, 1480 (1992).

[2] M. Gyulassy *et al.*, nucl-th/0302077; A. Kovner and U. A. Wiedemann, hep-ph/0304151; R. Baier *et al.*, Ann. Rev. Nucl. Part. Sci. **50**, 37 (2000).

[3] M. Gyulassy, P. Levai and I. Vitev, Nucl. Phys. B **594**, 371 (2001); Phys. Rev. Lett. **85**, 5535 (2000); Nucl. Phys. B **571**, 197 (2000); M. Djordjevic and M. Gyulassy, Nucl. Phys. A **733**, 265 (2004).

[4] T. Kinoshita, J. Math. Phys. **3**, 650 (1962); T. D. Lee, M. Nauenberg, Phys. Rev. **133**, B1549 (1964).

[5] I. Vitev, hep-ph/0501255.

[6] J. F. Owens, Rev. Mod. Phys. **59**, 465 (1987).

[7] Y. L. Dokshitzer, V. A. Khoze and S. I. Troian, Phys. Rev. D **53**, 89 (1996).

[8] R. Thomas, B. Kampfer and G. Soff, hep-ph/0405189.

[9] I. Vitev, M. Gyulassy, Phys. Rev. Lett. **89**, 252301 (2002); I. Vitev, J. Phys. G **30**, S791 (2004), references therein.

[10] M. Gluck, E. Reya, A. Vogt, Eur. Phys. J. C **5**, 461 (1998).

[11] J. Binnewies, B. A. Kniehl, G. Kramer, Z. Phys. C **65**, 471 (1995).

[12] J. W. Qiu, I. Vitev, hep-ph/0405068.

[13] I. Vitev, Phys. Lett. B **606**, 303 (2005)

[14] A. Adil and M. Gyulassy, Phys. Lett. B **602**, 52 (2004); X. N. Wang, Phys. Rev. C **70**, 031901 (2004).

[15] V. Greco, C. M. Ko and I. Vitev, Phys. Rev. C in press, nucl-th/0412043.

[16] P. Levai *et al.*, Nucl. Phys. A **698**, 631 (2002).

[17] K. Adcox *et al.*, Phys. Rev. Lett. **88**, 022301 (2002); S. S. Adler *et al.*, Phys. Rev. Lett. **91**, 072301 (2003).

[18] J. Adams *et al.*, Phys. Rev. Lett. **91**, 072304 (2003).

[19] Z. b. Xu, nucl-ex/0411001; H. Busching, nucl-ex/0410002.

[20] C. Adler *et al.*, Phys. Rev. Lett. **90**, 082302 (2003). J. Adams *et al.*, nucl-ex/0501016;

[21] J. Rak, J. Phys. G **30**, S1309 (2004).

[22] J. W. Qiu, I. Vitev, Phys. Lett. B **570**, 161 (2003); P. Levai, G. Fai and G. Papp, hep-ph/0502238.

[23] S. Pal, S. Pratt, Phys. Lett. B **574**, 21 (2003).

[24] J. W. Qiu and I. Vitev, Phys. Rev. Lett. **93**, 262301 (2004); Phys. Lett. B **587**, 52 (2004)

[25] I. Arsene *et al.*, nucl-ex/0403005.

[26] Y. Zhang, *et al.* Phys. Rev. C **65**, 034903 (2002); I. Vitev, Phys. Lett. B **562**, 36 (2003).

[27] I. Arsene *et al.*, Phys. Rev. Lett. **91**, 072305 (2003); S. S. Adler *et al.*, Phys. Rev. Lett. **91**, 072303 (2003); B. B. Back, Phys. Rev. Lett. **91**, 072302 (2003).

[28] J. F. Gunion and G. Bertsch, Phys. Rev. D **25**, 746 (1982).

[29] B. Z. Kopeliovich *et al.*, hep-ph/0501260.

Institute of Physics Publishing Journal of Physics: Conference Series **50** (2006) 127–133
doi:10.1088/1742-6596/50/1/015 5th International Conference on Physics and Astrophysics of Quark Gluon Plasma

Recent high $p_{\rm T}$ measurements in STAR

André Mischke *for the STAR Collaboration*

NIKHEF, Amsterdam, The Netherlands,
present address: Department of Subatomic Physics, Utrecht University, Princetonplein 5, 3584 CC Utrecht, The Netherlands.

E-mail: a.mischke@phys.uu.nl

Abstract. After five years of data taking, the STAR experiment at the Relativistic Heavy Ion Collider (RHIC) at Brookhaven National Laboratory provides precise measurements of particle production at high transverse momentum in p-p, d-Au, and Au-Au collisions at $\sqrt{s_{\rm NN}}$ = 200 GeV. We review recent results on the flavor dependence of high $p_{\rm T}$ particle suppression and hadron particle spectra at $\sqrt{s_{\rm NN}}$ = 62.4 GeV. New results on two-particle angular correlations for identified trigger particles and for low momentum associated charged hadrons in p-p and Au-Au as well as near-side $\Delta\eta$ correlations will be presented and discussed.

1. Introduction

The RHIC facility at Brookhaven National Laboratory provides Au-Au collisions at the highest energy presently available of $\sqrt{s_{\rm NN}}$ = 200 GeV. The high center-of-mass energy opens up the hard scattering regime which is accessed by the measurements of particle production at large transverse momentum and where pQCD is applicable. These high momentum particles originate from parton fragmentation in the early stage of the collisions. The scattered partons can be used to probe the produced medium of strongly interacting matter [1, 2].

A significant suppression of high $p_{\rm T}$ hadron production relative to a simple binary collision scaling from proton-proton collisions has been observed in central Au-Au collisions at RHIC [3, 4]. Additionally, it was found that jetlike correlations opposite to trigger jets disappear and that the elliptic anisotropy in hadron emission is large and saturates at very high transverse momenta [5, 6, 7]. All these findings are consistent with the picture of parton energy loss (induced gluon radiation) of the scattered partons in the extremely dense medium (jet quenching). In contrast, no suppression effects were seen in d-Au collisions [8], which provide an important control measurement for the effects in cold nuclear matter. From the d-Au measurements it was concluded that the observations made in Au-Au are due to final state interactions in the high density medium produced in such collisions.

In this paper, recent high $p_{\rm T}$ results from the STAR experiment [9] are presented. From the observations so far, particle production can be described dividing the transverse momentum axis into three ranges. The first range, $0 \lesssim p_{\rm T} \lesssim 2$ GeV/c, particles are produced via soft processes. These particles are usually called bulk matter. In the intermediate range, $2 \lesssim p_{\rm T} \lesssim 6$ GeV/c, the interplay between the probe and the medium can be studied. Initial state nuclear effects (e.g. Cronin) also contribute significantly in this range. The $p_{\rm T}$ range above $\sim$ 6 GeV/c seems to be the "clean" jet fragmentation regime where pQCD processes become dominant. Inclusive hadron spectra from p-p interactions are well described by pQCD. In the next sections,

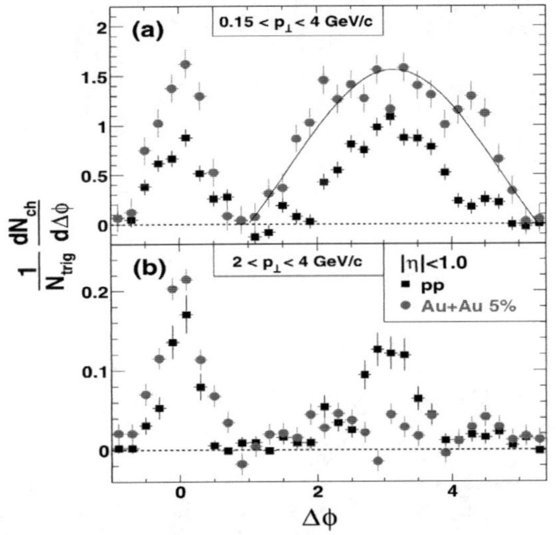

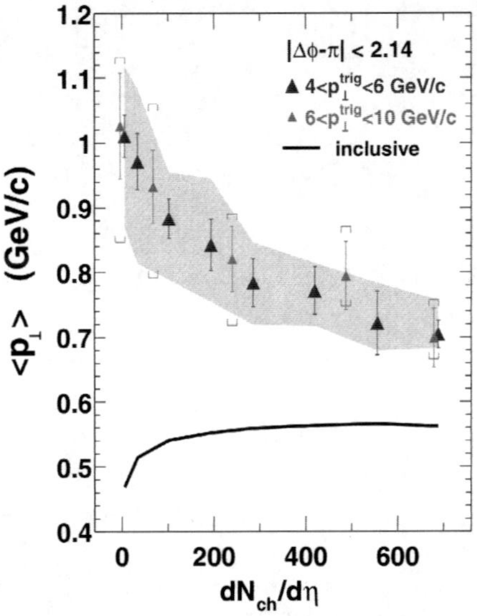

Figure 1. Background subtracted dihadron azimuthal angular distribution in p-p (solid circles) and 5% most central Au-Au collisions (solid squares) for $4 < p_T^{trig} < 6$ GeV/c and two associated p_T ranges $0.15 < p_T^{assoc} < 4$ GeV/c ($2 < p_T^{assoc} < 4$ GeV/c) upper (lower) plot ($|\eta| < 1$). The curve in the upper plot shows the shape of an $[A - B\cos(\Delta\phi)]$ function.

Figure 2. The $\langle p_T \rangle$ for the away-side associated charged hadrons selected in $4 < p_T^{trig} < 6$ GeV/c (systematic errors are indicated by the shaded band) and $6 < p_T^{trig} < 10$ GeV/c (systematic errors in caps) in Au-Au collisions. The leftmost data points are from p-p reaction. The solid line indicates the $\langle p_T \rangle$ for inclusive hadron production.

recent measurements of the STAR collaboration concerning inclusive particle spectra and particle correlations in Au-Au collisions at 200 and 62.4 GeV will be discussed using this classification.

2. Low transverse momentum range

Although the subject of this paper is high p_T, low transverse momentum particle production is discussed for parton fragmentation products appearing in this kinematical range. Due to full azimuthal coverage, the STAR detector has good capabilities to study leading particle correlations. The correlation contribution from azimuthal anisotropy (elliptic flow) is measured with high precision [6, 7]. It was observed that jetlike azimuthal angular ($\Delta\phi$) correlations opposite to trigger jets in the p_T range $4 < p_T^{trig} < 6$ GeV/c are suppressed for associate particle selected in $2 < p_T^{assoc} < 4$ GeV/c in central Au-Au collisions, whereas no suppression was measured in p-p, d-Au and peripheral collisions [8]. The method of leading particle correlation most likely favors the selection of high p_T trigger particles near the medium surface. Therefore, the back-to-back parton has to go through the produced medium and can probe it. The suppression exhibits a continuous centrality dependence [5]. Dihadron $\Delta\phi$ correlation studies with respect to the reaction plane in the 20–60% most central collisions have shown [7] that the suppression depends on the path length traversed, which is in line with the jet-quenching scenario.

Because of momentum conservation the fragmentation products of the away-side jet must balance the momentum of the trigger jet. In Figure 1, upper panel, the background (flow plus pedestal) subtracted $\Delta\phi$ distribution for p-p and the 5% most central Au-Au collisions are shown for the same trigger p_T, $4 < p_T^{trig} < 6$ GeV/c, but for an associate p_T range of $0.15 < p_T^{assoc} < 4$ GeV/c [10]. One observes a clear back-to-back correlation peak which is

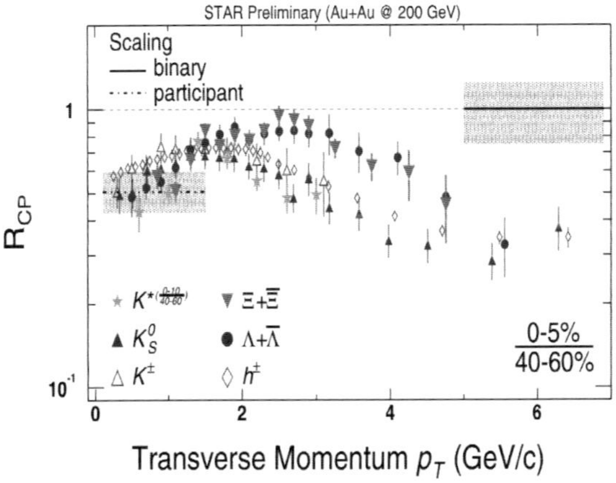

Figure 3. Nuclear modification factor $R_{CP}(p_{\rm T})$ for strange mesons and baryons using the centrality intervals 0–5% vs. 40–60% of the Au-Au collision cross section at $\sqrt{s_{\rm NN}} = 200$ GeV.

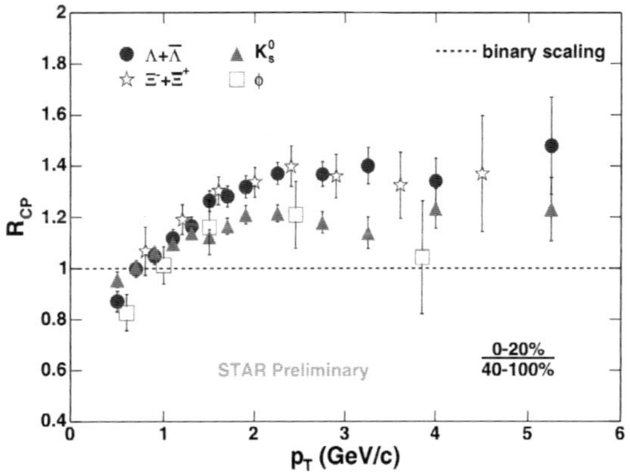

Figure 4. $R_{CP}(p_{\rm T})$ for hyperons (Λ, Ξ), K^0_s and ϕ mesons in d-Au collisions. Here, the ratio is calculated using the centrality intervals 0–20% vs. 40–100%.

widely dispersed and shows a momentum balance shape. Similar results have been found by the NA45/CERES collaboration for pions with $p_{\rm T} > 1.2$ GeV/c at CERN-SPS energies [11]. The away-side peak in Figure 1 exhibits two maxima symmetrically around the the peak center. Recently, it was suggested [12, 13] that these maxima can be expected from in-medium conical flow, caused by the energy deposition of partons traversing the produced matter faster than the speed of sound in the medium. Similar theoretical calculations for Mach shock waves in nuclear reactions have been performed by [14]. The resulting shock wave may give rise to increased particle production at well-defined angles with respect to the parton direction ($\Delta\phi_{\rm flow\ max.} \approx \pi \pm 1.$). Further analysis with higher statistics will clarify the significance of the away-side structure.

In addition, it was found that the mean $p_{\rm T}$ of the back-to-back associated charged hadrons in the trigger $p_{\rm T}$ ranges $4 < p_{\rm T}^{\rm trig} < 6$ GeV/c and $6 < p_{\rm T}^{\rm trig} < 10$ GeV/c decreases with centrality in Au-Au collisions as illustrated in Figure 2. The solid line in the figure reflects the $\langle p_{\rm T}\rangle$ for inclusive hadron production. For the most central events, the associate particles approach this line indicating thermal equilibration of the away-side fragmentation products with the bulk medium.

3. Intermediate transverse momentum range

Inclusive hadron suppression is quantified using the nuclear modification factor $R_{CP}(p_{\rm T})$, the ratio of the inclusive yields in central to peripheral collisions scaled by the number of binary collisions to take into account geometric effects. R_{CP} is expected to be one for pointlike incoherent processes (absence of nuclear effects). Deviations from unity indicate nuclear modifications such as the Cronin effect, shadowing and modifications of the fragmentation function in the produced medium.

In central Au-Au collisions, a suppression of high $p_{\rm T}$ particle production is observed. Moreover, R_{CP} exhibits a significant meson-baryon pattern in the intermediate $p_{\rm T}$ range ($2 \lesssim p_{\rm T} \lesssim 6$ GeV/c) [15, 16]. This pattern is plotted in Figure 3 for charged hadrons, kaons, hyperons (Λ, Ξ) and the K* resonance. The ϕ follows the meson trend as well (not shown). This pattern suggests a collective hadron production mechanism such as quark recombination or coalescence [17].

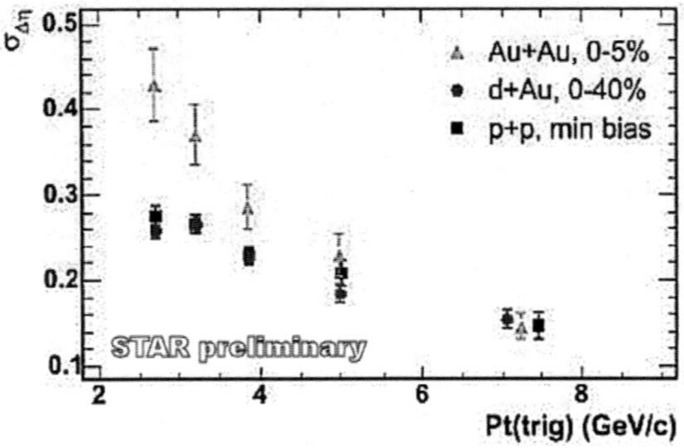

Figure 5. Near-side $\Delta\eta$ correlation width of charged hadrons as a function of $p_{\rm T}^{\rm trig}$ for p-p (square), d-Au (circle) and central Au-Au collisions (triangle).

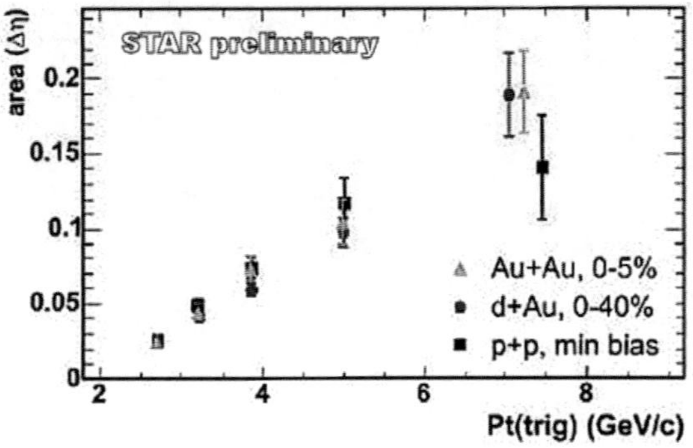

Figure 6. Trigger $p_{\rm T}$ dependence of the dihadron near-side $\Delta\eta$ correlation strength for p-p (square), d-Au (circle) and central Au-Au collisions.

R_{CP} was also studied by the STAR collaboration for ϕ mesons and hyperons in d-Au collisions [18] as illustrated in Figure 4. Here, the yields are enhanced for $p_{\rm T} > 1$ GeV/c likely due to multiple soft scatterings before the hard interaction (similar to the Cronin effect). The meson-baryon pattern is also present with a much smaller magnitude than in the Au-Au case ($\frac{\rm baryon}{\rm meson} \approx 1.15$). In d-Au collisions, the intermediate $p_{\rm T}$ range should be dominated by jet fragmentation and one would expect one curve for all particles. Hence, the different behavior for mesons and baryons is a possible hint of the quark content dependence of the Cronin effect (initial or final). Future high statistic analysis have to show whether the curves for mesons and baryons merge at even higher $p_{\rm T}$ for d-Au and Au-Au. For Au-Au collisions, one already notices a first hint around 6 GeV/c (cf. Figure 3).

Remarkably, the integrated yields of the $\Delta\phi$ correlation peaks for identified K_s^0 and Λ as leading particle do not show a significant flavor dependence at intermediate $p_{\rm T}$ [19]. This result does not favor the thermal quark coalescence as the dominant production source. Instead, it is more likely that one sees here the interplay between parton fragmentation and the produced bulk matter.

3.1. Pseudo-rapidity asymmetry in d-Au collisions

STAR has performed systematic studies of the pseudo-rapidity asymmetry and the centrality dependence of charged hadron production in d-Au collisions [20]. One observes a significant η dependence at the highest centrality bin (0–20% most central events) where the yields in the gold direction are higher than in the deuteron direction. The $p_{\rm T}$ dependence of the back(gold)-to-forward(deuteron) asymmetry ratio is in qualitative agreement with the saturation model [21, 22] but cannot be explained by initial state incoherent multiple scattering of the partons [23].

3.2. $\Delta\eta$ near-side dihadron correlations

Earlier STAR correlation analysis of soft hadrons ($p_{\rm T} < 2$ GeV/c) in central Au-Au collisions have shown [24] that near-side correlations have a jetlike peak in $\Delta\phi$ ($x - y$ plane) and $\Delta\eta$ ($y - z$ plane) with a similar width as measured in p-p reactions. This peak sits on top of a very broad, nearly flat distribution which can be attributed to long-range correlations. A new analysis [25] has been performed using high $p_{\rm T}$ particles which are originating from jet fragmentation. Figure 5 shows the peak width of the jetlike $\Delta\eta$ near-side correlation as a function of the trigger $p_{\rm T}$. This width is similar for p-p, d-Au and central Au-Au collisions at $p_{\rm T}^{\rm trig} > 6$ GeV/c but becomes broader for Au-Au when going to lower $p_{\rm T}^{\rm trig}$. However, the correlation strength is similar for all three collisions systems as shown in Figure 6. These

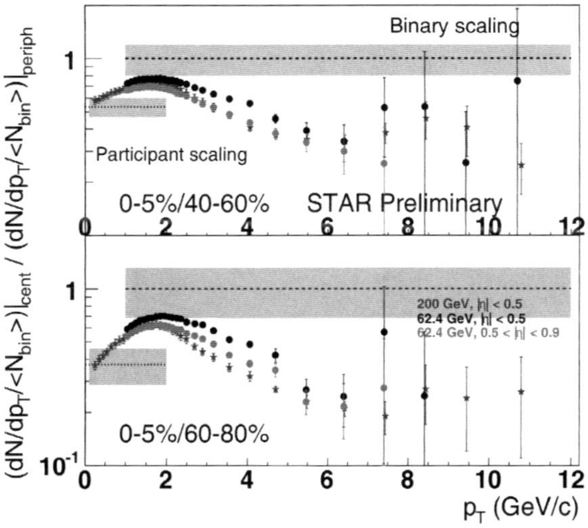

Figure 7. Charged hadron nuclear modification factor $R_{CP}(p_T)$ at cms energy of 62.4 (black and red points) and 200 GeV (blue points) for two different centrality interval ratios.

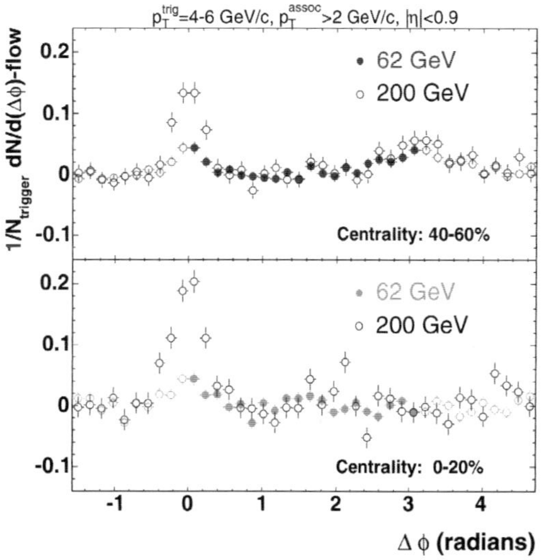

Figure 8. Dihadron azimuthal angular correlation distribution at $\sqrt{s_{NN}} = 62.4$ (full blue and orange points) and 200 GeV (open black points) for the kinematic cuts: $4 < p_T^{trig} < 6$ GeV, $p_T^{assoc} > 2$ GeV/c and $|\eta| < 0.9$.

findings might likely be interpreted by the coupling of the medium-induced gluon radiation to the collective longitudinal flow [26, 27].

3.3. High p_T measurements at $\sqrt{s_{NN}} = 62.4$ GeV

High p_T particle production is suppressed by a factor of about 5 in central Au-Au collisions. It has also been shown [28] that the nuclear modification factor is $R_{AA} \sim 1$ at top CERN-SPS energy. In 2004, the experiments at RHIC took Au-Au data at $\sqrt{s_{NN}} = 62.4$ GeV to probe the onset of suppression. In the jet-quenching picture, one would naively expect that the suppression is smaller at this energy because a higher fraction of quark jets are produced and the energy loss in the medium dependence on the color charged of the partons. At top RHIC energies, the number of quark and gluon jets should be about the same [29]. On the other hand, the smaller initial energy density leads to less parton energy loss whereas the steeper partonic p_T spectrum generates larger leading hadron suppression for the same magnitude energy loss.

In Figure 7, the $R_{CP}(p_T)$ is shown for two different centrality interval ratios for $\sqrt{s_{NN}} =$ 62.4 and 200 GeV. Within the errors, the high p_T suppression is similar above $p_T > 6$ GeV/c for the two collision energies. Consequently, the suppression seems to be driven by the nuclear geometry. The difference of the curves at the intermediate p_T can be attributed to the Cronin effect.

Two-particle azimuthal angular correlation for two different centrality bins at 62.4 and 200 GeV is shown in Figure 8. In comparison to 200 GeV, the near-side correlation strength is a factor ~ 3 smaller at 62.4 GeV due to the steeper underlying partonic spectrum. The away-side suppression is similar for both energies.

4. High transverse momentum range

The large acceptance ($0 \leq \phi \leq 2\pi$ and $-1 \leq \eta \leq 1$) of the STAR Barrel Electromagnetic Calorimeter (BEMC) [30, 31] enables the study of identified particle spectra at even higher p_T to probe where pQCD becomes dominant above nuclear and coalescence effects. In addition,

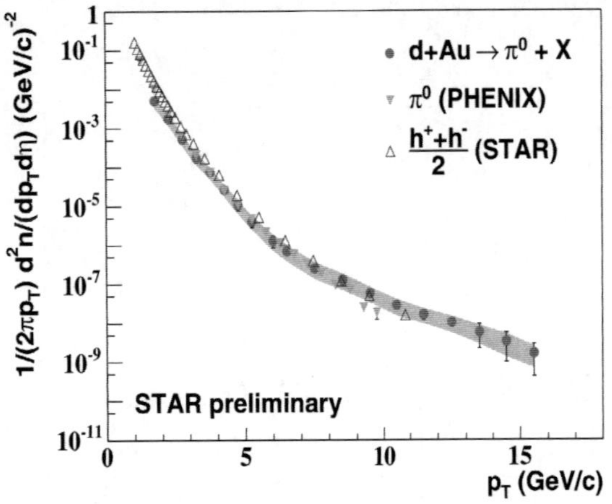

Figure 9. Inclusive neutral pion spectrum in d-Au collisions at $0 < \eta < 1$ (full circles). The error bars (shaded band) represent the statistical (systematic) uncertainties. Previous STAR measurements of the charged hadron cross section are shown by the open triangles, and the full triangles show π^0 results from PHENIX.

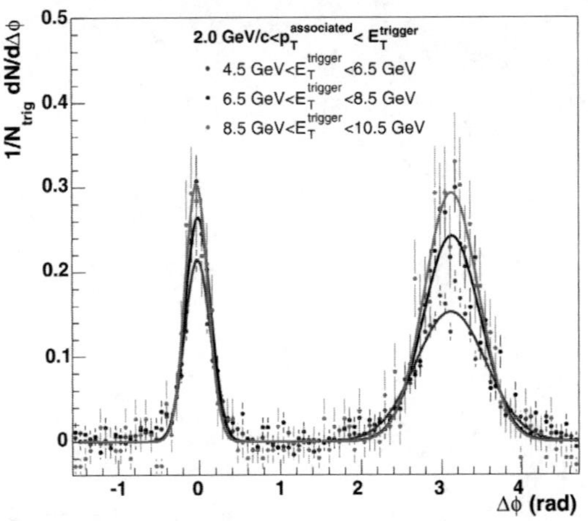

Figure 10. Pedestal background corrected photon-charged hadron $\Delta\phi$ correlation distribution for minbias d-Au collisions selected in the range $2.0\ \mathrm{GeV}/c < p_T^{\mathrm{assoc}} < E_T^{\mathrm{trig}}$ and three different E_T^{trig} ranges as indicated in the figure.

the high statistic 200 GeV Au-Au data set of Run IV will certainly extend the inclusive hadron spectra to higher p_T.

Preliminary analysis of neutral pion production in d-Au collisions has shown the capability of the BEMC [32], with the p_T spectrum measured up to ~ 15 GeV/c as plotted in Figure 9. The obtained spectrum shows reasonable agreement with previous STAR measurements on charged hadron cross sections [8] and the PHENIX results on π^0 production [33] as well as with NLO pQCD calculations [32]. Figure 10 shows the azimuthal angular photon - charged hadron correlation distribution in d-Au collisions at 200 GeV for different E_T^{trig} ranges [34]. The leading photon can be selected up to ~ 11 GeV/c. The obtained width and strength of the near- and away-side correlation peak are similar the the ones obtained from charged dihadron correlations [35].

5. Summary and future challenges

In this paper, recent high p_T results on inclusive particle spectra and two-particle $\Delta\phi$ correlation at $\sqrt{s_{\mathrm{NN}}} = 62.4$ and 200 GeV are presented. Similar behavior of the R_{CP} and the dihadron correlation are observed which can be interpreted as the nuclear geometry as the driving force of the suppression. Correlation analysis with lower p_T^{assoc} indicates thermal equilibration of the away-side fragmentation products with the bulk matter. The near-side $\Delta\eta$ correlation distribution can be interpreted as a coupling of the the parton energy loss to the collective flow. d-Au collisions provide essential information for the deeper understanding of the underlying mechanisms in Au-Au. The observed meson-baryon pattern can not be exclusively attributed to quark coalescence.

Further studies and measurements are needed to understand the evolution of the jet-quenching effect [36]. Minbias Cu-Cu collisions at $\sqrt{s_{\mathrm{NN}}} = 62.4$ and 200 GeV have been recorded this year. An additional low energy Au-Au run will be necessary to map the energy dependence of the suppression effect between top SPS ($\sqrt{s_{\mathrm{NN}}} = 17.4$ GeV) and 62.4 GeV. More details on the modification of the fragmentation function in the medium are needed which can be probed by direct photon tagged correlation studies because the hard scattered photons are not suffering from the surrounded medium. Heavy quark tagged correlations will offer the possibility to discriminate between quark and gluon jets since heavier quark are dominantly produced by gluon fragmentation.

Acknowledgments

We thank the RHIC Operations Group and RCF at BNL, and the NERSC Center at LBNL for their support. This work was supported in part by the HENP Divisions of the Office of Science of the U.S. DOE; the U.S. NSF; the BMBF of Germany; IN2P3, RA, RPL, and EMN of France; EPSRC of the United Kingdom; FAPESP of Brazil; the Russian Ministry of Science and Technology; the Ministry of Education and the NNSFC of China; Grant Agency of the Czech Republic, FOM and UU of the Netherlands, DAE, DST, and CSIR of the Government of India; Swiss NSF; and the Polish State Committee for Scientific Research.

References

[1] M. Gyulassy and M. Plümer. *Phys. Lett.*, B243:432, 1990.
[2] X. N. Wang and M. Gyulassy. *Phys. Rev. Lett.*, 68:1480, 1992.
[3] J. Adams *et al.* (STAR Collaboration). *Phys. Rev. Lett.*, 91:172302, 2003.
[4] J. Adams *et al.* (STAR Collaboration). *Phys. Rev. Lett.*, 89:202301, 2002.
[5] C. Adler *et al.* (STAR Collaboration). *Phys. Rev. Lett.*, 90:082302, 2003.
[6] C. Adler *et al.* (STAR Collaboration). *Phys. Rev. Lett.*, 90:032301, 2003.
[7] J. Adams *et al.* (STAR Collaboration). *Phys. Rev. Lett.*, 93:252301, 2004.
[8] J. Adams *et al.* (STAR Collaboration). *Phys. Rev. Lett.*, 91:072304, 2003.
[9] K. H. Ackermann *et al.* (STAR Collaboration). *Nucl. Instrum. Meth.*, A499:624–632, 2003.
[10] J. Adams *et al.* (STAR Collaboration). *submitted to Phys. Rev. Lett. (arXiv: nucl-ex/0501016)*, January 2005.
[11] G. Agakichiev *et al.* (CERES/NA45 Collaboration). *Phys. Rev. Lett.*, 92:032301, 2004.
[12] H. Stöcker. *Nucl. Phys.*, A750:121, 2005.
[13] J. Casalderrey-Solana, E. V. Shuryak and D. Teaney. arXiv: hep/ph-0411315, November 2004.
[14] J. Hofmann, H. Stöcker, U. W. Heinz, W. Scheid and W. Greiner. *Phys. Rev. Lett.*, 36:88–91, 1976.
[15] J. Adams *et al.* (STAR Collaboration). *Phys. Rev. Lett.*, 92:052302, 2004.
[16] M. A. C. Lamont *et al.* (STAR Collaboration). *J. Phys.*, G30:S963–S968, 2004.
[17] R. J. Fries. arXiv: nucl-th/0403036, March 2004.
[18] P. G. Jones *et al.* (STAR Collaboration). *J. Phys. G: Nucl. Part. Phys.*, 31:S399, 2005.
[19] Y. Guo. arXiv: nucl-th/0502015, February 2005.
[20] J. Adams *et al.* (STAR Collaboration). *Phys. Rev.*, C70:064907, 2004.
[21] D. Kharzeev, E. Levin and L. McLerran. *Phys. Lett.*, B561:93, 2003.
[22] D. Kharzeev, Y. V. Kovchegov and K. Tuchin. *Phys. Rev.*, D68:094013, 2003.
[23] X. N. Wang. *Phys. Lett.*, B565:116, 2003.
[24] J. Adams *et al.* (STAR Collaboration). *submitted to Phys. Rev. Lett. (arXiv: nucl-ex/0411003)*, January 2004.
[25] D. Magestro *et al.* (STAR Collaboration). *submitted to Eur. Phys. J.* **C**, January 2005.
[26] N. Armesto, C. A. Salgado and U. A. Wiedermann. *Phys. Rev. Lett.*, 93:242301, 2003.
[27] S. A. Voloshin. arXiv: nucl-th/0312065, December 2003.
[28] David d'Enterria. *Phys. Lett.*, B596:32, 2004.
[29] X. N. Wang. *Phys. Lett.*, B579:299, 2004.
[30] M. Beddo *et al.* (STAR Collaboration). *Nucl. Instrum. Meth.*, A499:725–739, 2003.
[31] J. Adams *et al.* (STAR Collaboration). *Phys. Rev.*, C70:054907, 2004.
[32] A. Mischke *et al.* (STAR Collaboration). *Nucl. Phys.*, A749:309c, 2005.
[33] S. S. Adler *et al.* (PHENIX Collaboration). *Phys. Rev. Lett.*, 91:072303, 2003.
[34] Subhasis Chattopadhyay *et al.* (STAR Collaboration). arXiv: nucl-ex/0403013, March 2004.
[35] J. Bielcikova *et al.* (STAR Collaboration). RIKEN BNL Research Center Workshop on "Jet Correlations at RHIC", to appear in RBRC proceedings volume, March 2005.
[36] J. Adams *et al.* (STAR Collaboration). *submitted to Nucl. Phys* **A** *(arXiv: nucl-ex/0501009)*, January 2005.

Stopping and Baryon Transport in Heavy Ion Reactions

F.Videbæk[1]

[1] Physics Department, Brookhaven National Laboratory, Upton, NY11973, USA

E-mail: `videbaek@bnl.gov`

Abstract. In this report I will give an experimental overview on nuclear stopping in hadron collisions, and relate observations to understanding of baryon transport. Baryon number transport is not only evidenced via net-proton distributions but also by the enhancement of strange baryons near mid-rapidity. Although the focus is on high-energy data obtained from pp and heavy ions from RHIC, relevant data from SPS and ISR will be considered. A discussion how the available data at higher energy relates and gives information on baryon junction, quark-diquark breaking will be made.

1. Introduction

A major goal for relativistic heavy ion reactions is to form hot nuclear matter at energy densities in clear excess over the value ($\approx 1\text{GeV/fm}^3$) predicted from Lattice QCD needed to make the phase transition or cross over to matter dominated by degrees of freedom of quarks and gluons rather than of hadrons. The next goal is to study and quantify the properties to confront those with properties of non-perturbative QCD. Such experiments has been carried out at RHIC with considerable progress and remarkable results from the first 3 years of experiments. Please refer to the white-papers from the 4 experiments in Refs.[1]. The deposited energy is essential to understand the formation of this medium. Because baryon number is conserved, and rapidity distributions are only slightly affected by rescattering in the later stages of the collisions, the measured baryon distribution retains information about energy loss and allows the degree of nuclear stopping to be determined. Such measurements may also help distinguish between different mechanism for transporting baryons to mid-rapidity.

At very low energies $1-15$ AGeV the hadrons preserve their identity with multiple collisions and excitation to resonances being important ingredient in the description of both stopping, transverse momentum spectra as well as strange particle production. At the higher energies partonic degrees become important, and many features can be described using a string picture. At these higher energies it have long be thought that the dominant mechanism for transport of baryon number is that of quark-diquark breaking (of the strings) where the baryon number is carried(associated) with the valence quarks. I.e. the distribution will reflect the distribution $q(x)-\bar{q}(x)$. Such a mechanism is not able to move the net-baryon number over a large range of x . These distribution are flat in x (e^{-y} in rapidity) for a single collision as observed at SPS energies. Already an analysis of the ISR data in pp collisions [2, 3], and later data from HERA that shows a non-zero baryon asymmetry of $\approx 8\%$ in γp reaction at more than 7 units of rapidity from the incident baryon [4] has demonstrated that additional mechanisms with a slower x and rapidity dependence are needed to describe the data. One such mechanism described in the afore mentioned publications is the baryon junction originally proposed in Ref. [5]. The baryon junction can be thought of as a final state where the incident quarks couple to a color decuplet state, or a topological structure where three gluons join in a junction carrying the baryon number with the valence quarks left in high rapidity mesons. This allows the baryon number to be carried to a much lower value of x through a diminished x (or $\sqrt{s_{NN}}$) dependence of the cross section. So even if the probability for such mechanism is small it may become important and possibly dominant at higher energies. It should be mentioned that other mechanism has been proposed in Ref. [6] (parton cascade) and Ref.[7] (diquark breaking) that may also be relevant for baryon transport. Clearly experimental data from both pp and AA are needed if we are to distinguish between these possibilities. In this paper I will review data in pp and AA collisions that sheds light on the issue of baryon transport and will compare data to predictions of models with and without the mechanism of baryon junction included. Some basic ideas comes from Refs.[8, 9]. See a recent paper[10] which has discussion as well as a wealth of references to models relevant to baryon dynamics.

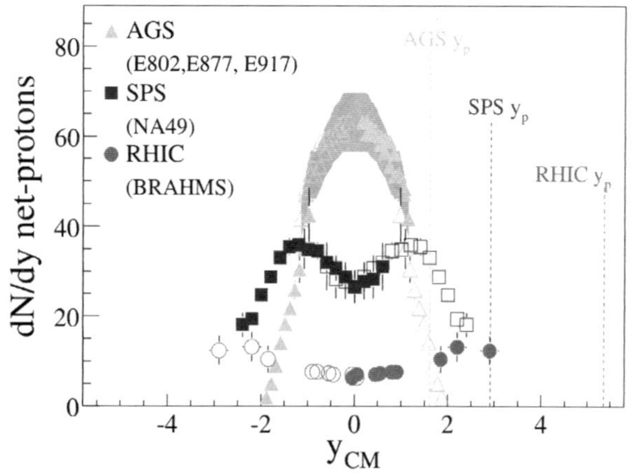

Figure 1. Rapidity density of net protons (i.e. number of protons minus number of antiprotons) measured at AGS, SPS, and RHIC for central collisions.

2. Experimental Considerations

The stopping in nuclear collisions can be estimated from the rapidity loss experienced by the baryons in the colliding nuclei. If incoming baryons have rapidity, y_b in the C.M. system and the average rapidity

$$< y >= \int_0^{y_b} y \frac{dN}{dy} dy / \int_0^{y_b} \frac{dN}{dy} dy \tag{1}$$

after the collision, the average rapidity loss is $\delta y = y_b - <y>$ [8, 9]. Here dN/dy denotes the number of net-baryons (number of baryons minus number of antibaryons) per unit of rapidity. In the case of full stopping δy approaches y_b. Thus, the distributions of dN/dy should be known from mid-rapidity to beam rapidity. Usually the measurements are for protons, in some cases for Λs while rarely have the neutrons been measured. To get the net baryon distributions corrections and extrapolations have to be made. At SIS energies rather detailed measurements have been obtained at 0.4 AGeV and 1.5 AGeV [11]. The clever use of medium mass beams with different

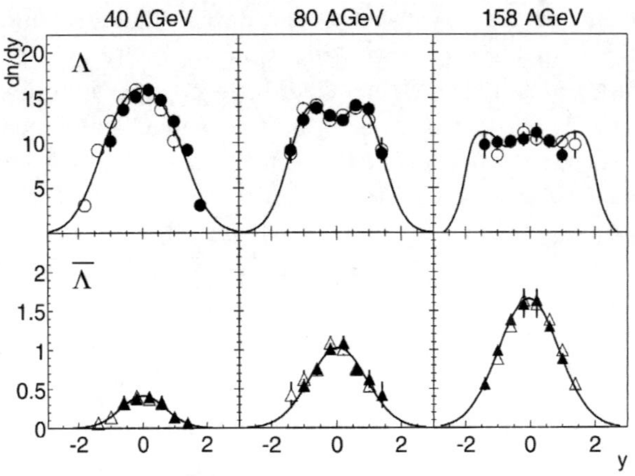

Figure 2. Rapidity distributions of Λ and $\bar{\Lambda}$ at 40, 80 , and 158 AGeV beam energy. The figure is from Ref.[16].

isospin content has shown that already at 1.5 AGeV the system is not fully stopped, but has a small degree of transparency.

At AGS energies the number of produced antiprotons is very small and the net-baryon distribution is similar to the proton distribution [12, 13, 14]. The net-proton rapidity distribution is centered around $y = 0$ and is rather narrow (Fig.1 The rapidity loss is about 1 for a beam rapidity of ≈ 1.6 . At CERN-SPS energies ($\sqrt{s_{NN}} = 17\ GeV, 158AGeV$ Pb+ Pb reactions) the rapidity loss is 1.75 for a beam rapidity of 2.9 [15], about the same relative rapidity loss as at the AGS.

At SPS another feature is visible (see Fig. 1). The net proton rapidity distribution shows a double 'hump' with a dip around $y = 0$. This shape results from the finite rapidity loss of the colliding nuclei and the finite width of each of the humps, which reflect the rapidity distributions of the protons after the collisions. This picture suggests that the reaction at the SPS is beginning to be transparent in the sense that fewer of the original baryons are found at midrapidity after the collisions, in contrast to the situation at lower energies. The net-Λ distributions on the other hand do not show this bump. The data from NA49 [16] displayed in Fig.2 show a rather flat distribution. This is certainly in part due to the higher inelasticity required to produced strange baryons. Also these data shows that at SPS the hyperon production is significant and must be considered for the net baryon distributions. A ratios of $\frac{\Lambda}{p}$ of ~ 0.4 is observed at SPS.

BRAHMS has measured the net proton rapidity distribution at RHIC in the interval $y = 0-3$ for central $(0 - 10\%)$ Au+Au collisions at $\sqrt{s_{NN}} = 200$GeV. Details of the analysis can be found in [9]. The results are displayed in Fig.1. The distribution measured at RHIC is both qualitatively and quantitatively very different from those at lower energies indicating a significantly different system is formed near mid-rapidity. At RHIC the Λ production is even more crucial for estimation of the net-baryon yield as well as correct for in the measured proton and anti-proton yields. A detailed discussion can be found in [9]. The ratio $\frac{\Lambda}{p}$ of ~ 0.9 was observed by STAR and PHENIX near midrapidity[17, 18]. The BRAHMS analysis presented assumes that the measured ratios at mid-rapidity is representative for all measured rapidities up to 3.

The net number of protons per unit of rapidity around $y = 0$ is only about 7 and the distribution is flat over at least the ± 1 unit of rapidity. The distribution rises in the rapidity range $y = 2 - 3$ to an average $dN/dy \approx 12$. Baryon conservation in the reactions can be exploited to set limits on the relative rapidity loss and the energy per baryon at RHIC. This is illustrated in Fig. 3, which in the insert shows two possible distributions whose integral areas correspond to the number of baryons present in the overlap between the colliding nuclei. From such distributions one can deduce a set of upper and lower limits for the rapidity loss at RHIC. Since not all baryons are measured with the bulk of these in the rapidity interval 3-5.4 these assumptions have to be made for these. The limits shown in the figure includes estimates of these effects [9] for different extrapolations. The conclusion is that the absolute rapidity loss at RHIC ($\delta y = 2.05 \pm 0.17$) is slightly larger than at SPS. The value is close to expectations from extrapolations of pA data at lower energies [19, 20]. In fact the relative rapidity loss is significantly reduced as compared to an extrapolation of the low energy systematic [8].

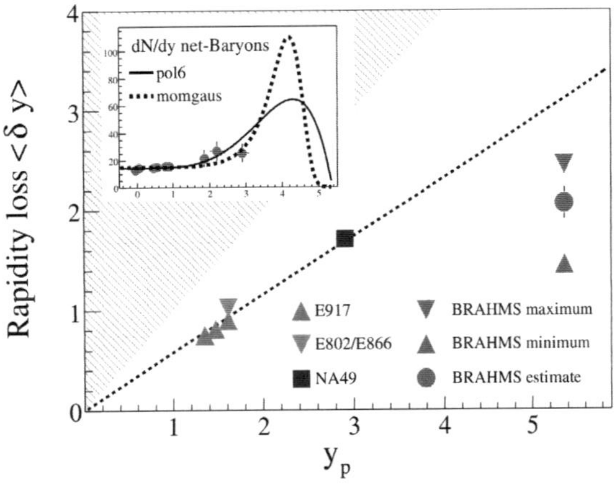

Figure 3. Average rapidity loss as deduced from net-proton distributions vs. beam rapidity. The straight line is the linear extrapolation from Ref.[8] for constant relative rapidity loss. Insert: two possible net-baryon distributions (Gaussian in p_T and 6'th order polynomial) respecting baryon number conservation.

Also from these distribution one can estimate the average energy loss of the colliding nuclei. We find this to be about 73 ± 6 GeV per nucleon, but still with a significant uncertainty if the extreme limits are assumed. The same limits on the relative rapidity loss gives a range of energy loss per baryon of 47 GeV $< E <$ 85 GeV.

3. Particle Ratios

Ratios of baryon to anti-baryons also give information on baryon transport albeit indirectly since it depends on both transport and baryon pair-production. In particular the fraction $p/\bar{p} - 1$ is the relative fraction of transported vs. produced protons. At RHIC several experiments have measured properties of the particle ratios. It is found that the centrality dependence of $\bar{p}/p$ is weak [21, 22, 23], the p_T-dependence up to several GeV/c is flat, and the ratios of anti-neutrons to neutrons was deduced from measurements of $\bar{d}/d$ [24] and found to be consistent with that of protons. In the following I will discuss results on $\bar{p}/p$ from PHOBOS on collision geometry and centrality dependence and from BRAHMS on rapidity dependence both being compared to models.

The particle ratios near mid-rapidity has been measured in d+Au, Au+Au and pp collisions by the PHOBOS collaboration[25, 26]. The $\bar{p}/p$ ratios in d-Au are very close to that observed in pp collisions ($\bar{p}/p \simeq 0.84$) while larger than what is seen in AuAu collisions ($\bar{p}/p \simeq 0.76$). This indicates that a larger fraction of baryon are transported to mid-rapidity in Au-Au collisions than in pp and d-Au collisions. This is consistent with the expectation of additional scattering in the heavy ion system. The surprising observation is that when the d-Au system is studied versus centrality, or rather the mean number of collisions $<\nu>$ estimated from a Glauber description is consistent with no dependence. This is shown in Fig.4 taken from their publication. This is in contrast to results from calculations of HIJING [27], AMPT[28] and RQMD[29] model that all predict a significant decrease of $\bar{p}/p$ with $<\nu>$. Such behavior arises naturally in the picture where multiple collisions cause increasing stopping and baryon transport, but is apparently not born out by the data from RHIC.

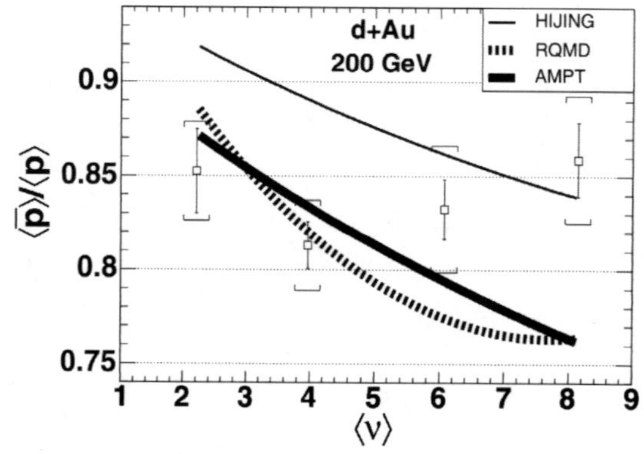

Figure 4. Ratios of anti-proton to proton as function of centrality in d-Au collisions at $\sqrt{s_{NN}} = 200$GeV. The figure is from the Phobos Collaboration [25]

BRAHMS has recently presented measurements and analysis of $p/\bar{p}$ ratios in pp collisions at 200 GeV as function of rapidity [30]. Figure 5 shows the resulting ratios of antiparticle to particle yields as a function of rapidity (left panel). For the ratios there is a clear midrapidity plateau and subsequent decrease with rapidity. This $\bar{p}/p$ ratio would implies that at midrapidity 12% of the protons carry baryon number that has been transported from the beam region at $y = 5.3$. It has been shown (see Ref.[31]) that one may need to correct for isospin effects before generalizing these results from $p+p$ to hadron–hadron collisions. At $y < 1.5$ the Au+Au ratios for the 20% most central collisions reported in [32] are noticeably similar to the present results. The kaon (not shown) and proton ratios remain consistent with the Au+Au results over our entire acceptance range. This is surprising in view of the different dynamics one might expect for the two systems. The ratio starts to decrease above $y = 1.5$, indicating a transition from the string breaking dominated regime at midrapidity to the fragmentation region. Though results for pp and AuAu looks similar in term of rapidity dependence there is a difference at mid-rapidity with the $\bar{p}/p$ values being lower in pp, as also shown by the PHOBOS results discussed above.

The right panel of Fig. 5 shows the present data and data from NA27 at $\sqrt{s} = 27.5$ GeV [34] (open triangles) shifted by the respective beam rapidities. Overlaying the two datasets we observe the ratios to be independent of the incident beam energy when viewed from the rest frame of one of the protons in the area where our rapidity region overlaps with he other experiment. This is consistent with the idea of limiting fragmentation that has also been observed for charged hadrons in nucleus–nucleus collisions [33, 35, 36]. We also note a transition in behavior at $y - y_b = -4$, indicative of a boundary between the midrapidity and fragmentation regions.

To interpret these results further, we confront predictions from theoretical models of hadron-hadron collisions with the data. The curves in the left panel of Figure 5 compare our results to the predictions of two such calculations, PYTHIA Ver. 6.303 [37] and HIJING/B [38], using the same p_T range as the present analysis. Both models give a good description of the pion data and for kaons at midrapidity. Also, PYTHIA clearly overestimates the $\bar{p}/p$ ratios. This is a well–known problem since PYTHIA employs only quark–diquark breaking of the initial protons.

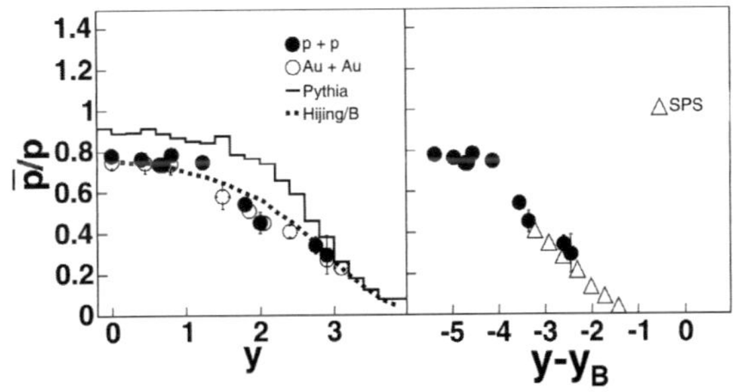

Figure 5. Left: $\bar{p}/p$ from p+p at $\sqrt{s_{NN}}$(solid) points compared with Au+Au [32] (open points), and predictions from PYTHIA [37] (solid histogram) and HIJING/B [38] (thick dashed line). Right ratios shifted by y_b compared with data from NA27 (triangles) at $\sqrt{s_{NN}}$=27.5 GeV [34].

The baryon junction scenario, incorporated as a model prediction in the HIJING/B event generator [38], is shown as the dashed lines in Fig. 5, exhibit a much better agreement with the data both in terms of overall magnitude and the width of the distribution. In Ref. [39] the author shows that baryon stopping in $p+p$ and Au+Au collisions at SPS and RHIC energies can be described using the same parameters for the baryon junction couplings, and predict that at RHIC the shapes of the rapidity distributions for $p+p$ and Au+Au will be similar for $|y| < 2$. The similarity of $\bar{p}/p$ in $p+p$ and $A+A$ up to $|y| < 3$ supports this prediction.

4. Model comparisons to Au-Au collisions at RHIC

In the following section comparisons to models at RHIC energies will be made. A large number of studies have been carried out both at lower energies and shows that the cascade mechanism via multiple collisions and resonances excitation is dominant at SIS and AGS energies. Though, already at SPS energies the stopping neither by cascade model nor by string models. In Ref.[38] the baryon junction mechanism was introduce to enhance stopping over what a conventional string description would give and achieved a satisfactory description of the NA49 net-proton data. In this paper there is also a prediction for RHIC which in fact over estimates the actual later measurements of protons and Λs.

In the left panel Fig. 6 the net-baryons measured by BRAHMS are compared to 3 models. The HIJING[27] model (full drawn curve) where the main mechanism for baryon transport is $q-qq$ string breaking result in a net-baryon yield at mid-rapidity slightly lower than the data, and with a mean rapidity loss of 1.7 being at the low end of the allowed range. The AMPT model (light line) on the other hand results in a much higher yield at mid-rapidity, still compatible with the data, but a mean rapidity loss at the upper end of the range. This model [28] includes the socalled popcorn mechanism to describe baryon - antibaryon production with parameters adjusted to described the NA49 data from SPS. The larger difference between the models are at the higher rapidities where the data are still lacking. The models results in significant different energy per baryon in the final state of 38 and 22 GeV, respectively. On the figure is also shown the calculation of the parton cascade description [6] (dashed points). This particular model should only be compared with data near mid-rapidity, since the spectator baryons that are left within a few units of beam rapidity are not explicitely dealt with in this model. In conclusion several models do in fact describe the measurements at RHIC energies, but the conventional string breaking description does underestimate the baryon transport to mid-rapidity.

Recently the HIJING/B model was improved in Ref.[10] by taking into account intrinsic k_T motion. The main idea is to see if the baryon junction mechanism is also able to account for the enhanced baryon over meson ratios seen in AA collisions in the intermediate p_T-range of 1-4 GeV/c. The model was also compared to the BRAHMS proton and net-proton distributions. The results are available in Ref.[10] as Figs. 4 and 6 and show an overall good description. The authors also calculated the mean rapidity loss to be 1.75 similar to that of the pure HIJING calculation shown in Fig.6 and a mean energy per baryon of 40 GeV. This number are within the experimental values of 2.0 ± 0.2 albeit on the lower range; as a result the mean energy is also considerable large than the value of 26 GeV quoted here albeit as pointed out this experimental has an considerable uncertainty. The model thus does transport additional baryons to mid-rapidity being a candidate for the correct description. Additionaly this transport is an important component in describing the enhancement of baryon over pions observed in the intermediate p_T range of 1-5 GeV/c. As stated by the authors it fails in describing the transverse spectra of kaons and Λs. All in all there are indications that the baryon junction picture are important for baryon dynamics at RHIC.

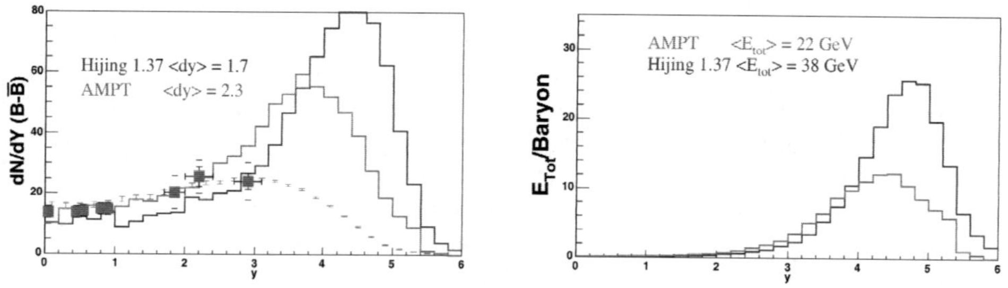

Figure 6. The left panel shows the rapidity density distributions compared to AMPT, HIJING and parton cascade model calculations. The right panels shows the rapidity distributions of energy per baryon.

5. Summary

At the highest energies so far available for heavy ion collisions, namely $\sqrt{s_{NN}} = 200$GeV at RHIC the mean rapidity loss in AA collisions seems to have reach an approximate saturated value of ≈ 2 units. In contract the measurements of net-proton distributions so far do not constrain the mean energy deposited in the collision as well, i.e. the energy that is available for particle production , longitudinal and transverse motion. Evaluation of the data, as well as comparison to models indicate that a likely range of energies are $25 - 37$ GeV/ net-baryon leaving $63 - 75\%$ for production. The analysis of the distributions as well as the $p/\bar{p}$ values < 1 in both pp and AA seem to require additional baryon transport mechanism over quark diquark breaking. The data still leaves open whether this is caused by the baryon junction mechanism, if it can be explained by other of the proposed mechanisms. Such mechanisms as these will in general not decrease the energy per baryon, since only baryon number is transported to mid-rapidity while the energy associated will reside at large rapidities in forward going mesons. Thus the direct connection between stopping of energy and rapidity loss of net-baryons is broken at the higher energies.

The outlook to have some of these questions clarified is encouraging. Not only are there more data to come from RHIC at both $\sqrt{s} = 62.4$ GeV and $\sqrt{s_{NN}} = 200$GeV, but also the upcoming ALICE experiment at LHC will have a distinct possibility to measure the baryon asymmetry in pp and AA collisions with a rapidity loss of up to$\sim 8 - 9.6$ units of rapidity which can help to disentangle the transport processes with different energy dependences.

6. Acknowledgments

This work was supported by the Division of Nuclear Physics of the of the U.S. Department of Energy under contract DE-AC02-98-CH10886. I also appreciate interesting discussions with S.Bass, Nu Xu, and H.Stocker on this subject, M.Murray and J.H.Lee for carefull reading of the manuscript, and will like to thank the organizers for an interesting conference in Kolkata.

References

[1] I. Arsene *et al.*, BRAHMS Collaboration, Nuc. Phys **A757**,1 (2005)
K.Adcox *et al.*,PHENIX CollaborationNuc. Phys A (in print) nucl-ex/0410003
B.B.Back *et al.*, PHOBOS Collaboration, Nuc. Phys A(in print) nucl-ex/04100220 (2004).
J. Adams *et al.*Nuc. Phys A (in print) nucl-ex/0501009

[2] D. Kharzeev Phys. Lett. B **378**, 238-246, (1996).

[3] B.Z. Kopeliovich and A. Zacharov Z.Physik**C43,**241(1998);

[4] B.Z. Kopeliovich and B. Vogh Phys.Lett**B446,**321(1999)

[5] G. C. Rossi and G. Veneziano, Nucl. Phys.**B123** ,507 (1977).

[6] S.A.Bass,B.Muller, and D.K.Srivastava, Phys.ReV.Lett. **91**,052302 (2003)

[7] A.Capella and B.Z.Kopeliovich Phys. Lett.**B381**,325 (1996).

[8] F. Videbæk and O. Hansen Phys.ReV**C52,** 2584 (1995)

[9] I. G. Bearden *et al.* Phys. Rev. Lett. **93,** 102301 (2004).

[10] V.Topor.Pop *et.al* Phys. Rev.**C70**,064906(2004).

[11] B. Hong (FOPI collaboration), Nucl.Phys.**A721,**317c (2003)

[12] B. B. Back *et al.*, Phys. Rev. Lett. **86**, 1970 (2001).

[13] L. Ahle *et al.*, Phys. Rev. **C60**, 064901 (1999).

[14] J. Barette *et al.*, Phys. Rev. **C62**, 024901 (2000).

[15] H. Appelshäuser *et.al* NA49 Collaboration. Phys.Rev.Lett.**82**,2471 (1999).

[16] T.Anticic *et.al* NA49 Collaboration. Phys.Rev.Lett.**93**,022302 (2004).

[17] S.S.Adler et.al (STAR collaboration) Phys. Rev. Lett. **89**,092301 (2002)

[18] K. Adcox et al.(PHENIX collaboration), Phys. Rev. Lett. **89**, 092302 (2002),

[19] W. Busza and A.S. Goldhaber Phys.Lett.**B139**,235 (1984)

[20] W. Busza and R. Ledoux. Ann.Rev.Nucl.Science **38**,150 (1988)

[21] S. S. Adler et al., Phys. Rev. **C69**, 034909 (2004).

[22] K. Adcox et al., Phys. Rev. **C69**, 024904 (2004).

[23] I. Arsene et al. BRAHMS Collaboration.nucl-ex/0503010

[24] K.Adcox.et.al (PHENIX collaboration). nucl-ex/0406004.

[25] B.B.Back *et al.*, PHOBOS Collaboration; Phys.ReV. **C70**, 011901 (2004)

[26] B.B.Back *et al.*, PHOBOS Collaboration; Phys.ReV. **C71**, 021901 (2005)

[27] X.N.Wang and M.Gyulassy, Phys.Rev. **44**, 3501 (1991)
M.Gyulassy and X.N.Wang, Comp.Phys.Comm.**83**,307 (1994).

[28] Z.W.Lin. *et al* Phys.ReV.**C64**,011902 (2001)
B.Zhang *et al* Phys.ReV.**C61**,067901(2000)

[29] H.Sorge Phys.ReV.**C52**,3291(1995)

[30] I.G.Bearden *et. al.* Phys.Lett.**B607**,42 (2005).

[31] H. G. Fischer, NA49 Collaboration, Nucl. Phys. **A715**,118 (2003).

[32] I.G. Bearden *et al.* BRAHMS Collaboration.Phys. Rev. Lett. **90**, 102301 (2003)

[33] I. G. Bearden *et al.*, BRAHMS Collaboration, Phys. Rev. Lett.**88**,202301(2002).

[34] M.Anguliar-Benitez,et.al Z.Phys. **C50** ,405 (1991)

[35] P. Deines-Jones *et al.*, Phys. Rev. C 62 (2000) 014903, hep-ex/9912008.

[36] B. B. Back *et al.*, PHOBOS Collaboration, Phys. Rev. Lett. **91**, 052303 (2003).

[37] T. Sjostrand, *et. al* Comput. Phys. Commun.**135**,238 (2001)

[38] S. E. Vance, M. Gyulassy and X. N. Wang, Phys. Lett. B 443 (1998) 45;

[39] A. Capella, Phys. Lett. **B542**,65 (2002)

Institute of Physics Publishing
doi:10.1088/1742-6596/50/1/017
Journal of Physics: Conference Series **50** (2006) 142–149
5th International Conference on Physics and Astrophysics of Quark Gluon Plasma

Electromagnetic measurements in high-energy heavy-ion collisions

Hideki Hamagaki

Center for Nuclear Study, Graduate School of Science, University of Tokyo, Hongo 7-3-1, Bunkyo-ku, Tokyo 113-0033 JAPAN

E-mail: `hamagaki@cns.s.u-tokyo.ac.jp`

Abstract. Electromagnetic measurements have been playing unique roles in the studies with high-energy heavy collisions. Many interesting results have been obtained in the recent RHIC experiments, which have been providing key information on the interpretation and understanding of the hot and dense nuclear matter created in the ultra-relativistic heavy ion collisions. In this article, some of the recent results are picked up, and implications of these results are discussed.

1. Introduction

Let me start with a little pedantic argument on the relations between electromagnetic probes and electromagnetic measurements. Electromagnetic probes are promptly produced leptons and photons without having intermediate long-lived composite states. Therefore, electromagnetic probes are penetrating probes by definition, which could provide direct information with minimum disturbance from when and where they are produced. Belonging to this category are; thermal and hard photons, and thermal and Drell-Yan lepton pairs.

Strictly speaking, all the other sources including leptonic and photonic decay of hadrons are background to the electromagnetic probes. It is, however, often the case that those 'background' sources are also the useful probes as well.

A good example is J/ψ. J/ψ, a charm and anti-charm bound state, is a hadronic composite with a long lifetime of $c\tau = 2264$ fm, and its yield has been considered to be a unique probe of deconfinement. Measurement of J/ψ via lepton-pairs, therefore, should be considered as a purely technical matter.

Situation is similar for π^0, η, and D and B mesons. Photons from π^0 and η are dominant background sources in view of measuring direct photons. High-p_T π^0 and η, however, reconstructed from two photons are very powerful tools to study jet production in p-p collisions and its medium modifications in p(d)-A and A-A collisions, although these measurements are not covered in this article. Leptonic decay of charm and bottom is a dominant source of single leptons in the medium p_T region in the collisions at the RHIC energies, and the leptons from decays of heavy quarks make it hard to access thermal and Drell-Yan pairs. Because of this dominance, however, heavy quark production and interactions with hot and dense media can be investigated by single leptons measured in p-p and A-A collisions.

The situation is a little different in case of light vector mesons, ρ, ω and ϕ. Their lifetimes, $c\tau(\rho) = 1.3$ fm, $c\tau(\omega) = 23.3$ fm, and $c\tau(\phi) = 46.2$ fm, are comparable to the lifetime of hot matter. Vector mesons measured with lepton-pairs may be considered as 'pseudo-penetrating' probes. A fraction of ω and ϕ would decay inside the medium. In case of ρ mesons, lifetime will be shorter than the lifetime of the sources. The yield and line-shape measured via ee-channel could be significantly different from the one measured via hadronic decay channel, since the mesons decayed inside the medium can only be detected without severe disturbance via ee-channel.

In this article, selected results of electromagnetic measurements from the RHIC experiments are introduced. At first, photon measurements in the high p_T region are picked up. Then, hot results on single lepton measurements are presented. Recent theoretical developments on J/ψ production are referred, and experimental status is provided. Current situation on low-mass pair measurements are reviewed and near-future plan is provided. Brief summary and outlook is provided in the end.

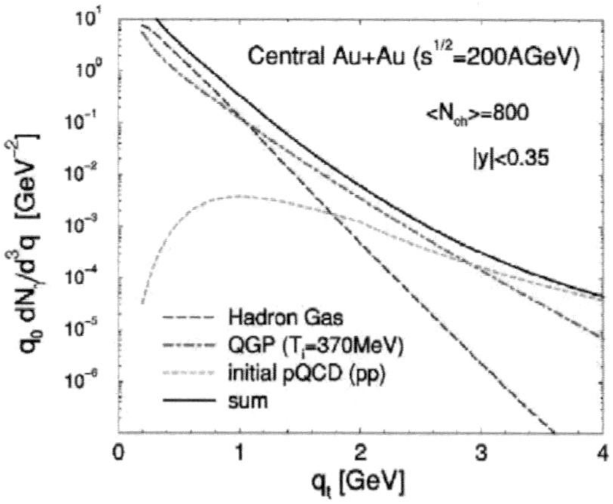

Figure 1. Prediction on the yield of direct photons from various stages; initial pQCD (dotted), QGP (dot-dashed) and hadron gas (dashed), of the central Au+Au collisions at $\sqrt{s_{NN}}$=200 GeV.

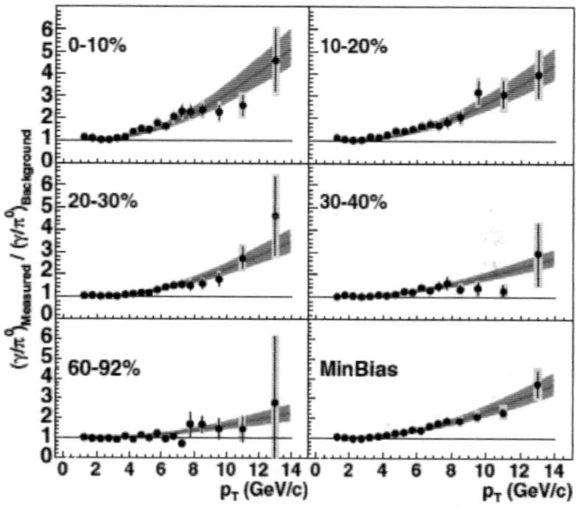

Figure 2. The ratio of $(\gamma/\pi^0)_{\text{measured}}$ to $(\gamma/\pi^0)_{\text{background}}$ as a function of p_T for several centrality bins of $\sqrt{s_{NN}}$ = 200 GeV Au+Au collisions. See text for details.

2. Direct photons

Direct photons come from various stages of collisions. They are in large divided into two categories; non-thermal and thermal. Non-thermal photons come from a hard process in the initial stage and from pre-equilibrium stage. Thermal photons are produced in the later stages; quark gluon plasma (QGP) phase and hadronic phase.

Although there are no explicit tags attached to photons where they are originated from, windows can be set so as to enrich photons from a certain stage. Figure 1 illustrates an expected photon yield from the three sources; a hadron gas phase, a QGP phase, and initial pQCD process[1]. Low-p_T region below $\sim$ 1 GeV/c is dominated by thermal photons from a hadron gas, and high-p_T region above $\sim$ 3 GeV/c is dominated by a pQCD hard process. A narrow window exist for photons from QGP phase in the p_T region from 1 to 3 GeV/c. It is to be added here that signal to background ratio in this momentum region is expected to be in the order of several %.

2.1. Hard pQCD photons from Au-Au collisions

The PHENIX experiment successfully measured the photons in the high pT region for Au-Au collisions at $\sqrt{s_{NN}}$ = 200 Gev[2]. Figure 2 shows a double ratio of measured $(\gamma/\pi^0)_{\text{Measured}}$

invariant yield ratio to the background decay $(\gamma/\pi^0)_{\text{Background}}$ ratio as a function of p_T for minimum bias and five centralities. Solid lines are the predictions with the NLO pQCD calculations, where the hard photon yield is calculated for p-p collisions and multiplied by N_{coll} assuming binary collision scaling. The shaded regions indicate the variation of the pQCD calculation for scale changes from $p_T/2$ to $2p_T$, plus $< N_{\text{coll}} >$ uncertainty.

Figure 3 shows the R_{AA} for prompt photons (close circles) and π^0 (open circles) at $p_T > 6.0$ GeV/c as a function of number of participant nucleons, where R_{AA} is the ratio of photon yield for Au-Au collisions scaled with N_{coll} relative to that for p-p collisions. Difference in the behavior between prompt photons and π^0 is clearly demonstrated. From these results, we can clearly conclude that the initial state effect, that is, modification of parton distribution, is not large, and π^0 suppression is dominantly the final state effect.

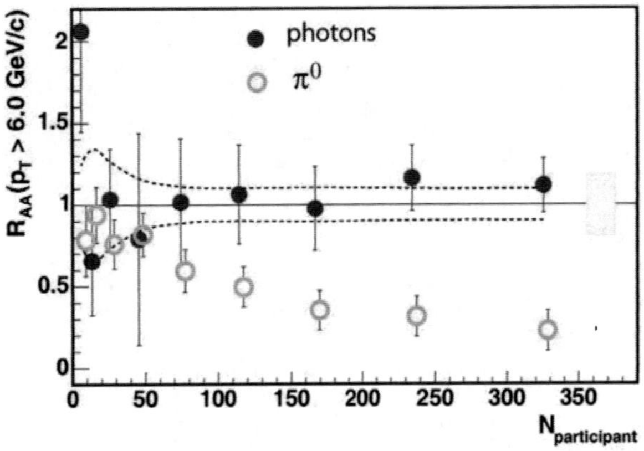

Figure 3. R_{AA} for prompt photons (closed circles) and π^0 (open circles) at $pT > 6.0$ GeV/c as a function of number of participant nucleons.

2.2. Search for thermal photons and beyond

Although photons at high p_T region are now getting under control, photon yield in the p_T region below ~ 4 GeV/c does not have any significance primarily because of statistics and large systematic uncertainty. It remains to be a big challenge.

Recently, a new mechanism to produce photons in the intermediate p_T region was introduced, where a photon is generated by an energetic quark via the pair-annihilation and gluon-Compton processes while passing through hot and dense medium[3, 4]. A related concept of modified photon bremsstrahlung of jets in the medium is also discussed[5]. There seems to be rich complexity in the photons in the intermediate momentum region.

3. Single lepton measurement

Only electrons are covered in this article. Single electrons come from many sources, which can be divided into two large categories; photonic and non-photonic. Photonic electrons are originated from photons and the typical sources are; Dalitz decay (internal conversion) of π^0, η, η', ω and ϕ, and external conversions of direct photons and decay mesons, π^0, η, and η'. Main sources of non-photonic electrons are; leptonic decay of D and B mesons, thermal pairs, lepton-pair decay of vector mesons, and Kaon decay $K_{e3} \rightarrow \pi^0 e\nu$.

3.1. Methods to extract the non-photonic components

There are two methods having been used to subtract a photonic component and extract a non-photonic one. One is a cocktail method, where all the background sources are cocktailed together to estimate the contributions using the Monte-Carlo simulations. The cocktail method was used in Ref. [6] for deducing single electron spectra in Au + Au collisions at $\sqrt{s_{NN}} = 130$ GeV.

Another is a converter method, which utilizes the data samples taken with and without an additional photon converter. The converter method was used in Ref. [7] for deducing non-photonic electrons in Au + Au collisions at $\sqrt{s_{NN}} = 200$ GeV. Advantage of this method over the cocktail method is a smaller systematic uncertainty, since the photonic contribution can be estimated directly from the data samples with and without the converter.

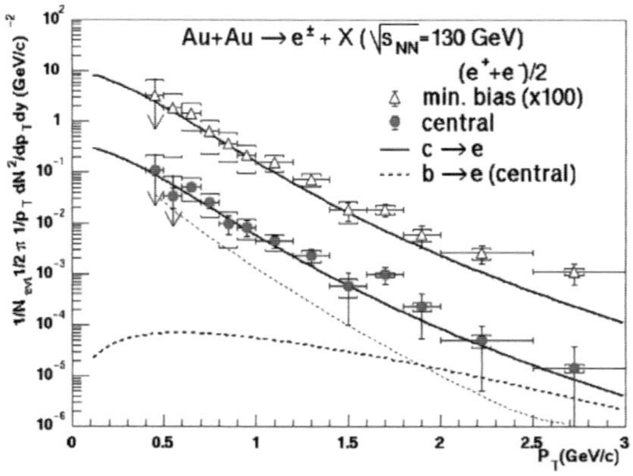

Figure 4. Transverse momentum spectra of non-photonic single electrons for minimum-bias and central 10 % $\sqrt{s_{NN}} = 130$ GeV Au + Au collisions.

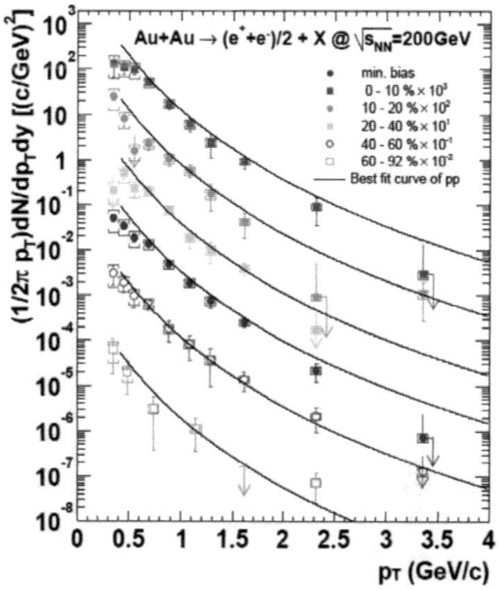

Figure 5. Transverse momentum spectra of non-photonic single electrons for various centrality bins in $\sqrt{s_{NN}} = 200$ GeV Au + Au collisions.

3.2. Electron spectra and yield

Figure 4 shows a momentum spectra of non-photonic single electrons for minimum-bias and central 10 % Au + Au collisions at $\sqrt{s_{NN}} = 130$ GeV[6]. As can be seen from the figure, major contribution comes from leptonic decay of charm and bottom quarks, as expected in the RHIC energies. Figure 5 shows invariant yields of non-photonic single electrons for several centrality bins in Au-Au collisions at $\sqrt{s_{NN}} = 200$ GeV[7]. Shown with the solid lines are the N_{coll} scaled momentum spectrum fitted for p-p collisions. Since the data points lie on the curves quite well, binary collision scaling seem to work quite well. To be more quantitative, N_{coll} dependence is obtained assuming the following dependence; $N_{AuAu}(p_{\mathrm{T}}; b) = N_{pp}(p_{\mathrm{T}}) \cdot N^{\alpha}_{\mathrm{coll}}$. The best fitted value for electron yield integrated over p_{T} from 0.8 to 4.0 GeV/c is, $\alpha = 0.938 \pm 0.075(stat) \pm 0.018(sys)$.

3.3. Is charm coupled weakly?

The results seem to suggest that charm quarks are weakly coupled with hot and dense matter. Theorists have argued about "dead cone" effect in order to account for the small energy loss of heavy quarks[8, 9, 10].

3.3.1. Charm energy loss: With higher statistics achieved using the cocktail method, detailed investigations in the high p_{T} region became possible. Figure 6 shows a PHENIX preliminary $R_{AA}(p_{\mathrm{T}})$ as a function of p_{T} for 0–10% central Au-Au collisions at $\sqrt{s_{NN}} = 200$ GeV. Larger hindrance is seen in the higher p_T region, the magnitude of which is comparable to that of π^0. This is a challenge to theorists, and will force them to re-investigate the energy loss mechanism.

3.3.2. Elliptic anisotropy of non-photonic electron: Elliptic anisotropy has been utilized as a tool to investigate the hydrodynamical behavior of the system. According to hydrodynamical calculations, large anisotropy is a consequence of fast thermalization and subsequent near-perfect hydrodynamical space-time evolution.

The PHENIX experiment reported the v_2 for non-photonic electrons in the p_T region below 1.5 GeV/c[11]. The result favors non-zero charm flow with a 90 % confidence level, when compared with a model calculation which is based on a quark coalescence model assuming the anisotropy of a D meson is the sum of the anisotropy from a light quark and a charm quark; $v_2(D) = v_2(\text{lightquark}) + v_2(\text{charm})$.

A recent result from the STAR experiment covers the higher p_T region[12], and the PHENIX + STAR combined result as shown in Fig. 7 strongly favor the calculation with charm flow.

Again, this should be a good exercise to theorists.

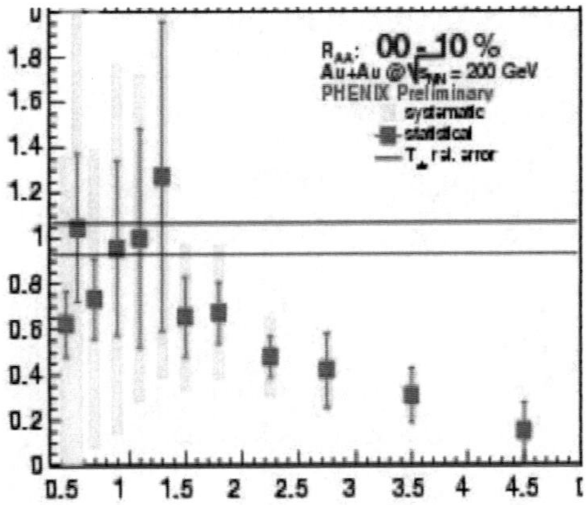

Figure 6. R_{AA} for non-photonic single electrons as a function of p_T for central 10 % of $\sqrt{s_{NN}}$ = 200 GeV Au + Au collisions.

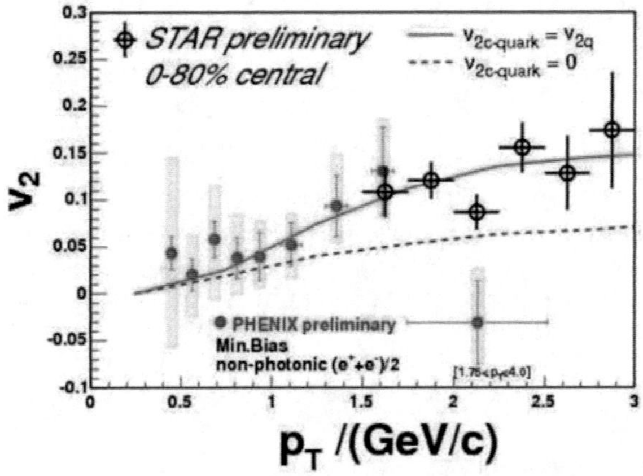

Figure 7. Magnitude v_2 of elliptic anisotropy of azimuthal angular distribution for non-photonic electron as a function of transverse momentum p_T. Preliminary results from both PHENIX (closed circles) and STAR (open circles) are shown.

3.3.3. Efforts in near future: Deduction of non-photonic electron is a complicated and difficult task. Since the results are so important, both PHENIX and STAR are working hard to re-visit this issue with the new analysis of Year-4 run. Year-4 run has much higher statistics, and full comparison between the cocktail method and the converter method will become possible. Higher statistics should also help to extend the p_T region.

4. J/ψ production in HI collisions

4.1. Experimental results at CERN-SPS

The history of experimental studies of J/ψ at CERN-SPS was not straight-forward. Figure 8 shows the yield ratio of J/ψ to Drell-Yan as a function of effective path length for various combinations of colliding systems, and provides a whole history at a glance[13]. When large suppression was observed in S+A collisions, it was considered to be what was predicted by

Matsui and Satz[14]. It turned out, however, that all the data points from S+A and p+A collisions lie on a common curve given by an empirical relation,

$$N(L) = N(0) \cdot \exp(-\sigma_{\rm abs}\rho L),$$

where $\sigma_{\rm abs}$ denoted the absorption cross section in nuclear medium, ρ is the nuclear density, and L denotes the path length. Large suppression strongly deviated from this trend, was finally observed in Pb+Pb collisions, which is considered to be a real consequence of deconfinement.

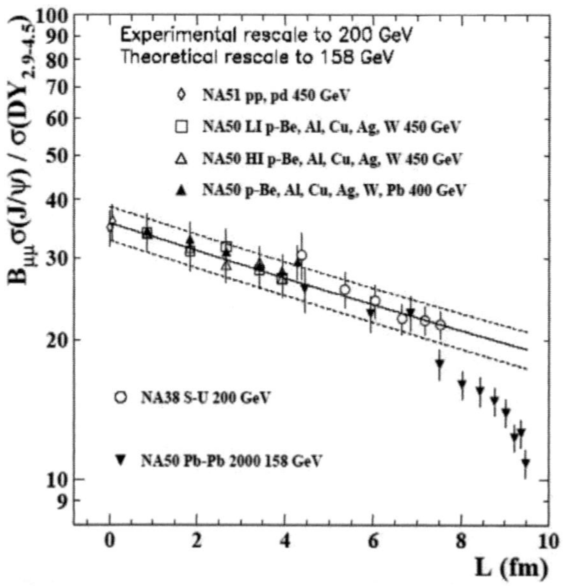

Figure 8. J/ψ over Drell-Yan ratio as a function of effect path length L in p+A, S+U and Pb+Pb collisions at CERN-SPS.

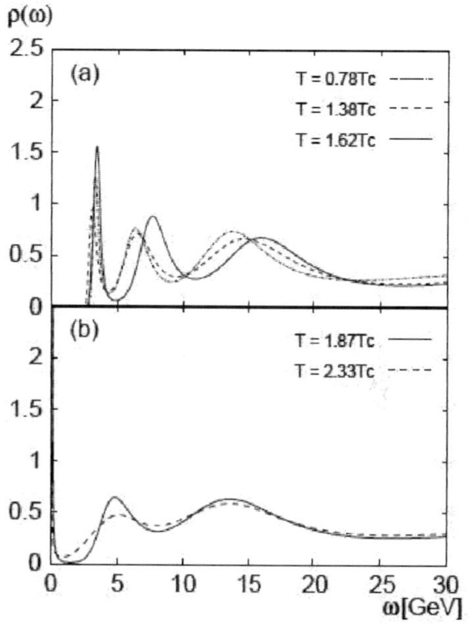

Figure 9. Spectral function of $c - \bar{c}$ system for several temperature below and above $T_{\rm C}$, calculated using the maximum entropy method in the lattice-QCD calculations.

4.2. Recent theoretical progress

4.2.1. Lattice-QCD calculations In the theoretical studies, lattice-QCD calculation is used more frequently as a tool to investigate properties of particles or states in media. A well-known example is the temperature dependence of potential of c-$\bar{c}$ system[15].

Recently, stability of J/ψ above $T_{\rm C}$ was studied by Asakawa and Hatsuda, using the maximum entropy method[16]. Figure 9 shows a spectral function of c-$\bar{c}$ system at several different temperature, where the lowest peak corresponds to J/ψ. The result indicates that the melting of J/ψ happens between $T = 1.62T_{\rm C}$ and $1.87T_{\rm C}$ The result need to be treated properly when constructing a scenario of J/ψ suppression.

4.2.2. J/ψ enhancement due to recombination of charms With increase of number density of charm and anti-charm quarks, probability increases of forming J/ψ's through recombination process. Probability is not high enough at the CERN-SPS energies, but may not be negligible at the RHIC energies. I was notified recently by Matsui that the idea of enhancement due to recombination is as old as the idea of suppression[17].

There are two branches of models, the one, categorized as statistical hadronization model, assumes the coalescence process only in the hadronization stage[19, 20], and the other, called kinetic formation model, adds J/ψ formation in the QGP in addition to the coalescence in the hadronization stage[21].

4.3. Current experimental situation at RHIC
The PHENIX experiment published a result on central dependence of J/ψ yield in Au+Au collisions[18]. Data quality is very poor, and precise comparison with the various models are not possible. Much improved result from the Year-4 run are expected to appear soon for central electron pairs and forward muon pairs.

5. Continuum and low-mass vector mesons
Still missing at RHIC are the continuum and low-mass vector mesons. Importance of these measurements cannot be overemphasized. Continuum, together with thermal photons, would provide direct information on the thermal properties and degree of freedom of the evolving matter. Low-mass vector meson is a sensitive probe of the chiral restoration and/or medium modification.

In both cases, subtraction of overwhelming combinatorial background is the challenge. Even if high statistical data is accumulated and systematic error can be maintained minimum, significant quality reduction cannot be avoided by subtraction of huge background.

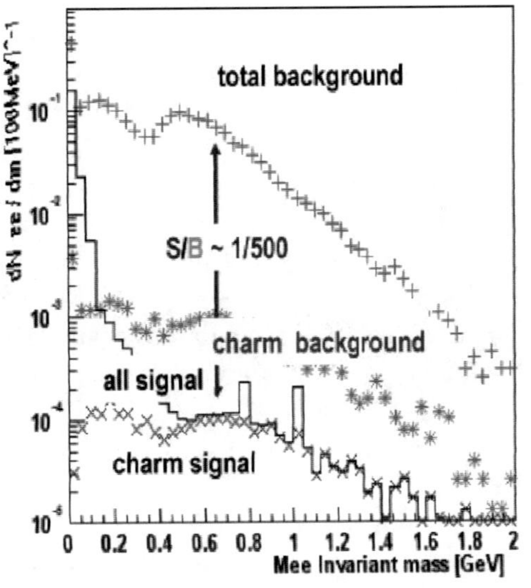

Figure 10. Invariant mass spectra of electron-positron pairs generated by a simulation, which indicate the magnitudes of signals and backgrounds in Au+Au collisions at RHIC.

5.1. Challenge to deduce combinatorials
Figure 10 shows invariant mass spectra of electron-pairs from a simulation which indicates the significant of combinatorial background in the low mass region in the PHENIX experiment for the Au+Au collisions.

Main source of combinatorial background is Dalitz pairs and external conversion of π^0. As was done properly in the CERES experiment at SPS, the combinatorial background can be reduced in principle, by utilizing the decay properties of π^0. Even after proper subtraction of the background, further complication is the large physics background from charm leptonic decay, as indicated in the figure. In the PHENIX experiment, an effort is being made to add Dalitz rejection capability, using a position sensitive Cherenkov detector called HBD (Hadron Blind Detector) placed in the near-field-free region near the beam pipe.

6. Summary and Outlook

Current status of electromagnetic measurements at RHIC was briefly overviewed. Mainly because of low yield and difficulty in measurement, the available results are rather limited at this moment, but it is expected that many results will be coming out soon from the data samples taken in RHIC Year-4 and Year-5 runs. The author is afraid that this article would become obsolete so quickly with a flood of new results.

In the end, remarks to some of electromagnetic measurements are provided.

Direct photon:
Successful detection of pQCD photons in Au-Au collisions is a big achievement, and the pQCD photon becomes the best reference of binary collision scaling. Measurement of photons in the low to intermediate p_T region is challenging but is quite important.

Single electron:
Charm and bottom are the dominant sources of single leptons, and total yield in the mid-rapidity region scales well with N_{coll}. Recent results for v_2 and R_{AA} suggest that a charm quark flows and has large energy loss, as light quarks. These results need re-evaluation with the new data set from the Year-4 run. Exclusive measurements of mesons with positive vertex determination is strongly favorable.

J/ψ:
Good theoretical progresses have been made both in lattice-QCD and phenomenological calculations. New experimental results with good statistics are definitely waited for.

Thermal pairs and low-mass vector mesons:
Reduction of combinatorial background.is definitely needed in order to gain accessibility to these physics quantities. The Hadron Blind Detector (HBD) seems to be the key device to serve for this purpose.

References

[1] Turbide S, Rapp R, and Gale C 2004 *Phys. Rev. C* **69** 014903.
[2] Adler S S *et al.* (PHENIX Collaboration) 2005 *Phys. Rev. Lett.* **94** 232301.
[3] Fries R J, Mueller B, and Srivastava D K 2003 *Phys. Rev. Lett.* **90** 132301.
[4] Fries R J, Mueller B, and Srivastava D K 2005 Preprint nucl-th/0507018.
[5] Zakharov B G 2004 *JTEP Lett* **80** 1 (Preprint hep-ph/0405101).
[6] Adcox K *et al.* (PHENIX Collaboration) 20052*Phys. Rev. Lett.* **88** 192303.
[7] Adler S S *et al.* (PHENIX Collaboration) 2005 *Phys. Rev. Lett.* **94** 082301.
[8] Dokshitzer Y L, and Kharzeev D E 2001 *Phys. Lett. B* **519** 199.
[9] Djordjevic M and Gyulassy M 2003 *Phys. Lett. B* **560** 37.
[10] Zhang B W, Wang E, and Wang X-N 2004 *Phys. Rev. Lett.* **93** 072301.
[11] Adler S S *et al.* (PHENIX Collaboration) Preprint nucl-ex/0502009, submitted to Phys. Rev. C.
[12] Laue F *et al.* (STAR Collaboration) 2005 *J. Phys.* **G31** S27 (Preprint hep-ex/0411007).
[13] Alessandro B *et al.* (NA50 Collaboration) 2005 *Eur. Phys. J.* **C39** 335 (Preprint hep-ex/0412036).
[14] Matsui T and Satz H 1986 *Phys. Lett.* **B178** 416.
[15] Karsch F, Peikert A, and Laermann E 2000 *Phys. Lett.* **478** 447.
[16] Asakawa M and Hatsuda T 2004 *Phys. Rev. Lett.* **92** 012001.
[17] Matsui T 1987 *Proceedings of the Second Workshop on Experiments and Detectors for a Relativistic Heavy Ion Collider (RHIC)*, eds. Ritter H G and Shor A, Lawrence Berkeley Laboratory, May 25-29, 1987, p251.
[18] Adler S S *et al.* (PHENIX Collaboration) 2004 *Phys. Rev. C* **69** 014901.
[19] Grandchamp L and Rapp R 2003 *Nucl. Phys.* **A715** 545.
[20] Andronic A, Braun-Munzinger P, Redlich K, and Stachel J 2003 *Phys. Lett.* **B571** 36.
[21] Thews R L, Schroedter M, and Rafelski J 2001 *Phys. Rev. C* **63** 054905.

Institute of Physics Publishing Journal of Physics: Conference Series **50** (2006) 150–157
doi:10.1088/1742-6596/50/1/018 5th International Conference on Physics and Astrophysics of Quark Gluon Plasma

Photons from Hot and Dense Hadronic Matter

Pradip Roy
Saha Institute of Nuclear Physics, Kolkata 700 064, INDIA

Abstract.

I. Introduction

Study of photon spectra emanating from hot and dense hadronic matter formed in ultra-relativistic heavy ion collisions is a field of considerable current interest. Electromagnetic probes have been proposed to be one of the most promising tools to characterize the initial state of the collisions [1]. Because of the very nature of their interactions with the constituents of the system they tend to leave the system without much change of their energy and momentum. In fact, photons (dilepton as well) can be used to determine the initial temperature, or equivalently the equilibration time. These are related to the final multiplicity of produced hadrons by isentropic expansion of the system formed due heavy ion collisions. By comparing the initial temperature with the transition temperature from lattice QCD, one can infer whether Quark Gluon Plasma (QGP) is formed or not. Furthermore, study of two photon interferometry could determine the source sizes at the very early stages of the collisions [2]. It has also been emphasized that to identify single photons from QGP, correct estimation from non-QGP sources is of utmost importance.

The hadronic matter produced in heavy ion collisions is usually considered to be a gas of the low lying mesons π, ρ, ω, and η, a_1 and nucleons. Reactions between these as well as the decays of the ρ and ω were considered to be the sources of thermal photons from hadronic matter [3, 4, 5]. In our quest for additional sources of thermal photons from hadronic matter we failed to find discussion on the role of baryons and *a priori* there is no reason for this. Hence, we consider it to be important and we shall see that these additional sources of photons contribute substantially in the thermal photon spectra.

In early nineties Kapusta et al. [4] has shown that photon emission rate from QGP is equal to that from hot hadronic matter at a fixed temperature [4]. Apparently this observation led to the conclusion that QGP photon might not be visible in a large hadronic background. Their calculation of photon emission rate from QGP was based on one-loop calculation. The next-to-leading order 2-loop calculation was done by Aurenche et al. [6] and it was turned out to be of the same order in α_s and of the same size or larger than the 1-loop result. So the hope of observing the QGP photons was again raised. However, this calculation is incomplete in the sense that it does not incorporate $O(1)$ suppression due to multiple scattering during the emission process. This point was later clarified by the same authors in a subsequent work [7]. The complete calculation of photon emission rate from QGP to order α_s has recently been completed by re-summing all the contributions from all orders in the multi-loop expansion [8].

In a hot and dense hadronic environment it is expected that the properties of hadrons change due to its interactions with the constituents of the medium. There are ample experimental evidences of this fact. The measurement of invariant mass spectra of electron-positron pairs in 12 GeV $p + A$ reactions shows a significant difference below the ω meson between $p + C$ and $p + Cu$ interactions[9]. This observation indicates that the spectral shape of the mesons is modified inside the nuclei. The enhancement of lepton pair yield in CERES data below the ρ-mass can only be explained by assuming the in-medium modifications of the ρ meson[10, 11].

We organize the paper as follows. In section II we shall discuss the formalism of photon production rates along with the results from QGP and hadronic matter. In particular, we shall focus on the the role of intermediary a_1 meson in the $\pi\,\rho \to a_1 \to \pi\,\gamma$. Section III will be devoted to discuss the in-medium effects in the photon production rates. Finally, we summarize in section IV.

II. Photon production rates

In relativistic heavy ion collisions one might expect that QGP is produced initially, it expands, cools and undergoes a phase transition when the temperature drops down to T_c. Another possible scenario is that the initial state consists of pure hadrons only. In both the cases hadronic matter appears inevitably. There are various sources of photons from relativistic heavy ion collisions: (i) Photons from QGP, (ii) photons from hadronic matter, (iii) photons from decay of $\pi^0(\eta) \to \gamma\,\gamma$ and (iv) hard photons $(A + B \to \gamma\,X)$. Decay photons are generally subtracted from the data and hence, we shall not discuss it here. Hard photon yield can be reliably calculated using perturbative quantum chromodynamics. Therefore, we shall discuss the thermal photon yield from hot hadronic matter in somewhat detail and shall be very brief while discussing photons from QGP.

a. Photons from Hadronic Matter

First we shall consider photon emission from reactions of the type $M\,M \to M\,\gamma$, where M generically denotes the low lying mesons. As mentioned earlier these type of reactions are thought to be the only sources of hadronic photons and a substantial amount of work has been done earlier [3]. Recently Turbide et al. [12] has claimed that ω-meson exchange t-channel exchange diagram in $\pi\,\rho \to \pi\,\gamma$ reaction is the single most dominant process of photon production for $E_\gamma > 2$ GeV. We have made a detail study of this process considering all possible diagrams involving π, ρ, ω, ϕ and a_1 in the intermediate state. For this purpose we have used different effective Lagrangians to obtain the relevant vertices. These include: (i) SU$(2)_L\times$SU$(2)_R$ Linear Sigma Model (LSM) by Kim et al. [13], (ii) A purely phenomenological approach by Xiong et al. [14], (iii) U$(2)_L\times$U$(2)_R$ chiral theory of mesons by B. A. Li [15], and (iv) SU$(3)_L\times$SU$(3)_R$ Non-linear Linear Sigma Model by Song et al. [5].

The emission rates of photons from the reaction $\pi\,\rho \to a_1 \to \pi\,\gamma$ for various phenomenological interactions mentioned above have been evaluated and the results are shown in fig. (1). We observe that the emission rates are different for these Lagrangians. This is primarily because these Lagrangians predict different values of the decay width ($\Gamma_{a_1\to\pi\gamma}$) for the process $a_1 \to \pi\,\gamma$. The experimental value of $\Gamma_{a_1\to\pi\gamma} = 640 \pm 246$ Kev. The $a_1\pi\gamma$ interaction of [13] predicts $\Gamma_{a_1\to\pi\gamma} = 670$ Kev. The interaction Lagrangians of Refs. [14], [15] predict the above decay width as 252 keV, 1420 keV respectively. These differences are clearly manifested in the photon emission rates shown in fig. (1).

Therefore, in our calculation we use the model of Kim et al. [13]. Individual contribution from different exchanges in the concerned reaction is shown in fig (2). It is seen that within our model the a_1 exchange contribution still remains dominant [16].

In fig (3) we compare the existing calculation with our new result. We find that the results are not interestingly different.

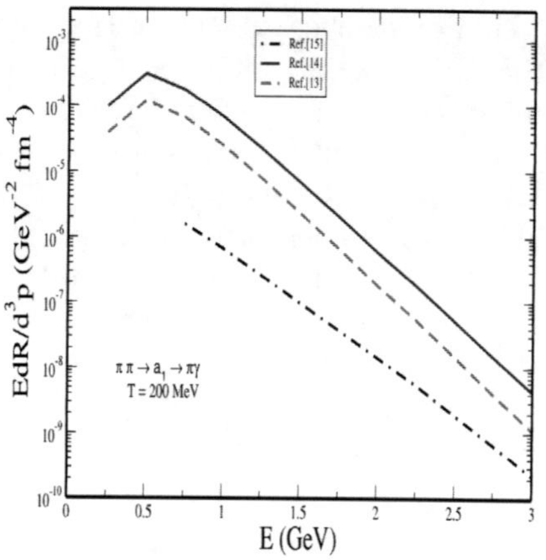

Figure 1. Photon production rate from $\pi\,\rho\,\rightarrow\,a_1\,\rightarrow\,\pi\,\gamma$ for various effective Lagrangians

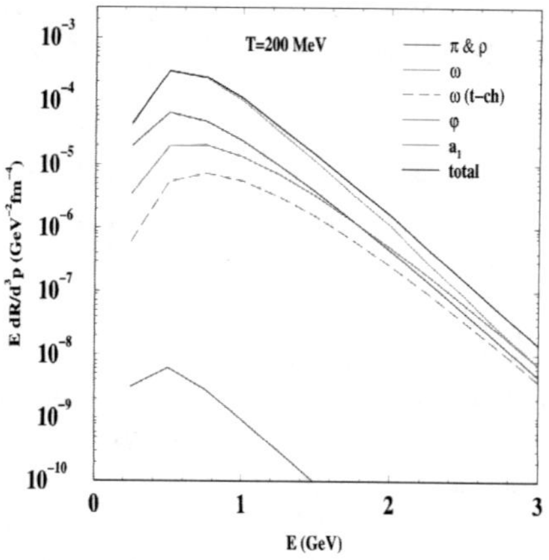

Figure 2. Individual contribution from different exchanges in the $\pi\,\rho\,\rightarrow\,\pi\,\gamma$

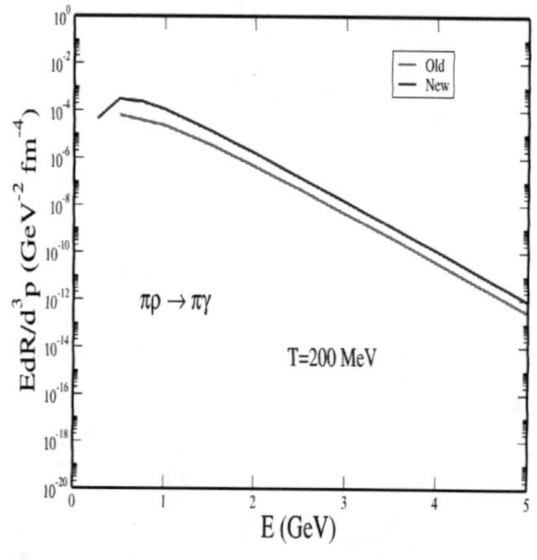

Figure 3. Comparison between old and new results of photon yield from $\pi\,\rho\,\rightarrow\,\pi\,\gamma$

The other important observation in this work is the photon production from reactions of the type [17] $P\,N(\bar{N}) \rightarrow \gamma\,N(\bar{N})$, $N\,\bar{N} \rightarrow \gamma\,P$, $V\,N(\bar{N}) \rightarrow \gamma\,N(\bar{N})$, $N\,\bar{N} \rightarrow \gamma\,V$, $A\,N(\bar{N}) \rightarrow \gamma\,N(\bar{N})$, $N\,\bar{N} \rightarrow \gamma\,A$, where P, A, V and N denote pseudo-scalar, axial vector, vector and nucleons respectively. As mentioned in the introduction that *a priori* there is no reason for not considering these type of reactions. We use a phenomenological Lagrangian to evaluate the photon yield from the above mentioned reactions.

$$\begin{aligned}
\mathcal{L}_{VNN} &= g_{\rho NN}\left[\bar{N}\gamma^\mu\tau N\cdot\rho_\mu\right.\\
&\quad\left.-\frac{\kappa}{2m_N}\bar{N}\sigma^{\mu\nu}\tau N\cdot\partial_\nu\rho_\mu\right]-g_{\omega NN}\,\bar{N}\gamma^\mu N\omega_\mu\\
\mathcal{L}_{ANN} &= \frac{g_{\pi NN}}{m_\pi}m_{a_1}\,\bar{N}\gamma^5\gamma^\mu\tau N\cdot\mathbf{a_1}_\mu\\
\mathcal{L}_{PNN} &= \frac{g_{\pi NN}}{m_\pi}\bar{N}\gamma^5\gamma^\mu\tau N\cdot\partial_\mu\pi+\frac{g_{\eta NN}}{m_\eta}\bar{N}\gamma^5\gamma^\mu N\partial_\mu\eta\\
\mathcal{L}_{em} &= eA^\mu\left[\bar{N}\gamma^\mu N-(\pi\times\partial_\mu\pi)_3\right.\\
&\quad\left.-[\rho^\nu\times(\partial_\nu\rho_\mu-\partial_\mu\rho_\nu)]_3\right]+\frac{e}{2}F^{\mu\nu}(\rho_\mu\times\rho_\nu)_3\\
&\quad+\frac{eg_\rho\kappa_6 f_\pi}{m_\rho^2}F^{\mu\nu}\left[(\partial_\mu\vec{a}_\nu-\partial_\nu\vec{a}_\mu)\times\vec{\pi}\right]_3
\end{aligned}\tag{1}$$

A monopole form factor of the type $(\Lambda^2 - M^2)/(\Lambda^2 - X^2)$, where Λ and M stand for the cutoff and the mass of the exchanged particle in the $X(= t/u)$ channels, respectively. The values and the coupling constants are taken from Ref. [18, 13]. It is to be noted here that the number of Feynman diagrams involved here is quite large and it requires too large a space to be presented here.

b. Photons from QGP

The photon emission rate from Compton ($q(\bar{q})\,g \rightarrow q(\bar{q})\,\gamma$) and annihilation ($q\,\bar{q} \rightarrow g\,\gamma$) processes has been calculated from the imaginary part of the photon self-energy by Kapusta et al. [4] in the 1-loop approximation. However, it has been shown by Aurenche et al. [6] that the two loop contribution is of the same order as the one loop due to the shielding of infra-red singularities. The complete calculation upto two loop was done by Arnold et al. [8] and the rate is given by

$$\frac{dN}{d^4x d^3k} = \frac{1}{(2\pi)^3}\mathcal{A}\left(\ln[T/m_q(T)] + \frac{1}{2}\ln(2E_\gamma/T) + C_{\rm tot}(E_\gamma/T)\right), \tag{2}$$

where $E_\gamma = k$ and $m_q^2(T) = 4\pi\alpha_s T^2/3$ and $\mathcal{A}$ is the leading log coefficient given by

$$\mathcal{A} = 2\,\alpha\,N_c \sum_i q_i^2\,\frac{m_q^2(T)}{E_\gamma}\,f_D(E_\gamma) \tag{3}$$

and

$$C_{\rm tot} = C_{2\to2}(E_\gamma/T) + C_{\rm brems}(E_\gamma/T)C_{\rm aws}(E_\gamma/T) \tag{4}$$

containing the dependence of the specific photon production processes. These are parametrized as follows:

$$\begin{aligned}
C_{2\to2} &= 0.04(E_\gamma/T)^{-1} - 0.3615 + 1.01\exp(-1.35E_\gamma/T)\\
C_{\rm brems} + C_{\rm aws} &= \sqrt{1+\frac{1}{6}N_f}\left(\frac{0.548\ln[12.28 + T/E_\gamma]}{(E_\gamma/T)^{3/2}}\right.
\end{aligned}$$

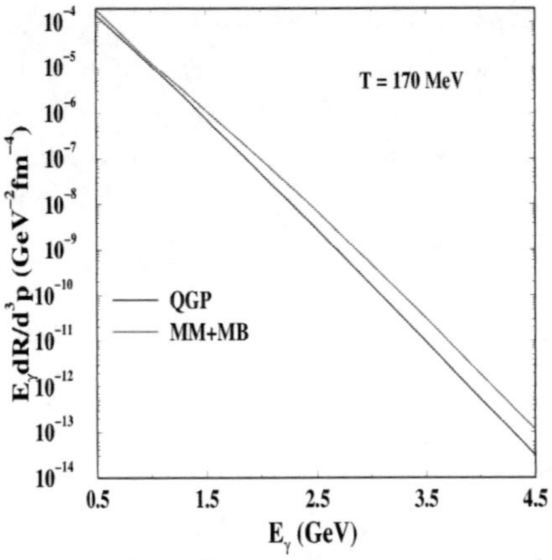

Figure 4. Comparison of static photon rates from QGP and hadronic matter at T=170 MeV

$$+\frac{0.133E_\gamma/T}{\sqrt{1+(E_\gamma/T)/16.27}}\Bigg) \tag{5}$$

In fig. (4) we compare the photon production rates from QGP and hadronic matter at a fixed temperature. In the hadronic yield we include all the possible sources of photon production from hadronic matter (except the hadronic decays). It is seen that the hadronic matter shines more brightly than the quark matter. However, in the hadronic sector we do not include the form factors in the strong vertices. Inclusion of the form factors may deplete the hadronic yield by 10 - 15 % [3, 4]. But as shall see in the next section the inclusion of hadronic in-medium modifications enhances the rate by a factor $\sim$ 2-3. Thus, with the incorporations of form factors and in-medium properties, the hadronic matter may still remain brighter.

III. In-medium properties of hadrons

The properties of hadrons are modified in a hot and/or dense medium. As mentioned earlier, there are enough experimental hints for this. For example, experimental data at KEK [9] indicate significant shape changes in e^+e^- invariant mass spectrum from $p+C$ to $P+Cu$ which means the spectral broadening of the decaying vector mesons. The Chaos Collaboration [19] also indicated lighter sigma meson in $\pi^+ A \to \pi^+\pi^\pm A'$ reactions. The dilepton measurement by CERES Collaboration in $Pb+Au$ collisions at CERN-SPS shows that there is enhancement in the yield below the rho meson mass. This observation could only be explained by assuming spectral modifications of hadrons in the medium (see fig. (5)) [11].

There are various theoretical models to estimate the spectral modifications of hadrons in the medium. We shall discuss Quantum Hadrodynamical Model (QHD) [20] in some what detail and briefly mention the final results of the other models. Let us first consider how nucleon properties are modified in a medium. In QHD the nucleons interact with each other by the exchange of a scalar σ and vector ω mesons (see fig. (6)).

The modified nucleon then enters in the ρ-self energy whose spectral function is given by

$$A_V = \frac{1}{\pi}\left[\frac{\mathrm{Im}\Pi}{(q^2-m_V^2+\mathrm{Re}\Pi)^2+(\mathrm{Im}\Pi)^2}\right] \tag{6}$$

The self energy Π is calculated using Walecka Model described by $\mathcal{L}_{VNN}$ in eq.(1).

The effective masses of hadrons in QHD can be parametrized as follows:

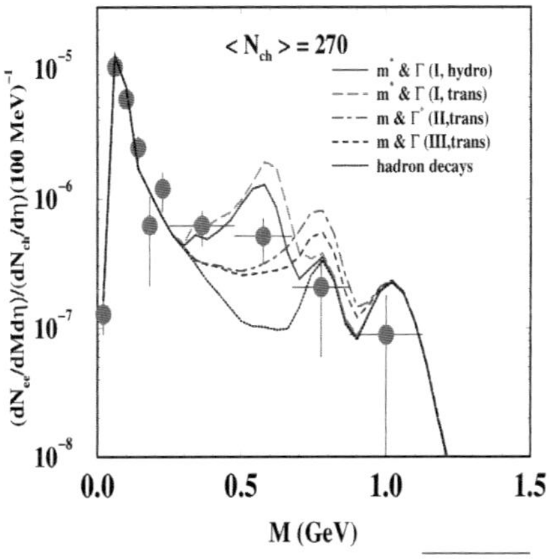

Figure 5. Dilepton spectra for $\langle N_{\rm ch} \rangle = 270$.

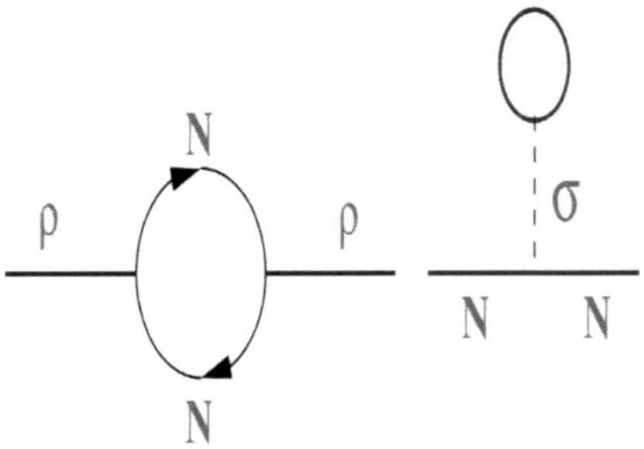

Figure 6. Feynman diagram for ρNN and σNN interactions. There is another diagram with σ replaced by ω.

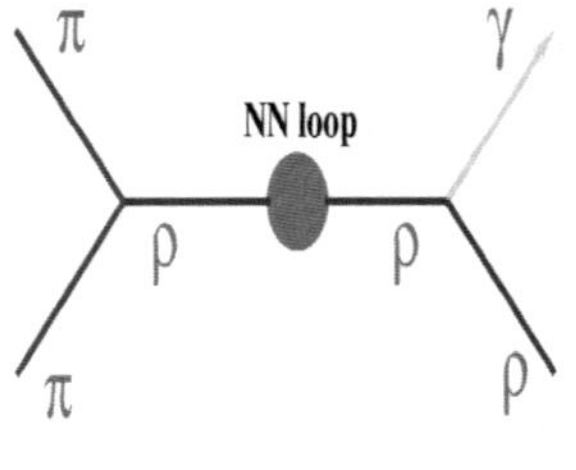

Figure 7. A typical Feynman diagram for in-medium effects in photon production

$$\frac{M_N^*}{M_N} = 1 + \sum_{j=0,1,2,3} a_j \left(\frac{T}{T_c}\right)^{j+1} \tag{7}$$

$a_0 = 0.56$, $a_1 =$ -2.306, $a_2 =$ 2.96, $a_3 =$ -1.277. and

$$\frac{m_\rho^*(\omega)}{m_\rho(\omega)} = 1 + \sum_{j=0,1,2,3} a_j \left(\frac{T}{T_c}\right)^{j+1} \tag{8}$$

$a_0 =$ 0.14(0.32), $a_1 =$ -0,609(-1.28), $a_2 =$ 0.958(1.745) $a_3 =$ -0.6157(-0.827) for ρ (ω) .

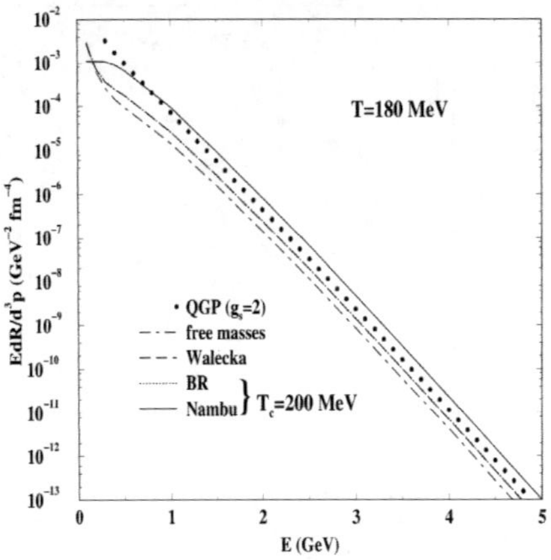

Figure 8. Photon rate with and without in-medium effects from $M\,M \rightarrow M\,\gamma$ reactions.

So far as the other scenarios are concerned, in the Universal scaling scenario (USS), the effective mass of hadrons (except pseudo-scalar) is given by

$$m_H^* \;=\; m_H\left(1-\frac{T^2}{T_c^2}\right)^{\lambda} \tag{9}$$

where, $\lambda = 1/6$ (Brown-Rho) and $\lambda = 1/2$ (Nambu)

A typical diagram of photon production where the medium effects enters is shown in fig. (7). If we include the in-medium effects in the photon production we see an enhancement by a factor of 2-3 in the yield (see fig. (8)). In the total rate from hadronic matter this enhancement will perhaps, be balanced by the decrease due to the inclusion of form factor in the hadronic vertices.

IV. Summary and Discussions

We have calculated the photon emission rate from hot and/or dense hadronic matter from all possible sources of photons. In fact, we have considered a new set of reactions (in addition to the existing sources of hadronic thermal photons) which were previously ignored. We find that the new set of reactions are equally important. The total rate from hadronic matter is then compared with that from quark matter and one finds that hadronic photons are brighter than that from QGP. This result implies that the observation of QGP photons seems to be difficult because of the large hadronic backgrounds.

The role of intermediary a_1 meson in the $\pi\,\rho \rightarrow \pi\,\gamma$ is re-visited. The observation of Turbide et al. [12] regarding the importance of ω meson t-channel diagram in the above mentioned reaction is no longer true within the ambit of the present model. In our model we find that the a_1 exchange diagram still constitutes the dominant contribution [16].

We have incorporated the in-medium effects in the photon producing reactions of the type $M\,M \rightarrow M\,\gamma$ and it is found that the photon yield increases by a factor of 2-3. However, we did not include the same in the new set of reactions involving nucleons. A detail study including all these effects along with space time evolution is in progress.

Acknowledgment: The author is grateful to J. Alam, S. Sarkar, A. K. Dutt-Mazumder, B. Dutta-Roy, B. Sinha, T. Hatsuda and B. mohanty as this work was done in collaboration with them.

[1] J. Alam, S. Raha and B. Sinha, Phys. Rep. **273** (1996) 243.

[2] J. Alam, B. Mohanty, P. Roy, S. Sarkar, and B. Sinha, Phys. rev **C67** (2003) 054902; S. Bass, B. Muller, and D. K. Srivastava, Phys. Rev. Lett. **93** (2004) 162301.

[3] J. Alam, S. Sarkar, P. Roy, T. Hatsuda, and B. Sinha, Ann. Phys. **286** (2000) 159; S. Sarkar, J. Alam, P. Roy, A. K. Dutt-Mazumder, B. Dutta-Roy, and B. Sinha, Nucl. Phys. **A634** (1998) 206; P. Roy, S. Sarkar, J. Alam, and B. Sinha, Nucl. Phys. **A653** (1999) 277.

[4] J. Kapusta, P. Lichard, and D. Seibert, Phys. Rev. **D44** (1991) 2774.

[5] C. Song, Phys. Rev. **C47** (1993) 2861.

[6] P. Auranche, F. Gelis, H. Zaraket, and R. Kobes, Phys. Rev. **D58** (1998) 085003

[7] P. Auranche, F. Gelis, and H. Zaraket, Phys. Rev. **D61** (2000) 116001; P. Auranche, F. Gelis, and H. Zaraket, Phys. Rev. **D62** (2000) 096012.

[8] P. Arnld, G. D. Moore, and L. G. Yaffe, J. High Energy Phys. **11** (2001) 057; P. Arnld, G. D. Moore, and L. G. Yaffe, J. High Energy Phys. **12** (2001) 009; P. Arnld, G. D. Moore, and L. G. Yaffe, J. High Energy Phys. **06** (2002) 030; T. renk, Phys. Rev. **C67** (2003) 064901.

[9] K. Ozawa et al. Phys. Rev. Lett. **86** (2001) 5019.

[10] G. Agakichiev et al., Phys. Lett. **B422** (1998) 405.

[11] J. Alam, P. Roy, S. Sarkar, and B. Sinha, Phys. Rev. **C 67** (2003) 054901; S. Sarkar, J. Alam, and T. Hatsuda, J.phys. **G30** (2004) 607; R. Rapp and J. Wambach, Adv. Nucl. Phys. **25** (2000) 1.

[12] S. Turbide, R. Rapp, and C. Gale, Phys. Rev. **C69** (2004) 014903.

[13] J. K. Kim, P. Ko, K. Y. Lee, and S. Rudaz, Phys. Rev. D53 (1996) 4787.

[14] L. Xiong, E. Shuryak, and G. E. Brown, Phys. Rev. D46 (1992) 3798.

[15] B. A. Li, Phys. Rev. D52 (1995) 5165.

[16] J. Alam, P. Roy, and S. Sarkar, "Comment on Hadronic production of thermal photons", Phys. rev. **C**, in press.

[17] J. Alam, P. Roy, S. Sarkar, Phys. Rev. **C68** (2003) 031901(R)

[18] O. Krehl, C. Hanhart, S. Krewald, and J. Speth, PRC 62 (2000) 025207.

[19] F. Bonutti et al, Nucl. Phys. **A677** (2000) 213.

[20] B. D. Serot and J. D. Walecka, "Advances in Nuclear Physics", Vol. 16, Plenum Press, New York (1986).

Institute of Physics Publishing Journal of Physics: Conference Series **50** (2006) 158–167
doi:10.1088/1742-6596/50/1/019 5th International Conference on Physics and Astrophysics of Quark Gluon Plasma

Peeking though the Colored Looking Glass: A Perspective on New Directions

L. McLerran

RIKEN BNL Research Center, Brookhaven National Laboratory, Upton, NY 11973, USA
and
Physics Department, Brookhaven National Laboratory, Upton, NY 11973, USA

Abstract. I discuss the Color Glass Condensate as a media. I argue that Pomerons, Odderons and Reggeons are the small fluctuation excitations of this media. I argue that understanding the effects of Pomeron loops leads to the idea that this media has a duality symmetry. I discuss the implications of the Color Glass Condensate for the initial conditions at RHIC

1. Introduction

In the last several years, an understanding of the wee partons which control the interaction of high energy particles has emerged.[1] These partons are largely gluons, and they control the properties of hadrons when the longitudinal momentum of the gluon scaled by that of the high energy hadron,

$$x = p^+_{gluon}/P^+_{hadron} \tag{1.1}$$

is $x \leq 10^{-2}$. In this region, the number density of gluons is so large that the typical separation between gluons is much less than the size of the hadron. The total number of gluons is large. This is the limit in which this gluonic matter can be treated as bulk matter.

This high energy density gluonic matter is called the Color Glass Condensate[2, 3, 4, 5, 6, 7, 8, 8, 9, 10]. The name arises because: The gluons are colored. The matter is ultimately made from the very high energy, large x, constituents of the hadron, and in making the low x matter, the time scale of evolution, which is Lorentz dilated relative to natural time scales, is transferred to the low x gluons. Therefore the low x gluons evolve on time scales enormous relative to their natural time scale. This is the property of a glass. Finally the gluons condense inside the hadron because of a negative potential energy produced by the high x gluons. This is stabilized by repulsive interactions among the gluons, which occurs when $1/\alpha_S$ gluons fill each phase space cell, so that

$$\alpha \frac{dN}{dyd^2p_Td^2r_T} \sim 1 \tag{1.2}$$

When phase space densities are large, the quantum occupation number per state is large, and the system is a condensate, with properties similar to traditional Bose-Einstein condensates. This means the condensate is highly coherent, and there are large classical chromodynamic

Figure 1. Alice and the looking glass.

fields present. Also, the coupling constant is weak, since the typical separation between gluons is small.

This Color Glass Condensate is a media. In this talk, I will argue that the small amplitude excitations of this media are the Pomerons, Odderons and Reggeons of S-matrix theory.[11, 12] These excitations can form quantum virtual fluctuations, which in the language of Feynman diagrams appear as loops. I will argue that understanding these loops leads to a theory with a duality symmetry.[13, 14, 15, 16, 17, 18, 19, 20, 21, 22, 23]

In the last section, I discuss the implications of the CGC for the formation of the Quark Gluon Plasma.[24, 25, 26, 27, 28, 29] There has been much excitement recently concerning the rapid thermalization of the QGP which has been observed at RHIC. This has led to the idea that in the early stages of the collision, the interactions are quite strong. One possibility is that this is due to intrinsic strong self coupling of the system. The other possibility, suggested by the CGC is that this is due to a weakly coupled system which interacts coherently.

Coherent interactions are well known to amplify the strength of intrinsically weak interactions. An example of this build up of interaction strength is gravity. The strongest force we experience daily is that of gravity, which arises from the coherent interaction with all of the constituents of the earth. This is in spite of the fact that gravity is intrinsically the weakest force between elementary particles.

I would like to begin this talk with an advertisement: The LHC heavy ion program will be largely about the Color Glass Condensate. The values of x and transverse momenta at LHC are ideally suited for the study of the CGC, and the order one effects which appear in collisions will be largely determined by CGC effects. Hints of the CGC have been seen in HERA and in RHIC experiments, but a real program can be developed at LHC which can test specific predictions. Alice staring at a colored looking glass is shown in Fig. 1.

2. Reggeons, Pomerons and Odderons

Reggeons are mathematical objects which were originally proposed as an explanation for Regge poles which appear in S-matrix theory. (In Fig. 2, you see Prof. Regge with Mikhail Gorbachev when he won the Pomranchuk Prize awarded at the Institute for Theoretical and Experimental Physics in Moscow.) Regge poles arise when one does an angular momentum decomposition of scattering amplitudes. The series representation for these amplitudes are then continued by dispersion theory into an integral in the complex angular momentum plane. It is found that a few simple poles in the complex angular momentum plane control most of the properties of these analytically continued integrals.

Figure 2. Prof. Regge winning the Pomeranchuk prize. Michail Gorbachev is in the background.

The most famous of these poles is that associated with the Pomeron, named after Dr. Pomeranchuk, shown in Fig. 3. This pole controls the high energy limit of scattering amplitudes. It was proven in the early days of Regge theory that the Pomeron leads to universal high energy behaviour for cross sections of strongly interacting particles. It was also argued, most notably by Gribov, that the Pomerons will interact among themselves, and behave like the elementary excitations of an effective field theory.[30, 31, 32, 33] In effect, Gribov showed that Pomerons were the quasi-particles of some as yet to be defined media, since their self interactions has all of these characteristics.

The odderon is the Pomeron's younger brother, see Fig. 4.[34] The Pomeron is the Reggeon responsible for the high energy limit of the real part of the S matrix. The Odderon controls the imaginary part. The Pomeron corresponds to charge conjugation even exchange of quantum numbers in high energy collisions, and the Odderon to C odd. Recall that two gluons make a C even color singlet state and the three gluons correspond to C odd. Therefore, insofar as the Pomeron and the Odderon can be thought of as composed of gluons, the Pomeron is the two gluon configuration and the odderon is the three gluon.

The Odderon has never been seen experimentally, whereas there is very good evidence for the Pomeron. This odd state of affairs is similar to living in a world where we knew of the existence of mesons (2 quark states) but had never seen baryons (2 quark states).

3. The Pomeron: A Modern Perspective

In spite of many years of experimental and theoretical work on strong interactions physics, some of the simplest and most common phenomena are only recently becoming understood. For example, the total cross section as a function of energy is shown in Fig. 5 The total cross section is a slowly varying function of energy, and it is widely believed that it saturates the unitarity limit,

$$\sigma \sim ln^2(E/E_0) \tag{3.3}$$

Figure 3. Dr. Pomeranchuk, the inventor of the Pomeron.

Figure 4. The Odderon Icon.

where E is the center of mass energy. The pomeron was originally introduced to explain *constant* total cross sections, and the data which showed a rising cross section which originated in the Serpukhov experiments, was a big surprise.

Later, Gribov and colleagues suggested that there was a perturbation theory associated with Pomeron dynamics, and that the bare or lowest order Pomeron actually would predict a power law growth of the cross section with energy. This means that at high energies, the interactions of Pomerons have to become strong, since unitarity requires that these interactions must decrease the rise of the cross section to $ln^2(E/E_)$. This led to the invention of Gribov's Reggeon calculus.

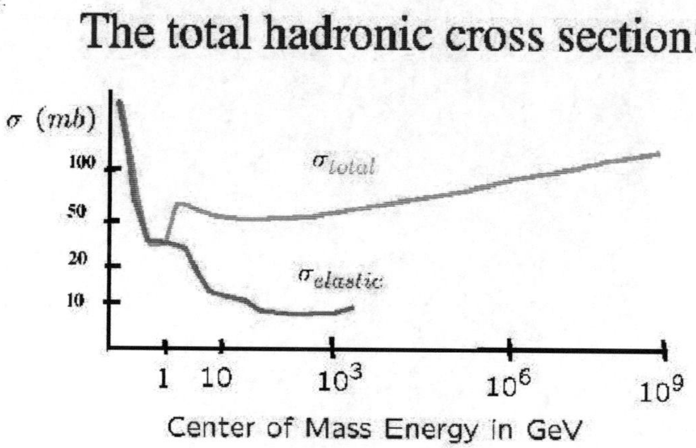

Figure 5. The total corss section as a function of energy.

In modern language, we can think about the bare Pomeron as being composed of gluons.[11] The rise with energy of the bare Pomeron occurs because the density of gluons is rising at small x like a power $1/x^\delta$. The phenomena of saturation of the gluon densities must occur because of the generation of a high density Pomeron (in old language) or gluon (in modern language) state. In this way of thinking about the problem, it is natural to imagine that the Pomeron is a dilute quasi-particle excitation of the Color Glass Condensate. This might be the case at for example low energies, or in some limited range of relatively large transverse momentum at high energy.

Using these simple ideas, we can understand how the cross section reaches its unitarity bound.[35, 36] Assume we have a distribution of gluons inside a hadron's transverse space,

$$\frac{dN}{dyd^2r_T} = Q^2_{sat}(y)e^{-2m_\pi r_T} \tag{3.4}$$

The exponentially falling behaviour in transverse space follows from general conditions for an isosinglet distribution of gluons. The factor of $Q^2_{sat}(y)$ is the total number of gluons per unit rapidity, and grows like $e^{\delta y}$. In scattering, a probe of a hadron will interact after it passes through the hadron and on the average some fixed number of gluons,

$$e^{\delta y}e^{-2m_\pi r_T} \sim constant \tag{3.5}$$

The cross section is therefore

$$\sigma \sim r_T^2 \sim y^2 \sim ln^2(E/\Lambda_{QCD}) \tag{3.6}$$

and saturates the Froissart bound. Saturation of the Froissart bound arises due to a tradeoff of the exponential growth of the number of gluons as a function of rapidity compared to the exponentially falling transverse profile.

4. The Pomeron and the Color Glass Condensate

The Color Glass Condensate description of gluonic matter is given by a renormalization group. The object of computation is the function which controls the fluctuations in the number distribution of gluons $F[\rho]$. If we have a longitudinal momentum cutoff Λ^+, the theory is described by Yang-Mills fields, sources of glue ρ, the Yang-Mills action and $F[\rho]$. The density matrix is

$$\int_{\Lambda^+} [dA][d\rho]\; e^{iS[A,\rho]}e^{-F[\rho]} \tag{4.7}$$

If we define

$$Z[\rho] = e^{-F[\rho]} \tag{4.8}$$

and

$$\eta = ln(1/x) \tag{4.9}$$

then Z satisfies the renormalization group equation

$$\frac{d}{d\eta}Z = -H(\rho, d/d\rho)Z \tag{4.10}$$

This is the equation for Euclidean time evolution of a Hamiltonian system. In this equation, η plays the role of a Euclidean time. For weak sources and small derivatives, this Hamiltonian is

$$\begin{aligned} H &= \kappa\alpha_s \int [dx][dy][dz]\frac{(x-y)^2}{(x-z)^2(y-z)^2} \\ &\frac{d}{d\rho(x)}\left(\rho(x)-\rho(z)\right)\left(\rho(y)-\rho(z)\right)\frac{d}{d\rho(y)} \end{aligned} \tag{4.11}$$

where κ is a numerical constant.

Correlation functions involving the Pomeron are the simplest operators one can construct

$$\rho(x)\cdot\rho(y) \tag{4.12}$$

5. The Pomeron and Saturation

The operators which describe the Pomeron can be generalized to an operator which describes the pomeron in a highly dense and coherent media. It is the so called dipole operator[37, 38]

$$O(x,y) = 1 - \frac{1}{N_c} < tr(U(x)U^\dagger(y)) > \tag{5.13}$$

where

$$U(x) = Pexp\left(i\int dx^- \frac{1}{-\nabla_T^2}\rho(x)\right) \tag{5.14}$$

where $x^\pm = (t \pm z)/\sqrt{2}$. These dipole operators exist only in the transverse plane. They describe a pomeron and arise from considering a dipole pair of quark-antiqaurk in the color glass field. They solve the Balitsky-JIMWLK equation[39, 7, 8, 40, 9]

$$\begin{aligned} \frac{d}{d\eta}O(x,y) &= \int [dz]\frac{(x-y)^2}{(x-z)^2(y-z)^2} \\ &(O(x,z)+O(z,y)-O(x,y)-O(x,z)O(z,y)) \end{aligned} \tag{5.15}$$

The real part of O is the Pomeron amplitude. This equation is the Balitsky-Kovchegov equation in mean field approximation and has been the subject of much study as a paradigm for saturation.

The Balitsky-Kovcehgov equation give exponential growth in y for the Pomeron amplitude at transverse momentum scales larger than that of the saturation momentum, $Q_{sat}(y)$. There is power law growth in y at smaller transverse momenta. The saturation momenta never saturates and grows exponentially in y.

6. The Odderon

The Odderon is the C-odd partner of the pomeron. It is made of three gluons, and in weak field is written as

$$d_{abc}\rho^a(x)\rho^b(y)\rho^c(z) \tag{6.16}$$

In strong field for coupling to mesons, it can be taken to be the imaginary part of the dipole operator O. Recall that the Pomeron is the real part of this operator. For coupling to baryons when the fields are strong in SU(3) gauge theory, it is given by the imaginary part of[12]

$$\epsilon_{\alpha\beta\gamma}\epsilon^{\alpha'\beta'\gamma'}U^{\alpha\alpha'}U^{\beta\beta'}U^{\gamma\gamma'} \tag{6.17}$$

Note that the real part of the baryon amplitude is the topological operator for baryon number. At high energies, the amplitude for the Odderon can be shown to behave as

$$\mid A_{odderon} \mid^2 \sim constant \tag{6.18}$$

The Pomeron is the simplest excitation of the Color Glass Condensate. The Odderon is a closely related, and is the second simplest quasi-particle excitation. The Odderon has never been experimentally probed.

7. Reggeons

There are of course many more quasi-particle excitations of the Color Glass Condensate. Reggeons are object made of N gluons:[41, 42, 43]

$$\rho_1 \cdots \rho_N \tag{7.19}$$

Computing the effects of Reggeons involves solving a N body problem in QCD. This problem may be simpler in the limit of a large number of colors N_c. In large N_c, the Reggeons form a chain with only nearest neighbor interactions. If one further assumes that this amplitude vanishes whenever any two coordinates of the chain are equal, then this JIMWLK equation which describes this evolution is conformally spearable. It has been conjectured by Lipatov-Fadeev-Korchemsky that this system may be exactly solvable and is a conformal field theory. It is not known what happens when the conformal spearability ansatz is relaxed, nor how to go beyond large N_c, nor how to deal with reggeons made of quarks.

8. Ploops

If one has quasi-particle excitations, these excitations may interact among themselves. They may have quantum interactions and generate loops. Loops made of Pomerons, we will call ploops.[44, 45, 46, 13, 14, 15, 16, 17, 18, 19, 20, 21, 22, 23]

The JIMWLK equation has no ploop. It describes tree graphs and has the structure that a j-gluon amplitude describing a pomeron evolves as

$$\frac{d}{dy}A_J = \sum_{j=k}^{\infty} c_k A_k \tag{8.20}$$

The evolution always involves more or equal number gluons than was in the original amplitude. This structure will always result when one considers strong fields, since in this limit the amplitude with larger numbers of gluons in them are of increasing importance.

To generate ploops, one needs to have terms with $k \leq j$ in the sum. These terms are clearly only important in the weak field limit.

We see that the introduction of ploops, introduces the idea of duality. For strong field evolution, the theory involves non-linearities in ρ since the multi-field amplitudes above are generated by inserting more powers of ρ in the JIMWLK Hamiltonian. On the other hand, the weak field limit is generated by powers of $d/d\rho$. A full theory with ploops must include effects of both non-linearities in ρ and non-linearities in $d/d\rho$. There should be some duality relationship between ρ and $d/d\rho$. (After this lecture was prepared, there have appeared many conjectures that ploop theory is self-dual.)

9. The Strongly Interacting QGP and the CGC

There is now compelling evidence that a Quark Gluon Plasma has been produced at RHIC.[24, 25] This has been discussed in several talks at this meeting, and I shall not repeat this discussion here. Suffice it to say, that one the most surprising pieces of information arising from the RHIC experiments is that the matter produced interacts with itself very strongly very early in the collision. This has led many of us to call the matter produced at RHIC as the sQGP: The strongly interacting Quark Gluon Plasma.

The Color Glass Condensate provides a theory of how the Quark Gluon Plasma originates in heavy ion collisions, as shown in Fig. 6. The coherent gluon fields of the Color Glass Condensate's of the two nuclei become time dependent when the nuclei pass through one another. As time passes, these time dependent fields become diluted by expansion and evolve into gluon radiation fields.

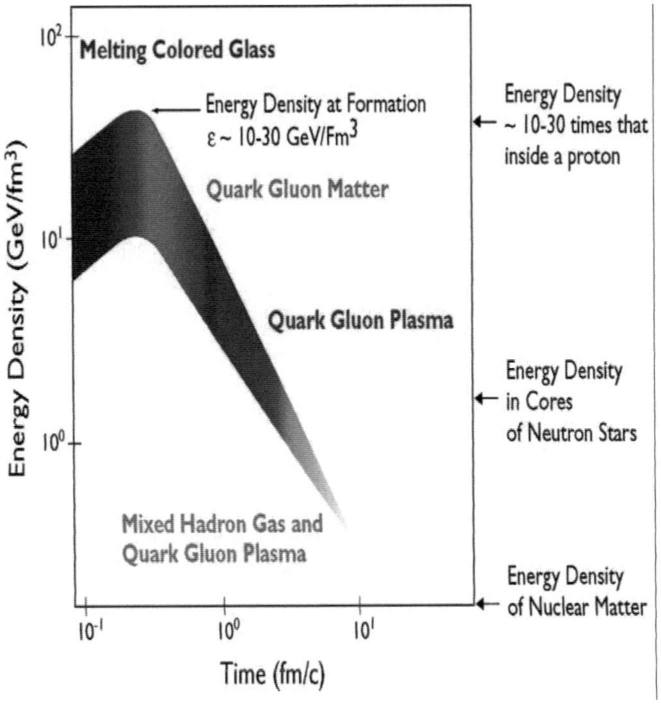

Figure 6. The evolution of matter produced in heavy ion collisions.

The details of this process are interesting. One has highly coherent fields below some momentum scale. These fields have a strength $1/g$ so that a particle interacting with them always has strength $g \times 1/g \sim 1$. When the typical momentum scale associated with these soft fields becomes less than or of the order of the typical momentum scale of all the degrees of freedom, then this strong interaction is of decreasing importance.

The bottom line is, however, that the CGC predicts very strong interactions at early time. The issue is how these fields thermalize. Recently, it has been proposed that these fields evolve into a strongly interacting plasma. This plasma is probably not correctly thought of as thermalized, yet longitudinal plasma instabilities induce a isotropic system. Therefore the magneto-hydrodynamic equation will predict behaviour very similar to that of a system in thermal equilibrium.[27, 28, 29]

This description contrasts with that proposed by Shuryak and collaborators where the system is thermalized by a complicated configuration of resonant states near T_c.[26] The test of these pictures is simple: At LHC where the initial energy per particle is significantly higher, Shuryak's scenario would have less early thermalization than is the case due to plasma instabilities.

10. Perspective for the LHC

The LHC is a machine which will test our ideas of the Color Glass Condensate. In the central region of such collisions, typical $x \sim 10^{-4}$ One can probe much smaller x values and high p_T by measuring produced particles in the forward region. There will be truly excellent test of the ideas of universality and evolution equations.

This is good for the CGC, but will make life a little more complicated for the studies of the QGP. Because we expect that the gluon distribution functions in nuclei will be much modified relative to A times that in a proton, they will have to be measured directly before QGP tests become unambiguous. This requires good pA or dA and pp measurements, or alternatively, good $\gamma - jet$ measurement with realistic extrapolation to Q^2 values of order a GeV. With this information in hand, one has potentially very interesting information about jet quenching due to the QGP or jet modification due to the CGC.

11. Acknowledgments

I have enjoyed enlightening conversations on the subject of this talk with Jean-Paul Blaizot, Yoshi Hatta, Edmond Iancu, Kazu Itakura, Genya Levin, Misha Lublinsky, Al Mueller and Anna Stasto. This manuscript has been authorized by Brookhaven Science Associates, LLC under Contract No. DE-AC02-98CH10886 with the U.S. Department of Energy. The United States Government retains, and the publisher, by accepting the article for publication, acknowledges, a world-wide license to publish or reproduce the published form of this manuscript, or allow others to do so, for the United States Government purposes.

References

[1] E. Iancu, A. Leonidov and L. McLerran, *"The Colour Glass Condensate: An Introduction"*, arXiv:hep-ph/0202270. Published in *QCD Perspectives on Hot and Dense Matter*, Eds. J.-P. Blaizot and E. Iancu, NATO Science Series, Kluwer, 2002;
E. Iancu and R. Venugopalan, *"The Color Glass Condensate and High Energy Scattering in QCD"*, arXiv:hep-ph/0303204. Published in *Quark-Gluon Plasma 3*, Eds. R.C. Hwa and X.-N. Wang, World Scientific, 2003;
H. Weigert, *"Evolution at small x_{bj}: The Color Glass Condensate"*, arXiv:hep-ph/0501087.

[2] L.V. Gribov, E.M. Levin, and M.G. Ryskin, *Phys. Rept.* **100** (1983) 1.

[3] A.H. Mueller and J. Qiu, *Nucl. Phys.* **B268** (1986) 427.

[4] J.-P. Blaizot and A.H. Mueller, *Nucl. Phys.* **B289** (1987) 847.

[5] L. McLerran and R. Venugopalan, *Phys. Rev.* **D49** (1994) 2233; *ibid.* **49** (1994) 3352; *ibid.* **50** (1994) 2225.

[6] Yu.V. Kovchegov, *Phys. Rev.* **D54** (1996) 5463; *Phys. Rev.* **D55** (1997) 5445.

[7] J. Jalilian-Marian, A. Kovner, L. McLerran and H. Weigert, *Phys. Rev.* **D55** (1997) 5414.

[8] J. Jalilian-Marian, A. Kovner, A. Leonidov and H. Weigert, *Nucl. Phys.* **B504** (1997) 415; *Phys. Rev.* **D59** (1999) 014014.

[9] E. Iancu, A. Leonidov and L. McLerran, *Nucl. Phys.* **A692** (2001) 583; *Phys. Lett.* **B510** (2001) 133; E. Ferreiro, E. Iancu, A. Leonidov and L. McLerran, *Nucl. Phys.* **A703** (2002) 489.
[10] E. Iancu and L. McLerran, *Phys. Lett.* **B510** (2001) 145.
[11] L. N. Lipatov, JETP **63** (1986) 904.
[12] Y. Hatta, E. Iancu, K. Itakura and L. McLerran, *"Odderon in the Color Glass Condensate"*, arXiv:hep-ph/0501171.
[13] E. Iancu and D.N. Triantafyllopoulos, *"A Langevin equation for high energy evolution with Pomeron Loops"*, arXiv:hep-ph/0411405 [to appear in *Nucl. Phys.* **A**].
[14] A.H. Mueller, A.I. Shoshi and S.M.H. Wong, *"Extension of the JIMWLK equation in the low gluon density region*, arXiv:hep-ph/0501088.
[15] A.H. Mueller, A.I. Shoshi and S.M.H. Wong, *in preparation.*
[16] E. Levin and M. Lublinsky, *"Towards a symmetric approach to high energy evolution: generating functional with Pomeron loops"*, arXiv:hep-ph/0501173.
[17] E. Levin, *"High energy amplitude in the dipole approach with Pomeron loops: asymptotic solution"*, arXiv:hep-ph/0502243.
[18] E. Iancu and D.N. Triantafyllopoulos, *Phys. Lett.* **B610** (2005) 253.
[19] J.-P. Blaizot, E. Iancu, K. Itakura and D.N. Triantafyllopoulos, *"Duality and Pomeron effective theory for QCD at high energy and large Nc"*, arXiv:hep-ph/0502221 [to appear in *Phys. Lett.* **B**].
[20] A. Kovner and M. Lublinsky, *Phys. Rev.* **D71** (2005) 085004.
[21] Y. Hatta, E. Iancu, L. McLerran, A.M. Stasto, and D.N. Triantafyllopoulos, *"Effective Hamiltonian for QCD evolution at high energy"*, arXiv:hep-ph/0504182.
[22] A. Kovner and M. Lublinsky, *"From target to projectile and back again: selfduality of high energy evolution"*, arXiv:hep-ph/0502119.
[23] A. Kovner and M. Lublinsky, *"Dense-Dilute Duality at work: dipoles of the target"*, arXiv:hep-ph/0503155.
[24] For a review of the theory of the Quark Gluon Plasma as seen at RHIC see the articles in *Nucl. Phys.* **A750** (2005).
[25] For a review of the experimental status of the Quark Gluon Plasma as seen at RHIC see the articles to be published in Nuclear Physics A: Results from the First Three Years at RHIC.
[26] E. V. Shuryak and I. Zahed, PLASMA *Phys. Rev.* **C70** (2004) 021901.
[27] Peter Arnold, Jonathan Lenaghan, Guy D. Moore, *JHEP* (2003) 0308 002.
[28] Peter Arnold, Jonathan Lenaghan, Guy D. Moore, Laurence, G. Yaffe, *Phys. Rev. Lett.* **94** (2005) 072302.
[29] Peter Arnold, Guy D. Moore, Laurence G. Yaffe, hep-ph/0505212.
[30] V.N. Gribov, I.Ya. Pomeranchuk, K.A. Ter-Martirosian, *Phys. Rev.* **139** (1965) 184.
[31] V.N. Gribov, *Sov. Phys. JETP* **26** (1968) 414; *Zh. Eksp. Teor. Fiz.* **53** (1967) 654.
[32] V.N. Gribov, A.A. Migdal, *Sov. J. Nucl. Phys.* **8** (1969) 583; *Yad. Fiz.* **8** (1968) 1002.
[33] V.A. Abramovsky, V.N. Gribov, O.V. Kancheli, *Yad. Fiz.* **18** (1973) 595; *Sov. J. Nucl. Phys.* **18** (1974) 308.
[34] C. Ewerz, *" The odderon in quantum chromodynamics"*, arXiv:hep-ph/0306137.
[35] Alexander Kovner and Urs Achim Wiedemann, *Phys. Lett.* **B551** (2003) 311.
[36] Elena Ferreiro, Edmond Iancu, Kazunori Itakura, Larry McLerran, *Nucl. Phys.* **A710** (2002) 373.
[37] A.H. Mueller, *Nucl. Phys.* **B415** (1994) 373; A.H. Mueller and B. Patel, *Nucl. Phys.* **B425** (1994) 471.
[38] A.H. Mueller, *Nucl. Phys.* **B437** (1995) 107.
[39] I. Balitsky, *Nucl. Phys.* **B463** (1996) 99; *Phys. Lett.* **B518** (2001) 235; *"High-energy QCD and Wilson lines"*, arXiv:hep-ph/0101042.
[40] J. Jalilian-Marian, A. Kovner and H. Weigert, *Phys. Rev.* **D59** (1999) 014015; A. Kovner, J. G. Milhano and H. Weigert, *Phys. Rev.* **D62** (2000) 114005.
[41] L.N. Lipatov, *Nucl. Phys.* **B452** (1995) 369.
[42] L.N. Lipatov, *JETP Lett.* **59** (1994) 596 *Zh. Eksp. Teor. Fiz.* **59** (1994) 571.
[43] L.D. Faddeev and G.P. Korchemsky, MODEL. *Phys. Lett.* **B342** (1995) 311.
[44] A.H. Mueller and A.I. Shoshi, *Nucl. Phys.* **B692** (2004) 175.
[45] E. Iancu, A.H. Mueller and S. Munier, *Phys. Lett.* **B606** (2005) 342.
[46] E. Iancu, K. Itakura, and D.N. Triantafyllopoulos, *Nucl. Phys.* **A742** (2004) 182.

Institute of Physics Publishing Journal of Physics: Conference Series **50** (2006) 168–175
doi:10.1088/1742-6596/50/1/020 5th International Conference on Physics and Astrophysics of Quark Gluon Plasma

J/ψ hadron interaction in vacuum and in QGP

Su Houng Lee[1] **and Taesoo Song**[2]

Institute of Physics and Applied Physics, Yonsei University, Seoul 120-749, Korea

Abstract. I first discuss the NLO QCD calculation for the dissociation cross section of J/ψ into open charm, which shows a large second order correction near the threshold. The large correction softens when effective thermal masses in the order of the binding energy are introduced for the quarks and gluons. Hence, with the thermal masses, the method is reliably applied to calculate the dissociation cross section of J/ψ, which according to recent lattice results seems to remain bound even up to $1.6 \times T_c$. The dominant contribution to the dissociation comes from the thermal gluons, while the effects due to thermal quarks are suppressed due to suppressed phase space. The relevance of this result is discussed in the context of RHIC and signature of QGP.

1. Quarkonium hadron interaction in QCD

Ever since J/ψ suppression was suggested as a signature of quark gluon plasma formation in the early stages of a relativistic heavy-ion collision[1], much work have been performed in estimating the quarkonium hadron interaction in QCD. This is so because J/ψ suppression is also expected to occur in the hadronic environment due to the absorption by the nucleons and by the comoving hadrons. Moreover, recent quenched lattice calculations show that the J/ψ remains bound even up to 1.6 to 2 $\times T_c$[2, 3, 4, 5]. Therefore, calculating the dissociation cross section of bound J/ψ by thermal quarks and gluons became a necessity because such information are crucial in estimating the hitherto neglected effects of the dynamical quarks on the bound state from lattice calculation, and in quantitatively obtaining the survival probability of the bound state in the QGP state. To address all of these questions in QCD, the basic ingredients are the dissociation cross sections of quarkonium by partons. The hadronic cross sections are then obtained by folding the elementary partonic cross section by the parton distribution amplitudes inside the hadrons, while the thermal cross section are obtained by folding it with the thermal distributions inside the QGP.

Many model calculations have been performed to calculate the dissociation cross section by hadrons. As can be seen from fig.1 for the dissociation cross section of J/ψ by pion, the calculations from each model vary greatly in their energy dependence and magnitude near the threshold. To understand the discrepancies and to obtain a consistent result, it is crucial to probe each model calculations further so as to determine the uncertainty and the valid energy range of each model calculations. The LO QCD approach was first performed by Peskin more than 25 years ago[6, 7]. The NLO calculation was performed only recently by making use of the Bethe-Salpeter amplitude and QCD factorization theorem[8]. As I will discuss in the next section, the second order correction becomes very large near threshold. This suggests that the QCD results are not reliable near threshold, and hence partly explains the differences with other model calculations in this energy region. The large second order correction in QCD can be traced down to a logarithmic behavior of the energy of the outgoing partons scaled with the binding energy. Therefore, for the charmonium case, where the binding energy is less than 1 GeV,

[1] Email: suhoung@yonsei.ac.kr.
[2] Email: songtsoo@yonsei.ac.kr.

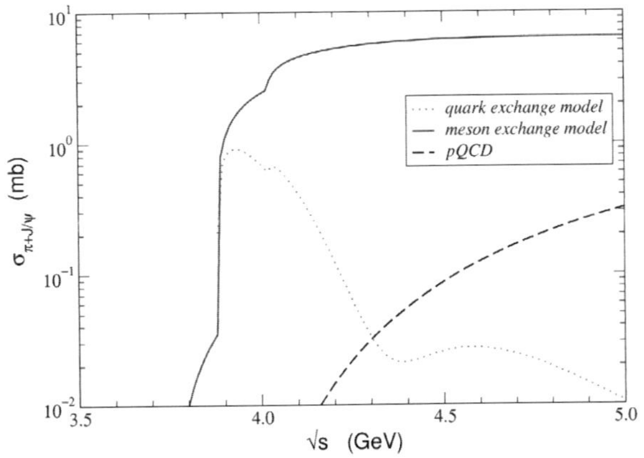

Figure 1. $\sigma_{J/\psi+\pi}$ cross section from meson exchange model [9], quark exchange model [10] and perturbative QCD [7] calculations.

introducing an effective thermal mass of few hundred MeV will substantially reduce the large fluctuation appearing near the threshold, which leads to a much more reliable QCD prediction. This fact enables us to use perturbative methods to calculate the dissociation cross section of J/ψ by the thermal gluons and quarks above the critical temperature T_C, and to estimate the effective width of the surviving bound states.

2. Quarkonium hadron interaction in QCD

2.1. Heavy quark propagation

Let us begin with some introduction on the propagation of heavy quarks in the QCD vacuum. The propagation of a heavy quark can be approximated by a perturbative quark propagation with a perturbative gluon insertion, which probes the non perturbative gluon field configuration in the QCD vacuum. Hence, the full heavy quark propagator is,

$$iS^A(q) = iS(q) + iS(q)(-ig\not{A})iS(q)\cdots, \tag{1}$$

where, $iS(q) = i/(\not{q} - m)$ and m is the heavy quark mass. The description in Eq.(1) is valid even for $q \to 0$, because $m \gg A \sim \Lambda_{QCD}$, where in the end only gauge invariant combination of the gauge field A will remain after taking the vacuum expectation value.

2.2. System with two heavy quarks

The propagation of a system composed of a heavy quark and an antiquark can also be approximated by combined perturbative heavy quark propagator with gluon insertions. However, since there a two heavy quarks involved, based on the operator product expansion, the propagation can be typically written in the following form.

$$\Pi(q) = ... + \int_0^1 dx \frac{F(q^2, x)}{(4m^2 - q^2 - (2x-1)^2 q^2)^n} \langle G^n \rangle \cdots, \tag{2}$$

where, $F(q^2, x)$ is a function depending on the structure of the two quark system and $\langle G^n \rangle \sim \Lambda_{QCD}^{2n}$ denotes the typical gauge invariant expectation value of gluonic operator of dimension $2n$.

Table 1. Physical processes involving two heavy quarks where $4m^2 - q^2 \gg \Lambda_{QCD}^2$ and where a perturbative treatment such as Eq.(2) are possible.

q^2	Process	Expansion parameter
0	Photo production of open charm	$\frac{\Lambda_{QCD}^2}{4m^2}$
$-Q^2 < 0$	QCD sum rules for heavy quark system	$\frac{\Lambda_{QCD}^2}{4m^2+Q^2}$
$m_{J/\psi}^2 > 0$	Dissociation cross section of bound states	$\frac{\Lambda_{QCD}^2}{4m^2-m_{J/\psi}^2}$

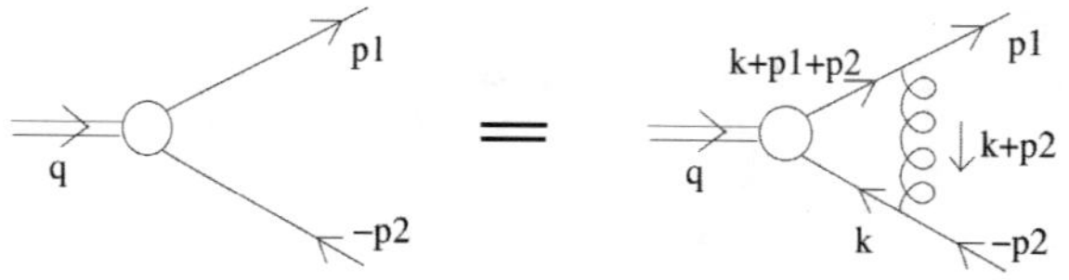

Figure 2. The Bethe-Salpeter equation for quarkonium.

The integration variable x can be thought of as the momentum fraction carried by one of the heavy quark. Here, one notes that such perturbative expansion is valid when $4m^2 - q^2 \gg \Lambda_{QCD}^2$. There are several cases where this condition is satisfied and perturbative QCD treatments are possible. Those cases are summarized in Table 1.

In the last line of Table 1, $4m^2 - m_{J/\psi}^2 \approx (2m + m_{J/\psi})\epsilon_0$, where ϵ_0 is the binding energy of the J/ψ. In QCD if $m \to \infty$, the bound state becomes Coulombic and $\epsilon_0 = m[N_c g^2/(16\pi)]^2 \gg \Lambda_{QCD}$ in the large N_c limit. Therefore, the expansion parameter becomes small and the dissociation cross can be calculated using perturbative QCD.

2.3. Historical perspective on LO calculation

The LO calculation for the dissociation of quarkonium by hadrons have been performed by Peskin[6] and applied to the charmonium system by Bhanot and Peskin[7] more than 25 years ago. The result have been rederived by Kharzeev and Satz[11] and by Arleo *et al.*[12], and used in the analysis of J/ψ suppression in heavy ion collision. All the derivation were based on calculating the real part of the correlation function using operator product expansion. Therefore, the derivation was rather time consuming and difficult to apply for a NLO calculation. Recently, we have derived anew the leading order pQCD result using the QCD factorization theorem [13]. Here, we will sketch the derivation. As in the original derivation by Peskin, one should note that the time scale of interaction to form the bound state, should be much smaller than that between the dipole and the hadrons. In momentum scale, it means that the bound state are produced by momentum scales of order $p \sim O(mg^2)$, while the typical scale for the interaction between the bound state and external hadron should be of order $\epsilon_0 \sim O(mg^4)$, which will be the separation scale in the OPE. Since the momentum scale of the bound state are larger than the separation scale, the bound state should be treated perturbatively, as in the Wilson coefficient. Hence, the bound state will be introduced as a vertex using the Bethe-Salpeter equation.

The Bethe-Salpeter equation, represented diagrammatically in Fig. 2, can be written as

$$\Gamma_\mu(p_1, -p_2) = -ig^2 C_F \int \frac{d^4K}{(2\pi)^4} \gamma^\alpha i\Delta(K + p_1 + p_2)\Gamma_\mu(K + p_1 + p_2, K)$$

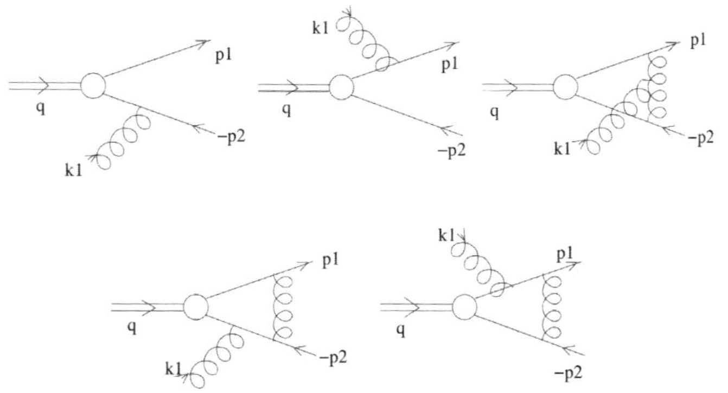

Figure 3. Leading order diagrams for $\Phi + g \to Q + \bar{Q}$. The two lower diagrams are suppressed in the large N_c limit.

$$\times i\Delta(K)\gamma_\alpha V(K + p_2), \tag{3}$$

where $C_F = (N_c^2 - 1)/N_c$, $i\Delta(K)$ is the quark propagator, and $iV(K)$ the gluon propagator. Changing variables to $q = p_1 + p_2$, $p \equiv (p_1 - p_2)/2$ and working in the rest frame of the bound state Φ, the BS amplitude reduces to the Schördinger equation with a relative bound state wave function $\psi(|\vec{p}|)$. Then the BS vertex can be written as,

$$\Gamma_\mu\left(\frac{q}{2} + p, -\frac{q}{2} + p\right) = \left(\epsilon_o + \frac{\vec{p}^2}{m}\right)\psi(|\vec{p}|)\sqrt{\frac{m_\Phi}{N_c}}\frac{1+\gamma^0}{2}\gamma_i g^i_\mu \frac{1-\gamma^0}{2}. \tag{4}$$

where $\epsilon_0 = 2m - m_\phi$. Using this BS vertex, one can obtain the LO matrix element for the LO graph given in Fig.3 by observing the following counting scheme.

$$|\vec{p_1}| \sim |\vec{p_2}| \sim O(mg^2), \qquad k^0 = |\vec{k}| \sim O(mg^4). \tag{5}$$

These are obtained from noting that the binding energy $\epsilon_o = m(N_c g^2/16\pi)^2 \sim O(mg^4)$ and that the three momentum of the heavy quarks are of $O(mg^2)$, and also using the non-relativistic energy conservation $m_\Phi + k_0 = 2m + |\vec{p_1}|^2/2m + |\vec{p_2}|^2/2m$.

3. Next to leading order result

The NLO result has been obtained in Ref.[8]. At this order, there are contributions with three body final states. These are $\Phi + q(g) \to Q + \bar{Q} + q(g)$. The counting scheme is similar to that of the LO with the incoming and outgoing parton momentum all of $O(mg^4)$. That is,

$$|\vec{p_1}| \sim |\vec{p_2}| \sim O(mg^2), \qquad k_1^0 = |\vec{k_1}| \sim k_2^0 = |\vec{k_2}| \sim O(mg^4), \tag{6}$$

where k_1 is the incoming quark (or gluon) momentum, and k_2 is the outgoing quark (or gluon) momentum. The full calculation have been performed recently by T. Song and S. H. Lee[8]. Here, I will only highlight some findings.

3.1. Divergences

There are divergences that appear in the intermediate stages of the calculation. As an example, Fig.4 is the lowest order diagram that involves an incoming quark. If the quarks and gluons are massless, the cross section coming from Fig.4 has collinear divergence when the outgoing

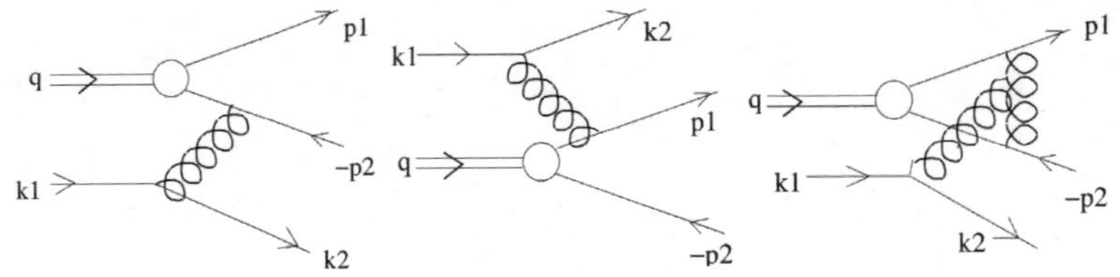

Figure 4. Next to leading order diagram with initial quark $\Phi + q \rightarrow Q + \bar{Q} + q$.

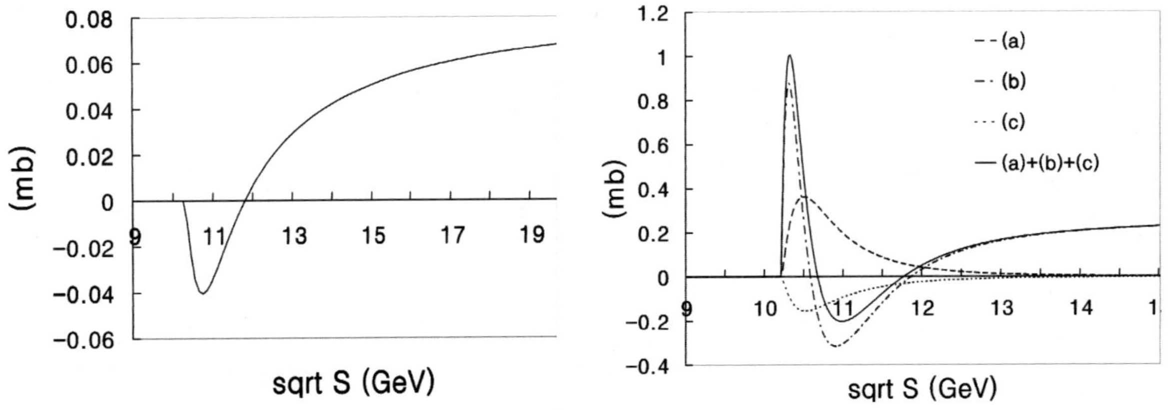

Figure 5. The left figure is the elementary cross section for $\Phi + q \rightarrow Q + \bar{Q} + q$ and the right figure is that for $\Phi + g \rightarrow Q + \bar{Q} + (g)$. The meaning of each label in the right figure is explained in [8].

quark momentum k_2 is parallel to k_1. These divergences disappear after mass factorization, which is a procedure to subtract out the contributions from the radiating gluons with transverse momentum smaller than the separation scale, and putting those into the gluon distribution function.

There are other soft and mixed soft-collinear divergences appearing in the intermediate stages of the calculations. However, those can be shown to disappear in the final answer[8].

3.2. Application to Upsilon dissociation cross section

Here we show the result applied to the upsilon dissociation cross section. The two independent parameters of the theory is determined by fitting the physical masses of $m_{\Upsilon(1S)}$ and $m_{\Upsilon(2S)}$ to the energies of the Coulomb bound states. Specifically, from the relation $m_{\Upsilon(2S)} - m_{\Upsilon(1S)} = 3/4\epsilon_o$, the binding energy is found to be $750 MeV$. Also, the bottom quark mass is found to be 5.1 GeV from equating it to $(m_{\Upsilon(1S)} + \epsilon_o)/2$. The coupling constant g is then found to be 2.53 from $g^2 = 16\pi/N_c\sqrt{\epsilon_o/m}$.

Fig.5 is the plot for the elementary total cross section of $\Phi+q \rightarrow Q+\bar{Q}+q$ and $\Phi+g \rightarrow Q+\bar{Q}+g$ as a function of the incoming energy. In the right figure, the different lines represent different parts of the diagrams explained in [8]. In both figures, there are some energy regions where the cross sections are negative. The negative values come from mass factorization, because finite parts of the cross sections have been subtracted out and placed in the definition of the distribution function. Therefore, the cross section becomes physical only after folding the elementary cross sections with the parton distribution functions (PDF), and adding those to the LO contribution. Nonetheless, it is clear that there are large fluctuation near the threshold

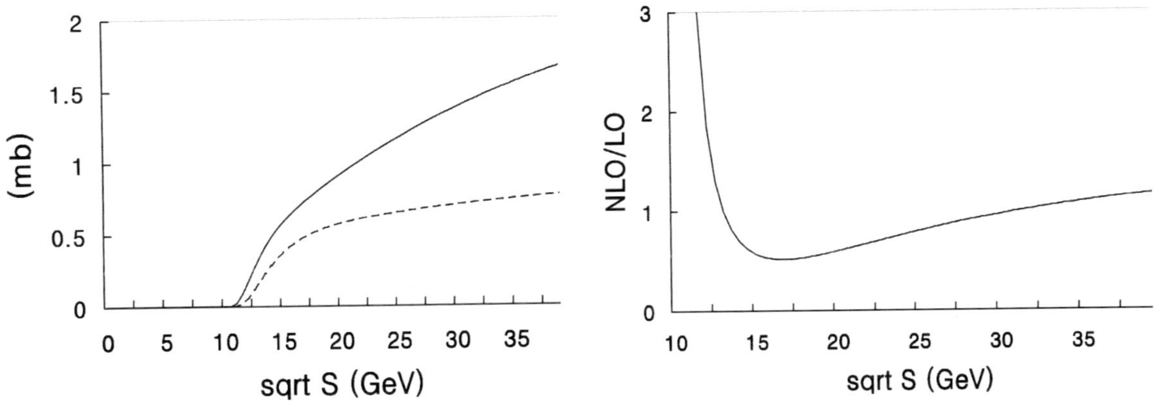

Figure 6. The left figure is the $\Upsilon(1S)$+nucleon total cross section to LO (dashed line) and to NLO (solid line). The right figure is the corresponding ratio between NLO and LO results.

and that the contributions from the gluon induced reactions are much larger than that from the quark.

The left figure of Fig.6 shows the LO and the NLO total dissociation cross section, while the right figure shows their ratio. As can be seen in the figure, the perturbative QCD approach is acceptable only in some limited energy region, and has large higher order corrections in the threshold region.

4. Lessons to learn

4.1. Gluon vs quark induced dissociation

As can be seen from Fig.5, the contribution from gluons are an order of magnitude larger than that from the quarks. The reason for such difference is not restricted to the case of Upsilon system, but is a general result coming from the following facts.

- There are many more gluon induced diagrams.
- Even for the diagrams in Fig.4, the contribution is larger when the incoming and outgoing quarks are replaced by gluons.

4.2. Origin of large correction

The reason for large fluctuation of the cross section near threshold can be traced down to the logarithmic behavior of the energy of the outgoing massless parton. The explicit form for Fig.4 is given by,

$$\log\left[\frac{2k_{2,0}}{\epsilon_0}\right] \tag{7}$$

where $k_{2,0}$ is the energy of the outgoing parton and ϵ_0 is the binding energy of the bound state. This term varies from -∞ to 0, as $k_{2,0}$ changes from 0 to $\epsilon_0/2$, which corresponds to the threshold region.

For the J/ψ, $\epsilon_0 \approx$ 700MeV. Therefore, an effective thermal mass of few hundred MeV is large enough for the quarks and gluons to tame the fluctuation and to make the higher order corrections convergent near the threshold. Such thermal mass occur in the QGP state at temperatures higher than the critical temperature $T > T_c$.

5. Dissociation of J/ψ due to thermal gluons and quarks

The previous discussions justifies the use of perturbative QCD method to estimate the dissociation of J/ψ by thermal quarks and gluons.

5.1. Gluon vs quark effects

As discussed in the previous section, a thermal mass for the quark and gluon of a few hundred MeV is enough to make the perturbative calculation meaningful. Moreover, from general grounds, the order α^2 dissociation induced by quarks are much smaller than that from gluons. Hence the thermal width acquired by J/ψ from quarks will be much smaller than that from the gluons.

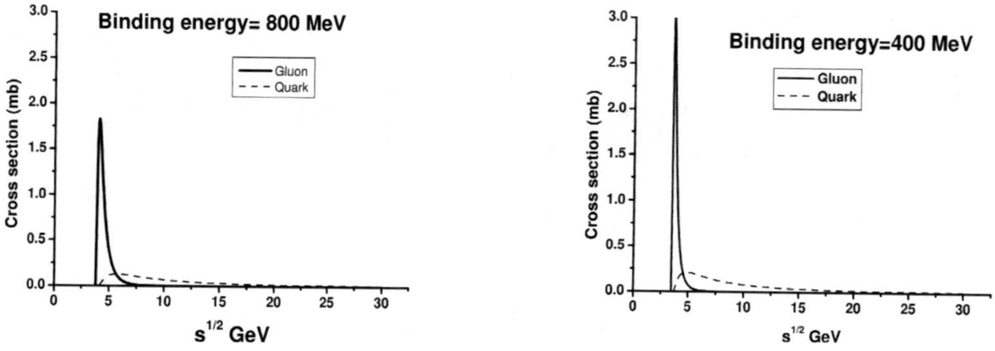

Figure 7. Lowest order dissociation cross section of J/ψ by gluons and quark.

Fig.7 shows the elementary dissociation cross section with an effective thermal mass of 200 MeV for the quarks and gluons. The Solid (dashed) line represents the lowest order dissociation due to a gluon (quark). For the gluon, the lowest order appears at order α, while for the quarks, it does at order α^2. The parameters have been fitted to the charmonium mass assuming a Coulomb bound state. Such procedure gives 800 MeV for the binding, but the value have been artificially changed to 400 MeV in producing the cross sections shown in the right figure of Fig.7.

Fig.8 represents the effective thermal cross section of gluons and quarks calculated from the following formula.

$$\langle\sigma\rangle = \frac{\int \sigma(p) p^2 f(p) dp}{\int p^2 f(p) dp}, \tag{8}$$

where $f(p)$ is the thermal distribution of the parton.

Finally, Fig.9 represents the effective thermal width calculated from the following simple formula.

$$\Gamma_{eff} = deg \cdot \int \sigma(p) f(p) p^2 dp, \tag{9}$$

where *deg* is the degrees of freedom of the parton, and we have assumed the relative velocity to be *c*. We can conclude the following.

(i) The effects of quarks to the thermal width of the J/ψ are negligible compared to that of gluons. Therefore, the conclusions about the fate of J/ψ up to $2 \times T_c$ from quenched lattice result, will not be modified by the introduction of dynamical fermions.

(ii) The thermal width becomes in the order of 1 GeV at temperatures of above 600 MeV.

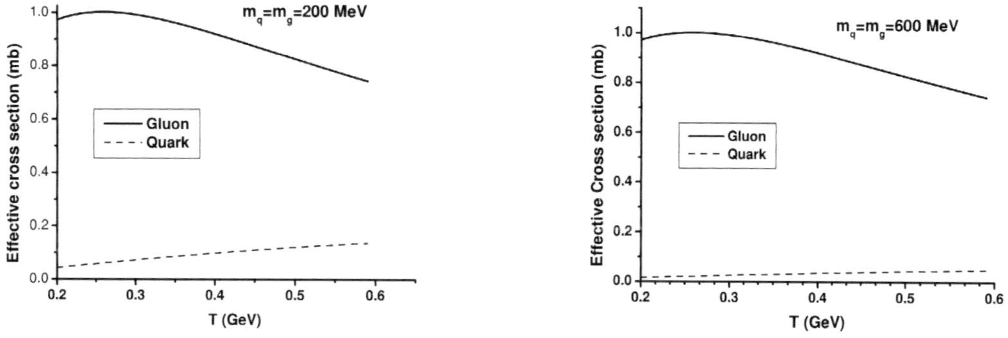

Figure 8. Effective thermal cross sections.

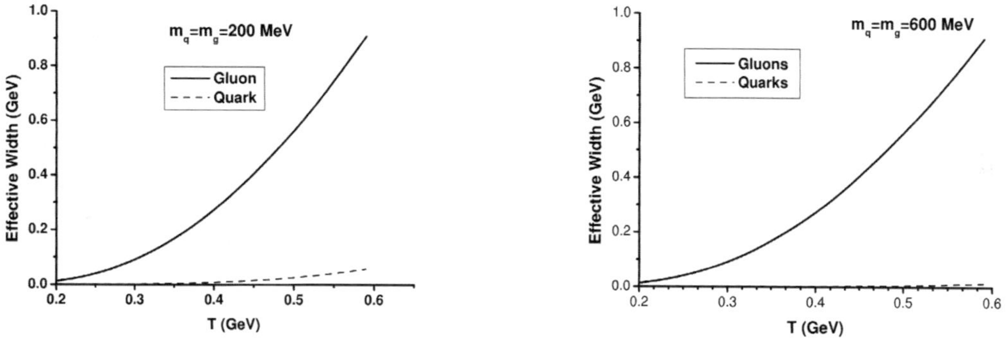

Figure 9. Effective thermal width of J/ψ at finite temperature.

Acknowledgments
I would like to thank the organizers for inviting me to this exciting conference.

References
[1] Matsui T and Satz H 1986 *Phys. Lett.* B **178** 416
[2] Asakawa M, Hatsuda T and Nakahara Y 2003 *Nucl. Phys.* A **715** 863
[3] Asakawa M and Hatsuda T 2004 *Phys. Rev. Lett.* **92** 012001
[4] Datta S, Karsch F, Petreczky P and Wetzorke I 2004 *Phys. Rev.* D **69** 094507
[5] Datta S, Karsch F, Petreczky P and Wetzorke I 2005 *J. Phys.* G **31** S351
[6] Peskin M E 1979 *Nucl. Phys.* **B156** 365
[7] Bhanot G and Peskin M E 1979 *Nucl. Phys.* **B156** 391
[8] Song T and Lee S H hep-ph/0501252
[9] Matinyan S G and Müller B 1998 *Phys. Rev.* C **58** 2994. Haglin K L 2000 *Phys. Rev.* **61** 031902. Lin Z and Ko C M 2000 *Phys. Rev.* **62** 034903. Oh Y, Song T and Lee S H 2001 *Phys. Rev.* **63** 034901. Sibirtsev A, Tsushima K and Thomas A W 2001 *Phys. Rev.* **63** 044906. Liu W, Ko C M and Lin Z W 2001
[10] Wong C-Y, Swanson E S and Barnes T E 2000 *Phys. Rev.* C **62** 045201. Wong C-Y, Swanson E S and Barnes T E 2002 *Phys. Rev.* C **65** 014903.
[11] Kharzeev D and Satz H 1994 *Phys. Lett.* B **334** 155
Kharzeev D, Satz H, Syamtomov A and Zinovjev G 1996 *Phys. Lett.* B **389** 595
[12] Arleo F, Gossiaux P-B, Gousset T and Aichelin J 2002 *Phys. Rev.* D **65** 014005
[13] Lee S H, Oh Y 2002 *J. Phys.* G **28** 1903. Oh Y S, Kim S and Lee Su Houng 2002 *Phys. Rev.* C **65** 067901

Strangeness and the Discovery of Quark-Gluon Plasma

Johann Rafelski[1], **Jean Letessier**[2]

[1]Department of Physics, University of Arizona, TUCSON, AZ 85718, USA

E-mail: `Rafelski@Physics.Arizona.EDU`

[2]Laboratoire de Physique Théorique et Hautes Energies
Université Paris 7, 2 place Jussieu, 75251 Cedex 05, France

E-mail: `JLetes@LPTHE.Jussieu.FR`

Abstract. Strangeness flavor yield s and the entropy yield S are the observables of the deconfined quark-gluon state of matter which can be studied in the entire available experimental energy range at AGS, SPS, RHIC, and, in near future, at the LHC energy range. We present here a comprehensive analysis of strange, soft hadron production as function of energy and reaction volume. We discuss the physical properties of the final state and argue how evidence about the primordial QGP emerges.

1. Introduction

The deconfined interacting quark–gluon plasma phase (QGP) is the *equilibrium* state of matter at high temperature and/or density. It is believed that this state has been present in the early Universe, 10-20μs into its evolution. The question is if, in the short time, 10^{-22}–10^{-23} s, available in a laboratory heavy ion collision experiment, the color frozen nuclear phase can melt and turn into the QGP state of matter. There is no valid first principles answer to this question available today, nor as it seems, will a first principles simulation of the dynamic heavy ion environment become available in the foreseeable future. To address this issue we study QGP experimentally, which requires development of laboratory experiments and suitable observables.

To form QGP in the laboratory we perform relativistic heavy ion collisions in which a domain of (space, time) much larger than normal hadron size is formed, in which color-charged quarks and gluons are propagating constrained by external 'frozen vacuum', which abhors color [1]. We expect a pronounced boundary in temperature and baryon density between confined and deconfined phases of matter, irrespective of the question if there is, or not, a true phase transition. We search for a boundary between phases considering the size of the interacting region and the magnitude of the reaction energy.

Detailed study of the properties of the deconfined state shows that QGP is rich in entropy and strangeness. The enhancement of entropy S arises because the color bonds are broken and gluons can be created. Enhancement of strangeness s arises because the mass threshold for strangeness excitation is considerably lower in QGP than in hadron matter. Moreover there are new mechanisms of strangeness formation in QGP involving reactions between (thermal) gluons. Thus S and s are the two elementary observables which are explored with soft hadronic probes, for further theoretical details and historical developments see our book [2]. The numerical work presented here was carried out with the public package of programs SHARE [3]. This report is a self contained summary of our recent results, see [4, 5, 6, 7].

Entropy enhancement, observed in terms of enhanced hadron multiplicity per net charge, has been among the first indications of new physics reach of CERN-SPS experimental heavy ion program [8]. The enhancement of strange hadron production both as function of the number of participating baryons, and reaction energy has been explored in several experiments at at BNL-AGS, CERN-SPS and and BNL-RHIC. We refrain from extensive historical survey of these results and present perhaps the latest, STAR-RHIC result in Fig. 1 [9]. In this presentation one sees the yield per participant $N_{\rm part}$ divided by a reference yield obtained in pp reactions. We observe that the enhancement rises both with the strangeness content in the hadron and with the size of the reaction region, indicating that the cause of this enhancement is a increased yield of strange quarks in the source, a qualitative expectation we will address in our quantitative analysis below. The gradual increase of the enhancement over the range of $N_{\rm part}$ is an important indicator of the physics mechanisms at work. This behavior agrees with our studies of kinetic strangeness production and strangeness yield increasing with the size of the reaction region. This enhancement of strange antibaryons which demonstrates that a novel strangeness production mechanism is present has been extensively studied at SPS energy range, where it was originally discovered [10, 11].

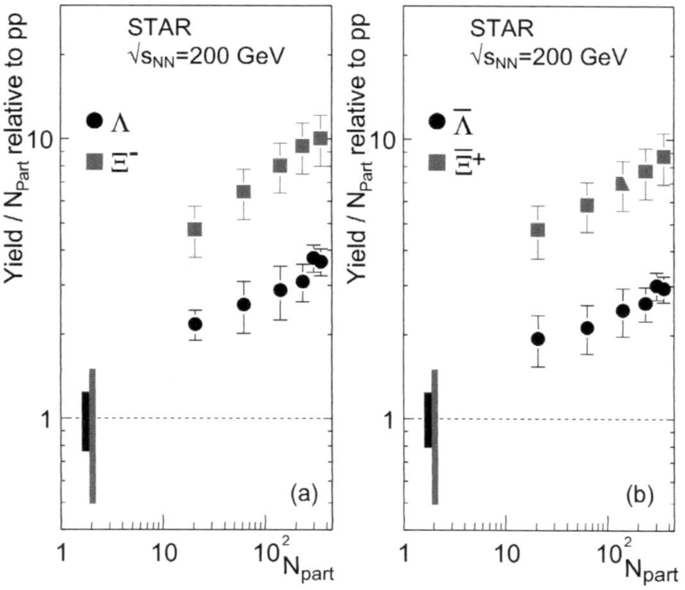

Figure 1. Yields per participant $N_{\rm part}$ relative to pp of Λ and Ξ^- on left and $\bar{\Lambda}$ and $\bar{\Xi}^+$ on right in Au+Au collisions at $\sqrt{s_{\rm NN}} = 200$ GeV Error bars are statistical. Ranges for pp reference data at $N_{\rm part} = 2$ indicate the systematic uncertainty.

Further evidence for parton dynamics prior to final state hadronization is obtained from the study of strange hadron transverse energy spectrum. The identity of hyperon and antihyperon spectra in particular $\Lambda, \overline{\Lambda}$ and $\Xi, \overline{\Xi}$ implies that both strange matter and antimatter must have been produced from a common source by the same fundamental mechanism. Furthermore they were not subject to interactions in their passage through the baryon rich hadron gas present at the SPS energy range.

The $\overline{\Lambda}, \overline{\Xi}$ annihilation in baryon-rich hadron gas is strongly momentum dependent. This should deform the shape of the antihyperon spectra as compared to the spectra of hyperons. Thus symmetry of the hyperon-antihyperon spectra also implies that there was no appreciable annihilation of the $\overline{\Lambda}, \overline{\Xi}$ after their formation. The working hypothesis is therefore that hadronization of the QGP deconfined phase formed in high energy nuclear collision is direct, fast (sudden) and occurs without significant sequel interactions. That can be further tested in a study of yields and spectra of unstable resonances [12].

At RHIC the parton level dynamics is convincingly demonstrated by quark content scaling of azimuthal asymmetry of the collective flow v_2. Further evidence is derived from the consideration of quark recombination formation of hadrons from QGP. We refer to the recent comprehensive survey of the RHIC result [1], and presentations at the meeting addressing these very interesting and recent developments. We will assume in this report that the case of quark-parton dynamics prior to hadronization is convincing, and will use our analysis of soft hadron production to find the thresholds of the onset of deconfinement and to determine the properties of the deconfined fireball at the time of its breakup into hadrons.

In next section 2 we introduce the statistical hadronization method of analysis of hadron production. We discuss data analysis as function of impact parameter and energy dependence of soft hadron production in section 3. We will present both the systematics of statistical model parameters, and the associated physical properties. In section 4, we address the physical QGP signatures indicating presence of a phase boundary, giving particular attention to the explanation of the 'horn' in the K^+/π^+ ratio, and the strangeness to entropy relative yield. We discuss, in section 5, the role these results play in understanding of the phase boundary separating QGP from normal confined matter. We also discuss briefly possible new soft hadron physics at LHC.

2. Statistical hadronization model

To describe the yields of particle produced we employ the statistical hadronization model (SHM). SHM is by definition a model of particle production in which the birth process of each particle fully saturates (maximizes) the quantum mechanical probability amplitude, and thus, the relative yields are determined solely by the appropriate integrals of the accessible phase space. For a system subject to global dynamical evolution such as collective flow, this is understood to apply within each local co-moving frame.

When particles are produced in hadronization, we speak of chemical freeze-out. Hadron formation from QGP phase has to absorb the high entropy content of QGP which originates in broken color bonds. The lightest hadron is pion and most entropy per energy is consumed in hadronization by producing these particles abundantly. It is thus important to free the yield of these particles from the chemical equilibrium constraint.

The normalization of the particle yields is, aside of the freeze-out temperature T, directly controlled by the particle fugacity $\Upsilon_i \equiv e^{\sigma_i/T}$, where σ_i is particle 'i' chemical potential. Since for each related particle and antiparticle pair, we need two chemical potentials, it has become convenient to choose parameters such that we can control the difference, and sum of these separately. For example, for nucleons N, and, respectively, antinucleons $\overline{N}$ the two chemical factors are:

$$\sigma_N \equiv \mu_b + T \ln \gamma_N, \qquad \sigma_{\overline{N}} \equiv -\mu_b + T \ln \gamma_N, \tag{1}$$

$$\Upsilon_N = \gamma_N e^{\mu_b/T}, \qquad \Upsilon_{\overline{N}} = \gamma_N e^{-\mu_b/T}. \tag{2}$$

The (baryo)chemical potential μ_b, controls the baryon number, arising from the particle difference. γ_N, the phase space occupancy, regulates the number of nucleon–antinucleon pairs present. There are many different hadrons, and in principle, we could assign to each a chemical potential and then look for chemical reactions which relate these chemical potentials. However, a more direct way to accomplish the same objective consists in characterizing each particle by the valance quark content. The relation between quark based fugacity and chemical potentials ($\lambda_{q,s} = e^{\mu_{q,s}/T}$) and the two principal hadron based chemical potentials of baryon number and hadron strangeness μ_i, $i = b, \mathrm{S}$ is:

$$\mu_b = 3\mu_q \qquad \mu_s = \frac{1}{3}\mu_b - \mu_\mathrm{S}, \qquad \lambda_s = \frac{\lambda_q}{\lambda_\mathrm{S}}. \tag{3}$$

An important (historical) anomaly is the negative S-strangeness in s-carrying-baryons. We will in general follow quark flavor and use quark chemical factors to minimize the confusion arising. In the local rest frame, the particle yields are proportional to the momentum integrals of the quantum distribution. As example, for the yield of pions π, nucleons N and antinucleons $\overline{N}$ we have:

$$\pi = Vg_\pi \int \frac{d^3p}{(2\pi)^3} \frac{1}{\gamma_q^{-2} e^{E_\pi/T} - 1}, \qquad E_i = \sqrt{m_i^2 + p^2}, \quad \gamma_q^2 < e^{m_\pi/T} \tag{4}$$

$$N = Vg_N \int \frac{d^3p}{(2\pi)^3} \frac{1}{\gamma_q^{-3} \lambda_q^{-3} e^{E_N/T} + 1}, \qquad \overline{N} = Vg_N \int \frac{d^3p}{(2\pi)^3} \frac{1}{\gamma_q^{-3} \lambda_q^{+3} e^{E_{\overline{N}}/T} + 1}. \tag{5}$$

There are two types of chemical factors γ_i and μ_i, and thus two types of chemical equilibriums. These are shown in table 1. The absolute equilibrium is reached when the phase space occupancy approaches unity, $\gamma_i \to 1$. The distribution of flavor (strangeness) among many hadrons is governed by the relative chemical equilibrium.

Table 1. Four quarks s, $\bar{s}$, q, and $\bar{q}$ require four chemical parameters; right: name of the associated chemical equilibrium

γ_i	controls overall abundance of quark ($i = q, s$) pairs	Absolute chemical equilibrium
λ_i	controls difference between quarks and antiquarks ($i = q, s$)	Relative chemical equilibrium

In order to arrive at the full particle yield, one has to be sure to include all the hadronic resonances which decay feeding into the yield considered, *e.g.*, the decay $K^* \to K + \pi$ feeds into K and π yields. The contribution is sensitive to temperature at which these particles are formed. Inclusion of the numerous resonances constitutes a book keeping challenge in study of particle multiplicities, since decays are contributing at the 50% level to practically all particle yields. A public statistical hadronization program, SHARE (Statistical HAdronization with REsonances) has simplified this task considerably [3].

The resonance decay contribution is dominant for the case for the pion yield. This happens even though each resonance contributes relatively little in the final count. However, the large number of resonances which contribute compensates and the sum of small contributions competes with the direct pion yield. For the more heavy hadrons, generally there is a dominant contribution from just a few, or even from a single resonance. The exception are the $\Omega, \overline{\Omega}$ which have no known low mass resonances, and also ϕ – the known resonances are very heavy and very few.

A straightforward test of sudden hadronization and the SHM is that within a particle 'family', particle yields with same valance quark content are in relation to each other well described by integrals of relativistic phase space. The relative yield of, *e.g.*, $K^*(\bar{s}q)$ and $K(\bar{s}q)$ or Δ and N are controlled by the particle masses m_i, statistical weights (degeneracy) g_i and the hadronization temperature T. In the Boltzmann limit one has (star denotes the resonance):

$$\frac{N^*}{N} = \frac{g^* {m^*}^2 K_2(m^*/T)}{g\, m^2 K_2(m/T)}. \tag{6}$$

Validity of this relation implies insensitivity of the quantum matrix element governing the coalescence-fragmentation production of particles to intrinsic structure (parity, spin, isospin), and particle mass. The measurement of the relative yield of hadron resonances is a sensitive test of the statistical hadronization hypothesis and lays the foundation to the application of the method in data analysis.

The method available to measure resonance yields depends in its accuracy significantly on the the precise nature of the hadronization process: the resonance yield is derived by reconstruction of the invariant mass of the resonance from decay products energies E_i and momenta p_i. Should the decay products of resonances after formation rescatter on other particles, then often their energies and momenta will change enough for the invariant mass to fail the acceptance conditions set in the experimental analysis. Generally, the rescattering effect depletes more strongly the yields of shorter lived resonances , as a greater fraction of these will decay shortly after formation, when elastic scattering of decay products on other produced particles is possible.

We further hear often the argument that the general scattering process of hadrons in matter can form additional resonance states. In our opinion, the loss of observability (caused by *any* scattering of *any* of the decay products is considerably greater than a possible production gain. The loss of resonance yield provides additional valuable information about the freeze-out conditions (temperature and time duration)[12].

3. Phase thresholds: volume and energy dependence

3.1. Statistical model parameters

In order to explore the properties of the fireball at hadronization as function of the volume at the top RHIC energy $\sqrt{s_{NN}} = 200\,\mathrm{GeV}$ Au–Au, we study the 11 centrality bins in which the $\pi^{\pm}, \mathrm{K}^{\pm}, p$ and $\bar{p}$ rapidity yields dN/dy for $y_{\mathrm{CM}} = 0$ have been recently presented, see table I and table VIII in Ref. [13]. These 6 particle yields and their ratios change rapidly. On the other hand, the additional two experimental results, the STAR $\mathrm{K}^*(892)/\mathrm{K}^-$ [14], and ϕ/K^- [15] show little centrality dependence.

In addition, three supplemental constraints help to determine the best fit:
A) strangeness conservation, *i.e.*, the (grand canonical) count of s quarks in all hadrons equals such $\bar{s}$ count for each rapidity unit;
B) the electrical charge to net baryon ratio in the final state is the same as in the initial state;
C) the ratio $\pi^+/\pi^- = 1. \pm 0.02$, which helps constrain the isospin asymmetry.
This last ratio appears redundant, as we already independently use the yields of π^+ and π^-. These yields have a large systematic error and do not constrain their ratio well, and thus the supplemental constraint is introduced, since SHARE allows for the isospin asymmetry effect. The 7 SHM parameters (volume per unit of rapidity dV/dy, temperature T, four chemical parameters $\lambda_q, \lambda_s, \gamma_q, \gamma_s$ and the isospin factor λ_{I3} are in this case studied in a systematic fashion as function of impact parameter using 11 yields and/or ratios and/or constraints, containing one (pion ratio) redundancy.

Although the number of degrees of freedom in such analysis is small, the χ^2 minimization yielding good significance is easily accomplished, showing good consistency of the data sample. The resulting statistical parameters are shown in Fig. 2, as function of participant number. We show on left results for the full non-equilibrium model allowing $\gamma_q \neq 1, \gamma_s \neq 1$ (full circles, blue) and semi-equilibrium setting $\gamma_s = 1$ (open circles, red). From top to bottom the (chemical) freeze-out temperature T, the occupancy factors γ_q, γ_s/γ_q and together in the bottom panel the

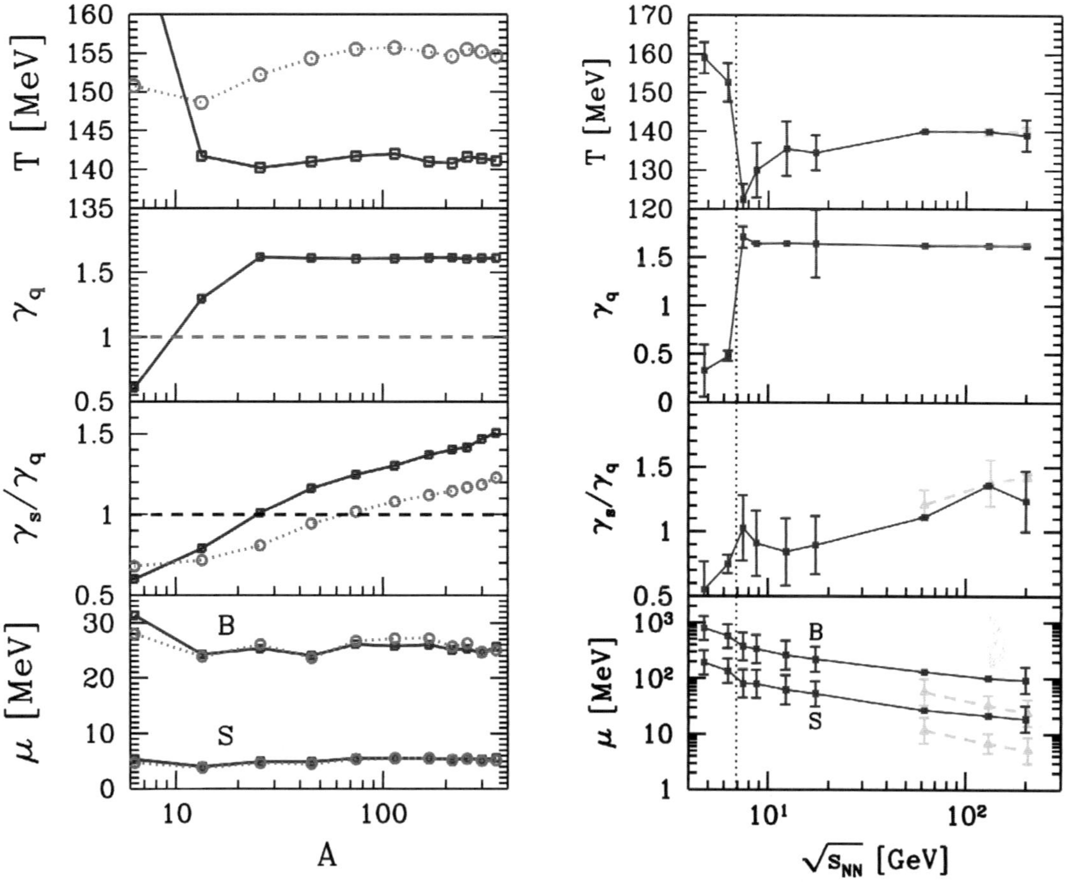

Figure 2. From top to bottom: temperature T, light quark phase space occupancy γ_q, the ratio of strange to light quark phase space occupancies γ_s/γ_q and the chemical potentials (B for baryochemical μ_B and S for strangeness μ_S) The lines guide the eye. Left: as function of centrality *i.e.* impact parameter dependence presented as dependence on inelastic (wounded) participant number A. Right: energy $\sqrt{s_{\rm NN}}$-dependence. Note that the results at RHIC (left side and the two highest energies in green on right) apply to central rapidity region dN/dy study, while the AGS/SPS energy results are obtained from total yields and are extrapolated to the top Brahms result at RHIC.

baryon μ_B and hyperon μ_S chemical potentials. On left, in Fig. 2, we present the results for the centrality dependence, as function of participant number, on right as function of energy.

To study the energy dependence, we must assemble several different experimental results from different facilities and experiments [6]. The results of this extensive analysis are shown on right hand side in Fig. 2. We note that when we are able to consider the total particle abundances, the number of participating nucleons is a tacit fit parameter. The lowest energy result is from our AGS study [4], the SPS data we used are from NA49 energy dependence exploration at the CERN-SPS [11, 16]. These results are for the total particle yields. The highest energy two results are based on studies of RHIC data at 130 and 200 GeV at central rapidity and address the dN/dy particle yields, with the highest point corresponding to the results presented at greatest centrality on the left hand side in Fig. 2.

There are several relevant features in Fig. 2. We see on left, that, for $A > 20$, there is no centrality dependence in freeze-out temperature T and chemical potentials $\mu_{B,S}$ (up to the variation which can be associated with fluctuation in the data sample). However, there is a change in values of the chemical potentials with reaction energy. This is result of rapidly decreasing baryon density, which due to reduced stopping is distributed over a wider range of rapidity as the reaction energy increases.

The middle sections, in Fig. 2, address the phase space occupancies which were obtained in terms of hadron particle yields. The quark-side occupancy parameters could be considerably different, indeed as model studies show, a factor 2 lower at the discussed conditions [2]: the hadron side phase space size is in general different from the quark-side phase space, since the particle degeneracies, particle spectra are quite different. We see similar behavior of γ_q as of T both for volume (that is, 'wounded' participant number A) and energy dependence seen in Fig. 2: the two lowest energy bins (top AGS and lowest SPS energy) deviate from the behavior seen at all other energies, as do the bins with $A < 20$. γ_s/γ_q as function of centrality rises steadily indicating longer lifespan of the fireball with increasing size. As function of energy, γ_s/γ_q reaches a plateau at 30 AGeV, with further rise only seen in the central rapidity results at RHIC. We note that, in our analysis, there is no saturation of the γ_s as we approach the most central reactions. This is inherent in the data we consider which includes the yields of ϕ, and K^*. This result is consistent with the implication that strangeness is not fully saturated in the QGP source, though it appears over-saturated when measured in the hadron phase space.

The deviation at the most peripheral centrality bins and at lowest reaction energy from trends set by other results could be an indication of the change in the reaction mechanism. As a threshold of centrality and/or energy are crossed a relatively small value of $\gamma_q \simeq 0.5$ grows rapidly to the maximum allowed by pion condensation condition, $\gamma_q \simeq e^{m_\pi/2T} \simeq 1.6$. This behavior signals a transformation of a chemically under-saturated phase of matter into something novel, where chemical equilibration is easy, and results in hadronization in a over-saturated phase of matter.

Independent of the chemical (non-)equilibrium assumption, the baryochemical potential $\mu_B = 25 \pm 1$ MeV is seen across 10 centrality bins. Similarly, we find strangeness chemical potential $\mu_S = 5.5 \pm 0.5$ MeV (related to strange quark chemical potential $\mu_s = \mu_B/3 - \mu_S$). The most notable variation, in Fig. 2, is the gradual increase in strangeness phase space occupancy γ_s/γ_q and thus strangeness yield with collision centrality. This effect was predicted and originates in an increasing lifespan of the fireball [17]. The over-saturation of the phase space has been also expected due to both, the dynamics of expansion [18], and/or reduction in phase space size as a parton based matter turns into HG [19]. This latter effect is also held responsible for the saturation of light quark phase space $\gamma_q \to e^{m_\pi/2T}$. A systematic increase of γ_s with collision centrality has been reported for several reaction energies [20].

3.2. Hadronization Temperature

Let us now discuss more in depth the magnitude of the hadronization temperature which at high energy/central collisions we find at $T = 140$ MeV. Some prefer the statistical hadronization to occur at higher temperature, perhaps as high as$T = 175$ MeV, a point argued at this meeting in great detail by Dr. Peter Braun-Münzinger. In his presentation we heard that he believes that the lattice results will reach to such high temperature near to $mu_{\rm B} = 0$ and this is where one should expect to see hadronization We disagree with both claims. For one,

We note that *chemical equilibrium QCD lattice* [21] results are mature and yield $T = 163 \pm 2$ MeV when extrapolated to physical quark mass scale. This result is in a very good agreement with the prior work on 2 and 3 flavor QCD which brackets this result by $T_{n_f=2,3} = T \pm 10$ MeV [22]. Moreover, the heavy ion collisions present a highly dynamical environment and one has to pay tribute to this especially regarding the value of hadronization temperature. For this reason we do not expect to find that the observed hadronization condition $T(\mu_B)$ will line up with the phase boundary curve obtained in study of a statistical system in thermodynamic limit on the lattice:

a) Dependence on parton collective flow:
A widely discussed effect which displaces the hadronization condition from the phase boundary is the expansion dynamics of the fireball. When the collective flow occurs at parton level, the color charge flow, like a wind, pushes out the vacuum [23], adding to thermal pressure a dynamical component. This can in general lead to supercooling and a sudden breakup of the fireball. We find that this can reduce the effective hadronization temperature by up to 20 MeV [24].

b) Dependence on quark chemical equilibration:
Lattice result have been discussed for 2-flavor lattice QCD at corresponding to $\gamma_q = 1, \gamma_s = 0$ (called) and for 2+1 flavor, corresponding to $\gamma_q = \gamma_s = 1$. While the precise nature of the phase limit is still under study it appears that for 2-flavor case the phase boundary temperature rises by about 7-10 MeV compared to the 2+1 case. We refer to the recent review of lattice QCD for further details [25]. Similarly the phase limit in pure gauge case corresponding in loose sense to $\gamma_q = \gamma_s = 0$ was seen near or even above $T = 200$ MeV.

These results do suggest that presence and the number of quarks matters regarding the precise location of the phase boundary and its nature. Its importance could be greatly enhanced, should the over-saturation of quark phase space have the same effect as would additional quark degrees of freedom. These are known to cause even for $\mu_B = 0$ the conversion of the phase crossover into a 1st-order phase transition which would, with these additional degrees of freedom, be expected at just the temperature we find in the SHM analysis.

We can be nearly sure that the chemical conditions matter and can displace the transition temperature. Because the degree of equilibration in the QGP depends on the collision energy, as does the collective expansion velocity, we cannot at all expect a simple hadronization scheme appropriate for the hadronization of nearly adiabatically expanding Universe.

Leaving this issue we note that, in the data analysis assuming chemical equilibrium, we find $T = 155 \pm 8$ MeV for the chemical equilibrium and strangeness non-equilibrium freeze-out, see Fig. 2. The error is our estimate of the propagation of the systematic data error, combined with the fit uncertainty; the reader should note that the error comparing centrality to centrality is negligible. The freeze-out temperature is for the semi-equilibrium and equilibrium model about 10% greater than the full chemical non-equilibrium freeze-out.

This result for T, in the equilibrium case, is in mild disagreement (1.5 s.d.) with earlier equilibrium fits [26, 27]. This, we believe, is due to some differences in data sample used, specifically, the hadron resonance production results used provide a very strong constraint for the fitted temperature, and more complete treatment by SHARE of hadron mass spectrum.

Interestingly, it seems that the general consensus about the chemical equilibrium best analysis result is in gross disagreement with the results advanced at this meeting by Dr. Peter Braun-Münzinger.

3.3. Physical Properties of the Fireball

We now turn our attention to the physical properties of the hadronizing fireball obtained summing individual contributions made by made of each of the hadronic particles produced. Often particles observed experimentally dominate (*e.g.* pions dominate pressure, kaon strangeness yield etc). However, it is important to include in this yields of particles predicted in terms of the SHM fit to the available results. Again, on left in Fig. 3. we show the behavior as function of impact parameter and on right as function of energy. From top to bottom we show the pressure P, energy density ϵ, entropy density σ, and the dimensionless ratio $\epsilon/T\sigma = E/TS$. All contributions are evaluated using relativistic expression, see [2].

When we fitted particle rapidity yields, the global fitted yield normalization factor is

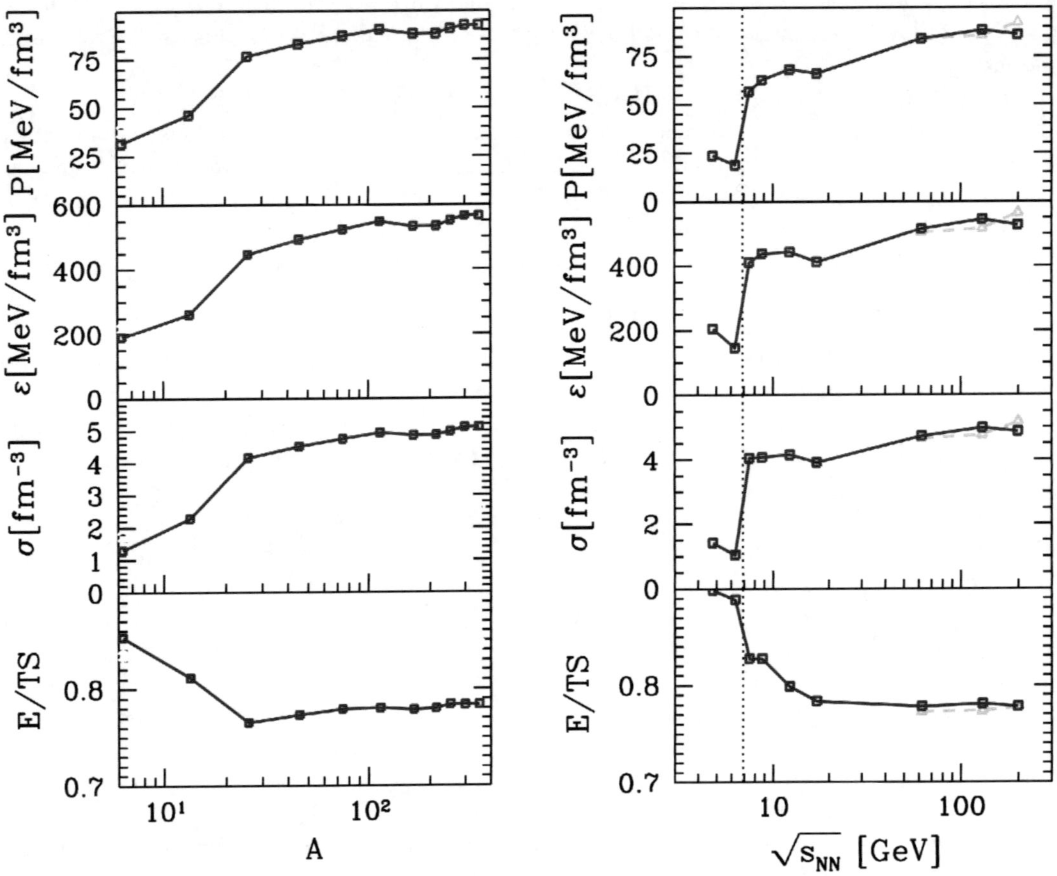

Figure 3. From top to bottom: pressure P, energy density $\epsilon = E/V$, entropy density S/V and E/TS; Left: as a function of centrality, Right as function of reaction energy. The results on left are for central rapidity, and on right are for the full phase space coverage, except for the high energy RHIC results (open triangles, green) which are for central rapidity. Open circles on left apply to the semi-non-equilibrium analysis with $\gamma_q = 1, \gamma_s \neq 1$, all other results are for the chemical nonequilibrium analysis.

$dV(A)/dy$, for total particle yields it is V which is also a function of centrality trigger. When we consider ratios of two bulk properties e.g. E/TS, the results are in general smoother indicating cancellation in the error propagating from the fit. The overall error is of the magnitude of the particle yield, for example pressure is dominated by pions and hence its precision is limited by this error. However, on left, the point-to-point error is minimal as the systematic error is common. On right, the absolute error matters as the fluctuations in the results presented show.

We note, in Fig. 3, that as the reaction energy passes the volume threshold $A = 20$ and even more so, the energy threshold $6.26\,\text{GeV} < \sqrt{s_{\text{NN}}} < 7.61\,\text{GeV}$ the hadronizing fireball becomes much denser. The entropy density jumps by factor 3–4, and the energy and baryon number density by a factor 2–3. The hadron pressure jumps up from $P = 25\,\text{MeV/fm}^3$ initially by factor 2 and ultimately more than factor 3. There is a gradual increase of $P/\epsilon = 0.115$ at low reaction energy to 0.165 at the top available energy. It is important to note that exactly the same behavior of the fireball physical properties arises both as function of reaction volume and reaction energy. This is the case for both the physical properties and the statistical parameters.

We believe that this shows a common change in the physical state created as function of energy available and reaction volume.

4. Search for an Energy Threshold

4.1. Kaon to pion ratio

One of the most interesting questions is if there is an energy threshold for the formation of a new state of matter. An important observable of the deconfined phase of matter is aside of strangeness, the high entropy content, which is arising from broken color bonds. The observable for both is the $\mathrm{K}^+/\pi^+ \propto \bar{s}/\bar{d}$ yield ratio [28]. This ratio has been studied experimentally [11] and a pronounced horn structure arises. We can describe this structure in our study of the particle yields only within the chemical non-equilibrium model. Although this change is associated a rather sudden modification of chemical conditions in the dense matter fireball, this effect is caused by two distinct phenomena: the rapid rise in strangeness $\bar{s}$ production below, and a rise in the antiquark $\bar{d}$ yield above a energy reaction threshold.

The measured K^+/π^+ ratio by NA49 is shown at top left of Fig. 4, where for comparison we also show the *pp* results. On right top, we present our results reduced to the correct relative scale, both for the total yield ratio for the AGS–RHIC energy range, and for the central rapidity results from RHIC. The solid line guides the eye to the fit results we obtained. To show that the K^+/π^+ ratio drop is due to a decrease in baryon density which leads to rise in the $\bar{d}$ yield, we show in bottom section of Fig. 4, the nearly baryon density independent K/π double ratio ratio Eq. (7), on left as function of $\sqrt{s_{\mathrm{NN}}}$, and on right as function of centrality of the reaction for $\sqrt{s_{\mathrm{NN}}} = 200$ GeV:

$$\frac{K}{\pi} = \sqrt{\frac{K^+}{\pi^+}\frac{K^-}{\pi^-}}. \tag{7}$$

Both upper and lower portion of Fig. 4 are also drawn on same relative scale.

Seen how the horn specifically arises in the one K^+/π^+, one can wonder if this is really a physical effect and how, in qualitative terms, a parameter γ_q, which controls the light quark yield, can help explain the horn structure seen in top of Fig. 8. We observe that this horn structure in the K^+/π^+ ratio traces out the final state valance quark ratio $\bar{s}/\bar{d}$, and in language of quark phase space occupancies γ_i and fugacities λ_i, we have:

$$\frac{\mathrm{K}^+}{\pi^+} \to \frac{\bar{s}}{\bar{d}} \propto F(T)\left(\frac{\lambda_s}{\lambda_d}\right)^{-1}\frac{\gamma_s}{\gamma_d} = F(T)\left(\sqrt{\lambda_{I3}}\frac{\lambda_s}{\lambda_q}\right)^{-1}\frac{\gamma_s}{\gamma_q}. \tag{8}$$

In chemical equilibrium models $\gamma_s/\gamma_q = 1$, and the K^+/π^+ ratio and its horn must arise solely from the variation in the ratio λ_s/λ_q and the change in temperature T which both are usually smooth function of reaction energy. The isospin factor λ_{I3} is insignificant in this consideration.

As collision energy is increased, increased hadron yield leads to a decreasing $\lambda_q = e^{\mu_{\mathrm{B}}/3T}$. We recall the smooth decrease of μ_{B} with reaction energy seen in bottom panel in Fig. 2. The two chemical fugacities λ_s and λ_q are coupled by the condition that the strangeness is conserved. This leads to a smooth λ_s/λ_q. The chemical potential effect is suggesting a smooth increase in the K^+/π^+ ratio.

For the interesting range of freeze-out temperature, $F(T)$ is a smooth function of T. Normally, one expects that T increases with collision energy, hence on this ground as well we expect an monotonic increase in the K^+/π^+ ratio as function of reaction energy. With considerable effort, one can arrange the chemical equilibrium fits to bend over to a flat behavior at $\sqrt{s_{\mathrm{NN}}^{\mathrm{cr}}}$. It is quasi impossible to generate the horn with chemical equilibrium model. To accomplish this, an additional parameter appears necessary, capable to change rapidly when hadronization conditions change. This is γ_q. It presence also allows T to vary in non-monotonic fashion, as is seen in Fig. 2.

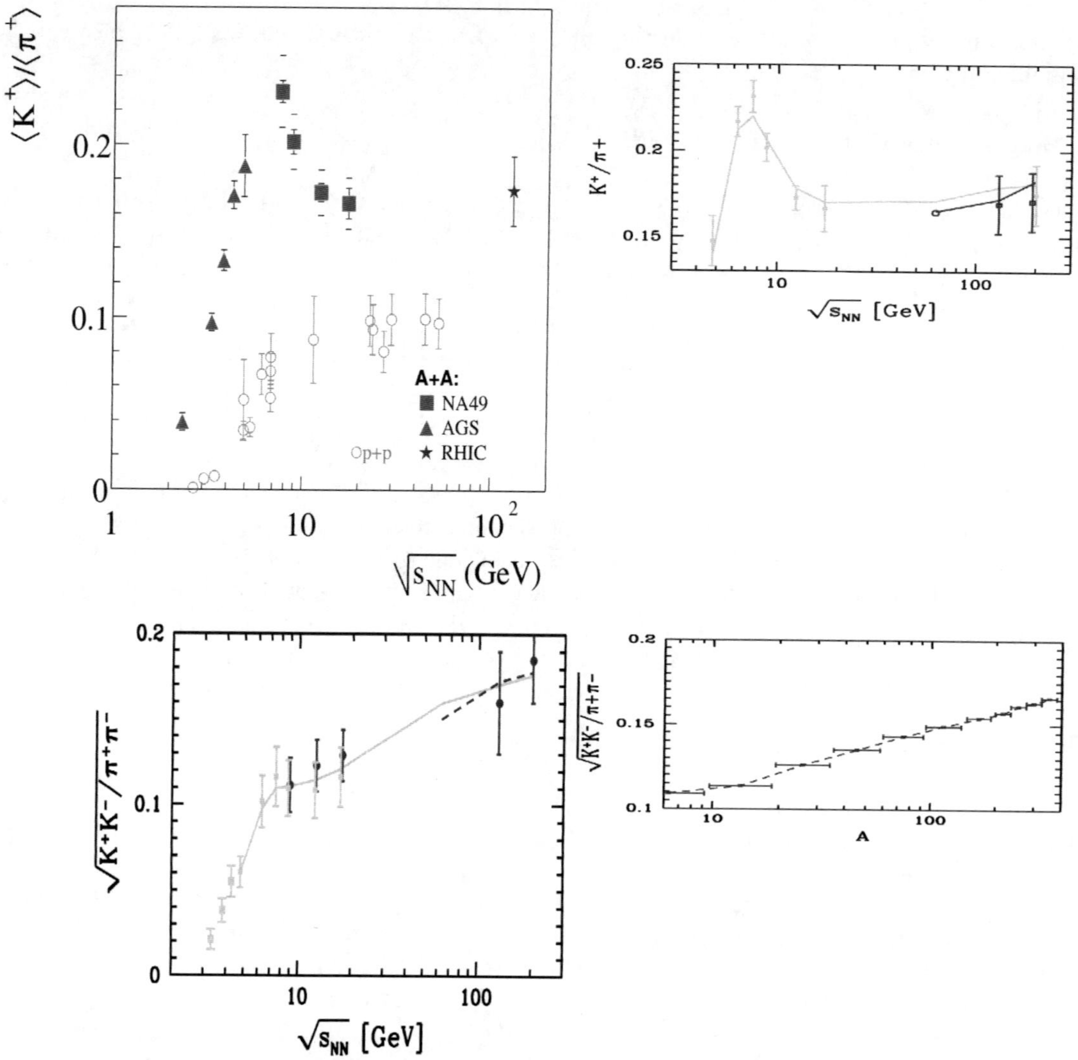

Figure 4. Top: Total yield-ratio of yields of K^+/π^+ for nuclear (filled symbols, green) and elementary interactions (open symbols, violet); (courtesy of NA49 collaboration [29]). On right: a fit to this K^+/π^+ ratio using SHM, at high energy also for central rapidity RHIC, line guides the eye between the fitted values. Bottom: the K/π ratio for nuclear collisions, squares on left (green): full phase space on left, circles (blue) are central rapidity results, left as function of collision energy and right as function of participant number at $\sqrt{s_{\rm NN}} = 200$ GeV.

4.2. QGP degrees of freedom and s/S ratio

The full SHM is capable to describe the data, and we now show that it can pinpoint the properties of the phase of matter that was created early on in the reaction. To see this we consider what we learn from the final state data about strangeness and entropy production. For this purpose we consider both the specific per baryon and per entropy yield of strangeness. In addition, we look at the cost in thermal energy to make strangeness. All these quantities are nearly independent of the dynamics of hadronization, since they are related to processes occurring early on, 'deep' inside the collision region, and long before hadronization.

In the QGP, the dominant entropy production occurs during the initial glue thermalization, and the thermal strangeness production occurs in parallel and/or just a short time later [30].

The entropy production occurs predominantly early on in the collision during the thermalization phase. Strangeness production by gluon fusion is most effective in the early, high temperature environment, however it continues to act during the evolution of the hot deconfined phase until hadronization [31]. Both strangeness and entropy are nearly conserved in the evolution towards hadronization and thus the final state hadronic yield analysis value for s/S is closely related to the thermal processes in the fireball at $\tau \simeq$ 1–2 fm/c. We believe that for reactions in which the system approaches strangeness equilibrium in the QGP phase, one can expect a prescribed ratio of strangeness per entropy, the value is basically the ratio of the QGP degrees of freedom.

We estimate the magnitude of s/S deep in the QGP phase, considering the hot stage of the reaction. For an equilibrated non-interacting QGP phase with perturbative properties:

$$\frac{s}{S} \equiv \frac{\rho_\mathrm{s}}{\sigma} = \frac{(3/\pi^2)T^3(m_s/T)^2 K_2(m_s/T)}{(32\pi^2/45)T^3 + n_\mathrm{f}[(7\pi^2/15)T^3 + \mu_q^2 T]} = \frac{0.027}{1+0.054(\ln\lambda_q)^2}\,. \tag{9}$$

Here, we used for the number of flavors $n_\mathrm{f} = 2.5$ and $m_s/T = 1$. We see that the result is a slowly changing function of λ_q; for large $\lambda_q \simeq 4$, we find at modest SPS energies, the value of s/S is reduced by 10%. Considering the slow dependence on $x = m_s/T \simeq 1$ of $W(x) = x^2 K_2(x)$ there is minor dependence on the much less variable temperature T.

The dependence on the degree of chemical equilibration which dominates is easily obtained separating the different degrees of freedom:

$$\frac{s}{S} = 0.027\frac{\gamma_s^\mathrm{QGP}}{0.38\gamma_\mathrm{G}^\mathrm{QGP} + 0.12\gamma_s^\mathrm{QGP} + 0.5\gamma_q^\mathrm{QGP} + 0.054\gamma_q^\mathrm{QGP}(\ln\lambda_q)^2}\,. \tag{10}$$

We assume that the interaction effects are at this level of the discussion canceling. Seen Eq. (10) we expect to see a gradual increase in s/S as the QGP source of particles approaches chemical equilibrium with increasing collision energy and/or increasing volume.

We repeat that it is important to keep in mind that this ratio s/S is established early on in the reaction, the above relations, and the associated chemical conditions we considered, apply to the early hot phase of the fireball. At hadron freeze-out the QGP picture used above does not apply. Gluons are likely to freeze faster than quarks and both are subject to much more complex non-perturbative behavior. However, the value of s/S is nearly preserved from the hot QGP to the final state of the reaction.

How does this simple prediction compare to experiment? Given the statistical parameters, we can evaluate the yields of particles not yet measured and obtain the rapidity yields of entropy, net baryon number, net strangeness, and thermal energy, both for the total reaction system and also for the central rapidity condition, also as function of centrality. In passing, we note that, in the most central reaction bin at RHIC-200, $dB/dy \simeq 15$ baryons per unit rapidity interval, implying a rather large baryon stopping in the central rapidity domain.

The rise of strangeness yield with centrality is faster than the rise of baryon number yield: $(ds/dy)/(dB/dy) \equiv s/B$ is seen in the top left panel in Fig. 5. The solid (blue) lines are for the chemical nonequilibrium central rapidity yields of particles at RHIC-200. Solid (green) lines, on right, are for total hadron yields and thus total yields of the considered quantities, *e.g.* strangeness, entropy. For the most central head-on reactions, we reach at RHIC-200 $s/B = 9.6 \pm 1$.

In the middle panel in Fig. 5 we compare strangeness with entropy production s/S, which we just evaluated theoretically. Again, on left, as function of participant number A at RHIC-200 and, on right, for the total reaction system for the most central 5-7% reactions. On left, we see a smooth transition from a flat peripheral behavior where $s/S \lesssim 0.02$ to smoothly increasing s/S reaching $s/S \simeq 0.028$ in most central reactions. This indicates that even at RHIC-200 for the more central reactions some new mechanisms of strangeness production becomes activated. On

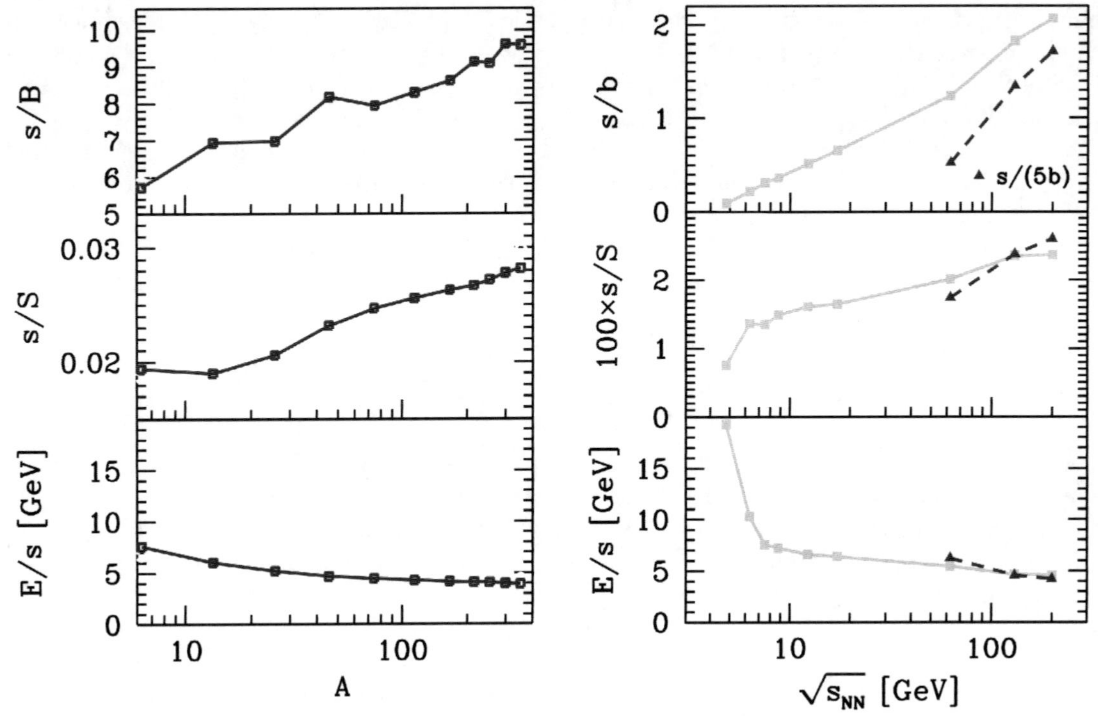

Figure 5. Strangeness per net baryon s/B, strangeness per entropy s/S, and $E_{\rm th}/s$ the thermal energy cost to make strangeness. Left: as a function of centrality, Right as function of reaction energy. The results on right include the central rapidity conditions at RHIC energies (dashed, blue) lines. The actual results are the symbols, the lines guide the eye.

right, we see that the change on s/S is much more drastic as function of reaction energy. After an initial rapid rise the further increase occurs beyond the threshold energy at slower pace.

In the bottom panel, on left, we see the thermal energy cost $E_{\rm th}/s$ of producing a pair of strange quarks as function of the size of the participating volume (*i.e.* A) This quantity shows a smooth drop which can be associated with transfer of thermal energy into collective transverse expansion after strangeness is produced. Thus, it seems that the cost of strangeness production is independent of reaction centrality. The result is different when we consider $\sqrt{s_{\rm NN}}$ dependence of this quantity, see bottom panel on right. There is a very clear change in the energy efficiency of making strangeness at the threshold energy. We will return to discuss possible reaction mechanisms below.

5. Final remarks

5.1. Phase boundary and hadronization conditions

The chemical freeze-out conditions we have determined presents, in the T–$\mu_{\rm B}$ plane, a more complex picture than naively expected, see Fig. 6. Considering results shown in Fig. 2, we are able to assign to each point in the T–$\mu_{\rm B}$ plane the associate value of $\sqrt{s_{\rm NN}}$. The RHIC dN/dy results are to outer left. They are followed by RHIC and SPS $N_{4\pi}$ results. The dip corresponds to the 30 and 40 AGeV SPS results. The top right is the lowest 20 AGeV SPS and top 11.6 AGeV AGS energy range. To guide the eye, we have added two lines connecting the fit results. We see that the chemical freeze-out temperature T rises for the two lowest reaction energies 11.6

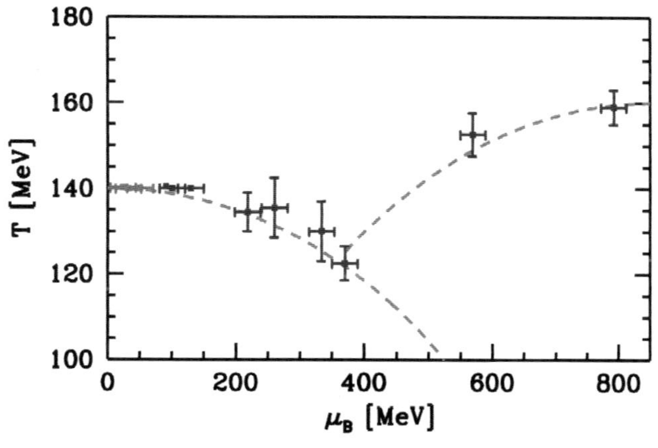

Figure 6. T–μ_B plane with points obtained in the SHM fit.

and 20 A GeV to near the Hagedorn temperature, $T = 160$ MeV, of boiling hadron matter.

Once the chemical non-equilibrium is allowed for, the data fit turns to be much more precise, and the picture of the phase boundary with smaller 'measurement' error reveal a much more complex structure, and contains physics details prior analysis based on a rudimentary model could not uncover. The shape of the hadronization boundary, shown in Fig. 6 in the T–μ_B plane, is the result of a complex interplay between the dynamics of heavy ion reaction, and the properties of both phases of matter, the inside of the fireball, and the hadron phase we observe. The dynamical effect, capable to shift the location in temperature of the expected phase boundary is due to the expansion dynamics of the fireball which occurs at parton level, and effects of chemical nonequilibrium, see subsection 3.2 for full discussion.

Possibly, not only the location, but also the *nature* of the phase boundary can be modified by variation of γ_i. We recall that for the 2+1 flavor case, there is a critical point at finite baryochemical potential with $\mu_B \simeq 350$ MeV [21]. For the case of 3 massless flavors, there can be a 1st order transition at all μ_B [22, 33, 34]. Considering a classical particle system, one easily sees that an over-saturated phase space, *e.g.*, with $\gamma_q = 1.6$, $\gamma_s \geq \gamma_q$ for the purpose of the study of the phase transition acts as being equivalent to a system with 3.2 light quarks and 1.6 massive (strange) quarks present in the confined hadron phase.

Considering the sudden nature of the fireball breakup seen in several observables [1], we conjecture that the hadronizing fireball leading to $\gamma_s > \gamma_q = 1.6$ super-cools and experiences a true 1st order phase boundary corresponding also at small μ_B. The system we observe in the final state prior to hadronization is mainly a quark–antiquark system with gluons frozen in prior expansion cooling of the QCD deconfined parton fluid. The quark dominance is necessary to understand *e.g.* how the azimuthal asymmetry v_2 varies for different particles [35].

These quarks and antiquarks have, in principle, at that stage a significant thermal mass. The evidence for this is seen in Fig. 3 in its two bottom panels, showing the dimensionless variable E/TS. The energy end entropy per particle of non-relativistic and semi-relativistic classical particle gas comprising both quarks and antiquarks is (see section 10, [2]):

$$\frac{E}{N} \simeq m + 3/2T + \ldots, \qquad \frac{S}{N} \simeq 5/2 + m/T + \ldots, \qquad \frac{E}{TS} \simeq \frac{m/T + 3/2}{m/T + 5/2}\,. \tag{11}$$

It is thus possible to interpret the fitted value $E/TS \to 0.78$ in terms of a quark matter made of particles with $m \propto aT$, $a = 2$ which is close to what is expected based on thermal QCD [36].

5.2. Looking forward to LHC

We expect considerably more violent transverse expansion of the fireball of matter created at LHC. The kinetic energy of this transverse motion must be taken from the thermal energy of the expanding matter, and ultimately this leads to local cooling and thus a reduction in the number of quarks and gluons. The local entropy density decreases, but the expansion assures that the total entropy still increases. Primarily, gluons are feeding the expansion dynamics, while strange quark pair yield, being weaker coupled remain least influenced by this dynamics. Model calculations show that this expansion yields an increase in the final QGP phase strangeness occupancy γ_s [32]. This mechanism, along with the required depletion of the non-strange degrees of freedom, in the feeding of the expansion, assures an increase in the K/π ratio given the nearly 30-fold increase of collision energy.

Depending on what we believe to be a valid hadronization temperature for a fast transversely expanding fireball, the possible maximal enhancement in the K/π ratio may be in the range of a factor 2–3. Perhaps even more interesting than the K/π ratio enhancement would be the enhancement anomaly in strange (antibaryon) yields. With $\gamma_s \gg 1$, we find that the more strange baryons and antibaryons are more abundant than the more 'normal' species. Specifically of interest would be $(\Omega^- + \overline{\Omega}^+)/(h^+ + h^-)$, $(\Xi^- + \overline{\Xi}^+)/(h^+ + h^-)$, and $2\phi/(h^+ + h^-)$ which should show a major, up to an order of magnitude shift in relative production strength. Detailed predictions for the yields of these particles require considerable extrapolation of physics conditions from the RHIC to LHC domain and this work is in progress [37].

Ultimately, strange s, and $\bar{s}$ quarks can exceed in abundance the light quark component, in which case we would need to rethink in much more detail the distribution of global particle yield. The ratios of neutral and charged hadrons would undergo serious change.

Regarding charm, we note that situation will not become as severe. Given the large charm quark mass, we expect that most of charm quark yield is due to first hard interactions of primary partons. For this reason, the yield of strange and light quarks, at time of hadronization, exceeds by about a factor 100 or more that of charm at central rapidity. Thus, even though charm phase space occupancy at hadronization may largely exceed the chemical equilibrium value, seen the low hadronization temperature, *e.g.*, $m_c/T \simeq 10$, it takes a factor $\gamma_c \simeq e^{10}/10^{1.5} = 700$ to compensate hadron yield suppression due to the high charm mass. Said differently, while strange quarks can compete in abundance with light quarks for $m_s/T \simeq 1$, charm (and heavier) flavor(s) will remain suppressed, in absolute yield, at the temperatures we can make presently in laboratory experiments.

5.3. Highlights

We have shown that strangeness, and entropy, at SPS and RHIC are well developed tools allowing the detailed study of hot QGP phase. Our detailed discussion of hadronization analysis results has further shown that a systematic study of strange hadrons fingerprints the properties of a new state of matter at point of its breakup into final state hadrons.

We have shown that it is possible to describe the 'horn' in the K^+/π^+ hadron ratio within the chemical non-equilibrium statistical hadronization model. We have shown that appearance of this structure is related to a rapid change in the properties of the hadronizing matter. Of most theoretical relevance and interest are the implications of non-equilibrium hadronization on the possible change in the location and *nature* of the phase boundary.

In summary, we have presented interpretation of the experimental soft hadron data production and discussed the production of strangeness and entropy that this analysis infers. We have seen, in quantitative way, how the relative strangeness and entropy production in most central high energy heavy ion collisions agrees with quark-gluon degree of freedom counting in hot primordial matter where the values of these quantities have been established.

Acknowledgments
Work supported in part by a grant from: the U.S. Department of Energy DE-FG02-04ER4131. LPTHE, Univ. Paris 6 et 7 is: Unité mixte de Recherche du CNRS, UMR7589. JR thanks Bikash Sinha and Jan-e Alam and the organizers of the 5th International Conference on Physics and Astrophysics of Quark Gluon Plasma, February 8 — 12, 2005 Salt Lake City, Kolkata, India for their kind hospitality. Dedicated to Professor Bikash Sinha on occasion of his 60th anniversary.

References

[1] Assessments by the RHIC experimental collaborations: *Hunting the Quark Gluon Plasma: results from the first 3 years at RHIC* BNL-73847-2005 Formal Report, April 18, 2005, to appear in Nucl. Phys. A (2005).
[2] J. Letessier and J. Rafelski, *Hadrons and quark - gluon plasma* Cambridge Monogr. Part. Phys. Nucl. Phys. Cosmol. **18** 1-397 (2002) (Cambridge, UK, 2002). Available on line to read freely with WIN and MAC platforms at http://site.ebrary.com/pub/cambridgepress/Doc?isbn=0521385369
[3] G. Torrieri, W. Broniowski, W. Florkowski, J. Letessier and J. Rafelski, [arXiv:nucl-th/0404083], Computer Physics Communications in press, see: www.physics.arizona.edu/~torrieri/SHARE/share.html
[4] J. Letessier, J. Rafelski and G. Torrieri, "Deconfinement energy threshold: Analysis of hadron yields at 11.6-A-GeV," arXiv:nucl-th/0411047.
[5] J. Rafelski, J. Letessier and G. Torrieri, "Centrality dependence of bulk fireball properties at RHIC," arXiv:nucl-th/0412072.
[6] J. Letessier and J. Rafelski, "Hadron production and phase changes in relativistic heavy ion collisions," arXiv:nucl-th/0504028.
[7] J. Rafelski and J. Letessier, Acta Phys. Polon. B **34**, 5791 (2003) [arXiv:hep-ph/0309030]; and J. Phys. G **30** (2004) S1 [arXiv:hep-ph/0305284].
[8] J. Letessier, A. Tounsi, U. W. Heinz, J. Sollfrank and J. Rafelski, Phys. Rev. Lett. **70**, 3530 (1993).
[9] H. Caines [STAR Collaboration], J. Phys. G **31**, S1057 (2005).
[10] G. E. Bruno [NA57 Collaboration], J. Phys. G **30**, S717 (2004) [arXiv:nucl-ex/0403036].
[11] M. Gazdzicki *et al.* [NA49 Collaboration], J. Phys. G **30**, S701 (2004) [arXiv:nucl-ex/0403023].
[12] J. Rafelski, J. Letessier and G. Torrieri, Phys. Rev. C **64**, 054907 (2001) [Erratum-ibid. C **65**, 069902 (2002)].
[13] S. S. Adler *et al.* [PHENIX Collaboration], Phys. Rev. C **69**, 034909 (2004) [arXiv:nucl-ex/0307022].
[14] H. B. Zhang [STAR Collaboration], "Delta, K* and rho resonance production and their probing of freeze-out dynamics at RHIC,", poster presentation at QM2004, Oakland, January 2004 [arXiv:nucl-ex/0403010]; J. Adams [STAR Collaboration], [arXiv:nucl-ex/0412019], Phys. Rev. C. (2005) in press.
[15] J. Adams *et al.* [STAR Collaboration], Phys. Lett. B **612**, 181 (2005) [arXiv:nucl-ex/0406003].
[16] M. Gazdzicki, Commented compilation of NA49 results, private communication.
[17] J. Letessier, A. Tounsi and J. Rafelski, Phys. Lett. B **389** (1996) 586.
[18] J. Rafelski and J. Letessier, Phys. Lett. B **469**, 12 (1999) [arXiv:nucl-th/9908024].
[19] J. Rafelski and J. Letessier, Nucl. Phys. A **702**, 304 (2002) [arXiv:hep-ph/0112027].
[20] B. Kampfer, J. Cleymans, P. Steinberg and S. Wheaton, Heavy Ion Phys. **21**, 207 (2004).
[21] Z. Fodor and S. D. Katz, JHEP **0404**, 050 (2004) [arXiv:hep-lat/0402006].
[22] F. Karsch, Nucl. Phys. A **698**, 199 (2002) [arXiv:hep-ph/0103314].
[23] T. Csorgo and J. Zimanyi, Heavy Ion Phys. **17**, 281 (2003) [arXiv:nucl-th/0206051].
[24] J. Rafelski and J. Letessier, Phys. Rev. Lett. **85**, 4695 (2000) [arXiv:hep-ph/0006200].
[25] F. Karsch and E. Laermann, arXiv:hep-lat/0305025. In R.C. Hwa, et al.: Quark gluon plasma III (2004) pp 1-59 (World Scientific, Singapore).
[26] P. Braun-Munzinger, K. Redlich and J. Stachel, [arXiv:nucl-th/0304013], and references therein.
[27] W. Broniowski, W. Florkowski and B. Hiller, Phys. Rev. C **68**, 034911 (2003) [arXiv:nucl-th/0306034].
[28] N. K. Glendenning and J. Rafelski, Phys. Rev. C **31**, 823 (1985).
[29] M. vanLeeuwen, Compilation of NA49 results as function of collision energy. Private communication (2003).
[30] J. Alam, B. Sinha and S. Raha, Phys. Rev. Lett. **73**, 1895 (1994).
[31] J. Rafelski and B. Müller, *Phys. Rev. Lett* **48**, 1066 (1982); **56**, 2334E (1986).
[32] J. Rafelski and J. Letessier, *Phys. Lett.* B **469**, 12 (1999).
[33] A. Peikert, F. Karsch, E. Laermann and B. Sturm, Nucl. Phys. Proc. Suppl. **73**, 468 (1999) .
[34] C. Bernard *et al.* [MILC Collaboration], Phys. Rev. D **71**, 034504 (2005) .
[35] H. Z. Huang and J. Rafelski, AIP Conf. Proc. **756**, 210 (2005) [arXiv:hep-ph/0501187].
[36] P. Petreczky, F. Karsch, E. Laermann, S. Stickan and I. Wetzorke, Nucl. Phys. Proc. Suppl. **106**, 513 (2002).
[37] J. Rafelski and J. Letessier, "Soft hadron relative multiplicites at LHC" (in preparation).

Institute of Physics Publishing Journal of Physics: Conference Series **50** (2006) 192–200
doi:10.1088/1742-6596/50/1/022 5th International Conference on Physics and Astrophysics of Quark Gluon Plasma

Recent Results on Strangeness and Heavy Flavour at RHIC

M A C Lamont† for the STAR Collaboration

† Physics Department, Yale University, New Haven, CT 06520, USA

E-mail: `Matthew.Lamont@Yale.edu`

Abstract.

The advent of RHIC has opened up the possibilities for a greater understanding of the hadronization process in relativistic heavy ion collisions. The large volume of data collected, together with high collisions energies, allows us to address both strangeness and charm production in great detail. This paper will present results on bulk strangeness production in Au+Au collisions at $\sqrt{s_{NN}}$ = 200 GeV, in particular on strangeness enhancement and strange quark scaling. Also presented are results on strange particle ratios and elliptic flow measurements in the intermediate p_T region, together with comparisons with theoretical models. An excitation function of strangeness yields and ratios are also presented with reference to the newly acquired Au+Au data at $\sqrt{s_{NN}}$ = 62.4 GeV.

Also presented in this paper are the first results on charm production measured in STAR. Utilising the methods of reconstructing D mesons through their leptonic channels, the cross-section for charm production in d+Au collisions at $\sqrt{s_{NN}}$ = 200 GeV is presented. Also shown is a first measurement of the elliptic flow of charmed mesons.

1. Strangeness Production at RHIC

1.1. Strangeness enhancement

The enhancement of strange particles in A+A collisions, with respect to p+p collisions, has long been predicted as a signature of a de-confined state of matter [1]. When calculating expected values for this enhancement, care has to be taken with the differences in the size of the system. This is achieved by utilising a canonical partition function for the smaller systems, where relevant quantum numbers must be conserved explicitly. For the larger volume A+A collisions though, a grand canonical partition function may be used which allows for the conservation of quantum numbers on the average. Utilising these models, predictions for strangeness enhancement as a function of strangeness, centrality and centre of mass energy have been obtained [2]. It is expected that the enhancement factors increase with strangeness content and centrality and are greatest at the lower SPS energies.

The $<p_T>$ of strange baryons in A+A collisions is approximately 1.1 GeV/c and hence the production of strange particles is predominantly in the p_T region below 2 GeV/c. Therefore, when comparisons are made between different collision systems regarding strangeness enhancement, it is typically this low p_T region which is being measured. STAR has good capabilities in measuring strange hadrons via their decay topology in the TPC in this p_T range, with acceptance down to a few hundred MeV [3]. Good coverage at low p_T is important as it allows for the constraint of fits, used to determine the yields ($dN/d\mathbf{y}$), as the data turns over at low p_T in A+A collisions (when plotted versus $m_T - m_0$) due to transverse radial flow.

The dependence of strangeness enhancement has been studied at the SPS in some detail for a number of lower energies than RHIC is capable of. Data on strangeness enhancement in Pb+Pb collisions are shown in the left two panels of Figure 1 [4]. While these data exhibit the expected ordering with respect to strangeness content of the hyperon, there appears to be no dependence on beam energy for both hyperons and anti-hyperons. The enhancement factors for the $\Lambda(\overline{\Lambda})$ and $\Xi^-(\overline{\Xi}^+)$ have also been calculated for Au+Au collisions at 200 GeV in STAR and is shown in the right panel of Figure 1. Note that there is no data from the $(\Omega^- + \overline{\Omega}^+)$ in the STAR figure. This comes from the lack of p+p reference data at 200 GeV on $(\Omega^- + \overline{\Omega}^+)$ production.

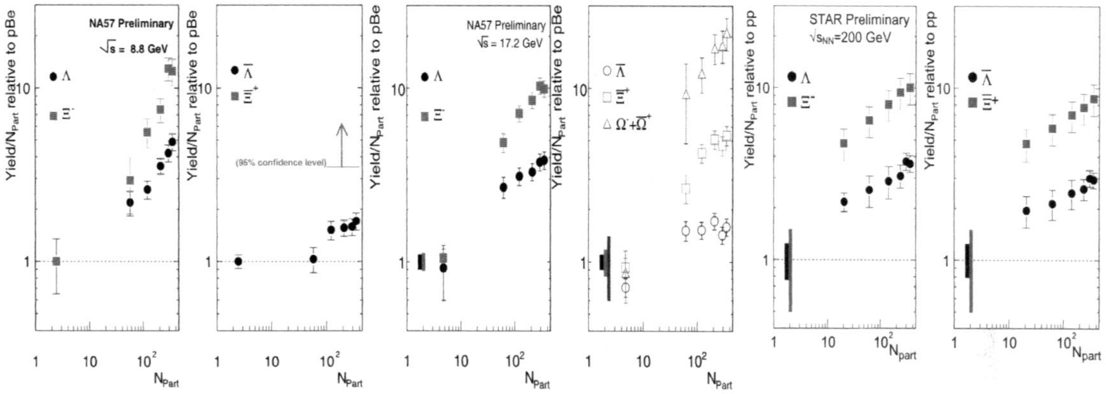

Figure 1. The enhancement factors for hyperons and anti-hyperons as a function of centrality for data from Pb+Pb collisions at $\sqrt{s_{NN}} = 8.8$ GeV (left panel), Pb+Pb collisions at $\sqrt{s_{NN}} = 17.2$ GeV (centre panel) and Au+Au collisions at $\sqrt{s_{NN}} = 8.8$ GeV (right panel).

It is evident that the predicted energy dependence of strangeness enhancement is not observed in the data. Although there is a dependence on energy of the strangeness yields in A+A collisions, the enhancement factors are affected by the increase in available energy for production in the reference systems. One other point of note from Figure 1 is that at the lower energies, the enhancement of the $\overline{\Lambda}$ appears to be constant with centrality (defined by the number of participants in the collision, N_{part}), whereas at the higher energy, no saturation is observed.

This observation initiated work on the fact that the volume may not scale linearly with N_{part}. It was found that when the valence quarks of a hadron were scaled with N_{part} for the light quarks and with the number of binary collisions (N_{bin}) for the strange quarks, then the yield of particles from peripheral to central A+A collisions appears to be constant [5].

1.2. Strangeness production at intermediate p_T

Also presented at this conference were results from the intermediate p_T region (defined in this paper as $2 < p_T < 6$ GeV/c). There has recently been a lot of interest in this range of the particle spectra, in particular for identified particles. This was initiated by results noting differences between $\overline{p}$ and π spectra measured by PHENIX [6]. PHENIX showed that in the most central dataset, the ratio of $\overline{p}/\pi$ increased with p_T up to 3 GeV/c where the statistics end. This is well above the ratio measured in elementary collisions [7]. In peripheral collisions, however, the situation is different and the ratio does follow that observed in elementary collisions.

One explanation of these phenomena is that the baryons and mesons are simply produced by the coalescence/recombination of thermal partons from the plasma. If that were the case, then this intuitively leads to a rising baryon/meson ratio. Various coalescence models produced post-dictions for the baryon/meson ratio which agreed well with the $\overline{p}/\pi$ measurements from PHENIX. All models showed that the ratio peaks in this intermediate region, and actually starts to decrease after $p_T \approx 4$ GeV/c [8–10].

In order to test the models further, it was required to extend the particle identification to higher transverse momenta. Using Λ and K^0_S measured in STAR, it is possible to reproduce the baryon/meson ratio and extend the measurement to higher p_T as the upper limit in p_T of Λ and K^0_S reconstruction is only limited by statistics. In the 200 GeV Au+Au run in 2001, it was possible to measure the Λ and K^0_S out to 6 GeV/c in p_T in the most central data, and out to 4.5 GeV/c in the most peripheral centrality bin. The Λ/K^0_S ratio is illustrated in the left panel of Figure 2 as a function of centrality in Au+Au collisions, together with data from p+p collisions at $\sqrt{s} = 200$ GeV.

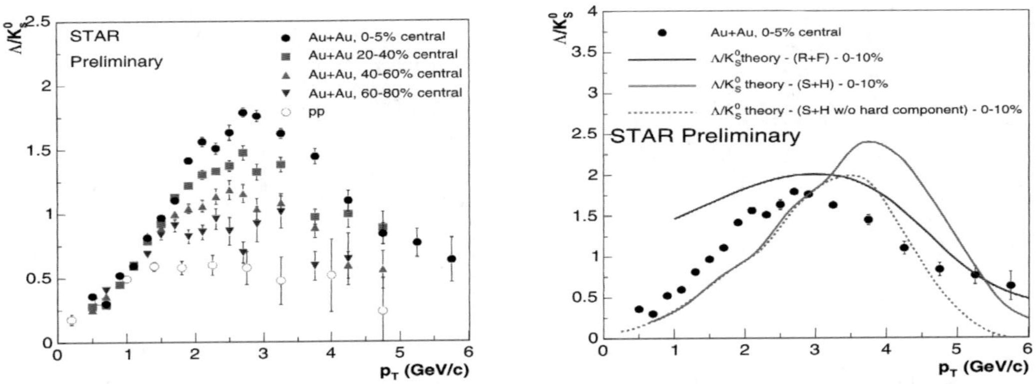

Figure 2. Left panel : The Λ/K^0_S ratios as a function of p_T and centrality for Au+Au and p+p collisions at $\sqrt{s_{NN}} = 200$ GeV. Right panel : The Λ/K^0_S ratio for the most central Au+Au bin together with theoretical predictions.

This data clearly exhibit a strong centrality and p_T dependence of the ratio. The ratio rises smoothly with increasing centrality, reaching a maximum at around $p_T \approx 3$ GeV/c before turning over and reducing towards the values obtained in the peripheral collisions. Due to the limited statistics with this dataset, it is not yet possible to determine at what value of p_T the ratio returns to the value from p+p collisions.

A comparison of this ratio to different coalescence models has been performed and is shown in the right panel of Figure 2. The first model (R+F) uses just recombination of thermal quarks and fragmentation at higher p_T [8] whereas the second model (S+H) allows the coalescence of soft and hard partons [9]. Though the models do not reproduce the data quantitatively, they do reproduce the qualitative details of the ratio well. One important prediction that is not yet available from these models is that of the centrality dependence of the ratios.

Another measurement using strange particles in the intermediate p_T region is that of the elliptic flow (v_2) of particles with respect to the reaction plane of the collision. This measurement is presented in the left panel of Figure 3 for Au+Au collisions at $\sqrt{s_{NN}} = 200$ GeV, averaged over all centralities. The results shown in the left panel of Figure 3 are important as they show that not only do the light particles flow, but the Ξ and Ω also flow with the same magnitude as the lighter hadrons. It has been shown previously that the Ξ and Ω have a smaller radial flow component than the lighter particles, which has been attributed to their presumably low hadronic cross-sections [11]. Though the v_2 is built up throughout the lifetime of the collision, the greatest contribution to the v_2 comes when the nuclear volume is most asymmetric in the early stages. As both the Ω and Ξ exhibit elliptic flow of the same magnitude as the Λ and K^0_S, it appears that this may be partonic in origin. To extend this hypothesis further, the measured v_2 of the different particles is divided by the number of valence quarks and plotted against the p_T similarly scaled by the number of valence quarks in the right panel of Figure 3.

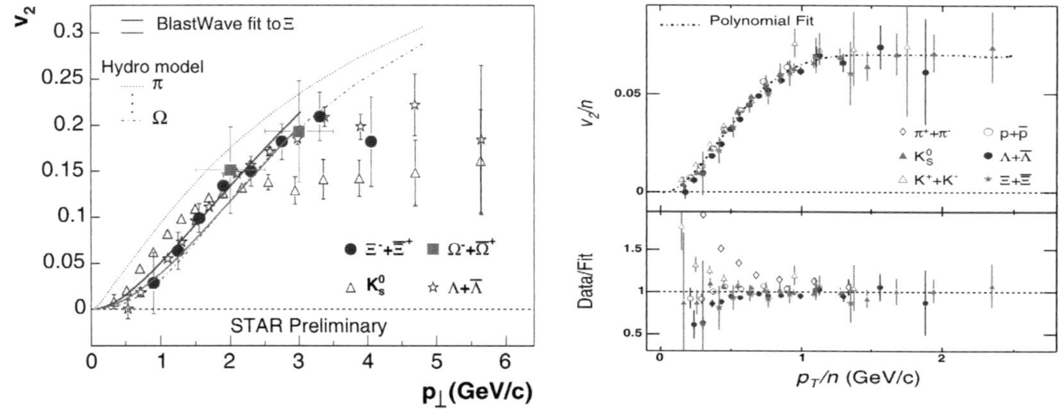

Figure 3. Left panel : The measured v_2 for strange particles at 200 GeV as a function of p_T plotted together with predictions from a hydrodynamical model and a blastwave fit to the pion v_2. Right panel : The elliptic flow at 200 GeV scaled by the number of valence quarks, plotted versus p_T scaled by the same number.

Together with the K^0_S, Λ and Ξ from STAR, data for pions, protons and charged kaons from PHENIX are also shown. The lower panel represents the deviation from a fit to the data. With the exception of the pions, the particles appear to scale well, with the largest deviations occurring at low p_T/n, in a region presumably dominated by hydrodynamics. This scaling is naturally reproduced in the coalescence models where the elliptic flow is built up when the collision is in the deconfined state.

Although there are a number of outstanding questions pertaining to the validity of the use of coalescence models, such as entropy conservation, the inclusion of higher Fock states (though this has been addressed recently [12]) and the use of a thermal mass for the constituent quarks of $\approx$ 300 MeV/c^2, it is clear that they can reproduce a number of features of the data. In all the different implementations, the coalescence mechanism dominates over fragmentation up to transverse momenta of about 6 GeV/c and it is therefore possible to divide the particle production into a number of different regimes : hydrodynamics, recombination, modified fragmentation and pure fragmentation.

Also shown at this conference were first results on strangeness production from Au+Au collisions at $\sqrt{s_{NN}}$ = 62.4 GeV. Excitation functions of both the yield and ratios of the singly and multiply strange particles showed little deviation from a common trend observed between the lower energy SPS data and the full energy RHIC data, mapping out the net baryon density dependence of the energy of the collision [13]. It was also found that the valence quark scaling of the v_2, as discussed earlier, also holds for the data at 62 GeV.

2. Heavy Flavour Production

2.1. Charm production cross section

The measurement of heavy quarks in relativistic heavy ion collisions is important as, due to their large masses, their production can be calculated in pQCD, even at low momenta. The calculation of the charm production cross section at RHIC energies is important in order to map out the area between $\sqrt{s} \approx 60$ GeV and $\sqrt{s} \approx 900$ GeV where there are existing measurements [14].

The reconstruction of charm is different to the well established topological analysis used for strange particles described in the previous section. In STAR, there are two methods used to reconstruct D mesons which are explained below.

Due to their short lifetimes, direct reconstruction of the decay vertex is not possible. Therefore a technique which takes all kaons and pions, identified through their specific ionisation in the TPC, and calculates an invariant mass distribution is used. The background is estimated by performing the same calculation on kaons and pions from different events, a technique which removes all possible correlations. Though this technique allows direct identification of D mesons, the branching ratio into this channel is very small at ≈ 4 %. The left panel of Figure 4 shows a mass plot before and after background subtraction. One setback of this technique is that D^0s are only reconstructed in the low p_T region.

In order to extend the reconstruction of D mesons to higher p_T, they are measured via their semi-leptonic decay channels. Electrons are initially identified in the TPC through their specific ionisation in the momentum range of 1.5 to 8 GeV/c where their dE/dx is higher than that of hadrons. This technique provides an e/h discrimination power of approximately 500. At low momentum ($0.2 < p_T < 3$ GeV/c), electrons can be identified in the STAR Time of Flight system, combining the measured velocity with the dE/dx information from the TPC. A cut on $|1/\beta - 1| \leq 0.003$ eliminates the hadrons crossing the electron band at low momentum. At higher momenta, electron candidates which pass the dE/dx cuts are extrapolated to the Barrel Electro-magnetic Calorimeter (EMC). The p/E_{tower} (where E_{tower} is the energy of the tower) of the candidates is then calculated as shown in the right panel of Figure 4. The electrons, due to their low mass, show a peak at ≈ 1, given by the filled in region in the plot whilst hadrons have a much wider distribution. Energy leakage to neighbouring towers means that the peak is not centred at 1, but this is expected and understood using GEANT simulations.

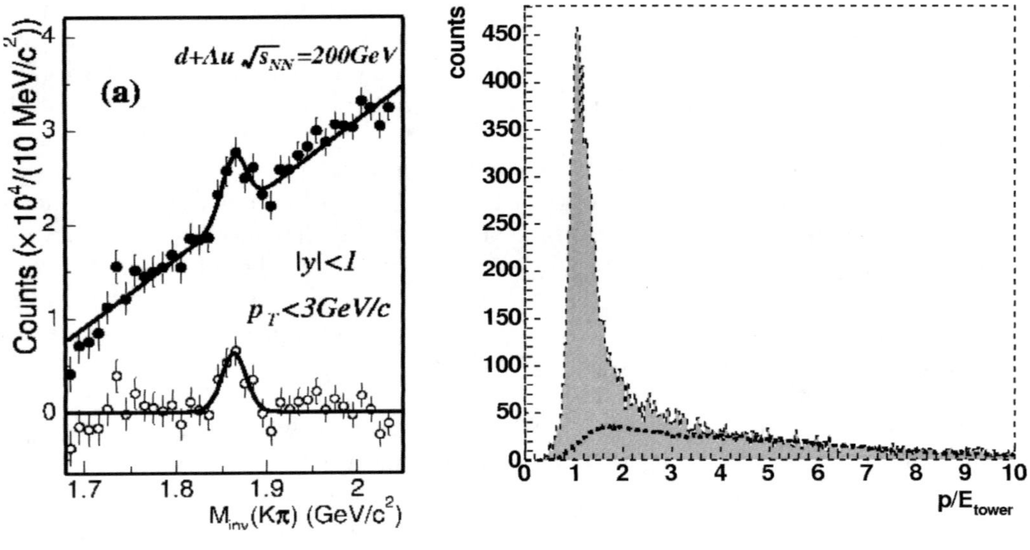

Figure 4. Left panel : Invariant mass distributions for the D^0 before and after background subtraction. Right panel : The measured p/E_{tower} for particles in the electro-magnetic calorimeter.

Another important detector is the Shower Max Detector (SMD) which provides for greater segmentation than the EMC alone. This enables the differentiation between electronic and hadronic signals through their profile: electronic showers tend to be more developed compared to hadronic showers. Whereas the background to the direct reconstruction of D mesons through the hadronic channel mainly comes from combinatorial effects, the semi-leptonic channel also has background sources of electrons. The majority of the background comes from photonic sources, such as π^0 Dalitz decays and photon conversions in the detector material. These can be reduced by calculating the invariant mass of di-electron pairs, and removing those pairs whose $m^2_{inv} <$ 0.02 (GeV/c^2)^2. Non-photonic sources, such as misidentified hadrons, leptons from heavy quarks and Drell-Yan, also contribute to the background but at a much smaller level and are estimated through simulation.

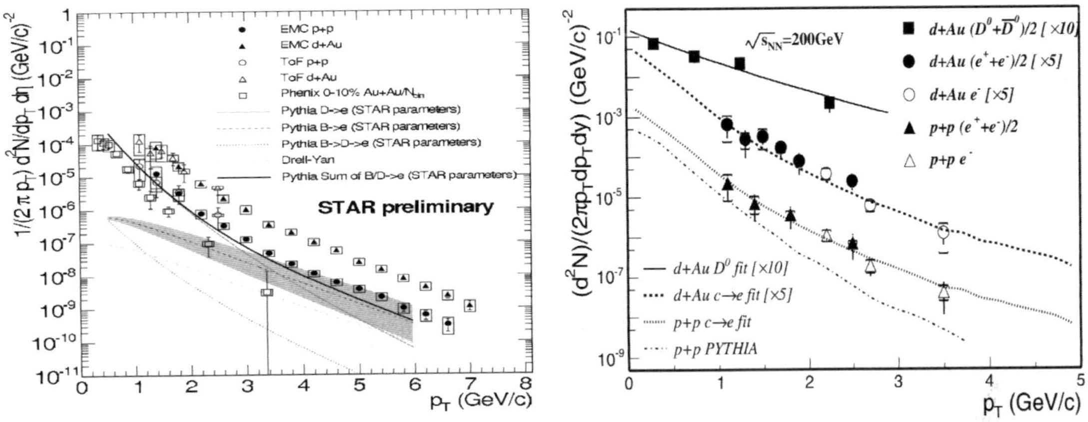

Figure 5. Left panel : STAR electron spectra from the Time Of Flight and Electro-magnetic Calorimeter detectors, together with contributions from a PYTHIA parameterisation. Right panel : Fits to the STAR electron and D^0 spectra.

The left panel in Figure 5 shows the electron spectra in both p+p and d+Au collisions over the full range in p_T using both the TOF and EMC detectors. Also shown are the various contributions to the spectrum from D and B decays, obtained by running Pythia with the parameters outlined elsewhere [15]. A simultaneous fit to the STAR electron data (from dE/dx and TOF only) together with the directly reconstructed D^0s was performed and is presented in the right panel of Figure 5.

The total cross section of charm production per nucleon-nucleon interaction in d+Au collisions from the combined fit is 1.4±0.2±0.4 mb which is in agreement with that obtained from fitting the D^0 spectra separately [16]. This is compared with other measurements as a function of energy in Figure 6, together with LO and NLO pQCD calculations. Although the STAR measurement appears to be high when compared to the PHENIX measurement and pQCD predictions, it is not inconsistent with measurements from cosmic ray experiments at $\sqrt{s}$ = 300 GeV.

2.2. Charm flow

One of the highlights from the strangeness section of this paper is that the measured v_2 of the strange hadrons appears to scale well with the number of valence quarks, which in turn led to the proposal of a number of coalescence/recombination models. It was proposed that if RHIC is hot and dense enough to produce a deconfined medium in Au+Au collisions, then charm quarks may be thermalised [17]. If this is the case, then they may also exhibit finite elliptic flow.

In Au+Au collisions, it is not possible to reconstruct D^0 mesons explicitly due to the large combinatorial background. We therefore must rely upon the technique which reconstructs them through their semi-leptonic decay channel. One important question is whether the elliptic flow of the measured electrons is representative of the meson flow? This is answered via simulations

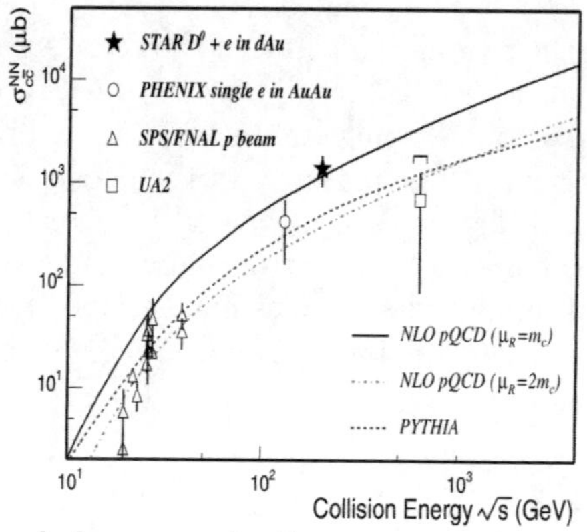

Figure 6. The measured charm production cross section as a function of energy.

and is shown in the left panel of Figure 7. It is evident that if the electron momentum is in the range 2 -3 GeV/c, there is a very strong corrrelation between the emission direction of the D meson and the decay direction of the resulting electron. Only data from the low momentum region is available due to the limited scale of the available calorimeter in the first Au+Au run at $\sqrt{s_{NN}}$ = 200 GeV.

The data used in this analysis are the 80% most central Au+Au collisions, constituting approximately 2 M events from the 2001 data taking period. The same technique as described in the previous section was used to remove background from the data sample. The electrons which still pass the various cuts are estimated to be mainly from D-meson decays (63%), the remainder ((37%) are from γ conversions and π^0 Dalitz decays. The azimuthal angles of the remaining electrons are then correlated with the reaction plane angles and the $\cos(2[\phi\text{-}\psi_{RP}])$ distribution calculated, as described elsewhere [18].

The azimuthal distribution of the background is simulated and then scaled using the π^0 single particle spectrum obtained from PHENIX [19]. This distribution is then subtracted from the total electron spectrum and the v_2 of the remainder is the calculated. This is done as a function of p_T and the result is shown in the right panel of Figure 7, where the error bars are statistical only and do not include an estimated 25% systematic error.

Also shown are data at lower p_T from PHENIX [20] together with predictions from the MP parton cascade model for the v_2 for D mesons themselves, and also the electrons coming from D mesons [21]. There are two scenarios, one where the charm quark has no flow (but the light quarks do), and one where the charm quark has as much flow as the light quarks. Within this model, the measured v_2 of the electrons in this p_T range agrees with the prediction where the charm quark flows as the light quarks, though the large systematical errors, still under investigation, do not rule out the first scenario.

3. Summary

In summary, this paper presented results on strangeness and charm production at RHIC. It was shown that understanding the enhancement of strangeness is a complicated process and a full understanding of the production of strangeness in p+p collisions as a function of energy is still required. In the intermediate p_T region, it was shown that there is a very large difference between the production of strange baryons and mesons which is not reproduced in elementary particle collisions. However, it can be explained by models which invoke the coalescence/recombination of constituent quarks. This leads to a natural explanation of the baryon/meson ratio and the common v_2 dependence with p_T when scaling both variables by the number of valence quarks in

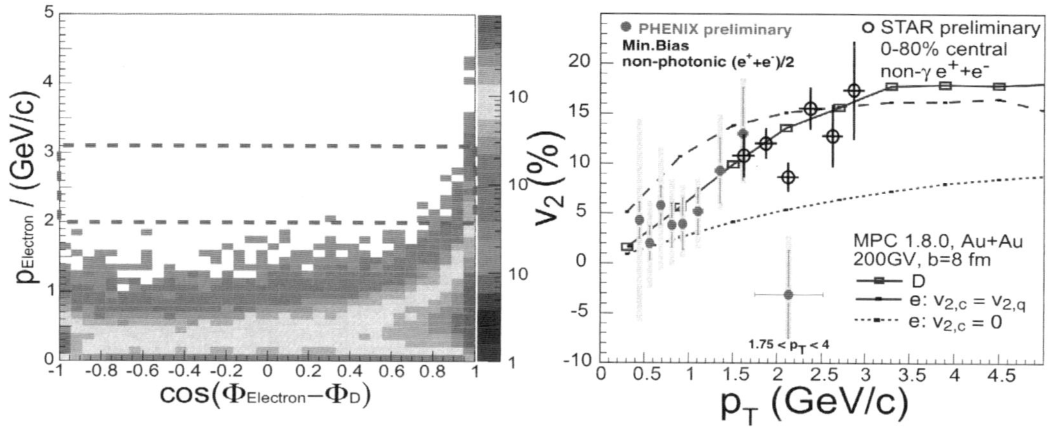

Figure 7. Left panel : The simulated relative angles in azimuth between the electron and the D meson. Right panel : The measured electron v_2 measured in STAR, together with preliminary data from PHENIX at lower p_T, and a theoretical model prediction.

the relevant hadron. The nucleon-nucleon production cross section of charm in d+Au collisions was also presented, obtained through both the hadronic and semi-leptonic decay channels. It was found to be slightly higher than model predictions though is in agreement with the trend of cosmic ray measurements at higher energies. A measurement of the elliptic flow (v_2) of electrons in STAR was presented, under the assumption that they originate from D mesons. The v_2 measurement extended the reach of an earlier PHENIX measurement. The data indicate that the charm quark flows as strongly as the light quarks when compared to a coalescence model, though the systematic errors on the measurement do not allow a definitive statement.

4. Acknowledgements

We thank the RHIC Operations Group and RCF at BNL, and the NERSC Center at LBNL for their support. This work was supported in part by the HENP Divisions of the Office of Science of the U.S. DOE; the U.S. NSF; the BMBF of Germany; IN2P3, RA, RPL, and EMN of France; EPSRC of the United Kingdom; FAPESP of Brazil; the Russian Ministry of Science and Technology; the Ministry of Education and the NNSFC of China; Grant Agency of the Czech Republic, FOM of the Netherlands, DAE, DST, and CSIR of the Government of India; Swiss NSF; the Polish State Committee for Scientific Research; and the STAA of Slovakia

5. References

[1] Rafelski J and Muller B 1982, *Phys. Rev. Lett.* **48** 1066
[2] Tounsi A, Mischke A and Redlich K, 2003, *Nucl. Phys.* A**715** 565
[3] Adler C *et al* (STAR Collaboration) 2002, *Phys. Rev. Lett.* **89** 092301
Adams J *et al* (STAR Collaboration) 2004, *Phys. Rev. Lett.* **92** 182301
[4] Elia D for the NA57 Collaboration 2005, *J. Phys. G: Nucl. Phys.* **31** S135
[5] Caines H (for the STAR Collaboration) 2004, *Preprint* nucl-ex/0412015
[6] Adcox K *et al* (PHENIX Collaboration) 2004, *Phys. Rev.* C **69** 024904
[7] Abreu P *et al* (DELPHI Collaboration) 2000, *Euro. Phys. J.* C **17** 207
[8] Fries R J, Muller B, Nonaka C, Bass S A 2003, *Phys. Rev.* C **68** 044902
[9] Greco V, Ko C M and Levai P 2003, *Phys. Rev.* C **68**, 034904
[10] Hwa R C and Yang C B 2003, *Phys. Rev.* C **67** 034902
[11] Adams J *et al* (STAR Collaboration) 2004, *Phys. Rev. Lett.* **92** 182301
[12] Fries R J, Muller B, Nonaka C, Bass S A 2005,*Preprint* nucl-th/0503003
[13] See proceedings of Strange Quark Matter 2004 to be published in *J. Phys. G: Nucl. Phys.*
[14] Rakobolskaya I N, Roganova T M and Sveshnikova L G 2003, *Nucl. Phys. B – ProceedingsSupplements* **122** 353

[15] Suaide A A P(for the STAR Collaboration) 2004, *J. Phys. G: Nucl. Phys.* **30** S1179
[16] Adams J *et al* (STAR Collaboration) 2005, *Phys. Rev. Lett.* **94** 062301
[17] Greco V, Ko C M and Rapp R 2004, *Phys. Lett.* B**595** 202
[18] Laue F (for the STAR Collaboration) 2005, *J. Phys. G: Nucl. Phys.* **31** S27
[19] Adler S S *et al* 2003, *Phys. Rev. Lett.* **91** 0702301
[20] Kaneta M *et al* 2004, *J. Phys. G: Nucl. Phys.* **30** S1217
[21] Molnar D 2005, *J. Phys. G: Nucl. Phys.* **31** S421

Institute of Physics Publishing Journal of Physics: Conference Series **50** (2006) 201–207
doi:10.1088/1742-6596/50/1/023 5th International Conference on Physics and Astrophysics of Quark Gluon Plasma

Structure of Vacuum, Matter and Antimatter: A Mechanism for Cold Compression

Walter Greiner

Frankfurt Institute for Advanced Studies,
J. W. Goethe-Universität,
Max-von-Laue-Str. 1,
D–60438 Frankfurt/Main, Germany

E-mail: greiner@fias.uni-frankfurt.de

Abstract. I will first present the vacuum for the e^+-e^- field of QED and show how it is modified for baryons in nuclear environment. Then I discuss the possibility of producing new types of nuclear systems by implanting an antibaryon into ordinary nuclei. The structure of nuclei containing one antiproton or antilambda is investigated within the framework of a relativistic mean-field model. Self-consistent calculations predict an enhanced binding and considerable compression in such systems as compared with normal nuclei. I present arguments that the life time of such nuclei with respect to the antibaryon annihilation might be long enough for their observation. This yields a mechanism for cold compression.

1. Introduction

It is generally accepted that physical vacuum has nontrivial structure. This conclusion was first made by Dirac on the basis of his famous equation for a fermion field which describes simultaneously particles and antiparticles. The Dirac equation in the vacuum has a simple form

$$(i\gamma^\mu \partial_\mu - m)\Psi(x) = 0 \ , \tag{1}$$

where $\gamma^\mu = (\gamma^0, \boldsymbol{\gamma})$ are Dirac matrices, m is the fermion mass and $\Psi(x)$ is a 4-component spinor field. For a plane wave solution $\Psi(x) = e^{-ipx}u_p$ this equation is written as

$$(\widehat{p} - m)u_p = 0 \ , \tag{2}$$

where $\widehat{p} = \gamma^0 E - \boldsymbol{\gamma}\mathbf{p}$. Multiplying by $(\widehat{p}+m)$ and requiring that $u_p \neq 0$ one obtains the equation $E^2 - \mathbf{p}^2 - m^2 = 0$ which has two solutions

$$E^{\pm}(\mathbf{p}) = \pm\sqrt{\mathbf{p}^2 + m^2} \ . \tag{3}$$

Here the + sign corresponds to particles with positive energy $E_N(\mathbf{p}) = E^+(\mathbf{p})$, while the − sign corresponds to solutions with negative energy. To ensure stability of the physical vacuum Dirac has assumed that these negative-energy states are occupied forming what is called now the Dirac sea. Then the second solution of eq. (3) receives natural interpretation: it describes holes in the Dirac sea. These holes are identified with antiparticles. Their energies are obviously given by

$E_{\overline{N}}(\mathbf{p}) = -E^{-}(-\mathbf{p}) = \sqrt{\mathbf{p}^2 + m^2}$. Unfortunately, the Dirac sea brings divergent contributions to physical quantities such as energy density, and one should introduce a proper regularization scheme to get rid off these divergences. This picture has received numerous confirmations in quantum electrodynamics and other fields.

One of the most fascinating aspects is the structure of the vacuum in QED and its change into charged vacuum states under the influence of strong (supercritical) electric fields [1], which we briefly remind of.

Fig. 1 shows the diving of the deeply bound states into the lower energy continuum of the Dirac equation. In the supercritical case the dived state is degenerate with the (occupied)

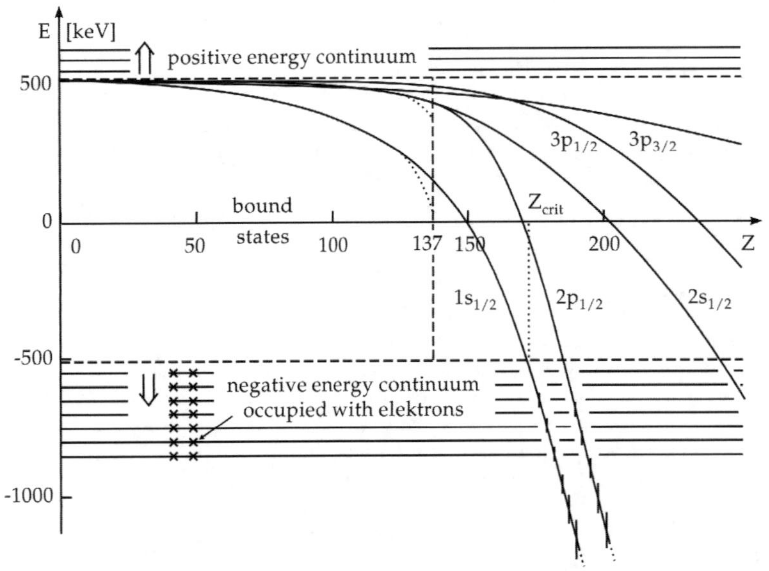

Figure 1. Lowest bound states of the Dirac equation for nuclei with charge Z. While the Sommerfeld fine-structure energies (dashed line) for $\xi = 1$ (s states) end at $Z = 137$, the solutions for extended Coulomb potentials (full line) can be traced down to the negative-energy continuum reached at the critical charge $Z_{\rm cr}$ for the $1s$ state. The bound states entering the continuum obtain a spreading witdth as indicated.

negative electron states. Hence spontaneous e^+e^- *paircreation* becomes possible, where an electron from the Dirac sea occupies the additional state, leaving a hole in the sea which escapes as a positron while the electron's charge remains near the source. This is a fundamentally new process, whereby the neutral vacuum of QED becomes unstable in supercritical electrical fields. It decays within about $10^{-19}s$ into a charged vacuum. The charged vacuum is now stable due to the Pauli principle, that is the number of emitted particles remains finite. The vacuum is first charged twice because two electrons with opposite spins can occupy the $1s$ shell. After the $2p_{1/2}$ shell has dived beyond $Z_{cr} = 185$, the vacuum is charged four times, etc. This change of the vacuum structure is not a perturbative effect, as are the radiative QED effects (vacuum polarization, self-energy, etc.). It is hoped to observe it in narrow energy windows of very heavy ions (e.g. U + Cm) near the Coulomb barrier, where "sticking" of the nuclei by about 10^{-19} s could be possible. Such sticking is important for observing a line-structure in the positron spectrum.

2. The vacuum structure in nuclear physics

It has been noticed already many years ago (see e. g. ref. [2]) that nuclear physics may provide a unique laboratory for investigating the Dirac picture of vacuum. The basis for this is given by relativistic mean-field models which are widely used now for describing nuclear matter and finite nuclei. Within this approach nucleons are described by the Dirac equation coupled to scalar and vector meson fields. Scalar S and vector V potentials generated by these fields modify plane-wave solutions of the Dirac equation as follows

$$E^{\pm}(\mathbf{p}) = V \pm \sqrt{\mathbf{p}^2 + (m-S)^2}\,. \tag{4}$$

Again, the + sign corresponds to nucleons with positive energy $E_N(\mathbf{p}) = V + \sqrt{\mathbf{p}^2 + (m-S)^2}$, and the − sign corresponds to antinucleons with energy $E_{\bar{N}}(\mathbf{p}) = -E^{-}(-\mathbf{p}) = -V + \sqrt{\mathbf{p}^2 + (m-S)^2}$. It is remarkable that changing sign of the vector potential for antinucleons is exactly what is expected from the G-parity transformation of the nucleon potential. As follows from eq. (4), in nuclear environment the spectrum of single-particle states of the Dirac equation is modified in two ways. First, the mass gap between positive- and negative-energy states, $2(m-S)$, is reduced due to the scalar potential and second, all states are shifted upwards due to the vector potential. These changes are illustrated in Fig. 2.

It is well known from nuclear phenomenology that a good description of nuclear ground state is achieved with $S \simeq 350$ MeV and $V \simeq 300$ MeV so that the net potential for nucleons is $V - S \simeq -50$ MeV. Using the same values one obtains for antinucleons very a deep potential, $-V - S \simeq -650$ MeV. Such a potential would produce many strongly bound states in the Dirac sea. However, if these states are occupied they are hidden from the direct observation. Only creating a hole in this sea, i.e. inserting a real antibaryon into the nucleus, would produce an observable effect. If this picture is correct one should expect the existence of strongly bound states of antinucleons with nuclei. Below we report on our recent studies of antibaryon-doped nuclear systems [3, 4].

3. Systems made of baryons and antibaryons

Unlike some previous works, we take into account the rearrangement of nuclear structure due to the presence of a real antibaryon. The structure of such systems is calculated using several versions of the relativistic mean–field model (RMF): TM1 [5], NL3 and NL-Z2 [6]. Their parameters were found by fitting binding energies and charge form-factors of spherical nuclei from ^{16}O to ^{208}Pb. The general Lagrangian of the RMF model is written as

$$\begin{aligned}
\mathcal{L} = & \sum_{j=B,\overline{B}} \overline{\psi}_j \left(i\gamma^\mu \partial_\mu - m_j\right)\psi_j \\
& + \frac{1}{2}\partial^\mu\sigma\partial_\mu\sigma - \frac{1}{2}m_\sigma^2\sigma^2 - \frac{b}{3}\sigma^3 - \frac{c}{4}\sigma^4 \\
& - \frac{1}{4}\omega^{\mu\nu}\omega_{\mu\nu} + \frac{1}{2}m_\omega^2\omega^\mu\omega_\mu + \frac{d}{4}(\omega^\mu\omega_\mu)^2 \\
& - \frac{1}{4}\vec{\rho}^{\mu\nu}\vec{\rho}_{\mu\nu} + \frac{1}{2}m_\rho^2\vec{\rho}^\mu\vec{\rho}_\mu \\
& + \sum_{j=B,\overline{B}} \overline{\psi}_j \left(g_{\sigma j}\sigma + g_{\omega j}\omega^\mu\gamma_\mu + g_{\rho j}\vec{\rho}^\mu\gamma_\mu\vec{\tau}_j\right)\psi_j \\
& + \text{Coulomb part}
\end{aligned} \tag{5}$$

Here summation includes valence baryons B, in fact the nucleons forming a nucleus, and valence antibaryons $\overline{B}$ inserted in the nucleus. They are treated as Dirac particles coupled to the scalar-isoscalar (σ), vector-isoscalar (ω) and vector-isovector ($\vec{\rho}$) meson fields. The calculations are

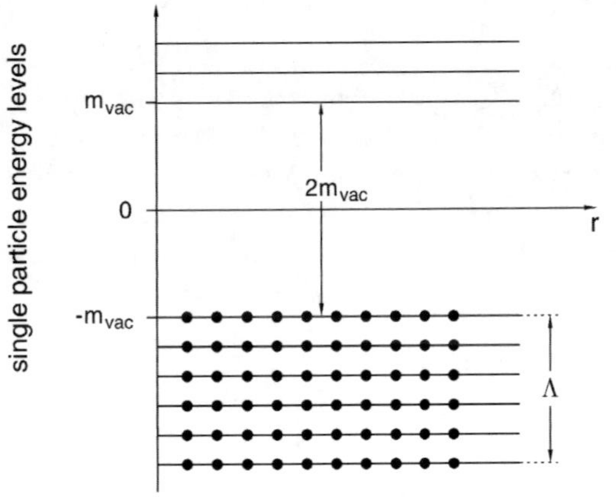

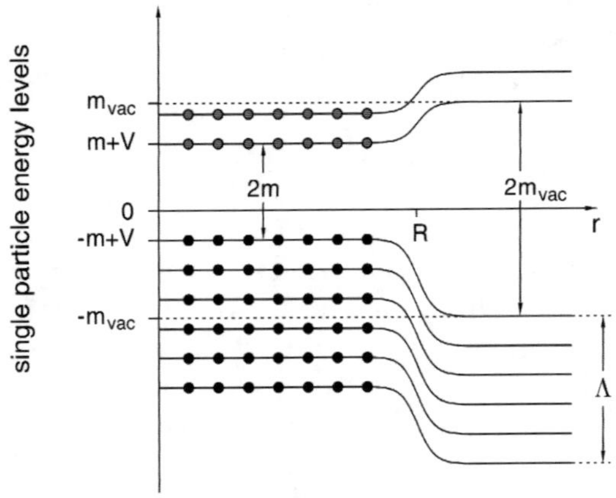

Figure 2. Schematic spectrum of Dirac equation in vacuum (upper panel) and in a nucleus of radius R (lower panel). A divergent contribution of negative-energy states is often regularized by introducing a cut-off momentum Λ

carried out within the mean-field approximation where the meson fields are replaced by their expectation values. Also a "no-sea" approximation is used. This implies that all occupied states of the Dirac sea are "integrated out" so that they do not appear explicitly. It is assumed that their effect is taken into account by nonlinear terms in the meson Lagrangian. Most calculations are done with antibaryon coupling constants which are given by the G-parity transformation ($g_{\sigma\overline{N}} = g_{\sigma N}$, $g_{\omega\overline{N}} = -g_{\omega N}$) and $SU(3)$ flavor symmetry ($g_{\sigma\overline{\Lambda}} = \frac{2}{3}g_{\sigma\overline{N}}$, $g_{\omega\overline{\Lambda}} = \frac{2}{3}g_{\omega\overline{N}}$). In isosymmetric static systems the scalar and vector potentials for nucleons are expressed as $S = g_{\sigma N}\sigma$ and $V = g_{\omega N}\omega^0$.

Following the procedure suggested in Ref. [7] and assuming the axial symmetry of the nuclear system, we solve effective Schrödinger equations for nucleons and an antibaryon together with differential equations for mean meson and Coulomb fields. We explicitly take into account the antibaryon contributions to the scalar and vector densities. It is important that antibaryons give a negative contribution to the vector density, while a positive contribution to the scalar density. This leads to increased attraction and decreased repulsion for surrounding nucleons. To maximize attraction, nucleons move to the center of the nucleus, where the antiproton has its largest occupation probability. This gives rise to a strong local compression of the nucleus

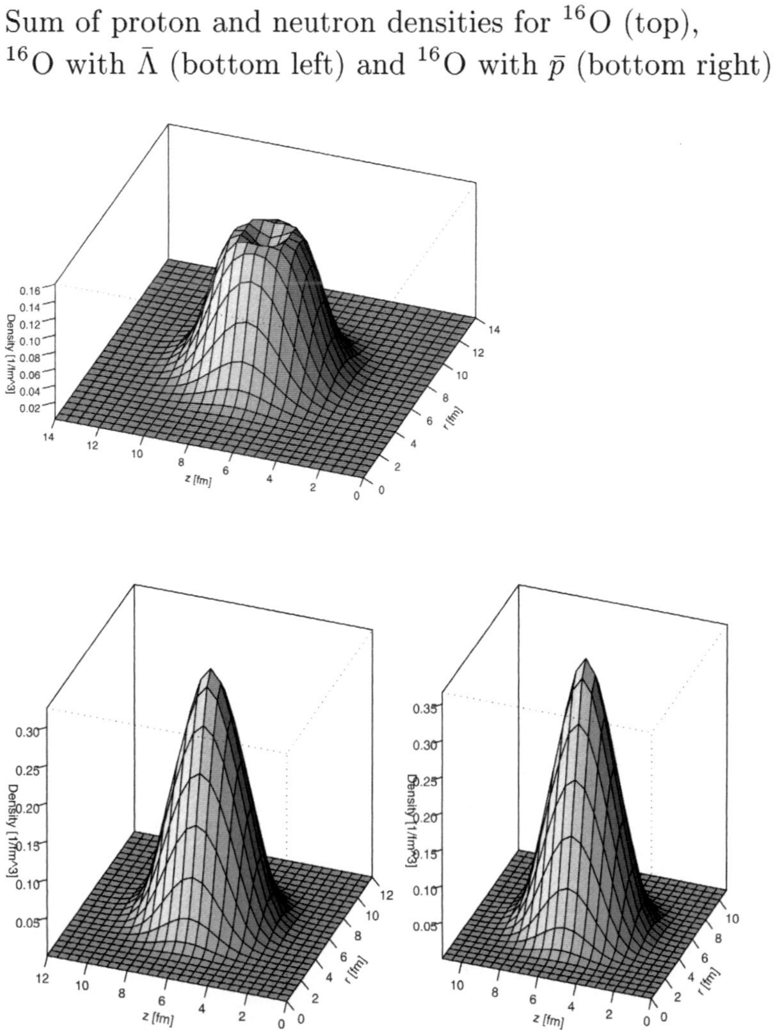

Figure 3. Sum of neutron and proton densities for ^{16}O (top), ^{16}O with $\overline{p}$ (bottom right) and ^{16}O with $\overline{\Lambda}$ (bottom left) calculated with the parametrization NL-Z2.

and leads to a dramatic rearrangement of its structure.

Results for the ^{16}O nucleus are presented in Fig. 3 which shows 3d plots of nucleon density distributions. The calculations show that inserting an antiproton into the ^{16}O nucleus leads to the increase of central nucleon density by a factor 2–4 depending on the parametrization. Due to a very deep antiproton potential the binding energy of the whole system is increased significantly as compared with 130 MeV for normal ^{16}O. The calculated binding energies of the $\overline{p}-^{16}$O system are 830, 1050 and 1160 MeV for the NL-Z2, NL3 and TM1, respectively. Due to this anomalous binding we call such systems super bound nuclei (SBN). In the case of antilambdas we rescale the coupling constants with a factor 2/3 that leads to the binding energy of 560÷700 MeV for the $\overline{\Lambda}-^{16}$O system. As a second example, we investigate the effect of a single antiproton inserted into the ^{8}Be nucleus. The normal ^{8}Be nucleus is not spherical, exhibiting a clearly visible 2α structure with the ground state deformation $\beta_2 \simeq 1.20$. As seen in Fig. 4, inserting an antiproton in ^{8}Be results in a much less elongated shape ($\beta_2 \simeq 0.23$) and disappearance of its cluster structure. The binding energy increases from 53 MeV to about 700 MeV. Similar, but weaker effects have been predicted [8] for the K^- bound state in the ^{8}Be

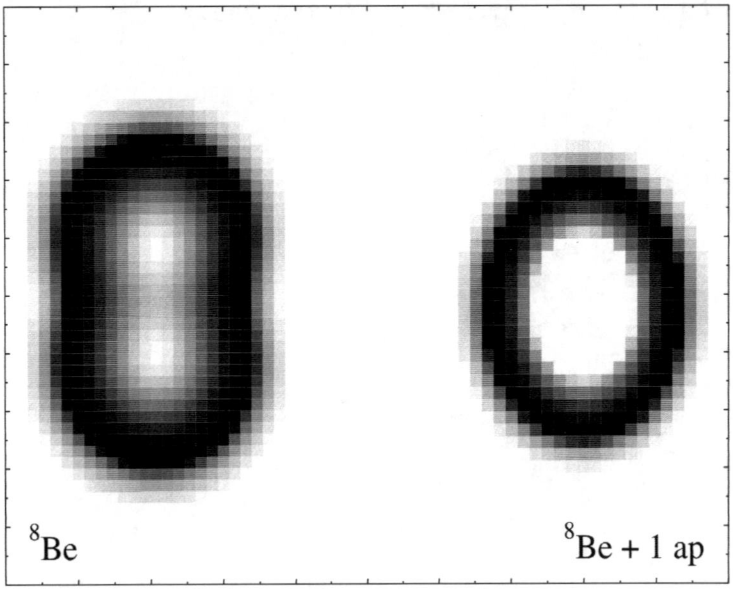

Figure 4. Contour plot of nucleon densities for ^{8}Be without (left) and with (right) antiproton calculated with the parametrization NL3.

nucleus.

The calculations have been performed also with reduced antinucleon coupling constants as compared to the G-parity prescription. We have found that the main conclusions about enhanced binding and considerable compression of $\bar{p}$-doped nuclei remain valid even when coupling constants are reduced by factor 3 or so.

4. Life-time estimates

The crucial question concerning possible observation of the super-bound nuclei (SBNs) is their life time. The main decay channel for such states is the annihilation of antibaryons on surrounding nucleons. The energy available for annihilation of a bound antinucleon equals $Q = 2m_N - B_N - B_{\overline{N}}$, where B_N and $B_{\overline{N}}$ are the corresponding binding energies. In our case this energy is at least by a factor 2 smaller as compared with the vacuum value of $2m_N$. This should lead to a significant suppression of the available phase space and thus to a reduced annihilation rate in medium. We have performed detailed calculations assuming that the annihilation rates into different channels are proportional to the available phase space. All intermediate states with heavy mesons like ρ, ω, η as well as multi-pion channels have been considered. Our conclusion is that decreasing the Q value from 2 GeV to 1 GeV may lead to the reduction of total annihilation rate by factor 20÷30. Then we estimate the SBN life times on the level of 5-25 fm/c which makes their observation feasible. This large margin in the life times is mainly caused by uncertainties in the overlap integral between antinucleon and nucleon scalar densities. Longer life times may be expected for SBNs containing antihyperons. The reason is that instead of pions more heavy kaons must be produced in this case. We have also analyzed multi-nucleon annihilation channels (Pontecorvo-like reactions) and have found their contribution to be less than 40% of the single-nucleon annihilation [3, 4].

We believe that such exotic nuclear states can be produced by using antiproton beams of multi-GeV energy, e.g. at the future GSI facility. It is well known that low-energy antiprotons annihilate on the nuclear periphery (at about 5% of the normal density). Since the annihilation cross section drops significantly with energy, a high-energy antiproton can penetrate deeper into the nuclear interior. Then it can be stopped there in an inelastic collision with a nucleon, e. g. via the reaction $A(\bar{p}, N\pi)_{\bar{p}}A'$, leading to the formation of a $\bar{p}$-doped nucleus. Reactions like $A(\bar{p}, \Lambda)_{\overline{\Lambda}}A'$ can be used to produce a $\overline{\Lambda}$-doped nuclei. Fast nucleons or lambdas can be used for triggering such events. In order to be captured by a target nucleus final antibaryons must be slow in the lab frame. Rough estimates of the SBN formation probability in a central $\bar{p}A$ collision give the values $10^{-5} - 10^{-6}$. With the $\bar{p}$ beam luminocity of $2{\cdot}10^{32}$ $\mathrm{cm}^{-2}\mathrm{s}^{-1}$ planned at GSI this will correspond to the reaction rate of a few tens of desired events per second.

Several signatures of SBNs can be used for their experimental observation. First, annihilation of a bound antibaryon can proceed via emission of a single photon, pion or kaon with an energy of about 1 GeV (such annihilation channels are forbidden in vacuum). So one may search for relatively sharp lines, with width of 10÷40 MeV, around this energy, emitted isotropically in the SBN rest frame. Another signal may come from explosive disintegration of the compressed nucleus after the antibaryon annihilation. This can be observed by measuring radial collective velocities of nuclear fragments. Such systems serve as a tool for creating cold compression of nuclear matter, in contrast to nuclear compression in high energy nucleus-nucleus encounters where nuclear shock waves – due to the Rankine-Hugoniot relation – yield compression and heating closely connected. For more details on all the points discussed here I refer to our comprehensive paper by Mishustin et al. [4].

I am grateful to T. J. Bürvenich, I.N. Mishustin and L.M. Satarov for fruitful discussions and help in the preparation of this talk.

References

[1] W. Greiner, B. Müller, J. Rafelksi, *Quantum electrodynamics of strong fields*, Springer Verlag, 2nd edition, December 1985

[2] N. Auerbach, A.S. Goldhaber, M.B. Johnson, L.D. Miller, A. Picklesimer, Phys. Lett. **B182** (1986) 221.

[3] T. Bürvenich, I.N. Mishustin, L.M. Satarov, J. A. Maruhn, H. Stöcker, W. Greiner, Phys. Lett. **B542** (2002) 261.

[4] I. N. Mishustin, L. M. Satarov, T. J. Bürvenich, H. Stöcker, and W. Greiner, Phys. Rev. C (2005), in print

[5] Y. Sugahara and H. Toki, Nucl. Phys. **A579** (1994) 557.

[6] M. Bender, K. Rutz, P.-G. Reinhard, J.A. Maruhn, and W. Greiner, Phys. Rev. **C60** (1999) 34304.

[7] G. Mao, H. Stöcker, and W. Greiner, Int. J. Mod. Phys. **E8** (1999) 389.

[8] Y. Akaishi and T. Yamazaki, Phys. Rev. **C65** (2002) 044005.

[9] I.N. Mishustin, L.M. Satarov, H. Stoecker, W. Greiner, Phys. Rev. **C 59** (1999) 3343.

Institute of Physics Publishing
doi:10.1088/1742-6596/50/1/024

Journal of Physics: Conference Series **50** (2006) 208–215
5th International Conference on Physics and Astrophysics of Quark Gluon Plasma

Quark deconfinement in neutron stars and gamma-ray bursts

Ignazio Bombaci

Dipartimento di Fisica "E. Fermi", Università di Pisa and INFN sezione di Pisa. Largo Bruno Pontecorvo 3, I-56127 Pisa, Italy.

Abstract. We discuss a model for the energy source of gamma-ray burst which is based on the "delayed conversion" of a metastable pure hadronic compact star ("neutron star") to a quark star (hybrid star or strange star).

1. Introduction

Due to the asymptotic freedom property of quantum chromodynamics (QCD), strong interacting matter at high temperature or/and density should be described in terms of quark and gluon degrees of freedom (Quark Gluon Plasma phase, QGP) rather than hadronic degrees of freedom (hadronic phase). A very schematic picture of the QCD phase diagram is shown in Fig. 1. Lattice calculations at zero baryon chemical potential (zero baryon density) suggest that at high temperature, for physical values of the quark masses, the transition to QGP is a "crossover" [1, 2] rather than a phase transition (which would be signaled by singularities in the thermodynamic observables). At high baryon chemical potential, many QCD inspired models suggest the transition to be a first order phase transition [3, 4]. In this region many possible color superconducting phases of quark matter are expected [5, 6, 7, 8, 9].

Neutron stars, the compact remnants of supernova (SN) explosions, can be viewed as a natural laboratory to test the low temperature and high baryon chemical potential region of the QCD phase diagram. In fact, neutron star structure calculations, based on a large variety of modern equations of state (EOS) of hadronic matter, predict a maximum stellar central density (the one for the maximum mass star configuration) in the range of $4-8$ times the saturation density ($\rho_0 \sim 2.8 \times 10^{14}$ g cm^{-3}) of nuclear matter. Thus the core of a neutron star is one of the best candidates in the universe where a deconfined phase of strong interacting matter could be found.

In this paper, we explore some of the most exciting consequences of the quark deconfinement phase transition in neutrons stars. Particularly, we show [10, 11, 12, 13] that above a threshold value of the stellar mass (stellar central pressure) a neutron star is metastable to the "decay" (conversion) to a compact star containing quark matter (a hybrid star or a strange star). The *mean lifetime* of the metastable configuration dramatically depends on the value of the stellar central pressure. The stellar conversion process liberates an enormous amount of energy (of the order of a few 10^{53} ergs [14]) which could cause a second delayed explosion [1]. Under favourable

[1] delayed with respect to the first explosion, *i.e.* the SN explosion, which formed the neutron star. In other words, we assume that quark matter is not formed during the stellar core collapse generating the SN explosion,

physical conditions this second explosion could be the energy source of a powerful gamma-ray burst (GRB) [11, 12, 13]. Thus this scenario is able to explain the inferred "delayed" connection between SN explosions and GRBs [16, 17, 18].

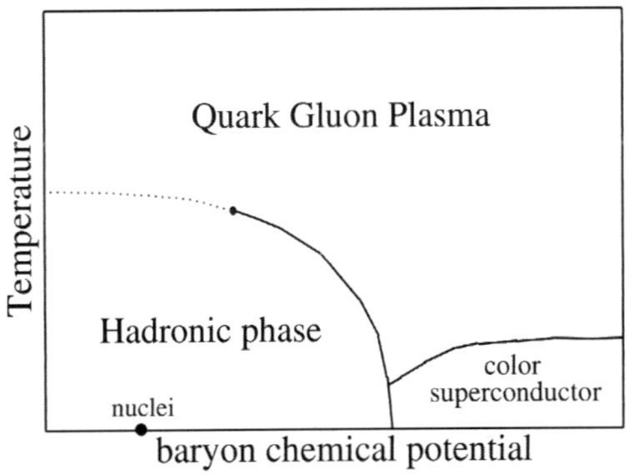

Figure 1. A schematic picture of the QCD phase diagram in the baryon chemical potential–temperature plane.

2. Classification of neutron stars

The possible presence of quark matter (QM) in neutron stars was realized soon after the introduction of quarks as the fundamentals constituents of hadrons [19], and since then this possibility has been widely explored (see *e.g.* [20, 21, 22, 23] and references therein quoted). Compact stars which possess a quark matter core, either as a mixed phase of deconfined quarks and hadrons or as a pure QM phase, are called *hybrid stars* (HySs). In the following of this paper, the more *traditional* neutron stars in which no fraction of QM is present, will be referred to as *pure hadronic stars* (HSs). A more exotic alternative to the existence of hybrid stars, is the conceivable existence of self-bound compact stars consisting completely of a mixture of *up* (u), *down* (d) and *strange* (s) quarks (together with an appropriate number of electrons to guarantee electrical neutrality) satisfying the Bodmer–Terazawa–Witten (BTW) hypothesis [24]. These stars have been called *strange stars* (SSs) [25], and their constituent matter as strange quark matter (SQM) [26]. Some indirect observational evidence for the possible existence of strange stars has been given in ref. [27, 28, 29, 30]. In the following, we will refer to hybrid stars and strange stars collectively as *quark stars* (QSs). Compact stars of each of the two classes (HSs and QSs) could be endowed with strong magnetic fields and could manifest their presence in the universe as pulsars or compact X-ray sources in binary systems.

3. Quantum nucleation of quark matter in hadronic stars

In the following we assume the quark deconfinement transition to be a first order phase transition [3, 4]. In our astrophysical scenario, we consider a pure HS whose central pressure (density) is increasing due to spin-down or due to mass accretion, *e.g.*, from the material left by the supernova explosion (see Fig. 5), or from a companion star. As the central pressure exceeds a threshold value P_0^*, the hadronic matter at the center of the star will be in a metastable (*super-compressed*) state, which could decay to the stable quark matter phase. This threshold point (static transition point) for the metastability of the hadronic phase can be determined using the Gibbs' criterion for phase equilibrium

$$\mu_H = \mu_{Q*} \equiv \mu_0^* \,, \qquad P_H(\mu_0^*) = P_{Q*}(\mu_0^*) \equiv P_0^* \,, \qquad T_H = T_{Q*} = 0 \tag{1}$$

or during the early stages of the evolution of the new born neutron star [15].

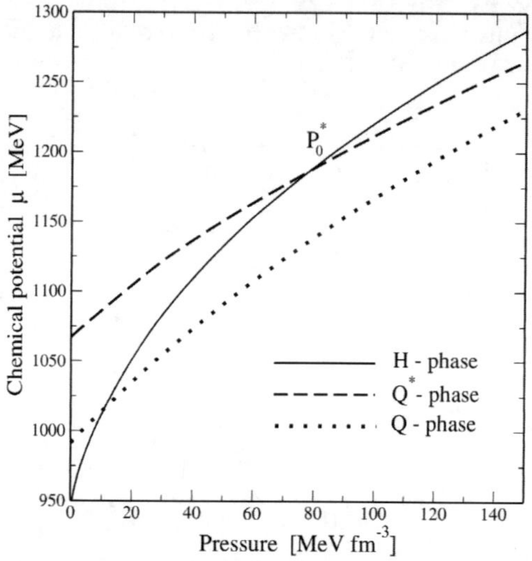

Figure 2. Phase chemical potential (Gibbs free energy per baryon) for the three phases (H, Q, and Q*) of matter discussed in the text as a function of the stellar central pressure. The H-phase is described with the GM3 model, for the Q- and Q*-phase we take $B = 98.83$ MeV/fm^3 within the MIT bag model.

between the the H- and the Q*-phase. Here μ_{Q*} and μ_H are the Gibbs free energies per baryon (*phase chemical potentials*) in the two matter phases, and are given by the relations

$$\mu_{Q*} = \frac{\varepsilon_{Q*} + P_{Q*}}{n_{b,Q*}}, \qquad \mu_H = \frac{\varepsilon_H + P_H}{n_{b,H}}, \tag{2}$$

ε_{Q*} (ε_H), P_{Q*} (P_H) and $n_{b,Q*}$ ($n_{b,H}$) denote respectively the total (*i.e.,* including leptonic contributions) energy density, the total pressure and baryon number density for the quark (hadron) phase.

The kinetic process that could form the new phase at zero temperature is quantum nucleation. In this process, the formation of a real spherical drop (with radius $\mathcal{R}$) of the new phase can be realized only as a results of the quantum penetration of the energy barrier

$$U(\mathcal{R}) = \frac{4}{3}\pi\mathcal{R}^3 n_{b,Q*}(\mu_{Q*} - \mu_H) + 4\pi\sigma\mathcal{R}^2 + 8\pi\gamma\mathcal{R}, \tag{3}$$

separating the metastable hadronic phase from the stable quark phase. The volume energy coefficient in Eq. (3) is proportional to the difference between the Gibbs free energy per baryon in the two phases at a fixed pressure $P_{Q*} = P_H$. The second term in Eq. (3) constitutes the drop surface energy, with σ being the surface tension for the surface separating the quark phase from the hadronic phase. The value of the surface tension σ is poorly known, and typical values used in the literature range within 10–50 MeV/fm^2 [31, 32]. Finally, the third term in Eq. (3), $E_{curv} = 8\pi\gamma\mathcal{R}$, is the so called curvature energy.

A virtual (sub-critical) droplet of deconfined QM moves back and forth in the potential energy well separating the two matter phases on a time scale $\nu_0^{-1} \sim 10^{-23}$ seconds, which is set by the strong interactions. This time scale is many orders of magnitude shorter than the typical time scale for the weak interactions, therefore quark flavor must be conserved during the early stages of the deconfinement transition. We will refer to this form of deconfined matter, in which the flavor content is equal to that of the β-stable hadronic system at the same pressure, as the

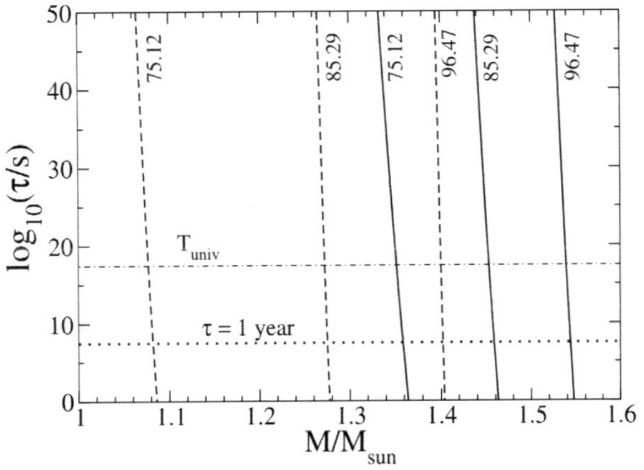

Figure 3. Nucleation time as a function of the gravitational mass of the hadronic star. Solid lines correspond to a value of $\sigma = 30$ MeV/fm^2 whereas dashed ones are for $\sigma = 10$ MeV/fm^2. In both cases $\gamma = 0$. The nucleation time corresponding to one year is shown by the dotted horizontal line. The different values of the bag constant (in units of MeV/fm^3) are plotted next to each curve. The hadronic phase is described with the GM1 model.

Q*-phase [12]. Soon afterwards a real (critical-size) drop of QM is formed, the weak interactions will have enough time to act, changing the quark flavor fraction of the deconfined droplet to lower its energy, and a droplet of β-stable SQM is formed (hereafter the Q-phase). Subsequently this drop of Q-matter will grow very rapidly and the original HS will be converted to a hybrid star or to a strange star, depending on the detail of the EOS for quark matter employed to model the phase transition.

To describe the hadronic phase, we use a model based on a relativistic lagrangian of hadrons interacting via the exchange of σ, ρ and ω mesons. Particularly, we use the Glendenning–Moszkowski (GM) EOS [33]. For the quark phase we use the phenomenological EOS of ref. [26] which is based on the MIT bag model for hadrons. The parameters here are: the quark masses (m_u, m_d, m_s), the bag constant B and the QCD structure constant α_s. In the present work, we take $m_u = m_d = 0$, $m_s = 150$ MeV and $\alpha_s = 0$

To calculate the nucleation time τ, *i.e.*, the time needed to form the first critical droplet of deconfined QM in the hadronic medium, one can use the relativistic quantum nucleation theory [32]. The nucleation time τ, can be calculated for different values of the stellar central pressure P_c. In Fig. 3, we plot τ as a function of the gravitational mass M_{HS} of the HS corresponding to the given value of the central pressure, as implied by the solution of the stellar structure equations in general relativity [34] for the pure HS sequences. Each curve refers to a different value of the bag constant and surface tension σ. As we can see, from the results in Fig. 3, a metastable hadronic star can have a mean-life time many orders of magnitude larger than the age of the universe $T_{univ} = (4.32 \pm 0.06) \times 10^{17}$ s. As the star accretes a small amount of mass (of the order of a few per cent of the mass of the sun), the consequential increase of the central pressure leads to a huge reduction of the nucleation time and, as a result, to a dramatic reduction of the HS *mean-life time*.

To summarize, pure hadronic stars having a central pressure larger than the static transition pressure P_0^* for the formation of the Q*-phase are metastable to the "decay" (conversion) to a more compact stellar configuration in which deconfined QM is present (HyS or SS). These metastable HS have a *mean-life time* which is related to the nucleation time to form the first

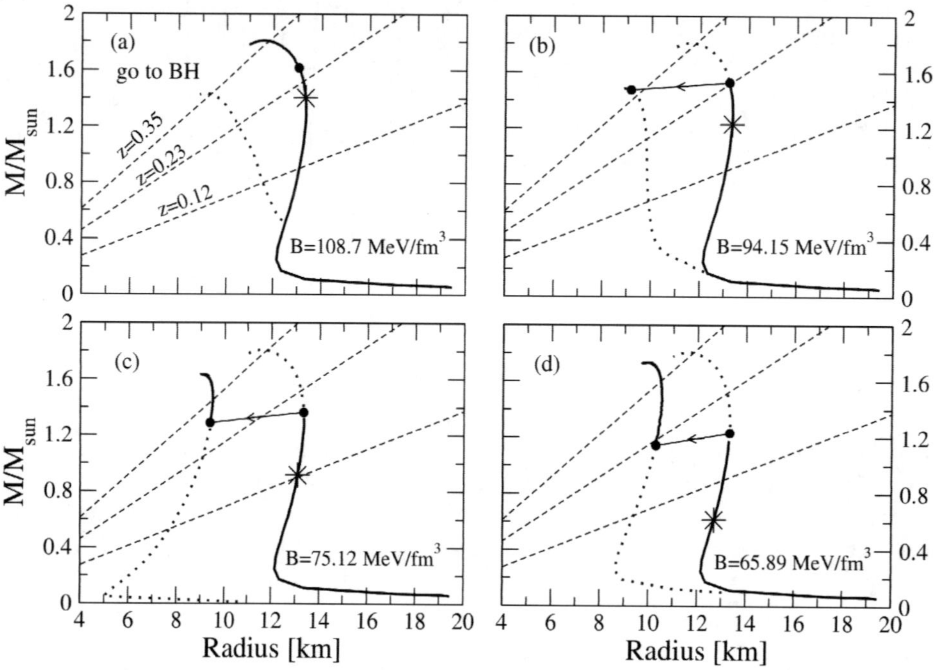

Figure 4. Mass-radius relation for pure HSs described within the GM1 model and for HyS or SS configurations for several values of the bag constant. The configuration marked with an asterisk represents the HS for which the central pressure is equal to P_0^* and $\tau = \infty$. The conversion process of the HS, with a mass equal to M_{cr}, into a final QS is denoted by the full circles connected by an arrow. In all the panels σ= 30 MeV/fm^2. The bifurcation point between the HyS and HS mass-radius curves, signals the onset of the mixed phase between the hadronic- and Q-phase (β-stable SQM). The dashed lines show the gravitational red shift deduced for the X-ray compact sources EXO 0748-676 ($z = 0.35$) and 1E 1207.4-5209 ($z = 0.12 - 0.23$).

critical-size drop of deconfined Q*-matter in their interior [2]. We define as *critical mass* M_{cr} of the metastable HS, the value of the gravitational mass for which the nucleation time is equal to one year: $M_{cr} \equiv M_{HS}(\tau{=}1\,\mathrm{yr})$. Pure hadronic stars with $M_H > M_{cr}$ are very unlikely to be observed. M_{cr} plays the role of an *effective maximum mass* for the hadronic branch of compact stars. While the Oppenheimer–Volkov maximum mass [34] $M_{HS,max}$ is determined by the overall stiffness of the EOS for hadronic matter, the value of M_{cr} will depend in addition on the bulk properties of the EOS for quark matter and on the properties at the interface between the confined and deconfined phases of matter (*e.g.*, the surface tension σ and the curvature coefficient γ).

In Fig. 4, we show the mass-radius (MR) curve for pure HSs within the GM1 model [33] for the EOS of the hadronic phase, and that for hybrid or strange stars for different values of the bag constant B. The configuration marked with an asterisk on the hadronic MR curves represents the HS for which the central pressure is equal to P_0^* and $\tau = \infty$. The full circle on the HS sequence represents the critical mass configuration, in the case σ = 30 MeV/fm^2 and $\gamma = 0$. The full circle on the HyS (SS) mass-radius curve represents the hybrid (strange) star

[2] The actual *mean-life time* of the HS will depend on the mass accretion rate or on the spin-down rate which modifies the nucleation time via an explicit time dependence of the stellar central pressure.

which is formed from the conversion of the hadronic star with M_{HS}=M_{cr}. We assume that during the stellar conversion process the total number of baryons in the star (*i.e.* the stellar baryonic mass) is conserved. Thus the total energy liberated in the stellar conversion is given [14] by the difference between the gravitational mass of the initial hadronic star ($M_{in} \equiv M_{cr}$) and that of the final hybrid or strange stellar configuration $M_{fin} \equiv M_{QS}(M^b_{cr})$ with the same baryonic mass: $E_{conv} = (M_{in} - M_{fin})c^2$.

The stellar conversion process, described so far, will start to populate the new branch of quark stars (the part of the QS sequence plotted as a continuous curve in Fig. 4). Long term accretion on the QS can next produce stars with masses up to the limiting mass $M_{QS,max}$ for the quark star configurations.

Figure 5. A schematic representation of the last stages of the evolution of a massive ($M \geq 8M_{sun}$) star leading to a delayed conversion of a pure hadronic star to a quark star (hybrid or strange star), and to the emission of a gamma-ray burst. Clockwise from the upper left corner of the figure: (i) nuclear burning stage of the progenitor star; (ii) supernova explosion and birth of a pure hadronic star ("neutron star"); (iii) mass accretion on the metastable hadronic star; (iv) conversion process of the hadronic star to a quark star (second "explosion") and GRB.

4. Quark Deconfinement Nova and GRBs

A mounting number of observational data [16, 17, 18, 35, 36, 37, 38] suggest a clear connection between supernova (SN) explosions and long-duration GRBs. In particular, the detection of X-ray spectral features in the X-ray afterglow of several GRBs, has given evidence for a possible time delay between the SN explosion and the associated GRB. For GRB 990705 the supernova explosion is evaluated [16] to have occurred a few years before the GRB, a few months before the burst [17] in the case of GRB 020813, while for GRB 011211 about four days before the burst [18].

These findings demand for a two-step explosion mechanism. The first event, in this scenario, is the SN explosion which forms a compact stellar remnant, *i.e.* a neutron star (NS); the second catastrophic event is associated with the NS and it is the energy source for the observed GRB.

These new observations, and the two-step scenario outlined above, poses severe problems for most of the current models for the central energy source of GRBs. The main difficulty of all these models is to understand the origin of the second "explosion", and to explain the long time delay between the two events.

In the *supranova* model [39] for GRBs the second catastrophic event is the collapse to a black hole of a *supramassive* neutron star, *i.e.* a fast rotating NS with a baryonic mass above the maximum baryonic mass for non-rotating configurations. In this model, the time delay between the SN explosion and the GRB is equal to the time needed by the fast rotating new born NS to get rid of angular momentum and to reach the limit for instability against quasi-radial modes where the collapse to a black hole occurs [40].

The delayed stellar conversion process, described so far, represents the second "explosion" – the *Quark Deconfinement Nova* (QDN) – in the two-step scenario proposed by the authors of ref. [11] to explain the delayed SN-GRB connection. The total energy (E_{conv}) liberated during the stellar conversion is in the range [12] 0.5–1.7$\times 10^{53}$ erg. This huge amount of energy will be mainly carried out by the neutrinos produced during the stellar conversion process. It has been shown [41] that near the surface of a compact stellar object, due to general relativity effects, the efficiency of the neutrino-antineutrino annihilation into e^+e^- pairs is strongly enhanced with respect to the Newtonian case, and it could be as high as 10%. The total energy deposited into the electron-photon plasma can therefore be of the order of 10^{51}–10^{52} erg.

The strong magnetic field of the compact star will affect the motion of the electrons and positrons, and in turn could generate an anisotropic γ-ray emission along the stellar magnetic axis. This picture is strongly supported by the analysis [42] of the early optical afterglow for GRB 990123 and GRB 021211. Moreover, it has been recently shown [43] that the stellar magnetic field could influence the velocity of the "burning front" of hadronic matter into quark matter. This results in a strong geometrical asymmetry of the forming quark matter core along the direction of the stellar magnetic axis, thus providing a suitable mechanism to produce a collimated GRB. Other anisotropies in the GRB could be generated by the rotation of the star.

In summary, the present scenario [10, 11, 12, 13] implies, as a natural consequence a two step-process which is able to explain a delayed connection between SN explosions and GRBs, giving also the correct energy to power GRBs. It is important to stress that the second explosion (QDN) take place in a "baryon-clean" environment due to the previous SN explosion. Moreover, it is possible to have different time delays between the two events since the *mean-life time* of the metastable hadronic star depends on the value of the stellar central pressure. Thus the present model is able to interpret a time delay of a few years (as inferred [16] for GRB 990705), of a few months (as in the case [17] of GRB 020813), of a few days (as deduced [18] for GRB 011211), or the nearly simultaneity of the two events (as in the case [38] of SN2003dh and GRB 030329).

Acknowledgments

I am grateful to my collaborators: Z. Berezhiani, A. Drago, F. Frontera, A. Lavagno, I. Parenti and I. Vidaña, with whom the results presented in this paper have been obtained. Special thanks to Luca Cesaro (INFN) for preparing Fig. 5.

[1] Z. Fodor and S.D. Katz, Prog. Theor. Suppl. 153 (2004) 86.
[2] F. Karsch, (2004) hep-lat/0412038.
[3] S.D.H. Hsu and M. Schwetz, Phys. Lett. B 432 (1998) 2003.
[4] O. Scavenius, A. Mocsy, I.N. Mishustin and D.H. Rischke, Phys. Rev. C 64 (2001) 045202.
[5] D. Balin and A. Love, Phys. Rep. 107 432 (1984) 325.
[6] M. Alford, K. Rajagopal and F. Wilczek, Phys. Lett. B 422 (1998) 2003.
[7] R. Rapp, T. Schäfer, E.V. Shuryak and M. Velkovsky, Phys. Rev. Lett. 81 (1998) 53.
[8] D. Blaschke and C. D. Roberts, Nucl. Phys. A 642 (1998) 197.
[9] M. Buballa, Phys. Rep. 407 (2005) 205.

[10] Z. Berezhiani, I. Bombaci, A. Drago, F. Frontera and A. Lavagno, Nucl. Phys. B (Proc. Suppl.) 113 (2002) 268.
[11] Z. Berezhiani, I. Bombaci, A. Drago, F. Frontera and A. Lavagno, Astrophys. J. 586 (2003) 1250.
[12] I. Bombaci, I. Parenti and I. Vidaña, Astrophys. J. 614 (2004) 314.
[13] A. Drago, A. Lavagno, and G. Pagliara, Phys. Rev. D 69 (2004) 057505.
[14] I. Bombaci, and B. Datta, Astrophys. J. 530 (2000) L72.
[15] M. Prakash, I. Bombaci, M. Prakash, P.J. Ellis, J.M. Lattimer, and R. Knorren, Phys. Rep. 280 (1997) 1.
[16] L. Amati, et al., Science, 290 (2000) 953; D. Lazzati, et al., Astrophys. J. 556 (2001) 471.
[17] N.R. Butler, et al., Astrophys. J. 597 (2003) 1010.
[18] J.N. Reeves, et al., Nature 414 (2002) 512.
[19] D. Ivanenko and D. F. Kurdgelaidze, Nuovo Cimento 2 (1969) 13; N. Itho, Prog. Theor. Phys. 44 (1970) 291; F. Iachello, W. D. Langer and A. Lande, Nucl. Phys. A 219 (1974) 612; J. C. Collins and M. J. Perry, Phys. Rev. Lett. 34 (1975) 1353.
[20] N. K. Glendenning, Compact Stars, Springer Verlag (1996).
[21] F. Weber, Pulsars as Astrophysical Laboratories for Nuclear and Particle Physics, IoP Pub. (1999).
[22] F. Weber, Prog. Part. Nucl. Phys. 54 (2005) 193.
[23] D. Blaschke, et al. (eds.), Lecture Notes in Physics, Vol. 578, Springer Verlag (2001).
[24] A.R. Bodmer, Phys. Rev. D 4 (1971) 1601; H. Terazawa, INS Rep. 336 (1979) Univ.of Tokyo, INS; J. Phys. Soc.Japan, 58 (1989) 3555; E. Witten, Phys. Rev. D 30 (1984) 272.
[25] C. Alcock, E. Farhi, and Olinto, Astrophys. J. 310 (1986) 261; P. Haensel, J.L. Zdunik, and R. Schaefer, Astron. and Astrophys. 160 (1986) 121.
[26] E. Farhi and R. L. Jaffe, Phys. Rev. D 30 (1984) 2379.
[27] I. Bombaci, Phys. Rev. C 55 (1997) 1587.
[28] K.S. Cheng, Z.G. Dai, D.M. Wai, and T. Lu, 1 Science, 280 (1998) 407.
[29] X.–D. Li, I. Bombaci, M. Dey, J. Dey, and E.P.J. van den Heuvel, Phys. Rev. Lett. 83 (1999) 3776.
[30] X.–D. Li, S. Ray, J. Dey, M. Dey, and I. Bombaci, Astrophys. J. 527 (1999) L51.
[31] H. Heiselberg, C. J. Pethick, and E.F. Staubo, Phys. Rev. Lett. 70 (1993) 1355.
[32] K. Iida and K. Sato, Phys. Rev. C 58 (1997) 2538.
[33] N. K. Glendenning and S. A. Moszkowski, Phys. Rev. Lett. 67 (1991) 2414.
[34] J.R. Oppenheimer, and G.M. Volkoff, Phys. Rev. 55 (1939) 374.
[35] J.S. Bloom et al., Nature 401 (1999) 453.
[36] L.A. Antonelli et al., Astrophys. J. 545 (2000) L39.
[37] L. Piro et al., Science 290 (2000) 955.
[38] J. Hjorth et al, Nature 423 (2003) 847.
[39] M. Vietri and L. Stella, Astrophys. J. 507, L45 (1998) L45.
[40] B. Datta, A.V. Thampan and I. Bombaci, Astron. and Astrophys. 334 (1998) 943.
[41] J.D. Salmonson and J.R. Wilson, Astrophys. J. 517 (1999) 859.
[42] B. Zhang, S. Kobayashi and P. Meszaros, Astrophys. J. 595 (2003) 950.
[43] G. Lugones, et al., Astrophys. J. 581 (2002) L101.

Institute of Physics Publishing Journal of Physics: Conference Series **50** (2006) 216–222
doi:10.1088/1742-6596/50/1/025 5th International Conference on Physics and Astrophysics of Quark Gluon Plasma

Exotica in rotating compact stars

Debarati Chatterjee, Sarmistha Banik, and Debades Bandyopadhyay
Saha Institute of Nuclear Physics, 1/AF Bidhannagar, Kolkata 700 064, India

Abstract. The determination of mass and radius of a single neutron star EXO 0748-676 has been reported recently. Also, the estimate of radius from the measurement of moment of inertia of pulsar A in double pulsar system PSR J0737-3039 would be possible in near future. Here we construct models of static and uniformly rotating neutron stars involving exotic matter and compare our theoretical calculations with the recent findings from observations to probe dense matter in neutron stars.

1. Introduction

Neutron stars are ideal astrophysical laboratories to study properties of cold and dense matter [1, 2]. The investigation of dense and cold matter in neutron star core has entered into a new phase. Observatories such as Chandra X-ray observatory, X-ray Multi Mirror (XMM)-Newton, Rossi X-ray Timing Explorer (RXTE) are pouring in many new and interesting data on compact stars. It is now possible to estimate mass and radius of a single neutron star from observations. These informations may shed light on the composition and equation of state (EoS) of matter at low temperature and high baryon chemical potential.

A lot of interest has been generated in thermonuclear x-ray bursts from the low mass x-ray binary EXO 0748-676. Already Cottam et al. [3] estimated the gravitational redshift $z = 0.35$ of three absorption lines in x-ray bursts from EXO 0748-676 with XMM-Newton. They also reported the narrow widths of those lines. Mass to radius ratio is immediately known from the measurement of $z = (1 - 2GM/c^2R)^{-1/2} - 1$, where M and R are neutron star mass and radius. On the other hand, the widths of absorption lines provide information about the surface rotational velocity (v_{rot}) which is related to radius and spin frequency $v_{rot} \propto \nu_{spin} R$. Villarreal and Strohmayer [4] reported the neutron star spin frequency (ν_{spin}) of 45 Hz in the x-ray burst oscillations of EXO 0748-676 with RXTE. These observations lead to a radius of 9.5-15 km and mass 1.5-2.3$M_\odot$ for EXO 0748-676 [4]. For the first time, it has been possible to determine mass and radius of a single neutron star.

Recently discovered double pulsar system PSR J0737-3039 [5] provides another opportunity for the determination of mass and radius of same neutron star. Two pulsars in this double pulsar system are denoted by pulsar A and pulsar B. Pulsar A has a spin period of $\sim$ 23 ms and mass 1.34$M_\odot$ whereas those of pulsar B are 2.8 s and 1.25$M_\odot$. Double neutron star binaries provide unique laboratories for testing relativistic gravity. Soon it will be possible to measure spin-orbit coupling of pulsar A from pulsar timing data [6, 7]. The moment of inertia for pulsar A could be determined from the measurement of spin-orbit coupling. Moment of inertia is dimensionally proportional to mass times radius squared. Already, the mass of pulsar A is known. Therefore, the measurement of moment of inertia will provide an estimate of radius of the compact object and put important constrain on the properties of neutron star matter.

Now the observed mass-radius (M-R) relationship can be directly compared with that of theoretical calculation. Consequently, the composition and EoS of dense matter in neutron stars may be probed. Here we discuss exotic forms of matter and their influence on various properties of static and rotating compact stars. In section 2, we describe the models of static and uniformly rotating compact stars and EoS adopted for this calculation. Results are discussed in section 3. And section 4 is devoted to summary and conclusions.

2. Model

Models of static and rotating compact stars are constructed by solving Einstein's field equations. The equilibrium structures of non-rotating stars are obtained from Tolman-Oppenheimer-Volkoff (TOV) equations. For rotating neutron stars, we consider stationary and axisymmetric equilibrium configurations. In this case, the metric is taken in the form [8, 9]

$$ds^2 = -e^{\gamma+\rho}dt^2 + e^{2\alpha}(dr^2 + r^2 d\theta^2) + e^{\gamma-\rho}r^2 sin^2\theta(d\phi - \omega dt)^2 \ , \tag{1}$$

where ω is the angular velocity of the local inertial frame. The metric functions in the line element depend on the radial coordinate r and the polar angle θ.

We model the neutron star matter as a perfect fluid and the stress-energy tensor is given by

$$T^{\mu\nu} = (\varepsilon + P)u^\mu u^\nu - P\, g^{\mu\nu} \tag{2}$$

where ε and P are the total energy density and pressure of the matter respectively and u^a is the fluid's four velocity. The moment of inertia of the star is defined by $I = J/\Omega$; angular momentum (J) is,

$$J = \int d^3x \sqrt{(-g)}(\epsilon + P)u^t(g_{\phi\phi}u^\phi + g_{\phi t}u^t) \tag{3}$$

where, u^ϕ and u^t are components of fluid four velocity and Ω is angular velocity of the star. For uniformly rotating neutron stars, we adopt a numerical code based on Komatsu-Eriguchi-Hachisu [9, 10] method.

Composition and structure of neutron stars depend on the nature of strong interaction. Matter density could exceed by a few times normal nuclear matter density in a neutron star interior. Consequently, the baryon and lepton chemical potential increase rapidly with baryon density. This might lead to the formation of exotic matter with a large strangeness fraction such as hyperon matter, Bose-Einstein condensate of K^- mesons and quark matter. Here we discuss equations of state with various compositions of dense matter in neutron star cores.

We construct hadronic EoS using relativistic field theoretical models [11, 12]. Here baryon-baryon interaction is mediated by exchange of mesons. The hadronic phase is composed of all members of the baryon octet, electrons and muons. The hadronic phase is to satisfy beta-equilibrium and charge neutrality conditions. In compact stars interior, the generalised β-decay processes may be written in the form $B_1 \longrightarrow B_2 + l + \bar{\nu}_l$ and $B_2 + l \longrightarrow B_1 + \nu_l$ where B_1 and B_2 are baryons and l represents leptons. The generic equation for chemical equilibrium condition is

$$\mu_i = b_i \mu_n - q_i \mu_e \ , \tag{4}$$

where μ_n, μ_e and μ_i are respectively the chemical potentials of neutrons, electrons and i-th baryon and b_i and q_i are baryon number and electric charge of i-th baryon respectively. The charge neutrality in hadronic phase is given by $Q^h = \sum_B q_B n_B^h - n_e - n_\mu = 0$, where n_B^h is the number density of baryon B in pure hadronic phase and n_e and n_μ are number densities of electrons and muons respectively.

Next we consider Bose-Einstein condensate of K^- mesons. Experimental results from heavy ion collisions suggest that the in-medium kaon-nucleon interaction is repulsive whereas it is attractive for antikaon-nucleon [13, 14]. Also, informations about antikaon-nucleon interaction in medium may be obtained studying the K^--atomic data. The analysis of K^--atomic data using a phenomenological density dependent potential showed that the real part of antikaon optical potential could be as large as -180 ± 20 MeV [15].

It was first demonstrated by Kaplan and Nelson within a chiral $SU(3)_L \times SU(3)_R$ model that K^- meson may undergo Bose-Einstein condensation in dense matter formed in heavy ion collisions [16]. The effective mass of antikaons decreases with increasing density because of the strongly attractive K^--baryon interaction in dense matter. Consequently, the in-medium energy of K^- mesons in the zero-momentum state also decreases with density. The s-wave K^- condensation sets in when the energy of K^- mesons equals to its chemical potential.

We consider K^- condensation as a first order phase transition. Here we adopt a relativistic field theoretical model to describe (anti)kaon-baryon interaction [11, 12, 17]. In this case, (anti)kaon-baryon interaction is treated in the same footing as baryon-baryon interaction. The pure K^- condensed phase is composed of members of the baryon octet embedded in the condensate, electrons and muons and maintains charge neutrality and beta-equilibrium conditions. The charge neutrality condition is $Q^{\bar{K}} = \sum_B q_B n_B^{\bar{K}} - n_{K^-} - n_e - n_\mu = 0$, where $n_B^{\bar{K}}$ is the number density of baryon B in pure antikaon condensed phase and n_{K^-} is the number density of antikaons.

Strangeness changing processes such as, $n \rightarrow p + K^-$ and $e^- \rightarrow K^- + \nu_e$, in a neutron star interior are responsible for the onset of K^- condensation. The requirement of chemical equilibrium yields

$$\mu_n - \mu_p = \mu_{K^-} = \mu_e \; , \tag{5}$$

(6)

where μ_{K^-} is the chemical potentials of K^- mesons. The above condition determines the onset of K^- condensation.

The deconfinement phase transition from hadronic to quark matter is a possibility in a neutron star interior. In this calculation, we also consider a first order phase transition from hadronic to quark phase. The pure quark matter is composed of u, d and s quarks. The EoS of pure quark matter is described in the MIT bag model [18].

As we are considering first order phase transitions, the mixed phase of two pure phases is governed by Gibbs phase rules and global conservation laws [1]. The Gibbs phase rules read, $P^I = P^{II}$ and $\mu_B^I = \mu_B^{II}$ where P^I, P^{II} and μ_B^I, μ_B^{II} are pressure and chemical potentials of baryon B in phase I and II respectively. The conditions for global charge neutrality and baryon number conservation are respectively $(1-\chi)Q^I + \chi Q^{II} = 0$ and $n_B = (1-\chi)n_B^I + \chi n_B^{II}$ where χ is the volume fraction in phase II. The total energy density in the mixed phase is given by $\epsilon = (1-\chi)\epsilon^I + \chi\epsilon^{II}$.

3. Results & Discussions

We adopt GM1 parameter set [11, 12, 19] for this calculation where nucleon-meson coupling constants are determined from the nuclear matter saturation properties. The vector meson coupling constants for (anti)kaons and hyperons are determined from the quark model [20] whereas the scalar meson coupling constants for hyperons and antikaons are obtained from the potential depths of hyperons and antikaons in normal nuclear matter [12, 20]. As the phenomenological fit to the K^- atomic data yielded a very strong real part of antikaon potential U_{K^-} at normal nuclear matter density, we perform this calculation with an antikaon optical potential of -160 MeV at normal nuclear matter density ($n_0 = 0.153 fm^{-3}$). The coupling constants for strange mesons with hyperons and (anti)kaons are taken from Ref.[12, 20]. Also, we consider a bag constant $B^{1/4} = 200$ MeV and strange quark mass $m_s = 150$ MeV for quark matter EoS.

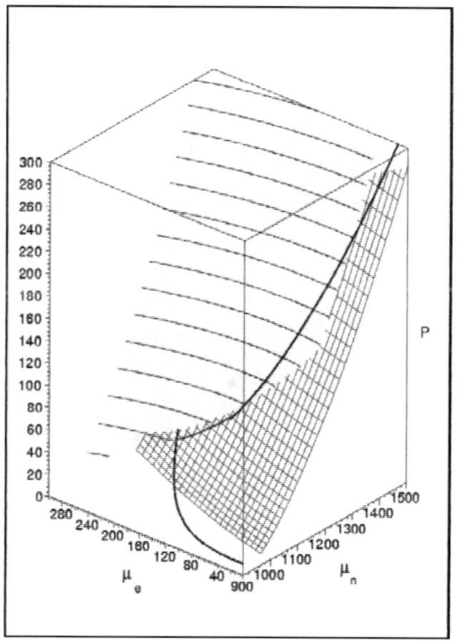

Figure 1. Pressure P (MeV fm^{-3}) as a function of μ_n (MeV) and μ_e (MeV) for pure hadronic phase (wired surface) and K^- condensed phase (shaded surface). The solid line is the charge neutrality line in pure hadronic phase, the mixed phase and pure K^- condensed phase.

We compute equations of state for four different compositions of neutron star matter. The EoS with n, p, e and μ is denoted by "np". With further inclusion of hyperons, it becomes "npH". The EoS undergoing a first order phase transition from nuclear to quark matter is represented by "npQ". In the last case, the EoS involves first order phase transitions from nuclear to K^- condensed matter and then from K^- condensed matter to quark matter and is referred to as "npKQ" [21].

It is evident from Eq. (4) that neutron and electron chemical potentials are two independent quantities. One can express energy density and pressure of neutron star matter using μ_n and μ_e. As an example, we demonstrate here the construction of EoS involving a first order phase transition from nuclear to K^- condensed matter. In Figure 1, pressure is plotted as a function of neutron and electron chemical potentials for pure hadronic and K^- condensed phase. Here shaded surface denotes the antikaon condensed phase whereas the wired surface represents the hadronic phase. It is noted that two surfaces intersect along a common line. This implies that two phases having same pressure along the intersection line could co-exist. In this figure, the solid line represents the EoS of charge neutral and beta-equilibrated matter. It has three parts. Below P=42 MeV fm^{-3}, the pressure in pure hadronic phase is higher than that of the antikaon condensed phase making the hadronic phase physically preferred one. Therefore, the solid line below pressure 42 MeV fm^{-3} denotes the charge neutrality line in pure hadronic phase. This phase is mainly composed of neutrons, protons, electrons and muons. Two surfaces first meet at $2.23n_0$ and P=42 MeV fm^{-3} and this is the beginning of the mixed phase. After the onset of the phase transition, the mixed phase represented by the solid line follows the intersection part of two surfaces. The Gibbs conditions of phase equilibrium and global charge neutrality are satisfied along the intersection line of two surfaces. We find that the electron chemical potential μ_e decreases in the mixed phase. This is attributed to the fact that with the formation of K^- condensate, electrons are replaced by K^- mesons. The phase transition ends at $3.51n_0$ and pressure 70 MeV fm^{-3}. The solid line above pressure 70 MeV fm^{-3} represents the charge

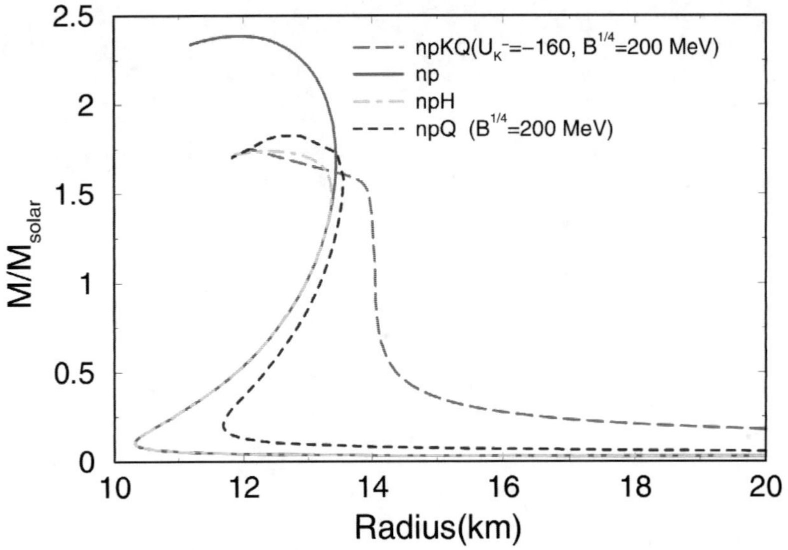

Figure 2. Mass-radius relationship of non-rotating stars calculated with different equations of state.

neutrality line in pure antikaon condensed phase.

In Figure 2, the mass-radius relationship of non-rotating neutron stars is displayed for different EoS. For np case, the maximum neutron star mass is 2.39 $M_\odot$ and the corresponding radius is 11.94 km. With the inclusion of softening components such as hyperons, Bose-Einstein condensate of K^- mesons or quarks, the maximum neutron star masses are reduced. For npH case, the maximum mass and the corresponding radius are 1.74$M_\odot$ and 12.41 km respectively. Maximum masses and the corresponding radii are 1.83 $M_\odot$ and 12.88 km for npQ case and 1.75$M_\odot$ and 12.10 km for npKQ case, respectively. Among all cases considered here, the maximum neutron star mass for np case is the largest because the EoS in this case is the stiffest. Now we compare our results with the recent findings from EXO 0748-676. Villarreal and Strohmayer [4] have predicted that the best fit values of mass and radius for EXO 0748-676 are M$\sim$ 1.8 $M_\odot$ and R$\sim$ 11.5 km respectively. It is found that theoretical M-R relationships shown in Figure 2 are consistent with the observed values. It shows that there might be room for exotic matter in neutron star interior.

Now we focus on the estimate of radius from the calculation of moment of inertia of pulsar A in the double pulsar binary PSR J0737-3039. In this calculation, we adopt some of EoS used for explaining EXO 0748-676 data. Also, we use the inputs gravitational mass M=1.337$M_\odot$ and angular velocity Ω = 276.8 s^{-1} corresponding to a spin period P=$2\pi/\Omega$=22.7 ms to calculate the moment of inertia of the uniformly rotating star. It follows from the calculation that for np, npH and npQ cases the central energy density corresponding to pulsar A is $\sim 5.4 \times 10^{14}$ g cm^{-3}. We also note that the moment of inertia and radius of pulsar A are I $\sim 4.90 \times 10^{45}$ g cm^2 and R $\sim$ 14 km for all these cases studied here. It exhibits that hyperons or quark matter does not change the moment of inertia and radius appreciably.

In Figure 3, we have shown the moment of inertia in dimensionless unit as a function of compactness parameter (M/R) of the neutron star for different EoS. Already, it has been noted that these EoS give rise to maximum neutron star masses > 1.7 $M_\odot$. Therefore, I/MR^2 versus

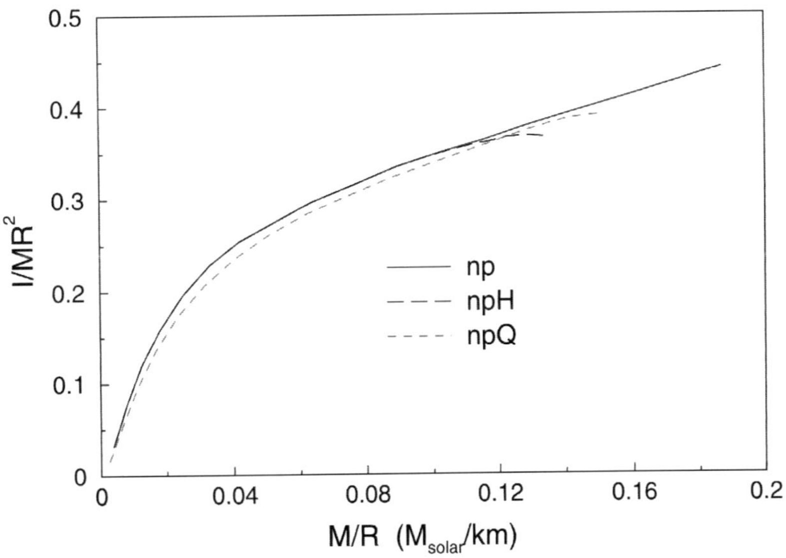

Figure 3. Moment of inertia (I) is plotted with compactness parameter (M/R) for three different EoS.

M/R curves shown in Figure 3 can be approximated by [7]

$$I \simeq (0.237 \pm 0.008)\, MR^2 \left[1 + 4.2\,\frac{M\,km}{M_\odot\,R} + 90\left(\frac{M\,km}{M_\odot\,R}\right)^4\right]. \tag{7}$$

The values of moment of inertia that we have obtained from the rotating neutron star model for fixed angular frequency are comparable with the values calculated using Eq. (7). Thus from the measurements of moment of inertia and mass of pulsar A, the radius could be determined by inversion of Eq. (7).

It has been shown that the moment of inertia of a neutron star depends quite strongly on the EoS. The predictions for neutron star moment of inertia and radius are quite different in non-relativistic and relativistic models [6]. Therefore, a measurement of moment of inertia with 10 percent uncertainty could impose new constraints on the EoS and could rule out some classes of equations of state that currently exist.

4. Summary and conclusions

We have computed various properties of static and uniformly rotating compact stars using EoS with first order nuclear to K^- condensed matter phase transition and also K^- condensed matter to quark matter phase transition. Mass and radius of EXO 0748-676 have been measured. These informations put stringent constaints on EoS. Our results are consistent with the mass and radius of EXO 0748-676. This, in turn, shows that bizarre matter might exist in this star. We have also calculated the moment of inertia for pulsar A of double pulsar binary PSR J0737-3039. It is found that the values of moment of inertia and radius obtained with relativistic EoS are quite different from that of the nonrelativistic model of Akmal and Pandharipande [6]. Soon the measurement of moment of inertia for pulsar A would be possible. This would lead to an estimate of the radius of pulsar A. Therefore, it would be possible to rule out some EoS with the help of the measured value of moment of inertia and mass of pulsar A. Lattimer and

Prakash showed that there was an empirical correlation between the radius and matter pressure in the density region n_0 to $2n_0$ [22]. Therefore, the determination of radius of pulsar A having a mass $M=1.337M_\odot$ could give an estimate of matter pressure around saturation density. It is interesting to note that we have informations about mass and radius of a single neutron star. More events like this would be available in future. Therefore, it would be worth investigating the dense matter EoS from the knowledge of masses and radii of neutron stars and using a numerical inversion of neutron star structure equation [23].

Acknowledgments:

Two authors (SB and DB) acknowledge the support for this work by the Department of Science and Technology (DST), Government of India and German Academic Exchange Service (DAAD), Germany.

References

[1] Glendenning N K 1997 Compact stars (New York, Springer)
[2] Weber F 1999 Pulsars as Astrophysical Laboratories for Nuclear and Particle Physics (Bristol and Philadelphia, Institute of Physics Publishing)
[3] Cottam J, Paerels F and Mendez M 2002 Nature **420** 51
Miller C 2002 Nature **420** 31
[4] Villarreal A R and Strohmayer T E 2004 astro-ph/0409584
[5] Lyne A G et al. 2004 Science **303** 1153
[6] Morrison I A, Baumgarte T W, Shapiro S L and Pandharipande V R 2004 astro-ph/0411353
[7] Lattimer J M and Schutz B F 2004 astro-ph/0411470
[8] Cook G B, Shapiro S L and Teukolsky S A 1994 Astrophys. J. **422** 227
[9] Stergioulas N and Friedman J L 1995 Astrophys. J. **444** 306
Stergioulas N 2003 Living Rev. Rel. **6** 3
[10] Komatsu H, Eriguchi Y and Hachisu I 1989 MNRAS **237** 355
Komatsu H, Eriguchi Y and Hachisu I 1989 MNRAS **239** 153
[11] Banik S and Bandyopadhyay D 2001 Phys. Rev. C**63** 035802
[12] Banik S and Bandyopadhyay D 2001 Phys. Rev. C**64** 055805
[13] Pal S, Ko C M, Lin Z and Zhang B 2000 Phys. Rev. C**62** 061903(R)
[14] Li G Q, Lee C -H and Brown G E 1997 Phys. Rev. Lett. **79** 5214
[15] Friedman E, Gal A, Mareš J and Cieplý A 1999 Phys. Rev. C**60** 024314
[16] Kaplan D B and Nelson A E 1986 Phys. Lett. B**175** (1986) 57
Nelson A E and Kaplan D B 1987 Phys. Lett. B **192** 193
[17] Glendenning N K and Schaffner-Bielich J 1998 Phys. Rev. Lett. **81** 4564
[18] Farhi E and Jaffe R L 1984 Phys. Rev. D **30** 2379
[19] Glendenning N K and Moszkowski S A 1991 Phys. Rev. Lett. **67** 2414
[20] Schaffner J and Mishustin I N 1996 Phys. Rev. C **53** 1416
[21] Banik S, Hanauske M, Bandyopadhyay D and Greiner W 2004 Phys. Rev. D**70** 123004
[22] Lattimer J M and Prakash M 2001 Astrophys. J. C**550** 426
[23] Lindblom L 1992 Astrophys. J. C**398** 569

Institute of Physics Publishing Journal of Physics: Conference Series **50** (2006) 223–229
doi:10.1088/1742-6596/50/1/026 5th International Conference on Physics and Astrophysics of Quark Gluon Plasma

Color superconducting strange quark matter at finite temperature

Amruta Mishra[1], **Hiranmaya Mishra**[2]

[1]Department of Physics, Indian Institute of Technology, New Delhi- 110016, India
[2]Theory Division, Physical Research Laboratory, Navrangpura, Ahmedabad 380 009, India

E-mail: [1]`amruta@physics.iitd.ac.in`, [2]`hm@prl.ernet.in`

Abstract. We investigate the phase diagram of strange quark matter in beta equllibrium where the lighter up and down quarks form the two flavor superconducting matter whereas the strange quark remains unpaired. This is studied wihin a Nambu-Jona-Lasinio model. The variational method as used here allows us to investigate simultaneous formation of condensates in quark–antiquark as well as in diquark channels. Color and electric charge neutrality conditions are imposed in the calculation of the thermodynamic potential. Medium dependance of strange quark mass plays a sensitive role in maintaining charge neutrality conditions. At zero temperature the system goes from the gapless phase to the usual BCS phase through an intermediate normal phase as density is increased. However, at higher temperature the gapless to BCS transition becomes a smooth transition. The gapless modes show a smooth behaviour with respect to temperature vanishing above a critical temperature which is larger than the BCS transition temperature.

1. Introduction

Color superconductivity has become a compelling topic in QCD during the last few years. At sufficiently high baryon densities, when nucleons get converted to quark matter, the resulting quark matter is in one kind or the other of the many different possible color superconducting phases at low enough temperatures [1]. The rich phase structure is essentially due to the fact that the quark quark interaction is not only strong and attractive in many channels but also many degrees of freedom are possible for quarks like color, flavor and spin so that various kinds of BCS pairing are possible. Although such phases seems to be almost unlikely to be created in present hevay ion collision experiments, perhaps, it is natural to expect some color suprconducting phase to exist in the core of compact stars where the densities are above nuclear matter densities and temperatures are of the order of tens of keV. However, to consider quark matter for neutron stars, color and electrical charge neutrality conditions need to be imposed for the bulk quark matter. Recently it was observed that imposition of neutrality conditions lead to pairing of quarks with different fermi momenta giving rise to gapless modes [2, 3]. In the present work we generalise the variational approach of Ref. [3] to study the color superconductivity as well as the chiral symmetry breaking to include the effects of temperatures. This will be particularly relevant for the physics of proto neutron stars. Here, we compute the thermodynamic potential at finite temperatures for charge neutral ud-s matter where the two lighter quraks take part in diquark condensation. We also self consistently detrmine the quark antiquark condensates and the diquark condensates- the former being related to chiral symmetry breaking. We thus in the prsent calculation keep all orders in the strange quark mass while computing the thermodynamic potential. The temperature dependence and the nature of the gapless modes as a function of the baryon density shall be particularly relevant for cooling of the neutron stars through neutrino emission [5].

2. An ansatz for the ground state and the thermodynamic potential

We here write down the ansatz for the variational state as a squeezed coherent state involving quark antiquark as well as diquark condensates as given by [3, 4]

$$|\Omega\rangle = \mathcal{U}_d|vac\rangle = \mathcal{U}_d\mathcal{U}_Q|0\rangle, \tag{1}$$

Here, $\mathcal{U}_Q$ and $\mathcal{U}_d$ are unitary operators creating quark–antiquark and diquark pairs respectively. Explicitly,

$$\mathcal{U}_Q = \exp\left(\int q^{0i}(\mathbf{k})^\dagger(\boldsymbol{\sigma}\cdot\mathbf{k})h_i(\mathbf{k})\tilde{q}^{0i}(\mathbf{k})d\mathbf{k} - h.c.\right). \tag{2}$$

In the above, $h_i(\mathbf{k})$ is a real function of $|\mathbf{k}|$ which describes vacuum realignment for chiral symmetry breaking for quarks of a given flavor i. Similarly, the unitary operator $\mathcal{U}_d$ describing diquark condensates is given as

$$\mathcal{U}_d = \exp\left(B_d^\dagger - B_d\right)\int\left[q_r^{ia}(\mathbf{k})^\dagger r f^{ia}(\mathbf{k})q_{-r}^{jb}(-\mathbf{k})^\dagger\epsilon_{ij}\epsilon_{3ab} + \tilde{q}_r^{ia}(\mathbf{k})r f_1^{ia}(\mathbf{k})\tilde{q}_{-r}^{jb}(-\mathbf{k})\epsilon_{ij}\epsilon_{3ab}\right]d\mathbf{k}. - h.c.\bigg). \tag{3}$$

Thus, we have only the quarks of colors red and green (a=1,2) and flavors u,d (i=1,2) taking part in diquark condensation . We have also introduced here (color, flavor dependent) functions $f^{ia}(\mathbf{k})$ and $f_1^{ia}(\mathbf{k})$ respectively for the diquark and diantiquark channels. As may be noted the state constructed in Eq.(3) is spin singlet and is antisymmetric in both color and flavor. To include the effects of temperature and density we write down the state at finite temperature and density $|\Omega(\beta,\mu)\rangle$ taking a thermal Bogoliubov transformation over the state $|\Omega\rangle$ using thermofield dynamics (TFD) as described in Ref.s [6, 7]. We then have,

$$|\Omega(\beta,\mu)\rangle = \mathcal{U}_{\beta,\mu}|\Omega\rangle = \mathcal{U}_{\beta,\mu}\mathcal{U}_d\mathcal{U}_Q|0\rangle. \tag{4}$$

where $\mathcal{U}_{\beta,\mu}$ is given as

$$\mathcal{U}_{\beta,\mu} = e^{\mathcal{B}^\dagger(\beta,\mu)-\mathcal{B}(\beta,\mu)}, \tag{5}$$

with

$$\mathcal{B}^\dagger(\beta,\mu) = \int\left[q_I'(\mathbf{k})^\dagger\theta_-(\mathbf{k},\beta,\mu)\underline{q}_I'(\mathbf{k})^\dagger + \tilde{q}_I'(\mathbf{k})\theta_+(\mathbf{k},\beta,\mu)\underline{\tilde{q}}_I'(\mathbf{k})\right]d\mathbf{k}. \tag{6}$$

In Eq.(6) the ansatz functions $\theta_\pm(\mathbf{k},\beta,\mu)$ will be related to quark and antiquark distributions and the underlined operators are the operators in the extended Hilbert space associated with thermal doubling in TFD method. In Eq.(6) we have suppressed the color and flavor indices on the quarks as well as the functions $\theta(\mathbf{k},\beta,\mu)$. All the functions in the ansatz in Eq.(4) are to be obtained by minimising the thermodynamic potential.

Since we shall be dealing with non–asymptotic densities, we take here a simple model for quark interactioni, namely the Nambu Jona Lasino model, with four fermion point interaction explicitly given as

$$\mathcal{H} = \sum_{i,a}\psi^{ia\dagger}(-i\boldsymbol{\alpha}\cdot\boldsymbol{\nabla} + \gamma^0 m_i)\psi^{ia} - G_s\sum_{A=0}^{8}\left[(\bar{\psi}\lambda^A\psi)^2 - (\bar{\psi}\gamma^5\lambda^A\psi)^2\right] - G_D(\bar{\psi}\gamma^5\epsilon\epsilon^b\psi^C)(\bar{\psi}^C\gamma^5\epsilon\epsilon^b\psi) \tag{7}$$

We shall here assume isospin symmetry with m_u=m_d. In Eq.(7) λ^A, $A = 1, \cdots 8$ denote the Gellman matrices acting in the flavor space and $\lambda^0 = \sqrt{\frac{2}{3}}\,\mathbb{1}_f$, $\mathbb{1}_f$ as the unit matrix in the flavor space. To calculate the thermodynamic potential we shall also have to specify the chemical potentials relevant for the system. Here we shall be interested in the form of quark matter that might be present in compact stars older than few minutes so that chemical equilibriation under weak interaction is there. The chemical potential here is then a matrix in the color and the flavor space and is given by $\mu_{ij,ab} = (\mu\delta_{ij} + Q_{ij}\mu_E)\delta_{ab} + (Q_{3ab} + Q_{8ab}\mu_8)\delta_{ij}$, where, Q, Q_3 and Q_8 are generators of $U(1)_{em}$ of electromagnetism, and $U(1)_3$,$U(1)_8$ subgroups of the color gauge group. Here, i, j are flavor indices and a, b are color indices. One can then compute the expectaion value of the Hamiltonian with respect to the state $|\Omega(\beta,\mu)\rangle$ defined in Eq.(4) and evaluate the thermodynamic potential as a functional of the various condensate functions ($h_i(\mathbf{k})$, $f^{ia}(\mathbf{k})$ as well as the thermal functions $\theta_\pm(\mathbf{k},\beta)$. It is possible to determine these functions variationally by functional minimisation of the thermodynamic potential.

The details of these minimisation can be found in Ref. [?] and here we quote the final results for the thermodynamic potential which is given as

$$\Omega = \Omega_{ud} + \Omega_s + \Omega_e, \tag{8}$$

where, Ω_{ud} is the contribution from the light quarks and is given as

$$\begin{aligned}
\Omega_{ud} &= \frac{8}{(2\pi)^3}\int d\mathbf{k}\left[\sqrt{\mathbf{k}^2+m^2} - \frac{1}{2}(\bar\omega_- + \bar\omega_+)\right] \\
&- \frac{4}{\beta(2\pi)^3}\sum_{i=1,2}\int d\mathbf{k}\left[\log(1+\exp(-\beta\omega_{-i}) + \log(1+\exp(-\beta\omega_{+i})\right] \\
&+ \frac{\Delta^2}{4G_D} + \sum_{i=1,2}\frac{(M^i-m)^2}{8G_s} \\
&+ \frac{4}{(2\pi)^3}\int d\mathbf{k}\left[\sqrt{\mathbf{k}^2+m^2} - \frac{1}{2}(\epsilon_1+\epsilon_2)\right] \\
&- \frac{2}{\beta(2\pi)^3}\sum_{i=1,2}\int d\mathbf{k}\left[\log(1+\exp(-\beta(\epsilon_i - \mu^{i3})) + \log(1+\exp(-\beta(\epsilon_i + \mu^{i3}))\right].
\end{aligned} \tag{9}$$

Similarly the contribution from the strange quarks to the thermodynamic potential, Ω_s is given as

$$\begin{aligned}
\Omega_s &= \frac{6}{(2\pi)^3}\int d\mathbf{k}\left[\sqrt{\mathbf{k}^2+m_s^2} - \sqrt{\mathbf{k}^2+M_s^2}\right] \\
&- \frac{2}{\beta(2\pi)^3}\sum_{a=1,3}\int d\mathbf{k}\left[\log(1+\exp(-\beta(\epsilon_3 - \mu^{3a})) + \log(1+\exp(-\beta(\epsilon_3 + \mu^{3a}))\right] + \frac{(M_s-m_s)^2}{8G_s}.
\end{aligned} \tag{10}$$

Finally, the contribution of the electron to the total thermodynamic potentail is given as $\Omega_e = -\mu_E^4/12\pi^2\,(1 + 2\pi^2T^2/\mu_E^2)$ The first three lines in Eq.(9) correspond to the contribution from the quarks taking part in the condensation while the fourth and fifth lines corrspond to the contribution from the two light quarks with the blue color. The quasi particle energies are given by $\omega_{-1} = \bar\omega_- + \delta_\epsilon - \delta_\mu$ and $\omega_{-2} = \bar\omega_- - \delta_\epsilon + \delta_\mu$, for u and d quarks respectively. Similarly for the antiparticles the energies are given as $\omega_{+1} = \bar\omega_+ + \delta_\epsilon + \delta_\mu$ and $\omega_{+2} = \bar\omega_+ + \delta_\epsilon - \delta_\mu$, for u and d quarks. Here $\bar\omega_\pm = \sqrt{\Delta^2 + (\bar\epsilon \pm \bar{m}u)^2}$, with $\bar\epsilon$ and $\bar\mu$ being the average energy and average chemical potential of the two quarks that condense. We do not write down here the gap equations for the superconducting gap as well as the three mass gaps as they are not very illuminating for our discussions that follow. Interested reader however can find the same in Ref.[?]. Thus the thermodynamic potential is a function of three parameters: the two mass gaps and a superconducting gap which need to be minimised subjected to the conditions of electrical and color charge neutrality namely $Q_E = 2/3\rho^1 - 1/3\rho^2 - 1/3\rho^3 - \rho_e = 0$, and, $Q_8 = 1/\sqrt{3}\sum_i(\rho^{i1} + \rho^{i2} - 2\rho^{i3}) = 0$. Here we have denoted $\rho^{ia} = \langle\psi^{ia\dagger}\psi^{ia}\rangle$ and$\rho^i = \sum_a \rho^{ia}$– the two indices in the superscript rering to flavor and color indices respectively. the results of such a constrained minimisation are discussed in the follwing section.

3. Results and discussions

For numerical calculations we have taken the values of the parameters of NJL model as follows. The cutoff Λ and the scalar coupling G_s are chosen by fitting the pion decay constant $f_\pi = 93$ MeV and the chiral condensate $\langle\bar{u}u\rangle^{1/3} = -250$ MeV$=\langle\bar{d}d\rangle^{1/3}$. We also take the current quark masses of u and d quarks as zero. This leads to the coupling constant G_s and the cut off Λ as $G_s = 5.0163$ GeV^{-2} and $\Lambda = 0.6533$ GeV. Similar to Ref.[9] we take the current quark mass of strange quarks as $m_s = 120$ MeV as a typical value giving "reasonable" vacuum properties. With this choice of parameters, the constituent quark masses at zero temperature and density are given as $M_1 = 0.313$ GeV$=M_2$, and for strange quark $M_3 = 0.541$ GeV. These values are similar to those obtained in Ref.[10], where the parameters have been fixed by fitting vacuum masses and decay constants of pesudoscalar mesons. In Fig.1 we show the gap parameters as a function of density without imposition of charge neutrality condition. At $\mu_B = 960$ MeV, the system goes from vacuum solution to a color superconducting phase while the masses of the light quarks drop to zero from their vacuum values.

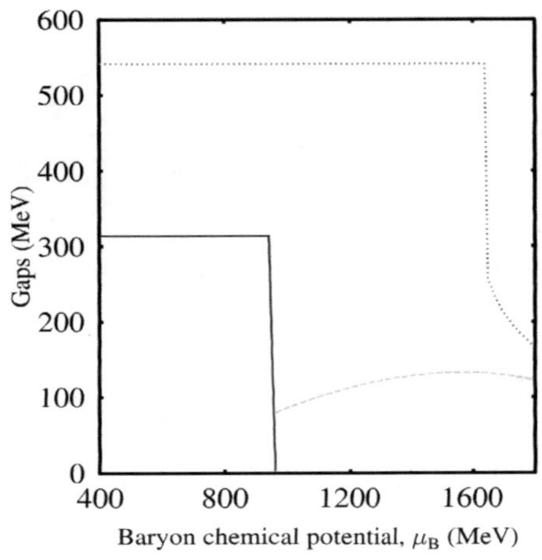

Figure 1. Gap parameters when charge neutrality conditions are not imposed. Solid curve refers to masses of u and d quarks, dotted curve refrs to mass of strange quarks and the dashed curve refers to the superconducting gap

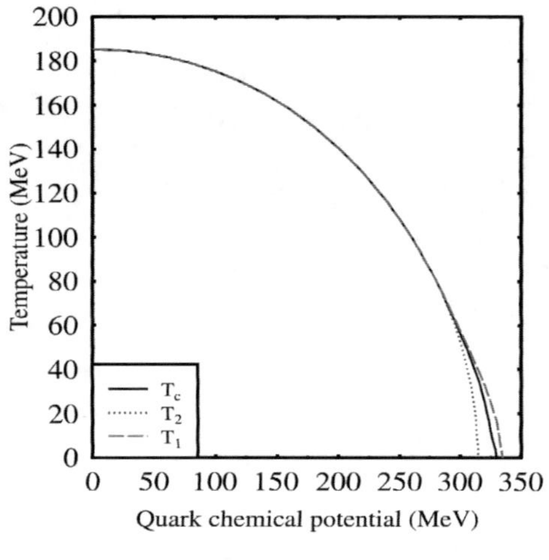

Figure 2. Phase diagram in the (μ,T) plane. Middle curve is the critical line and the outer lines are the lower and upper spinodals

In Fig.2 we show the resulting phase diagram for the chiral transition in the temperature and quark chemical potential ($\mu_B/3$) plane in the present model when charge neutrality conditions are not imposed. The middle line is the critical line in this plane. We also show here the upper (T_1) and the lower (T_2) spinodial lines constraining the region of spinodal instability region. In this region there are solutions of the mass gap equations which are metastable and fluctuations in this region is important. The tricritical point (T_c, μ_c) turns out to be (74.9,285) MeV beyond which the transition is second order.

We next extend our discussions to the case where the charge neutrality conditions (μ_E=0=μ_8) are imposed. At lower baryon densities charge neutrality condtition forces the d-quark density to be larger than u- quarks. This makes the d-quark masses vanishing much earlier than that of the u- quarks as density is increased. At higher chemical potential the strange quarks help in maintaining the charge neutrality condition. The superconducting gap is shown in Fig.s3

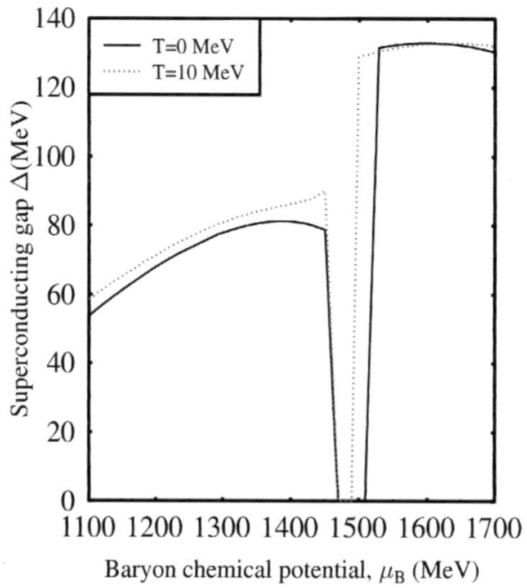

Figure 3. Supeconducting gap as a function of density for charge neutral matter

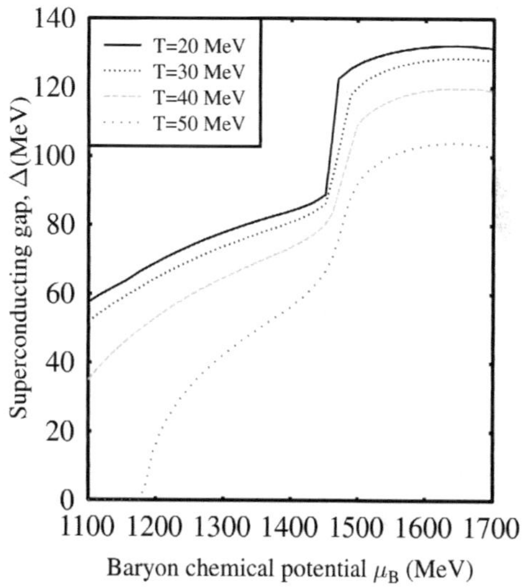

Figure 4. Superconducting gap as a function of baryon chemical potential for different temperatures for charge neutral matter.

and Fig.4 for different temperatures. As may be seen for smaller values of the baryon chemical potential ($\mu_B < 1470$ MeV) we have smaller values of the gap which reaches a maximum about 80 MeV at $\mu_B = 1390$ MeV at zero temperature. Then it decreases and vanishes at $\mu_B = 1470$ MeV=μ_2. The gap remains zero till $\mu_B = \mu_3 = 1530$ MeV. At μ_3 the system jumps from the normal phase to the BCS phase with a gap about 132 MeV. The baryon number density jumps from $\rho_B = 0.96\text{fm}^{-3}$ to 1.54 fm^{-3} i.e. from 6 to 9.5 times the nuclear matter density. The interval in the baryon chemical potential between the gapless and BCS phase within which it is normal quark matter decreases with increase in temperature and then disappears as may be noted in Fig.s4. The sharp transition between gapless phase to the BCS phase as a function of baryon chemical potential at smaller temperature becomes a smooth transition as temperature increases as may be seen from Fig.4.

Let us next discuss now some of the characteristics of the gapless modes which occur for smaller values of the baryon chemical potential ($\mu_B < \mu_2$). At zero temperature, gapless modes occur when the gap is less than half the difference of the chemical potential $\delta\mu$ of the

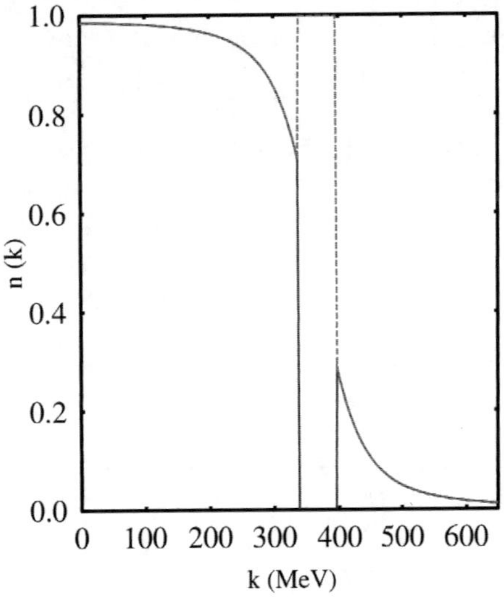

Figure 5. Occupation number as a function of momentum for u quarks. Solid line corresponds to u quarks and dotted line corresponds to d quarks

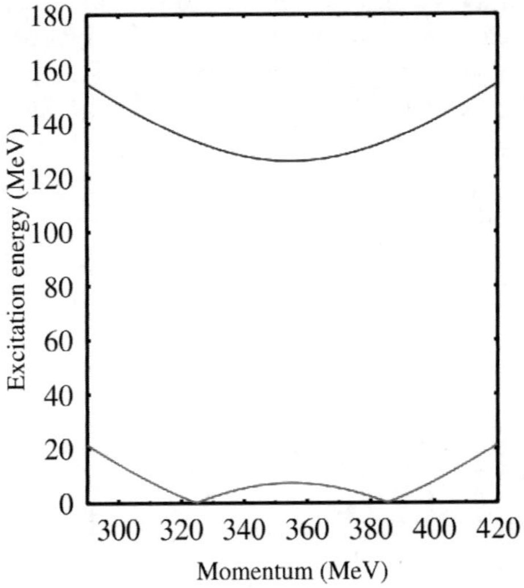

Figure 6. Excitation energies as functions of chemical potential

two condensing quarks. It is easy to show also that the excitation energy ω_2 of the d-quark vanishes at momenta $\mu_- = \bar{\mu} - \sqrt{\delta\mu^2 - \Delta^2}$ and $\mu_+ = \bar{\mu} + \sqrt{\delta\mu^2 - \Delta^2}$ at zero temperature. In Fig.6 we plot the dispersion relations - the excitation energies of the quasi particles as functions of momentum for quark chemical potential $\mu_q = (1/3)\mu_B \simeq 379$MeV and at zero temperature. The gap in this case turns out to be $\Delta = 59$ MeV and the difference of the chemical potential turns out to be $\delta_\mu \simeq -66$ MeV at zero temperature. Far from the pairing region $\bar{\mu} \simeq 354$ MeV, the spectrum looks like that of usual BCS type dispersion relations. Of the two excitation energies, ω_1 shows a minimum at the average fermi momentum $\bar{\mu}$ with a value $\omega_1(|\mathbf{k}| = \bar{\mu}) = \Delta + \delta_\mu \simeq 125$MeV. On the other hand, ω_2 becomes gapless ($\omega_2 = 0$)at momenta μ_- and μ_+. In this 'breached' pairing region [11] we have only unpaired down quarks and no up quarks. In this phase the number densities of the condensing u and d quarks are not the same. The occupation numbers in the momentum space for the u and d quarks are plotted in Fig.5. In the region between μ_- and μ_+, the occupation numbers resemble like that of normal matter as ρ_u becomes almost zero but for the vanishingly small antiparticle contribution. On the other hand, in the same region, the occupation number for d- quarks becomes unity in the same limit for the antiparticle contributions. This gives rise to difference in number densities of the u and d quarks that take part in diquark condensation and in fact is given as This is infact given by $\delta\rho_{sc}(T = 0) = 2/(3\pi^2)(\mu_+^3 - \mu_-^3)$. We also note here that the ratio of T_C to the gap $/Delta(T = 0)$ for the gapless phase is not the BCS value of 0.567 but can be as large as 0.81 for the chemical potential considered here.

As the chemical potential is increased beyond 1530 MeV at zero temperature the solution for the gap jumps to a higher value of about 130 MeV almost similar to the case when charge neutrality condition is not imposed. This corresponds to the usual BCS solution. In this case, the numbers of u quarks and d quarks participating in condensation are the same. The charge neutrality condition however is maintained by the third color, the electron as well as the strange quarks. One essential effect of including strange quarks is that the electron density starts decreasing for chemical potentials greater than the strange quark mass as the strange quark can carry the negative charge to maintain electric charge neutrality condition. This has the effect that the lowest excitation energy $\omega_2 = \bar{\omega} + \delta_\mu$ in the BCS pairing case becomes large due to both the large value of the gap as well as the smaller magnitude of the electron chemical potential.

4. Summary

We have analysed here in NJL model the structure of vacuum in terms of quark–antiquark as well as diquark pairing at finite temperature. The methodology uses an explicit variational construct of the trial state. The gap function as well as the thermal distribution functions are also determined variationally. To consider neutron star matter, we have imposed the color and electric charge neutrality conditions through the introduction of appropriate chemical potentials.

We observe that the mass of the strange quark plays a sensitive role in maintaining charge neutrality condition. Further, we observed that solutions for the gap quations which may not be free energtically preferble solutions, can be the preferred solutions when constraints of charge neutrality conditions are imposed. We also find that with increase in chemical potential, the quark matter has a transition from a gapless phase to the BCS phase through an intermediate normal quark matter phase and this interval decreases and finally vanishes at higher temperature. The sharp transition at zero temperature becomes a smooth transition at higher temperatures as density is increased.

We have focussed our attention here to the two flavor superconducting phase with strange quarks. The variational method adopted can be directly generalised to include color flavor locked phase and one can then make a free energy comparison regarding the possibility of which phase would be thermodynamicaly favourable at what density. This will be particularly interesting for cooling of neutron stars with a CFL core through neutrino emission.

[1] For reviews see Rajagopal, K. and Wilczek,F. arXiv:hep-ph/0011333; Hong, D.K. Acta Phys. Polon. B32,1253 (2001); Alford, M.G., Ann. Rev. Nucl. Part. Sci 51, 131 (2001); Nardulli,G., Riv. Nuovo Cim. 25N3, 1 (2002); Reddy, S., Acta Phys Polon.B33, 4101(2002); Schaefer,T., arXiv:hep-ph/0304281; Rischke,D.K., Prog. Part. Nucl. Phys. 52, 197 (2004); Ren,H.C., arXiv:hep-ph/0404074; Huang,M. arXiv: hep-ph/0409167; Shovkovy, I.,arXiv:nucl-th/0410191.

[2] Shovkovy Igor, Huang Mei, Phys. Lett. B **564**, 205 (2003), Huang Mei and Shovkovy Igor,Nucl. Phys. **A729**, 835 (2)003.

[3] Amruta Mishra and Hiranmaya Mishra, Phys. Rev. D **69**, 014014 (2004).

[4] H. Mishra and J.C. Parikh, Nucl. Phys. **A679**, 597 (2001).

[5] Jaikumar P. and Prakash, M.Phys. Lett. B **516**, 345 (2001).

[6] H. Umezawa, H. Matsumoto and M. Tachiki *Thermofield dynamics and condensed states* (North Holland, Amsterdam, 1982) ; P.A. Henning, Phys. Rep.253, 235 (1995).

[7] Amruta Mishra and Hiranmaya Mishra, *J. Phys. G: Nucl. Phys.* 231431997.

[8] Mishra,A. and Mishra, H., arXiv: hep-ph/0412213.

[9] Buballa M.and Oertel M., Nucl. Phys. **A703**, 770 (2002).

[10] P. Rehberg, S.P. Klevansky and J. Huefner, Phys. Rev. C **53**, 410 (1996).

[11] W.V. Liu and F. Wilczek,*Phys. Rev. Lett.* 900470022003,E. Gubankova, W.V. Liu and F. Wilczek, *Phys. Rev. Lett.* 910320012003.

Institute of Physics Publishing
doi:10.1088/1742-6596/50/1/027

Journal of Physics: Conference Series **50** (2006) 230–237
5th International Conference on Physics and Astrophysics of Quark Gluon Plasma

Equation of State and Collective Dynamics

Ulrich Heinz[1]

Department of Physics, The Ohio State Univesity, Columbus, OH 43210, USA

Abstract. This talk summarizes the present status of a program to quantitatively relate data from the Relativistic Heavy Ion Collider (RHIC) on collective expansion flow to the Equation of State (EOS) of hot and dense strongly interacting matter, including the quark-gluon plasma and the quark-hadron phase transition. The limits reached with the present state of the art and the next steps required to make further progress will both be discussed.

1. The QCD Equation of State and ideal fluid dynamics

With relativistic heavy-ion collisions one explores the phase diagram of strongly interacting bulk matter in the regime of high energy density and temperature. Lattice QCD (LQCD) tells us [1] that for zero net baryon density QCD matter undergoes a phase transition at $T_{\rm cr} = 173 \pm 15\,\rm MeV$ from a color-confined hadron resonance gas (HG) to a color-deconfined quark-gluon plasma (QGP). The critical energy density $e_{\rm cr} \simeq 0.7\,\rm GeV/fm^3$ [1] corresponds roughly to that in the center of a proton. At the phase transition, the normalized energy density e/T^4 rises rapidly by about an order of magnitude over a narrow temperature interval $\Delta T \lesssim 15 - 20\,\rm MeV$, whereas the pressure p/T^4 (which is proportional to the grand canonical thermodynamic potential) is continuous and rises more gradually (Fig. 1). Both seem to saturate at about 80-85% of the Stefan-Boltzmann value for an ideal gas of noninteracting quarks and gluons, the energy density more quickly (at about $1.2\,T_{\rm cr}$), the pressure more slowly. Above about $2\,T_{\rm cr}$, the lattice data follow the Equation of State of an ideal gas of massless particles, $e = 3p$.

For many years this observation has been interpreted as lattice QCD support for the hypothesis of a weakly interacting, perturbative QGP. The recent RHIC data taught us (as I will show) that this interpretation was quite wrong.

It was recognized over 3 decades ago (see review [2]) that information about the EOS of

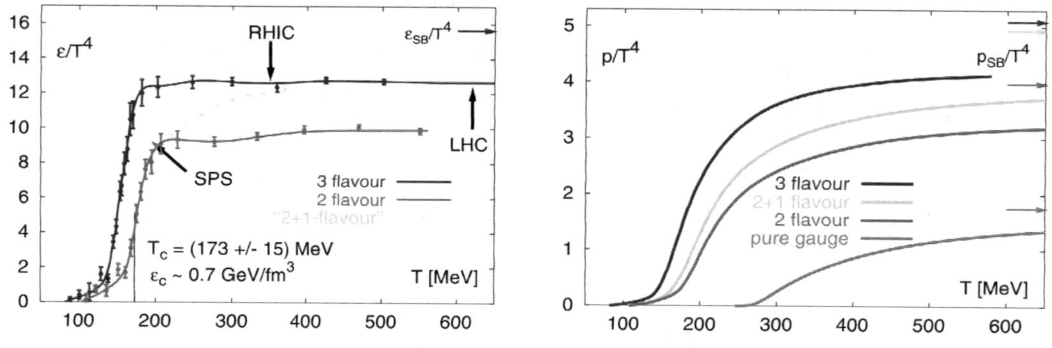

Figure 1. The normalized energy density e/T^4 (left) and pressure p/T^4 (right) from lattice QCD [1] for 0, 2 and 3 light quark flavors, as well as for 2 light + 1 heavier (strange) quark flavors. Horizontal arrows on the right indicate the corresponding Stefan-Boltzmann values for a non-interacting quark-gluon gas.

[1] Email: `heinz.9@osu.edu`. Work supported by the U.S. Department of Energy, grant DE-FG02-01ER41190.

strongly interacting matter can be extracted by studying the collective dynamics of relativistic heavy-ion collisions. This connection is particularly direct in the framework of *ideal fluid dynamics* which becomes applicable if the matter formed in the collision approaches *local thermal equilibrium*. The latter requires sufficiently strong interactions in the medium that local relaxation time scales are shorter than the macroscopic evolution time scale. In this limit the local conservation laws for the baryon number, energy and momentum currents, $\partial_\mu j^\mu_B(x) = 0$ and $\partial_\mu T^{\mu\nu} = 0$, can be rewritten as the relativistc Euler equations for ideal fluid motion:

$$\dot{n}_B = -n_B\,(\partial\cdot u), \qquad \dot{e} = -(e+p)\,(\partial\cdot u), \tag{1}$$

$$\dot{u}^\mu = \frac{\nabla^\mu p}{e+p} = \frac{c_s^2}{1+c_s^2}\,\nabla^\mu \ln\left(\frac{e}{e_0}\right). \tag{2}$$

The dot denotes the time derivative in the local fluid rest frame ($\dot{f} = u\cdot\partial f$) and ∇^μ the gradient in the directions tranverse to the fluid 4-velocity u^μ. The first line describes the dilution of baryon and energy density due to the local expansion rate $\partial\cdot u$, which itself is driven according to (2) by the pressure or energy density gradients providing the fluid acceleration. The absolute value of the energy density e is locally irrelevant: the initial maximal energy density e_0 only matters by setting the overall time scale between the beginning of hydrodynamic expansion and final decoupling, thereby controlling how much flow can develop globally. The details of the flow pattern are thus entirely controlled by the temperature dependent speed of sound $c_s^2 = \frac{\partial p}{\partial e}$.

According to the LQCD data, the latter is $c_s^2 \approx \frac{1}{3}$ for $T > 2\,T_{\rm cr}$, then drops steeply near $T \approx T_{\rm cr}$ to values near $c_s^2 \approx \frac{1}{20}$ (the "softest point", see Fig. 11 in [1]), before rising again in the hadron resonance gas phase to $c_s^2 \approx 0.15$ [3]. A key goal of flow studies in relativistic heavy ion collisions is to find traces of this "softest point" in the data.

2. "Flavors" of transverse flow in heavy ion collisions

Experimentally one studies flow by analyzing the transverse momentum spectra of the emitted hadrons. In central ($b=0$) collisions between spherical nuclei, the flow is azimuthally symmetric about the beam axis. This "radial flow" integrates over the entire pressure history of the collision from initial thermalization to final decoupling ("freeze-out"), due to persistent pressure gradients. In noncentral ($b \neq 0$) collisions, or central collisions between deformed nuclei such as uranium [4], the nuclear reaction zone is spatially deformed, and anisotropies of the transverse pressure gradients result in transverse flow anisotropies. These can be quantified by Fourier expanding the measured final momentum spectrum $dN/(dy\,p_\perp dp_\perp\,d\phi_p)$ with respect to the azimuthal angle ϕ_p. For collisions between equal nuclei, the first non-vanishing Fourier coefficient at midrapidity is the *elliptic flow* $v_2(p_\perp, b)$. Since v_2 is driven by pressure *anisotropies* and the spatial deformation of the reaction zone creating such anisotropies quickly decreases as time proceeds, the elliptic flow is sensitive to the EOS only during the early expansion stage [5] (the first $\sim 5\,\text{fm}/c$ in semicentral Au+Au collisions [6]), until the spatial deformation has disappeared.

Depending on the initial energy density (i.e. beam energy), the hot expanding fireball spends this crucial time either entirely in the QGP phase, or mostly near the quark-hadron phase transition, or predominantly in the hadron resonance gas phase [6], thereby probing different effective values of the sound speed c_s. To the extent that ideal fluid dynamics is valid in all these cases, an excitation function of the elliptic flow v_2 should thus allow to map the temperature dependence of the speed of sound and identify the quark-hadron phase transition, via a minimum in the function $v_2(\sqrt{s})$ [6]. This will be further discussed below (see Section 5.2 and Fig. 6).

3. Model parameters and predictive power of hydrodynamics

The hydrodynamic model requires *initial conditions* at the earliest time at which the assumption of local thermal equilibrium is applicable, and a *"freeze-out prescription"* at the end when the system becomes too dilute to maintain local thermal equilibrium. Both are described in detail elsewhere [7]. Different approaches to freeze-out invoke either the Cooper-Frye algorithm [8] (used by us), in which chemical freeze-out of the hadron abundances at $T_{\rm cr}$ [9] must be implemented by hand by introducing non-equilibrium chemical potentials below $T_{\rm cr}$ [10, 11, 12], or a hybrid approach [13, 14] that switches from a hydrodynamic description to a microscopic hadron cascade at the quark-hadron transition, letting the cascade handle the chemical and thermal freeze-out kinetics. While the radial flow patterns from the two freeze-out procedures don't differ much, the elliptic flow can be quite different if the spatial deformation of the fireball is still significant during the hadronic stage of the expansion, as I will discuss in Sec. 5.2.

We have solved the relativistic equations for ideal hydrodynamics under the simplifying assumption of boost-invariant longitudinal expansion (see [6, 15] for details). This is adequate near midrapidity (the region which most RHIC experiments cover best), but not sufficient to describe the rapidity distribution of emitted hadrons and of their transverse flow pattern which require a (3+1)-dimensional hydrodynamic code such as the one by Hirano [10].

The initial and final conditions for the hydrodynamic evolution are fixed by fitting the pion and proton spectra at midrapidity in *central* ($b=0$) collisions; additionally, we use the centrality dependence of the total charged multiplicity $dN_{\rm ch}/dy$. I stress that this is the *only* information used from $b\neq 0$ collisions, and it is necessary to fix the ratio of soft to hard collision processes in the initial entropy production. An upper limit for the initial thermalization time $\tau_0 \leq 0.6\,{\rm fm}/c$, the initial entropy density $s_0 = 110\,{\rm fm}^{-3}$ in the fireball center (corresponding to an initial maximal energy density $e_0 \approx 30\,{\rm GeV/fm}^3$ and an initial central fireball temperature $T_0 \approx 360\,{\rm MeV} \approx 2\,T_{\rm cr}$), the baryon to entropy ratio, and the freeze-out energy density $e_{\rm dec} = 0.075\,{\rm GeV/fm}^3$ are all fixed from the $b=0$ pion and proton spectra (these numbers refer to 200 A GeV Au+Au collisions at RHIC [12]). The initial entropy density profile is calculated for all b from the collision geometry, using the Glauber model with soft/hard ratio as fixed above. Except for pions and protons, all other hadron spectra in $b=0$ collisions and all spectra for $b\neq 0$ collisions (including all flow anisotropies such as v_2 which vanish at $b=0$) are then parameter-free predictions of the model. Note that all calculated hadron spectra include feeddown from decays of unstable hadron resonances.

4. Successes of ideal fluid dynamics at RHIC

4.1. Hadron momentum spectra and radial flow

Figure 2 shows the single particle $p_\perp$-spectra for pions, kaons and antiprotons (left panel) as well as Ω baryons (right panel) measured in Au+Au collisions at RHIC, together with hydrodynamical results [12]. In order to illustrate the effect of additional radial flow generated in the late hadronic stage below $T_{\rm cr}$, two sets of curves are shown: the lower (blue) bands correspond to kinetic decoupling at $T_{\rm cr} = 165\,{\rm MeV}$, whereas the upper (red) bands assume decoupling at $T_{\rm dec} = 100\,{\rm MeV}$. The width of the bands indicates the sensitivity of the calculated spectra to

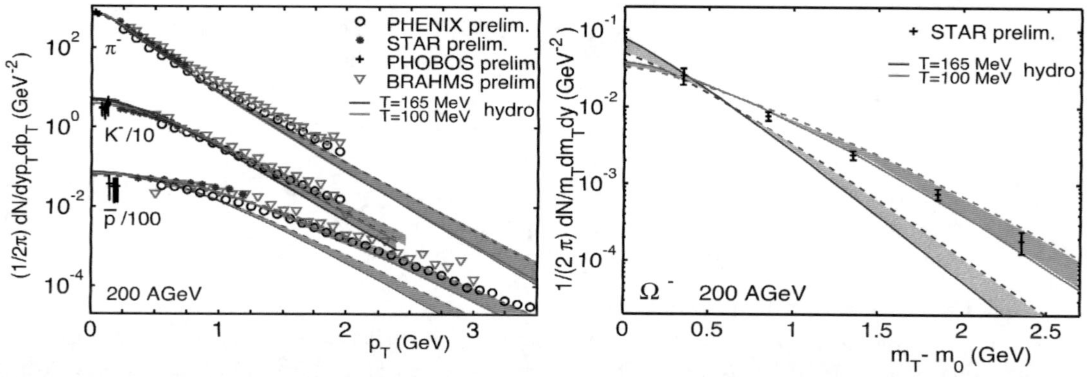

Figure 2. Negative pion, kaon, antiproton, and Ω spectra from central Au+Au collisions at $\sqrt{s} = 200\,A\,{\rm GeV}$ measured at RHIC [16]. The curves show hydrodynamical calculations [12] (see text).

an initial transverse flow of the fireball already at the time of thermalization [12]. In the hydrodynamic simulation it takes about 9-10 fm/c until most of the fireball becomes sufficiently dilute to convert to hadronic matter and another 7-8 fm/c to fully decouple [6]. Figure 2 shows that by the time of hadronization the dynamics has not yet generated enough radial flow to reproduce the measured $\bar{p}$ and Ω spectra; these heavy hadrons, which are particularly sensitive to radial flow, require the additional collective "push" created by resonant quasi-elastic interactions during the fairly long-lived hadronic rescattering stage. The flattening of the $\bar{p}$ spectra by radial flow provides a natural explanation for the (initially puzzling) experimental observation that for $p_\perp > 2$ GeV/c antiprotons become more abundant than pions [15].

As shown elsewhere (see Fig. 1 in [17]), the model describes these and all other hadron spectra not only in central, but also in *peripheral* collisions, up to impact parameters of about 10 fm, and with similar quality. No additional parameters enter at non-zero impact parameter.

4.2. Elliptic flow

Figure 3 shows the predictions for the elliptic flow coefficient v_2 from Au+Au collisions at RHIC, together with the data [18, 19]. For impact parameters $b \leq 7$ fm ($n_{\rm ch}/n_{\rm max} \geq 0.5$) and transverse

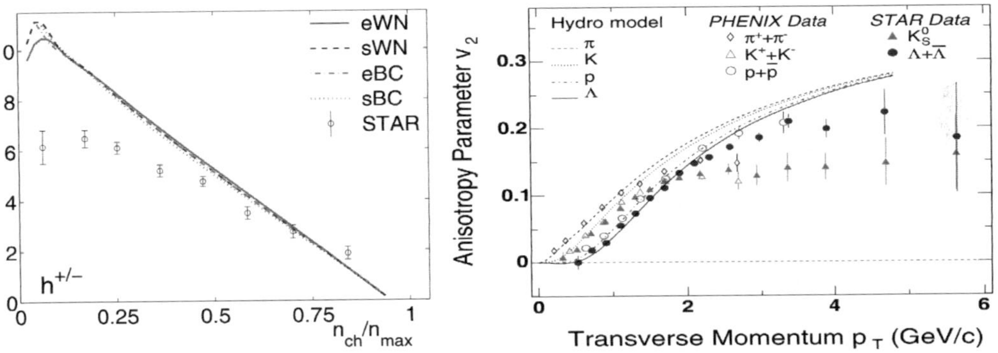

Figure 3. Left: $p_\perp$-averaged elliptic flow for all charged hadrons from 130 A GeV Au+Au collisions, as a function of collision centrality ($n_{\rm ch}$ is the charged multiplicity at $y = 0$). The curves are hydrodynamic calculations with different choices for the initial energy density profile (see [20]). Right: Differential elliptic flow $v_2(p_\perp)$ for identified hadrons from minimum bias Au+Au collisions at 200 A GeV [18, 19, 21], together with hydrodynamic curves from [22].

momenta $p_\perp \lesssim 1.5$ GeV/c the data are seen (left panel of Fig. 3) to exhaust the upper limit for v_2 obtained from the hydrodynamic calculations. For larger impact parameters $b > 7$ fm the $p_\perp$-averaged elliptic flow v_2 increasingly lags behind the hydrodynamic prediction; this will be discussed in detail in Sec. 5.2. As a function of $p_\perp$ (right panel of Fig. 3) the elliptic flow of all hadrons measured so far is very well described by hydrodynamics, for $p_\perp \lesssim 1.5 - 2$ GeV/c. In particular the hydrodynamically predicted *mass splitting* of v_2 at low $p_\perp$ is perfectly reproduced by the data. This mass splitting depends on the EOS [22], and the EOS including a quark-hadron phase transition used here describes the data better than one without phase transition (see Fig. 2 in [17]). Ideal fluid dynamics with a QGP EOS thus gives an excellent and very detailed description of *all* hadron spectra below $p_\perp = 1.5$ GeV/c. Since this $p_\perp$-range includes more than 99% of all produced hadrons, it is fair to say that *the bulk of the fireball matter formed in Au+Au collisions at RHIC behaves very much like a perfect fluid.*

4.3. Final source eccentricity in coordinate space

While spectra and elliptic flow reflect the *momentum* structure of the hadron emitting source, its *spatial* deformation can be tested with 2-pion Hanbury Brown-Twiss (HBT) correlations measured as a function of the azimuthal emission angle [23, 24]. Even though the initial spatial deformation of the reaction zone in non-central Au+Au collisions at RHIC is finally completely gone, many pions are already emitted from earlier times when the spatial deformation is still significant. For Au+Au collisions at $b = 7\,\mathrm{fm}$, the spatial eccentricity of the time-integrated hydrodynamic pion emission function is $\langle\varepsilon_x\rangle = 0.14$ (or 56% of its initial value $\varepsilon_x(\tau_0) = 0.25$) [23]. Using azimuthally sensitive pion HBT measurements, the STAR Collaboration has measured in the corresponding impact parameter bin [25] $\langle\varepsilon_x\rangle = 0.11 \pm 0.035$ (or $45 \pm 15\%$ of the initial deformation). This agreement can be counted as another success for the hydrodynamic model.

5. Viscous effects at RHIC

5.1. QGP viscosity

As evident in the right panel of Fig. 3, the hydrodynamic prediction for $v_2(p_\perp)$ gradually breaks down above $p_\perp \gtrsim 1.5\,\mathrm{GeV}/c$ for mesons and above $p_\perp \gtrsim 2.2\,\mathrm{GeV}/c$ for baryons. The empirical fact [21] that both the $p_\perp$-values, where this break from hydrodynamics sets in, and the saturation values for v_2 at high $p_\perp$ for baryons and mesons are always (*i.e. for all collision centralities!*) related by 3 : 2 (i.e. by their ratios of valence quark numbers), independent of their masses, tells its own interesting story (see e.g. [26]): It strongly suggests that in this $p_\perp$ region hadrons are formed by coalescence of color-deconfined quarks, and that the elliptic flow is of partonic origin (i.e. generated before hadronization), with a $p_\perp$-shape that follows hydrodynamics at low $p_\perp$ up to about 750 MeV and then gradually breaks away [26].

Since $v_2(p_\perp)$ is a measure for the relatively small differences between the in-plane and out-of-plane slopes of the $p_\perp$ spectra, it is more sensitive to deviations from ideal fluid dynamic behaviour than the angle-averaged slopes. Two model studies [27, 28] showed that v_2 reacts particularly strongly to shear viscosity. As the mean free path of the plasma constituents (and thus the fluid's viscosity) goes to zero, v_2 approaches the ideal fluid limit from below [29] (see Fig. 4a). At higher transverse momenta it does so more slowly than at low $p_\perp$ [29], approaching a constant saturation value at high $p_\perp$. The increasing deviation from the ideal fluid limit for

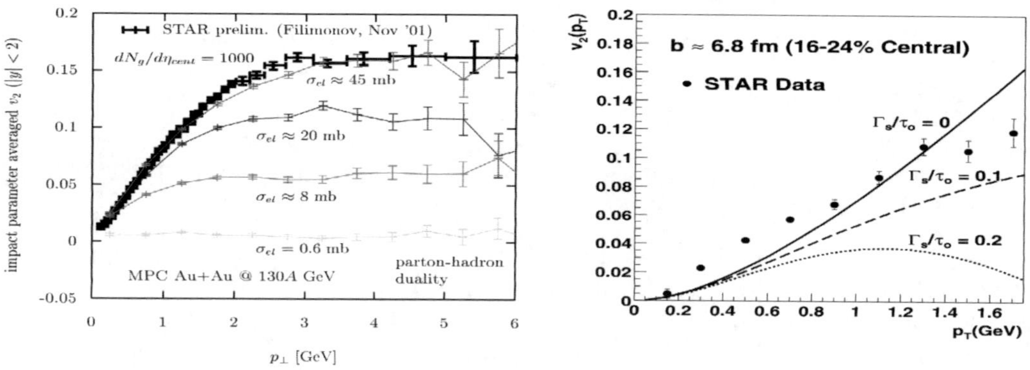

Figure 4. Left: Elliptic flow from a parton cascade [29], compared with STAR data, for different parton-parton scattering cross sections. Larger cross sections lead to smaller mean free paths. Right: Perturbative effects of shear viscosity on the elliptic flow $v_2(p_\perp)$ [28] (see text).

growing $p_\perp$ is qualitatively consistent with the expected influence of shear viscosity: Teaney [28] showed that the lowest order viscous corrections to the local equilibrium distribution increase quadratically with $p_\perp$ so that v_2 remains increasingly below the ideal fluid limit as $p_\perp$ grows (see Fig. 3b). From Fig. 4b Teaney concluded that at RHIC the normalized sound attenuation length $\frac{\Gamma_s}{\tau} = \frac{4}{3T\tau}\frac{\eta}{s}$ (where η is the shear viscosity, T the temperature and s the entropy density) cannot be much larger than about 0.1. This puts a stringent limit on the dimensionless ratio η/s, bringing it close to the recently conjectured absolute lower limit for the viscosity of $\eta/s = \hbar/(4\pi)$ [30]. This would make the quark-gluon plasma the most ideal fluid ever observed [30].

These arguments show that deviations from ideal fluid dynamics at high $p_\perp$ must be expected, and that they can be large even for fluids with very low viscosity. At which $p_\perp$ non-ideal effects begin to become visible in $v_2(p_\perp)$ can be taken as a measure for the fluid's viscosity. To answer the *quantitative* question what the RHIC data on partonic elliptic flow and its increasing deviation from ideal fluid behaviour above $p_\perp \gtrsim 750\,\mathrm{MeV}/c$ imply for the *value* of the QGP shear viscosity η requires a numerical algorithm for solving viscous relativistic hydrodynamics (see [31] for an update).

5.2. Viscosity of the hadron resonance gas

Ideal fluid dynamics also fails to describe the elliptic flow v_2 in more peripheral Au+Au collisions at RHIC and in central and peripheral collisions at lower energies (see Fig. 5a), as well as at forward rapidities in minimum bias Au+Au collisions at RHIC [32]. Whereas hydrodynamics predicts a non-monotonic beam energy dependence of v_2 (Fig. 6a [6]), with largest values at upper AGS and lower SPS energies, somewhat lower values at RHIC and again larger values at the LHC, the data seem to increase monotonically with $\sqrt{s}$.

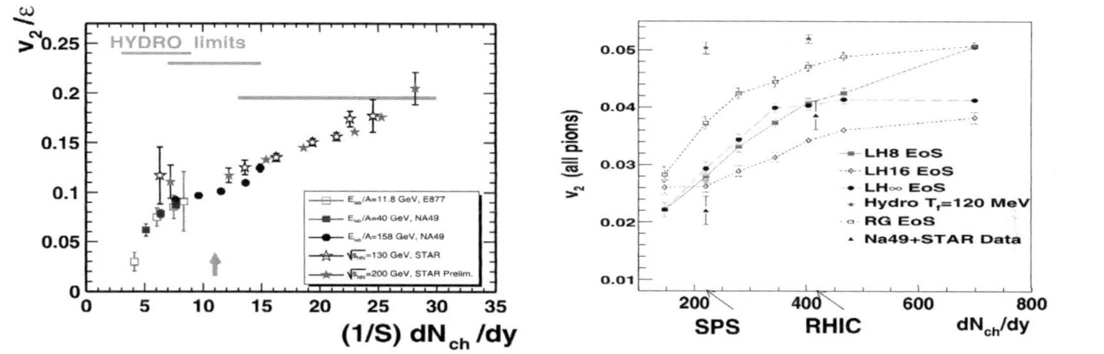

Figure 5. Left: Scaled elliptic flow v_2/ε (where ε denotes the initial spatial eccentricity) vs. the charged multiplicity density per unit initial transverse overlap area S [33]. Right: Elliptic flow v_2 for minimum bias Au+Au collisions at various collision energies, parametrized by the final charged multiplicity density at midrapidity [14]. See text for discussion.

Such a monotonic rise is consistent with "hybrid" calculations by Teaney [14] (Fig. 5b) where the fireball undergoes ideal fluid dynamic evolution only while in the QGP, followed by hadronic *kinetic* evolution using RQMD after hadronization. Figure 5b shows several curves corresponding to different equations of state during the hydrodynamic evolution (see [14]), with LH8 being closest to the data. The difference between the points labelled LH8 and the hydrodynamic values at the top of the figure is due to the different evolution during the late hadronic stage. Obviously, at lower collision energies and for impact parameters $b \sim 7\,\mathrm{fm}$ (which dominate v_2 in minimum bias collisions), ideal fluid dynamics continues to build additional elliptic flow during the hadronic stage, but RQMD does not. The initial energy densities are smaller than at RHIC and the fireball does not spend enough time in the QGP phase for the spatial eccentricity ε to fully disappear before entering the hadron resonance gas phase. Anisotropic pressure gradients thus still exist in the hadronic phase, and ideal fluid dynamics reacts to them according to the stiffness of the hadron resonance gas EOS ($p \approx 0.15e$). Teaney's calculations [14] show that RQMD responds to these remaining anisotropies much more weakly, building very little if any additional elliptic flow during the hadronic phase. The hadron resonance gas is a highly viscous medium, unable to maintain local thermal equilibrium. The failure of the hydrodynamic model in situations where the initial energy density is less than about $10\,\mathrm{GeV/fm}^3$ [34] is therefore likely *not* the result of viscous effects in the early QGP fluid, but rather caused by the highly viscous late hadronic stage which is unable to efficiently respond to any remaining spatial fireball eccentricity. Similar arguments hold at forward rapidities at RHIC [34] where the initial energy densities are also significantly smaller than at midrapidity while the initial spatial eccentricities are similar.

The large hadron gas viscosity spoils one of the clearest experimental signatures for the quark-hadron phase transition, the predicted [6] non-monotonic beam energy dependence of v_2 which was already described in Sec. 2 and is shown in Fig. 6a. As one comes down from infinite beam energy, v_2 is predicted to first decrease (due to the softening of the EOS in the phase transition region) and then recover somewhat in the moderately stiff hadron gas phase. The hadron gas viscosity spoils this recovery, leading to an apparently monotonous decrease of v_2 with falling beam energy (Fig. 5). However, recent PHENIX data from Au+Au collisions at $\sqrt{s}=62\,A\,\mathrm{GeV}$ [35] indicate that this decrease may not be quite as monotonous as suggested by Fig. 5a. Figure 6b shows that, at fixed $p_\perp$, the elliptic flow $v_2(p_\perp)$ is essentially constant over the entire energy range explored at RHIC (from 62 to 200 A GeV), decreasing only when going further down to SPS and AGS energies. When integrated over $p_\perp$, this turns into a monotonous behaviour as in Fig. 5a, due to the steepening $p_\perp$ spectra at lower $\sqrt{s}$. While Fig. 6b does not confirm the hydrodynamically predicted *rise* of v_2 towards lower $\sqrt{s}$, it may still reflect this predicted non-monotonic structure in the elliptic flow excitation function, after strong dilution by hadronic viscosity effects. Obviously, many and more systematic hybrid calculations of the type pioneered by Teaney [14] are necessary to explore to what extent we can eventually prove the existence of a QCD phase transition using elliptic flow data in the SPS-RHIC energy domain.

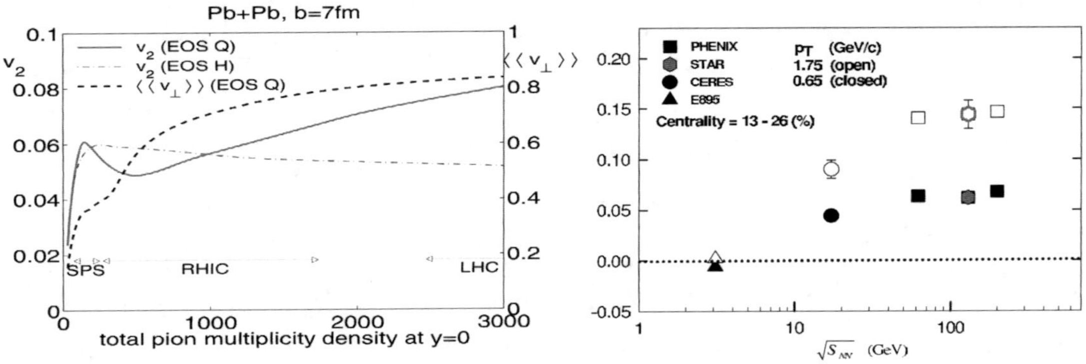

Figure 6. Left: Excitation function of radial and $p_\perp$-integrated elliptic flow for $b=7$ fm Pb+Pb or Au+Au collisions [6]. The horizontal axis gives $dN_\pi/dy(b{=}7\,\mathrm{fm})$, and the horizontal arrows indicate how dN_π/dy was expected to correspond to the beam energy ranges covered by SPS, RHIC and LHC before the RHIC data were available. Right: Elliptic flow at fixed $p_\perp=0.65$ and $1.75\,\mathrm{GeV}/c$ from A+A collisions with $A\approx 200$ at a variety of beam energies [35].

Obviously, it will be important to confirm non-viscous fluid behaviour at higher initial energy densities than so far explored, by extending Fig. 5a to the right and verifying that v_2/ε settles on the hydrodynamic curve. This can be done with Pb+Pb collisions at the LHC, or with full-overlap U+U collisions at RHIC [4]. In addition, the large spatial source deformations achievable in full-overlap side-on-side U+U collisions allow for decisive systematic studies of the non-linear path-length dependence of QCD radiative energy loss of fast partons. In [4] we give quantitative arguments why a U+U collision program should be seriously considered at RHIC.

6. Conclusions

The collective flow patterns observed at RHIC provide strong evidence for fast thermalization at less than 1 fm/c after impact and at energy densities more than an order of magnitude above the critical value for color deconfinement. The thus created thermalized QGP is a strongly coupled plasma which behaves like an almost ideal fluid. These features are first brought out in heavy-ion collisions at RHIC because only there the initial energy densities and QGP life times are large enough for the ideal fluid character of the QGP to really manifest itself, in the form of fully saturated hydrodynamic elliptic flow, undiluted by late non-equilibrium effects from the highly viscous hadron resonance gas which dominates the expansion at lower energies.

We are now ready for a systematic experimental and theoretical program to quantitatively extract the EOS, thermalization time and transport properties of QGP and hot hadronic matter. This requires more statistics and a wider systematic range for soft hadron production data, but more importantly a wide range of systematic simulation studies with the "hydro-hadro" hybrid algorithms and, above all, a (3+1)-dimensional viscous relativistic hydrodynamic code (see [31] for more on that).

References

[1] Karsch F and Laermann E 2004 *Quark-Gluon Plasma 3*, ed R C Hwa and X-N Wang (Singapore: World Scientific) p 1

[2] Stöcker H and Greiner W 1986 *Phys. Rept.* **137** 277

[3] Sollfrank J *et al* 1997 *Phys. Rev.* C **55** 392

[4] Heinz U and Kuhlman A 2005 *Phys. Rev. Lett.* **94**, in press (*Preprint* nucl-th/0411054)

[5] Sorge H 1997 *Phys. Rev. Lett.* **78** 2309

[6] Kolb P F, Sollfrank J, and Heinz U 2000 *Phys. Rev.* C **62** 054909.

[7] Heinz U 2004 *AIP Conf. Proc.* **739** 163

[8] Cooper F and Frye G 1974 *Phys. Rev.* D **10** 186

[9] Braun-Munzinger P, Magestro D, Redlich K, and Stachel J 2001 *Phys. Lett.* B **518** 41

[10] Hirano T and Tsuda K 2002 *Phys. Rev.* C **66** 054905

[11] Rapp R 2002 *Phys. Rev.* C **66** 017901; Teaney D 2002 *Preprint* nucl-th/0204023

[12] Kolb P F and Rapp R 2003 *Phys. Rev.* C **67** 044903

[13] Bass S A and Dumitru A 2000 *Phys. Rev.* C **61** 064909

[14] Teaney D, Lauret J, and Shuryak E V 2001 *Phys. Rev. Lett.* **86** 4783 and *Preprint* nucl-th/0110037

[15] Kolb P F and Heinz U 2004 *Quark-Gluon Plasma 3*, ed R C Hwa and X-N Wang (Singapore: World Scientific) p 634

[16] PHENIX Collaboration, Chujo T *et al* 2003 *Nucl. Phys.* **A715** 151c;
STAR Collaboration, Barannikova O, Wang F *et al* 2003 *Nucl. Phys.* **A715** 458c;
PHOBOS Collaboration, Wosiek B *et al* 2003 *Nucl. Phys.* **A715** 510c;
BRAHMS Collaboration, Ouerdane D *et al* 2003 *Nucl. Phys.* **A715** 478c;
STAR Collaboration, Suire C *et al* 2003 *Nucl. Phys.* **A715** 470c

[17] Heinz U and Kolb P F 2002 *Proc. 18th Winter Workshop on Nuclear Dynamics*, ed. R. Bellwied, J. Harris, and W. Bauer (Debrecen, Hungary: EP Systema) p.205 (*Preprint* hep-ph/0204061)

[18] STAR Collaboration, Ackermann K H *et al* 2001 *Phys. Rev. Lett.* **86** 402;
STAR Collaboration, Adler C *et al* 2001 *Phys. Rev. Lett.* **87** 182301;
STAR Collaboration, Adler C *et al* 2002 *Phys. Rev. Lett.* **89** 132301 and *Phys. Rev.* C **66** 034904;
STAR Collaboration, Adler C *et al* 2003 *Phys. Rev. Lett.* **90** 032301;
STAR Collaboration, Adams J *et al* 2004 *Phys. Rev. Lett.* **92** 052302

[19] PHENIX Collaboration, Adcox K *et al* 2002 *Phys. Rev. Lett.* **89** 212301;
PHENIX Collaboration, Adler S S *et al* 2003 *Phys. Rev. Lett.* **91** 182301

[20] Kolb P F, Heinz U, Huovinen P, Eskola K J, and Tuominen K 2001 *Nucl. Phys.* **A696** 175

[21] Sorensen P R 2003 Kaon and Lambda production at intermediate p(T): Insights into the hadronization of the bulk partonic matter created in Au + Au collisions at RHIC, Ph.D. thesis (*Preprint* nucl-ex/0309003)

[22] Huovinen P, Kolb P F, Heinz U, Ruuskanen P V and Voloshin S A 2001 *Phys. Lett* B **503** 58

[23] Heinz U and Kolb P 2002 *Phys. Lett.* B **542** 216

[24] Retière F and Lisa M A 2004 *Phys. Rev.* C **70** 044907

[25] STAR Collaboration, Adams J *et al* 2004 *Phys. Rev. Lett.* **93** 012301

[26] Müller B 2005 *Nucl. Phys.* **A750** 84; Molnar D and Voloshin S A 2003 *Phys. Rev. Lett.* **91** 092301

[27] Heinz U and Wong S M H 2002 *Phys. Rev.* C **66** 014907

[28] Teaney D 2003 *Phys. Rev.* C **68** 034913 and private communication

[29] Molnar D and Gyulassy M 2002 *Nucl. Phys.* **A697** 495 (Erratum *ibid.* **A703** 893);

[30] Kovtun P, Son D T, and Starinets A O 2004 *Preprint* hep-th/0405231

[31] Chaudhuri A K and Heinz U 2005 these proceedings

[32] Hirano T 2002 *Phys. Rev.* C **65** 011901

[33] NA49 Collaboration, Alt C *et al* 2003 *Phys. Rev.* C **68** 034903

[34] Heinz U and Kolb P F 2004 *J. Phys. G: Nucl. Phys.* **30** S1229

[35] PHENIX Collaboration, Adler S S *et al* 2004 *Preprint* nucl-ex/0411040

doi:10.1088/1742-6596/50/1/028 5th International Conference on Physics and Astrophysics of Quark Gluon Plasma

Hadron production in ultra-relativistic nuclear collisions and the QCD phase boundary.

Peter Braun-Munzinger
GSI and Technical University,
Darmstadt,
Germany

Abstract. We update briefly our understanding of hadron production in relativistic nucleus-nucleus collisions in terms of statistical models with emphasis on the relation of the data to the QCD phase boundary and on a puzzle in the beam energy dependence.

1. Hadron yields, statistical description, and the phase boundary to the QGP.

Hadron yields or more specifically yield ratios observed in central nuclear collisions at AGS, SPS, and RHIC energies can be described with high precision within a hadro-chemical equilibrium approach [1, 2, 3, 4, 5, 6, 7, 8, 9], governed by a chemical freeze-out temperature T_{ch} and baryo-chemical potential μ. A recent review, found in [10], provides a wealth of information on the subject. The data at SPS and RHIC energy comprise multi-strange hadrons including the Ξ, $\bar{\Xi}$, Ω and $\bar{\Omega}$ baryons. Their yields ratios (to pions, e.g.) agree particularly well with the chemical equilibrium calculation and are enhanced by more than an order of magnitude as compared to observations in pp collisions. We present as an example the results for central Au-Au collisions at RHIC energy in Fig. 1.

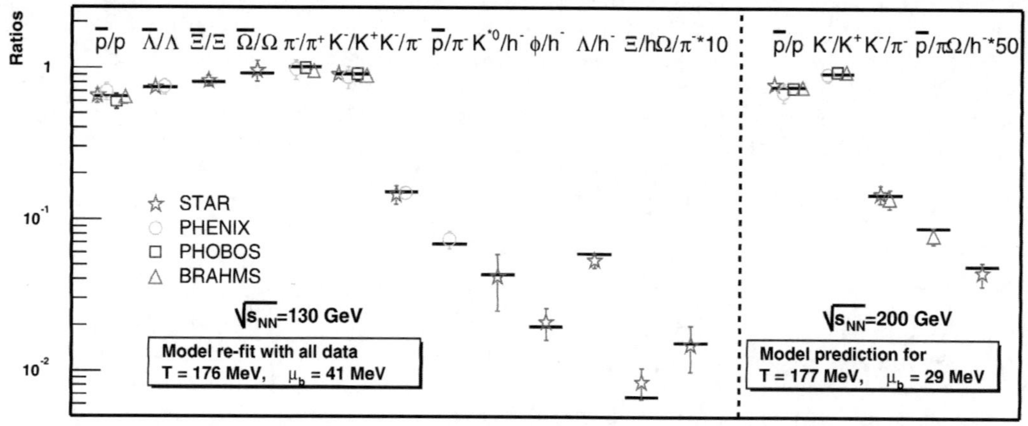

Figure 1. Fit of particle ratios for Au-Au collisions measured at RHIC energies. The measurements are the symbols, the thermal model values are the lines [6, 10]

A remarkable exception is observed for strongly decaying, wide resonances. In particular, the Δ/p ratio is overpredicted while the ρ/π ratio, both determined in (rather peripheral, though) Au-Au collisions by the STAR collaboration [11, 12], is underpredicted by the model [10]. For an interesting and provocative speculation as to where this might come from see [13]. Clearly precision data for central collisions and also at other beam energies are needed to settle this issue.

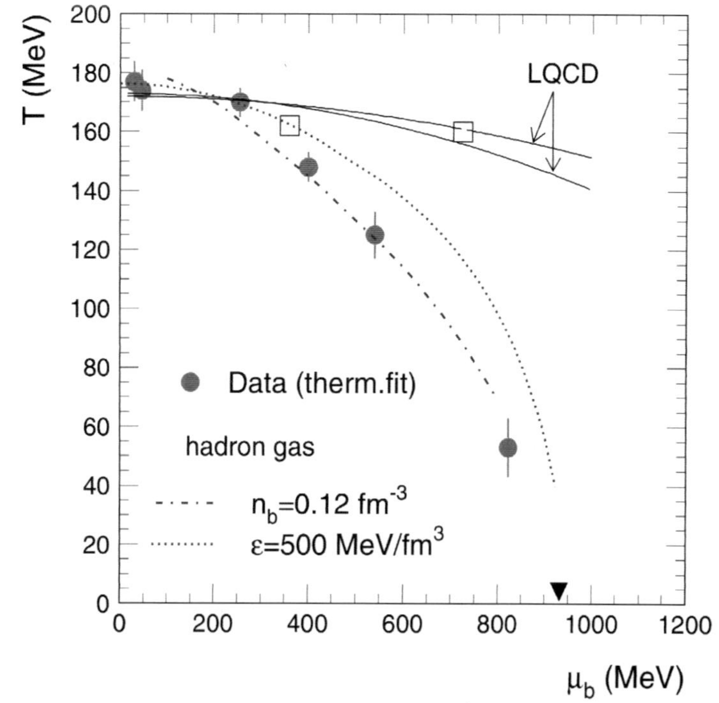

Figure 2. Phase diagram of hadronic matter and chemical freeze-out points [14]. The open squares represents recent estimates [15, 16] of the position of the tri-critical point (see text).

The chemical parameters T_{ch} and μ determined from fits to data at all available energies are plotted in the QCD phase diagram shown in Fig. 2, taken (slightly modified) from [14]. The phase transition lines in this figure are obtained from recent analyses within the framework of lattice QCD [15, 16] and include recent estimates [15, 16] of the position of the tri-critical point[1]. Also included in this figure are lines of constant total baryon density and lines of constant energy density, computed within the framework of the hadronic gas model of [3]. The line of constant energy density $\epsilon = 500$ MeV/fm^3 should in our view be a reasonable phenomenological description of the phase boundary. Interestingly, it is, at $\mu > 500$ MeV, much closer to the chemical freeze-out points than to the phase lines from lattice QCD. Apparently, the phase boundary lines obtained from state-of-the-art calculations within the framework of lattice QCD are not at all lines of constant energy density. Rather, the energy density grows by more than a factor of 2 when going from 0 to 500 MeV in μ. This rather surprizing result deserves further attention.

An important observation, already made in[1], concerning the results of the thermal model

[1] Position and even the existence of such a tri-critical point are currently hotly debated. For a critical discussion see [17].

calculations is that, for top SPS energy and above, the chemical parameters determined from the measured hadron yields coincide within the uncertainties of about ±10 MeV with the phase boundary as determined from lattice QCD calculations. A natural question arises: is this coincidence accidental and, if not, what enforces equilibration at the phase boundary? Considerations about collisional rates and timescales of the hadronic fireball expansion [18] imply that at SPS and RHIC energies the equilibrium cannot be established in a purely hadronic medium and that is is the phase transition which drives the particles densities and ensures chemical equilibrium.

In [18] it is further shown that many body collisions become important close to T_c and provide a mechanism for rapid equilibration. Because of the strong density increase near the QCD phase transition such multi-particle collisions provide a natural explanation for the observation of chemical equilibration at RHIC energies and lead to T_{ch}=T_c within an accuracy of a few MeV. Any scenario with T_{ch} substantially smaller than T_c would require that either multi-particle interactions dominate even much below T_c or that the two-particle cross sections are larger than in the vacuum by a high factor. Both of the latter hypotheses seem unlikely in view of the rapid density decrease in the hadronic phase of the expanding fireball. The critical temperature determined from RHIC for $T_{ch} \approx T_c$ coincides well with recent lattice estimates [19] for $\mu = 0$. The same arguments as discussed here for RHIC energy also hold for SPS energies: it is likely that also there the phase transition drives the particle densities and insures chemical equilibration.

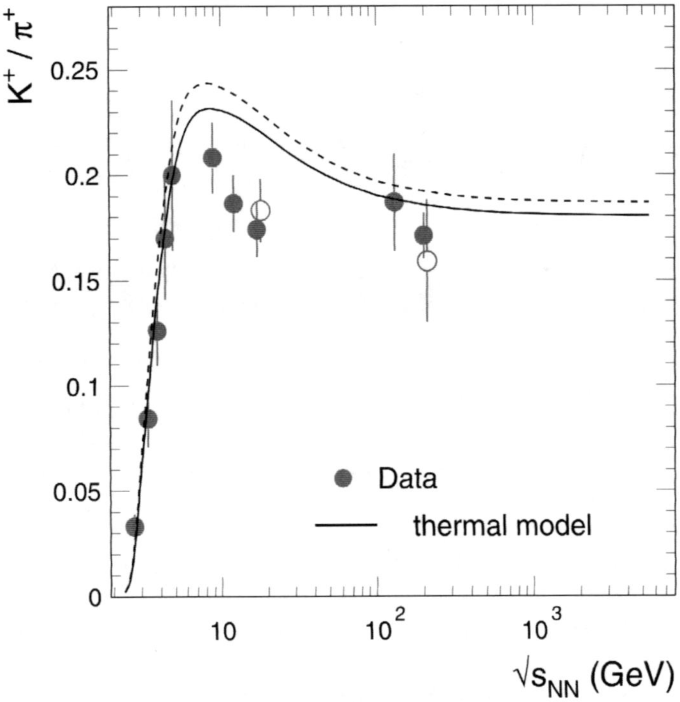

Figure 3. The K^+/π^+ ratio from NA49 [25] and thermal model predictions from [27]. The dashed and solid lines correspond to calculations without and with account of the resonance widths.

It was alternatively proposed [20, 22, 21] that the observed hadron abundances arise from a direct production of strange (and non-strange) particles by hadronization. How this happens microscopically is unclear. To escape the above argument that $T_{ch}=T_c$ one would have to argue that the particle yields are established without hadronic rescattering. This is unlikely since the abundances are determined by hadronic properties (masses) with high precision. Second, one may question if the "chemical temperature" extracted from the abundances is a universal temperature which also governs the local kinetic aspects and can be associated with the critical temperature of a phase transition in equilibrium. Indeed, in a prethermalization process, different equilibrium properties are realized at different time scales. Nevertheless, all experience shows that kinetic equilibration occurs before chemical equilibration. It seems hard to imagine that chemical equilibrium abundances are realized at a time when the kinematic distributions are not yet close to their equilibrium values.

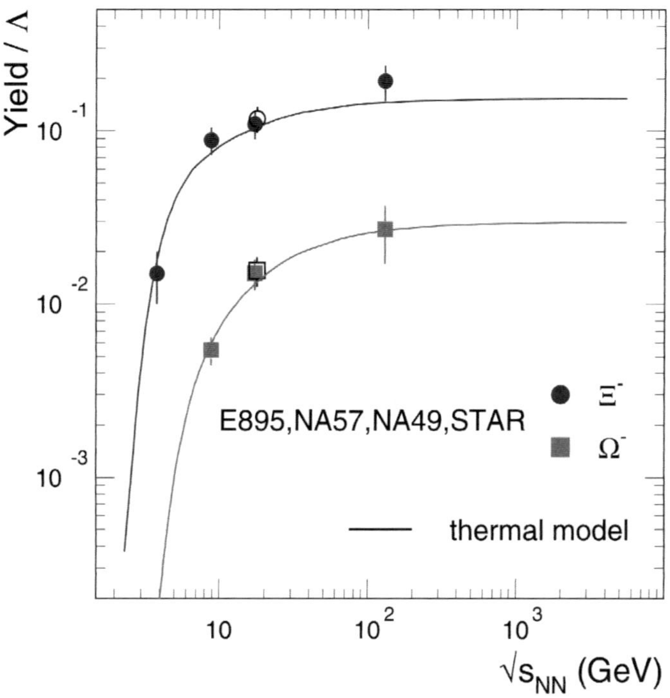

Figure 4. Energy dependence of particle rations involving strange baryons compared to thermal model calculations of [27].

The chemical equilibrium curve and the QCD phase boundary as obtained from lattice QCD calculations begin to differ for baryo-chemical potential values of $\mu > 400$ MeV. Despite that multi-strange baryons are produced with yields very close to what is obtained in a full chemical equilibration scenario (see below). This could be an indication for new, as yet unidentified equilibration processes, different from those advocated in [18] and also incompatible with the "phase space filling" scenario through hadronization put forward by [20, 21, 22], in a dense, baryon-rich medium. Considering the discussion above, a rather radical but appealing possibility is that the chemical freeze-out curve defines the phase boundary also at large μ values up to $\mu \approx 600$ MeV. We note in this context that the chemical freeze-out points determined from measurements at SIS energy should still be considered with caution as no multi-strange baryons have been measured there.

The relatively smooth energy dependence of the above discussed thermal parameters is to be contrasted with recent observation by the NA49 collaboration of an anomaly in the K^+/π^+ ratio near $\sqrt{s_{nn}} = 8$ GeV [25, 26]. These results look rather striking, and indeed, the narrowness of the observed structure is missed by the thermal model predictions, as is visible in Fig. 3.

We note, however, that the anomaly seems to be confined to ratios involving light mesons. For strange baryons, the situation is quite different, as is demonstrated in Fig. 4. Here, the observed energy dependence is very well reproduced by the calculations and no anomaly is visible.

The situation is even more puzzling considering that the energy averaging inherent in ultra-relativistic nucleus-nucleus collisions leads to a characteristic smearing of all possible intrinsic structures. The effect can be estimated [28] using simple geometrical arguments and is of the order 10 % of the cm. energy per nucleon at SPS energy. The corresponding width is very close to the width of the structure observed by the NA49 collaboration and would imply that the strength of the intrinsic structure is even much larger than visible in Fig. 3. Further research is necessary to shed light on this puzzle.

Acknowledgments. Discussions with Anton Andronic and Johanna Stachel are gratefully acknowledged.

[1] P. Braun-Munzinger, J. Stachel, J.P. Wessels and N. Xu, Phys. Lett. **B344** (1995) 43 and **B365** (1996) 1.
[2] J. Cleymans, D. Elliott, H. Satz, and R.L. Thews, Z. Phys. **C74** (1997) 319.
[3] P. Braun-Munzinger, I. Heppe and J. Stachel, Phys. Lett. **B465** (1999) 15.
[4] J. Cleymans and K. Redlich, Phys. Rev. **C60** (1999) 054908.
[5] F. Becattini, J. Cleymans, A. Keranen, E. Suhonen, and K. Redlich, Phys. Rev. **C64** (2001) 024901.
[6] P. Braun-Munzinger, D. Magestro, K. Redlich, and J. Stachel, Phys. Lett. **B518** (2001) 41.
[7] N. Xu and M. Kaneta, Nucl. Phys. **A698** (2002) 306c.
[8] F. Becattini, J. Phys. **G28** (2002) 1553.
[9] R. Rapp and E. Shuryak, Phys. Rev. Lett. **86** (2001) 2980.
[10] P. Braun-Munzinger, K. Redlich, and J. Stachel, nucl-th/0304013, invited review in Quark Gluon Plasma 3, eds. R.C. Hwa and X.N. Wang, (World Scientific Publishing, 2004).
[11] P. Fachini et al., STAR collaboration, nucl-ex/0403026, J. Phys. **G30** (2004) S735.
[12] J. Adams et al., STAR collaboration, Phys. Rev. Lett. **92** (2004) 092301.
[13] G.E. Brown, Chang-Hwan Lee, and M. Rho, nucl-th/0507073.
[14] A. Andronic and P. Braun-Munzinger, hep-ph/0402291.
[15] S. Ejiri et al., hep-lat/0312006.
[16] Z. Fodor and S. Katz, JHEP **0404** (2004) 050.
[17] O. Philipsen, Proc. Lattice05 conf., Dublin, Ireland, July 2005, to be published. See also http://www.maths.tcd.ie/lat05/plenary-talks/Philipsen.pdf.
[18] P. Braun-Munzinger, J. Stachel, C. Wetterich, Phys. Lett. **B596** (2004) 61.
[19] F. Karsch, E. Laermann, A. Peikert, Nucl. Phys. **A605** (2001) 579.
[20] R. Stock, Phys. Lett. **B465** (1999) 277.
[21] U. Heinz, Nucl. Phys. **A638** (1998) 357.
[22] U. Heinz, Nucl. Phys. **A685** (2001) 414.
[23] J. Adams et al., STAR coll., Phys. Rev. Lett. **92** (2004) 182301, nucl-ex/0307024.
[24] F. Becattini, Z. Physik **C69** (1996) 485, F. Becattini and U. Heinz, Z. Physik **C76** (1997) 269.
[25] S.V. Afanasiev et al., NA49 collaboration, Phys. Rev. **C66** (2002) 054902.
[26] C. Blume et al., NA49 collaboration, J. Phys. **G31** (2005) S685.
[27] A. Andronic, P. Braun-Munzinger, J. Stachel, manuscript in preparation.
[28] P. Braun-Munzinger, D. Miskowiec, J. Stachel, manuscript in preparation.

Institute of Physics Publishing
doi:10.1088/1742-6596/50/1/029
Journal of Physics: Conference Series **50** (2006) 243–250
5th International Conference on Physics and Astrophysics of Quark Gluon Plasma

Partonic collectivity in high-energy nuclear collisions

Nu Xu
Nuclear Science Division
Lawrence Berkeley National Laboratory, 1 Cyclotron Road, CA 94720, USA
E-mail: NXu@lbl.gov

Abstract. After a brief introduction to the physics of high-energy nuclear collisions, we will present recent experimental results that are closely connected to the properties of the matter produced in Au+Au collisions at RHIC. Collective motion with parton degrees of freedom is called partonic collectivity. We will focus on collective observables such as transverse radial flow and elliptic flow. With experimental observations, we will demonstrate that, at RHIC, collectivity has already been developed prior to the hadronic stage. Finally, we will develop a plan for future measurements that are needed for characterization of the partonic Equation of State in high-energy nuclear collisions.

1. Introduction

This is my first trip to India. I am very grateful to the conference organizers. India, as I was taught when I was young, is a great land with a long history of civilization. One of the few quotations I recite often is this: *"Each soul is potentially divine. The goal is to manifest this divinity within by controlling nature: external and internal. Do this either by work, or worship, or psychic control, or philosophy - by one, or more, or all of these - and be free." – S. Vivekananda.* I do believe in *work*. I believe this short trip is the beginning of a long journey toward happiness.

The purpose of the high-energy nuclear collision program at both Brookhaven National Laboratory (BNL) and CERN is to probe strongly interacting matter under extreme conditions, *i.e.* at high densities and temperatures. Naturally the search for the existence of a new form of matter - the matter with partonic equation of state (EOS) that colloquially called quark-gluon plasma - is the experimental focus of the program. It has been demonstrated that the flow measurement is one of the most powerful methods for studying partonic EOS in high-energy nuclear collisions. For recent reviews, see [1, 2] and references therein.

The term flow has two important aspects: (i) collectivity of produced hadrons and (ii) the local thermalization among these hadrons [3]. As long as there are interactions among constituents, collectivity of the matter will be developed provided that the distribution of matter density is inhomogeneous. When the interactions last long enough the system will eventually approach local equilibrium and hence develops hydrodynamic type flow. At the early stage of a high-energy nuclear collision, both the matter density and its gradient are large, therefore we expect the development of partonic collectivity - the collective motion of partons. The issues of partonic local equilibrium can be addressed by studying heavy-flavor (*c*-, *b*-quarks) collectivity.

This is because the collisions that generate the collective motion for heavy-quarks will likely lead to thermalization among the light quarks (u-, d-, s-quarks). It is important to note that collectivity is cumulative through the expansion phase and is not affected by the details of the hadronization process.

This paper is organized as follows: First we will discuss the results of hadron transverse momentum spectra. The freeze-out thermodynamic parameters (T_{fo}, β_T) as a function of collision centrality will be discussed. Comparing with the results of π, K, and p, we will show that multi-strange hadrons like the ϕ and Ω freeze-out at a relatively earlier time with a higher temperature and smaller collective velocity. In section 3, we will discuss the event anisotropy parameter v_2. Since the multi-strange hadrons Ξ and Ω have demonstrated the $v_2(p_T)$ is similar to other baryon's $v_2(p_T)$, we will argue that the collectivity has been developed before hadronization. The first result of the open-charm v_2, as observed in the electron decayed channel, will be presented in section 4 and a brief summary will be given in section 5.

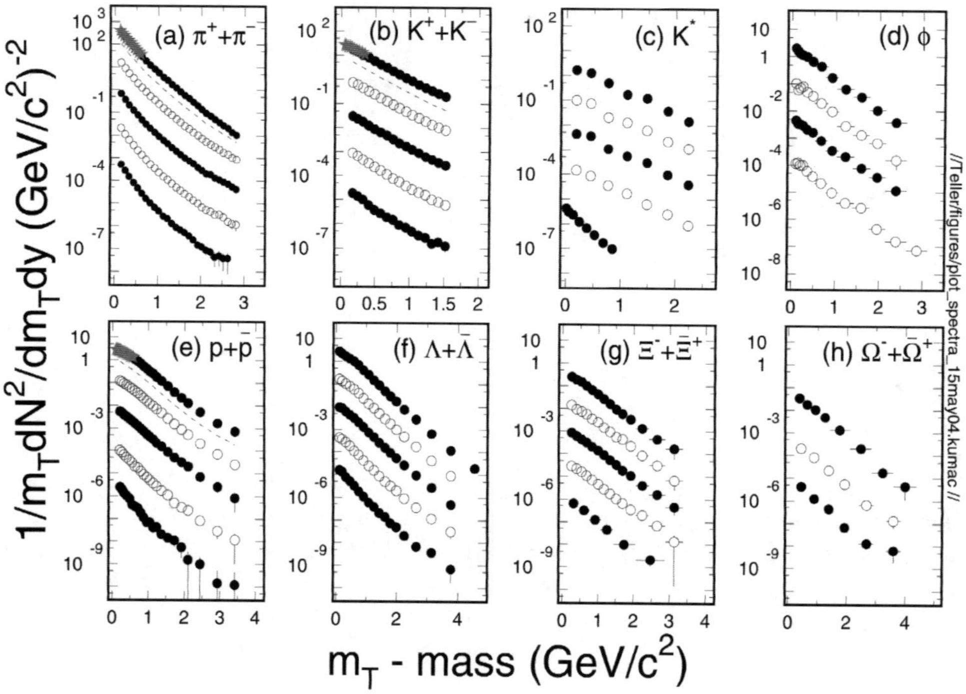

Figure 1. Mid-rapidity hadron spectra from $\sqrt{s_{\rm NN}} = 200$ GeV $Au + Au$ and $p + p$ collisions. Most and least central collisions are listed from top-down. For pions (a), kaons (b), and protons (e) [4], the centrality bins are: 0-5%, 20-30% (10^{-1}), 40-50% (10^{-2}), 60-70% (10^{-3}), and 80-92% (10^{-4}). The star symbols (0-5%) are data from [5]. The dashed-lines represent the spectra from minimum biased collisions. For K^* (c) [6], from top to bottom, respectively, the centralities are 0-10%, 10-30%, 30-50%, 50-80% for $Au + Au$ collisions and $p + p$ collisions. For ϕ-meson (d) [7], the centralities are: 0-5%, 10-30% (10^{-1}), 30-50% (10^{-2}), and 50-80% (10^{-3}). For Λ (f) and Ξ (g) [8], the centrality bins are: 0-5%, 10-20% (10^{-1}), 20-40% (10^{-2}), 40-60% (10^{-3}), 60-80% (10^{-4}). For the Ω baryon (h) [8], the centralities are: 0-10%, 20-40% (2×10^{-2}), and 40-60% (10^{-3}).

2. Transverse momentum distributions - collective expansion

In Figure 1, the mid-rapidity transverse momentum distributions (the invariant spectra are plotted as a function of $m_T -$ mass) for pions, kaons, protons [4], K^* [6], ϕ [7], Λ, Ξ, and Ω

[5, 8], from $\sqrt{s_{\mathrm{NN}}} = 200$ GeV $Au + Au$ and $p + p$ collisions, are shown. While pion spectra show a p_T−power-law shape, most of the hadron spectra are m_T−exponential, especially the strange hadrons like K, ϕ, Ξ and , Ω. In order to characterize the transverse motion, an exponential fit or a power-law fit is often used for the measured spectra. In addition, for spectra extracted from heavy ion collisions, hydrodynamics motivated fits [9] are applied.

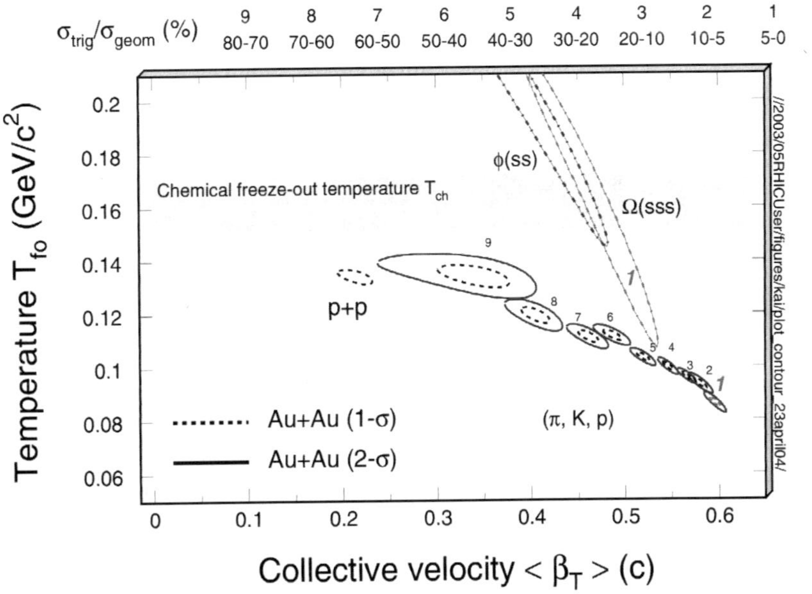

Figure 2. χ^2 contours, extracted from the thermal+radial flow fits, for copiously produced hadrons π, K and protons and multi-strange hadrons ϕ and Ω. On the top of the plot, the numerical labels indicate the centrality selection. For π, K and protons, 9 centrality bins (from top 5% to 70-80%) were used from 200 GeV Au+Au collisions. The results from 200GeV $p+p$ collisions are also shown. For ϕ and Ω, only the most central results are presented. Dashed- and solid-lines are the 1-σ and 2-σ contours, respectively.

The thermal fits allow us to construct common freeze-out contours, for a given group of hadrons, in the (T_{fo}, $\langle\beta_T\rangle$) plane. In this approach, the thermal random motion and ordered collective motion are separated [9, 10]. The χ^2 contours are extracted from the thermal + radial flow fits and presented in the temperature-velocity space. Figure 2 shows a qualitatively different behavior, in the temperature versus collective velocity plane, between the relatively rarely produced multi-strange particles and copiously produced particles π, K and p. On top of the plot, the numerical labels indicate the centrality selection. For π, K and protons, 9 centrality bins were used from the 200 GeV $Au + Au$ collisions and $p + p$ collisions [5]. For ϕ and Ω, only the most central results are shown. Dashed- and solid-lines are the 1-σ and 2-σ contours, respectively.

As the centrality increases from peripheral to central collisions, for copiously produced hadrons, π, K, p, the values of the temperature parameter decreases and the velocity parameter increases. At the most central collisions, the velocity becomes as high as 60% of the speed of light and the thermal freeze-out temperature is at $\sim$ 100 MeV. On the other hand, the fit results indicate that the minima for multi-strange hadrons are not sensitive to collision centrality and they are all close to a temperature of T $\sim$ 160 - 170 MeV and an average velocity of $\langle\beta_T\rangle \sim 0.4c$. Note that this temperature is the same as the chemical freeze-out temperature [11, 12, 13, 14] and it is close to the value of the phase transition temperature [15]. Due to relatively small total hadronic cross sections [16, 17, 18, 19] and their heavy mass, multi-strange particles do not participate in the evolution of the system during the hadronic phase. As a result, they de-associate from the system near the hadronization point with $T_{fo} \sim 170$ MeV and $\beta_T \sim 0.4c$. The finite value of the collective velocity must have been developed before the hadronization - the partonic collectivity.

Two short notes regarding to the contour plot Fig. 2 are in order:

(1) The direct measurements of the cross section from the multi-strange hadron and nucleon interactions are scarce. Early study of photo-production of ϕ-meson has concluded a relatively small cross section of $\phi + p$ interaction, although the result is model dependent [20]. Recent report from the $\Xi^- + p$ elastic and inelastic scatterings concluded a relatively a small value of cross section [21]. In addition, I would argue that the Fig. 2 is a measure of the effective interaction cross sections in heavy ion collisions. Since there is either no or small number of high state of resonances for most of the multi-strange hadrons and the mass of the ϕ-meson is very close to that of the protons, the observation of the different freeze-out time between ϕ-meson and protons, for example, has to come from the difference in the strength of the interaction. The claim [22] of a common hydrodynamic model fit to all particle species from Au+Au collisions at RHIC, including the Ω baryons, is not supported by the experimental observations. Besides other problems, the treatment of freeze-out in the hydrodynamic model alone is not adequate.

(2) As one can see in the figure that there is finite velocity in $p+p$ collisions at $\sqrt{s_{NN}}$ =200 GeV. Thermal model fit tests show that it is caused by the jet production [23] in the high energy collisions. In principle, such jetiness-collectivity, caused by large momentum transfer, is different from the hydrodynamic type of collectivity discussed above. In central $Au + Au$ collisions, the collective effect arise from the jet production is probably small due to the severe energy loss in the medium [24].

3. Elliptic flow v_2 – Evidence for partonic collectivity at RHIC

The particle azimuthal distribution with respect to the reaction plane at a given rapidity window can be de-convoluted by the expansion:

$$\frac{dN}{d\phi} \approx v_0(1 + 2v_1 \cos(\Delta\phi) + 2\mathrm{v}_2 \cos(2\Delta\phi)). \tag{1}$$

The first and second Fourier coefficients, v_1 and v_2, are connected to direct flow and elliptic flow, respectively. The coefficient v_0 is a normalization constant and $\Delta\phi$ is defined as the azimuthal angle difference between the particle and the event reaction plane. At a given rapidity the coefficients are: $v_1 = \langle\cos(\Delta\phi)\rangle$ and $v_2 = \langle\cos(2\Delta\phi)\rangle$. Because the rescattering induced expansion naturally reduces the spatial anisotropy, the early information of the collision dynamics can be learned through measuring v_1, v_2 and higher order harmonics [25, 26, 27, 28].

The measured elliptic flow v_2 from the minimum bias $Au + Au$ collisions at $\sqrt{s_{NN}} = 200$ GeV for π, K_0^S, p, Λ [29, 30] are shown in Figure 3 (a). Respectively, from top to bottom the dashed-lines represent the elliptic flow of $\pi, K, p, \Lambda, \Xi, \Omega$ from hydrodynamic calculations [31]. In the low p_T region, the trends of v_2 are well reproduced by the hydrodynamic calculations. At higher p_T, the v_2 is found to be saturated and hydrodynamic results over-predict the data. While the baryons saturate at $p_T \geq 3$ GeV/c with $v_2 \sim 0.2$, mesons saturation starts earlier at lower values of v_2.

Figure 3 (b) and (c) show v_2 for the multi-strange baryons $\Xi^- + \overline{\Xi}^+$ and $\Omega^- + \overline{\Omega}^+$ [8, 32], respectively. Although they tend to suffer much less rescatterings during the later hadronic stage of the collisions [16], the values of v_2 for strange-baryons, within the experimental uncertainties, are found to be as high as other hadrons at a given p_T . Therefore, the v_2 must have been developed at an earlier stage prior to hadronization. This measurement and the early freeze-out

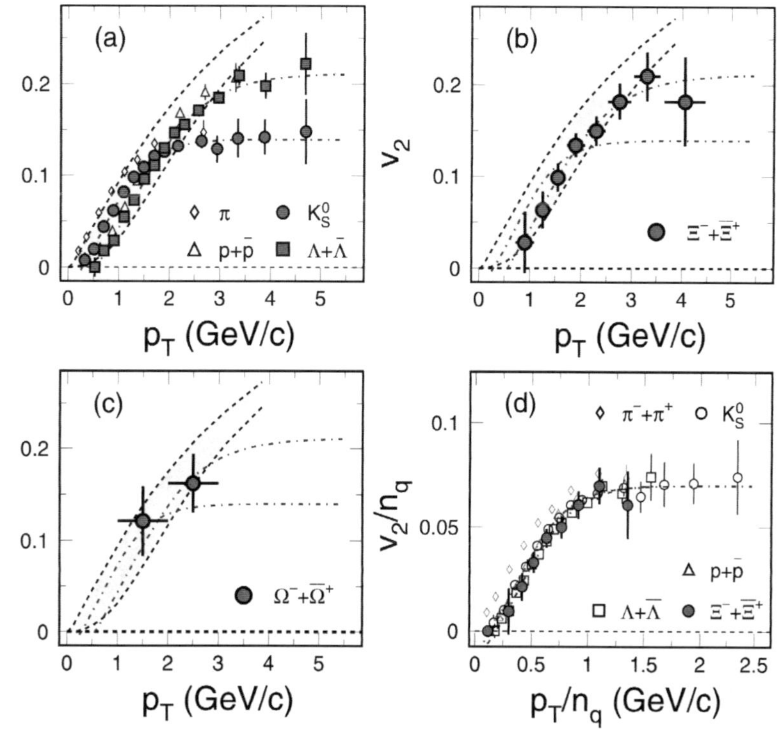

Figure 3. (a) Experimental results of the transverse momentum dependence of the event anisotropy parameters for π, K_S^0, $p+\overline{p}$, $\Lambda+\overline{\Lambda}$ [30, 29]. Hydrodynamic calculations are shown as thick-dashed-lines. From top to bottom are the results for π, K, p, Λ, $\Xi^-+\overline{\Xi}^+$, and $\Omega^-+\overline{\Omega}^+$; Multi-strange baryon elliptic flow v_2 are shown in (b) for Ξ and (c) for Ω. Plots (b) and (c) are from preliminary STAR results [8, 32]. (d) Number of constituent quark (n_q) scaled v_2/n_q versus scaled p_T/n_q. All curves are from [33].

(see Fig. 2) for the multi-strange hadrons constitute a clear piece of evidence for the development of partonic collectivity in $Au+Au$ collisions at RHIC.

The measured v_2 distributions have been fitted with the equation given as [33]:

$$f_{v_2} = \frac{a}{1+exp(-(x-b)/c)} - d, \tag{2}$$

where parameters a, b, c and d are fixed from the fit. The fit results to K_S^0 and Λ are shown as dot-dashed lines in Fig. 3. According to coalescence approaches [34], after scaling both values of v_2 and p_T with the number of the constituent quarks (NCQ) of the corresponding hadron, all particles should fall onto one single curve. Fig. 3(d) shows the scaled v_2 versus the scaled p_T. Indeed, with the exception of pions, all particles follow a single curve. This observation implies that due to high parton density and intense interactions partons have already developed collective motion in heavy ion collisions at RHIC.

In the NCQ scaled plot, Fig. 3(d), the fit results are shown as dashed lines. For kaon, proton and lambda, the scaling seems to be working within $p_T/n_q \leq 2.5$ GeV/c. Pions (open triangles) do not follow the scaling, because of a large fraction of hadrons produced through resonance decays. This is particularly true for pions in high-energy heavy ion collisions [35, 36]. At mid-rapidity, in collisions at RHIC, as high as $\sim 80\%$ of pions are from resonance decays and the dominant sources for pion production are ρ, ω and baryon resonances like Λ and Δ.

4. Heavy-flavor collectivity - light-flavor thermalization at RHIC

In the previous sections, we established the evidence of partonic collectivity for light-flavor (u, d, s) hadrons. In order to access the partonic EOS, one must address the issue of early excitation/thermalization [37] of the system. Because the masses of heavy-flavors are much larger than the highest possible excitation of the system reached in $Au + Au$ collisions at RHIC, a large number of rescatteings would be required in order to develop any collective motion. Furthermore, since the number density of light-flavors is much higher than that of heavy-flavors, the intensity of rescattering should be even stronger among the light-flavors. As expected for any physical system, intensive rescattering among constituents will eventually lead to thermalization. Therefore, the heavy-flavor collectivity could be used to probe the intensity of partonic rescattering and therefore the degree of light-flavor thermalization. In practice, this means that we should measure the v_2 of charm-hadrons [33, 38, 39, 40, 41].

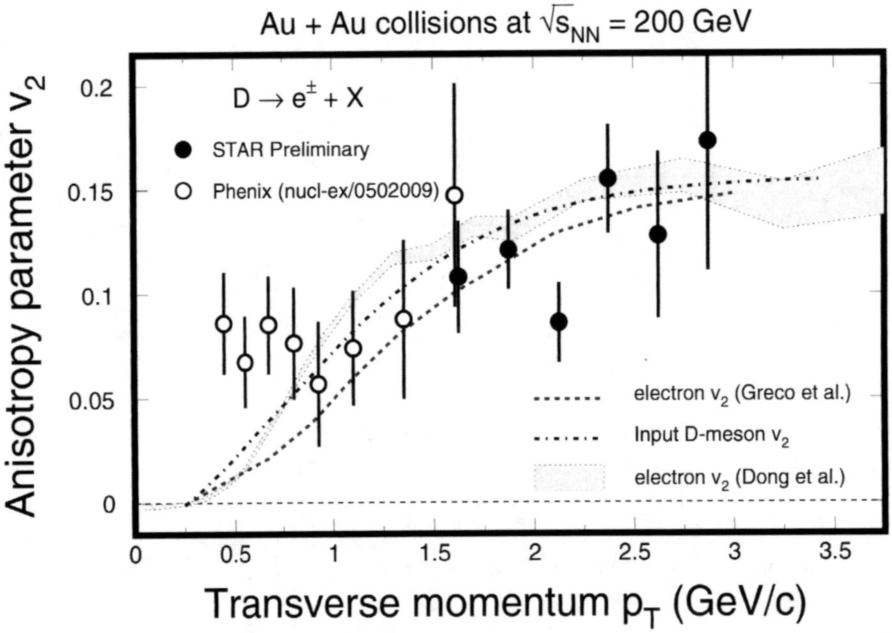

Figure 4. Non-photonic electron elliptic flow $v_2(p_T)$ from minimum bias $Au + Au$ collisions at $\sqrt{s_{NN}} = 200$ GeV. Data from PHENIX [43] and STAR [45] experiments are shown as open- and filled-circles, respectively. Only statistical errors are shown. Dashed-line and the hatched band are charm-hadron decayed electron $v_2(p_T)$ from refs. [33, 38].

The heavy-flavor program has just started at RHIC [42] and there is no measurement of v_2 with reconstructed charm-hadron yet. However, as discussed in [33, 38], the charm-hadron v_2 can be inferred by their decayed electron v_2. The results of the non-phonic electron v_2 from minimum bias $Au+Au$ collisions at $\sqrt{s_{NN}}$ =200 GeV [43, 44, 45] are shown in Figure 4. These non-photonic electrons are presumably from the decays of charm-hadrons. Only statistical errors are shown in the figure and the systematic errors are in the order of 25%. The dashed-line and hatched-band are charm-hadron decayed electrons from model calculations [38, 33]. The dot-dashed-line [33] is the D-meson v_2 assuming NCQ-scaling. Since there are several resonance states for charm-hadrons, hadronic interactions could also develop non-vanishing v_2 [39, 40]. A recent study based on the HSD model indeed shows the value of maximum v_2 in the order of 2-3% [39] in the b=7 fm $Au + Au$ collisions at RHIC. Due to the large statistical uncertainties in data, both model results are consistent with the measurements. The observation indicates, if confirmed with (high statistics) reconstructed charm-hadron measurements, the onset of collectivity for the heavy-flavors and the light-flavor thermalization at RHIC.

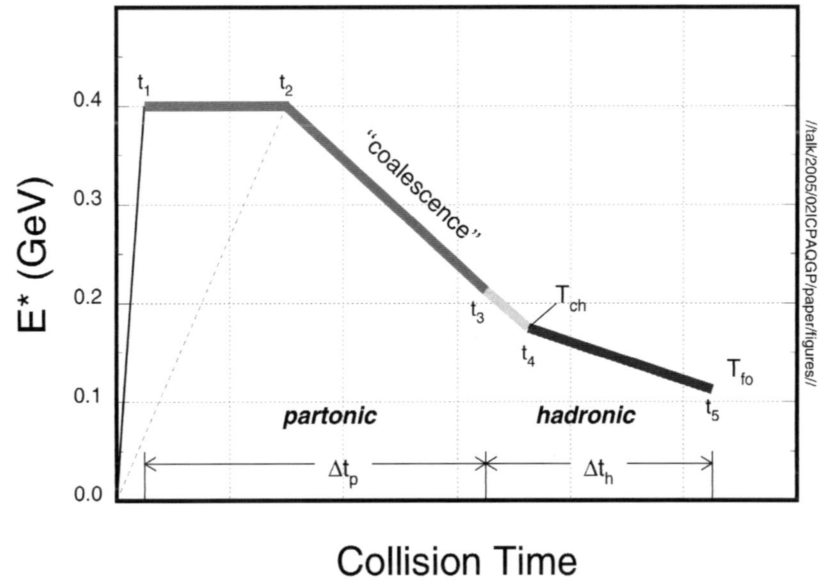

Figure 5. Excitation of high-energy nuclear collision as a function of time. In case of thermalization, E^* has the meaning of a temperature T.

Assuming the system has reached thermalization at the early stage, now the questions are: What is the temperature? How long does the system stay at the equilibrium? More importantly: How does the transformation from a partonic system to a hadronic system occur? What are the dynamics for hadron mass formation?

Figure 5 is a schematic time evolution of the system in high-energy nuclear collisions. The important realization here is that, in both partonic and hadronic periods, collective expansion occurred. During the partonic period (Δt_p) interactions are dominated by quarks and gluons. During t_2 and t_3, the temperature T is in the order of T_c and colored bound states [46, 47] are coalesced. Since gluons are disappearing and giving their way to massive quarks, the degrees of freedom are gradually reduced. At the same time, both quarks and the colored bound-states are acquiring mass. In order to avoid the shrinking of entropy, the system is expanding rapidly – further enhancing the partonic collectivity. Near the end of the period, colorless hadrons are formed. The duration between t_3 and t_4 is expected to be small and most of hadrons are born in a state of chemical equilibrium. Later, the hadronic interactions during Δt_h lead to further expansion and cooling of the system until the kinetic freeze-out. Hadronic observables have allowed us to extract information as early as $t \leq t_3$. However, earlier dynamic information of the system, which answers the first two questions, one has to extract from photon and low mass di-lepton measurements. RHIC detector upgrades are necessary in order to perform these measurements with high precision.

—

Acknowledgments: I am thankful for many enlightening discussions with Drs. P. Braun-Munzinger, J.E. Gonzalez, M. Gyulassy, U. Heinz, P. Huovinen, F. Liu, M. Oldenburg, A. Poskanzer, H.G. Ritter, K. Schweda, P. Sorensen, E.V. Shuryak, Z. Xu, and P.F. Zhuang. This work has been supported by the U.S. Department of Energy under Contract No. DE-AC03-

76SF00098.

[1] H.G. Ritter and X.N. Wang, proceedings of *"Quark Matter 2004"*, J. Phys. **G30**, (2004) and references therein.
[2] RHIC experiments whitepapers: nucl-ex/0410020 (BRAHAMS); nucl-ex/0410003 (PHENIX); nucl-ex/0410022 (PHOBOS); nucl-ex/0501009 (STAR).
[3] W. Reisdorf and H.G. Ritter, Ann. Rev. Nucl. Part. Sci. **47**, 663(1997).
[4] S.S. Adler *et al.*, (PHENIX Collaboration), Phys. Rev. **C69**, 034909(2004).
[5] J. Adams *et al.*, (STAR Collaboration), Phys. Rev. Lett. **92**, 112301(2004).
[6] J. Adams *et al.*, (STAR Collaboration), subm. to Phys. Rev. **C**, December, 2004. nucl-ex/0412019.
[7] J. Adams *et al.*, (STAR Collaboration), nucl-ex/0406003 and E. Yamamoto, (STAR Collaboration), Nucl. Phys. **A715**, 466c(2003).
[8] K. Schweda *et al.*, (STAR Collaboration), J. Phys. G: Nucl. Part. Phys. **30**, S693(2004).
[9] E. Schnedermann, J. Sollfrank, and U. Heinz, Phys. Rev. **C48**, 2462(1993).
[10] I. G. Bearden *et al.*, (NA44 Collaboration), Phys. Rev. Lett **78**, 2080(1997).
[11] P. Braun-Munzinger, J. Stachel, J. Wessels, and N. Xu, Phys. Lett. **B344**, 43(1995); P. Braun-Munzinger, I. Heppe, and J. Stachel, Phys. Lett. **B465**, 15(1999).
[12] P. Braun-Munzinger, these proceedings.
[13] J. Cleymans and K. Redlich, Phys. Rev. Lett. **81**, 5284(1998).
[14] N. Xu and M. Kaneta, Nucl. Phys. **A698**, 306c(2001).
[15] F. Karsch, Nucl. Phys. **A698**, 199c(2002).
[16] H. van Hecke, H. Sorge, and N. Xu, Phys. Rev. Lett. **81**, 5764(1998).
[17] Y. Cheng, F. Liu, Z. Liu, K. Schweda, and N. Xu, Phys. Rev. **C68**, 034901(2003).
[18] C. Alt *et al.*, (NA49 Collaboration), sub. Phys.Rev.Lett., Sept. 2004; nucl-ex/0409004
[19] M. Bleicher, these proceedings.
[20] H.-J Behrend *et al.*, Phys. Lett. **B56**, 408(1975).
[21] J.K. Ahn ıet al., sub. to Phys. Lett. **B**, nucl-ex/0502010.
[22] U. Heinz, these proceedings.
[23] Zhixu Liu, in preparation, February 2005.
[24] C. Adler *et al.*, (STAR Collaboration) Phys. Rev. Lett. **90**, 082302(2003).
[25] H. Sorge, Phys. Lett. **B402**, 251(1997).
[26] J.-Y. Ollitrault, Phys. Rev. **D46**, 229(1992).
[27] S. Voloshin, Nucl. Phys. **A715**, 379c(2003).
[28] J. Adams *et al.*, (STAR Collaboration) Phys. Rev. Lett. **92**, 062301(2004).
[29] J. Adams, *et al.*, (STAR Collaboration), Phys. Rev. Lett. **92**, 052302(2004).
[30] S.S. Adler *et al.*, (PHENIX Collaboration), Phys. Rev. Lett. **91**, 182301(2003).
[31] P. Huovinen, P. Kolb, U. Heinz, P.V. Ruuskanen, and S. Voloshin, Phys. Lett. **B503**, 58(2001).
[32] J. Castillo *et al.*, (STAR Collaboration), J. Phys. G: Nucl. Part. Phys. **30**, S1207(2004).
[33] X. Dong, S. Esumi, P. Sorensen, N. Xu, and Z. Xu, Phys. Lett. **B597**, 328(2004).
[34] Z. Lin and C. Ko, Phys. Rev. Lett. **89**, 202302(2002); R.J. Fries, B. Mueller, C. Nonaka, S.A. Bass, Phys. Rev. Lett. **90**, 202303(2003); D. Molnar and S. Voloshin, Phys. Rev. Lett. **91**, 092301(2003).
[35] Zhang-bu Xu, nucl-ex/0307014, (2003).
[36] R.J. Fries, B. Mueller, C. Nonaka, S.A. Bass, Phys. Rev. **C68** 044902(2003); C. Nonaka,B. Mueller,M. Asakawa, S.A. Bass, and R.J. Fries, Phys. Rev. **C69**, 031902(2004).
[37] H. Stöcker, Nucl. Phys. **A750**, 121(2005) and these proceedings.
[38] V. Greco, C. M. Ko, R. Rapp, Phys. Lett. **B595**, 202(2004).
[39] E.L. Bratkovskaya, W. Cassing, H. Stöcker, N. Xu, Phys.Rev. **C**, in print; nucl-th/0409047.
[40] H. van Hees and R. Rapp, Phys. Rev. **C71**, 034907(2005).
[41] X.L. Zhu, P.F. Zhuang and N. Xu, Phys. Lett. **B607**, 107(2005).
[42] J. Adams, *et al.*, (STAR Collaboration), Phys. Rev. Lett. **94**, 062301(2004).
[43] M. Kaneta *et al.* (PHENIX Collaboration), J. Phys. **G30**, S1217(2004); S.S. Adler *et al.* (PHENIX Collaboration), nucl-ex/0502009.
[44] H. Hamagaki, these proceedings.
[45] F. Laue *et al.* (STAR Collaboration), nucl-ex/0411007.
[46] E. Shuryak and I. Zahed, Phys. Rev. **C70**, 021901(2004); Phys. Rev. **D69**, 046005(2004); these proceedings.
[47] Hong-Jo Park, Chang-Hwan Lee, and Gerald E. Brown, hep-ph/0503016.

Institute of Physics Publishing Journal of Physics: Conference Series **50** (2006) 251–258
doi:10.1088/1742-6596/50/1/030 5th International Conference on Physics and Astrophysics of Quark Gluon Plasma

Hydrodynamical evolution of dissipative QGP fluid

A. K. Chaudhuri[1] **and U. Heinz**[2]

[1]Variable Energy Cyclotron Centre, 1/AF,Bidhan Nagar, Kolkata - 700 064
[2]Department of Physics, The Ohio State University Columbus, OH 43210

E-mail: akc@veccal.ernet.in; heinz@mps.ohio-state.edu

Abstract. In the framework of the Mueller-Israel-Stewart theory of dissipative fluid dynamics, we have studied the space-time evolution of a QGP fluid. For simplicity, we have considered shear viscosity only and neglected bulk viscosity and heat conductivity. Shear viscosity opposes the expansion and cooling of the fluid. As a result, the lifetime of the fluid is extended. We also find that the parton p_T distribution is considerably flattened.

1. Introduction

One of the primary goals of relativistic heavy ion collisions is the creation and detection of the (lattice QCD [1]) predicted deconfined phase of quarks and gluons. Experiments at RHIC, BNL, gave strong indications of QGP formation in central Au+Au collisions [2]. Hydrodynamics provides a simple, intuitive dynamic description of relativistic heavy ion collisions. Most of the experimental data on soft hadron production at RHIC are well described by ideal hydrodynamics. However, some problems remain. For example, one of the most important signals of collectivity, the elliptic flow, saturates beyond $p_T \sim 1.5{-}2$ GeV/c while ideal hydrodynamics predicts a continuous increase with p_T [3].

Ideal hydrodynamic may not give an accurate description of the early stage of the collision. In the early stages of collision, particle momenta are predominantly in the beam direction, while ideal hydrodynamic assumes a locally isotropic distribution. The anisotropy in the momentum distribution will give rise to shear viscous stress. Other dissipative effects, e.g. bulk viscosity and heat conduction can also affect the hydrodynamic evolution of QGP. Unlike ideal flow, non-ideal flow generates entropy. The space-time evolution is also changed.

Relativistic theories for dissipative fluid dynamics were formulated long ago [4, 5, 6]. In early formulations, called 1st order theories [4, 5], the entropy 4-current contained terms of first order in the dissipative fluxes; these had the undesirable feature of violating causality. This is corrected in so-called 2nd order theories, due to Grad, Mueller and Israel and Stewart [6]. In 2nd order theories, the entropy 4-current contains terms of 2nd order in the dissipative fluxes. The space of thermodynamic variables is extended to include the dissipative fluxes, and relaxation equations for these dissipative fluxes are obtained from the positivity condition for entropy production $\partial_\mu s^\mu \geq 0$.

Up to now, only a few authors have considered the effects of dissipation in relativistic heavy ion collisions [7, 8, 9, 10, 11, 12, 13]. Most of these studies considered one dimensional systems using 1st order theories. Only recently 2nd order theories have been implemented numerically [10, 12, 13]. In the present paper, using the causal 2nd order theory of dissipative fluid dynamics, we study the effect of shear viscosity on the space-time evolution of a QGP fluid. The paper is organized as follows: In Section II, we present the equations for 2nd order dissipative hydrodynamics. Our choice of equation of state, transport coefficients and initial conditions is described in Section III. Results for the space-time evolution of the QGP fluid, with and without dissipation, are presented Section IV. Conclusions are drawn in Section VI.

2. Equations for causal dissipative hydrodynamics

We consider a QGP fluid in the central rapidity region, with zero net baryon density and chemical potential, $n_B = 0$, $\mu_B = 0$. We neglect the effects of heat conduction ($\mu_B = 0$) and bulk viscosity (massless particles) and account only for shear viscosity. We work in the Landau-Lifshitz energy frame.

The energy-momentum tensor, including the shear viscous pressure tensor $\pi^{\mu\nu}$, is written as [6]

$$T^{\mu\nu} = (\varepsilon + p)u^\mu u^\nu - pg^{\mu\nu} + \pi^{\mu\nu} \tag{1}$$

where ε is the energy density, p is the hydrostatic pressure, and u is the hydrodynamic 4-velocity, normalized by $u^\mu u_\mu = 1$. $T^{\mu\nu}$ satisfies the energy-momentum conservation law

$$\partial_\mu T^{\mu\nu} = 0. \tag{2}$$

In the Israel-Stewart theory of dissipative fluids [6], the dissipative fluxes are treated as independent thermodynamic variables which satisfy kinetic relaxation equations. The transport (relaxation) equations for the shear viscous pressure read

$$\beta_2 D\pi^{\mu\nu} = -\frac{1}{2\eta}\pi^{\mu\nu} + \nabla^{\langle\mu}u^{\nu\rangle} \tag{3}$$

where $D = u^\mu\partial_\mu$ is the convective time derivative. η is the shear viscosity coefficient. β_2 is related to relaxation time by $\tau_\pi = 2\eta\beta_2$. The angular bracket in $\nabla^{\langle\mu}u^{\nu\rangle}$ is defined as

$$\nabla^{\langle\mu}u^{\nu\rangle} = \frac{1}{2}[\nabla^\mu u^\nu + \nabla^\nu u^\mu] - \frac{1}{3}\Delta^{\mu\nu}\partial_\sigma u^\sigma, \tag{4}$$

where $\nabla^\mu = \partial^\mu - u^\mu D$ is the transverse gradient operator and $\Delta^{\mu\nu} = g^{\mu\nu} - u^\mu u^\nu$ is the projector orthogonal to the flow velocity u^μ.

The viscous pressure tensor $\pi^{\mu\nu}$ is symmetric and traceless $\pi^{\mu\nu} = \pi^{\nu\mu}$ and $\pi^\mu_\mu = 0$. Furthermore, it is transverse to the hydrodynamic 4-velocity, $u_\mu\pi^{\mu\nu} = 0$. Thus it has 5 independent components.

There are 10 unknowns (ε, p, three components of the hydrodynamic velocity u, and 5 viscous pressure components) and 9 equations (4 energy-momentum conservation equations and 5 transport equations for the independent components of $\pi^{\mu\nu}$. The system is closed by the equation of state $p = p(\varepsilon)$. We assume longitudinal boost-invariance and cylindrical symmetry in the transverse direction. These additional symmetries reduce the independent variables to four (energy density ε, radial velocity v_r and two components, say π^{rr} and $\pi^{\phi\phi}$, of the viscous pressure tensor). In (τ, r, ϕ, η) coordinates, the hydrodynamic and relaxation equations read (see appendix)

$$\partial_\tau \tilde{T}^{\tau\tau} + \partial_r\left(v_s\tilde{T}^{\tau\tau}\right) = -\frac{\tilde{p} - \tilde{\pi}^{rr}/\gamma^2 - r^2\tilde{\pi}^{\phi\phi}}{\tau}, \tag{5}$$

$$\partial_\tau \tilde{T}^{\tau r} + \partial_r\left(v_r\tilde{T}^{\tau r}\right) = \frac{\tilde{p} + r^2\pi^{\phi\phi}}{r} - \partial_r\left(\tilde{p} + \frac{\tilde{\pi}^{rr}}{\gamma^2}\right), \tag{6}$$

$$\partial_\tau\pi^{rr} + v_r\partial_r\pi^{rr} = -\frac{1}{\tau_\pi\gamma}\left(\pi^{rr} - 2\eta\nabla^{\langle r}u^{r\rangle}\right), \tag{7}$$

$$\partial_\tau\pi^{\phi\phi} + v_r\partial_r\pi^{\phi\phi} = -\frac{1}{\tau_\pi\gamma}\left(\pi^{\phi\phi} - 2\eta\nabla^{\langle\phi}u^{\phi\rangle}\right). \tag{8}$$

where $\tilde{A}^{\mu\nu} = r\tau A^{\mu\nu}$, $\tilde{p} = r\tau p$ and $v_s = T^{\tau r}/T^{\tau\tau}$.

The four equations are solved simultaneously using the SHASTA-FCT algorithm [14]. At each time step, we obtain the energy density ε and radial velocity v_r from $T^{\tau\tau}$, $T^{\tau r}$, and π^{rr} by iteratively solving (in analogy to the ideal fluid case [15])

$$v_r = \frac{T^{\tau r}}{T^{\tau\tau} + p(\varepsilon) + \pi^{rr}/\gamma^2}, \qquad \varepsilon = T^{\tau\tau} - v_r T^{\tau r}. \tag{9}$$

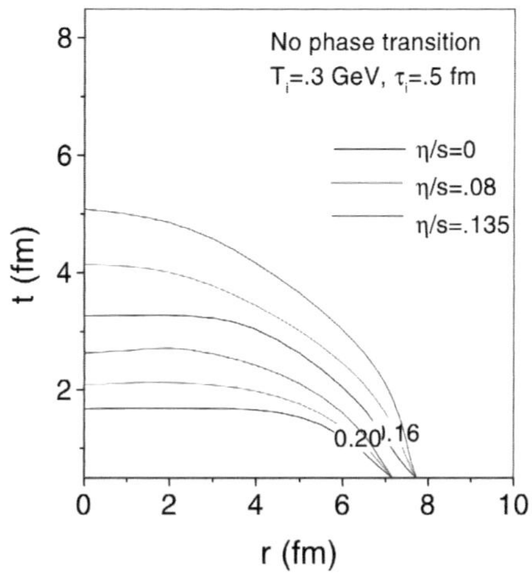

Figure 1. Contours of constant temperature in the r–τ plane for ideal and non-ideal hydrodynamic evolution, using $T_i = 0.3$ GeV at $\tau_i = 0.5\,\text{fm}/c$.

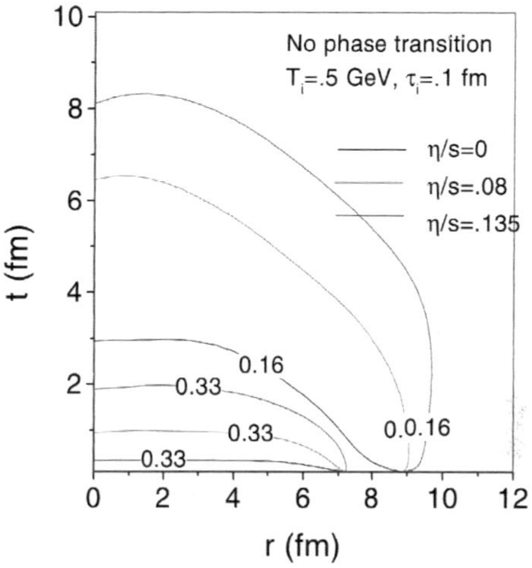

Figure 2. Contours of constant temperature in the r–τ plane for ideal and non-ideal hydrodynamic evolution, using T_i=0.5 GeV at $\tau_i = 0.1\,\text{fm}/c$.

3. Equation of state, transport coefficient and initial conditions

The equation of state is one of the important inputs for hydrodynamics. It is the link between the macroscopic and microscopic world. In this exploratory work, we have used a simple equation of state, $p = \frac{1}{3}\varepsilon$, with $\varepsilon = aT^4$, $a = (16 + \frac{21}{2}N_f)\frac{\pi^2}{30}$.

In classical kinetic theory, explicit expressions can be obtained for the viscosity coefficient η and the relaxation time τ_π in terms of the collision term. For a strongly coupled QGP, neither η or τ_π are known. We treat them as phenomenological parameters. For guidance, we use perturbative [17, 16] and AdS/CFT [18] estimates for η, respectively, and a kinetic theory estimate [6] for τ_π.

The shear viscosity coefficient η for hot QCD was determined perturbatively to leading logarithmic accuracy in [16, 17]. For $\alpha_s \approx 0.5$ the result in [17] gives

$$\frac{\eta}{s} = 0.135. \tag{10}$$

Recently the shear viscosity was also evaluated in a strongly coupled gauge theory ($N = 4$ SUSY YM theory), using the AdS/CFT correspondence [18]. The corresponding AdS/CFT estimate of shear viscosity is

$$\frac{\eta}{s} = 0.08 \tag{11}$$

In kinetic theory, in the Boltzmann gas approximation, the relaxation time is estimated as $\tau_\pi = 2\eta\beta_2 = 2\eta\frac{3}{4p}$ [6]..

For the initial energy density distribution in the transverse plane, we use the Woods-Saxon parameterisation:

$$\varepsilon(r) = \frac{\varepsilon_0}{1 + e^{\frac{r-R}{a}}}, \tag{12}$$

with $R = 6.4$ fm, $a = 0.54$ fm. This is not very realistic, but facilitates comparison with the results of [13]. ($\varepsilon_0 = aT_i^4$ is the central energy density at initial time $\tau = \tau_i$.) We have considered two sets of initial conditions: (i) $T_i = 0.3$ GeV and $\tau_i = 0.5$ fm/c, and (ii) $T_i = 0.5$ GeV and $\tau_i = 0.1$ fm/c. Both sets have similar total initial entropy content.

We also assume that at the initial time τ_i the radial velocity vanishes ($v_r(r) = 0$).

For the non-ideal fluid, initial viscous pressures, π^{rr} and $\pi^{\phi\phi}$ are required. Even though v_r and its derivatives are zero initially, due to the Bjorken longitudinal motion the stress tensor is not zero: $\nabla^{\langle r}u^{r\rangle >} = r^2\nabla^{\langle\phi}u^{\phi\rangle >} = \frac{1}{3\tau_i}$. We assume that at initial time τ_i, the viscous pressure components are fully relaxed to the Bjorken scaling expansion values,

$$\pi^{rr} = r^2\pi^{\phi\phi} = \frac{2\eta}{3\tau_i}. \tag{13}$$

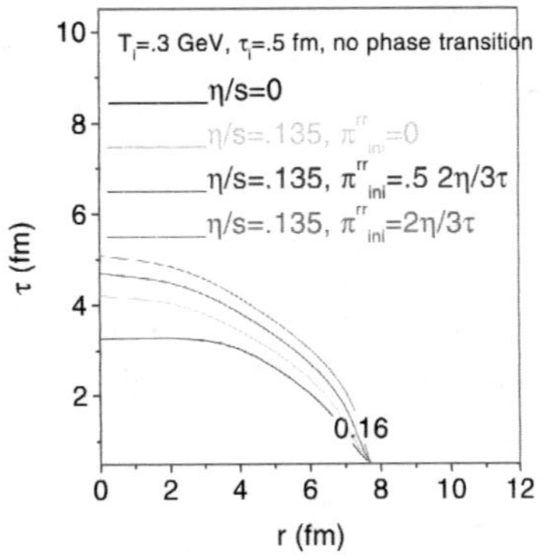

Figure 3. Freeze-out surface at T_f=160 MeV for different initial viscous pressures.

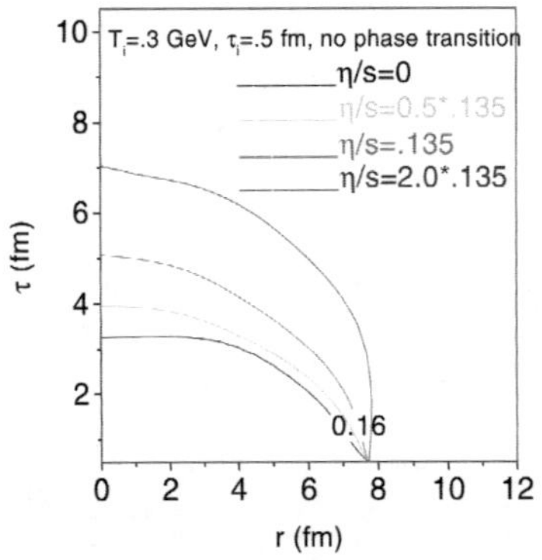

Figure 4. Freeze-out surface at T_f=160 MeV for different shear viscosities η.

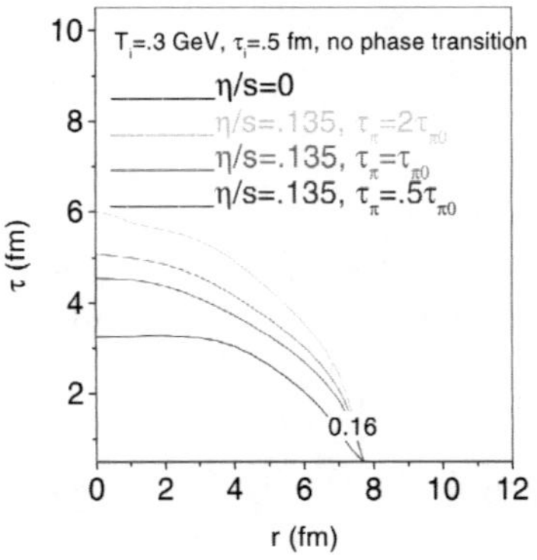

Figure 5. Freeze-out surface at T_f=160 MeV for different relaxation times τ_π.

4. Results

4.1. Space-time evolution of ideal and non-ideal QGP

With have solved the hydrodynamic equation for ideal and non-ideal fluids, with identical initial conditions. We do not invoke a phase transition. The initial QGP evolves without undergoing phase transition until it freezes out at a an assumed freeze-out temperature $T_f = 160$ MeV.

In Figs. 1 and 2, we show contours of constant temperature in the $r\!-\!\tau$ plane. The black lines are for the ideal fluid, while the red and blue lines are for non-ideal fluids with η/s=0.08 and 0.135, respectively. For ideal hydrodynamics the almost vanishing slope of the isotherms near $r = 0$ indicates that the transverse rarefaction wave has not reached the fireball center before the system decouples at $T_f = 0.16$ GeV. The cooling at $r = 0$ is controlled by the longitudinal scaling expansion. Consequently, the Bjorken cooling law, $T^3\tau$=constant, is well maintained at $r = 0$. For non-ideal flow, the Bjorken cooling law is violated. Viscosity opposes expansion and cooling. Thus the fluid cools more slowly and the lifetime of the fluid is extended. For a fluid with $T_i = 0.3$ GeV at $\tau_i = 0.5$ fm/c, the lifetime at $r = 0$ is extended by 30% and 55% for η/s=0.08 and 0.135, respectively. For the higher initial temperature ($T_i = 0.5$ GeV at $\tau_i = 0.1$ fm/c), the viscosity effects are even more prominent: in the fireball center the fluid life time is enhanced by more than 120% and 170% for η/s=0.08 and 0.135, respectively. One also observes an increased transverse expansion.

The space-time evolution of non-ideal fluids depends sensitively on the (i) initial viscous pressure, (ii) viscosity coefficient and (iii) relaxation time. In Fig. 3 the freeze-out surface at $T_f = 0.16$ GeV is shown for different initial viscous pressures, $\pi^{rr}(= r^2\pi^{\phi\phi})$=0, $\frac{\eta}{3\tau_i}$ and $\frac{4\eta}{3\tau_i}$, respectively, using η/s=0.135. The higher the initial viscous pressure, the more extended is the freeze-out surface. The life time of the dissipative QGP is extended by 20% if the initial viscous pressure is increased from zero to $\frac{4\eta}{3\tau_i}$. The freeze-out surface also depends sensitively on the value of the viscosity coefficient (Fig. 4). As the viscosity decreases, departure of the freeze-out surface from ideal behavior also decreases. In Fig. 5 we show the freeze-out surface for different relaxation times, $\tau_\pi = 0.5\,\tau_{\rm kin}$, $\tau_{\rm kin}$ and $2\,\tau_{\rm kin}$ (where $\tau_{\rm kin} = \frac{3\eta}{2p}$), for fixed viscosity $\eta/s = 0.135$. As the relaxation time is increased by a factor 4, the freeze-out time in the fireball center is decreased by 25%.

The evolution of the radial expansion velocity is also changed in viscous fluids. In Fig. 6 we show lines of constant radial flow v_r in the $r-\tau$ plane, for the ideal fluid (black line) and non-ideal fluids (red and blue lines). The radial velocity grow faster in the non-ideal fluid.

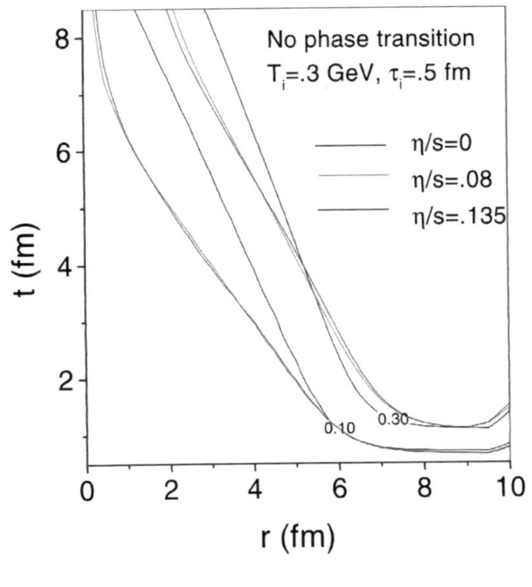

Figure 6. Contours of constant radial velocity in the $r-\tau$ plane for ideal and non-ideal hydrodynamic evolution, using T_i=0.3 GeV at τ_i=0.5 fm/c.

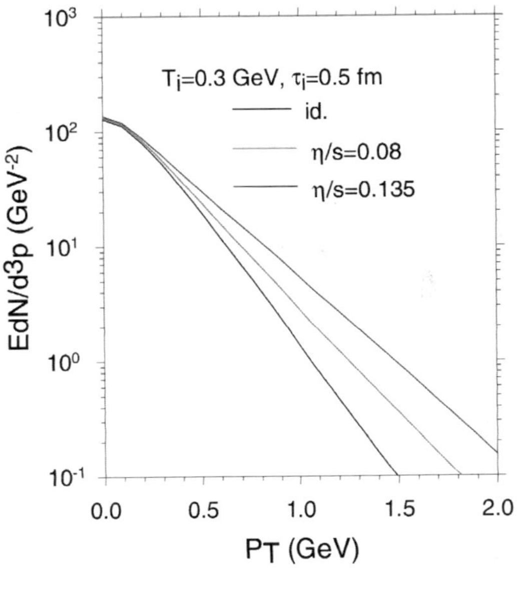

Figure 7. The final parton p_T distribution at freeze-out ($T_f = 160$ MeV). Initial temperature and time are $T_i = 0.5$ GeV and $\tau_i = 0.1$ fm/c.

4.2. Viscous corrections to the distribution function

The changed space-time evolution of the dissipative QGP may have significant effects on particle production. Although we cannot compute the p_T spectra of final hadrons, since we did not include the hadronization phase transition, we can calculate the p_T spectra of partons on the $T = T_f$ freeze-out surface.

When viscous corrections are included, the distribution function is modified from its local thermal equilibrium form [6, 11, 13]. When the viscous corrections are small, the distribution function can be written as

$$f(x,p) = f_{\rm eq}(x,p)(1+\Delta f) \tag{14}$$

with the equilibrium distribution function

$$f_{\rm eq}(x,p) = \frac{g}{2\pi^2}\frac{1}{e^{u\cdot p/T}\pm 1} \approx \frac{g}{2\pi^2}e^{-u\cdot p/T}.$$

If Δf is restricted to 2nd order in p^μ, its form can be determined as [6, 11, 13]

$$\Delta f = C p_\mu p_\nu \pi^{\mu\nu} \tag{15}$$

where $C = \frac{1}{2(\varepsilon+p)T^2}$.

In Fig. 7, we show the parton p_T spectra at $T_f = 160$ MeV. The black line is for the ideal fluid. The red and blue lines are for non-ideal fluids with η/s=0.08 and 0.135, respectively. Viscous flow is seen to enhance high p_T production by a factor 5–10, depending on viscosity. However, since $\pi^{\mu\nu} \neq 0$ already at initial time τ_i, already the initial p_T spectra for the viscous fluids are flatter than for the ideal fluid. To obtain the same measured final p_T-spectrum in the viscous case thus requires a retuning of initial conditions. This is presently under investigation.

5. Summary and conclusions

We have studied the effects of shear viscosity on the space-time evolution of a relativistic QGP fluid. In the 2nd order theory of dissipative fluid dynamics, the dissipative fluxes are treated as thermodynamic variables which follow kinetic relaxation equations. These are solved simultaneously with the energy-momentum conservation equations. Viscosity opposes expansion and cooling. Consequently, for a fixed freeze-out temperature, the freeze-out time (fluid lifetime) is increased. The evolution of the radial flow velocity is also changed – the viscous fluid generates more transverse flow. Viscous effects flatten the p_T distribution and require a retuning of the initial conditions if the final spectra are to be kept fixed at their measured values.

Acknowledgement: The work of U.H. was supported by the U.S. Department of Energy under contract DE-FG02-01ER41190.

References

[1] Karsch F, Laermann E, Petreczky P, Stickan S and Wetzorke I, 2001 *Proccedings of NIC Symposium* (Ed. H. Rollnik and D. Wolf, John von Neumann Institute for Computing, Jülich, NIC Series, vol.9, ISBN 3-00-009055-X, pp.173-82,2002.)
[2] see 2004 *Proccedings of 17th International Conference on Ultra Relativistic Nucleus-Nucleus Collisions (Quark Matter 2004)*, Oakland, California, 11-17 Jan 2004. Published in 2004 *J. Phys. G* **30**
[3] Kolb P E and Heinz U *Preprint* nucl-th/0305084
[4] Eckart C 1940 *Phys. Rev.* **58** 919
[5] Landau L D and Lifshitz E M 1959 *Fluid Mechanics* (Pergamon, MewYork)
[6] Grad H, 1949 Commun. *Pure Appl. Math.* **2** 331; Mueller I 1967 *Z. Phys.* **198** 329; Israel W 1976 *Ann. Phys. (N.Y.)* **100** 310; Stewart J M 1977 *Proc. Roy. Soc. A* **357** 59; Israel W and Stewart J M 1979 *Ann. Phys. (N.Y.)* **118** 341
[7] Kajantie K 1984 *Nucl. Phys. A* **418** 41c; Baym G 1984 *Nucl. Phys. A* **418** 525c; Hosoya A and Kajantie K 1985 *Nucl. Phys. B* **250** 666
[8] Danielewicz P and Gyulassy M 1985 *Phys. Rev. D* bf 31 666
[9] Chaudhuri A K 1995 *Phys.Rev. C*bf 512889, 2000 *J. Phys. G* **26** 1433, 2000 *Phys. Scripta* **61** 311
[10] Muronga A 2004 *Phys. Rev. C* **69** 044901 (*Preprint* nucl-th/0309055)
[11] Teaney D 2003 *Phys. Rev. C* **68** 033913

[12] Teaney D 2004 *J. Phys. G: Nucl. Part. Phys.* **30** S1247
[13] Muronga A and Rischke D H *Preprint* nucl-th/0407114
[14] Boris J P and Book D L 1976 *Methods in computational Physics* **16** 85
[15] Rischke D H, Bernard S and Maruhn J A 1995 *Nucl. Phys.A* **595** 346
[16] Baym G, Monien H, Pethick C J and Ravenhall D G 1990 *Phys. Rev. Lett* **64** 1867
[17] Arnold P, Moore G D and Yaffe L G 2000 *J. High Energy. Phys.* **0011** 001 (for a complete leading order calculation see P. Arnold, G. D. Moore and L. G. Yaffe, 2003 *J. High Energy. Phys.* **0305** 051)
[18] Policastro G, Son D T and Starinets A O 2001 *Phys. Rev. Lett.***87** 081601; 2002 *J. High Energy Phys.* **09** 043 (*Preprint* hep-th/0210220)

Appendix A. Coordinate transformations

Instead of Cartesian coordinates $x^\mu = (t, x, y, z)$ we use curvilinear cylindrical coordinates in longitudinal proper time and rapidity, $\bar{x}^m = (\tau, r, \phi, \eta)$:

$$t = \tau \cosh \eta; \qquad \tau = \sqrt{t^2 - z^2} \tag{A.1}$$

$$x = r \cos \phi; \qquad r = \sqrt{x^2 + y^2} \tag{A.2}$$

$$y = r \sin \phi; \qquad \phi = \tan^{-1}\left(\frac{y}{x}\right) \tag{A.3}$$

$$z = \tau \sinh \eta; \qquad \eta = \frac{1}{2} \ln \frac{t+z}{t-z}. \tag{A.4}$$

The differentials

$$dt = d\tau \cosh \eta + d\eta\, \tau \sinh \eta, \tag{A.5}$$

$$dx = dr \, \cos \phi - d\phi\, r \sin \phi, \tag{A.6}$$

$$dy = dr \, \sin \phi + d\phi\, r \cos \phi, \tag{A.7}$$

$$dz = d\tau \sinh \eta + d\eta\, \tau \cosh \eta, \tag{A.8}$$

and the metric tensor is easily read off from

$$\begin{aligned} ds^2 &= g_{\mu\nu} dx^\mu dx^\nu = dt^2 - dx^2 - dy^2 - dz^2 \\ &= \bar{g}_{mn} d\bar{x}^m d\bar{x}^n = d\tau^2 - dr^2 - r^2 d\phi^2 - \tau^2 d\eta^2, \end{aligned} \tag{A.9}$$

namely

$$\bar{g}_{mn} = \begin{pmatrix} 1 & 0 & 0 & 0 \\ 0 & -1 & 0 & 0 \\ 0 & 0 & -r^2 & 0 \\ 0 & 0 & 0 & -\tau^2 \end{pmatrix}, \qquad \bar{g}^{mn} = \begin{pmatrix} 1 & 0 & 0 & 0 \\ 0 & -1 & 0 & 0 \\ 0 & 0 & -1/r^2 & 0 \\ 0 & 0 & 0 & -1/\tau^2 \end{pmatrix} \tag{A.10}$$

In curvilinear coordinates we must replace the partial derivatives with respect to x^μ by covariant derivatives (denoted by a semicolon) with respect to $\bar{x}^m$:

$$\bar{T}^{ik}_{;p} = \frac{\partial \bar{T}^{ik}}{\partial \bar{x}^p} + \Gamma^i_{pm} \bar{T}^{mk} + \bar{T}^{im} \Gamma^k_{mp}.$$

The only non-vanishing Christoffel symbols are

$$\Gamma^\tau_{\eta\eta} = \tau; \quad \Gamma^\eta_{\tau\eta} = \Gamma^\eta_{\eta\tau} = 1/\tau; \quad \Gamma^r_{\phi\phi} = -r; \quad \Gamma^\phi_{r\phi} = \Gamma^\phi_{\phi r} = 1/r. \tag{A.11}$$

The hydrodynamic 4-velocity $u^\mu = \gamma(1, v_x, v_y, v_z)$ is transformed to $\bar{u}^m = \gamma(1, v_r, 0, 0)$, with $\gamma = 1/\sqrt{1 - v_r^2}$. From here on, we drop the bars over tensor components in $\bar{x}$-coordinates for simplicity.

Appendix B. Relaxation equations for the viscous pressure tensor

Being symmetric and traceless, the viscous pressure tensor $\pi^{\mu\nu}$ has 9 independent components. The assumption of boost invariance reduces this number by 3 ($\nabla^{\langle m}u^{\eta\rangle}=0,\ m\neq\eta$). The additional assumption of cylindrical symmetry eliminates two more components ($\nabla^{\langle m}u^{\phi\rangle}=0,\ \eta\neq m\neq\phi$. The transversality condition $u_m\pi^{mn}=0$ eliminates another two components (u_ϕ and u_η vanish and thus yield no constraints). Thus, with boost-invariance and cylindrical symmetry, the viscous pressure tensor has only two independent components which we here choose as π^{rr} and $\pi^{\phi\phi}$.

The viscous pressure tensor relaxes on a short kinetic time scale τ_π to 2η times the shear tensor $\sigma^{\mu\nu}=\nabla^{\langle\mu}u^{\nu\rangle}$ [6]. The relaxation equations for π^{rr} and $\pi^{\phi\phi}$ are

$$\partial_\tau\pi^{rr}+v_r\partial_r\pi^{rr} = -\frac{1}{\tau_\pi\gamma}\left(\pi^{rr}-2\eta\nabla^{\langle r}u^{r\rangle}\right), \tag{B.1}$$

$$\partial_\tau\pi^{\phi\phi}+v_r\partial_r\pi^{\phi\phi} = -\frac{1}{\tau_\pi\gamma}\left(\pi^{\phi\phi}-2\eta\nabla^{\langle\mu}u^{\phi\rangle}\right). \tag{B.2}$$

The (rr) and $(\phi\phi)$ components of the shear tensor can be written as

$$\nabla^{\langle r}u^{r\rangle} = -\partial_r u^r-u^r Du^r-\frac{1}{3}\Delta^{rr}\partial\cdot u=\gamma^2\left(\frac{\partial\cdot u}{3}-\partial_\tau\gamma-\partial_r(\gamma v_r)\right), \tag{B.3}$$

$$\nabla^{\langle\phi}u^{\phi\rangle} = -\frac{1}{r^2}\frac{u^r}{r}-\frac{1}{3}\Delta^{\phi\phi}\partial\cdot u=\frac{1}{r^2}\left(\frac{\partial\cdot u}{3}-\frac{\gamma v_r}{r}\right), \tag{B.4}$$

with the scalar expansion rate

$$\partial\cdot u=\partial_\tau u^\tau+\partial_r u^r+\frac{u^\tau}{\tau}+\frac{u^r}{r}=\left(\partial_\tau+\frac{1}{\tau}\right)\gamma+\left(\partial_r+\frac{1}{r}\right)(\gamma v_r) \tag{B.5}$$

and the convective detivative

$$D=u\cdot\partial=\gamma(\partial_\tau+v_r\partial_r). \tag{B.6}$$

The projectors Δ^{rr} and $\Delta^{\phi\phi}$ are,

$$\Delta^{rr} = g^{rr}-u^r u^r=-1-(u^r)^2=-\gamma^2$$

$$\Delta^{\phi\phi} = g^{\phi\phi}-u^\phi u^\phi=-\frac{1}{r^2}.$$

The remaining non-vanishing components of the viscous pressure tensor $\pi^{\tau\tau}$, $\pi^{\tau r}$, and $\pi^{\eta\eta}$ are eliminated using the constraints $u_m\pi^{mn}=0$ and $g_{mn}\pi^{mn}=0$, yielding

$$\pi^{\tau\tau}=v_r\pi^{\tau r}=v_r^2\pi^{rr},\quad \pi^{\eta\eta}=-\frac{1}{\tau^2}\left(-\frac{\pi^{rr}}{\gamma^2}+r^2\pi^{\phi\phi}\right). \tag{B.7}$$

Appendix C. Energy-momentum conservation

With longitudinal boost-invariance and cylindrical symmetry, the energy-momentum conservation equations $T^{mn}{}_{;n}=0$ yield

$$\partial_\tau\tilde{T}^{\tau\tau}+\partial_r\left(v_s\tilde{T}^{\tau\tau}\right) = -\frac{\tilde{p}-\tilde{\pi}^{rr}/\gamma^2-r^2\tilde{\pi}^{\phi\phi}}{\tau}, \tag{C.1}$$

$$\partial_\tau\tilde{T}^{\tau r}+\partial_r\left(v_r\tilde{T}^{\tau r}\right) = \frac{\tilde{p}+r^2\pi^{\phi\phi}}{r}-\partial_r\left(\tilde{p}+\frac{\tilde{\pi}^{rr}}{\gamma^2}\right). \tag{C.2}$$

where $\tilde{A}^{mn}\equiv r\tau A^{mn}$, $\tilde{p}\equiv r\tau p$, and $v_s\equiv T^{\tau r}/T^{\tau\tau}$.

Institute of Physics Publishing
doi:10.1088/1742-6596/50/1/031

Review of HBT or Bose-Einstein correlations in high energy heavy ion collisions

T. Csörgő
MTA KFKI RMKI, H - 1525 Budapest 114, P.O.Box 49, Hungary

Abstract. A brief review is given on the discovery and the first five decades of the Hanbury Brown - Twiss effect and its generalized applications in high energy nuclear and particle physics, that includes a meta-review. Interesting and inspiring new directions are also highlighted, including for example source imaging, lepton and photon interferometry, non-Gaussian shape analysis as well as many other new directions. Existing models are compared to two-particle correlation measurements and the so-called RHIC HBT puzzle is resolved. Evidence for a (directional) Hubble flow is presented and the conclusion is confirmed by a successful description of the pseudorapidity dependence of the elliptic flow as measured in Au+Au collisions by the PHOBOS Collaboration.

Child, how happy you are sitting in the dust, playing with a broken twig all the morning.
I smile at your play with that little bit of a broken twig.
I am busy with my accounts, adding up figures by the hour.
Perhaps you glance at me and think, "What a stupid game to spoil your morning with!"
Child, I have forgotten the art of being absorbed in sticks and mud-pies.
I seek out costly playthings, and gather lumps of gold and silver.
With whatever you find you create your glad games,
I spend both my time and my strength over things I never can obtain.
In my frail canoe I struggle to cross the sea of desire, and forget that I too am playing a game.
(R. Tagore: Playthings)

1. Introduction

First I attempted to give a brief summary of the numerous achievements of particle interferometry from 1955 to 2005, to celebrate the Golden Jubilee: the 50 year anniversary of the Hanbury Brown - Twiss effect. As it is clearly impossible to summarize 50 years of efforts and discoveries of the order of one thousand scientific papers in a few pages, I ended up in creating a meta-review. I had to be very brief in reviewing - see my talk in ref. [1] for more details, historical remarks and animated illustrations.

2. The Hanbury Brown - Twiss effect

One of the most inspiring books that I have read recently was R. Hanbury Brown's brief autobiography – Boffin. I recommend the reading of this book for all students all who enter the field of particle interferometry in high energy particle and nuclear physics [2].

As this note is contributed to the ICPA-QGP 2005 conference organized in India, I should mention that Robert Hanbury Brown was born here in Aruvankadu, Nilgiri Hills, India in 1916 – he passed away in Andover, England, in 2002. Similarly to him Richard Q. Twiss was also born in India and they both had loved this country.

Hanbury Brown had been trained as an engineer and he was marked with an ingenious talent as well as a subtle sense of humor. Let me quote his book on how he reflects on his most famous discovery: "*I was a long way from being able to calculate, whether it would be sensitive enough to measure a star. To do that one has to be familiar with photons and as an engineer my education in physics had stopped far short of the quantum theory. Perhaps just as well, otherwise like most physicists I would have come to the conclusion that the thing would not work – ignorance is sometimes a bliss in science." ... "In fact to a surprising number of people the idea that the arrival of photons at two separated detectors can ever be correlated was not only heretical but patently absurd, and they told us so in no uncertain terms, in person, by letter, in print, and by publishing the results of laboratory experiments, which claimed to show that we were wrong ...* " . Opponents cited also from sacred books of Quantum Mechanics, for example quoting Dirac [3]: "*Interference between two different photons can never occur.*" Vigorous objections were based on two laboratory experiments, the first of which was performed by the group of Jánossy at KFKI, my home institute in Budapest, Hungary. Their paper concluded [4], that " *... in agreement with quantum theory, the photons of two coherent light beams are independent from each other .*" However, Hanbury Brown and Twiss analyzed this and other experiments, and pointed out [5] that to observe the HBT effect one needs a very intensive source of light with a very narrow bandwidth, such as an isotope lamp that neither of these counter-experiments had.

The first stellar intensity interferometer, the pilot model was set up at Jodrell Bank in England 50 years ago, in 1955. This equipment was used to make the first measurement of the angular diameter of a main sequence star, Sirius [6]. This pilot experiment demonstrated the value of the method of intensity correlations and was the precursor to the construction of the Stellar Intensity Interferometry in Narrabi, Australia, completed in 1963. The giant reflectors of the Narrabi Observatory had parabolic surfaces of 22 feet diameter and a focal length of 36 feet, mounted on a circular railway truck, that had 618 feet diameter, or 73 % of the 843 feet diameter of the AGS accelerator of Brookhaven National Laboratory in Upton, LI, USA.

During the last 50 years, the study of intensity correlations became a broad field not only in quantum optics, solid state physics and astronomy, but extended to include the investigations of correlations among various type and number of particles in high energy particle and nuclear physics, [7–15].

3. Particle interferometry from 1955 to 2005 - a review of reviews

The field of particle interferometry or femtoscopy in high energy particle and heavy ion physics is firmly established: there is even a book written about the topic [16]. A collection of the early articles on Bose-Einstein correlations in high energy physics is also available [17]. Here I collected some of the most inspiring review papers of the field.

For newcomers, I recommend to start with the summary of Lörstad from 1989 [18], written at a period when preliminary correlation data indicated signatures for a first order QGP to hadron transitions in O + Au collisions at CERN SPS energies. Boal, Gelbke and Jennings [19] summarized boson and fermion intensity correlations in subatomic physics, with emphasis on final state interactions. The review of Bauer, Gelbke and Pratt [20] is recommended for techniques applied in low and intermediate energy heavy ion collisions in particular for the noting of the energy dependence of the effective source sizes in nucleon-nucleon intensity correlations. A pedagogical work aimed at the identification of experimental difficulties in two-boson interferometry was compiled by Zajc [21], a must for every pedestrian entering this field. Schutz [22] summarized hard photon interferometry in heavy ion physics. Harris and Müller overviewed the suggested signatures for the production of a quark-gluon plasma, including Bose-Einstein correlations [23]. Light particle correlation data in heavy ion collisions at intermediate and low energies was compiled by Ardouin [24]. Data and theory of correlations in relativistic heavy ion collisions were summarized by Heinz and Jacak [28]. The multivariate Gaussian approach to the parameterization of particle interferometry in relativistic heavy ion collisions was highlighted by Wiedemann and Heinz [25]. The invariant Buda-Lund particle interferometry method was summarized by me and Lörstad [26]. The review of Weiner [27] highlighted the applications of quantum optical methods in boson interferometry in high energy particle and nuclear physics. Bose-Einstein correlations in electron-positron annihilations at LEP were introduced and reviewed by Kittel [29], summarized at the Z fragmentation [30], and at the W^+W^- decays [31]. Model independent methods, similarities between interferometry results in electron-positron annihilation data and in relativistic heavy ion collision data, critical review of the resonance and Coulomb effects, limitations of the Gaussian approach and the Bertsch-Pratt and Yano-Koonin parameterizations were discussed [32] together with applications of source image reconstruction and quantum optical methods like squeezed states and pion lasers. Tomasik and Wiedemann reviewed central and non-central particle interferometry data after the first results at RHIC became accessible [33]. Alexander [34] reviewed Bose-Einstein and Fermi-Dirac correlations in particle physics, with focus on the mass and transverse mass dependence of the radius parameters of these correlations in electron-positron annihilation at LEP. Heinz [35] reviewed the concepts applied in high energy heavy ion physics, including particle interferometry. The review of Padula [36] starts with an excellent historical overview of interferometry in the 1960's and 1970's, covering the current formalism and squeezing of mass-modified bosons and fermions. These reviews well summarize the first 50 years of particle interferometry.

4. Interesting new directions

Improvement on the two-particle Coulomb corrections. If $\lambda < 1$, either partial coherence develops in the source, or some fraction of the pions comes from the extended, halo part of the source [32, 95]. This influences, how the relative separations of the particle pairs are distributed, hence the weight assigned to various parts of the two-particle Coulomb wave function. CERES [37] was the first experiment to apply this averaging self-consistently. By now, STAR [38], PHENIX [11] and PHOBOS [39] have also utilized this method suggested by Bowler [40], Sinyukov and collaborators [41].

Many-body Coulomb effects for finite sources. The core-halo type of problem exists not only in the construction of the Bose-Einstein correlation function of two charged particles, but in the determination of the Bose-Einstein correlation function of three or more charged particles. The many-body Coulomb corrections are notoriously difficult, but the circumstances are favorable in high energy heavy ion and particle physics for the application of an approximate, cluster expansion method, that is based on an asymptotically correct form of the multiparticle Coulomb wave function [42].

Intensity correlations of non-identical particle pairs are new and promising tools in heavy ion physics, given the large possible combinations of various type of particles. This method can not only be used to learn about the temporal sequence of particle emission, but also to learn about the strong final state interactions between various type of particle pairs [14, 43].

Intensity correlations of penetrating probes provides information not only on the phase-space distribution of the decoupling system at the time of freeze-out but also adds the information about the temporal evolution of the fireball.

a) Photon interferometry attracted a strong theoretical activity in order to utilize correlations of direct photons to search for quark gluon plasma phase suggested first by Makhlin [44]. This field is pioneered by Srivastava and collaborators, [46–48, 52] as well as by the organizers of this conference [51]. Earlier papers highlighted to photon interferometry at the CERN SPS energy domain [45, 49, 50], where the first experimental results on photon interferometry were reported recently by WA98 in Pb+Pb collisions [53].

b) Lepton interferometry is also based on penetrating probes that can escape from the hadronic fireball during its time evolution, hence lepton correlations can give information on the volumes of homogeneity at any stage of the expansion, at least in principle. The first case study has just been published by Alam, Mohanty and collaborators [54] using a 3+1 dimensional, relativistic hydrodynamical model with spin dependent invariant amplitudes and a bag model equation of state.

The key experimental question for photon and lepton interferometry is if one can find a transverse momentum window where photons or leptons from a quark gluon plasma phase dominate the single particle spectrum, overcoming contributions e.g. gammas from the π^0 decays, or in case of electrons, from Dalitz and open charm decays and Drell-Yan processes.

Q-boson interferometry has been investigated theoretically by Zhang, Padula, Anchiskin and collaborators in refs. [55–57]. It remains to be seen if q-boson correlations can ever be observed in experiments of high energy physics.

Similarities between Bose-Einstein correlations in different reactions might indicate an important limitation of our understanding of the particle correlations in $e^+ + e^-$, hadron + hadron, d+A and A+B reactions. For example, the mass and the transverse mass dependencies of the Bertsch - Pratt radius parameters in these reactions seems to be very similar, but the scales and the reaction mechanisms are completely different [29–32, 34].

Continuous emission and escaping probabilities has to be taken into account, if particle emission is not confined to a narrow freeze-out hypersurface, but is continuous during the time evolution of the system. This scenario has been considered by Grassi, Hama and collaborators, who developed first the continuous emission model [58]. This model was further improved with the help of the introduction of the escaping probabilities and a self-consistent theoretical formulation of the particle emission from hydrodynamical sources during the whole period of expansion [59].

Initial conditions that fluctuate event by event are seen in realistic Monte-Carlo calculations. Averaging over the final states of hydrodynamics started from fluctuating initial conditions yields HBT radii closer to the measured values as compared to the case when the initial conditions are first averaged over and the hydrodynamic evolution starts from the smoothed initial conditions, as shown with the help of a 3+1 dimensional relativistic hydrodynamical codes developed by the SPHERIO collaboration [60].

Rapidity dependence of the HBT radii was explored by PHOBOS at RHIC and NA49 at CERN SPS. For the side, out and longitudinal components only weak rapidity dependence was found in both cases. The Yano-Koonin-Podgoretskii fit revealed that the longitudinal flow profile is nearly boost invariant in both reactions [39, 61].

Azimuthally sensitive HBT measurements push the multi-variate Gaussian parameterization to its limits. The transverse momentum and the azimuthal angle dependence of the 3 by 3 symmetric radius matrix provide important constraints for the dynamics of the effective source in these collisions, pioneered by M. Lisa and collaborators [62]. STAR data on asHBT indicate [38], that the effective source is extended in the direction perpendicular to the impact parameter, similarly to the geometry of the nuclear overlap.

Beyond the Gaussian approximation: a) The Edgeworth and Laguerre expansions utilize complete sets of polynomials that are orthogonal with respect to given weight functions. The method was worked out by S. Hegyi and the present author [63]. The Edgeworth expansion has been applied to quantify the non-Gaussian features of Bose-Einstein correlations by the L3 and the STAR collaborations [13, 64, 65].

b) The Lévy index of stability, $0 < \alpha \leq 2$ is a new parameter of the two- and three-particle Bose-Einstein correlation functions [66–68]. It characterizes the power-law tails of Lévy stable distributions, which appear in physical systems where the final distribution is obtained as convolution of many elementary random steps. If the distribution functions of the elementary steps are characterized by finite means and variances, central limit theorems determine that the limit distribution has to be a Gaussian and $\alpha = 2$. Generalized central limit theorems state that for certain elementary processes with infinite variance or infinite mean a limiting distribution exist, but the limit distribution develops a power-law tail and $\alpha < 2$.

c) Imaging methods were developed by Brown and Danielewicz[69] to reconstruct the relative coordinate distribution of the source by inverting the two-particle correlation function with a kernel, based on known final state interaction and symmetrization effects. This method was applied first at the AGS energies [70]. Preliminary PHENIX results indicate [71] the existence of a long tail in the relative coordinate distributions in d+Au and Au+Au collisions at the maximal RHIC energies.

Multi-boson symmetrization effects and pion lasers are explored with the pion-laser model of S. Pratt [72], that I could understand only from ref. [73], which lead to the analytical solution of the model with Zimányi [74, 75]. Among others we found that in the rare gas limit two and three-particle symmetrization can be performed perturbatively, within Poisson distributed clusters. More theoretical and experimental work is needed to investigate these effects under conditions of RHIC and LHC reactions.

In-medium hadronic mass modifications can signal partial $U_A(1)$ symmetry restoration, measurable with two-pion Bose-Einstein correlations [76]. Current STAR Au+Au HBT radii were shown in agreement with a model assuming chiral symmetry restoration [77]. The quantum freeze-out problem of in-medium mass shifted bosons has been solved [78] correcting the idea of Andreev and Weiner [79]. It was extended to the case of fermions [80]. A liquid of mass-modified hadrons is thus signalled by back-to-back correlations between the detected particle - antiparticle pairs.

Let me add the following important comment here. The most recent studies of the phase structure of strongly interacting matter with the help of lattice QCD calculations suggest that the transition from the hadronic phase to a this form of matter, the quark-gluon plasma, happens at a critical temperature of $T_c = 164 \pm 2$ MeV and the form of the transition is a cross-over, where hadrons might exist above the critical temperature in a broad range of baryo-chemical potentials up to $\mu_B \approx 350$ MeV, where the transition changes to a second order phase transition. At even larger baryochemical potentials, a first order phase transition is seen, and the critical temperature decreases slightly with increasing μ_B-s, see ref. [86] for further details. If indeed the degrees of freedom in a fluid are dominated by non-hadronic states, and if hadronization and freeze-out happens simultaneously in a time-like deflagration as suggested in ref. [81], then the back-to-back particle-antiparticle correlations have to vanish. Hence the excitation function of the these correlations changes drastically at the transition from a hadronic to a non-hadronic liquid. Hence it is sensitive to the creation of a quark gluon plasma, which underlines the importance of search for squeezed states in ultra-relativistic heavy ion collisions.

For advanced studies, I recommend the following works and important contributions to particle interferometry in high energy physics from its first decade, [87]-[90] its second [91]-[98], third [99]-[117], fourth [118]-[130] and fifth decade [131]-[149].

5. The "RHIC HBT puzzle" and its resolution

Before contrasting the experimental results at RHIC with theoretical expectations and fits, let me go back first to the theory of particle interferometry for expanding sources formulated by Pratt [107, 108]. The directional dependence of the HBT correlation function was investigated by Hama and Padula as early as in 1987 [109], who observed that the longitudinal effective radius parameter is proportional to the proper-time of freeze-out, and proportional to the square root of temperature over energy. Their eq. (7) of ref. [109] can be considered as one of the first precursors to the studies of the ratios or the differences between R_{out} and R_{side}. Makhlin and Sinyukov derived a famous and experimentally well tested formula for boost-invariant particle emitting sources: $R_{long} = \tau_f \sqrt{T_f/m_t}$, ref. [110]. Bertsch introduced [111] the naming convention (out,side,longitudinal) for the components of the relative momentum that are parallel with the mean transverse momentum, perpendicular to the longitudinal and the mean momentum, and parallel with the longitudinal direction, respectively, with an influential graphical illustration, supported by Monte-Carlo cascade simulations by Bertsch, Gong and Tohyama [112]. Using a Gaussian approximation, the popular fit parameters thus became λ, R_{out}, R_{side}, and R_{long}.

5.1. HBT result at CERN SPS energies

The first preliminary results from the NA35 experiment in O+Pb reactions at CERN SPS energies indicated a signal for the first order QGP-hadron transition, R_{out} was found to be much larger, than R_{side}, a three sigma effect [113, 114]. These data were described by Padula and Gyulassy in terms of a quark gluon plasma model but also in terms of a conventional hadronic resonance gas model [115, 116]. Kaon interferometry has been suggested to increase the selectivity of these correlation measurements [117].

The preliminary NA35 data were changed and the difference between the out and the side HBT radius parameters in S+Pb collisions was found to vanish within errors [118, 119]. This observation was published by the second generation experiment NA44, which attempted to determine precisely all the three radius components at mid-rapidity in a broad transverse mass interval [120], valid not only for pions but also for kaons [121, 122]. The results indicated an approximate equality and simultaneous scaling of the radius parameters of the correlation functions, namely that these radius parameters within rather small experimental errors turned out to be rather similar, obeying a common dependence on the transverse mass of the pair. At the same time the NA44 collaboration observed the scaling of the single particle spectra, indicating that the slope parameter increases with increasing transverse mass of the particles [123]. Both of these effects were explained in terms of a three-dimensionally expanding, axially symmetric, suddenly decaying fireball [124, 125]. The existence of scaling limiting cases in certain domain of the parameter space was a focal point for the formulation and subsequent development of this Buda-Lund hydro model.

5.2. HBT predictions for RHIC energies

Soon after the observation of the approximate equality and transverse mass scaling of the radius parameters of the Bose-Einstein correlation function (HBT radii for short), Gyulassy and Rischke predicted that the HBT radius in the out direction will exceed the HBT radius in the side direction, predicting $R_{out}/R_{side} > 1$, values reaching up to 4 or even 20 in case of an ideal first order QGP to hadron phase transition, [126, 127]. These detailed predictions about a slowly burning log of QGP obtained a lot of experimental attention, in contrast to the earlier and less well-known but in fact more successful model of ref. [81], that predicted $R_{out} \approx R_{side} \approx R_{long}$ in a broad transverse mass interval at RHIC for a QGP which has reduced entropy density and harmonizes suddenly at every location from a supercooled state. Presently neither of the four experimental predictions made in ref. [81] are in disagreement with the known features of particle production in Au+Au collisions at RHIC as far as I know, hence the picture of a suddenly frozen, supercooled QGP has not yet been invalidated by experimental observations at RHIC.

5.3. Less Unpromising Models and the Au+Au HBT data at RHIC

The first detailed measurements of Bose-Einstein or HBT correlations by STAR at RHIC [10] have excluded the validity of the slowly burning quark-gluon plasma picture predicted in ref. [126, 127]. These measurements were confirmed by PHENIX [11] and the approximate equality of the HBT radii was established in a large transverse momentum interval. Recently, PHOBOS has extended these measurements to a broad rapidity interval without changing the overall picture [132].

Of the order of 50 models were shown to be unable to describe these observables at RHIC. The difficulty faced here by some of the theoretical approaches to describe the data is frequently referred to as the "RHIC HBT puzzle" [131].

With the help of my students, M. Csanád and A. Ster, we have scanned the literature and attempted to determine the list of less unpromising models, that cannot be excluded with the tools of mathematical statistics, using HBT data at RHIC. The following models were found to pass this HBT test:

- The multi-phase transport model (AMPT) of Lin, Ko and Pal [133].
- The hadronic cascade model calculation of Humanic [134].
- The Buda-Lund hydro model [135–138].
- The Cracow single freeze-out thermal model [139].
- The Blast-wave model as implemented by Retiere and Lisa [140, 141].
- The parameterized time-dependent expanding souce model of Renk [142, 143].
- Ref. [77] that investigates the onset of the chiral phase transition and its effects on the HBT radii. Effectively, this model also leads to a Buda-Lund type of parameterization.

As the number of models that pass the HBT test is fairly sizeable at present, additional criteria are needed to select the most relevant models.

6. Evidence for a Hubble flow in Au+Au at RHIC

To constrain the class of the not obviously wrong models, additional observables are needed. Let me chose here the identified single particle spectra of PHENIX, the pseudorapidity distribution measured by BRAHMS and PHOBOS, and the transverse mass dependence of the HBT radius parameters as measured by PHENIX and STAR. Humanic's cascade, the blastwave model, the Buda-Lund hydro model and the Cracow hydro model pass these additional criteria as well. In my talk I have illustrated [1], how a transverse and a longitudinal Hubble constant can be extracted from the blastwave, the Buda-Lund and the Cracow model fits to these data. Within errors, all of these three models are agreement with an approximately direction independent Hubble type of flow profile, where the transverse and the longitudinal Hubble constants are within errors the same, $u_{long} = H_{long} r_{long}$ and $u_{transv} = H_{transv} r_{transv}$, with $H_{long} = H_{transv} = 0.13 \pm 0.02\ fm^{-1}$. This result is well illustrated by the upper plots in Fig.1.

These observations are confirmed by a perfect description of the rapidity dependence of the elliptic flow with the help of the ellipsoidal generalization of the Buda-Lund hydro model, as illustrated on the lower plots of Figure 1. These data by the PHOBOS collaboration were not reproduced with the help of hydrodynamical models earlier. The lower right panel of Figure 1 illustrates, that the PHOBOS data are not only perfectly described with the ellipsoidally symmetric generalization of the Buda-Lund hydro model, but also that the PHOBOS data can be scaled to the theoretically predicted form of the scaling function.

This ellipsoidally symmetric Buda-Lund model is based on direction dependent Hubble constants [85]. The asymptotic emergence of a direction independent Hubble flow is seen analytically in the new families of exact analytic solutions of non-relativistic hydrodynamics that are the basis this Buda-Lund hydro model [144–146].

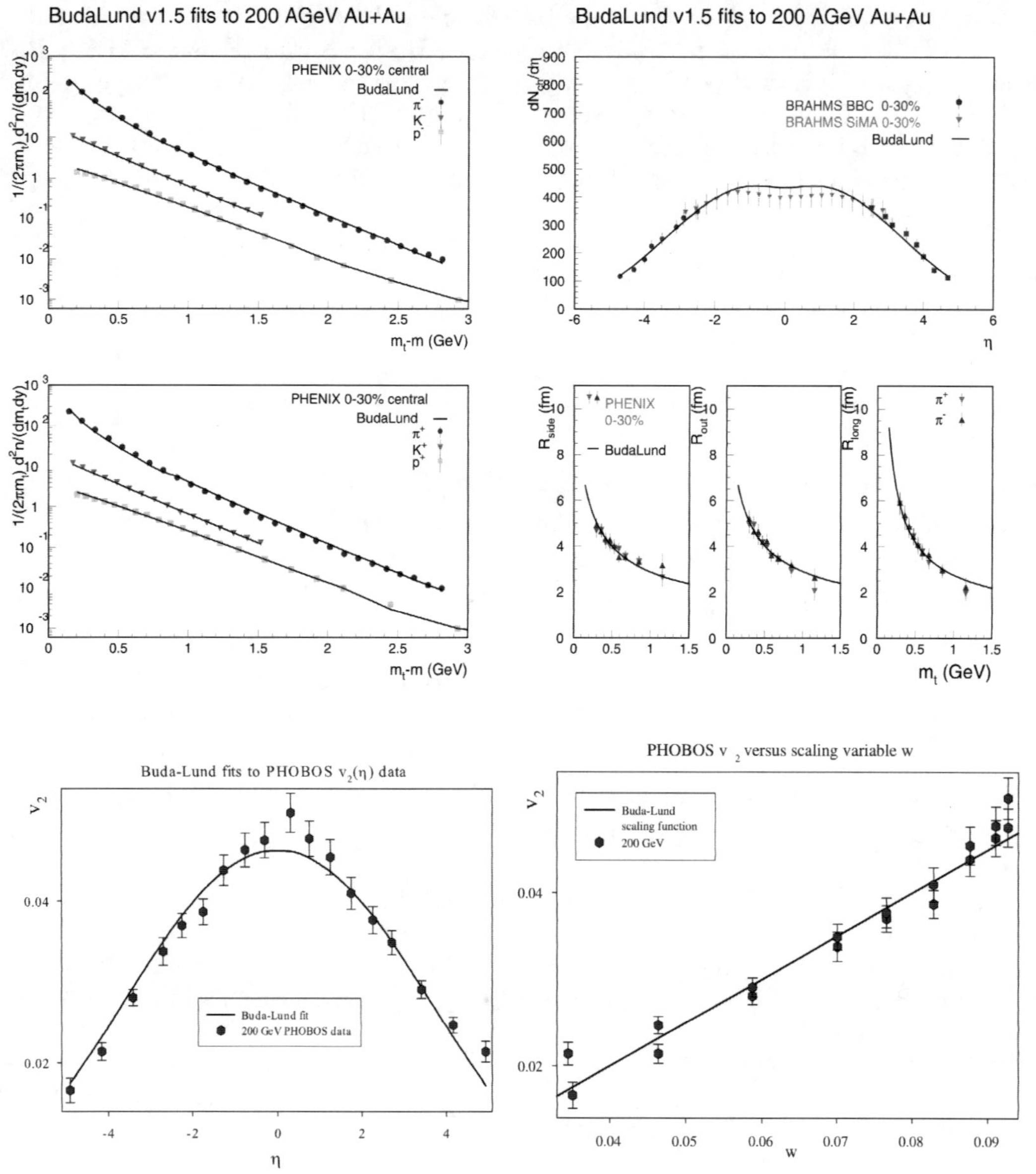

Figure 1. The upper panels show a simultaneous Buda-Lund fit to final Au+Au data at $\sqrt{s_{NN}} = 200$ GeV from refs. [82], [11, 83]. The fit parameters are summarized in Table 1 of ref.[138]. The ellipsoidally symmetric Buda-Lund hydro model is fitted [150] to preliminary PHOBOS $v_2(\eta)$ data at 200 GeV on the lower left panel [84]. In the lower right panel, these PHOBOS data on $v_2(\eta)$ are re-scaled and compared to the scaling function $v_2(w) = I_1(w)/I_0(w)$ predicted by the Buda-Lund hydro model [85].

7. Conclusions

After a historic introduction and a review of reviews I have highlighted some of the interesting new directions, including interferometry with penetrating probes like photons and leptons and the quantification of the non-Gaussian behaviour of the correlation functions. It may well be that a non-trivial energy dependence of these correlations reveals itself not in the scale parameters like the HBT radii but in the shape parameters like the Lévy index of stability α. Indeed, for p+p reactions, this parameter has been linked [68] to the anomalous dimension of QCD which follows a non-trivial energy dependence due to the running of the strong coupling constant. I also noted that the disappearance of the particle-antiparticle back-to-back correlations at a certain colliding energy can be a new signal for a transition from a hadronic fluid to a non-hadronic fluid, hence to the onset of the deconfinement transition. More experimental efforts are needed to look for squeezed states and chiral symmetry restoration at the RHIC energies.

The HBT data at RHIC provide important and selective constraints for the model builders and indicate that a large number of models implemented the space-time evolution of the source incorrectly. Many models failed to describe these observables, but this is not a puzzle. Mysteries surround the HBT effect from the very beginning of its discovery. I noted, that there was a successful prediction [81] in 1994 for the simultaneous equality of the HBT radii in Au+Au reactions at RHIC in a broad transverse mass interval, and all the qualitative predictions of this model are in agreement with the present data. Hence currently it cannot be excluded that a suddenly hadronizing supercooled QGP is present in these reactions. I presented the list of the less unpromising models - theoretical descriptions that pass the HBT test in Au+Au collisions at RHIC. At present 7 such models were found in the literature. Then I suggested to try to determine the common part of these models and to increase the sensitivity of the test by comparing these models to the single particle spectra, the HBT radii and to the elliptic flow measurements at RHIC. I have argued that the hadronic cascade model of Humanic, the Buda-Lund hydro model, the blastwave model and the Cracow model passes all these tests. In an attempt to determine the common features of these models, the transverse and the longitudinal Hubble constants were found to be the same within errors in the best fits to the data by the Buda-Lund, the Blastwave and the Cracow models, providing an evidence for a nearly fully developed Hubble flow and an indication of deconfinement temperatures reached in central Au+Au collisions at RHIC [138, 151]. These observations are confirmed by a perfect description of the rapidity dependence of the elliptic flow with the help of the Buda-Lund hydro model. Based on a forthcoming manuscript [150] I have shown that the PHOBOS rapidity dependent elliptic flow data, previously resisting a description in a hydrodynamic picture, actually agree with the theoretically predicted scaling function if the proper scaling variable w is determined. The solution of the HBT puzzle at RHIC thus naturally provides a solution to the puzzling rapidity dependence of the elliptic flow at RHIC as well.

Acknowledgments:
I would like to express my sincere thanks to the Organizers of this meeting for creating an excellent scientific atmosphere for ICPA-QGP 2005. This work was supported by following grants: OTKA T034269, T038406, T049466, the OTKA-MTA-NSF grant INT0089462, the NATO PST.CLG.980086 grant and the exchange program of the Hungarian and the Polish Academy of Sciences.

[1] T. Csörgő, *"Review of Bose-Einstein or Hanbury Brown - Twiss correlations"*, invited talk at the ICPA-QGP 2005 conference, Kolkata, India, `http://www.veccal.ernet.in/~icpaqgp/prog.html`

[2] R. Hanbury Brown: *"BOFFIN - A personal story of the Early Days of Radar, Radio Astronomy and Quantum Optics "* Published under the Adam Hilger imprint by IOP Publishing Ltd, Briston - New York and Philadelphia, ISBN 0 - 7503 - 0130 - 9

[3] P. A. M. Dirac, *"The Principles of Quantum Mechanics"* Oxford University Press, 1930, p. 14-15.

[4] A. Ádám, L. Jánossy and P. Varga, Acta Phys. Hungarica 4 (1955) 301.

[5] R. Hanbury Brown and R. Q. Twiss: Nature **178** (1956) 1447 - 1448.

[6] R. Hanbury Brown and R. Q. Twiss, Nature **178**, 1046 (1956).

[7] N. M. Agababyan *et al.* [EHS-NA22 Collaboration], Phys. Lett. B **332** (1994) 458.
[8] N. Neumeister *et al.* [UA1-Minimum Bias-Collaboration], Phys. Lett. B **275** (1992) 186.
[9] I. G. Bearden *et al.* [the NA44 Collaboration], Phys. Rev. Lett. **87** (2001) 112301
[10] C. Adler *et al.* [STAR Collaboration], Phys. Rev. Lett. **87** (2001) 082301
[11] S. S. Adler *et al.* [PHENIX Collaboration], Phys. Rev. Lett. **93** (2004) 152302
[12] M. Acciarri *et al.* [L3 Collaboration], Phys. Lett. B **493** (2000) 233 [arXiv:hep-ex/0008020].
[13] P. Achard *et al.* [L3 Collaboration], Phys. Lett. B **540** (2002) 185 [arXiv:hep-ex/0206051].
[14] A. Kisiel [STAR Collaboration], J. Phys. G **30**, S1059 (2004) [arXiv:nucl-ex/0403042].
[15] J. Adams *et al.* [STAR Collaboration], Phys. Rev. Lett. **91**, 262301 (2003) [arXiv:nucl-ex/0306028].
[16] R. M. Weiner, *Introduction to Bose-Einstein Correlations and Subatomic Interferometry*, John Wiley and Sons; 1st edition (2000) 232 pages, ISBN 0 471 96922 2
[17] R. M. Weiner, *Bose-Einstein Correlations in Particle and Nuclear Physics: A Collection of Preprints*, John Wiley and Sons; 1st edition (1997) 483 pages, ISBN 0 471 96979 6
[18] B. Lörstad, Int. J. Mod. Phys. A **4**, 2861 (1989).
[19] D. H. Boal, C. K. Gelbke and B. K. Jennings, Rev. Mod. Phys. **62**, 553 (1990).
[20] W. Bauer, C. K. Gelbke and S. Pratt, Ann. Rev. Nucl. Part. Sci. **42**, 77 (1992).
[21] W. A. Zajc, *"A pedestrian's guide to interferometry," Given at NATO Advanced Study Institute on Particle Production in Highly Excited Matter, Castelvecchio Pascoli, Italy, 12-24 Jul 1992* Proc. NATO ASI series, Il Ciocco, 1992, 435-459
[22] Y. Schutz, Acta Phys. Polon. B **25**, 485 (1994).
[23] J. W. Harris and B. Muller, Ann. Rev. Nucl. Part. Sci. **46**, 71 (1996) [arXiv:hep-ph/9602235].
[24] D. Ardouin, Int. J. Mod. Phys. E **6**, 391 (1997).
[25] U. A. Wiedemann and U. W. Heinz, Phys. Rept. **319** (1999) 145 [arXiv:nucl-th/9901094].
[26] T. Csörgő and B. Lörstad, arXiv:hep-ph/9901272.
[27] R. M. Weiner, Phys. Rept. **327**, 249 (2000) [arXiv:hep-ph/9904389].
[28] U. W. Heinz and B. V. Jacak, Ann. Rev. Nucl. Part. Sci. **49**, 529 (1999) [arXiv:nucl-th/9902020].
[29] W. Kittel, arXiv:hep-ph/0111462.
[30] W. Kittel, Acta Phys. Polon. B **32**, 3927 (2001) [arXiv:hep-ph/0110088].
[31] W. Kittel, arXiv:hep-ph/9905394.
[32] T. Csörgő, Heavy Ion Phys. **15**, 1 (2002) [arXiv:hep-ph/0001233].
[33] B. Tomasik and U. A. Wiedemann, arXiv:hep-ph/0210250. In Hwa, R.C. and X.-N. Wang (eds.): Quark gluon plasma 3, World Scientific, Singapore, pp. 715-777
[34] G. Alexander, Rept. Prog. Phys. **66**, 481 (2003) [arXiv:hep-ph/0302130].
[35] U. W. Heinz, arXiv:hep-ph/0407360.
[36] S. S. Padula, Braz. J. Phys. **35**, 70 (2005) [arXiv:nucl-th/0412103].
[37] D. Adamova *et al.* [CERES collaboration], Nucl. Phys. A **714** (2003) 124 [arXiv:nucl-ex/0207005].
[38] J. Adams *et al.* [STAR Collaboration], Phys. Rev. Lett. **93**, 012301 (2004) [arXiv:nucl-ex/0312009].
[39] B. B. Back *et al.* [PHOBOS Collaboration], arXiv:nucl-ex/0409001.
[40] M. G. Bowler, Phys. Lett. B **270**, 69 (1991).
[41] Y. Sinyukov, R. Lednicky, S. V. Akkelin, J. Pluta and B. Erazmus, Phys. Lett. B **432** (1998) 248.
[42] E. O. Alt, T. Csörgő, B. Lörstad and J. Schmidt-Sorensen, Eur. Phys. J. C **13** (2000) 663
[43] S. Voloshin, R. Lednicky, S. Panitkin and N. Xu, Phys. Rev. Lett. **79**, 4766 (1997)
[44] A. N. Makhlin, Sov. J. Nucl. Phys. **49**, 151 (1989) [Yad. Fiz. **49**, 238 (1989)].
[45] R. Ostendorf *et al.*, Phys. Rev. Lett. **73**, 34 (1994).
[46] D. K. Srivastava and J. I. Kapusta, Phys. Lett. B **307**, 1 (1993).
[47] D. K. Srivastava and J. I. Kapusta, Phys. Rev. C **48**, 1335 (1993).
[48] D. K. Srivastava, Phys. Rev. D **49**, 4523 (1994).
[49] A. Timmermann, M. Plümer, L. Razumov and R. M. Weiner, Phys. Rev. C **50**, 3060 (1994)
[50] J. Pisut, N. Pisutova and B. Tomasik, Z. Phys. C **71**, 139 (1996).
[51] J. E. Alam, B. Mohanty, P. Roy, S. Sarkar and B. Sinha, Phys. Rev. C **67**, 054902 (2003)
[52] S. A. Bass, B. Muller and D. K. Srivastava, Phys. Rev. Lett. **93**, 162301 (2004)
[53] M. M. Aggarwal *et al.* [WA98 Collaboration], Phys. Rev. Lett. **93** (2004) 022301
[54] J. e. Alam, B. Mohanty, A. Rahaman, S. Sarkar and B. Sinha, Phys. Rev. C **70**, 054901 (2004)
[55] Q. H. Zhang and S. S. Padula, Phys. Rev. C **69**, 024907 (2004)
[56] D. V. Anchishkin, A. M. Gavrilik and N. Z. Iorgov, Eur. Phys. J. A **7** (2000) 229
[57] D. V. Anchishkin, A. M. Gavrilik and N. Z. Iorgov, Mod. Phys. Lett. A **15** (2000) 1637
[58] F. Grassi, Y. Hama, S. S. Padula and O. J. Socolowski, Phys. Rev. C **62**, 044904 (2000)
[59] Y. M. Sinyukov, S. V. Akkelin and Y. Hama, Phys. Rev. Lett. **89**, 052301 (2002)
[60] O. . J. Socolowski, F. Grassi, Y. Hama and T. Kodama, Phys. Rev. Lett. **93**, 182301 (2004)

[61] S. Kniege *et al.* [NA49 Collaboration], J. Phys. G **30**, S1073 (2004)
[62] M. A. Lisa, U. W. Heinz and U. A. Wiedemann, Phys. Lett. B **489**, 287 (2000)
[63] T. Csörgő and S. Hegyi, Phys. Lett. B **489**, 15 (2000).
[64] P. Achard *et al.* [L3 Collaboration], Phys. Lett. B **547** (2002) 139
[65] J. Adams *et al.* [STAR Collaboration], arXiv:nucl-ex/0411036.
[66] T. Csörgő, S. Hegyi and W. A. Zajc, arXiv:nucl-th/0402035.
[67] T. Csörgő, S. Hegyi and W. A. Zajc, Eur. Phys. J. C **36**, 67 (2004)
[68] T. Csörgő, S. Hegyi, T. Novák and W. A. Zajc, Acta Phys. Polon. B **36**, 329 (2005)
[69] D. A. Brown and P. Danielewicz, Phys. Lett. B **398** (1997) 252
[70] S. Y. Panitkin *et al.* [E895 Collaboration], Phys. Rev. Lett. **87**, 112304 (2001)
[71] P. Chung *et al.*, PHENIX Collaboration, Nucl. Phys. A **749** (2005) 275.
[72] S. Pratt, Phys. Lett. B **301** (1993) 159.
[73] W. Q. Chao, C. S. Gao and Q. H. Zhang, J. Phys. G **21** (1995) 847
[74] T. Csörgő and J. Zimányi, Phys. Rev. Lett. **80** (1998) 916
[75] J. Zimányi and T. Csörgő, Heavy Ion Phys. **9** (1999) 241
[76] S. E. Vance, T. Csörgő and D. Kharzeev, Phys. Rev. Lett. **81**, 2205 (1998)
[77] J. G. Cramer, G. A. Miller, J. M. S. Wu and J. H. S. Yoon, Phys. Rev. Lett. **94**, 102302 (2005)
[78] M. Asakawa, T. Csörgő and M. Gyulassy, Phys. Rev. Lett. **83**, 4013 (1999)
[79] I. V. Andreev and R. M. Weiner, Phys. Lett. B **373** (1996) 159
[80] P. K. Panda, T. Csörgő, Y. Hama, G. Krein and S. S. Padula, Phys. Lett. B **512** (2001) 49
[81] T. Csörgő and L. P. Csernai, Phys. Lett. B **333**, 494 (1994)
[82] I. G. Bearden *et al.* [BRAHMS Collaboration], Phys. Rev. Lett. **88**, 202301 (2002)
[83] S. S. Adler *et al.* [PHENIX Collaboration], Phys. Rev. C **69**, 034909 (2004)
[84] S. Manly *et al.* [PHOBOS Coll.], Nucl. Phys. A**715**, 611 (2003)
[85] M. Csanád, T. Csörgő and B. Lörstad, Nucl. Phys. A **742**, 80 (2004)
[86] Z. Fodor and S. D. Katz, JHEP **0404** (2004) 050 [arXiv:hep-lat/0402006].
[87] G. Goldhaber et al., Phys. Rev. Letters **3** (1959) 181 .
[88] G. Goldhaber, S. Goldhaber, W. Y. Lee and A. Pais, Phys. Rev. **120** (1960) 300.
[89] R. J. Glauber, Phys. Rev. **130** (1963) 2529.
[90] R. J. Glauber, Phys. Rev. **131** (1963) 2766.
[91] G. I. Kopylov, Phys. Lett. B **50** (1974) 472.
[92] E. V. Shuryak, Phys. Lett. B **44** (1973) 387.
[93] E. V. Shuryak, Yad. Fiz. **18** (1973) 1302.
[94] G. Cocconi, Phys. Lett. B **49**, 459 (1974).
[95] P. Grassberger, Nucl. Phys. B **120**, 231 (1977).
[96] G. N. Fowler and R. M. Weiner, Phys. Rev. D **17** (1978) 3118 [Erratum-ibid. D **18** (1978) 4358].
[97] M. Gyulassy, S. K. Kauffmann and L. W. Wilson, Phys. Rev. C **20** (1979) 2267.
[98] F. B. Yano and S. E. Koonin, Phys. Lett. B **78** (1978) 556.
[99] M. I. Podgoretsky, Sov. J. Nucl. Phys. **37** (1983) 272 [Yad. Fiz. **37** (1983) 455].
[100] M. Gyulassy, Phys. Rev. Lett. **48**, 454 (1982).
[101] M. Deutschmann *et al.* Nucl. Phys. B **204**, 333 (1982).
[102] A. Vourdas and R. M. Weiner, Phys. Rev. D **38**, 2209 (1988).
[103] S. Barshay, Phys. Lett. B **130**, 220 (1983).
[104] T. Akesson *et al.* [Axial Field Spectrometer Collaboration], Phys. Lett. B **187** (1987) 420.
[105] B. Andersson and W. Hofmann, Phys. Lett. B **169** (1986) 364.
[106] X. Artru and M. G. Bowler, Z. Phys. C **37** (1988) 293.
[107] S. Pratt, Phys. Rev. Lett. **53** (1984) 1219.
[108] S. Pratt, Phys. Rev. D **33** (1986) 1314.
[109] Y. Hama and S. S. Padula, Phys. Rev. D **37** (1988) 3237.
[110] A. N. Makhlin and Y. M. Sinyukov, Z. Phys. C **39** (1988) 69.
[111] G. F. Bertsch, Nucl. Phys. A **498** (1989) 173C.
[112] G. Bertsch, M. Gong and M. Tohyama, Phys. Rev. C **37** (1988) 1896.
[113] T. J. Humanic *et al.* [NA35 Coll.], Z. Phys. C **38** (1988) 79.
[114] A. Bamberger *et al.* [NA35 Coll.], Phys. Lett. B **203** (1988) 320.
[115] S. S. Padula and M. Gyulassy, Nucl. Phys. A **498** (1989) 555C.
[116] M. Gyulassy and S. S. Padula, Phys. Lett. B **217** (1989) 181.
[117] M. Gyulassy and S. S. Padula, Phys. Rev. C **41** (1990) 21.
[118] J. Baechler *et al.* [NA35 Collaboration], Nucl. Phys. A **525** (1991) 327C.
[119] T. Alber [NA35 and NA49 Collaborations], Nucl. Phys. A **590** (1995) 453C.

[120] H. Beker *et al.*, Phys. Rev. Lett. **74** (1995) 3340.
[121] H. Beker *et al.*, Z. Phys. C **64** (1994) 209.
[122] I. G. Bearden *et al.* [NA44 Collaboration], Nucl. Phys. A **661** (1999) 435.
[123] I. G. Bearden *et al.* [NA44 Collaboration], Phys. Rev. Lett. **78**, 2080 (1997).
[124] T. Csörgő and B. Lörstad, Phys. Rev. C **54**, 1390 (1996) [arXiv:hep-ph/9509213].
[125] T. Csörgő and B. Lörstad, Nucl. Phys. A **590**, 465C (1995) [arXiv:hep-ph/9503494].
[126] D. H. Rischke and M. Gyulassy, Nucl. Phys. A **597**, 701 (1996) [arXiv:nucl-th/9509040].
[127] D. H. Rischke and M. Gyulassy, Nucl. Phys. A **608**, 479 (1996) [arXiv:nucl-th/9606039].
[128] S. Chapman, P. Scotto and U. W. Heinz, Phys. Rev. Lett. **74**, 4400 (1995)
[129] S. Chapman, P. Scotto and U. W. Heinz, Heavy Ion Phys. **1**, 1 (1995) [arXiv:hep-ph/9409349].
[130] I. G. Bearden *et al.*, Phys. Rev. C **58** (1998) 1656.
[131] M. Gyulassy, Lect. Notes Phys. **583** (2002) 37 [arXiv:nucl-th/0106072].
[132] B. Holzman *et al.* [PHOBOS Collaboration], J. Phys. G **30** (2004) S1049 [arXiv:nucl-ex/0406027].
[133] C. M. Ko, Z. W. Lin and S. Pal, Heavy Ion Phys. **17**, 219 (2003) [arXiv:nucl-th/0205056].
[134] T. J. Humanic, Nucl. Phys. A **715**, 641 (2003) [arXiv:nucl-th/0205053].
[135] T. Csörgő and A. Ster, Heavy Ion Phys. **17**, 295 (2003)
[136] M. Csanád, T. Csörgő, B. Lörstad and A. Ster, Acta Phys. Polon. B **35**, 191 (2004)
[137] M. Csanád, T. Csörgő, B. Lörstad and A. Ster, Nukleonika 2004; 49: S49-S55
[138] M. Csanád, T. Csörgő, B. Lörstad and A. Ster, J. Phys. G **30**, S1079 (2004)
[139] W. Broniowski, A. Baran and W. Florkowski, AIP Conf. Proc. **660**, 185 (2003)
[140] F. Retiere, J. Phys. G **30**, S335 (2004) [arXiv:nucl-ex/0307026].
[141] F. Retiere and M. A. Lisa, Phys. Rev. C **70** (2004) 044907 [arXiv:nucl-th/0312024].
[142] T. Renk, Phys. Rev. C **70**, 021903 (2004) [arXiv:hep-ph/0404140].
[143] T. Renk, Phys. Rev. C **69** (2004) 044902 [arXiv:hep-ph/0310346].
[144] T. Csörgő, Central European Journal of Physics 2(4) 2004 556-565 [arXiv:nucl-th/9809011].
[145] T. Csörgő *et al.*, Phys. Rev. C **67** (2003) 034904
[146] T. Csörgő and J. Zimányi, Heavy Ion Phys. **17** (2003) 281 [arXiv:nucl-th/0206051].
[147] M. A. Lisa *et al.* [E895 Collaboration], Phys. Lett. B **496** (2000) 1 [arXiv:nucl-ex/0007022].
[148] U. W. Heinz, A. Hummel, M. A. Lisa and U. A. Wiedemann, Phys. Rev. C **66** (2002) 044903
[149] M. A. Lisa [the STAR Collaboration], Acta Phys. Polon. B **35** (2004) 37 [arXiv:nucl-ex/0312012].
[150] M. Csanád, private communication. M. Csanád et al, manuscript in preparation.
[151] M. Csanád for the PHENIX Collaboration, arXiv:nucl-ex/0505001.

Institute of Physics Publishing
doi:10.1088/1742-6596/50/1/032

Journal of Physics: Conference Series **50** (2006) 271–278
5th International Conference on Physics and Astrophysics of Quark Gluon Plasma

Correlations in STAR: interferometry and event structure

Mikhail Kopytine for the STAR Collaboration

Department of Physics, Kent State University, Kent, Ohio, USA

E-mail: kopytin@bnl.gov

Abstract. STAR observes a complex picture of RHIC collisions where correlation effects of different origins – initial state geometry, semi-hard scattering, hadronization, as well as final state interactions such as quantum intensity interference – coexist. Presenting the measurements of flow, mini-jet deformation, modified hadronization, and the Hanbury Brown and Twiss effect, we trace the history of the system from the initial to the final state. The resulting picture is discussed in the context of identifying the relevant degrees of freedom and the likely equilibration mechanism.

1. Introduction

This talk is an overview of recent STAR results in the field known as "correlations and fluctuations". Out of multiple reasons to analyze correlations and fluctuations at RHIC known before the data became available, there is one that seems to dominate now: by measuring correlations and their evolution we learn about *equilibration* of the system. Is it taking place? What is the mechanism? And what is equilibrating? Novel and advanced data analysis techniques have been created lately to address these issues. I will focus on the developments since Quark Matter 2004.

Our evidence for equilibration includes elliptic flow, medium modification of mini-jets and of charge-dependent correlations. Related issues are the azimuthally-sensitive HBT and the blast wave picture.

2. Elliptic and directed flow as a particle number correlation

In non-central collisions the initial spatial configuration is anisotropic. But the configuration space anisotropy does not produce momentum space anisotropy unless there is another mechanism involved, such as local equilibration via rescattering. If the latter is occuring, then it is sensible to talk about pressure gradients, and the azimuthal variation thereof. This picture naturally calls for analysis of correlations in velocity between local elements of expanding matter, but experimentally one is limited to final state particles. The flow analysis in this field began with diagonalization of the (*momenta*-based) sphericity tensor [1]. In experiments with limited acceptance in rapidity at higher energies (AGS, SPS and RHIC) a different technique [2] became popular which utilizes the observed azimuthal anisotropy by expanding the particle *number*

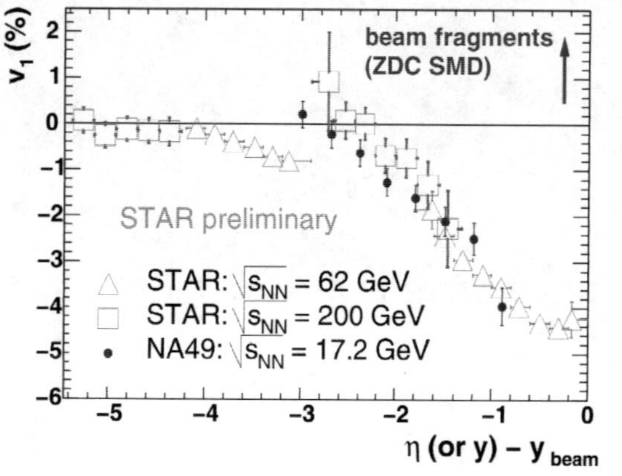

Figure 1. Directed flow coefficient as a function of pseudorapidity relative to the projectile beam in STAR and NA49

distribution with respect to the reaction plane angle Ψ_r into a Fourier series

$$E\frac{d^3N}{d^3p} = \frac{1}{2\pi}\frac{d^2N}{p_t\,dp_t\,dy}\{1 + \sum_{m=1}^{\infty} 2v_m \cos[m(\phi - \Psi_r)]\}, \tag{1}$$

with v_2 typically being the largest component. Cumulant-based analyses, developed later [3], distinguish between two-particle and higher order correlations and separate flow (a multiparticle correlation) from the lower-order "non-flow".

STAR's recent measurements of directed flow (v_1), shown in Fig.1, use 3-particle cumulants to minimize the "non-flow" effects. In the limiting fragmentation picture [4], what matters for the fragmenting system is the relative rapidity with the nearby spectators, hence the choice of the abscissa. The data from different energies are consistent with limiting fragmentation. The v_1 term requires knowledge of Ψ_r between 0 and 2π and thus an azimuthal direction rather than orientation of the reaction plane. This direction, given by the spectators, is reconstructed event-by-event by means of a position-sensitive zero-degree shower-maximum detector (ZDC SMD) which selectively detects fragmentation neutrons behind a RHIC dipole magnet. These measurements indicate that the directed flow at RHIC is opposite in azimuthal angle to the direction of the spectators (an "anti-flow"). We will return to the velocity aspect of the elliptic flow in Sections 3 and 6.

3. Final state geometry

While v_m characterize the particle distribution in momentum space, observation of the corresponding structure in the configuration space rests with the Hanbury Brown-Twiss (HBT) interferometry. Herein a correlation function is defined as the ratio of two-particle probability density to the uncorrelated reference, which in STAR is obtained by forming mixed pairs. In the Cartesian parameterization,

$$C_{\rm fit}(\vec{q}) = 1 + \lambda \exp(-q_o^2 R_o^2 - q_s^2 R_s^2 - q_l^2 R_l^2), \tag{2}$$

the momenta of a pair are projected onto the transverse plane where one identifies directions of their sum and difference. Projections of the momentum difference on those directions become, respectively, outward q_o and sideward q_s components. The longitudinal component q_l is orthogonal to both of them. Corresponding to q are the radii parameters R. Fig.2 shows a $\sqrt{s_{NN}}$ dependence of the quantities representing the best fits of the parameters in Eq.2 to the experimental data. The bottom panel presents density of π^+ multiplicity per unit of rapidity, which displays a steady logarithmic increase with $\sqrt{s_{NN}}$. The latter may be contrasted with

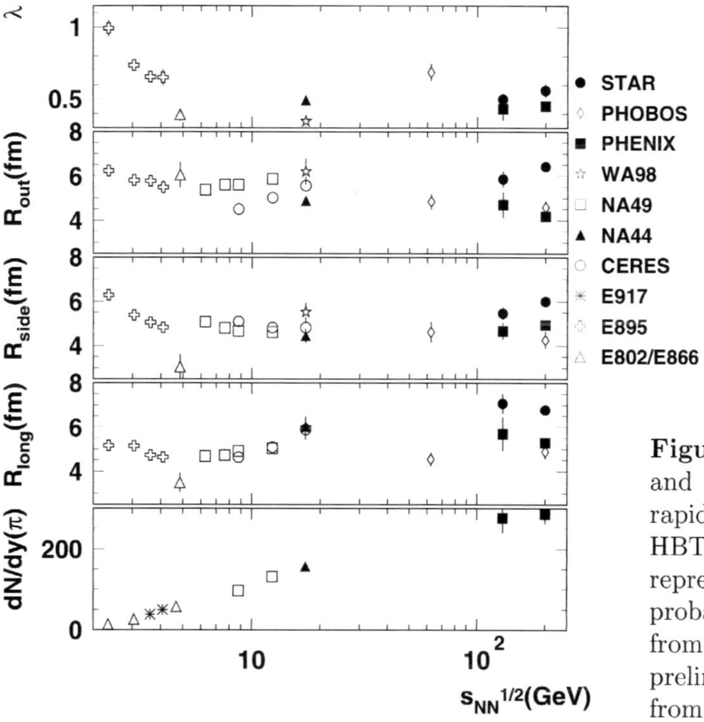

Figure 2. Fit parameters for $\pi^-\pi^-$ and dN/dy of π^+ at or near mid-rapidity as a function of $\sqrt{s_{NN}}$. The HBT data are either p_t integrals or represent a p_t bin closest to the most probable inclusive p_t and are taken from [7]. The NA49 HBT data are preliminary. The dN/dy data are from [8].

the general lack of energy dependence of the radius parameters which underscores the non-triviality of interpreting the HBT results. In the hydrodynamical framework the concept of *length of homogeneity*[1] has been introduced [5], and it is known that the measured radii reflect those lengths[2] (or rather, according to [6], a combined effect of the correlation lengths in the *primordial correlator* between the *currents* emitting the pion radiation and the source size). These lengths are smaller than the geometrical source sizes in an expanding (due to pressure gradients) and thus inhomogeneous system. A more subtle issue is that of the little relative difference between R_o and R_s, which is believed to increase with longer emission.

The quantitative understanding of this information remains a model-dependent problem. In particular, one may assume a local, thermally-equilibrated system and thus ascribe all the sources of inhomogeneity (dynamical correlations) to hydrodynamic expansion. Taking this approach and fitting the data (including single-particle spectra) with the modified Blast Wave model, Retiere and Lisa obtained a set of parameters [10]. Remarkably, the azimuthally-varying component ρ_2 in their description of the transverse velocity field appears to be non-zero and well constrained by the fit. This represents an indirect measurement of the transverse-velocity aspect of elliptic flow which complements the particle-number-related v_2.

The azimuthally-dependent radii recently published by STAR [9] show an effective source elongated perpendicularly to the reaction plane. This indicates shorter homogeneity lengths (and thus larger gradients) in-plane than out-of-plane, which is consistent with the geometry of the early stage of the collision. The early emission is consistent with short emission time.

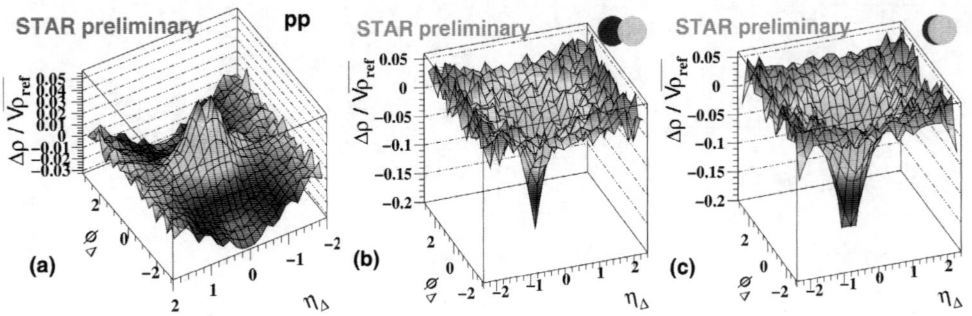

Figure 3. Comparison of CD correlations in η and ϕ: (a) pp collisions at $\sqrt{s} = 200$ GeV, (b) peripheral AuAu collisions, (c) central AuAu collisions. (b) and (c) are at $\sqrt{s_{NN}} = 130$ GeV.

4. Charge-dependent number correlations– modified hadronization in the medium

The charge-dependent (CD) number correlation is defined as a difference between the like-sign and the unlike-sign correlations. (In this and subsequent sections we discuss results obtained without particle identification). The correlations are defined by comparing same-event pairs with mixed-event pairs, obtained from mixing similar events. To characterize and normalize correlations, we use

$$\frac{\Delta\rho}{\sqrt{\rho_{\text{ref}}}} = \frac{\rho_{a,b} - \rho_a\rho_b}{\sqrt{\rho_a\rho_b}}, \tag{3}$$

where a and b index particles (or in the discretized case, bins in the kinematic space occupied by particles), $\rho_{a,b}$ is pair density distribution, while ρ_a and ρ_b are the corresponding single-particle distributions, whose product forms the reference density ρ_{ref}. Experimentally one deals with fluctuating histogram bin contents n_a and n_b and Eq.3 becomes (see Appendix A)

$$\frac{\Delta\rho}{\sqrt{\rho_{\text{ref}}}} \to \frac{\overline{n_a n_b} - \overline{n_a} \times \overline{n_b}}{\sqrt{\overline{n_a} \times \overline{n_b}}} = \frac{\text{Cov}[n_a, n_b]}{\sqrt{\overline{n_a} \times \overline{n_b}}} \tag{4}$$

If the variance in the bin population n is dominated by the Poissonian process, $\text{Var}[n] = \bar{n}$, and if in presence of correlations $\text{Cov}[n_a, n_b] \propto \text{Var}[n]$ (number of correlation sources is proportional to multiplicity) this normalization eliminates the trivial multiplicity dependence of the correlation amplitude. Eq.3 represents a *per-particle* measure of correlations. [3]

In paper [15], CD correlations in η and ϕ have been analyzed simultaneously by using *joint autocorrelation*. Fig.3 shows a comparison of CD correlations in pp collisions (panel (a)) with peripheral and central AuAu collisions. Kinematic cuts on the track separation are applied to eliminate the contribution of HBT and Coulomb effects in AuAu collisions (panels (b) and (c)) [15]. A much broader HBT correlation for pp collisions contributes the gaussian peak at the origin in (a). Transverse momentum conservation also contributes to the structure at the origin (the ridge beneath the gaussian peak). The main feature, the negative gaussian peak on η_Δ, is attributed to local charge conservation during hadronization. If the process happens on a longitudinally expanding string (as e.g. in Lund model), the strict alternation of positive and negative hadrons in rapidity holds (modulo interspersed neutrals). This implies that η_Δ for an unlike-sign pair tends to be shorter than for the like sign, regardless of ϕ_Δ, creating a groove in the CD plot, clearly seen in Fig.3(a). The ϕ_Δ-independent groove appears to diminish in

[1] The term "effective length" was used in the original article.

[2] This result was derived for locally thermally equilibrated systems.

[3] In this terminology, the correlation function commonly used in HBT represents a *per-particle-pair*-normalized quantity. The choice of normalization depends on the correlation mechanism under study.

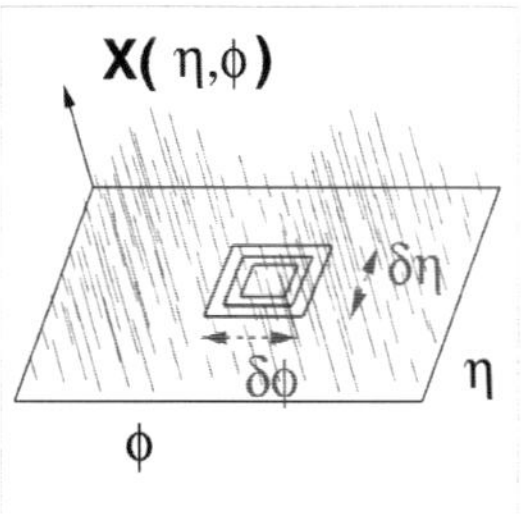

Figure 4. A fantasy event. Positions of spikes indicate η, ϕ kinematics of individual particles, their heights – magnitude of quantity X assigned to a particle, which could be p_t or a derived quantity. This event has azimuthal anisotropy, however its v_2 may be 0, illustrating the difference between number and p_t flow. The analysis proceeds by selecting a bin of size $\delta\eta, \delta\phi$, integrating the *random field* X within the bin, finding the variance of that integral in an event sample, then feeding this information into an integral equation to yield the correlation of X, as discussed in Appendix B.

amplitude from (a) and (b) to (c). In addition, the negative peak at 0 relative angle changes its shape, becoming narrower from peripheral to central AuAu collisions and acquiring more η-ϕ symmetry. We hypothesize that this reflects a change in the hadronization geometry from one dominated by independent string break-ups in peripheral collisions to a bulk hadronization process in which individual strings are no longer relevant.

5. Charge-independent number correlations – longitudinal minijet deformation

The charge-independent (CI) number correlation sums the like-sign and the unlike-sign correlations. In AuAu collisions, these correlations have been analyzed by STAR in the space of pseudorapidity η and azimuthal angle ϕ [11, 13] and in the transverse momentum space [12].

Our discrete-wavelet technique [11] extracts information about correlation structure by measuring the power spectra of local fluctuations in the density of charged hadrons with respect to a mixed-event reference (the so-called "dynamic texture" of the event). As discussed in [14], this observable is a measure of the gradient of the two-particle correlation function, but takes less computing. [4]

The experiment [11] reveals a reduction in the dynamic texture (and thus in the correlation function gradient) along the η-direction at $p_t > 0.6$ GeV/c in the central AuAu collisions, compared to an expectation based on peripheral events. HIJING simulations [11] point to minijets as the likely source of correlations of this scale and magnitude. The possible mechanism of the reduction is a coupling between the minijet fragments and the longitudinally expanding bulk medium. In paper [13], the same correlation structure is investigated in the joint η-ϕ autocorrelation technique. It is found that the correlations are broadened in η_Δ with centrality. A reduction in the correlation gradient ("dynamic texture") and the elongation of the correlation function along η_Δ are consistent descriptions of the effect.

6. p_t correlations –minijets and elliptic flow

We require a direct measurement of correlations in the velocity field, independent of particle number correlations. To form an independent measurement, these have to be separated from particle number correlations. For a random field (such as particle number or velocity distributed in η and ϕ, see Fig.4) variance of the integrated content of a bin can be related via an integral equation to the correlation function of the field [16, 17]. The correlation function can be reconstructed from the measured variance by solving the integral equation derived in [16, 17]. This process offers a computational advantage since the bin integration involves $\mathcal{O}(N)$ computations, whereas two-particle correlations require $\mathcal{O}(N^2)$ (N is the event multiplicity used). By varying the bin size one varies the range of the difference variable in the correlation function.

[4] The number of computations required is $\mathcal{O}(N)$, in contrast with $\mathcal{O}(N^2)$ required for direct construction of the correlation function.

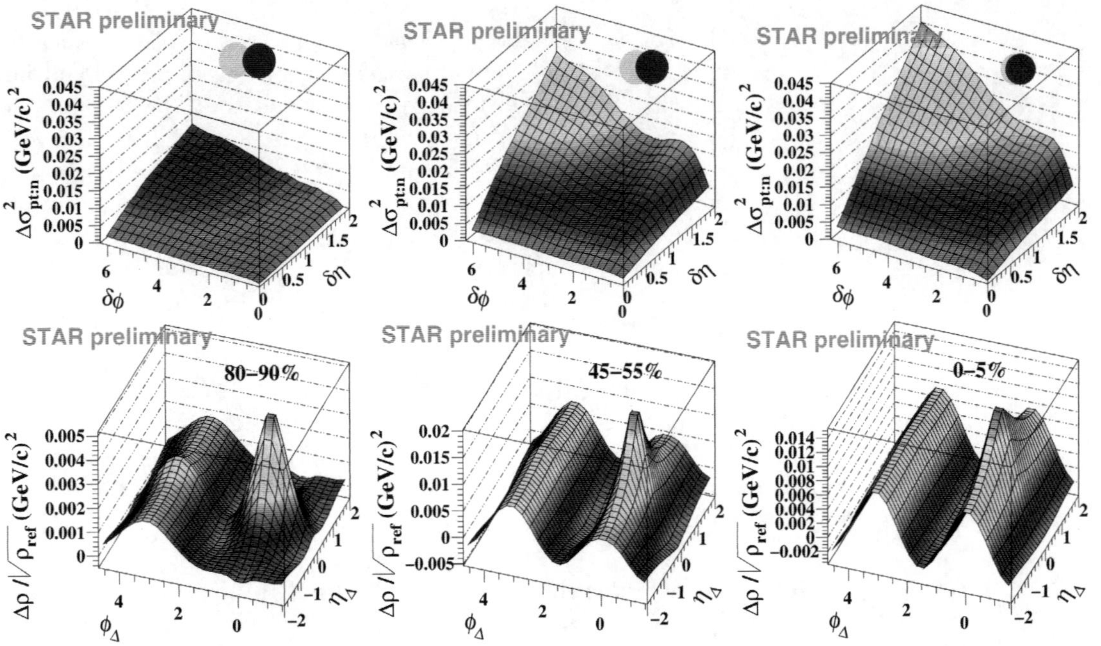

Figure 5. Top: scale dependence of the reference-subtracted variance $\Delta\sigma^2_{p_t:n}$. Bottom: reference-subtracted normalized correlation functions reconstructed from $\Delta\sigma^2_{p_t:n}$ by inversion. The percentage intervals indicate centrality for each column-wise pair of plots.

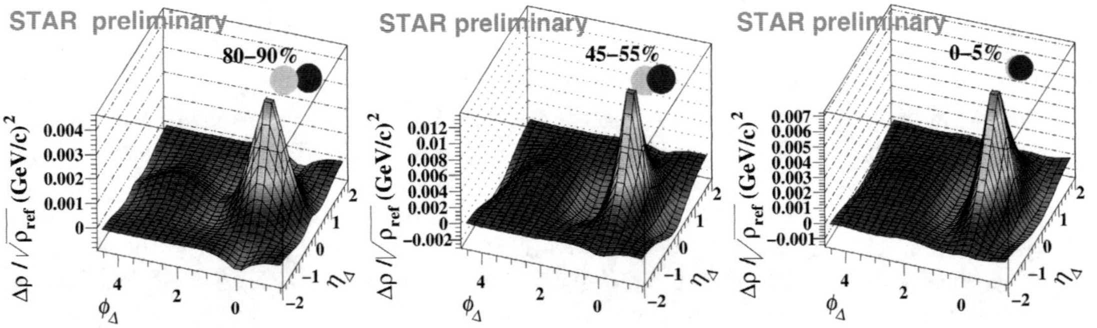

Figure 6. Same data as in Fig.5 with subtracted azimuthal harmonics v_1 and v_2.

In order to extract the p_t correlation alone, we need to disentangle the effect of n from that of $p_{t,i}$ in the variance of $p_t \equiv \sum_{i\in(\eta,\phi)\mathrm{bin}} p_{t,i}$, and from statistical fluctuations. We denote the desired variance as $\Delta\sigma^2_{p_t:n}$. As an estimator of $\Delta\sigma^2_{p_t:n}$, we use $\mathrm{Var}[p_t - n\hat{p}_t]/\bar{n} - \sigma^2_{\hat{p}_t}$, where the latter term is the inclusive per-hadron p_t variance. In the case when p_t in the bin, event-by-event, derives from the same parent distribution (not the case in Fig.4), $\Delta\sigma^2_{p_t:n}$ converges to 0, despite fluctuations (statistical or not) in n.

Top panels in Fig.5 represent scale dependence of the variance $\Delta\sigma^2_{p_t:n}$ for three centrality classes. The bottom panels, representing per-particle normalized correlations plotted on difference variables, are obtained from the respective top panels by solving the integral equation derived in [16, 17] (see Appendix B) which relates variance and correlation.

While the quadrupole wave in Fig.5(bottom row) is the reflection of the elliptic flow in the $p_t : n$ correlation function, the peak at 0 angle differences likely reflects minijets [19]. But unlike flow, the effect of minijets on the HBT radii at RHIC energy is not part of the standard (local equilibrium) picture. Therefore this effect may constitute an under-appreciated factor driving the effective radii (Fig.2) down at higher $\sqrt{s_{NN}}$.

The amplitude of the minijet peak has a non-monotonic behavior with increasing centrality, rising in peripheral and falling in central collisions. The peak is seen to be elongated longitudinally in central collisions, showing qualitatively the same behavior as seen with number correlations by using wavelet analysis and joint autocorrelation (Section5).

7. Conclusions

The small effective HBT radii at RHIC indicate that the system is highly inhomogeneous and decouples rapidly. STAR has evidence that semi-hard scattering leaves a trace in the "soft" p_t domain which is usually considered a safe realm for local equilibrium approximations, thus possibly explaining some of the inhomogeneity. Whether or not the minijet correlations present a challenge to the picture of local equilibrium which constitutes a basis for much of the HBT interpretation infrastructure, will depend on the effective final-state-particle number involved in a minijet, or in other words, on the *order* of those correlations (higher order being more hydro-like and thus probably less of a challenge). Therefore a study of higher order correlations and cumulants may be a promising future direction. The phenomenon of η-broadening of minijets at "soft" p_t (Section 5 and 6) can clarify the nature of the medium created in central AuAu collisions, provided that the coupling mechanism, apparently sensitive to that nature, is under theoretical control.

Appendix A. Notation

In our notation i is particle index, n is number of particles within a kinematic cut (bin), $\overline{(\ldots)}$ denotes an average over events, $\hat{p_t}$ is an inclusive mean p_t per particle, x_Δ (variants: η_Δ, ϕ_Δ) is difference variable $= x_i - x_{i'}$, δx is scale (range of local integration, see Fig.4), Δx is the upper limit on δx. Covariance of x and y is denoted as Cov$[x, y]$, variance of x – as Var[x].

Appendix B. From variance to correlation

The connection between variance and correlation function – the basis of analysis in Section6 – is discussed in detail using integral calculus in [16] and the algebra of discrete bins in [17]. For a random field $X(t)$, the *autocorrelation* function is defined [18] as

$$\rho(X, t_\Delta) \equiv \overline{X(t)X(t + t_\Delta)} \tag{B.1}$$

(assuming that correlations do not depend on position t).

The variance of X, integrated over bin of width $\delta\eta, \delta\phi$, centered at 0 (see Fig.4) is

$$\text{Var}[X; \delta\eta, \delta\phi] = \overline{\left(\int_{-\delta\eta/2}^{\delta\eta/2}\int_{-\delta\phi/2}^{\delta\phi/2} X \, d\eta \, d\phi\right)^2} - \left(\int_{-\delta\eta/2}^{\delta\eta/2}\int_{-\delta\phi/2}^{\delta\phi/2} \overline{X} \, d\eta \, d\phi\right)^2 = \tag{B.2}$$

$$\int_{-\delta\eta/2}^{\delta\eta/2} d\eta_1 \int_{-\delta\phi/2}^{\delta\phi/2} d\phi_1 \int_{-\delta\eta/2}^{\delta\eta/2} d\eta_2 \int_{-\delta\phi/2}^{\delta\phi/2} d\phi_2 [\overline{X(\eta_1, \phi_1)X(\eta_2, \phi_2)} - \overline{X(\eta_1, \phi_1)} \times \overline{X(\eta_2, \phi_2)}] \tag{B.3}$$

As we form the reference-subtracted variance $\Delta\sigma_X^2 \equiv \text{Var}[X; \delta\eta, \delta\phi] - \text{Var}[X_{\text{ref}}; \delta\eta, \delta\phi]$, the first term in the brackets above becomes the reference-subtracted correlation $\Delta\rho(X, \eta_\Delta, \phi_\Delta)$ and the second term cancels with the reference since $\overline{X} = \overline{X_{\text{ref}}}$.

$$\Delta\sigma_X^2 = \int_{-\delta\eta/2}^{\delta\eta/2} d\eta_1 \int_{-\delta\phi/2}^{\delta\phi/2} d\phi_1 \int_{-\delta\eta/2}^{\delta\eta/2} d\eta_2 \int_{-\delta\phi/2}^{\delta\phi/2} d\phi_2 \Delta\rho(X, \eta_1 - \eta_2, \phi_1 - \phi_2) \tag{B.4}$$

$$= 2\int_0^{\delta\eta} d\eta_\Delta 2\int_0^{\delta\phi} d\phi_\Delta (\delta\eta - \eta_\Delta)(\delta\phi - \phi_\Delta)\Delta\rho(X, \eta_\Delta, \phi_\Delta) \tag{B.5}$$

Eq.B.5 is obtained by making a change of the integration variables from η_1,η_2 to η_1 and $\eta_\Delta = \eta_1 - \eta_2$, integrating over η_1, and repeating the same for ϕ (see [16] for details). In the actual analysis, this integral equation is discretized by partitioning the acceptance into $m_\delta \times n_\delta$ *microbins* of size $\varepsilon_\eta \times \varepsilon_\phi$. X is replaced by $(p_t - n\hat{p_t})/\sqrt{\bar{n}}$. This denominator creates the reference number correlation $\rho_{\rm ref}(n)$ in the final expression B.6 which makes it a per-particle measure:

$$\Delta\sigma^2_{p_t:n}(m_\delta\varepsilon_\eta, n_\delta\varepsilon_\phi) = 4\sum_{k,l=1}^{m_\delta,n_\delta} \varepsilon_\eta\varepsilon_\phi K_{m_\delta n_\delta:kl}\frac{\Delta\rho(p_t : n; k\varepsilon_\eta, l\varepsilon_\phi)}{\sqrt{\rho_{\rm ref}(n; k\varepsilon_\eta, l\varepsilon_\phi)}} \tag{B.6}$$

Here the kernel K is a discrete representation of the continuous case :

$$(\delta\eta - \eta_\Delta)(\delta\phi - \phi_\Delta) \rightarrow \varepsilon_\eta\varepsilon_\phi K_{m_\delta n_\delta:kl} \equiv \varepsilon_\eta\varepsilon_\phi(m_\delta - k + \frac{1}{2})(n_\delta - l + \frac{1}{2}) \tag{B.7}$$

Knowing the computationally cheaper quantity $\Delta\sigma^2_{p_t:n}$, we solve the integral equation B.6 for $\Delta\rho/\sqrt{\rho_{\rm ref}}$ using standard numerical techniques [17].

References

[1] Gutbrod H H, Poskanzer A M and Ritter H G 1989 *Rept. Prog. Phys.* **52** 1267
[2] Voloshin S A and Zhang Y 1996 *Z. Phys.* C **70** 665
Poskanzer A M and Voloshin S A 1998 *Phys. Rev.* C **58** 1671
[3] STAR Collaboration 2002 *Phys. Rev.* C **66** 034904
[4] Benecke J, Chou T T, Yang C N and Yen E 1969 *Phys. Rev.* **188** 2159
[5] Makhlin A N and Sinyukov Yu M 1988 *Z. Phys.* C **39**
[6] Weiner R 2000 *Bose-Einstein Correlations and Subatomic Interferometry* (New York:Wiley)
[7] PHOBOS Collaboration 2004 *Preprint* nucl-ex/0409001,
STAR Collaboration 2004 *Preprint* nucl-ex/0411036,
PHENIX Collaboration 2002 *Phys. Rev. Lett.* **88** 192302
PHENIX Collaboration 2004 *Phys. Rev. Lette.* **93** 152302
E895 Collaboration 2000 *Phys. Rev. Lett.* **84** 2798,
NA44 Collaboration 1998 *Phys. Rev.* C **58** 1656,
Kniege S for the NA49 Collaboration 2004 *J. Phys. G* **30** S1073,
E802 Collaboration 2002 *Phys. Rev.* C **66** 054906,
CERES Collaboration 2002 *Nucl. Phys.* A **714** 124,
WA98 Collaboration 2000 *Eur. Phys. J.* C **16** 445,
[8] PHENIX Collaboration 2004 *Phys. Rev.* C **69** 024904,
PHENIX Collaboration 2004 *Phys. Rev.* C **69** 034909,
E866 Collaboration and E917 Collaboration 1999 *Phys. Lett.* B **476** 1,
NA44 Collaboration 1999 *Phys. Lett.* B **471** 6,
NA49 Collaboration 2002 *Phys. Rev.* C **66** 054902.
[9] STAR Collaboration 2004 *Phys. Rev. Lett.* **93** 012301
[10] Retiere F and Lisa M A 2004 *Phys. Rev.* C **70** 044907
[11] STAR Collaboration 2005 *Phys. Rev.* C **71** 031901(R)
[12] STAR Collaboration 2004 *Preprint* nucl-ex/0408012
[13] STAR Collaboration 2004 *Preprint* nucl-ex/0411003
[14] Kopytine M for the STAR Collaboration 2004 *Preprint* nucl-ex/0403011
[15] STAR Collaboration 2004 *Preprint* nucl-ex/0406035
[16] van Marcke E 1988 *Random Fields: Analysis and Synthesis* (Cambridrge:The MIT Press)
[17] Trainor T A, Porter R J and Prindle D J 2004 *Preprint* hep-ph/0410182
[18] Flandrin P 1999 *Time-Frequency/Time-Scale Analysis* (San Diego:Academic Press) p.37
[19] Liu Q, Trainor T A, and Prindle D J 2004 *Preprint* hep-ph/0410180

Institute of Physics Publishing
doi:10.1088/1742-6596/50/1/033

Journal of Physics: Conference Series **50** (2006) 279–288
5th International Conference on Physics and Astrophysics of Quark Gluon Plasma

Review of Parton Recombination Models

Steffen A. Bass[1,2]

[1] Department of Physics, Duke University, Durham, North Carolina 27708-0305, USA
[2] RIKEN BNL Research Center, Brookhaven National Laboratory, Upton, New York 11973, USA

E-mail: bass@phy.duke.edu

Abstract. Parton recombination models have been very successful in explaining data taken at RHIC on hadron spectra and emission patterns in Au+Au collisions at transverse momenta above 2 GeV/c, which have exhibited features which could not be understood in the framework of basic perturbative QCD. In this article I will review the current status on recombination models and outline which future challenges need to be addressed by this class of models.

1. Introduction

Collisions of heavy nuclei at relativistic energies are expected to lead to the formation of a deconfined phase of strongly interacting nuclear matter, often referred to as a Quark-Gluon-Plasma (QGP). Recent data from the Relativistic Heavy-Ion Collider (RHIC) at Brookhaven Lab have provided strong evidence for the existence of a transient QGP – among the most exciting findings are strong (hydrodynamic) collective flow [1, 2, 3, 4, 5, 6], the suppression of high-p_T particles [7, 8, 9, 10] and evidence for parton recombination as hadronization mechanism at intermediate transverse momenta [11, 12, 13, 14, 15].

The development of parton recombination as a key hadronization mechanism for hadrons with transverse momenta up to a couple of GeV/c was triggered by a series of experimental observations which could not be understood in a straightforward manner either in the framework of perturbative QCD or via regular soft physics, i.e. hydrodynamic scaling.

These observations, which in a broader context have often been referred to as the *baryon puzzle at RHIC* are in particular:

- **baryon/meson particle ratios**: It has been found that the ratio of protons over positively charged pions is equal or above one for $p_T > 1.5$ GeV/c and remains approximately constant up to 4 GeV/c [16, 17, 18, 19]. The rise of this ratio as a function of p_T can be well understood for transverse momenta between zero and roughly 1.5 GeV/c as a result of a hydrodynamic expansion with a strong radial flow component (which affects baryons more strongly than light mesons). However, for $p_T > 1.5$ GeV/c one would expect hadron productions to be described by perturbative QCD and the fragmentation of hard partons – here the proton/pion ratio would be given by the ratio of the respective fragmentation functions and be on the order of 0.1 – an order of magnitude below the observed value.
- **flavor dependence of the nuclear modification factor**: The strong nuclear suppression of the pion yield at transverse momenta larger than 2 GeV/c in central Au + Au collisions, compared to $p + p$ interactions is widely seen as the experimental confirmation of jet

quenching, the phenomenon that high energy partons lose energy when they travel through the hot medium created in a heavy ion collision [20, 21, 22], entailing a suppression of intermediate and high P_T hadrons. However, the amount of suppression seems to depend on the hadron species [24, 23]. In fact, in the production of protons and antiprotons between 2 and 4 GeV/c the suppression seems to be completely absent. Generally, mesons appear to suffer from a strong energy loss while baryons and antibaryons do not. This behavior is surprising, since the fragmentation of a hard parton into a hadron should first occur after it has traversed the deconfined medium – jet-quenching should therefore be flavor-blind and not exhibit a meson/baryon asymmetry.

- **constituent quark number scaling of the elliptic flow**: Measurement of the elliptic flow v_2 as a function of p_T shows a surprising species dependence above $p_T \approx 1$ GeV/c. For transverse momenta below 1 GeV/c the elliptic flow coefficient v_2 shows a monotonous increase as a function of p_T, which is well described in the framework of hydrodynamics and exhibits the expected mass-scaling. Above 1 GeV/c, however, the elliptic flow saturates in a species dependent fashion which cannot be described by hydrodynamics. A rescaling of the v_2 vs. p_T curve by the number of valence quarks contained in the respective hadron on both axes (i.e. plotting v_2/n vs. p_T/n with n the number of valence quarks) yields a *universal* curve for all hadron species [25, 14], strongly indicative of parton recombination.
- **two component shape of single particle spectra**: Transverse momentum spectra of identified hadrons exhibit a unique two-component shape as a function of p_T: exponential (i.e. thermal) for low and intermediate transverse momenta (up to 2-4 GeV/c, depending on the hadron species) and a power-law shape for higher transverse momenta. The power-law shape is indicative of pQCD driven particle production, the exponential hints at a thermalized system, possibly with hydrodynamic features. While the low and high p_T limits of the spectral shape do not come as surprise, the details of the transition from one regime to the other can neither be directly addressed in a thermal model nor by a pQCD mini-jet calculation.

While attempts have been made to describe the above observations with a variety of different physics concepts, e.g. gluon junctions in the case of the high p_T baryon excess [26], or a hydrodynamic plus pQCD jet-production approach for the spectral shape [27], only models based on parton recombination at intermediate p_T combined with hadron production via fragmentation at high p_T have been able to describe all of the above phenomena within one consistent framework.

2. Formalism of the Recombination plus Fragmentation approach

Inclusive hadron production at sufficiently large momentum transfer can be described by perturbative quantum chromodynamics (pQCD). The invariant cross section for a hadron h with momentum P can be given in factorized form [28]

$$E\frac{d\sigma_h}{d^3P} = \sum_a \int_0^1 \frac{dz}{z^2} D_{a\to h}(z) E_a \frac{d\sigma_a}{d^3P_a}. \tag{1}$$

The sum runs over all parton species a and σ_a is the cross section for the production of parton a with momentum $P_a = P/z$. Thus the parton production cross section has to be convoluted with the probability that parton a fragments into hadron h. The probabilities $D_{a\to h}(z)$ are called fragmentation functions [29]. Like parton distributions they are non-perturbative quantities. However, they are universal and once measured, e.g. in e^+e^- annihilations, they can be used to describe hadron production in other hard QCD processes.

Using (1) we can estimate the ratio of protons and pions. Taking, e.g., the common parametrization of Kniehl, Kramer and Pötter (KKP) [30], the ratio $D_{a\to p}/D_{a\to\pi^0}$ is always smaller than 0.2 for each parton a. This reflects the well known experimental fact that pions are much more abundant than protons in the domain where pQCD is applicable. The excess of pions over protons even holds down to very low P_T, smaller than 1 GeV, where perturbative calculations are no longer reliable. In that domain one can argue that the difference in mass, $M_p \gg M_\pi$, lays a huge penalty on proton production. The small value of the p/π^0 ratio predicted by these calculations over the entire range of P_T is the reason why the ratio $p/\pi^0 \sim 1$ measured at RHIC is so surprising.

It has been suggested that the fragmentation functions $D_{a\to h}(z)$ can be altered by the environment [31, 32]. The energy loss of the propagating parton in the surrounding medium leads, in first approximation, to a rescaling of the variable z. This would affect all produced hadrons in the same way, and thus cannot explain the observations at RHIC. In a picture with perturbative hadron production and jet quenching alone, the different behavior of hadrons can not be described by one consistent set of energy loss parameters. To save the validity of the purely perturbative approach, species dependent non-perturbative contributions to the fragmentation functions have to be introduced *ad hoc* to explain the data [33].

Perturbative hadron production consists of three steps: production of a parton in a hard scattering, propagation and interaction with a medium,and finally hadronization of the parton. Only modifications in our understanding of hadronization are able to provide an explanation of the experimental observations, since the other steps are blind to the hadron species that will eventually be created.

Recombination can be interpreted as the most "exclusive" form of hadronization, the endpoint of a hypothetical resummation of fragmentation processes to arbitrary twist. Comparing a simple model for recombination with single parton fragmentation, one finds that these two mechanisms of hadron production compete differently depending on the phase space density of partons. The competition between fragmentation and recombination is dominated by the slope and the absolute value of the phase space distribution of partons. Recombination always wins over fragmentation for an exponentially falling parton spectrum, but fragmentation takes over if the spectrum has the form of a power law, e.g. as provided by pQCD. This insight can be applied to hadron production in relativistic heavy ion collisions at midrapidity and transverse momenta of a few GeV/c where a densely populated phase space is expected. For the recombination of three quarks into a baryon the momenta of three partons have to be added up, but only two momenta in the case of a meson. Assuming an exponential parton spectrum this implies for a baryon a distribution $\sim [\exp(-P_T/3)]^3$ and for mesons $\sim [\exp(-P_T/2)]^2$, predicting e.g. a constant p/π^+ ratio where the value is determined by simple counting of quantum numbers [11].

The idea of quark recombination was proposed long ago to describe hadron production in the forward region of $p+p$ collisions [34]. This was later justified by the discovery of the leading particle effect, the phenomenon that, in the forward direction of a beam of hadrons colliding with a target, the production of hadrons sharing valence quarks with the beam hadrons are favored. E.g. in the Fermilab E791 fixed target experiment with a 500 GeV π^- beam [35] the asymmetry

$$\alpha(x_F) = \frac{d\sigma_{D^-}/dx_F - d\sigma_{D^+}/dx_F}{d\sigma_{D^-}/dx_F + d\sigma_{D^+}/dx_F} \tag{2}$$

between D^- and D^+ mesons grows nearly to unity when the Feynman variable x_F, measuring the longitudinal momentum relative to the beam momentum, approaches 1. Fragmentation would predict this asymmetry to be very close to zero. However, recombination of the $\bar{c}$ quark from a $c\bar{c}$ pair produced in a hard interaction with a d valence quark from the π^-, propagating in forward direction with large momentum, is highly favored compared to the recombination of the c with a $\bar{d}$ which is only a sea quark of the π^-. This leads to the enhancement of D^- over

D^+ mesons in the forward region.

The leading particle effect is a clear signature for the existence of recombination as a hadronization mechanism and has been addressed in several publications recently [36, 37]. In this case recombination is favored over fragmentation only in a certain kinematic situation (the very forward direction), which is only a small fraction of phase space.

In central heavy ion collisions many more partons are produced than in collisions of single hadrons. The idea that recombination may then be important for a wide range of rapidities – and at least up to moderate transverse momenta – was advocated before [39, 40, 41]. However, it was only recently that RHIC data indicated that recombination could indeed be a valid approach up to surprisingly high transverse momenta of a few GeV/c.

Charm and heavy hadron production have the advantage that the heavy quark mass provides a large scale that permits a more rigorous treatment of the recombination process [37]. The description of recombination into pions and protons seems to be theoretically less rigorous. However, a simple counting of quantum numbers in a picture where the structure of hadrons is dominated by their valence quarks often provides surprisingly good results. This has been pointed out for particle spectra and ratios [38, 11, 42] and for elliptic flow [47, 14, 48]. Recombination of D mesons in heavy ion collisions has been investigated in [43].

The recombination approach can be formulated in terms of Wigner functions. The yield of mesons M recombining from two partons a, b is given by [12]

$$\frac{dN_M}{d^3P} = \sum_{a,b} \int \frac{d^3R}{(2\pi)^3} \frac{d^3q d^3r}{(2\pi)^3} W_{ab}\left(R - \frac{r}{2}, \frac{P}{2} - q; R + \frac{r}{2}, \frac{P}{2} + q\right) \Phi_M(r,q). \tag{3}$$

W_{ab} and Φ_M are the Wigner functions of the partons and the meson respectively, P and R are the momentum and spatial coordinate of the meson and the sum runs over all possible combinations of quantum numbers, essentially leading to a degeneracy factor C_M. The generalization of this formula for baryons is straightforward [12]. The Wigner function for the partons is usually factorized into classical one-particle phase space distributions, $W_{ab}(r_a, p_a; r_b, p_b) = w_a(r_a, p_a) w_b(r_b, p_b)$. This assumes that the partons are completely uncorrelated before hadronization.

One of the interesting conclusions of this ansatz is that if the parton spectrum is exponential in transverse momentum p_T, $w = Ae^{-p_T/T}$, with a slope parameter $1/T$, then fragmentation and recombination would provide meson spectra

$$dN_{\text{frag}}/d^2P_T D \propto Ae^{-P_T/\langle z\rangle T}\langle D\rangle, \qquad dN_{\text{reco}}/d^2P_T \propto A^2 e^{-P_T/T} \tag{4}$$

respectively. Here $\langle D\rangle$ and $\langle z\rangle$ are average values of the fragmentation function and the scaling variable z. Since $\langle z\rangle < 1$, fragmentation is less effective than recombination on an exponential spectrum, as long as the normalization A is not too small. On the other hand, if the parton spectrum has power law form, $w = Bp_T^{-\alpha}$, the yields are $dN_{\text{frag}}/d^2P_T \sim P_T^{-\alpha}$ and $dN_{\text{reco}}/d^2P_T \sim P_T^{-2\alpha}$ for mesons. This implies that fragmentation will dominate at large P_T for power law spectra (which are predicted by perturbative QCD).

3. Implementations

Four groups have contributed to the the development of the recombination approach in the past years in a major fashion. These groups are:

- Duke/Minnesota/Osaka (Duke) [11, 44, 12, 45, 46]
- Texas A& M/Budapest (TAM) [13, 42, 52, 53]
- Oregon [38, 51]

- Ohio State/Wayne State (Ohio) [48, 50, 49]

While utilizing the same basic principles of the recombination approach, the groups differ in their particular implementation of the model. The Duke and Oregon groups chose a semi-analytic approach and try to evaluate Equation (3) using certain assumptions that essentially result in a convolution of one-particle distributions and hadron wave functions in longitudinal momentum space (longitudinal with respect to the hadron momentum). For example, the Duke group writes the meson spectrum from recombination as [12]

$$E\frac{dN_M}{d^3P} = C_M \int_\Sigma d\sigma \frac{P \cdot u}{(2\pi)^3)} \int dx\, w_a(\sigma, xP^+) w_b(\sigma, (1-x)P^+) \, |\phi_M(x)|^2 \tag{5}$$

where $d\sigma$ integrates over the hadronization hypersurface Σ, u is the four vector orthogonal to Σ, x is the momentum fraction of parton a in the meson, P^+ is the light cone momentum and ϕ_M the wave function of the meson. The parton phase that undergoes hadronization is assumed to have an exponential part at low p_T (soft partons) and a power law tail at high p_T (hard partons). For central Au+Au collisions at RHIC, the parameters chosen by the Duke group are those of a thermal distribution with temperature $T = 175$ MeV and average radial flow velocity $v = 0.55c$ for the soft partons, while the hard partons are taken from a minijet calculation [54] including energy loss [12]. The temperature is being kept fixed in these calculations, whereas the radial flow velocity and the energy loss coefficient have been fitted to data [12].

The Ohio and TAM groups utilize Monte Carlo implementations of the recombination process. These can be connected to microscopic transport models that prepare the parton state before hadronization. One of the main differences lies in the treatment of the connection between soft and hard partons. While the Duke group strictly separates soft and hard physics, allowing only the soft partons to recombine and only the hard partons to fragment, the TAM group includes additional coalescence of soft and hard partons [13, 42]. The Oregon group carries the approach even a step further and replaces fragmentation functions by a scenario where minijet partons develop a shower which subsequently recombines. In this approach, one is able to describe fragmentation functions reasonably well. Applied to heavy ion collisions, this allows for the recombination of soft partons with shower partons [51].

4. Current status

All four model implementations describe hadron data from RHIC at intermediate and large transverse momenta very well. The calculations for spectra reproduce the two-component shape of the spectra with the transition around $P_T = 4$ GeV/c for mesons and 6 GeV/c for baryons from soft (recombination) to hard (perturbative) particle production. Soft-hard coalescence can improve the fit to data points in the transition region, which is then extended to even higher P_T. The baryon/meson ratios in the recombination approach are essentially given by the ratio of degeneracy factors C_B/C_M, leading naturally to an enhanced proton/pion ratio. The most recent results can be found e.g. in [12, 42, 51].

It should be noted, however, that recombination plus fragmentation models are limited to providing a microscopic description of hadronization. They do not attempt to describe the full dynamics of a relativistic heavy-ion collision, but require the phase space distribution of partons at the time of hadronization as input (Duke and TAM approach) or can be used as a reverse filter to extract this distribution from the experimental data (Oregon approach).

The dependence on the parton distribution at hadronization limits the predictive power of the recombination models to such observables which exhibit a flavor-dependency (baryons vs. mesons) or allow for a comparison between mass scaling and constituent quark number scaling. Among the predictions of the recombination plus fragmentation approach, which were subsequently confirmed by experiments, were the nuclear suppression factor of the phi meson

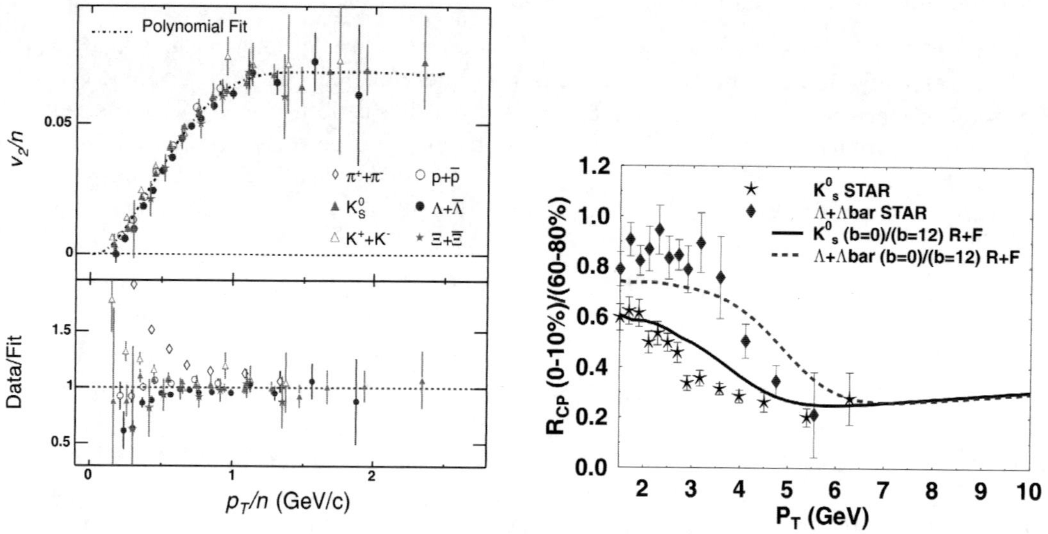

Figure 1. Left: scaled elliptic flow v_2/n vs. scaled transverse momentum p_t/n, with n the number of valence quarks in the respective hadron. The scaling law is strongly indicative of parton recombination (figure taken from [55]). Right: nuclear modification factor R_{CP} as a function of transverse momentum for Λ and K^0_S in the recombination+fragmentation approach compared to STAR data (figure taken from [12]).

[56, 57] (exhibiting the same behavior vs p_T as for kaons and pions) as well was the quark number scaling of the elliptic flow of the cascade and the omega baryon [58].

5. Challenges:

The recombination plus fragmentation model has been extremely successful in describing single particle observables such as spectra, ratios and elliptic flow. Recently, however, more sophisticated dynamical correlation observables have become available and questions have been posed concerning the range of applicability of the recombination approach as standard model of hadronization:

5.1. Jet-like two particle correlations

One of the biggest challenge for the recombination models has been the measurement of dynamic two-particle correlations. The picture of quarks recombining from a collectively flowing, deconfined thermal quark plasma appears to be at odds with the observation of "jet-like" correlations of hadrons observed in the same transverse momentum range of 2 to 5 GeV/c [59, 60, 61]. Triggering on a hadron, e.g., with transverse momentum 2.5 GeV/c $< p_T <$ 4 GeV/c, the data shows an enhancement of hadron emission in a narrow angular cone around the direction of the trigger hadron in a momentum window below 2.5 GeV/c. Can such correlations be reconciled with the claim that hadrons in this momentum range are mostly created by recombination of quarks?

Obviously, the observation is incompatible with any model which assumes that no correlations exist among the quarks before recombination. Such correlations require deviations from a global thermal equilibrium in the quark phase. One mechanism for correlations is already well established: elliptic flow.

In [62] it was shown that correlations among partons in a quark-gluon plasma naturally

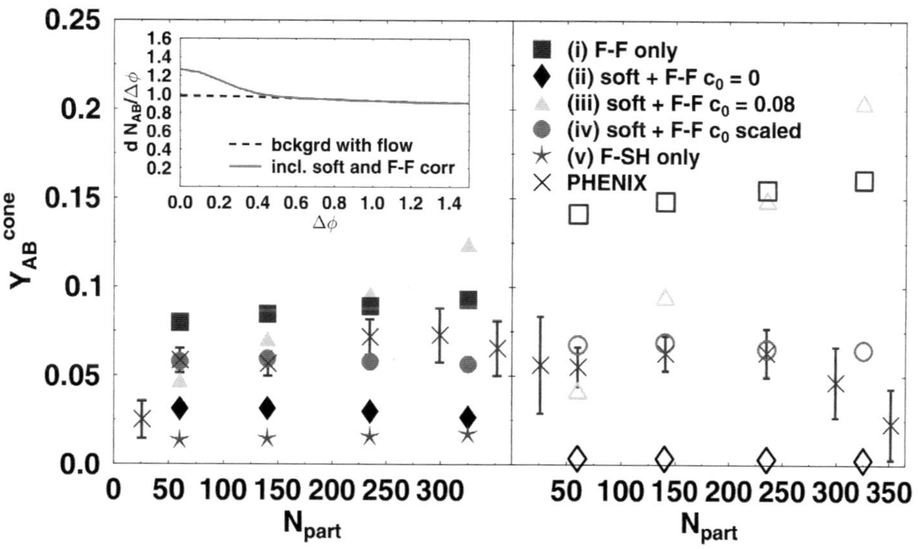

Figure 2. Associated particle yield per trigger hadron A. for meson (left panel) and baryon triggers (right panel) as a function of centrality. Different correlation scenarios in the recombination approach are compared to PHENIX data [60] (figure taken from [62]).

translate into correlations between hadrons formed by recombination of quarks. Correlations are even enhanced by by an amplification factor $Q = n_A n_B$ similar to the scaling of elliptic flow. The interaction of hard partons with the medium has been discussed as one plausible mechanism for the existence of such parton correlations, even though other scenarios for the creation of parton-parton correlations in the deconfined phase are possible. A numerical example that shows that two-parton correlations of order $\approx 10\%$ will be sufficient to explain hadron correlations as measured by the PHENIX collaboration. One may conclude that the existence of localized angular correlations among hadrons are not in contradiction with the recombination scenario.

5.2. Charged particle fluctuations:

Fluctuations of the net electric charge of all particles emitted into a specified rapidity window have been proposed as a possible signal for the formation of deconfined quark matter in relativistic heavy ion collisions [63, 64]. The argument at the basis of this proposal is that charge fluctuations in a quark-gluon plasma are expected to be significantly smaller (by a factor $3-4$) than in a hadronic gas. Because the net charge contained in a given volume is locally conserved and can only be changed by particle diffusion, thermal fluctuations generated within the deconfined phase could survive hadronization and final state interactions.

The most widely used measure for the entropy normalized net charge fluctuations is the D measure [64]:

$$D = 4\langle(\Delta Q)^2\rangle / N_{\mathrm{ch}}, \tag{6}$$

where $\langle(\Delta Q)^2\rangle$ denotes the event-by-event net charge fluctuation within a given rapidity window Δy, and N_{ch} is the total number of charged particles emitted in this window. For a free plasma of quarks and gluons $D \approx 1$, while for a free pion gas $D \approx 4$. Several experiments have measured net charge fluctuations in heavy ion collisions at the CERN Super-Proton Synchrotron (SPS) and at the Relativistic Heavy Ion Collider (RHIC) in Brookhaven [65, 66, 67, 68]. The results for D are generally somewhat smaller than 4, but much larger than the value predicted for a free quark-gluon gas. Applying the formalism of parton recombination to the charged particle fluctuation observables [69, 70] one finds, however, that within the present systematic uncertainties, parton recombination is compatible with the measured charged particle fluctuations. This finding is of significant importance, since the incompatibility of the charged particle fluctuation measurements with the deconfinement hypothesis has posed a serious problem to the emerging picture of Quark-Gluon-Plasma formation at RHIC.

5.3. Entropy balance between the two phases:
The recombination model has been developed for hadron production at intermediate transverse momenta of a few GeV/c. In that domain, one does not have to be concerned about the apparent violation of entropy conservation due to recombining two or three partons respectively into one meson or baryon, since the bulk matter at low transverse momenta can act as a heat-bath and compensate for the perceived reduction in entropy. It is tempting, however, to attempt the extension of the recombination concept to low P_T: once recombination is recognized to be the dominant hadronization mechanism at intermediate P_T from 2 to 5 GeV/c, it is quite reasonable to expect that this mechanism extends down to very small transverse momenta, where quarks are even more abundant. However, the theoretical description will only be on solid ground once the problems of energy and entropy conservation have been addressed properly.

One proposal to deal with the entropy balance in the recombination picture has been the explicit treatment of resonances [53]. While the decay of resonances will certainly decrease the mismatch in particle number between the deconfined and the confined phase, this line of argument alone might be insufficient to fully balance the entropies in both phases. A comprehensive analysis of the total entropy created at RHIC was performed in [70] and contrasted it with lattice-gauge predictions on the entropy-density of a deconfined system of quarks and gluons approaching T_C. The results of that analysis indicate that

(i) The entropy per particle in the hadronic gas, and therefore the entropy content of the hadronic phase at chemical freezeout, is considerably larger than often assumed (mostly due to large masses of hadronic resonances in the system).

(ii) The entropy density of the quark phase is significantly suppressed near T_c, most likely due to correlations among the quasiparticles caused by their strong interactions.

These two conclusions make the recombination picture of hadronization far more compatible with the entropy constraint. If the quark-gluon plasma at hadronization consists of strongly interacting quasi-particles (e.g. constituent quarks) with strong correlations, and if many of the hadrons created at hadronization are heavy, quark recombination and the concomitant particle number decrease could be reconciled with the second law of thermodynamics. At present, however, a more rigorous statement concerning the entropy balance cannot be made, since the state of the art does not allow a direct comparison of the entropy content of both phases, due to the volume at hadronization not being unambiguously known. Besides, lattice calculations do not provide the number of (quasi-)particles, because there is no lattice definition of particle density. Therefore, a more detailed comparison of the entropy content of the hadronic phase and the quark phase at hadronization remains as a problem for future investigations.

5.4. Resonances and elliptic flow scaling law violations:
The v_2 scaling law has been impressively confirmed by measurements at RHIC. Pions and kaons ($n = 2$), protons, Λ and Ξ ($n = 3$) [58] all fall on one universal curve, if v_2 and P_T are divided by the number n of valence quarks [25]. Slight deviations can be observed and are largest for pions. This can be due to its nature as a Goldstone boson – its mass is much smaller than the sum of its constituent quark masses – or because a large fraction of all pions are not created at hadronization but by the subsequent decay of hadron resonances [53]. This poses the interesting question how the existence of a hadronic phase generally affects hadron production

at intermediate and large momenta. In general the v_2 scaling law implies that the elliptic flow is *additive* for any type of composite particle with respect to the v_2 of its constituents. However, this implies that e.g. the v_2 of a K_0^* meson in the recombination domain will therefore be the sum of the v_2 contributions of the d and $\bar{s}$ quarks if it has been formed from a hadronizing QGP, or the sum of the contributions of the K^+ and π^- if it has been formed through the coalescence of a kaon and a pion in the hadron phase. Naively, if the kaon and pion themselves are formed by quark recombination, one would expect a scaling of the K_0^* with $n = 4$ in the hadronic phase, while it scales with $n = 2$, like the stable mesons, if it is created at the phase transition. The scaling could further be altered by the possibility that one or both of the coalescing hadrons could come from fragmentation, which breaks the scaling law. The interplay of these production mechanism may therefore result in scaling law violations for hadronic resonances. These scaling law violations expected for hadronic resonances in turn can be utilized to estimate the amount of hadronic rescattering [46].

6. Outlook

The recombination plus fragmentation approach is able to describe most available RHIC data on spectra, ratios, nuclear suppression and elliptic flow of hadrons, including their impact parameter dependence, for transverse momenta above 1–2 GeV/c – for v_2 even down to very low P_T – consistently with a very small number of globally adjusted parameters. The challenges described in section 5 are being addressed and improved model implementations are likely to succeed in describing the forthcoming data in those sectors. The biggest challenge in the long run will be to determine whether the recombination approach can be extended to the bulk of the matter at low transverse momenta – for this a far more detailed knowledge of the non-perturbative dynamics of QCD seems necessary. To date, the success of the recombination approach (and in particular the elliptic flow scaling law) is among the most compelling and direct pieces of evidence for the formation of a deconfined phase of partons at RHIC.

Acknowledgments

This work was supported in part by RIKEN, the Brookhaven National Laboratory,DOE grants DE-FG02-96ER40945 and DE-AC02-98CH10886, as well as a DOE Outstanding Junior Investigator Award. I would like to thank my collaborators Masayuki Asakawa, Rainer Fries, Berndt Mueller and Chiho Nonaka, without whom the research descibed in this article would not have been possible.

References

[1] P. Huovinen, P. F. Kolb, U. W. Heinz, P. V. Ruuskanen and S. A. Voloshin, Phys. Lett. B **503**, 58 (2001).
[2] D. Teaney, J. Lauret and E. V. Shuryak, [arXiv:nucl-th/0110037].
[3] S. S. Adler *et al.* [PHENIX Collaboration], hys. Rev. Lett. **91**, 182301 (2003).
[4] S. Esumi (for the PHENIX collaboration), Nucl. Phys. A**715**, 599 (2003).
[5] C. Adler *et al.* [STAR Collaboration], Phys. Rev. Lett. **90**, 032301 (2003); *ibid.* **89** 132301 (2002); *ibid.* **87** 182301 (2001).
[6] J. Adams *et al.* [STAR Collaboration], Phys. Rev. Lett. **92**, 052302 (2004).
[7] X. N. Wang, Phys. Rev. C **63**, 054902 (2001).
[8] M. Gyulassy, I. Vitev, X. N. Wang, Phys. Rev. Lett. **86**, 2537 (2001).
[9] K. Adcox *et al.* [PHENIX Collaboration], Phys. Rev. Lett. **88**, 022301 (2002).
[10] J. Adams *et al.* [STAR Collaboration], Phys. Rev. Lett. **91**, 172302 (2003).
[11] R. J. Fries, B. Müller, C. Nonaka and S. A. Bass, Phys. Rev. Lett. **90**, 202303 (2003).
[12] R. J. Fries, B. Müller, C. Nonaka and S. A. Bass, Phys. Rev. C **68**, 044902 (2003).
[13] V. Greco, C. M. Ko, and P. Lévai, Phys. Rev. Lett. **90**, 202303 (2003).
[14] S. A. Voloshin, Nucl. Phys. A **715** (2003) 379.
[15] C. Nonaka, R. J. Fries and S. A. Bass, Phys. Lett. B **583**, 73 (2004).
[16] K. Adcox *et al.* [PHENIX Collaboration], Phys. Rev. Lett. **88**, 242301 (2002).

[17] C. Adler *et al.* [STAR Collaboration], Phys. Rev. Lett. **86**, 4778 (2001).
[18] K. Adcox *et al.* [PHENIX Collaboration], Phys. Rev. C **69**, 024904 (2004).
[19] J. Adams *et al.* [STAR Collaboration], Phys. Rev. Lett. **92**, 112301 (2004).
[20] M. Gyulassy and X. Wang, Nucl. Phys. B **420**, 583 (1994); R. Baier, Y. L. Dokshitzer, A. H. Mueller, S. Peigne and D. Schiff, Nucl. Phys. B **484**, 265 (1997).
[21] R. Baier, Y. L. Dokshitzer, A. H. Mueller and D. Schiff, JHEP **0109**, 033 (2001);
[22] B. Müller, Phys. Rev. C *in print* (2003), nucl-th/0208038.
[23] S. S. Adler *et al.* [PHENIX Collaboration], Phys. Rev. Lett. **91**, 172301 (2003).
[24] J. Adams *et al.* [STAR Collaboration], Phys. Rev. Lett. **92**, 052302 (2004).
[25] P. Sorensen [STAR Collaboration], J. Phys. G **30**, S217 (2004).
[26] I. Vitev and M. Gyulassy, Nucl. Phys. A **715**, 779 (2003).
[27] T. Hirano and Y. Nara, Phys. Rev. C **69**, 034908 (2004).
[28] J. F. Owens, Rev. Mod. Phys. **59**, 465 (1987).
[29] J. C. Collins and D. E. Soper, Nucl. Phys. B **194**, 445 (1982).
[30] B. A. Kniehl, G. Kramer and B. Pötter, Nucl. Phys. B **582**, 514 (2000).
[31] X. F. Guo and X. N. Wang, Phys. Rev. Lett. **85**, 3591 (2000); Nucl. Phys. A **696**, 788 (2001).
[32] E. Wang and X. N. Wang, Phys. Rev. Lett. **89**, 162301 (2002).
[33] X. N. Wang, nucl-th/0305010.
[34] K. P. Das and R. C. Hwa, Phys. Lett. B **68**, 459 (1977); Erratum *ibid.* **73**, 504 (1978); R. G. Roberts, R. C. Hwa and S. Matsuda, J. Phys. G **5**, 1043 (1979).
[35] E. M. Aitala *et al.* [E791 Collaboration], Phys. Lett. B **371**, 157 (1996).
[36] J. C. Anjos, J. Magnin and G. Herrera, Phys. Lett. B **523**, 29 (2001).
[37] E. Braaten, Y. Jia and T. Mehen, Phys. Rev. Lett. **89**, 122002 (2002);
[38] R. C. Hwa and C. B. Yang, Phys. Rev. C **67**, 034902 (2003) [arXiv:nucl-th/0211010].
[39] C. Gupt, R. K. Shivpuri, N. S. Verma and A. P. Sharma, Nuovo Cim. A **75**, 408 (1983)
[40] T. Ochiai, Prog. Theor. Phys. **75**, 1184 (1986).
[41] T. S. Biro, P. Levai and J. Zimanyi, Phys. Lett. B **347**, 6 (1995); T. S. Biro, P. Levai and J. Zimanyi, J. Phys. G **28**, 1561 (2002).
[42] V. Greco, C. M. Ko and P. Levai, Phys. Rev. C **68**, 034904 (2003).
[43] R. Rapp and E. V. Shuryak, hep-ph/0301245.
[44] R. J. Fries, B. Muller, C. Nonaka and S. A. Bass, J. Phys. G **30**, S223 (2004).
[45] C. Nonaka, R. J. Fries and S. A. Bass, Phys. Lett. B **583**, 73 (2004).
[46] C. Nonaka, B. Muller, M. Asakawa, S. A. Bass and R. J. Fries, Phys. Rev. C **69**, 031902 (2004).
[47] Z. W. Lin and C. M. Ko, Phys. Rev. Lett. **89**, 202302 (2002).
[48] D. Molnár and S. A. Voloshin, Phys. Rev. Lett. **91**, 092301 (2003).
[49] D. Molnár, J. Phys. G **30**, S235 (2004).
[50] Z. w. Lin and D. Molnar, Phys. Rev. C **68**, 044901 (2003).
[51] R. C. Hwa and C. B. Yang, Phys. Rev. C **67**, 034902 (2003).
[52] V. Greco, C. M. Ko and R. Rapp, Phys. Lett. B **595**, 202 (2004).
[53] V. Greco and C. M. Ko, Phys. Rev. C **70**, 024901 (2004).
[54] R. J. Fries, B. Muller and D. K. Srivastava, Phys. Rev. Lett. **90**, 132301 (2003).
[55] M. D. Oldenburg [STAR Collaboration], J. Phys. G **31**, S437 (2005).
[56] J. Adams *et al.* [STAR Collaboration], arXiv:nucl-ex/0406003.
[57] S. S. Adler *et al.* [PHENIX Collaboration], arXiv:nucl-ex/0410012.
[58] J. Castillo [STAR Collaboration], J. Phys. G **30**, S1207 (2004).
[59] C. Adler *et al.* [STAR Collaboration], Phys. Rev. Lett. **90**, 082302 (2003)
[60] A. Sickles [PHENIX collaboration], J. Phys. G **30**, S1291 (2004).
[61] S. S. Adler *et al.* [PHENIX Collaboration], arXiv:nucl-ex/0408007.
[62] R. J. Fries, S. A. Bass and B. Muller, Phys. Rev. Lett. in print, arXiv:nucl-th/0407102.
[63] M. Asakawa, U. W. Heinz, and B. Müller, Phys. Rev. Lett. **85**, 2072 (2000).
[64] S. Jeon and V. Koch, Phys. Rev. Lett. **85**, 2076 (2000).
[65] J. Adams *et al.* [STAR Collaboration], Phys. Rev. C **68**, 044905 (2003); G. D. Westfall [STAR collaboration], J. Phys. G **30**, S1389 (2004).
[66] K. Adcox *et al.* [PHENIX Collaboration], Phys. Rev. Lett. **89**, 082301 (2002); J. Nystrand [PHENIX Collaboration], Nucl. Phys. A **715**, 603 (2003).
[67] H. Sako and H. Appelshäuser [CERES/NA45 Collaboration], J. Phys. G **30**, S1371 (2004).
[68] C. Alt *et al.* [NA49 Collaboration], [arXiv:nucl-ex/0406013].
[69] A. Białas, Phys. Lett. B **532**, 249 (2002).
[70] C. Nonaka, B. Muller, S. A. Bass and M. Asakawa, arXiv:nucl-th/0501028.

Institute of Physics Publishing
doi:10.1088/1742-6596/50/1/034

Journal of Physics: Conference Series **50** (2006) 289–292
5th International Conference on Physics and Astrophysics of Quark Gluon Plasma

Heavy-Ion Physics at LHC

Yves Schutz

CERN, CH-1211 Genève 23

Yves.schutz@in2p3.fr

Abstract. The new conditions which will be reached when LHC will collide lead ions are discussed together with the probes which will become available for studying the properties of the hottest matter ever formed in the laboratory. The experimental requirements and how the LHC experiments, in particular ALICE, will face the challenges are presented.

1. Introduction

Before the first data will be available from the LHC, which is expected to come into operation in 2007, it will remain impossible to predict what the properties of the matter formed in heavy ion collisions will be at such unprecedented high energies. Will matter have the properties of weakly interacting plasma of gluons and quarks carrying the small Chiral mass and thus provide the unique opportunity to study the true vacuum of QCD? Will matter instead be the same perfect partonic fluid or strongly interacting quark and gluon plasma which has been discovered at RHIC energies? Will the new state of matter characterized by a saturated gluon density and dubbed as the Color Glass Condensate play an essential role as the precursor state for the formation of hot nuclear matter? Will there be yet a new state of matter discovered at LHC? Nobody is really in a position today to make firm predictions. Therefore, experiments must be prepared for the unexpected and cannot afford to neglect any observable. The ALICE experiment has been conceived within this spirit and tries to be as complete as possible exploiting all presently known techniques to detect and identify all kinds of particles, which can be produced in nuclear interactions, over a very broad momentum range.

2. From SPS to LHC

Let us briefly review the recent history of the exploration of the properties of hot nuclear matter or the "search of the quark gluon plasma". In February 2000, CERN announced that a "new State of matter" had been created [1]. Compelling evidence for this new state of matter was provided by the results collected over about 10 years by seven experiments dedicated to measuring a number of crucial observables. The large energy densities reached in Pb-Pb collisions at $\sqrt{s_{NN}}$=17.3 GeV exceeded substantially the critical energy density at which the transition from normal hadronic matter toward deconfined partonic matter is predicted by theory. The measured strangeness enhancement and suppressed charmonia bound state production, as well as hints for thermal electromagnetic radiation, constituted a set of prominent evidences. However the interpretation in terms of the formation of quark gluon plasma was not unique enough to establish unambiguously the properties of the new state of matter.

Four years later, following a very successful start of the RHIC operation, the announcement was made that a strongly coupled QGP has been discovered at RHIC [2]. The claim for discovery has however not been made by the four RHIC experiments because the smoking gun of the QGP formation has not yet been observed. There is nevertheless a solid line of empirical evidence [3] for the observation of deconfined matter. In Au-Au collisions at $\sqrt{s_{NN}}$=200 GeV, matter is now formed largely exceeding the critical value. An unexpected large elliptic flow, reaching the saturation limit predicted by the hydrodynamic model, has been observed for all measured particle species and has been interpreted in terms of formation of an early collective system at the partonic level. Jets have been found to be measurable through the measurement of leading hadrons and hadron correlations and the results indicate that jet properties are modified by the medium traversed by hard scattered partons. It has been further demonstrated, through observation that similar modifications are absent in d-Au collisions, that the modifications are due to final state effects on the hard scattered partons traversing a color dense opaque medium. A consistent interpretation of the data is given in terms of the formation of a strongly coupled QGP rather than the nearly ideal gas predicted by the theory.

With the advent of LHC and the huge step in energy to reach the nominal $\sqrt{s_{NN}}$=5500 GeV, the study of heavy-ion collisions will enter a new and unexplored regime. Quantitatively matter is predicted to be formed at energy densities and temperatures largely exceeding (by a factor up to 4 for the temperature) the critical values at which QCD calculation on the lattice predict the deconfinement transition to occur and the Chiral symmetry to be restored. In this regime the coupling constant could be sufficiently reduced to approach the state of a weakly interacting gas of the QCD vacuum quanta. The system is expected to be thermalized within a tenth of an *fm/c* and to remain much longer in the deconfined phase (more than ten *fm/c*) enhancing the production of rare probes. At the LHC energies the collisions will also enter a qualitatively new regime of partons kinematics. The particle production from very forward to central rapidity will be dominated by hard processes involving partons with small Bjorken-x values, from, for example, $x=10^{-2}$ for Q=100 GeV transferred momentum at central rapidity to, $x=10^{-5}$ for the production of J/ψ (Q=10 GeV) at very forward rapidity (y=4). The saturation regime will therefore determine the property of the initial stage of the collision and the QGP formed at LHC might evolve from a Color Glass Condensate [4]. Hard processes, where the transferred momentum is much larger than the QCD scale or the thermal scale ($Q \gg \Lambda_{QCD}$, T), will contribute significantly (98%) to the total cross section. Particles produced in hard processes are thus expected to be abundant at LHC energies, providing a new and unique set of observables probing the medium at very early times.

3. Hard probes at LHC

They are produced in the early stage of the collision, because of their large virtuality the formation time is short allowing them to be decoupled from the medium, and their production cross section is calculable by perturbative QCD. Hard scattered partons appearing in the detector as a jet of hadrons, heavy flavor quarks and quarkonia, and weakly interacting probes such as photons or W/Z bosons belong to this category of probes.

The properties of partons created in hard scattering in the very early phase of the collision will be modified by the traversed medium. These modifications will be imprinted in the final hadronic state (or jet) of the parton and therefore provide a unique probe of the medium formed during the collision at an early stage. In the vacuum, a parton produced in some hard process with a high virtuality E_T, will lose its virtuality by parton splitting until it reaches the hadronic scale Q (~ 1 GeV). At this scale, the perturbative approximation ceases to apply and the partonic shower fragments into hadrons. The typical length of this hadronisation scales linearly with E_T, $L_{hadr} \sim E_T/Q_{hadr}^2$. The medium will modify these vacuum dynamics. While traversing the medium, high energy partons will more effectively lose the virtuality by gluon radiation. The total radiated energy depends on the properties of the medium which are characterized by a transport coefficient (energy loss by unit of distance) [5], $\boldsymbol{q}$, and scales with the square root of E_T, $L_{therm} \sim \sqrt{(E_T/\boldsymbol{q})}$. The competition between full thermalization (all the parton energy E_T is radiated) and hadronisation depends on the initial energy of the parton and consequently on the observed final-state hadron energy, p_T.

For sufficiently low parton energies so that $L_{therm} < L_{hadr} < L_{medium}$ (i.e., $p_T \leq 2$ GeV/*c*), the hard parton is thermalized in the heat bath and contributes to the bulk properties of the medium. This sets the ability to detect hadrons down to low p_T (~ 100 MeV/*c*) as a first experimental requirement. The bulk properties of the LHC medium will most likely follow a monotonic extrapolation from the properties of the RHIC medium. However, one has to be prepared for changes due, for example, to the difference in the expansion dynamics and in the thermodynamics by the inclusion of heavy quarks.

For parton energies for which all the scales are equivalent ($L_{therm} \sim L_{hadr} \sim L_{tmedium}$), or equivalently for hadrons with $p_T \sim$ 2-10 GeV/c, the hadronisation of the hard parton shower will occur inside the medium, so that the medium can be used as a control parameter to modify the non-perturbative hadronisation process. Thus one can expect to gain some insight into the dynamics of this process and learn about the nature of the pre-hadronic states. Medium modified fragmentation functions have indeed been observed at RHIC [6] in the measurement of proton to pion ratio enhanced by about a factor 10 with respect to the same ratio measured in e^+-e^- annihilations and with the ratio expected from perturbative calculations applying the factorization theorem. This sets the next experimental requirement, the necessity of excellent particle identification in a p_T range extending at least up to 10 GeV/c.

For more energetic partons satisfying $L_{therm} > L_{hadr} > L_{tmedium}$ (i.e., $p_T >$ 10 GeV/c), the hadronisation occurs in the vacuum, and the final hadronic state will carry information on the energy lost by the parton inside the medium. Thus, measuring the jet properties under these conditions will enable to derive information on the properties of the medium. The most straightforward measurement will be, as demonstrated by RHIC data, to probe the reduced energy of the initial parton with the reduced energy of the leading hadron. Hadron angular correlations measuring the shape of back to back correlations provide another tool to study medium induced modification of jet properties. Measuring high p_T hadrons with high momentum resolution over a broad range of momenta is the experimental requirement to perform such studies. This kind of measurement is however not very sensitive to the medium properties because the parton energy loss scales like $\boldsymbol{q}L^2$ [7], where L is the length of the traversed medium, rendering the interior of the medium opaque to these probes.

A better probe will be the modification of the jet topology [7], by measuring the hadron multiplicity in the jet or the heating of the jet (transverse momentum distribution of the hadrons with respect to the jet axis). These observables will be more sensitive to the medium properties as they scale as $\boldsymbol{q}L$, linearly with the distance traveled inside the medium. For these jet topology measurements one will take advantage of the copious production of energetic jets at LHC. Leading order perturbative QCD calculation predicts that one 10 GeV jet is produced in every Pb-Pb collision and up to 10^4 jets of 200 GeV transverse energy will be produced in one LHC heavy-ion run (effective running 10^6 seconds) at the nominal luminosity of $5\ 10^{26}$ $cm^{-2}s^{-1}$. These jets can be identified together with their topology, provided sufficiently high quality charged hadron and/or electromagnetic calorimetry measurements are available [8]. Because of the high production rate, the quality of jet measurement will be greatly improved by photon tagging. High resolution photon measurements and high quality photon identification are necessary experimental requirements.

At LHC, jet quenching measurements can be extended to heavy quarks to provide an additional constraint to the determination of the transport coefficient. Indeed heavy quarks lose energy less efficiently through medium induced gluon radiation than light quarks do. The measurement of heavy to light quark double nuclear modification ratio will thus be sensitive to the mass and color-charge dependence of medium induced energy loss [7]. The identification of heavy flavor D or B mesons will require, in addition to high quality particle tracking and particle identification (for detection in the hadronic channel), high resolution vertexing as well as tracking and identification of high momentum electrons (for detection in the semi-electronic channel).

4. Heavy-ion experiments

Among the three experiments at LHC, ALICE, ATLAS and CMS, which have an approved heavy-ion program, ALICE is the dedicated heavy-ion experiment. It implements all the requirements which have been mentioned in the previous section. The challenge is to do in one experiment what is being done in four at RHIC. Since predictions for particle density span a broad range of values, ALICE has been optimized for charged particle densities of the order of $dN_{ch}/d\eta$ = 4000, but can provide acceptable performances up to values as high as 8000. All known techniques for particle detection and identification are exploited in the design of the ALICE detectors, allowing to measure the flavor (hadrons, leptons and photons) content and the phase space distribution of heavy-ion events on an event by event basis. ALICE is equipped for jet physics which is going to dominate the LHC heavy-ion program. On the low p_T side, up to several tens of GeV/*c*, jets will be mostly characterized through hadron correlations and leading particle measurements. Jets with p_T around 100 GeV/*c* and above will be identified allowing detailed study of their topology. Fragmentation functions will be measured down to low fractional momenta for identified charged and neutral particles. ALICE is however still missing (only because of financial constrains) an electromagnetic calorimeter with large acceptance, which would greatly extend the jet physics reach of ALICE.

5. Conclusion

In 2007, the LHC experiments will be ready to take data. First *pp* beams are scheduled to collide, with reduced luminosity, by mid 2007. A first heavy-ion pilot run may take place already in 2007. The first full heavy-ion run is scheduled to take place in 2008, with reduced luminosity. Within the first 15 minutes at nominal luminosity, data could be collected with sufficient statistics to measure the particle multiplicity, the low energy particle spectra, and the particle production ratios. Within one month of data collection, the statistics would be sufficient to study rare high p_T processes such as jets, D and B mesons, quarkonia and direct photons. The following runs could then be devoted to the same measurements in *pA* collisions, and to heavy-ion mass and energy scans. Until then, the collaborations continue to invest a huge amount of effort to guarantee a timely completion of the detectors and the accelerator.

References

[1] *New State of Matter created at CERN*, CERN Press Relase, February 2000, http://info.web.cern.ch/Press/PressReleases/Releases2000/PR01.00EQuarkGluonMatter.html

[2] Miklos Gyulassy, *What have we learned at RHIC*, J. Phys. G: Nucl. Part. Phys. 30 (2004) S911-S918.

[3] *Proceedings of Quark Matter 2004*, Oakland , 11-17 January 2004, J. Phys. G: Nucl. Part. Phys. 30 (2004).

[4] *The Color Glass Condensate at RHIC and LHC*, L. Mc Lerran, in these proceedings.

[5] *Calculating Quenching Weights*, C.A. Salgado and U.A. Wiedemann, Phys. Rev. D68 014008 (2003).

[6] *Identified charged particle spectra and yields in Au+Au collisions at* $\sqrt{s_{NN}}$ *= 200 GeV*, S. Adler etal. (PHENIX collaboration), Phys. Rev. C 69 034909 (2004).

[7] *Modification of jet shapes and jet multiplicities,* C.A. Salgado and U.A. Wiedemann, Phys. Rev. Lett. 93 042301, (2004).

[8] *Jet physics in heavy-ion collisions at the LHC with the ALICE detector,* A. Morsch, J. Phys. G: Nucl. Part. Phys. 31 (2005) S597-S602.

[9] *Testing the color charge and mass dependence of parton energy loss with heavy-to-light ratios at RHIC and LHC*, N. Armesto, A. Dainese, C. Salgado, and U. Wiedemann, hep-ph/0501225 (2005).

Institute of Physics Publishing Journal of Physics: Conference Series **50** (2006) 293–299
doi:10.1088/1742-6596/50/1/035 5th International Conference on Physics and Astrophysics of Quark Gluon Plasma

Experimental Summary

Johanna Stachel
Physikalisches Institut der Universität Heidelberg,
Philosophenweg 12,
D69120 Heidelberg

1. Introduction

This paper gives highlights of the experimental results shown at this conference. As additional reference the four experimetal papers summarizing the results from the first 3 years of RHIC running will be used [1, 2, 3, 4]. This summary is a highly personal selection of topical results and does not claim to be of comprehensive nature. The author apologizes to all conference participants who contributed interesting results that are not mentioned here.

2. Global observables and initial condition

Transverse energy production was studied by PHENIX and STAR [5, 6] and measured as 600 GeV per unit pseudorapidity for the most central 5 % of Au + Au collisions (unless mentioned otherwise, here always results at the top RHIC energy of $\sqrt{s}$ = 200 GeV are used). Using the Bjorken estimate and, very conservatively, a value τ_0 = 1 fm/c gives for the initial energy density ϵ_0 = 5.5 GeV/fm^3. Choosing instead the saturation scale as the measure for $\tau_0 = 1/Q_0$ = 0.14 fm/c gives a value 7 times larger. Most likely these two values bracket the time scale of equilibration. In any case, the initial energy density significantly exceeds the critical energy density for the phase transition obtained by lattice QCD [7] as ϵ_c = 0.7 GeV/fm^3.

Charged particle production is addressed by all four experiments. The total charged particle multiplicity for the most central 5 % of Au + Au collisions at full RHIC energy was determined by BRAHMS [8] to 4600. The corresponding number measured by PHOBOS [9] is 5300 for the most central 6 %. This leads to a total hadron multiplicity of about 7500. As discussed by Roland [10] at this conference, the concept of limiting fragmentation holds very accurately for charged particle production over 3-4 units away from beam rapidity comparing SPS and RHIC energies. Mohanty [11] showed that photons (dominantly from π^0 decay) also exhibit this behavior. Assuming that this extends to LHC energy, Roland [10] estimated an upper limit of 1300 - 1800 for $dN_{ch}/d\eta$ for Pb + Pb collisions at LHC.

The centrality dependence of charged particle multiplicty distributions and, in particular, its logarithmic growth with number of participants have been taken as signs of gluon saturation and in that respect support the relevance of the concept of color glass condensate at RHIC energies. Recent calculations by Salgado [12] show that this approach works just as well for data at top SPS energy and this casts some doubt whether color glass condensate is indeed the correct underlying physical concept. In the talk of Hofmann the scaling of charged particle multiplicity with beam energy was compared to heavy ion data at lower beam energy as well as to pp and e^+e^- data [13, 3]. The heavy ion data from top SPS energy upwards are found to agree after participant scaling with pp data at twice the center of mass energy. This is indicative of the large degree of stopping in a heavy ion collisions. The energy deposition and therefore the energy available for particle production is comparable to an e^+e^- collision at the same c.m. energy. This is reflected quantitatively in the BRAHMS data on projectile nucleon rapidity shift as shown by Debbe [14, 15]. The best estimate for the mean beam nucleon rapidity shift is 2.0 $\pm$ 0.4 units, very close to the asymptotic value determined by Busza and Goldhaber [16] and later Busza and Ledoux [17] from central pA collisions. Of the incident beam energy 73 $\pm$ 6 % is available for heating of the fireball and particle production.

3. Hadron production

In the talk of Braun-Munzinger [18] it was shown, that RHIC data on hadron production at mid-rapidity can be very accurately reproduced as the particle yields of a chemically equilbrated system in terms of a grand canonical ensemble (see [19] for a complete set of references). A temperature of 177 $\pm$ 5 MeV is needed and this value is practically unchanged from $\sqrt{s} = 130$ and 17.3 GeV, while the baryon chemical potential drops continuously with increasing beam energy and has fallen, for top RHIC energy, to 29 $\pm$ 5 MeV. It appears that from top SPS energy on a constant temperature characterizes hadron yields, while at lower SPS energies the temperature is significantly lower; at $\sqrt{s} = 8.8$ GeV it is only 148 $\pm$ 5 MeV. A possible explanation how the system can achieve equilibrium apparently in a very short time even for hadrons with multiple (triple) strangeness and antinuclei was presented by Braun-Munzinger [20]; due to the rapid increase in density in the immediate vicinity of the phase transition multiparticle collisions become very important and can drive even Ω yields into equilibrium within a fraction of a fm/c.

The systematic behavior of hadro-chemical freeze-out parameters with beam energy as shown by Braun-Munzinger [19] leads to maxima of certain particle ratios for intermediate beam energies driven by the interplay of increasing temperature and decreasing baryon chemical potential; examples are the ratio K^+/π with a broad maximum around $\sqrt{s_{nn}} \approx 10$ GeV and the ratio Λ/π with a pronounced peak at $\sqrt{s_{nn}} = 5$ GeV. In the talk of Friese (Fig.4 in [21]) the measured excitation function of such ratios by NA49 was shown and put into context with lower and higher beam energy data. The observed ratio Λ/π follows the expected trend with the pronounced maximum with good precision. In the ratio K^+/π the data appear to show a sharper maximum and, in particular, the NA49 data points at 80 and 158 A GeV/c beam momentum appear to lie nearly two standard deviations below the thermal model prediction. While there has been much speculation about the physics behind this deviation at this stage it remains an experimental question requiring confirmation. It should be noted that the ratios K^-/π and K^0_s/h^- do not show anything unusual.

4. Azimuthal Anisotropies

In has been observed for already quite some time that the elliptic flow parameters v_2 as function of transverse momentum and for different particle species can be remarkably well reproduced by hydrodynamics calculations [22, 23, 24]. The data are described quantitatively up to about 2 GeV/c in transverse momentum. The elliptiv flow observable for more massive particles like protons or Lambdas even exhibits sensitivity to the equation of state favoring an equation of state with a QGP - hadronic matter phase transition [22]. As pointed out e.g. by McLerran at this conference, it requires strong interactions at short times for the hydrodynamic description to work so well. This could happen for instance if the initial state is a color-glass condensate. Considering the beam energy dependence of elliptic flow, the data show a continuous increase. When plotting the elliptic flow coefficient divided by the eccentrity as function of the charged particle rapidity density per cross sectional area, the data appear to scale and to increase about linearly (see Fig. 25 in [25]. The hydrodynamics results, conversely, show a weak drop with beam energy and at top RHIC energy data and calculations meet. The significant overprediction of data by hydrodynamics at SPS energy is linked to the equation of state used, which has a softest point such that there is a maximum in v_2 at SPS energy and a minimum at full RHIC energy [4].

As shown by Jacak (Fig.4 in [26]) the elliptic flow increases in data by about 50% from SPS energy to $\sqrt{s_{nn}} = 62.4$ GeV and then remains about constant for larger beam energies althoug the energy density there increases about 30 %, supporting a soft equation of state. In the talk by Jacak were also shown the results of various hydrodynamics calculations together with transverse momentum spectra and elliptic flow for protons and pions (Fig.5 in [26]). While qualitatively all calculations that incorporate a phase transition between QGP and hadronic phase reproduce the overall features shown by the data, there are significant deviations in detail. This has to do with how the hadronic phase and freeze-out are treated. Common to all calculations is a rapid equilibration on a time scale of 0.5-1 fm/c.

Another type of scaling has been discovered in the elliptic flow data at RHIC: when normalizing both the flow coefficient v_2 and the transverse momentum to the number of constituent quarks in a hadron, the data for different hadronic species scale for transverse momenta above about 2 GeV/c as shown by Xu (Fig.3 in [27]). This has been interpreted as partons in the QGP carrying the flow information and preserving it in coalescing [27].

5. Jet quenching

One of the highlights of the RHIC program is the observation of the suppression of production of hadrons at high transverse momentum. This was addressed in several presentations at this conference. When normalizing the transverse momentum spectra observed in Au + Au collisions to the corresponding pp spectra multiplied with the number of binary collisions, a suppression factor $R_{AA}(p_t)$ is obtained. Similarly central and peripheral Au + Au collisions can be compared. It is observed in all four experiments [1, 2, 3, 4] that for central Au + Au collisions production of pions/charged hadrons is suppressed by a factor 4-5 for transverse momenta in excess of 4 GeV/c. There is reasonable agreement between the experiments, although around 2.5 GeV/c a 20 % difference between the STAR data and those of the other experiments is noted (see e.g. Fig.2 of [28]). Since the suppression is not seen in d + Au collisions [1, 2, 3, 4], it is obviously an effect induced by the medium the parton or jet traverses.

Presumably the suppression is due to medium induced radiative energy loss of the parton before it fragments and hadronizes as was predicted [29, 30]. Calculations employing a large initial gluon rapidity density of about 1100 can account for the data [31]. The beam energy dependence of the R_{AA} factor was presented recently by d'Enterria [32] and it appears that the suppression evolves in a very smooth way from top SPS energy onwards. The $R_{AA}(p_t = 4\text{GeV/c})$ value of 1.0 measured at the SPS represents already a slight suppression as compared to the normal Cronin enhancement [32]. Going from $\sqrt{s_{nn}} = 17.3$ to 62.4 to 200 GeV the gluon rapidity density needed grows from 400 to 650 to 1100 [31]. Alternatively this suppression can be described by increasing and large opacities of the medium traversed [33].

These high initial gluon densities correspond to an initial temperature of about twice the critical temperature and to initial energy densities $\epsilon_0 = 14$ - 20 GeV/fm^3 well in line with the initial conditions needed for the hydrodynamics calculations to describe spectra and elliptic flow (see section 4) and bracketed by the estimates based on the Bjorken formula (see section 2).

It was observed since some time that the high p_t suppression pattern is different for different hadronic species. This is very cleanly demonstrated by recent STAR data on the ratio Λ/K^0_s which was shown by Lamont [34] as function of p_t for different centralities. For central collisions this ratio peaks at $p_t \approx 3$ GeV/c at a value of 1.6, close to the ratio 3/2 expected in quark coalescence models. However, such models at present give quantitatively only a rough description of the data as shown in Fig. 2 of [34]. In the talk of Jacak a similar rise of the ratio for proton over pion p_t spectra is shown for d + Au collisions (see Fig. 2 in [26]) and the question arises why (constituent) quark coalescence would be an appropriate description for a d + Au collision. To test the coalescence hypothesis jet partners have been looked for for a high p_t meson or baryon both on the near and on the away side. The fact that within present measurement accuracy a jet partner is equally likely for a trigger baryon or meson [35] (very similar data from STAR where shown by Mischke and Mironov at this conference) does not strengthen the valence quark coalescence scenario.

In d + Au collisions no suppression is seen in central collisions at mid-rapidity. Recent data by BRAHMS on R_{dAu} as function of pseudo-rapidity shown that a significant suppression sets in at forward rapidities [36]. This is supported by similar data from PHOBOS [37]. It has been suggested that this could be a manifestation of the color glass condensate visible as one probes smaller values of x and a corresponding analysis describes the data well [38]. On the other hand, the same data can be described equally well in terms of the parton coalescence picture [39]. In the talk by Vitev it was shown, that the same feature is exhibited by d + Au data at $\sqrt{s}$ = 19.4 GeV. This sheds doubt on both the color glass and the quark coalescence descriptions and points to alternative explanations such as put forward by Kopeliovich et al. [40] e.g. in terms of energy loss of the incoming parton.

Parton thermalization is displayed in a very clean way be recent results of the STAR collaboration [41, 42]. Evaluating the mean transverse momentum in a cone opposite to a high p_t trigger particle as a funtion of centrality, a gradual decrease for more central Au + Au collisions is observed and in the most central collisions a value very close to the inclusive mean p_t is reached (see Fig. 2 in [42]). To actually connect the high energy parton energy loss to the radiated gluons Vitev calculated di-hadron correlations (see Fig. 3 in [31]). Comparing calculations with and without the socalled gluon feed-back to STAR data [41] shows that quenching of the away-side jet at high p_t is accompanied by a simultaneous strong enhancement at p_t below 2 GeV/c.

In azimuthal correlations of two high p_t particles it was seen that the away-side peak disappears in central Au + Au collisions. Recently it was shown that the effect is very strong in case the away-side jet is emitted out of the reaction plane and much weaker for emission in the reaction plane [43]. This supports the strong correlation of the suppression with the length of matter traversed by the parton. When lowering the p_t cut on the correlated hadron, a very broad structure appears on the side opposite to the trigger particle. This was shown by Mischke for a cut on the correlated hadron of p_t = 0.15 - 4. GeV/c (see Fig. 1 in [42]). This calls to mind a similar observation at SPS energy by CERES [44] where for a condition $p_t \geq 1.2$ GeV/c also very strong broadening of the away-side structure with increasing collision centrality was observed. In the talk of Jacak a tantalizing observation was shown [45]: For a trigger particle p_t of 4-6 GeV/c and a correlated particle p_t of 1.0 - 2.5 GeV/c the away-side peak seen in peripheral Au + Au collisions develops actually into a hole at $\Delta\phi = \pi$ for more central collisions while a very broad peak appears with a maximum at $\Delta\phi = \pi - 1$ as can be seen in Fig. 8 of [26]. Could this be the Mach cone due to the sonic boom of the quenched jet? This goes back to an idea developed by Stöcker et al. for nucleons and nuclei emitted in nuclear reactions in the mid 1970ies (see references [14,15] in [46]) and was suggested for the current scenario of a parton traversing a quark-gluon plasma [47, 48]. If this could be established it would have far reaching consequences since it would be an observable linked directly to the speed of sounds of the quark-gluon plasma and thereby its equation of state. At present it remains an experimental challenge to establish an actual cone topology in two dimensions.

6. Open charm and charmonia

Open charm has been measured indirectly from the inclusive electron p_t spectra after subtracting known contributions from photon conversions and light hadron decays by PHENIX [49]. Recently similar measurements have become possible at high p_t for STAR using the calorimeter [34]. The spectrum remaining after subtraction is dominated [1] by open charm

[1] A possible contribution to the electron spectrum from the Drell-Yan production process cannot be ruled out at present, though.

and beauty contributions but the method is limited by the systematic error introduced by the subtraction. A significant step ahead will come in the future when silicon micro-vertex-detectors will be implemented. Nevertheless, very interesting results have come already from the present method. In the talk of Jacak results for an elliptic flow analysis of the open charm decay elecrons were shown (see Fig. 6 of [26]). There is a significant nonzero value in the transverse momentum range 0.4 - 1.6 GeV/c. At this meeting STAR data were shown by Lamont [34] that complement the PHENIX results and extend the overall transverse momentum coverage by adding the range p_t = 1.5 - 3.0 GeV/c. Together the data paint a consistent picture that indeed the electrons from open charm decay exhibit elliptic flow. What is more, the data are described quantitatively by a calculation [50] based on thermalization and elliptic flow of the charm quark (see Fig. 7 of [34]); the corresponding calculation with no charm quark flow (and hence only an effect from the flow of the light quark in a D meson) is a factor two below the data, altough with the presently still large error bars this is an effect of only a few standard deviations. The present data can be taken as an indication that the charm quark thermalizes to a significant degree. Note that this is a necessary prerequisite for any formation of charmed hadrons by statistical hadronization [51]. On the other hand, in that case also jet quenching should be observed for charmed hadrons. Indeed in the talks by Hamagaki and Tabaru [52] it was shown that electron spectra after the subtraction of contributions from conversion and light hadron decays show high p_t suppression for central collisions. The R_{AA} factor drops practically as low as for pions at p_t of 4 GeV/c, i.e. to values of about 0.2 (see Fig.7 in [26]). In a recent publication [53] the suppression for electrons from D meson decay was studied for different transport coefficients and in the talk of Jacak it was shown, that the present data would be consistent with a calculation using a transport coefficient of 14 GeV^2/fm (see Fig. 2 of [53]) at the upper end of the range needed to reproduce the data for pions. This is very surprizing, in particular also in view of the fact, that at p_t of about 4 GeV/c also the contribution of b-quarks to the electron spectrum should become sizeable.

The overall charm production cross section at RHIC energy has been measured indirectly by PHENIX [49] for $\sqrt{s}$ = 130 and 200 GeV from the inclusive electron spectra in the way described above for Au + Au collisions and at full RHIC energy also for d + Au and pp collisions [54]. It is found that the integrated charm cross section, when scaled with the number of binary collisions, agrees for all three collisions systems [52]. The value is about 30% above a NLO pQCD calculation (as shown e.g. by Lamont [34] in these proceedings) but agrees within errors. In STAR, D mesons have been reconstructed via their hadronic decay to $K\pi$ in d + Au collisions and a charm cross section per nucleon nucleon collision has been extracted [55]. It is twice as large as the PHENIX value by about two standard deviations. This obvious experimental discrepancy needs to be resolved; most likely this will come only with the future charm measurement using the displaced vertex feature.

For charmonia there were final results for NA50 at the CERN SPS presented at this meeting by Quintans [56]. Analysing all pA data, a cross section for normal nuclear absorption of 4.1±0.4 mb was extraced [57]. To this normal nuclear suppression all results from heavy ion collisions can be compared. It turns out that S + U data as well as data from peripheral Pb + Pb collisions agree with this normal nuclear absorption curve. For transverse energies above 40 GeV or a length of nuclear matter seen by the J/ψ of $L \geq 7$ fm the points from Pb + Pb collisions fall increasingly below this normal nuclear absorption curve. The range of L = 5.3 - 8 fm covered by the new NA60 experiment for In + In collisions is exactly in the interesting region where the deviation is starting. First data were shown by Seixas at this meeting (although not contained in these proceedings) and they confirm well the systematic trend seen in Pb + Pb collisions. Nothing has changed on the front of theoretical interpretation. The suppression can be explained by disappearance of the J/ψ (or possibly only the charmonia states that feed it) in a hot colored medium or by interaction with comovers, albeit with a very large density of more than $1/fm^3$, i.e. a value not deemed achievable for a hadron gas.

7. Direct photons

Direct photons in pp collisions have been measured at top RHIC energy by PHENIX [58]. It was shown at this conference by Cole and Bathe (see Fig. 1 in [59]) that they agree well with a NLO pQCD calculation by Gordon and Volgelsang [60]. The data are still statistics limited plus there is some remaining scale uncertainty. Here additional high luminosity pp running is desirable. For Au + Au collisions, the measurement by PHENIX [61] shown at this meeting by Bathe and Jacak agrees also well with the same NLO pQCD calculation. There is no definitive answer yet whether k_t broadening is needed. The centrality dependence is well reproduced over the entire range in p_t covered out to 13 GeV/c just by scaling with the number of collisions (see Figs. 2,3 in [59]). Comparing this to the observed high p_t suppression seen for hadron spectra (see above) confirms that this is indeed a final state effect due to the medium that should not and apparently does not affect photons. Comparing the measured γ/π^0 ratio to the expected one due to hadronic decays a direct photon component is apparent for p_t above 4 GeV/c. With systematic errors of 10 % this means that at the present stage there is unfortunately no sensitivity yet to thermal photons radiated by the plasma; they are expected to dominate over direct hard photons below about 3 GeV/c. To achieve the required sensitivity of a few % in γ/π^0 in the range p_t = 1-3 GeV/c requires a quantum jump in the quality of the experimental data. Nevertheless, this very difficult measurement is very desirable.

8. Fluctuations and correlations

For this interesting subject I refer to the authoritative review given by Voloshin in these proceedings [62]. Here only very few points are picked up. Transverse momentum fluctuations have been studied at the SPS and at RHIC and many different observables have been proposed and have been used to quantify these fluctuations. Nonstatistical fluctuations have been seen at all energies from $\sqrt{s_{nn}}$ = 8 to 200 GeV. Voloshin has proposed to use the momentum correlator as a measure. In this quantity all present measurements give a relative fluctuation of 1 % without any hint to increased fluctuations possibly indicating the vicinity of a critical point (see Fig. 4 of [62]). The question was raised by Koch at this meeting whether possibly the data cover not the optimal region in p_t . Another possibility is that the data integrated over opening angles average over too many different contributions to correlations which might lead to a loss in sensitivity [63]. Voloshin pointed out that, nevertheless, the current data are very useful because they are sensitive to velocity profiles for the expansion of the fireball as demonstrated in his Fig.5. This should indeed be exploited more to get quantitative results that then constrain the interpretation of transverse momentum spectra and two particle correlations.

A very tantalizing preliminary NA49 result on the charged kaon to pion multiplicity fluctuations was shown by Friese (see Fig.6 in [21]): at the smallest SPS energies the fluctuations are larger than expected in contrast to the fluctuations in the proton to pion ratio that look 'normal'. It would be extremely desirable to confirm this very difficult and, if confirmed, possibly far reaching measurement with better particle identification capability.

Acknowledgements. Valuable discussions with P. Braun-Munzinger are gratefully acknowledged. This work was supported in part by the HGF GSI virtual institute VH-VI-146.

[1] K. Adcox et al., PHENIX Collaboration, Nucl.Phys. **A757** (2005) 184, nucl-ex/0410003.
[2] I. Arsene et al., BRAHMS Collaboration, Nucl.Phys. **A757** (2005) 1, nucl-ex/0410020.
[3] B.B. Back et al., PHOBOS Collaboration, Nucl.Phys. **A757** (2005) 28, nucl-ex/0410022.
[4] J. Adams et al., STAR Collaboration, Nucl.Phys. **A757** (2005) 102, nucl-ex/0501009.
[5] S.S. Adler et al., PHENIX Collaboration, Phys.Rev. **C71** (2005) 034908; Erratum-ibid. **C71** (2005) 049901.
[6] J. Adams et al., STAR Collaboration, Phys.Rev. **C70** (2004) 054907.
[7] F. Karsch, Lect. Notes. Phys. **583** (2002) 209, hep-lat/0106019.
[8] I.G. Bearden et al., BRAHMS Collaboration, Phys. Lett. B **523** (2001) 227.

[9] B.B. Back et al., PHOBOS Collaboration, sub. Phys. Rev. Lett., nucl-ex/0301017.
[10] G. Roland for the PHOBOS Collaboration, these proceedings.
[11] B. Mohanty for the STAR Collaboration, these proceedings.
[12] J. L. Albacete, N. Armesto, J. G. Milhano, C. A. Salgado, and U. A. Wiedemann, Proceedings of "Hard Probes 2004", Eur. Phys. J. **C**, in print, hep-ph/0502167.
[13] D. Hofmann for the PHOBOS Collaboration, these proceedings.
[14] R. Debbe for the BRAHMS Collaboration, these proceedings.
[15] I.G. Bearden et al., BRAHMS Collaboration, Phys. Rev. Lett. **93** (2004) 102301.
[16] W. Busza and A.S. Goldhaber, Phys. Lett. B **139** (1984) 235.
[17] W. Busza and R. Ledoux, Ann. Rev. Nucl. Part. Sci. **38** (1988) 119.
[18] P. Braun-Munzinger, these proceedings.
[19] P. Braun-Munzinger, K. Redlich, and J. Stachel, in Quark-Gluon Plasma 3, eds. R.C. Hwa and X.N. Wang, (World Scientific, Singapore, 2004), 491.
[20] P. Braun-Munzinger, J. Stachel, and C. Wetterich, Phys. Lett. B **596** (2004) 61.
[21] V. Friese for the NA49 Collaboration, these proceedings.
[22] P. Huovinen, Nucl. Phys. A **715** (2003) 299c.
[23] D. Teaney, J. Lauret, and E.V. Shuryak, Nucl. Phys. A **698** (2002) 479c; and same authors in nucl-th/0110037.
[24] U. Heinz, J. Phys. **G31** (2005) S717.
[25] T. Alt et al., NA49 Collaboration, Phys. Rev. C **68** (2003) 034903.
[26] B. Jacak for the PHENIX Collaboration, these proceedings.
[27] N. Xu for the STAR Collaboration, these proceedings.
[28] T.K. Hemmick, J. Phys. **G30** (2004) S659.
[29] X.N. Wang and M. Gyulassy, Phys. Rev. Lett. **68** (1992) 1480.
[30] H. Baier, Y.L. Dokshitzer, A.H. Mueller, S. Peigne, and D. Schiff, Nucl. Phys. **B483** (1997) 291 and **B484** (1997) 265.
[31] I. Vitev, these proceedings.
[32] D. d'Enterria, Proceedings of "Hard Probes 2004", Eur. Phys. J. **C**, in print, nucl-ex/0504001.
[33] Q. Wang and X.N. Wang, Phys. Rev. C **71** (2005) 014903.
[34] M.A.C. Lamont for the STAR Collaboration, these proceedings.
[35] S.S. Adler et al., PHENIX Collaboration, Phys. Rev. C **71** (2005) 051902.
[36] I. Arsene et al., BRAHMS Collaboration, J. Phys. **G30** (2004) S759.
[37] B.B. Back et al., PHOBOS Collaboration, Phys. Rev. C **70** (2004) 061901.
[38] D. Kharzeev, Y.V. Kovchegov, and K. Tuchin, Phys. Lett. B **599** (2004) 23.
[39] R.C. Hwa, C.B. Yang, and R.J. Fries, Phys. Rev. C **71** (2005) 024902.
[40] B.Z. Kopeliovich, J. Nemchik, I.K. Potashnikova, M.B. Johnson, and I. Schmidt, hep-ph/0501260.
[41] J. Adams et al., STAR Collaboration, nucl-ex/0501016.
[42] A. Mischke for the STAR Collaboration, these proceedings.
[43] J. Adams et al., STAR Collaboration, Phys. Rev. Lett. **93** (2004) 252301.
[44] D. Adamova et al., CERES Collaboration, Phys. Rev. Lett. **92** (2004) 032301.
[45] S.S. Adler et al., PHENIX Collaboration, nucl-ex/0507004.
[46] H. Stöcker, these proceedings.
[47] H. Stöcker, Nucl. Phys. A **750** (2005) 121.
[48] J. Casalderrey-Solana, E. Shuryak, and D. Teaney, hep-ph/0411315.
[49] K. Adcox et al., PHENIX Collaboration, Phys. Rev. Lett. **88** (2002) 192303;
S.S. Adler et al., PHENIX Collaboration, Phys. Rev. Lett. **94** (2004) 082301.
[50] V. Greco, C.M. Ko, R. Rapp, Phys. Lett. B **595** (2004) 202.
[51] P. Braun-Munzinger, J. Stachel, Phys. Lett. B **490** (2000) 196.
[52] T. Tabaru for the PHENIX Collaboration, these proceedings.
[53] N. Armesto, A. Dainese, C.A. Salgado, U. Wiedemann, Phys. Rev. D **71** (2005) 054027.
[54] K. Adcox et al., PHENIX Collaboration, J. Phys. **G30** (2004) S1189.
[55] J. Adams et al., STAR Collaboration, Phys. Rev. Lett. **94** (2005) 062301.
[56] C. Quintans for the NA50 Collabortion, these proceedings.
[57] B. Alessandro et al., NA50 Collaboration, Eur. Phys. J. **C39** (2005) 335.
[58] S.S. Adler et al., PHENIX collaboration, Phys. Rev. D **71** (2005) 071102R.
[59] S. Bathe for the PHENIX Collaboration, theses proceedings.
[60] L.E. Gordon and W. Vogelsang, Phys. Rev. D **48** (1993) 3136.
[61] S.S. Adler et al., PHENIX collaboration, Phys. Rev. Lett. **94** (2005) 232301.
[62] S. Voloshin, these proceedings.
[63] D. Miskowiecz, CERES Collaboration, private communication.

Institute of Physics Publishing Journal of Physics: Conference Series **50** (2006) 300–310
doi:10.1088/1742-6596/50/1/036 5th International Conference on Physics and Astrophysics of Quark Gluon Plasma

Summary of theoretical contributions

H. Stöcker[1,2]

[1] Frankfurt Institute for Advanced Studies (FIAS), Johann Wolfgang Goethe University, Max-von-Laue-Str. 1, 60438 Frankfurt am Main, Germany
[2] Institut für Theoretische Physik, Johann Wolfgang Goethe Universität, Max-von-Laue-Str. 1, 60438 Frankfurt am Main, Germany

Abstract. Results from various theoretical approaches and ideas presented at this exciting meeting are reviewed. I also point towards future directions, in particular hydrodynamic behaviour induced by jets traveling through the quark-gluon plasma, which might be worth looking at in more detail.

1. Theoretical overview

We have witnessed an exciting conference with an excellent program, heated scientific discussions and lots of new data and theoretical ideas. Our topics ranged from astrophysics to field theory, from heavy-ion reaction phenomenology to big-bang cosmology.

My task, i.e. to review all in all about 30 theory talks, is combined in this paper with a cross-disciplinary analysis of experiments, which verify - pardon, falsify the theoretical conjectures in many cases - there is an old saying that a theory can never be verified: even if lots of data support the theory, at some point the theory will always go astray...

Let me rearrange the order of the theory talks on the topics of our meeting[1]:

- Equation of State
- Collective Dynamics
- Jets : Production and Quenching
- Results from $p+p$, $p(d)+A$ and $A+A$ collisions
- Signatures of Quark Gluon Plasma
- QCD at Finite Temperature and Density
- Multiparticle production, fluctuations and correlations
- Cosmological Implications of the QCD Phase Transition
- QCD Phenomenology
- Low x behaviour of QCD
- Strangeness and heavy flavor production

The common interest is given by the titel of the conference: "*Physics and astrophysics of the quark gluon plasma*"[2]

[1] Instead of giving explicit references I refer the reader to the electronic proceedings available on the Web
[2] Somewhat in this summary, also pardon the fact that some of the many interesting items have not been taken up here, because I did not witness the first days of the conference.

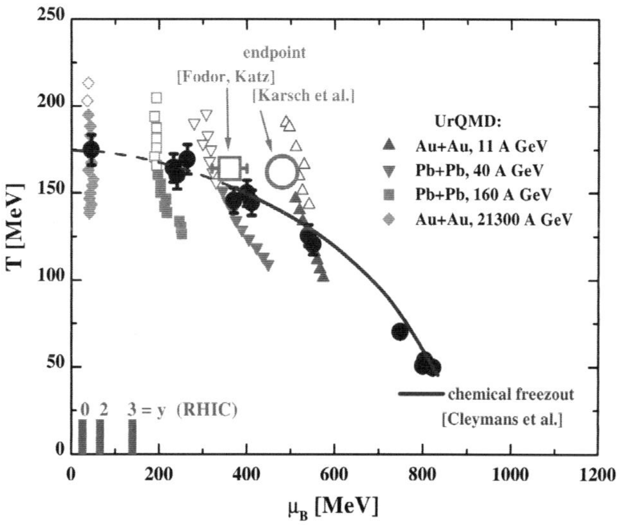

Figure 1. The phase diagram with the critical end point at $\mu_B \approx 400$ MeV, $T \approx 160$ MeV as predicted by the Swansea-Bielefeld and Wuppertal/Budapest collaborations [1]. In addition, the time evolution in the $T - \mu$-plane of a central cell in UrQMD calculations [2] is depicted for different bombarding energies. Note, that the calculations indicate that bombarding energies $E_{LAB} \lesssim 40$ A·GeV are needed to probe a first order phase transition as predicted by the Swansea-Bielefeld and Wuppertal-Budapest collaborations. At RHIC (see insert at the μ_B scale) this point is accessible in the fragmentation region only (taken from [3]). The new conjecture by Gavai and friends is that the critical endpoint is moving to the left to $T = 0.95\ T_c$, $\mu_B = 1.1 - 1.3T_c \approx$ 190–220 MeV. In this case the top SPS energy range would be best suited to explore the endpoint in central Au+Au collisions.

- Astrophysics
- Lattice
- Colored Glass
- Fluctuations & DCCs
- J/Psi & EM Probes
- Strangeness
- Transport Theory
- Hydro & Jets

John Ellis gave a beautiful survey of the common issues in both heavy-ion physics and the big bang cosmology: We do in both cases study a very fast expansion of dense/hot strongly interacting matter, and do have the task to reconcile whatever happened in the first few nanoseconds of the big bang from the sparse debris found nowadays. The connection to the matter-anti matter asymmetry problems is particularly exciting for future topical studies at the LHC. This is quite analogous to the transient 6-8 fm/c $\approx 2 - 2.5 \times 10^{-23}s$ timescale of the collision processes at RHIC.

The intense astrophysics discussions between Bombaci and Banyopadyhyay about the possible occurence of massive strange quark stars (SQS), the transition of neutron stars to

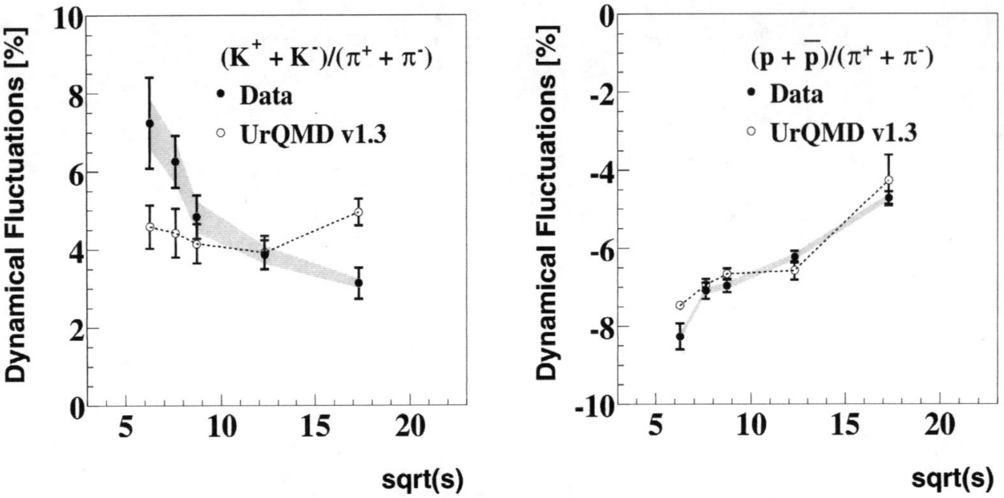

Figure 2. Energy dependence of the event-by-event fluctuation signal of the $(K^+ + K^-)/(\pi^+ + \pi^-)$ ratio (left panel) and the $(p + \bar{p})/(\pi^+ + \pi^-)$ ratio (right panel). The systematic errors of the measurements are shown as grey bands (from Ref. [4]).

strange hyperon-, hybrid- and quark stars, and the relation to the gamma ray bursts (quark-deconfinement nova-model) has been of particular interest - this transition is predicted to yield a radius-collapse of several kilometers.

The first observation of the "double delight pulsar" psr-j0737-3039 will enable us to pin down the mass-radius curves by the spin-orbit effect with high precision. D. Bandyopadyhyay showed that soft equations of state (EOS) are ruled out by EXO 0748-676. The connection of conjectured different color superconducting phases to the cooling curves of SQS have been pointed out in the paper by Mishra and Mishra.

The lattice-QCD (lQCD) discussions between Gavai and Laermann centered about the questions on the order of the phase transition and on the speed of sound. Laermann stated that there is no indication for criticality, while Gavai and friends showed that the critical endpoint is at $T = 0.95\ T_c$, $\mu_B/T = 1.1 - 1.3$, i.e. less than half of the μ_B=400 MeV values given by the Swansea-Bielefeld and Wuppertal-Budapest collaborations (cf. Fig 1). Gavai showed also that all the way up to $T = 2T_c$, the speed of sound is much less, $c_s^2 = 0.15$ at $T = 1.1\ T_c$, than that of a noninteracting ultrarelativistic (massless) gas, $c_s^2 = 1/3$.

In the colored glass condensate section, Venugopalan explored the demise of the structure function, in particular how the dipole and higher multipole operators may turn out to be the more relevant observables at high energies. Adding valence quark contributions, Kovchegov showed a quite satisfactory agreement of the Color-Glass-Condensate (CGC)-model to the observed rapidity dependence of the p_T-distributions. McLerran iterated the theme of the Color Glass Condensate as THE Medium: Pomerons, Odderons, Reggeons as Quasiparticle excitations of the CGC - does this mean that the CGC is the initial phase for the QGP? Is the strong Quark-Gluon Plasma (sQGP) really the CGC? Is rapid 'thermalization' due to the CGC? Does flow arise largely from the CGC? Well, definitely LHC is THE CGC machine – according to McLerran.

Fluctuations and Disordered Chiral Condensates (DCC's) were discussed by Koch, Csörgö, Chandrasekar and Randrup, among others. K/π fluctuations increase towards lower beam energy with a significant enhancement over the hadronic cascade model UrQMD [5] (cf. Fig. 2)! On the other hand, p/π fluctuations are negative – this indicates a strong contribution from resonance decays, as was shown by Koch in comparing NA49-data to UrQMD results.

Dileptons, J/Psi, and photons have been discussed by Lee, Mustafa and Koch (among others). Large corrections on the QCD NLO Quarkonium- Gluon/hadron dissociation cross section have been reported even for the Ypsilon system, especially near threshold. The thermal width of the J/Ψ should be $\sim$ 1 GeV at T=600 MeV according to Lee's estimates.

Strangegeness and equilibration has been the main topic of Rafelski, Cleymans, Braun-Munzinger and Bleicher. The structure in the K/π ratios reported by NA49 near $\sqrt(s) = 8$ GeV is not reproduced by any model (cf. Fig. 3), but Peter Braun-Munzinger notes: the natural smearing is 3 GeV near that energy - how can the 'horn' then be so steep? Hadron-string models work well globally, as Bleicher reports, but these models do NOT give MULTI-STRANGE BARYONs! Is the alternative a four parameter nonequilibrium thermal model, with $T, \mu, \Gamma_\mu, \Gamma_s$, by Rafelski et al.?

The extreme density/temperature dependence of the characteristic equilibration time, $\tau_{eq.} \sim T^{-60}$, was pointed out by Braun-Munzinger, which implies that all particles freeze out at about the hadronization time. According to Braun-Munzinger this might be due to Carsten Greiner's conjecture of Hagedorn states as intermediate doorway states.

Deeply bound $\bar{p}$ and K^- states as gateway to cold and dense matter were discussed by Walter Greiner: $\bar{p}$'s – due to G-parity in the strong interactions – and K^- can suppress repulsive vector fields, thus predicting discrete bound states with binding energies of several 100th MeV and 20 fm/c life times [8]. Formation of such cold and highly dense nuclear system at densities $\rho\, 3-5\rho_0$ will be studied in dense $\bar{p}$ - nuclear systems at FAIR (GSI)and the K-nucleus collisions at J-PARC.

Jacak, Shuryak, Heinz and Chauduri discussed applications of hydrodynamics to RHIC-collisions. The reasons why hydro does reasonably well fit both, radial and elliptic flow for a large number of hadron species (cf. Fig. 4), is still not fully settled. The question of early thermalization and the unsatisfactory rapidity distributions from ideal hydrodynamics remain open.

Bass showed in his talk, however, that the recombination/quark coalescence models (cf. Fig. 5) can help analyze the participant scaling and even the charm flow. However, as Bleicher showed, even the hadron/string model UrQMD may exhibit "recombination" and participant scaling.

Jacak showed the PHENIX jet-pair distributions, which clearly give a novel signal to the away-side jet suppression (cf. Fig. 6 for STAR results), i.e. the recent topic of Mach-cones induced by stopped jets in the quark-gluon liquid [12]. This is most important as an observable, because it links the parton dynamics and collective flow and the jet tomography to the measurement of the speed of sound in the medium - be it a weakly or strongly coupled plasma: the opening angle of the Mach-shock-wakes directly gives the speed of sound in the medium, which is linked to both, the appearance of vector potentials and the parton/constituent mass parameters.

2. Interlude on Mach shocks

Sideward peaks around the away-side jet have been predicted recently [12] as a signature of Mach shock waves created by stopping partonic jets propagating through a QGP formed in an ultrarelativistic heavy–ion collision. Analogous Mach shock waves were studied long ago for heavy-ion induced Mach shocks travelling through cold hadronic matter [14, 15] as well as in nuclear Fermi liquids [16, 17]. It has been argued that Mach–like motions of quark–gluon matter can appear via the excitation of collective plasmon waves by the moving color charge associated with the leading jet [12, 22].

Point–like perturbations (a small body, a hadron or parton etc.) moving with a supersonic speed in the spatially homogeneous ideal fluid produce the Mach region of the perturbed matter [18]. In the fluid rest frame (FRF) the Mach region has a conical shape (cf. Fig. 7) with an

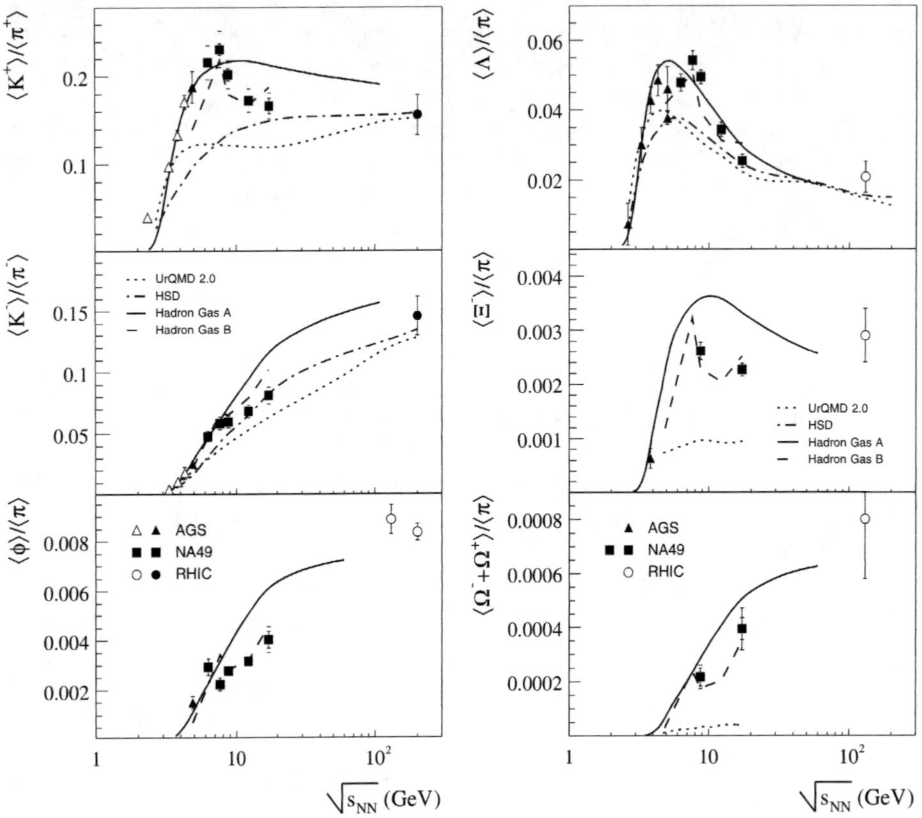

Figure 3. The energy dependence of the 4π-yields of strange hadrons, normalized to the pion yields, in central Pb+Pb/Au+Au collisions. The data are compared to string hadronic models UrQMD2.0 [5]: dotted lines; HSD[6]: dashed-dotted lines) and statistical hadron-gas models from Braun-Munzinger and Becattini and collaborators (with strageness undersaturation: dashed line, assuming full equilibrium: solid line). The figure is taken from [7].

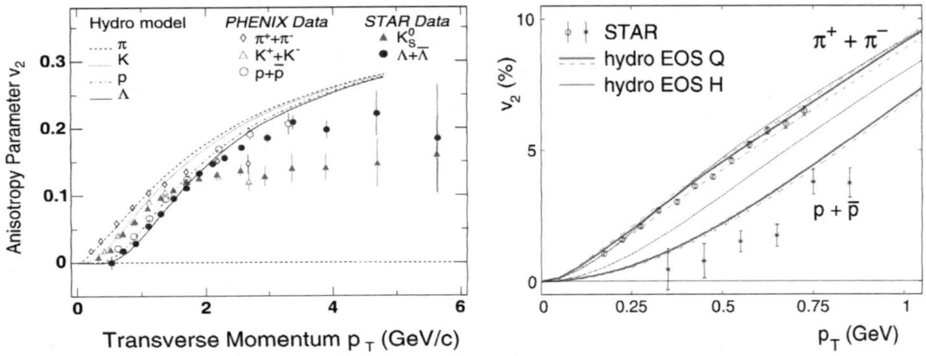

Figure 4. Differential elliptic flow $v_2(p_\perp)$ for several identified hadron species from minimum bias Au+Au collisions at $\sqrt{s}$= 130 GeV (right) and $\sqrt{s}$ = 200 GeV compared with hydrodynamic predictions from [9]. The figure is taken from [10].

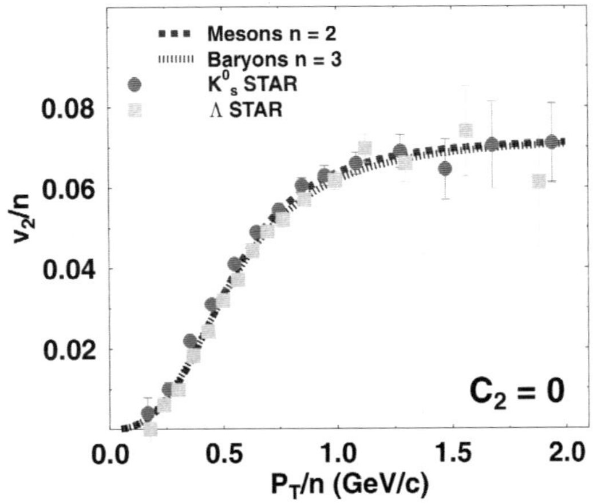

Figure 5. Scaled elliptic flow $v_2/n(p_\perp/n)$ of baryons and mesons as calculated from the quark flow compared to data for Λ-hyperons and kaons. The figure is taken from [26].

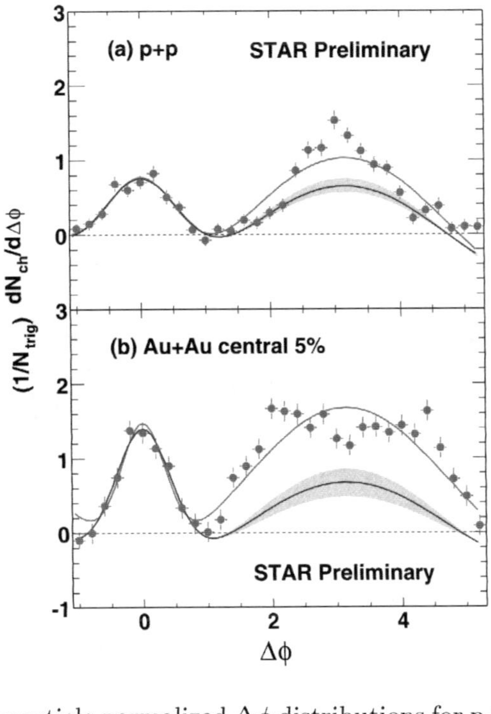

Figure 6. The per trigger particle normalized $\Delta\phi$ distributions for p+p (a) and 5% most central Au+Au collisions (b). The figure is taken from [24].

opening angle with respect to the direction of particle propagation given by[3]

$$\tilde{\theta}_M = \sin^{-1}\left(\frac{c_s}{\tilde{v}}\right), \qquad (1)$$

where c_s denotes the sound velocity of the unperturbed (upstream) fluid and $\tilde{\boldsymbol{v}}$ is the particle

[3] Here and below the quantities in the FRF are marked by a tilde.

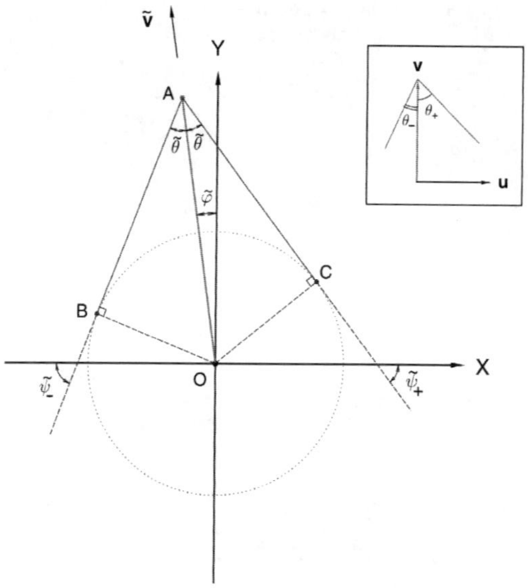

Figure 7. Mach region created by jet moving with velocity $\boldsymbol{v}$ orthogonal to the fluid velocity $\boldsymbol{u}$. Main plot and insert correspond to FRF and CMF, respectively. It is assumed that jet moves from O to A in FRF. Dotted circle represents the front of sound wave generated at point O.

velocity with respect to the fluid. In the FRF, trajectories of fluid elements (perpendicular to the surface of the Mach cone) are inclined at the angle $\Delta\theta = \pi/2 - \widetilde{\theta}_M$ with respect to $\widetilde{\boldsymbol{v}}$. Strictly speaking, formula (1) is applicable only for weak, sound–like perturbations and certainly not valid for space–time regions close to a leading particle. Nevertheless, it suffices for a qualitative analysis of flow effects. Following Refs. [12, 13, 19] one can estimate the angle of preferential emission of secondaries associated with a fast jet in the QGP. Substituting $\widetilde{v} = 1, c_s = 1/\sqrt{3}$ into Eq. (1) gives the value $\Delta\theta \simeq 0.96$ rad $= 61^o$This agrees well with positions of maxima of the away–side two–particle distributions observed by the STAR Collaboration (cf. 6) in central Au+Au collisions at RHIC energies (cf. also B. Jacak's talk).

Let us consider [13] the case when the away–side jet propagates with velocity $\boldsymbol{v}$ parallel to the matter flow velocity $\boldsymbol{u}$. Assuming that $\boldsymbol{u}$ does not change with space and time, and performing the Lorentz boost to the FRF, one sees that a weak Mach shock has a conical shape with the axis along $\boldsymbol{v}$. In this reference frame, the shock front angle $\widetilde{\theta}_M$ is given by (1). Transformation from the FRF to the c.m. frame (CMF) shows that the Mach region remains conical, but the Mach angle becomes smaller in the CMF:

$$\tan\theta_M = \frac{1}{\gamma_u}\tan\widetilde{\theta}_M\,, \tag{2}$$

where $\gamma_u \equiv (1-u^2)^{-1/2}$ is the Lorentz factor corresponding to the flow velocity $\mathbf{u}$. The resulting expression for the Mach angle in the CMF is

$$\theta_M = \tan^{-1}\left(c_s\sqrt{\frac{1-u^2}{\widetilde{v}^2 - c_s^2}}\right)\,, \tag{3}$$

where

$$\widetilde{v} = \frac{v \mp u}{1 \mp vu}\,, \tag{4}$$

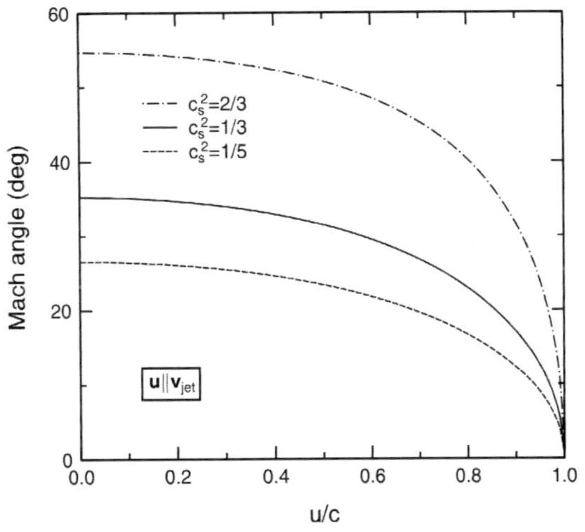

Figure 8. Mach cone angles for jet propagating collinearly to the matter flow as a function of fluid velocity u. Different curves correspond to different values of sound velocity c_s.

and upper (lower) sign corresponds to the jet's motion in (or opposite to) the direction of collective flow. For ultrarelativistic jets ($v \to 1$) one can take $\widetilde{v} \simeq 1$ which leads to a simpler expression

$$\theta_M \simeq \tan^{-1}\left(\frac{c_s\gamma_s}{\gamma_u}\right) = \sin^{-1}\left(c_s\sqrt{\frac{1-u^2}{1-u^2c_s^2}}\right), \tag{5}$$

where $\gamma_s = (1-c_s^2)^{-1/2}$. According to (5), in the ultrarelativistic limit θ_M does not depend on the direction of flow with respect to the jet. The Mach cone becomes more narrow as compared to jet propagation in static matter. This narrowing effect has a purely relativistic origin. Indeed, the difference between θ_M from (5) and the Mach angle in absence of flow ($\lim_{u\to 0}\theta_M = \sin^{-1} c_s$) is of second order in the collective velocity u. The Mach angle calculated from (5) is shown in Fig. 8 (from [13]) as a function of u for different sound velocities c_s. Following Ref. [19], the value $c_s^2 = 1/5$ is identified with the hadronic matter and $c_s^2 = 1/3$ with ideal QGP composed of massless quarks and gluons. The value $c_s^2 = 2/3$ may be chosen to represent a strongly coupled QGP [20]. We see that precise measurements will provide valuable information on the properties of the quark-gluon liquid [11, 12].

3. Future directions

I propose future correlation measurements which can yield spectroscopic information on the plasma:

(i) Measure the sound velocity of the expanding plasma by the emission pattern of the plasma particles traveling sideways with respect to the jet axis: The dispersive wave generated by the wake of the jet in the plasma yields preferential emission to an angle (relative to the jet axis) which is given by the ratio of the leading jet particles' velocity, devided by the sound velocity in the hot dense plasma rest frame. The speed of sound for a non-interacting gas of relativistic massless plasma particles is $c_s \approx \frac{c}{\sqrt{3}} \approx 57\%\, c$, while for a plasma with strong vector interactions, $c_s \approx c$, since strong shocks can yield larger speeds. They are also related – unlike the linearized sound waves – to strong matter flow with high flow velocities

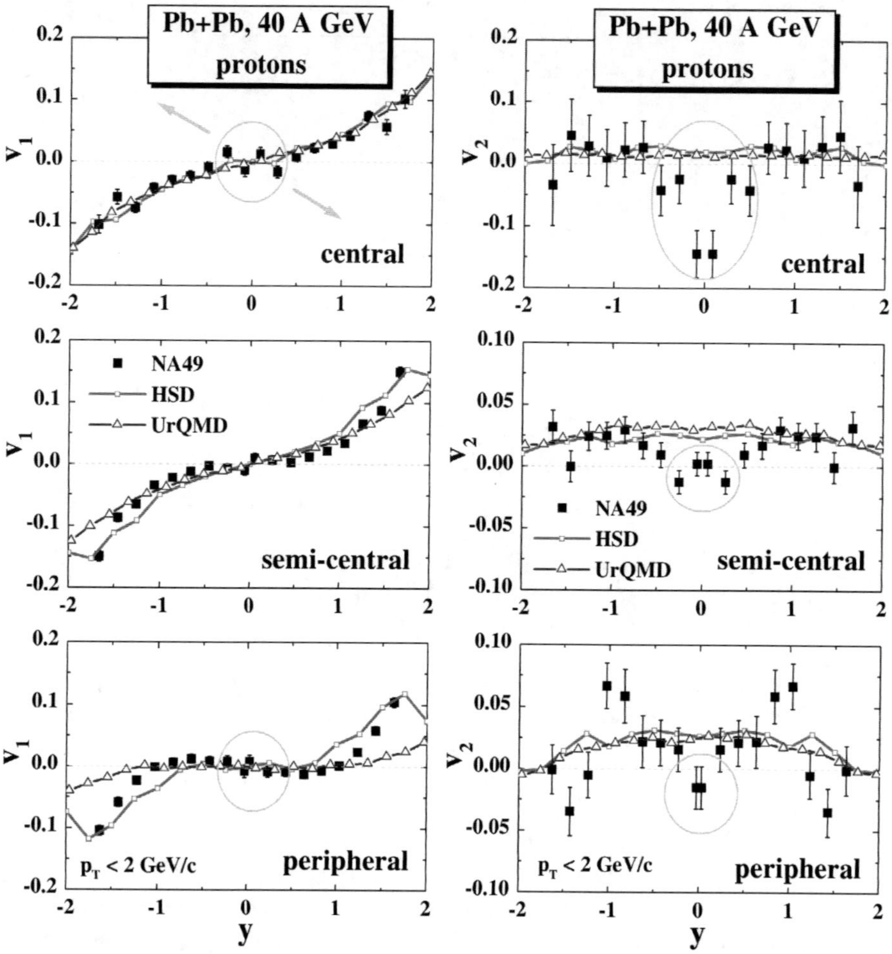

Figure 9. The flow v_1 and v_2 for protons from NA49 [27] for Pb+Pb at 40 A GeV in comparison to the results of the hadron/string models HSD (red lines) and UrQMD (blue lines). Note the large v_2 deviations (at y_{cm}) of the data from the best conventional hadronic transport theories. The figure is taken from [25].

v_f approaching the speed of light relative to the expanding medium. Hence, the emission angle measurement can yield information of the interactions in the plasma.

(ii) The NA49 collaboration has observed the collapse of both, v_1- and v_2-collective flow of protons (cf. Fig. 9), in Pb+Pb collisions at 40 A·GeV, which presents first evidence for a first order phase transition in baryon-rich dense matter. It should be possible to study the nature of this transition and the properties of the expected chirally restored and deconfined phase both at the forward fragmentation region at RHIC, with upgraded and/or second generation detectors, and at the new GSI facility FAIR.

(iii) A critical discussion of the use of collective flow as a barometer for the equation of state (EoS) of hot dense matter at RHIC showed that hadronic rescattering models can explain $< 30\%$ of the observed elliptic flow v_2 for $p_T > 2$ GeV/c [23, 28]. I interpret this as evidence for the production of superdense matter at RHIC with initial pressure way above hadronic pressure, $p > 1$ GeV/fm^3.

(iv) The fluctuations in the flow, v_1 and v_2, should be measured. Ideal Hydrodynamics predicts

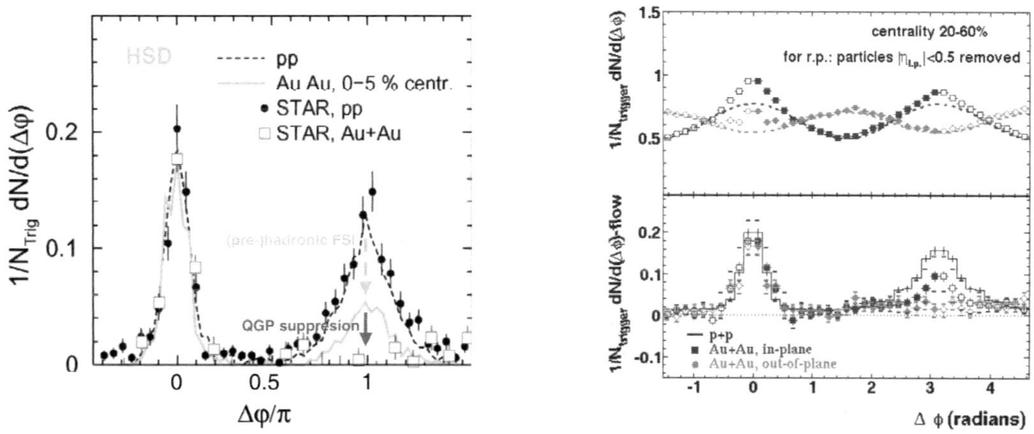

Figure 10. Left: STAR data on near-side and away-side jet correlations compared to the HSD model for p+p and central Au+Au collisions at midrapidity for $p_T^{Trig} = 4 \ldots 6\,\mathrm{GeV}/c$ and $p_T = 2\,\mathrm{GeV}/c \ldots p_T^{Trig}$ [28, 29]. Right: High p_T correlations: in-plane vs. out-of-plane correlations of the probe (jet+secondary jet fragments) with the bulk (v_2 of the plasma at $p_T > 2\,\mathrm{GeV/c}$), prove the existence of the initial plasma state (STAR-collaboration, preliminary).

that they are larger than 50 % due to initial state fluctuations. The QGP coefficient of viscosity may be determined experimentally from the fluctuations observed and proof the conjecture of Ref. [11].

(v) The connection of v_2 to jet suppression has proven experimentally that the collective flow is not faked by minijet fragmentation and theoretically that the away-side jet suppression can only partially ($< 50\%$) be due to pre-hadronic or hadronic rescattering [23] (cf. Fig. 10).

(vi) I propose upgrades and second generation experiments at RHIC, which inspect the first order phase transition in the fragmentation region, i.e. at $\mu_B \approx 200 - 400$ MeV ($y \approx 3 - 5$), where the collapse of the proton flow – analogous to the 40 A·GeV data – should be seen.

Let me finally express my birthday greetings to Bikash and thank him and his crew for decades of exciting physics conjectures, his strong involvement into our field and courage to built up such a great school of young successful scientists in India, which are highly competitive in the whole world.

Acknowledgments

This work is partially supported by GSI, BMFT, DFG, and DAAD. Let me thank Elena Bratkovskaya, Leonid Satarov and Igor Mishustin (from FIAS) for their contributions to this Summary.

References

[1] Z. Fodor and S. D. Katz, JHEP **0203** (2002) 014; JHEP **0404** (2004) 050.
[2] L. V. Bravina *et al.*, Phys. Rev. **C 60** (1999) 024904; Nucl. Phys. **A 698** (2002) 383.
[3] E. L. Bratkovskaya *et al.*, Phys. Rev. **C 69** (2004) 054907.
[4] Ch. Roland, J. Phys. G **31** (2005) S1075.
[5] S.A. Bass *et al.*, Prog. Part. Nucl. Phys. **42** (1998) 255; M. Bleicher *et al.*, J. Phys. G **25** (1999) 1859.
[6] W. Ehehalt and W. Cassing, Nucl. Phys. A **602** (1996) 449; W. Cassing and E. L. Bratkovskaya, Phys. Rep. **308** (1999) 65.

[7] Ch. Blume, J. Phys. G **31** (2005) S685.
[8] Y. Akaishi and T. Yamazaki , Phys. Rev. C65 (2002) 044005.
[9] P. Huovinen *et al.*, Phys. Lett. **B 503** (2001) 58.
[10] U. Heinz, J. Phys. G **31** (2005) S717.
[11] A. Peshier and W. Cassing, Phys. Rev. Lett. 94 (2005) 172301.
[12] H. Stöcker, Nucl. Phys. **A 750** (2005) 121; H. Stöcker, E. L. Bratkovskaya *et al.*, J. Phys. **G 31** (2005) S929.
[13] L.M. Satarov, H. Stöcker and I.N. Mishustin, hep-ph/0505245.
[14] J. Hofmann, H. Stöcker, W. Scheid, and W. Greiner, in report of Int. Workshop on BeV/nucleon Collisions of Heavy Ions: How and Why, Bear Mountain, NY, 1974 (report BNL–AUI, 1975);
H.G. Baumgardt, J.U. Schott, Y. Sakamoto, E. Schopper, H. Stöcker, J. Hofmann, W. Scheid, and W. Greiner, Z. Phys. **A 273** (1975) 359;
J. Hofmann, H. Stöcker, U. Heinz, W. Scheid, and W. Greiner, Phys. Rev. Lett. **36** (1976) 88.
[15] H. Stöcker, J. Hofmann, J.A. Maruhn, and W. Greiner, Prog. Part. Nucl. Phys. **4** (1980) 133.
[16] A.E. Glassgold, W. Heckrotte, and K.M. Watson, Ann. Phys. **6** (1959) 1.
[17] V.A. Khodel, N.N. Kurilkin, and I.N. Mishustin, Phys. Lett. **B 90** (1980) 37.
[18] L.D. Landau and E.M. Lifshitz, *Fluid Mechanics*, Pergamon Press, NY, 1959.
[19] J. Casalderrey–Solana, E.V. Shuryak, and D. Teaney, hep–ph/0411315.
[20] E.V. Shuryak and I. Zahed, Phys. Rev. **C 70** (2004) 021901.
[21] D. Kharzeev, private communication.
[22] W. Schäfer, H. Stöcker, B. Müller, and W. Greiner, Z. Phys. **A 288** (1978) 349.
[23] K. Gallmeister and W. Cassing, Nucl. Phys. **A 748** (2005) 241.
[24] F. Wang (STAR Collaboration), J. Phys. **G 30** (2004) S1299.
[25] H. Stöcker, E. L. Bratkovskaya, M. Bleicher, S. Soff, and X. Zhu, J. Phys. **G 31** (2005) S929.
[26] B. Müller, R. J. Fries, and S. A. Bass, nucl-th/0503003.
[27] C. Alt *et al.*, Phys. Rev. **C 68** (2003) 034903.
[28] W. Cassing, K. Gallmeister, and C. Greiner, Nucl. Phys. **A 735** (2004) 277.
[29] W. Cassing, K. Gallmeister, and C. Greiner. J. Phys. **G 30** (2004) S801.

Journal of Physics: Conference Series **50** (2006) 311–314
5th International Conference on Physics and Astrophysics of Quark Gluon Plasma

Studying nuclear matter created in p+p, d+Au and Au+Au collisions using charged kaons

Camelia Mironov for the STAR Collaboration
Department of Physics Kent State University
Kent, Ohio 44242, USA

E-mail: cmironov@bnl.gov

Abstract. Kaons, as the main carriers of strangeness, make for an important probe of the medium produced in relativistic heavy ion collisions. By topologically reconstructing the charged kaons, we reach high transverse momentum limits not presently accessible with the traditional reconstruction methods. We present recent results on nuclear modification factors involving charged kaons from three collision systems: p+p, d+A and Au+Au at 200GeV center of mass energy. A comparison to other strange mesons and baryons will be made and conclusions on medium properties, particle production mechanisms, initial and final state effects will be presented.

1. Introduction

Some of the questions that the RHIC program at Brookhaven National Laboratory tries to answer are: what are the properties (density, degrees of freedom etc) of the medium created after colliding relativistic heavy ions, what is the nature of the hadronization mechanism and is it a unique process independent of transverse momentum? The answers have to take into account the fact that the measured observables are influenced by both initial state (e.g. multiple scattering, gluon saturation) and final state (partonic or hadronic interactions) effects. In order to disentangle these effects, comparisons to simpler baseline systems (d+Au and p+p) are performed and some of them are presented in this paper.

A useful tool that has been applied to these studies is the *Nuclear modification factor* R^h_{AB}:

$$R^h_{AB} = s\frac{1}{< N^{AB}_{binary} >}\frac{d^2N^h_{AB}}{d^2N^h_{pp}} \tag{1}$$

where d^2N^h is the hadron differential yield and $< N^{AB}_{binary} >$ is the mean number of binary collisions. It is thought that for $p_T > 2$ GeV/c, particle production should scale with the number of binary collisions and hence $R^h_{AB} \approx 1$. An $R^h_{AB} < 1$ is indicative of in-medium energy loss whilst an $R^h_{AB} > 1$ is traditionally attributed to soft parton scattering prior to the hard collision (Cronin effect) [1]. An alternative measurement that has been applied to the data is the ratio of spectra in central to peripheral collisions, R_{CP}. The assumption made is that the information provided by either ratio is consistent, as long as the peripheral Au+Au, d+Au and p+p data are similar. This is a reasonable hypothesis as a dense medium is presumably not created in any of these systems, which could cause many hadronic or partonic re-scattering in the final state.

Previous results on unidentified spectra have shown $R^h_{AB} < 1$ in central Au+Au collisions and $R^h_{AB} > 1$ in minimum bias d+Au collisions [2]. Models based on LO pQCD calculations interpreted the data as medium induced jet quenching in central Au+Au collisions and initial state Cronin effect in d+Au data [3, 4]. These first inclusive hadron results pointed to (string) fragmentation as the hadronization mechanism for $p_T >$2GeV/c.

However, further studies using identified particle distributions revealed new phenomena. The first measurements of identified particle R_{CP} in Au+Au collisions at $\sqrt{s_{NN}} = 200$ GeV showed differences in the level of suppression between kaons and the Λ in the intermediate transverse momentum region between 2 and 6 GeV/c [9]. These measurements were inconsistent with a simple model of jet quenching in the medium and required further explanation. Further measurements of R_{CP} for other particles (K^*, Ξ and Ω) showed that this observed effect was not dependent on the particle mass, but rather on the particle species: mesons exhibited a different level of suppression to baryons [10].

In order to explain this baryon-meson anomaly a number of recombination models postulated that baryons are formed from three, and mesons from two, coalescing quarks respectively [8]. Therefore, assuming the same partonic p_T in a thermal partonic medium, the baryons will be pushed further in p_T compared to mesons and their suppression will occur later in transverse momentum.

In this paper we present the R_{AuAu} and R_{dAu} measurements for identified baryons and mesons as obtained by the STAR Collaboration.

2. Data analysis

The experimental results presented were obtained using events collected by the STAR experiment during the $\sqrt{s_{NN}} = 200$ GeV runs in 2001 (Au+Au and p+p) and 2003 (d+Au). The majority of the particles used are reconstructed via their decay products, either through topological cuts (Λ, $\overline{\Lambda}$, K^0_S, $K^{\pm}$, $\Omega, \overline{\Omega}$, Ξ and $\overline{\Xi}$) on tracks measured in the Time Projection Chamber (TPC) [11] or through an event mixing technique [12]. The proton R_{AuAu} is obtained by combining information from the Time of Flight (TOF) and the Cherenkov (RICH) detectors [11].

Charged kaon decays are detected through their characteristic 'kink' topology: $K^{\pm} \rightarrow \mu^{\pm}\nu(64\%)$, $K^{\pm} \rightarrow \pi^{\pm}\pi^0(21\%)$ [13]. The charged parent (kaon) decays to a neutral daughter (not detected) and a charged daughter with the same sign. Therefore, the kaon identification process reduces to the finding of 'bent' tracks (kinks) in the TPC. Geometrical cuts and kinematical constraints are applied to the candidates in order to reduce the background and isolate the signal. In these datasets, the transverse momentum limitation is dictated by statistics only: 3 GeV/c in p+p, 4 GeV/c in d+Au and 5 GeV/c in Au+Au.

3. Results

In Fig. 1 the R_{AuAu} for different baryons and mesons is presented as a function of p_T for central Au+Au collisions. The R^{meson}_{AuAu} is consistent with the R^{meson}_{CP}, and the R_{AuAu} and R_{CP} of the proton are also in agreement [5]. However, for the strange and multi-strange baryons, the R_{AuAu} shows the opposite effect to the R^{baryon}_{CP} results [10]: it is not reduced but increased with respect to binary scaling in the intermediate (2 to 6 GeV/c) p_T region. Moreover, there seems to exist an ordering, which follows the strange quark content: Ξ hyperons (strangeness S=-2) are enhanced more than Λ hyperons (S=-1) which are in turn enhanced more than protons (S=0).

A first attempt at determining R_{dAu} is shown in Figure 2. The present level of uncertainties in the d+Au analysis do not allow us to draw significant conclusions about a possible strangeness ordering also in d+Au. However, we can affirm that the baryon-meson difference observed in Au+Au is present also in d+Au but at a smaller level which can not account for the meson-baryon differences seen in R^{AuAu}_{CP}.

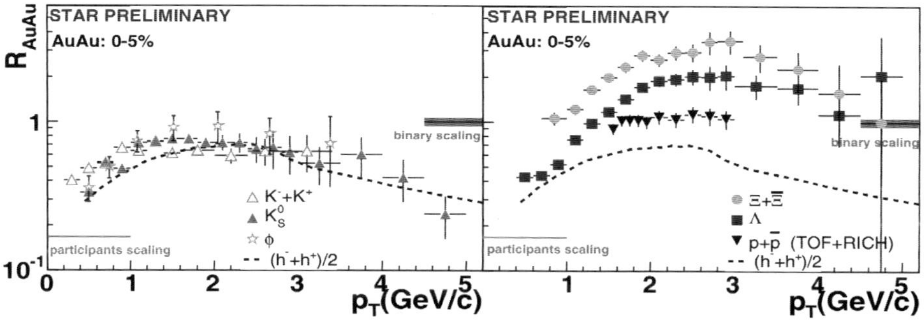

Figure 1. R_{AuAu} for mesons(left), baryons (right) and inclusive charged hadrons (dashed line). Statistical errors only are shown.

4. Discussion

Though the experimental results on R_{CP} point to the validity of the quark coalescence picture, several other models offer a different explanation to describe both the central to peripheral ratio and the central to p+p ratio [15, 14]. In particular, $HIJING/B\overline{B}$ v2.0 attributes the meson-baryon separation to a different p_T 'kick' for di-quarks than for quarks and to the presence of an additional baryon production mechanism (string junctions loops) which has an increasing contribution from p+p to d+Au to Au+Au collisions [14]. In the same string fragmentation scenario, the presence of 'strong color field' effects influence the strange baryon production and produces the strangeness ordering in the R_{AA} plot.

In addition to these hadronization scenarios, other effects might influence the Au+Au to p+p spectra ratios. In particular, if canonical suppression dominates the strange baryon production in p+p collisions but not in peripheral Au+Au collisions [16], then a plausible reason for the R_{AA}^{baryon} increase with respect to binary scaling is not only an enhancement of the yield in central Au+Au collisions, but a suppression of the yield in p+p collisions. It was shown that this suppression will increase as a function of strangeness content for baryons [16], and therefore it should lead to larger differences between R_{CP} and R_{AA} for multi-strange baryons than for singly strange or non-strange baryons.

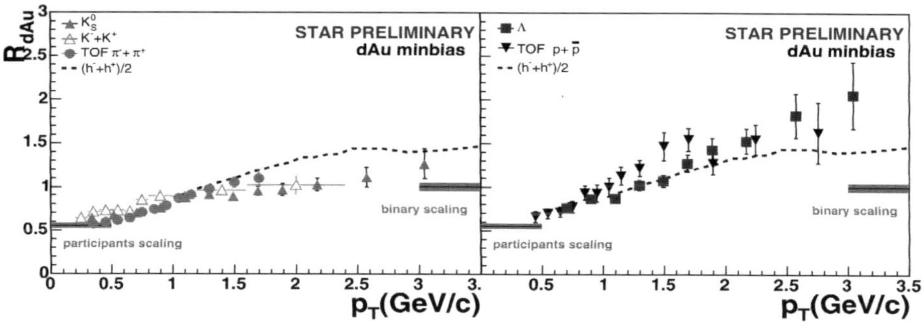

Figure 2. R_{dAu} for mesons(left), baryons (right) and inclusive charged hadrons (dashed line). Statistical errors only are shown.

5. Summary

We have presented nuclear modification factors R_{AuAu} and R_{dAu} for identified hadrons, as measured by the STAR Collaboration. A difference between baryons and mesons is observed, similar to the one measured in the Au+Au central to peripheral ratios (R_{CP}). However, the R_{CP}^{baryon} is reduced in comparison to binary scaling at all p_T in Au+Au collisions, whereas the R_{AuAu}^{baryon} in the intermediate p_T region is increased for strange baryons. In addition, the R_{AuAu}^{baryon} values show an ordering according to the baryon strangeness content, a feature which was not found in the R_{CP}^{baryon} values. Less of a difference is observed between R_{CP} and R_{AuAu} in the case of mesons (strange and non-strange) and non-strange baryons. The meson baryon differences are also present in R_{dAu} but at a smaller level which can not account for the meson-baryon differences seen in R_{CP}^{AuAu}.

References

[1] Antreasyan D *et al Phys. Rev. D* 19 (1979) 764
[2] Adams J *et al* (STAR Collaboration) *Phys. Rev. Lett.* 91 072304 (2003)
[3] Adams J *et al* (STAR Collaboration) *Phys. Rev. Lett.* 91 172302 (2003)
[4] Accardi A *Preprint* hep-ph/0212148
[5] Adler S *et al* (PHENIX Collaboration) *Phys. Rev. Lett.* 91 172301 (2003)
[6] Adams J *et al* (STAR Collaboration) *Phys. Rev. Lett.* 91 072304 (2003)
[7] Adams J *et al* (STAR Collaboration)*Preprint* nucl-ex/0408016
[8] Fries R J *Preprint* nucl-th/0403036
[9] Adams J *et al* (STAR Collaboration)*Phys. Rev. Lett.* 92 052302 (2004)
[10] Schweda K (STAR Collaboration) *J. Phys. G: Nucl. Phys.* 30 693 (2004)
[11] Ackermann K H *et al Nucl. Instrum. Meth.*, A 499 (2003) 624-632
[12] Adams J *et al* (STAR Collaboration) *Phys. Lett. B* 612 181 (2005)
[13] Deng W 2002 *Ph.D Thesis* Kent State University
[14] Topor-Pop V *et al Preprint* nucl-th/0407095
[15] Hirano T and Nara Y *Preprint* nucl-th/0307015
[16] Caines H (STAR Collaboration) *Preprint* nucl-ex/0412015

Institute of Physics Publishing
doi:10.1088/1742-6596/50/1/038
Journal of Physics: Conference Series **50** (2006) 315–318
5th International Conference on Physics and Astrophysics of Quark Gluon Plasma

Measurement of Δ^{++} resonance production in d+Au collisions at $\sqrt{s_{NN}} = 200$ GeV.

D. K. Mishra (for the STAR Collaboration)

Institute of Physics, Sachivalaya Marg, Bhubaneswar 751005, INDIA

E-mail: dmishra@iopb.res.in

Abstract. The production of Δ^{++} resonance in d+Au collisions at $\sqrt{s_{NN}} = 200$ GeV is investigated using the data taken by the STAR detector at RHIC. Systematics of Δ^{++} resonance production relative to p and $\overline{p}$ yields have been studied. The nuclear modification factor R_{CP}, defined in terms of the ratio of Δ^{++} yields for central to peripheral data shows a rise in the intermediate p_T region, in line with Cronin enhancement.

1. Introduction

Relativistic heavy ion collisions provide a unique opportunity to study nuclear matter in states of high temperature and energy density. Under these conditions, it is expected that nuclear matter would go through a phase transition from a (confined) hadronic matter to a deconfined phase or quark-gluon plasma (QGP). The study of hadronic resonances with extremely short life-times ($\sim$few fm/c), comparable with the lifetime of the fireball source, may provides a unique tool to probe the collision dynamics.

At high temperature and density, the hadron production and the collision dynamics can be studied through the resonance decays and regeneration. In heavy ion collision a hot and dense fireball source is created. During the expansion of the fireball source, the short lived resonances and their hadronic decay daughters may undergo rescattering with other particles inside the medium, which causes signal loss and modification in the observed momentum distribution of resonances. This introduces a mass shift while reconstructing the resonance from detected daughter particles. In some cases even the resonances can not be reconstructed from rescattered daughter particles in which case the corresponding primordial resonance yields can get reduced. On the other hand, regeneration of resonances through the interaction of daughter particles with the medium can lead to enhancement over primordial resonance yields. The measurement of the resonance yield, in particular the ratio with respect to the corresponding stable particles, its mass and momentum distribution, when compared to those for p+p collisons, can thus provide information about hadronization and the time evolution of source from chemical to kinetic freeze-out [1].

In this connection, it has been suggested that the properties of Δ resonances formed in heavy ion collisions may differ considerably from those of the free resonances [2, 3], possibly yielding new information about nuclear matter at high temperature and energy density. Further, it has been shown that the relative yields of resonances such as $N^*(1440)$ and $\Delta(1232)$ compared to the nucleon are smaller than that in the case of stable strange baryons [4]. This is because the widths are much larger ($\Gamma_{N^*} \simeq 300$ MeV, $\Gamma_{\Delta} \simeq 120$ MeV) and the final state nucleon has a greater scattering cross section in hadronic matter than the strangeness carrying stable hadron. Therefore, it was of interest to look at the $\Delta(1232)$ state whose detection might suggest whether or not chemical and kinetic freeze-out are indeed nearly coincident.

In the present note we report some results on the above mentioned study through the production of Δ^{++} resonace in $d + Au$ data as taken by the STAR experiment. The main idea behind this analysis has been to compare the results with the corresponding $p + p$ and $Au + Au$ data taken by the same experiment earlier to understand the in-medium effects in a consistent manner.

2. Analysis

The data set used for this analysis was taken in the 2002 - 2003 RHIC run for $d + Au$ collisions at $\sqrt{s_{NN}} = 200$ GeV. The main tracking device here is the Time Projection Chamber (TPC) [5], which provides, for charged particles, both momentum information and particle identification. Through the ionization energy loss (dE/dx) in the TPC, charged pions and kaons are identified with momenta upto about 0.75 GeV/c while protons and anti-protons are identified with momenta upto about 1.1 GeV/c. In the STAR detector, a minimum bias trigger for $d + Au$ collisions was defined by requiring at least one beam-rapidity neutron in the Zero Degree Calorimeter in the Au beam direction, which is assigned negative pseudorapidity (η) [6]. The events with z-vertices within $\pm$ 50 cm from the center of TPC were accepted for analysis. Centrality of the $d+Au$ collisions was determined by the uncorrected charged particle multiplicity (N_{ch}) within a pseudorapidity window of $-3.8 < \eta < -2.8$ as measured by the Forward Time Projection Chamber (FTPC) along the Au beam direction. The $d+Au$ events were divided into three collision centrality classes: $(0-20)\%$ ($N_{ch} \geq 17$), $(20-40)\%$ ($10 \leq N_{ch} < 17$) and $(40-100)\%$ ($N_{ch} < 10$) of the total inelastic cross section. In the present analysis, about 11.7M $d + Au$ minimum bias events have been used.

The Δ^{++} and $\overline{\Delta}^{--}$ signals were obtained by the invariant mass reconstruction from their potential daughter track combinations, substracting in both cases the combinatorial backgrounds obtained from mixed-events [7] via the decay channels $\Delta^{++}(1232) \rightarrow p\pi^+$ and $\overline{\Delta}^{--}(1232) \rightarrow \overline{p}\pi^-$. Fig. 1(a) shows the invariant mass distribution from the same-event pair superposed on results obtained from mixed-events. Since the Δ signal is only a small fraction ($\sim 1.5\%$) of the background in the Δ mass range one has to substract a huge combinatorial background obtained from mixed events to get the real Δ signal. Fig. 1(b) shows the invariant mass distribution after background substraction fitted to the function: $PS \times BW + RBG$ [8]. Where BW is the relativistic $p-$wave Breit-Wigner function, BW $= \frac{M_{p\pi}\Gamma M_0}{(M_{p\pi}^2 - M_0^2)^2 + M_0^2\Gamma^2}$, PS is the Boltzmann phase space factor, PS $= \frac{M_{p\pi}}{\sqrt{M_{p\pi}^2 + p_T^2}} \times exp(\frac{-\sqrt{M_{p\pi}^2 + p_T^2}}{T_{fo}})$ and RBG is the residual Gaussian background, RBG $= \exp(\frac{-1}{2}(\frac{M_0 - M_{p\pi}}{\Gamma})^2)$. Within these parametrization, T_{fo} is the temperature at which the resonance is emitted, M_0 is the Δ mass, Γ is the momentum dependent Δ width, p_T is the Δ transverse mementum.

From the fits to the invariant mass spectra at mid-rapidity $|y| < 0.5$ for each p_T bin, the raw resonance yields were extracted for minimum bias data sample . The raw yields were then corrected for acceptance and efficiency by using the corresponding numbers from Monte-Carlo simulations. The corrected yields as mentioned above are then used to get the resonance yield $((2\pi)^{-1}d^2N/p_T dp_T dy)$ as a function of transverse momentum, p_T.

The p_T spectra for Δ for minimum bias as well as for different collision centralities are presented in Fig.1(a). The dashed lines in Fig.1(a) represent the exponential fits to the data with the function $(2\pi)^{-1}d^2N/p_T dp_T dy = (dN/dy)/(2\pi T(T + m_0))exp(-(\sqrt{p_T^2 + m_0^2} - m_0)/T)$, where dN/dy is the mid-rapidity yield, T being the inverse slope parameter and m_0 is the Δ

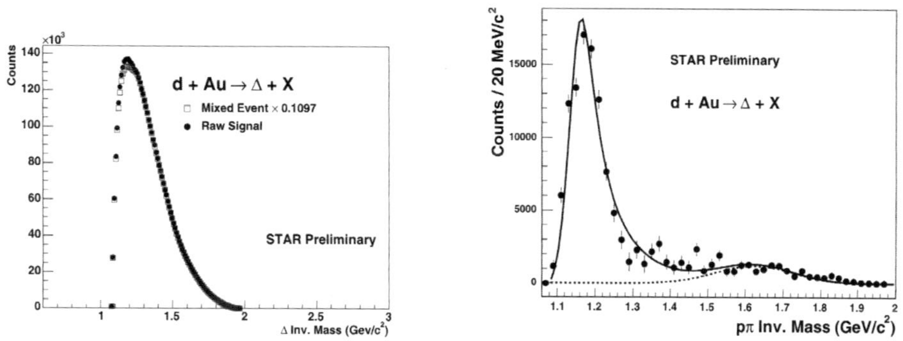

Figure 1. (a) The invariant mass distribution from the same event pair as well as from mixed-event pair (b) The invariant mass distribution for Δ^{++} after the mixed-event background substraction.

mass = 1232 MeV. From the fit, we can extract the integrated Δ^{++} yields at mid-rapidity and the inverse slope parameters T.

3. Results

Figure 2(b) shows the $\overline{\Delta}^{--}/\Delta^{++}$ ratio for different p_T bins alongwith the corresponding $\bar{p}/p$ ratio. It doesn't show any significant p_T dependence.

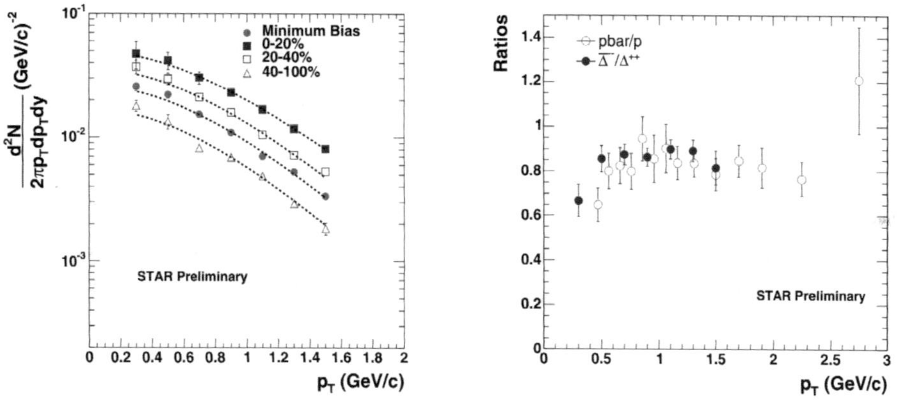

Figure 2. (a) The p_T spectra for minimum bias as well as for different centralities fitted with an exponential function. (b) The $\overline{\Delta}^{--}/\Delta^{++}$ and $\bar{p}/p$ ratios as a function of p_T. The error bars on the data are statistcal errors

The yields dN/dy for the stable charged hadrons ($\pi^{\pm}, K^{\pm}, p$ and $\bar{p}$) have also been measured in the STAR experiment for the $d+Au$ collisions at $\sqrt{s_{NN}} = 200$ GeV. Using those data on stable hadrons ($p, \bar{p}$) together with the present results one can calculate the ratio of resonance yield relative to corresponding value for $p, \bar{p}$. Figure 3(a) shows the $(\Delta^{++} + \overline{\Delta}^{--})/(p + \bar{p})$ ratio as a function of corresponding mean value of uncorrected charged hadron multiplicity at midrapidity $dN_{ch}/d\eta$ for different centrality bins. In the case of the Δ resonance, p-π total interaction cross section, which determines the regeneration effect, is comparable to the π-π total interaction cross section, which determines rescattering effect [9]. The over all effect is that the observed $(\Delta^{++} + \overline{\Delta}^{--})/(p + \bar{p})$ ratios are flat from peripheral $d + Au$ to central $d + Au$ collisions. In the same figure (Fig. 3(a)) the Δ^{++} dN/dy, has also been plotted as a function of number of charged hadrons for different collision centralities. It shows that, dN/dy for Δ increases as we go from peripheral to central collisions.

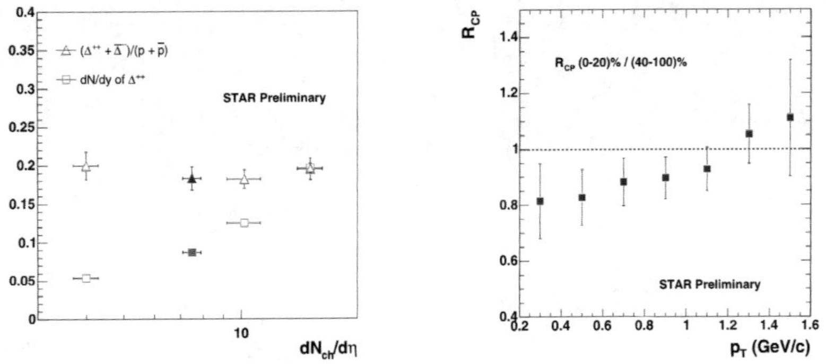

Figure 3. (a) The $(\Delta^{++} + \overline{\Delta}^{--})/(p + \overline{p})$ ratio and dN/dy for Δ as a function of centrality $(dN_{ch}/d\eta)$, closed symbols are for minimum bias, (b)The Δ^{++} nuclear modification factor R_{CP} as a function of p_T. The error bars on the data points are statistical errors.

The nuclear modification factor, as given by

$$R_{CP} = \frac{< N_{bin}^{peripheral} > d^2N^{central}/dp_Td\eta}{< N_{bin}^{central} > d^2N^{peripheral}/dp_Td\eta} \tag{1}$$

for Δ resonance has also been estimated in $d + Au$ collisions. Fig. 3(b) shows the R_{CP} values between central $((0-20)\%)$ and peripheral $((40-100)\%)$ collisons for the Δ production as a function of p_T. The ratio R_{CP} is seen to be lower than unity in the low p_T region. It is larger than unity for $p_T > 1$ GeV/c, which shows enhancement that can be explained by Cronin effect at the intermediate p_T region [11].

4. Conclusions

In this report, priliminary STAR results on $\Delta^{++}(1232)$ resonance production at mid-rapidity in $d + Au$ collisions at $\sqrt{s}_{NN}$= 200 GeV are presented. The ratio of Δ^{++} yield to that of p and $\overline{p}$ has been found to be independent of collision centrality. This indicates regeneration may be compensating for loss in the observed yield due to rescattering of the daughter particles. dN/dy of Δ is found to increase with collision centrality. The $\overline{\Delta}^{--}/\Delta^{++}$ ratio has been found to be close to 0.85, independent of p_T. This indicates that the mid-rapidity region is not net-baryon free. In the intermediate p_T region, the nuclear modification factor, R_{CP}, for Δ^{++} shows a rise indicating Cronin enhancement.

References

[1] M. Bleicher and J. Aichelin, Phys. Lett. B **530** 81-87 (2002).
[2] S. A. Bass *etal.*, Phys. Lett. B **335**, 289 (1994).
[3] J. Cugnon, D. Kinet, and J. Vandermeulen, Nucl. Phys. A **379**, 553 (1982).
[4] G. Torrieri and J. Rafelski, Phys. Lett. B **509**, 239 (2001).
[5] K. H. Ackermann *etal.* for STAR collaboration, Nucl. Phys. A **661**, 681 (1999).
[6] J. Adams *etal.* STAR Collaboration, Phys. Rev. Lett, **91**, 172304 (2003).
[7] C. Adler *etal.*, Phys. Rev. C **66**, 061901(R) (2002).
[8] J. Adams *etal.* STAR Collaboration, Phys. Rev. Lett, **92**, 092301 (2004).
[9] M. Bleicher *etal.*, hep-ph/9909407.
[10] J. Adams *etal.*, Phys. Rev. Lett. **92**, 052302 (2004).
[11] D. Kharzeev *etal.*, Phys. Lett. B**561**, 93 (2003); J. L. Albacete *etal.*, Phys. Rev. Lett. **92**, 082001 (2004); D. Kharzeev *etal.*, Phys. Rev. D **68**, 094013 (2003); R. Baier *etal.*, Phys. Rev. D **68**, 054009 (2003).

Institute of Physics Publishing
doi:10.1088/1742-6596/50/1/039
Journal of Physics: Conference Series **50** (2006) 319–322
5th International Conference on Physics and Astrophysics of Quark Gluon Plasma

First Results from Photon Multiplicity Detector at RHIC

B. Mohanty (for STAR Collaboration[1])
Variable Energy Cyclotron Centre, 1/AF, Bidhan Nagar, Kolkata - 700064
E-mail: bmohanty@veccal.ernet.in

Abstract. We present the first measurement of multiplicity and pseudorapidity distributions of photons in the pseudorapidity region $2.3 \leq \eta \leq 3.7$ for different centralities in Au + Au collisions at $\sqrt{s_{NN}}$ = 62.4 GeV. The pseudorapidity distribution of photons, dominated by neutral pion decays, has been compared to those of identified charged pions, photons, and inclusive charged particles from heavy ion and nucleon-nucleon collisions at various energies. Scaling of photon yield with number of participating nucleons and limiting fragmentation scenario for inclusive photon production has been studied.

1. Introduction

Inclusive charged particle multiplicity measurements at RHIC have so far revealed a lot of information on the nature and dynamics of particle production in heavy ion collisions [1, 2]. It has been observed that the charged particle pseuorapidity density at mid rapidity scales with the number of participating nucleons and the number of collisions. However, the total charged particle multiplicity is found to scale with number of participant nucleons only. Limiting fragmentation (LF) [3] behaviour of inclusive charged particles have been studied at RHIC. It has been observed that inclusive charged particles follow a energy independent and centrality dependent limiting fragmentation scenario [1, 2]. It will be interesting to study the above physics aspects with inclusive photon multiplicity measurements [4] at RHIC. In this paper we present the first photon multiplicity measurements in the forward rapidity at RHIC for the Au + Au collisions at $\sqrt{s_{NN}}$ = 62.4 GeV in the STAR experiment.

2. Experiment and data analysis

The photon multiplicity measurements were done in the STAR experiment [5] at RHIC by a highly granular gas based photon multiplicity detector (PMD) in pseudorapidity region $2.3 \leq \eta \leq 3.7$. The details of the construction and design of PMD can be found in the following Ref. [6]. The data presented here corresponds to 0 to 80% of Au+Au hadron cross section. The minimum bias trigger was obtained from the information of the following trigger detectors. An array of scintillator slats arranged in a barrel surrounding the Time Projection chamber (TPC), called central trigger barrel (CTB), which measures charged particles and two zero degree hadronic calorimeters (ZDCs) at ± 18 m from the detector center [7].The charged particles from TPC

[1] Complete author list can be found at the end of this proceeding

within $| \eta | < 0.5$ was used for the centrality selection for data and simulation in the present paper.

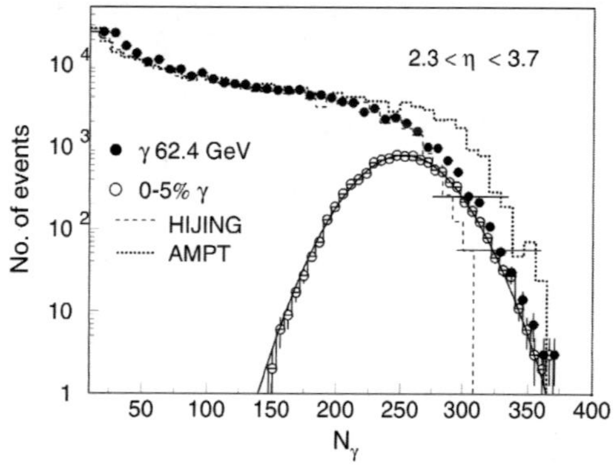

Figure 1. Minimum bias N_γ distribution. Comparison with HIJING and AMPT models are shown. The solid curve is the fit to a Gaussian.

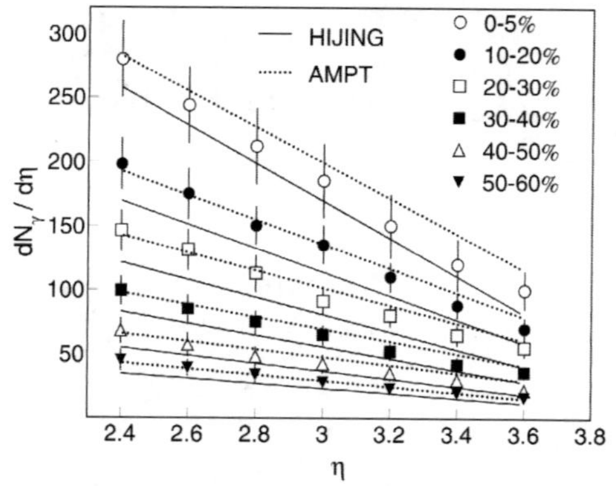

Figure 2. $\frac{dN_\gamma}{d\eta}$ for various event centrality classes compared to HIJING and AMPT model calculations.

In the present analysis, only the data from the preshower plane of the PMD has been used. The data analysis proceeded through the following steps: (a) Calibration of gain of all cells of PMD, (b) clustering of hits on PMD and (c) photon-hadron discrimination. Details of each of the above steps of analysis of can be found in Ref. [4].

3. Results

3.1. Multiplicity and Pseudorpatidity distributions

Fig. 1 shows the minimum bias distribution of N_γ along with results from HIJING [8] and AMPT [9] models. We observe that HIJING underpredicts the measured photon multiplicity. AMPT slightly overpredicts the total measured photon multiplicity for central collisions. However, within the systematic errors [4] it is difficult to make a firm conclusion. The top 5% central multiplicity distribution (open circles) is fitted to a Gaussian with a mean of 252.

Fig. 2 shows the pseudorapidity distribution of photons for various event centrality classes. The errors shown are systematic and statistical. The results from HIJING are systematically lower compared to data for mid-central and peripheral events. The results from AMPT compare well to the data.

3.2. Scaling of photon multiplicity with number of participating nucleons

Fig. 3 shows the variation of total number of photons per participant pair in the PMD coverage as a function of the number of participants. N_{part} is obtained from Glauber calculations [10]. We observe that the total number of photons per participant pair is approximately constant with centrality. The values from HIJING are lower compared to the data. The values from AMPT agree fairly well with those obtained from the data.

3.3. Limiting fragmentation scenario for photons

Fig. 4 compares the photon spectra in Au + Au collisions at $\sqrt{s_{NN}} = 62.4$ GeV, with the top SPS energy photon data for Pb + Pb collisions [11] and charged particle data at $\sqrt{s_{NN}} = 130$ and

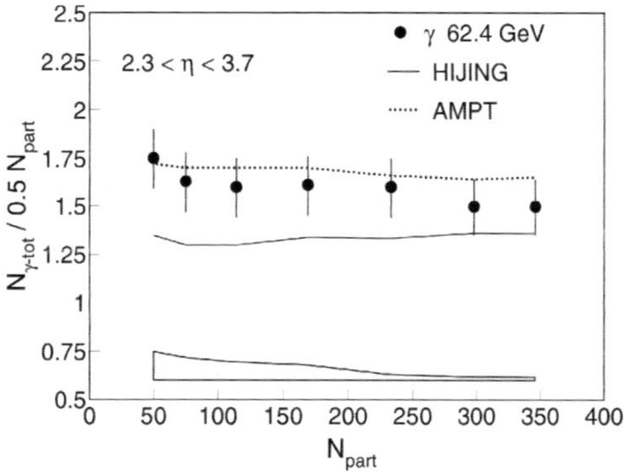

Figure 3. Variation of N_γ per participant pair in PMD coverage ($2.3 \leq \eta \leq 3.7$) as a function of N_{part}. The error bars shown are systematic errors and the lower band reflects uncertainties in N_{part} calculations.

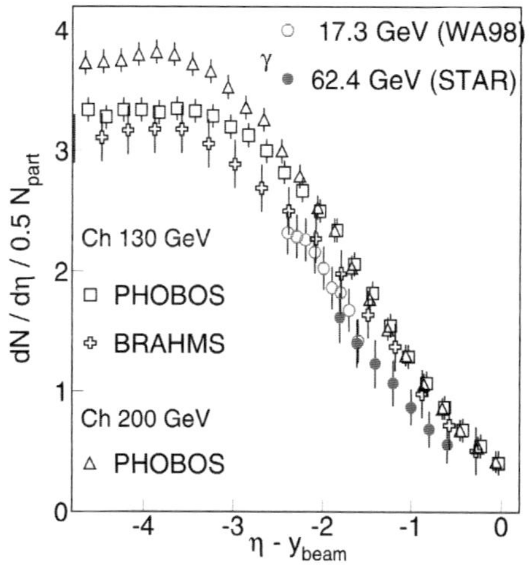

Figure 4. Variation of $\frac{dN_\gamma}{d\eta}$ and $\frac{dN_{ch}}{d\eta}$ normalized to N_{part} with η - y_{beam} for central collisions at various collision energies. The error bars shown are systematic errors.

200 GeV [1, 2] for Au+Au as a function of η - y_{beam} for central collisions. The SPS and RHIC photon results are consistent with each other, suggesting that photon production follows the LF behavior. However, the photon multiplicity values are lower compared to charged particles.

In Fig. 5 we study the centrality dependence of the LF behavior for inclusive photons and compare the photon spectra with those for charged particles. In the forward η region, the photon production cross section as a function of η - y_{beam} is independent of centrality. The dependence of LF spectra on the collision system is established by the comparison of charged particles spectra from pp and $p\bar{p}$ collisions at $\sqrt{s_{NN}}$ = 53 and 200 GeV, respectively, and γ at 546 GeV [12]. We observe that the photon results in the forward rapidity region from $p\bar{p}$ collisions at $\sqrt{s_{NN}}$ = 546 GeV are in close agreement with the measured photon yield in Au+Au collisions at $\sqrt{s_{NN}}$ = 62.4 GeV within the LF scenario.

Fig. 6 shows the total charged pion rapidity density from SPS [13] and scaled photon rapidity density at $\sqrt{s_{NN}}$ = 62.4 GeV as a function of y-y_{beam}. HIJING calculations indicate that about 93-96% of the photons are from π^0 decays. The photon results in Fig. 6 have been scaled down accordingly to reflect approximately twice the π^0 spectrum. The results show that pion production in heavy ion collisions in the fragmentation region agrees with the LF picture.

4. Summary

In summary, we have presented the first results of photon multiplicity measurements at RHIC in the pseudorapidity region $2.3 \leq \eta \leq 3.7$. The pseudorapidity distributions of photons have been obtained for various centrality classes. Photon production per participant pair is found to be approximately independent of centrality in this pseudorapidity region. Comparison with photon and charged pion data at RHIC and SPS energies shows, for the first time in heavy ion collisions, that photons and pions follow an energy independent limiting fragmentation behavior, as has been previously observed for inclusive charged particles. However photons, unlike charged particles, follow a centrality independent limiting fragmentation scenario.

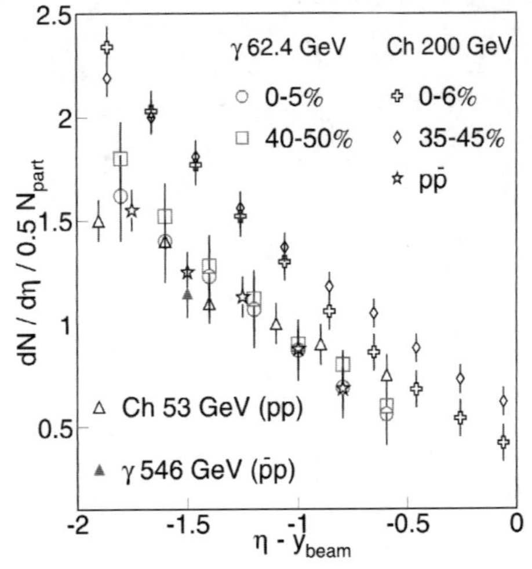

Figure 5. Centrality dependence of limiting fragmentation for inclusive photons and charged particles. Comparison to $p\bar{p}$ and pp collisions. The error bars shown are systematic errors.

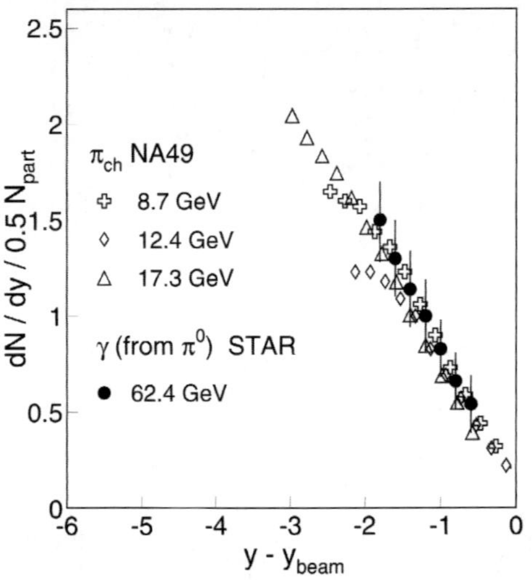

Figure 6. Variation of $\frac{dN_\gamma}{d\eta}$ scaled down to reflect π^0 contribution and $\frac{dN_{\pi^\pm}}{dy}$ normalized to N_{part} for central collisions at various collision energies with y - y_{beam}. The error bars include the systematic errors.

4.1. Acknowledgments

We thank the RHIC Operations Group and RCF at BNL, and the NERSC Center at LBNL for their support. This work was supported in part by the HENP Divisions of the Office of Science of the U.S. DOE; the U.S. NSF; the BMBF of Germany; IN2P3, RA, RPL, and EMN of France; EPSRC of the United Kingdom; FAPESP of Brazil; the Russian Ministry of Science and Technology; the Ministry of Education and the NNSFC of China; SFOM of the Czech Republic, FOM and UU of the Netherlands, DAE, DST, and CSIR of the Government of India; the Swiss NSF. We acknowledge the help of CERN for use of GASSIPLEX chips in the PMD readout.

5. References

[1] I.G. Bearden et al., (BRAHMS Collaboration), 2001 *Phys. Lett.* B **523** 227; 2002 *Phys. Rev. Lett.* **88** 202301.
[2] B.B. Back et al., (PHOBOS Collaboration), 2001 *Phys. Rev. Lett.* **87** 102303; 2003 *Phys. Rev. Lett.* **91** 052303; nucl-ex/0410022.
[3] J. Benecke, T.T. Chou, Chen-Ning Yang and E. Yen, 1969 *Phys. Rev.* **188** 2159; R. Beckmann, S. Raha, N. Stelte and R.M. Weiner, 1981 *Phys. Lett.* B **105** 411.
[4] J. Adams et al.,(STAR Collaboration), nucl-ex/0502008
[5] K. H. Ackermann, et al., 2003 *Nucl. Instr. Meth.* A **499** 624.
[6] M.M. Aggarwal et al., 2003 *Nucl. Instr. Meth.* A **499** 751; 2002 *Nucl. Instr. Meth.* A **488** 131.
[7] F.S. Bieser et al., 2003 *Nucl. Instrum. Meth.* A **499** 766.
[8] X-N. Wang and M. Gyulassy, 1991 *Phys. Rev.* D **44** 3501.
[9] B. Zhang, C.M. Ko, Bao-An Li and Zi-wei Lin, 2000 *Phys. Rev.* C **61** 067901.
[10] J. Adams et al. (STAR Collaboration), nucl-ex/0311017.
[11] M.M. Aggarwal et al., (WA98 Collaboration), 1999 *Phys. Lett.* B **458** 422.
[12] K. Alpgard et al., (UA5 Collaboration), 1982 *Phys. Lett.* B **115** 71; G.J. Alner et al., 1986 *Z. Phys.* C **33** 1.
[13] S.V. Afanasiev et al., (NA49 Collaboration), 2002 *Phys. Rev.* C **66** 054902.

Probing initial states at RHIC through particle production at forward and backward rapidities using PHENIX muon arms

Anuj K Purwar for PHENIX Collaboration

P-25 Subatomic Physics Group,
MS H846,
Los Alamos National Laboratory,
Los Alamos, NM 87545, USA

E-mail: anuj@lanl.gov

Abstract. Particle multiplicities have yielded some of the most interesting insights at RHIC (Relativistic Heavy Ion Collider). To study the initial state effects at RHIC, we need to look at the small-x region in the gold nucleus, which can be achieved by going to forward rapidities (in the deuteron going direction) for d+Au collisions at $\sqrt{s_{NN}} = 200$ GeV. We measure charged hadrons at forward and backward rapidities using the PHENIX north and south muon arms for d+Au collisions at $\sqrt{s_{NN}} = 200$ GeV. We observe a Cronin-like enhancement on the gold going direction and a suppression on the deuteron going direction. This is qualitatively consistent with the expectation of shadowing/saturation type effects in the small-x region, which is probed at forward rapidities. However the enhancement at backward rapidities has not been explained.

Data from the Relativistic Heavy Ion Collider (RHIC) for Au+Au collisions at $\sqrt{s_{NN}} = 200$ GeV at mid-rapidity indicated a suppression [1] of particle yields as compared to the expectation from naive scaling of p+p collisions. This was consistent either with energy loss in the deconfined medium or depletion of low-x partons due to gluon saturation processes. A control experiment was done by colliding deuteron and gold nuclei at the same energy to quantify whether or not this is due to initial state effects (gluon saturation) or final state effects (deconfined medium). Run 3 data with d + Au collisions at $\sqrt{s_{NN}} = 200$ GeV, showed an enhancement at mid-rapidity [2]. This was inconsistent with the gluon saturation hypothesis [3, 4]. By looking at particle multiplicities at forward rapidity (in the deuteron going direction) we can explore the small-x region of the Au nucleus, where the gluon saturation effects are thought to be more important.

In this paper we seek to determine the nuclear modification factor R_{cp} by looking at the particle multiplicities for d+Au collisions at forward and backward rapidities using the PHENIX muon arms. R_{cp} is defined as the ratio of the particle yield in central collisions to that in peripheral collisions, each normalized by the average number of nucleon-nucleon binary collisions ($\langle N_{coll}\rangle$):

$$R_{cp} = \frac{\langle \left(\frac{dN}{d\eta dp_T}\right)^{Central} \rangle / \langle N_{coll}^{Central} \rangle}{\langle \left(\frac{dN}{d\eta dp_T}\right)^{Peripheral} \rangle / \langle N_{coll}^{Peripheral} \rangle} \quad (1)$$

R_{cp} is what we get divided by what we expect. If particle production in heavy ion collisions was simply an incoherent sum of p+p collisions, then R_{cp} would be unity. Any deviation from unity would indicate a different kind of physics. Thus, we assume that peripheral collisions are similar to p+p collisions, allowing us to normalize R_{cp} independent of the p+p reference spectrum. As a result many systematics related to detector efficiencies cancel, leading to a relatively clean measurement with minimal systematics. The downside is that our most peripheral centrality bin of 60-88% still corresponds to 3-4 collisions and not all nuclear effects might be eliminated. By looking at R_{cp} at forward (and backward) rapidities we can probe momentum fraction x, which can be related to the rapidity (Y)and tranverse energy (M_T)of the particle as: $x \approx \frac{M_T}{\sqrt{s}} e^{-y}$. Hadrons with $p_T > 2$ GeV/c in the deuteron going direction (forward rapidity: $1.4 < \eta < 2.2$) can be measured by north arm and are are sensitive to low-x partons in the gold nucleus $0.001 < x < 0.03$. Whereas, hadrons in the gold going direction (backward rapidity: $-2.0 < \eta < -1.4$) can be measured by the south arm and are sensitive to high-x partons in the gold nucleus $0.04 < x < 0.5$.

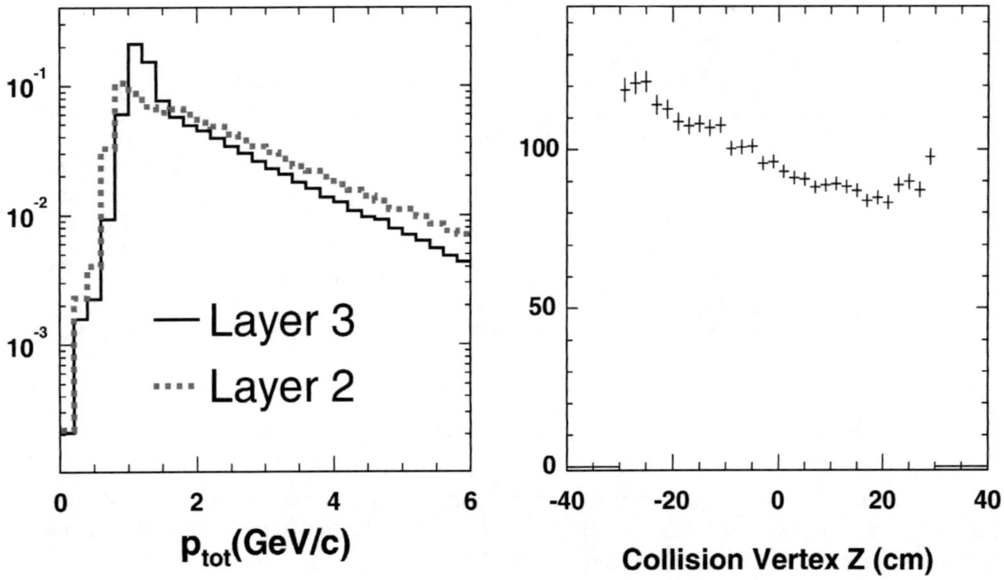

Figure 1. (Left) The total momentum p_{tot} measured in the MuTr for all charged tracks penetrating to MuID layers 2 (grey) and 3 (black). (Right) Collision vertex distribution for events with muons at forward rapidity, normalized by the minimum bias collision vertex distribution.

Each muon arm is a radial magnetic field spectrometer with a muon tracker (MuTr) and a muon identifier (MuID) consisting of layers of absorber and tracking. In addition, the pole tips of the Central Magnet (CM) made of steel and brass, act as hadron absorbers to absorb pions and kaons. The MuTr consists of three stations of cathode-strip readout tracking chambers mounted inside conical-shaped muon magnets. The MuID is segmented with 4 layers of steel absorber with gaps in between instrumented with Iarocci tubes arrayed in x and y directions. Muon tracks can be identified based on the depth of the MuID hit. Although the PHENIX muon arms [5] were designed to detect muons, we use them in this analysis to measure charged hadrons

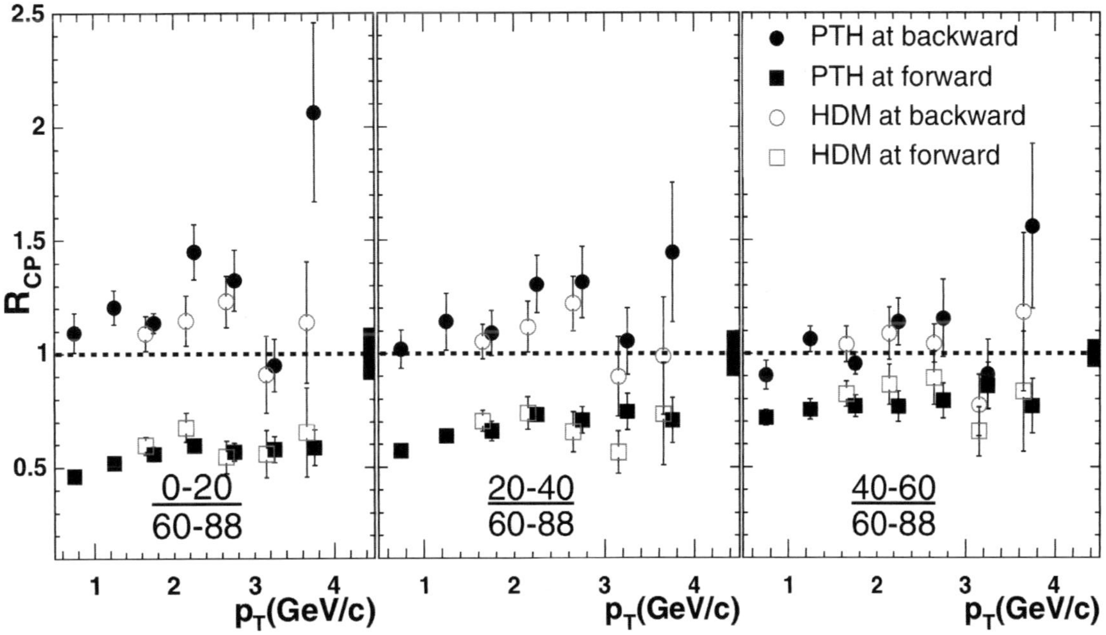

Figure 2. (color online). R_{cp} as a function of p_T at forward rapidity (squares) and backward rapidity (circles) for different centrality classes.

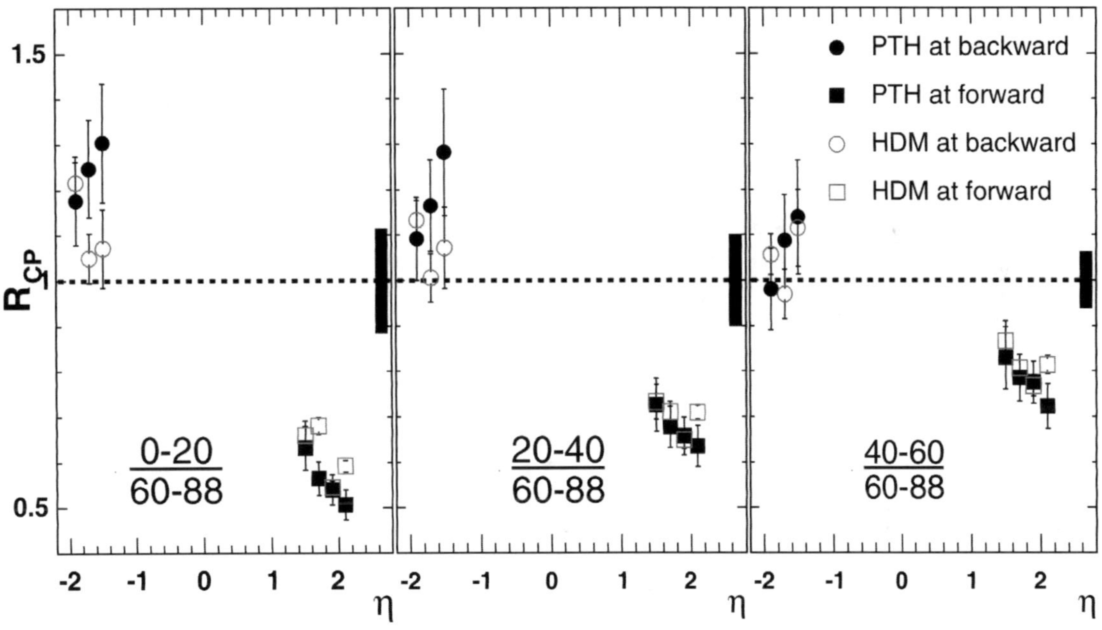

Figure 3. (color online). R_{cp} as a function of η for $1.5 < p_T < 4.0\,\mathrm{GeV}/c$ for different centrality classes.

in two ways. First, we can measure punch through hadrons by looking at the particles that stop in the shallow MuID layers because hadrons can lose a larger portion of their energy due to

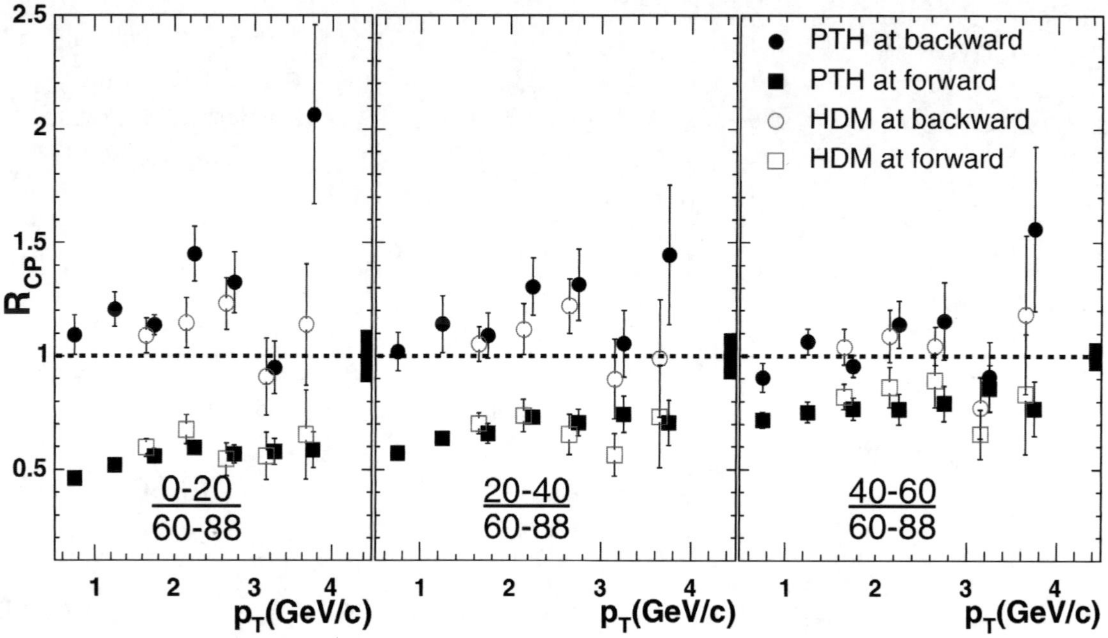

Figure 4. (color online). R_{cp} as a function of p_T at forward rapidity (squares) and backward rapidity (circles) for different centrality classes.

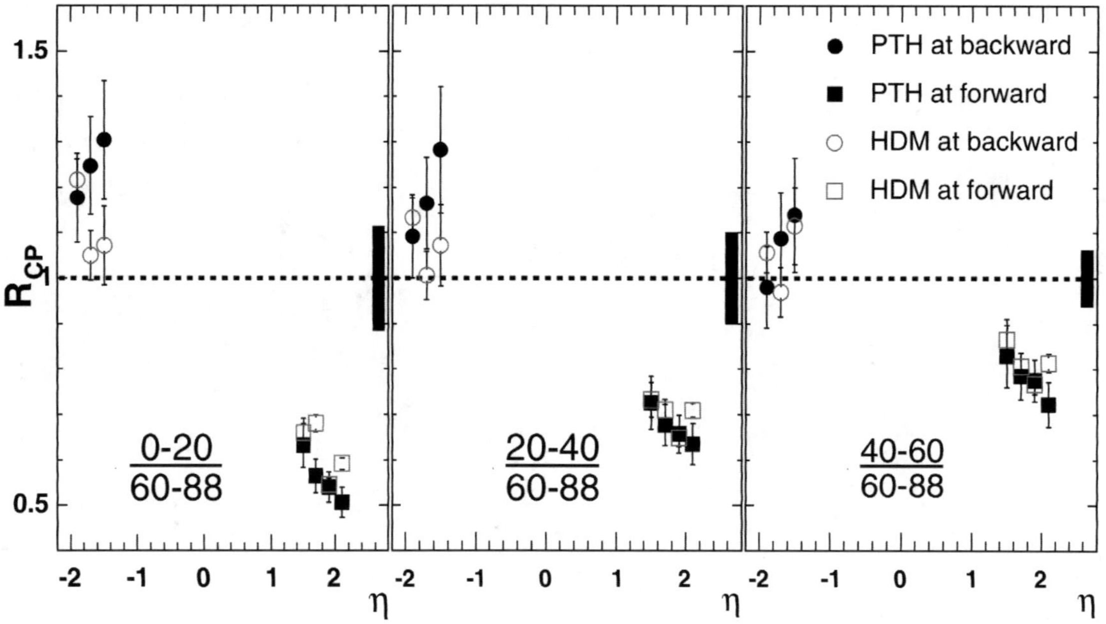

Figure 5. (color online). R_{cp} as a function of η for $1.5 < p_T < 4.0\,\text{GeV}/c$ for different centrality classes.

hadronic interactions. If we look at the distribution of the total momentum p_{tot} as measured at Station 1 stopped at given shallow MuID gaps , (Figure 1) we see that there is a peak of mostly slow muons around 1.3 GeV, and a longer hadronic tail. This allows us to select hadrons by demanding that the associated track stopped in the shallow MuID layers (2 & 3) and had a momentum 3σ above the stopped muon peak. Angular cuts were made to reduce backgrounds due to showering hadrons and from small-angle background. Second, we can measure muons identified by hits in the deep MuID layers that come from decays of light mesons: primarily pions and kaons. The decay probability is given by:

$$P_{decay}(p, L) = 1 - \exp\left(-\frac{L \cdot m}{\tau \cdot p}\right) \tag{2}$$

where $L \sim 41$ cm, is the distance from the collision vertex to the absorber; p, m and τ are the momentum, mass and proper lifetime of the parent particle. As a result, the further the collsion from the absorber, the greater the probability to get a muon from a light meson decay. In comparison, charmed hadrons have almost no collision-vertex dependence due to their short decay times. Thus by looking at the the collision-vertex distribution for events normalized by the minimum bias distribution (Figure 1), we can separate the muons that come from pions/kaons from other contributions.

This analysis was done using two data sets: 67 million minimum bias events which require at least one hit in the PHENIX Beam-Beam Counters (BBC) corresponding to $88 \pm 4\%$ of all inelastic $d + Au$ collisions [2] and 5.3 billion MuID triggered events. Using the particle multiplicity in the BBC, we can divide the events into four centrality classes: $60 - 88\%$ ($\langle N_{coll} \rangle = 3.1 \pm 0.3$), $40 - 60\%$ ($\langle N_{coll} \rangle = 7.0 \pm 0.6$), $20 - 40\%$ ($\langle N_{coll} \rangle = 10.6 \pm 0.7$),and $0 - 20\%$ ($\langle N_{coll} \rangle = 15.4 \pm 1.0$). We plot R_{cp} for both punch-through hadrons (PTH) and hadron decay muons (HDM) as a function of pseudorapidity η (Figure 5), and as a function of transverse momentum p_T (Figure 4) at forward and backward rapidities. R_{cp} vs η shows a suppression at forward rapidity that increases with rapidity and an enhancement at the backward rapidity. Moreover this suppression (enhancement) at forward (backward) direction is greatest for most central events and decreases for less central events. Similar behaviour is seen in the R_{cp} vs p_T plot, although the p_T dependence of R_{cp} is very weak. Systematic uncertainties in our analysis are shown by shaded bars. More details can be found in [6].

These results (as well as similar results from the BRAHMS experiment [7]) can be explained by several theoretical models including gluon saturation [3, 4], coherent multiple scattering [8] and recombination [9].

To summarize, we have measured R_{cp} for charged hadrons using the PHENIX muon arms for d+Au collisions. A suppression in hadron yields is seen at forward rapidity (deuteron going direction) and an enhancement is seen in the backward direction (gold going direction). Forward suppression is qualitatively consistent with several theories from shadowing/saturation type effects to initial state multiple scattering to recombination. However, the enhancement at backward rapidity has not been explained. We might be able to distinguish between different models by studies of charm R_{cp} at forward rapidities in d+Au collisions.

[1] K. Adcox, et al arXiv:nucl-ex/0410003.
[2] S. S. Adler *et al.*, Phys. Rev. Lett. **91**, 072303 (2003).
[3] L. McLerran and R. Venugopalan, Phys. Rev. D **49**, 2233 (1994); Phys. Rev. D **49**, 3352 (1994).
[4] D. Kharzeev, Y.V. Kovchegov and K. Tuchin, arXiv:hep-ph/0405045.
[5] K. Adcox *et al.*, Nucl. Instrum. Methods **A499**, 469 (2003) and references therein.
[6] S. S. Adler *et al.*, Phys. Rev. Lett. **94**, 082302 (2005).
[7] I. Arsene *et al.*, Phys. Lett. **B595**, 209 (2004).
[8] J. Qiu and I. Vitev, arXiv:hep-ph/0405068.
[9] R. Hwa, C.B. Yang and R.J. Fries, arXiv:nucl-th/0410111.

Institute of Physics Publishing — Journal of Physics: Conference Series **50** (2006) 328–331
doi:10.1088/1742-6596/50/1/041 — 5th International Conference on Physics and Astrophysics of Quark Gluon Plasma

Measurement of direct photons in $\sqrt{s_{NN}} = 200\,\mathrm{GeV}$ $p+p$ and Au+Au collisions with the PHENIX Experiment at RHIC

Stefan Bathe for the PHENIX Collaboration
University of California, Riverside, CA, USA
E-mail: bathe@bnl.gov

Abstract. The measurement of direct photons in $\sqrt{s_{NN}} = 200\,\mathrm{GeV}$ p+p and Au+Au collisions is presented. The signal is compared to NLO pQCD calculations, which, in case of Au+Au, are scaled with the number of underlying nucleon-nucleon collisions. The agreement of the calculation with the data in both cases confirms the scaling of hard processes with the number of nucleon-nucleon collisions and supports the explanation of the earlier-observed pion suppression as a final-state effect.

1. Introduction

Direct photons are a unique probe of the hot and dense matter created at RHIC: they allow access to the initial, thermalized state of the nuclear collision. Their measurement, however, is challenging. One has to cope with a large background from hadronic decays.

While several definitions of direct photons are used, we pragmatically attribute this term to all photons that are not decay photons. The main direct photon production processes are quark-antiquark annihilation and quark-gluon Compton scattering [1]. In addition, direct photons are produced through fragmentation of hard partons. These are also called bremsstrahlung photons.

Direct photon measurements in p+p collisions provide a superb test of pQCD. They probe the gluon distribution function. At RHIC they can also probe the spin gluon distribution function [2] in the polarized-proton program. In contrast to hadron measurements, their interpretation does not suffer much from uncertainties on the fragmentation function. Furthermore, they provide a constraint on the hard-scattering contribution to direct photon production in Au+Au collisions.

In heavy ion collisions, thermal direct photons allow, in principle, measurement of the temperature of the collision system in its hottest phase. Also, hard direct photons serve as a crucial baseline for the interpretation of the earlier observed high-p_T hadron suppression [3]. Interactions of hard partons with the medium provide an additional source of direct photons, either through annihilation and Compton scattering [4] or through medium-induced bremsstrahlung [5].

With its high-resolution, highly-segmented electromagnetic calorimeter PHENIX [6] has an excellent capability to measure photons. PHENIX has made precision measurements of neutral pions and η mesons up to transverse momenta of 15 GeV/c. With this crucial measurement of the direct photon background, PHENIX has extracted direct photons in $p+p$ and Au+Au collisions at $\sqrt{s_{NN}} = 200\,\mathrm{GeV}$.

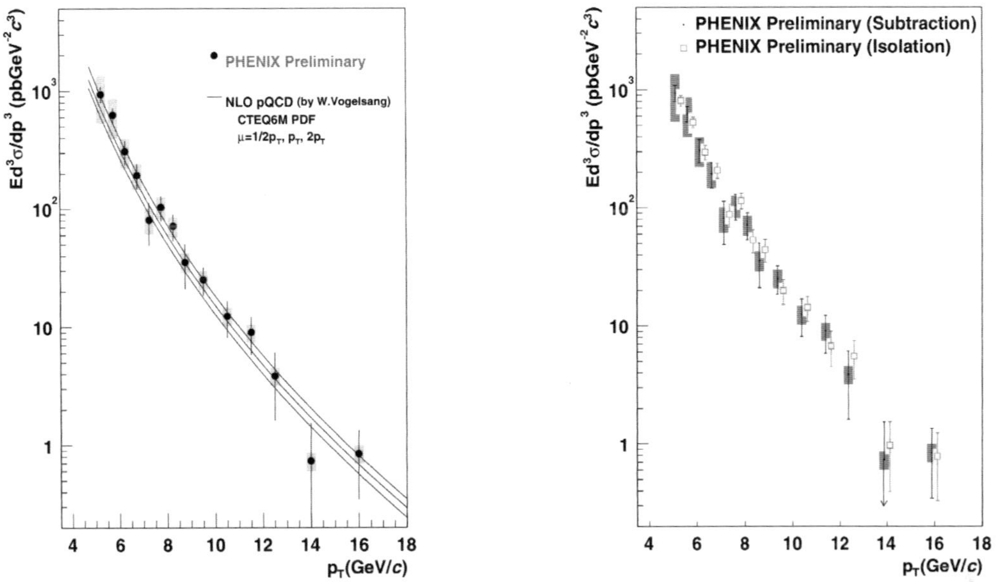

Figure 1. PHENIX preliminary direct photon invariant cross section [7] in $p+p$ collisions at $\sqrt{s}$ = 200 GeV . The left panel shows the result obtained with even-by-event subtraction of decay photons. The curves represent a NLO pQCD calculation for three different scales. The right panel compares this result to the result obtained by an isolation cut.

2. Direct Photon Analysis

The first step in the direct photon analysis in Au+Au is to get a clean inclusive photon sample. To achieve this, non-photon backgrounds, as from charged particles, have to be subtracted. The next step is to measure the p_T spectrum of π^0's and η's. From these spectra, the number of decay photons is determined in a Monte Carlo calculation, which also considers contributions from other hadronic decays. Finally, the decay photon spectrum is subtracted from the inclusive photon spectrum to obtain the direct photon spectrum. This method makes no attempt to identify direct photons on an event-by-event basis. Therefore it is called the subtraction method or statistical method. Further details can be found in [8].

In p+p collisions, where the multiplicity and the occupancy on the detector are small, it is possible to tag and subtract decay photons event by event. In addition, an isolation cut, *i. e.* a cut on the maximum energy allowed in a cone around a direct photon, allows further suppression of decay photons on an event-by-event basis. As an additional feature the isolation cut helps to identify fragmentation photons. Further details on the $p+p$ analysis can be found in [7].

For the measurement of direct photon production in $p+p$ collisions, an integrated luminosity of 266 nbarn^{-1} sampled in the 2003 RHIC $p+p$ run was analyzed. The Au+Au result is based on the 2002 run with an integrated luminosity of 24 μbarn^{-1}. Photons were detected by the electromagnetic calorimeter (EMCal), which is located in the two central arms of the PHENIX detector, each covering $2 * 90°$ in azimuth and a pseudorapidity (η) range of ± 0.35. For the $p+p$ measurement, only one arm was used. The energy calibration was checked by the position and width of the π^0 invariant-mass peak. The fraction of charged-particle contamination was determined with the central-arm tracking detectors which are located in front of the EMCal.

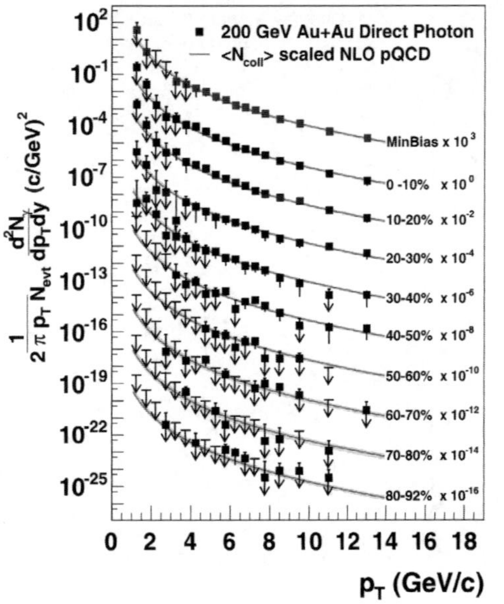

Figure 2. Direct photon invariant yields [8] as a function of p_T for 9 centrality selections and minimum bias Au+Au collisions at $\sqrt{s_{NN}}$ = 200 GeV. The error bars indicate the total error. Arrows indicate measurements consistent with zero yield with the tail of the arrow indicating the 90% confidence level upper limit. The solid curves are pQCD predictions.

3. Direct Photon Cross Section in $p + p$ Collisions

Figure 1 shows the PHENIX preliminary direct photon cross section measured in $\sqrt{s} = 200\,\text{GeV}$ $p + p$ collisions as a function of p_T [7]. The left panel shows the result obtained with event-by-event tagging and subtraction of decay photons. The curves represent a NLO pQCD calculation for three different scales [9]. The calculation agrees well with the measurement. The right panel of Fig. 1 compares this result to the result obtained by an isolation cut. No significant reduction of the direct photon yield by the isolation cut is observed. This suggests that either the contribution from fragmentation photons is small or that the efficiency of the isolation cut to discount fragmentation photons is low. These preliminary results substantially increase the statistical significance and the p_T reach of the published result from the 2002 run [10].

4. Direct Photon Yield in Au+Au Collisions

The direct photon spectra measured in $\sqrt{s_{NN}} = 200\,\text{GeV}$ Au+Au collisions [8] are shown in Fig. 2 for different centrality selections. They are compared to NLO pQCD calculations [9] scaled by the corresponding number of binary nucleon collisions. It can be seen that the pQCD calculations provide a good description of the measured direct photon spectra.

A common way to study possible medium effects in AA collisions is the *nuclear modification factor* R_{AA}, *i. e.* the ratio of the AA invariant yields to the NN-collision-scaled $p+p$ invariant yields [3]. The centrality dependence of R_{AA} for the integrated yield in $p_T > 6$ GeV/c for direct photons (closed circles) is now compared to that of π^0's (open circles) in Fig. 3. The direct photon $p + p$ yield is taken as the NLO pQCD calculation [9] as in the previous figure, while the π^0 $p + p$ yield is taken from the measured π^0 yield [11]. The R_{AA} trend confirms the observation from above in more detail: the direct photon production in Au+Au is consistent with the binary-scaled $p + p$ pQCD calculation. This is in sharp contrast [3] to the centrality dependence of the π^0 R_{AA}, indicating that the observed large suppression of high-p_T hadron production in central Au+Au collisions is dominantly a final-state effect due to parton energy loss in the dense produced medium, rather than an initial-state effect.

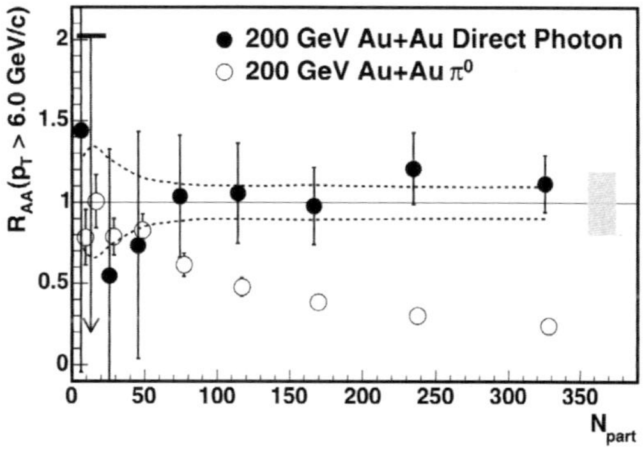

Figure 3. $R_{AA}(p_T > 6\text{GeV}/c)$ as a function of centrality (N_{part}) for direct photons (closed circles) and π^0's (open circles) [8]. The error bars indicate the total error excluding the error on $\langle N_{coll} \rangle$ shown by the dashed lines and the scale uncertainty of the NLO calculation shown by the shaded region at the right.

5. Summary

Direct photon production has been measured in $p+p$ and Au+Au collisions at $\sqrt{s_{NN}} = 200$ GeV. In $p+p$, NLO pQCD calculations agree well with the measurement. The equally good agreement between measurement and binary scaled pQCD calculations in Au+Au suggests that nuclear modifications at mid-rapidity are small. The result provides strong confirmation that the observed large suppression of high p_T hadron production in central Au+Au collisions is dominantly a final-state effect.

References

[1] Peitzmann T and Thoma M H. Direct photons from relativistic heavy-ion collisions. *Phys. Rept.*, **364**:175–246, 2002.

[2] Saito N et al. Spin physics with the PHENIX detector system. *Nucl. Phys.*, A**638**:575–578, 1998.

[3] Adler S S et al. Suppressed π^0 production at large transverse momentum in central Au + Au collisions at $\sqrt{s_{NN}} = 200$ GeV. *Phys. Rev. Lett.*, **91**:072301, 2003.

[4] Fries R J, Muller B and Srivastava D K. High energy photons from passage of jets through quark gluon plasma. *Phys. Rev. Lett.*, **90**:132301, 2003.

[5] Jeon S, Jalilian-Marian J and Sarcevic I. Prompt photon and inclusive π^0 production at RHIC and LHC. *Nucl. Phys.*, A**715**:795, 2003.

[6] Adcox K et al. PHENIX detector overview. *Nucl. Instrum. Meth.*, A**499**:469–479, 2003.

[7] Okada K. Measurement of prompt photons in $\sqrt{s} = 200$ GeV $p+p$ collisions, 2005. hep-ex/0501066. To appear in the proceedings of the 16th International Spin Physics Symposium (SPIN 2004), Trieste, Italy, October 10–16, 2004.

[8] Adler S S et al. Centrality dependence of direct photon production in $\sqrt{s_{NN}} = 200$ GeV Au + Au collisions. *Phys. Rev. Lett.*, **94**:232301, 2005.

[9] Gordon L E and Vogelsang W. Polarized and unpolarized prompt photon production beyond the leading order. *Phys. Rev.*, D**48**:3136–59, 1993.

[10] Adler S S et al. Mid-rapidity direct-photon production in p + p collisions at $\sqrt{s} = 200$ GeV. *Phys. Rev.*, D**71**:071102, 2005.

[11] Adler S S et al. Mid-rapidity neutral pion production in proton proton collisions at $\sqrt{s} = 200$ GeV. *Phys. Rev. Lett.*, **91**:241803, 2003.

Heavy quark and J/ψ production at RHIC/PHENIX

Tsuguchika Tabaru for the PHENIX Collaboration
RIKEN BNL Research Center, Building 510A, Brookhaven National Laboratory, Upton, NY 11973, USA

E-mail: tsugu@bnl.gov

Abstract. Single electrons from open heavy quarks and di-leptons from J/ψ mesons have been studied systematically at RHIC/PHENIX using data from p+p, d+Au and Au+Au collisions at $\sqrt{s_{NN}}$=62.4 GeV, 130 GeV and 200 GeV. From the single electron study, the charm quark yield is found to scale with the number of binary collisions. This scaling has recently been confirmed using Au+Au collisions at 62.4 GeV. A new analysis shows that a high p_{T} suppression of single electrons is observed in Au+Au collisions at 200 GeV. This suppression suggests that heavy quarks lose significant energy in the medium. A weak rapidity dependence is seen in J/ψ yields from d+Au data, which can be interpreted as a cold matter effect. We report also the results of a measurement of the spin alignment of the J/ψ in the p_{T} range around 1.5 GeV/c.

1. Introduction

There is growing evidence that a new state of hot and/or dense matter is formed in high energy heavy ion collisions at RHIC [1]. This new state of matter is thought to be made of de-confined quarks and gluons, and it is expected to behave differently from normal nuclear matter. Heavy quarks (charm and bottom) and quarkonia are produced via hard processes in the early stage of the collisions, and they travel through the medium of partons. Due to their heavy mass, heavy quarks can experience interaction with the medium in a different way from light quarks [2]. Therefore, measurements of open heavy quark yields in heavy ion collisions provide new information on the nature of the medium created at RHIC. Also, it is predicted that the J/ψ yield will be either suppressed [3] or enhanced [4] in the presence of de-confined quarks and gluons.

In high energy nuclear collisions, a large fraction of electrons with transverse momentum (p_{T}) larger than 1 GeV/c comes from decays of heavy quarks. Thus, one can measure heavy quark production via single electrons. PHENIX has measured open heavy quark yields using single electrons, and J/ψ yields using both di-electron and di-muon pairs, allowing a systematic study of the effects of nuclear matter on these heavy quark signals. In this article, the recent results of the heavy quark data at PHENIX are presented.

2. Open heavy quark measurement

The open heavy quark yields were studied through their electron decays. The electrons were measured in two central arms [5], each covering $\Delta\phi=\pi/2$ at mid-pseudo-rapidity ($|\eta|<0.35$). The charged tracks were reconstructed with drift chambers and pad chambers. In each arm the electrons were identified by a Ring Imaging Čerenkov counter (RICH) and an ElectroMagnetic Calorimeter (EMCal) in the p_{T} range of 0.5 to 5 GeV/c.

Yields of electrons coming from heavy quark decays are extracted by subtracting background electrons. The main components of the background electron yield, called "photonic electrons" come from the Dalitz decays of neutral pions and eta mesons, and from gamma conversions. All other electrons, including those from heavy quark decays, are called "non-photonic electrons". The backgrounds in non-photonic electrons are small, and come from di-electron decays and semi-leptonic decays of other mesons. PHENIX evaluates the backgrounds mainly by the following two methods. In the first method, the "converter subtraction" method, special data sets were taken with a thin brass cylinder (radiation length of 1.7%) installed around the collision vertex position. Since the brass converter increases only the yield of conversion electrons, the yield and p_T shapes of photonic electrons can be evaluated experimentally and precisely, especially in the low p_T region. The advantage of this method is that the systematic error in the result is small. The disadvantage is that the statistical error is dominated by the data taken with the converter, about 10% of all the data. In the second method, the "cocktail subtraction" method, all of the background electron yields are evaluated by simulation using as input the meson and photon distributions measured by the PHENIX detector. This method is good to study signal electrons at high p_T, where signal to background ratio is larger than unity. The results of the cocktail subtraction measurement are confirmed by those of the converter subtraction method.

To study the behavior of heavy quarks in the new state of matter systematically, various data were collected in p+p, d+Au, and Au+Au [6] collisions at $\sqrt{s_{NN}}$=200 GeV, and Au+Au at 62 GeV and 130 GeV [7]. Comparison between p+p and d+Au data [8], shows that the p_T distribution of single electrons scales with the number of binary collisions (N_{coll}), within the uncertainty of the measurement. The total yield of single electrons is also found to be proportional to N_{coll} in Au+Au collisions. Figure 1 shows the Au+Au results [6], with the signal electron yield normalized by N_{coll} as a function of N_{coll}. This was also studied at 62.4 GeV. Figure 2 shows the new result, the p_T spectra of signal electrons in Au+Au collisions at 62.4 GeV. There are also plotted the ISR single electron data taken in p+p collisions at the same energy, scaled by N_{coll} [9, 10, 11]. The plot shows that the single electron yield scales with N_{coll}. These analyses were done using the converter subtraction method.

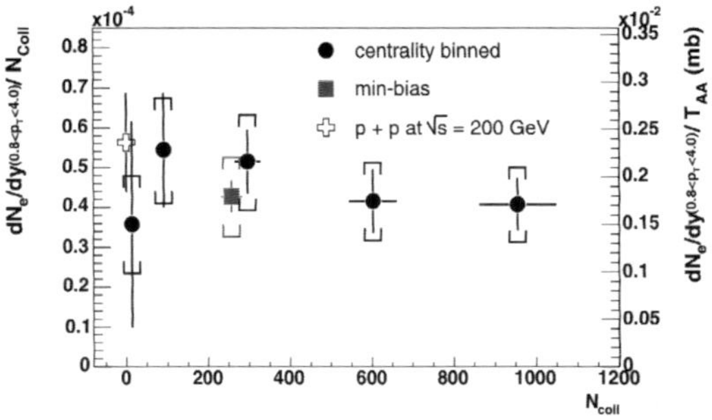

Figure 1. Non-photonic electron yield ($0.8<p_T<4.0$ GeV/c) measured in Au+Au collisions at $\sqrt{s_{NN}}$=200 GeV scaled with N_{coll} (left-hand scale) as a function of centrality given by N_{coll}. This electron yield translates to the electron cross section per NN collision in the above p_T range (right-hand scale).

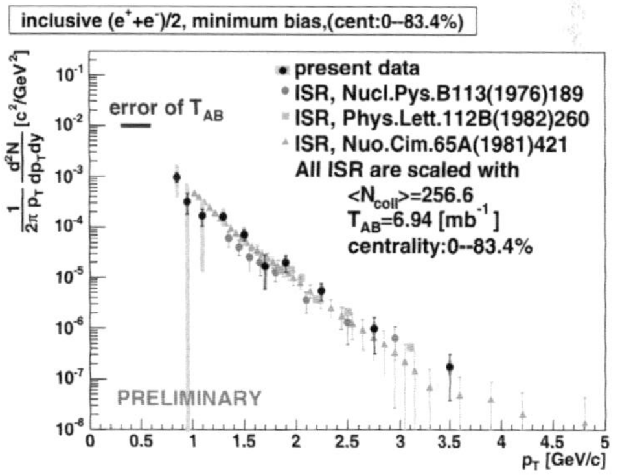

Figure 2. Single electron p_T spectra in Au+Au collisions (new data) and the ISR p+p data [9, 10, 11] at $\sqrt{s_{NN}}$=62 GeV. The ISR data are scaled by N_{coll}.

Figure 3 shows new results, namely the p_T distribution of single electrons from heavy quark decays in Au+Au collisions at $\sqrt{s_{NN}}$=200 GeV, and the nuclear modification factor in 2.5<p_T<5.0 GeV as a function of the number of participants. This analysis uses the same data set as that of Fig. 1. The statistics are greatly improved because the cocktail subtraction method was used. The new high statistics result clearly shows that the single electron spectra are suprresed at high p_T. This suppression suggests that heavy quarks lose significant energy in the media. The elliptic flow of charm mesons is also measured [12]. The measured eliptic flow of heavy flavor electron is nonzero with a 90 % confidence level. The new Au+Au data taken in 2004 should clarify these things because of their much higher statistics.

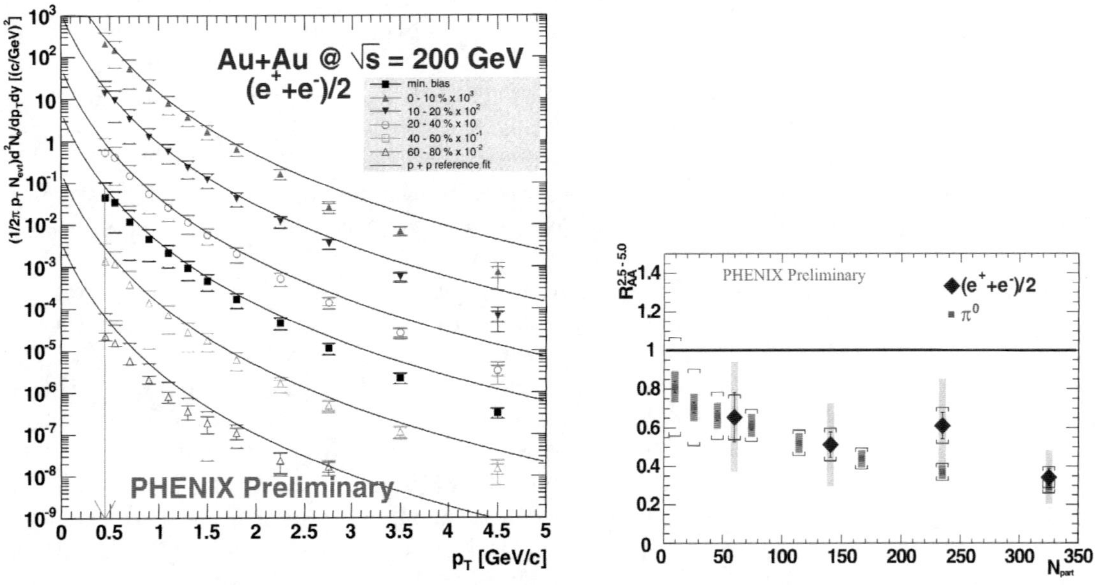

Figure 3. Left: single electron p_T spectra for each centrality bin in Au+Au collisions at $\sqrt{s_{NN}}$=200 GeV (points), compared with the p+p result scaled by number of binary collisions (solid lines). Right: nuclear modification factor (R_{AA}) of single electron (diamond) in 2.5<p_T<5.0 GeV as a function of the number of participants. Those of neutral pions are also plotted by square points as reference data.

3. J/ψ measurement

The J/ψ meson yields were measured via the di-electron decay channel using the central arms and via the di-muon decay channel using the muon arms at forward (1.2<η<2.4) and backward pseudo-rapidity (-2.2<η<-1.2). The production cross sections were measured in p+p [13] (Fig. 4-(a)), d+Au and Au+Au [14] collisions. The nuclear modification factor for d+Au, the ratio of the binary collision scaled d+Au cross section to the corresponding p+p cross section, is shown versus rapidity in (Fig. 4-(b)). There is a weak rapidity dependence that suggests effects of nuclear absorption or nuclear (anti-) shadowing. We expect about 600 $J/\psi \rightarrow e^+e^-$ events and several thousand $J/\psi \rightarrow \mu^+\mu^-$ events from the Au+Au data set taken in 2004, which is still being analyzed. This should allow us to make a plot similar to (Fig. 4-(b)) for Au+Au.

The production mechanism of J/ψ mesons was also studied with J/ψ spin alignment in d+Au data at mid-rapidity in the p_T around 1.5 GeV/c. The spin alignment was measured by the positron angular distribution, $1 + \lambda \cos^2 \theta$, where θ is the angle between the positron momentum and the momentum direction of the J/ψ in the rest frame. Figure 5 shows the measured λ as a function of p_T.

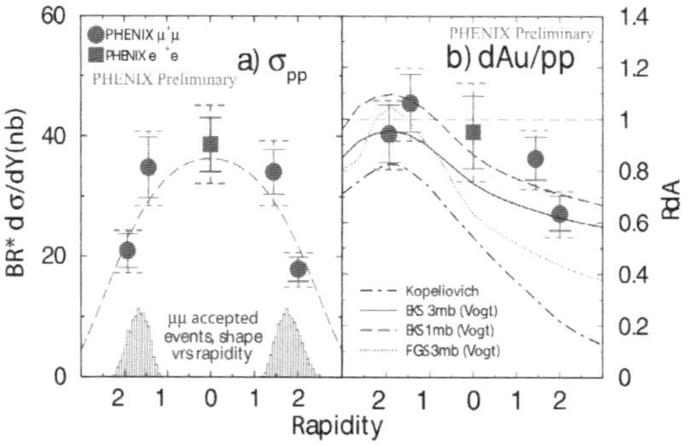

Figure 4. The left panel shows the J/ψ production cross section as a function of rapidity in p+p collisions at $\sqrt{s_{NN}}$=200 GeV. The right panel shows the nuclear modification factor for d+Au collisions (R_{dA}) as a function of rapidity.

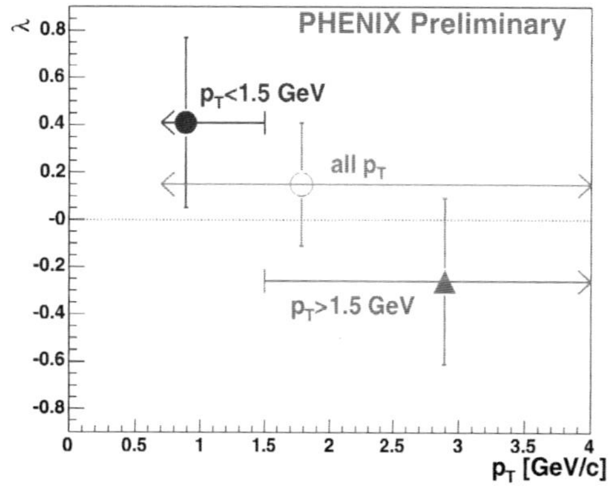

Figure 5. The J/ψ polarization parameter as a function of p_{T} in d+Au collisions at $\sqrt{s_{NN}}$=200 GeV. The filled circle and the triangle points show the λ in p_{T} below and above 1.5 GeV/c, respectively. The open circle shows the polarization parameter for all p_{T}.

4. Summary

Open heavy quark yields were measured by PHENIX through their decay electrons at $\sqrt{s_{NN}}$=200 GeV in p+p, d+Au and Au+Au collisions, and at 62.4 GeV and 130 GeV in Au+Au collisions. The electrons p_{T} spectra from open heavy flavor in d+Au scale with N_{coll} at mid rapidity. In Au+Au collisions at 200 GeV, the total yield of single electrons scales with the number of binary collisions, while the single electron yields are suppressed at high p_{T}.

The J/ψ meson production cross sections were measured in p+p, d+Au and Au+Au collisions at 200 GeV. A weak rapidity dependence is seen in the d+Au nuclear modification factor. No significant spin alignment of J/ψ is seen at the mid-rapidity in the p_{T} range around 1.5 GeV/c.

References

[1] Adcox K *et al.* *Preprint* nucl-ex/0410003
[2] Dokshitzer Y L and Kharzeev D E 2001 *Phys. Lett.* B **519** 199
[3] Matsui T and Satz H 1986 *Phys. Lett.* B **178** 416
[4] Thews R L, Schroedter M and Rafelski J 2001 *Phys. Rev.* C **63** 054905
[5] Adcox K *et al.* 2003 *Nucl. Instrum. Methods Phys. Res.* A **499** 469
[6] Adler S S *et al.* 2005 *Phys. Rev. Lett.* **94** 082301
[7] Adcox K *et al.* 2002 *Phys. Rev. Lett.* **88** 192303
[8] Kelly S *et al.* 2004 *J. Phys.* G **30** S1189
[9] Büsser F W *et al.* 1976 *Nucl. Phys.* B **113** 189
[10] Perez P *et al.* 1982 *Phys. Lett.* B **112** 260
[11] Basile M *et al.* 1981 *Nuovo Cimento* A **65** 421
[12] Adler S S *et al.* *Preprint* nucl-ex/0502009
[13] Adler S S *et al.* 2004 *Phys. Rev. Lett.* **94** 051802
[14] Adler S S *et al.* 2004 *Phys. Rev.* C **69** 014901

Institute of Physics Publishing Journal of Physics: Conference Series **50** (2006) 336–339
doi:10.1088/1742-6596/50/1/043 5th International Conference on Physics and Astrophysics of Quark Gluon Plasma

Study of jet fragmentation via azimuthal correlations: a probe for the sQGP at RHIC

N N Ajitanand

For the PHENIX Collaboration

ajit@mail.chem.sunysb.edu

Abstract. Two particle azimuthal correlation functions are used to study flow and jet fragmentation in the hot QCD matter believed to be created in Au+Au collisions at RHIC ($\sqrt{s_{NN}}$ =200GeV). Detailed results on jet topologies and yields, and their flavor dependence are obtained via a novel technique for decomposition of the combined correlations from jets and flow in Au+Au collisions. The results indicate significant broadening of the away-side jet attributable to important interactions between the flowing medium and the scattered partons which fragment into jets.

1. Introduction

The RHIC experiments have been designed to recreate the first moments after the Big Bang when the universe existed as a plasma of quarks and gluons (QGP). This phase of nuclear matter is widely anticipated in the most energetic Au+Au collisions at RHIC. In fact, recent estimates[1] of the energy density achieved in relatively central Au+Au collisions at $\sqrt{s_{NN}}$ =200GeV far exceeds that required for creating the QGP according to Lattice QCD calculations[2]. The concomitant high matter density is expected to give rise to large pressure gradients which is evidenced in the observed large azimuthal anisotropy (v_2) of particle emission from the collision zone. The value of this anisotropy is close to the predictions of the hydrodynamic model which in turn implies the creation of a strongly interacting thermalized medium. In addition to the dominant soft processes giving rise to the formation of the medium, there are relatively rare hard parton-parton collisions. The scattered partons propagate through the medium radiating gluons and interacting with the medium till they finally fragment into jets. Thus jets are an ideal probe of the medium provided one can de-convolute the jet signal from the collective flow effects. Possible medium associated modifications of the jet topology are a conical emission akin to a “sonic boom”[3] and distortions induced via interactions between scattered and flowing partons[4]. To study jet modification it is necessary to first build two particle azimuthal correlations.

2. Azimuthal correlations

Two particle azimuthal correlations are built by pairing a leading hadron in the range 2.5<pt<4.0 with an associated hadron in the range 1.0<pt<2.5. The correlation function C is given by :

$$C(\Delta\phi) = \frac{N_{\mathrm{Real}}(\Delta\phi)}{N_{mix}(\Delta\phi)}$$

Here $\Delta\phi$ is the difference of azimuthal angles of the pair. The real distribution is built from pair members belonging to the same event and mixed distributions are made of pair members belonging to different events. Thus the correlation function is free of geometric acceptance effects and carries only the physics effects i.e. long range correlations from flow and short range correlations from jets.

2. Deconvolution Method

It is necessary to decompose the correlation function to obtain the jet function. A two source model gives[5]

$$C(\Delta\phi) = a_0\left[H(\Delta\phi)+J(H(\Delta\phi))\right],$$

where, H is a second harmonic function characterized by $p_2 = v_2^{lo} \times v_2^{hi}$.

The values v_2^{lo} and v_2^{hi} are simply the average of $\cos(2(\phi-\Phi_{RP}))$ for the lower and higher p_T ranges and Φ_{RP} is the reaction plane azimuth obtained from the Beam-Beam counters placed at large η. To obtain a_0 the ZYAM assumption is made i.e. the minimum of the Jet function J is zero. The following plots demonstrate the effectiveness of the de-convolution method using 3-dimensional simulations with two types of away side jets (both giving similar jet functions) in a flowing background of particles. The solid squares represent the extracted function and the solid line the input jet function. Panels (b), (c) and (d) show results for the case where the leading particle is unconstrained, constrained in the reaction plane and constrained perpendicular to the reaction plane respectively. A comparison of the input and output jet-pair distribution indicates that decomposition is robust.

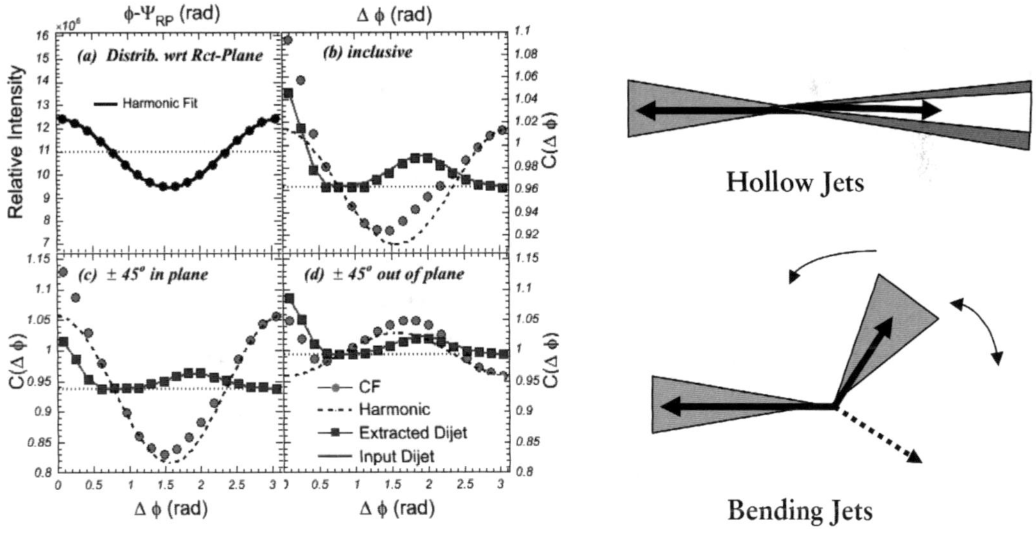

Figure 1: Decomposition of simulated data. Panel (a) shows the harmonic contribution. Panels, (b), (c) and (d) show the combined correlation function (solid circles) and harmonic contribution (dashed line) and the extracted jet-pair distribution (filled squares)for inclusive and leading particle constrained in and out of plane respectively. Schematic jet shapes are indicated to the right of the figure.

4. Results

Figure 2 show the extracted jet functions following decomposition of the measured correlation functions for 200 Gev Au+Au collisions. The points represent nominal values. The lines indicate the results of one and two sigma systematic error variation on the v_2 value used for decomposition. For the most peripheral collisions (panel (f)) the characteristic distribution for a di-jet is evident. For all other centralities, there is a clear broadening of the away-side jet and possibly a minimum at ~ 180^0. The latter may be the result of the predicted interaction of the away-side jet with the medium[1,4].

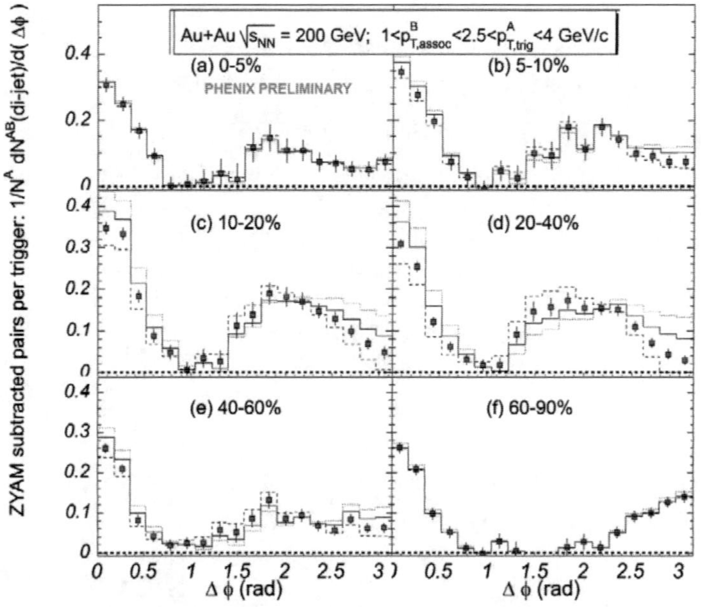

Figure 2: Extracted jet-pair distributions for Au+Au collisions. Results are shown for trigger particle with $2.5 < p_T < 4$ and associated particle with $1.0 < p_T < 2.5$ GeV/c for several centralities as indicated.

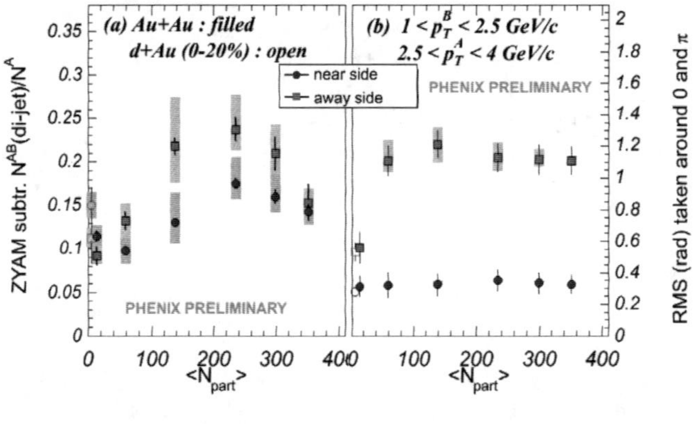

Figure 3: Extracted yields (a) and jet widths (b) as a function of centrality. Results are shown for the near and away-side jets.

Figure 3 shows the integrated near and away side yields and second moments of the jet widths as a function of centrality. The away side yield is generally higher than the near side yield.

The near side widths are independent of centrality while the away side values rises quickly and then remain relatively constant for mid-central to central collisions.

Similar jet extractions have been performed for flavor identified associated particles in the same pt range (1.0 < pt < 2.5 GeV/c). Figure 4 shows the ratio of baryonic yield to the mesonic yield for the near and away side jets. The away side is generally more baryon rich than the near side.

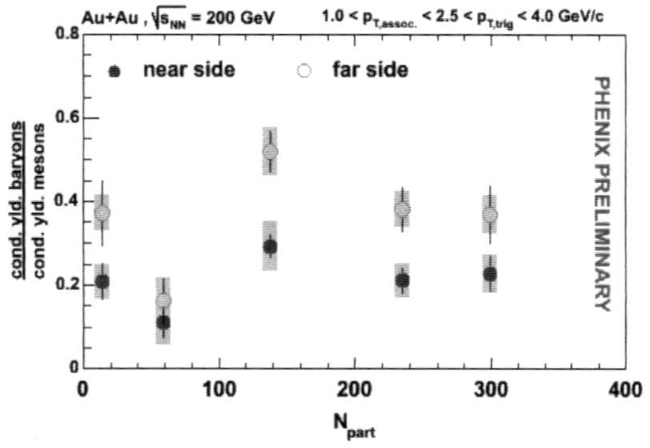

Figure 4: Centrality dependence of the extracted ratio of associated Baryon to Meson jet yields. Results are shown for the near and away-side jets.

3. Conclusions

A robust procedure for the decomposition of correlation functions has been used to extract di-jet correlations in Au+Au collisions. The observed characteristics of the near-side jet are similar to that observed in p+p and d+Au collisions. However, significant distortions of the away-side jet are observed. The flavor dependence of jet yields indicates a larger baryonic content for the away-side jet. Such information may provide more detailed information on the mechanism/s for medium induced modification to jets which traverse the strongly interacting high energy density matter produced at RHIC.

References

[1] S.S. Adler et al (PHENIX collaboration) nucl-ex/0409015
[2] F. Karsch, Nucl. Phys. A698, 199 (2002)
[3] Casalderrey-Solana et al hep-ph/0411315, (2005); Stocker et al. nucl-th/0406018 (2004)
[4] Armesto,Salgado,Wiedemann hep-ph/0411341
[5] Ajitanand et al. nucl-ex/0501025

Institute of Physics Publishing
Journal of Physics: Conference Series **50** (2006) 340–343
doi:10.1088/1742-6596/50/1/044
5th International Conference on Physics and Astrophysics of Quark Gluon Plasma

Event by Event fluctuation in K/π ratio at RHIC

Supriya Das (for the STAR Collaboration [1])
Experimental Physics Division, Variable Energy Cyclotron Center,
1/AF Bidhan Nagar, Kolkata -700064, INDIA

E-mail: supriya@veccal.ernet.in

Abstract. We present the preliminary results of event by event fluctuation in K/π ratio in Au+Au collision at $\sqrt{s}_{NN}$ = 200 GeV and at 62.4 GeV using STAR detector at RHIC. Two different methods have been used to extract the dynamical fluctuation. The results from the study of energy and centrality dependence of the dynamical fluctuation are presented. From the excitation function it is seen that at two RHIC energies the measure of dynamical fluctuation is constant with values very close to that at 12.3 GeV at SPS. The dynamical fluctuation is found to be positive and decreasing with increasing centrality at RHIC. The results are compared with those from HIJING model calculation with jets. Results from HIJING are found to be very close to data from central collisions whereas it over predicts the data for peripheral events.

1. Introduction

A key question of the heavy ion program at the Relativistic Heavy Ion Collider (RHIC) is to understand whether the hot and dense matter produced in the midst of the relativistic heavy ion collisions undergoes a transition to and from a quark gluon plasma (QGP) phase before it hadronizes. The study of fluctuations of specific observables on event by event basis, can provide comprehensive insight in this matter [1-7]. Several recent experimental studies at the SPS and RHIC have focused on the study of fluctuations in relativistic heavy ion collisions. Results from NA49 experiment at SPS [8, 9] show interesting behavior of both K/π ratio as well as the event by event dynamical fluctuation in that quantity. In this paper we present preliminary results from studies on the event by event fluctuation in the K/π ratio for Au+Au collisions at $\sqrt{s}_{NN}$ = 200 GeV and at 62.4 GeV using the STAR detector at RHIC.

2. STAR experiment

STAR experiment at RHIC measures charged hadrons over a wide range of pseudorapidity ($|\eta| < 1$) and azimuthal angle ($-\pi < \phi < \pi$), using the large Time Projection Chamber [10]. An overview of the experiment can be found in [11]. The kaons and pions are identified based on their relative energy deposition (dE/dx) in the TPC within the momentum range of 100 MeV - 600 MeV. This wide acceptance of the detector enhances the possibility to study event by event physics. Specially the wide momentum range available for particle identification enables to study event by event fluctuation in K/π ratio.

[1] Complete author list can be found at the end of this proceedings

3. Analysis

We have used the dataset from RHIC run II for Au+Au 200 GeV and RHIC run IV for Au+Au at 62.4 GeV. Around 300K minimum bias events, corresponding to 0 - 80% of Au+Au hadronic interaction cross-section, are analysed for this work. In minimum bias mode the events were triggered by two ZDCs and a minimum signal at the CTB. Only events for which the z-vertex is within ±25 cm of the center of the TPC are accepted. The global tracks with at least 15 hits in TPC and z component of Distant of Closest Approach (DCA) of the track with the primary vertex is less than 3 cm are used in this analysis. Two separate methods of analysis are used, one to obtain the overall measure of the dynamical fluctuation in central events and the other one to obtain the centrality dependence of the dynamical fluctuation.

3.1. Dynamical fluctuation and its excitation function

The kaons and pions in each event are identified from their relative energy deposition in TPC. Then the ratio of multiplicities of kaons and pions is obtained on event by event basis. The width (σ_{data}) of the distribution of this ratio for sufficiently large number of events, gives the fluctuation in the ratio for real data. In order to isolate the strength of non-statistical fluctuation, we need to establish a reference which tells us the contribution to fluctuation from finite number of particles detected in the detector, experimental resolution in dE/dx but not containing any further correlations between the particles. This statistical limit corresponding to uncorrelated kaon and pion production, is estimated based on the ratio of kaon and pion multiplicities obtained from pseudo events constructed using a mixed event technique (σ_{mixed}). This mixing of the events has been done picking up one track from every event randomly to construct one mixed event while the multiplicity of the mixed events follow that of the real events. The dynamical fluctuation in the quantity has been obtained by the formula

$$\sigma_{dyn} = \sqrt{\sigma^2_{data} - \sigma^2_{mixed}} \qquad (1)$$

This approach was used by NA49 for central events at various SPS energies. We have therefore used this to compare the results from SPS.

3.2. Results

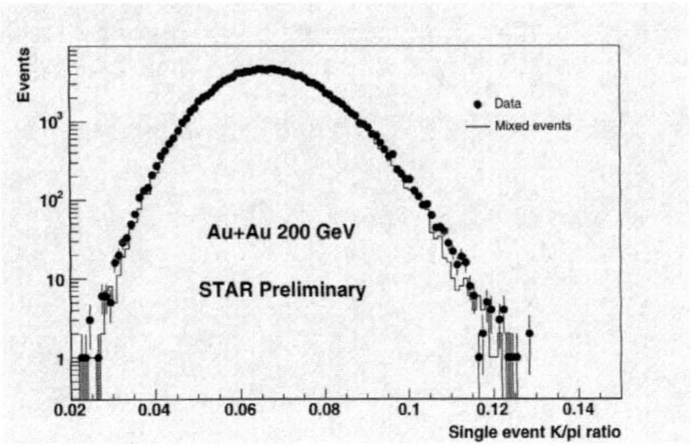

Figure 1. Distribution of K/π ratio for real data and mixed events for Au+Au at 200 GeV

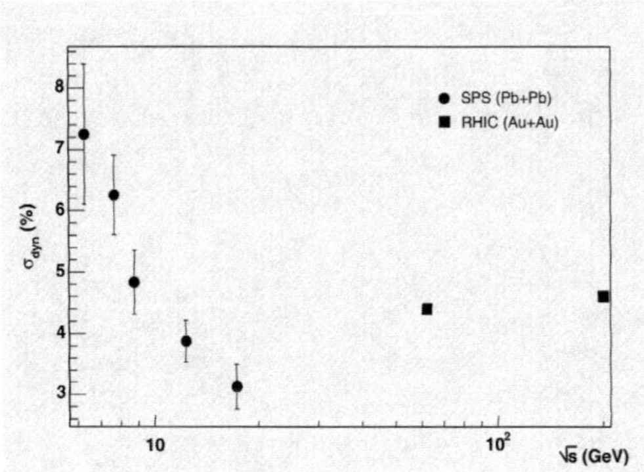

Figure 2. Excitation function for σ_{dyn}

Fig. 1 shows the distribution of K/π ratio for top 5% central events in Au+Au collisions at 200 GeV. The distributions obtained from real and mixed events are superimposed to help comparing the widths of the distributions. The values of σ_{dyn} in different cases are tabulated in the table below.

In Fig. 2 the energy dependence (excitation function) of σ_{dyn} is plotted. The data points up to 17 GeV are obtained from NA49 collaboration [9]. At two RHIC energies, the measure of dynamical fluctuation appears to be constant having values close to SPS data at 12.3 GeV.

Table 1. Values of σ_{dyn}

Experiment	System	Ratio type	σ_{data}	σ_{mixed}	σ_{dyn}
NA49	Pb+Pb at 17.3 GeV	K/π	-	-	3.12(±0.37)%
STAR	Au+Au at 200 GeV	K/π	17.78%	17.23%	4.6(±0.025)%
STAR	Au+Au at 200 GeV	K^+/π^+	24.29%	24.10%	3.06(±0.066)%
STAR	Au+Au at 200 GeV	K^-/π^-	24.81%	24.55%	3.61(±0.055)%
STAR	Au+Au at 62.4 GeV	K/π	20.7%	20.3%	4.44(±0.01)%

3.3. Centrality dependence of dynamical fluctuation

To study the centrality dependence of the dynamical fluctuation we have used the variance ν_{dyn}, first suggested in Ref. [12], which is derived from single and two-particle distribution functions. In this method the dynamical fluctuation has been expressed as the contribution from three terms representing the correlation strength of kaons, pions and the cross correlation between them [13] :

$$R = R_{KK} + R_{\pi\pi} - 2R_{K\pi} \tag{2}$$

This approach has been used in several cases where the fluctuation of the ratio of two quantities is measured [14].

Experimentally this quantity is obtained from the following relation:

$$\nu_{dyn,K/\pi}(M) = \frac{\langle K(K-1)\rangle_M}{\langle K\rangle_M^2} + \frac{\langle \pi(\pi-1)\rangle_M}{\langle \pi\rangle_M^2} - 2\frac{\langle K\pi\rangle_M}{\langle K\rangle\langle \pi\rangle_M} \tag{3}$$

The notation $\langle O\rangle_M$ is used to indicate the average of the quantity O for all events with uncorrected multiplicity M of charged hadrons within pseudorapidity range $|\eta| < 0.5$. Centrality is defined dividing this whole range of multiplicity in nine contiguous bins.

3.4. Results

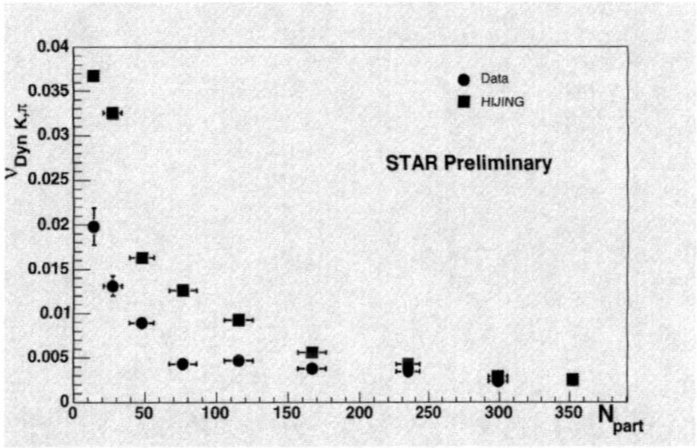

Figure 3. Centrality dependence of $\nu_{dyn,K/\pi}$ for Au+Au collisions at $\sqrt{s_{NN}} = 200$ GeV.

Fig. 3 shows the variation of $\nu_{dyn,K/\pi}$ with centrality where along X-axis plotted is N_{part} which is the number of participants in Au+Au collisions calculated by optical glauber model calculation. Error on the data points correspond to statistical errors only. Results from HIJING[15] calculation is shown to compare with the data. 10K HIJING events for each centrality bin have been used for this purpose.

4. Summary

We have presented the preliminary results of the event by event fluctuation in K/π ratio measured at RHIC using the STAR detector. Two different methods were used to extract the dynamical fluctuation. In one method the distribution of K/π ratio from data was compared with that obtained from mixed events which provides the statistical limit of this fluctuation, corresponding to uncorrelated kaon and pion production. The results from this method show the dynamical fluctuation in K/π for top 5% central Au+Au at 200 GeV as 4.6%, for K^+/π^+ as 3.06%, for K^-/π^- as 3.61% and for K/π for top 5% central events for Au+Au at 62.4 GeV as 4.44%. The excitation function of dynamical fluctuation has also been studied combining RHIC data with data from SPS. At two RHIC energies, the measure of dynamical fluctuation appears to be constant having values close to SPS data at 12.3 GeV. Another method based on correlation studies were used to study the centrality dependence of this dynamical fluctuation was used. The result from this shows the dynamical fluctuation for Au+Au 200 GeV is positive and decreases with increasing centrality. These results were compared with those obtained from HIJING and the model calculation results are very close for the central collisions whereas for peripheral events the model overpredicts the data.

5. Acknowledgments

We thank the RHIC Operations Group and RCF at BNL, and the NERSC Center at LBNL for their support. This work was supported in part by the HENP Divisions of the Office of Science of the U.S. DOE; the U.S. NSF; the BMBF of Germany; IN2P3, RA, RPL, and EMN of France; EPSRC of the United Kingdom; FAPESP of Brazil; the Russian Ministry of Science and Technology; the Ministry of Education and the NNSFC of China; Grant Agency of the Czech Republic, FOM of the Netherlands, DAE, DST, and CSIR of the Government of India; Swiss NSF; the Polish State Committee for Scientific Research; and the STAA of Slovakia.

6. References

[1] Stephanov M, Rajagopal K and Shuryak E 1998 *Phys. Rev. Lett.***81** 4816.
[2] Stephanov M, Rajagopal K and Shuryak E 1999 *Phys. Rev.* D **60** 114028.
[3] Jeon S and Koch V 1999 *Phys. Rev. Lett.* **83** 5435.
[4] Jeon S and Koch V 2000 *Phys. Rev. Lett.* **85** 2076.
[5] Asakawa M, Heinz U and Müller B 2000 *Phys. Rev. Lett.* **85** 2072.
[6] Heiselberg H 2001 *Phys. Rep.* **351** 161.
[7] Heiselberg H and Jackson AD 2001 *Phys. Rev.* C **63** 064904.
[8] Afanasiev *et al.* for NA49 Collaboration 2000 *Phys. Rev. Lett.* **86** 1965.
[9] Roland G for NA49 Collaboration 2004 *J. Phys.* G **30** S1381.
[10] Ackermann K H *et al.* for STAR Collaboration 1999 *Nucl.Phys.* A **661** 681
[11] Ackermann K H *et al.* for STAR Collaboration 2003 *Nucl. Instrum. Meth.* A **499** 624
[12] Voloshin S 2002 *Proc. International Nuclear Physics Conference* **610** 591.
[13] Pruneau C, Gavin S and Voloshin S 2002 *Phys. Rev.* C **66** 044904.
[14] J Adams *et al.* for STAR Collaboration 2003 *Phys. Rev.* C **68** 044905.
[15] Wang X N and Gyulassy M 1992 *Phys. Rev.* D **44** 844.

Institute of Physics Publishing Journal of Physics: Conference Series **50** (2006) 344–348
doi:10.1088/1742-6596/50/1/045 5th International Conference on Physics and Astrophysics of Quark Gluon Plasma

Indications of the onset of deconfinement at low SPS energies

Volker Friese (for the NA49 collaboration)

Gesellschaft für Schwerionenforschung, Planckstr. 1, 64291 Darmstadt, Germany

E-mail: v.friese@gsi.de

Abstract.
We present evidence for an onset of deconfinement at low SPS energies obtained by the NA49 experiment in central Pb+Pb collisions in an energy scan from 20 to 158 AGeV. As indicators for such an onset, the independence of mean m_t on beam energy, a sharp peak in the strangeness-to-pion ratio and dynamical fluctuations in the event-by-event K/π ratio are discussed.

1. Introduction

According to lattice QCD calculations, hadronic matter undergoes a transition to a deconfined phase once a critical energy density is surpassed [1]. High-energy collisions of heavy ions provide a tool to study the properties of strongly interacting matter under extreme conditions and, in particular, the phase transition and the deconfined phase. The NA49 collaboration has investigated central Pb+Pb collisions at beam energies between 20 and 158 AGeV ($\sqrt{s_{NN}} = 6.3 - 17.3$ GeV) in a search for signals of the onset of deconfinement.

The production of strange particles is one of the observables proposed as a signal of a deconfined state [2]. Therefore, one of the prime interests of NA49 is to measure spectra and yields of a multitude of strange particle species. In addition, event-by-event fluctuations possibly connected to a critical point in the phase diagram of strongly interacting matter have recently attracted attention. In the following, we present results on these observables obtained in the context of the NA49 energy scan programme, which indicate that indeed the transition from confined to deconfined matter occurs at low SPS energies.

NA49 [3] is a fixed target experiment operating at the CERN-SPS. Its basic components are four large-volume TPCs for the detection of the major part of the charged particles produced in the collisions. The large rapidity coverage allows the determination of rapidity-integrated yields. Particle identification is provided by the energy loss measurement in the TPCs. The PID capabilities are augmented by TOF detectors covering kaons at mid-rapidity.

For the study of the energy dependence of the relevant observables, the NA49 results are compared to data obtained at AGS and RHIC. A compilation of references to these data can be found in [4].

2. Transverse dynamics

The transverse-mass spectra of the particle species measured by NA49 at mid-rapidity for the beam energies 20 and 30 AGeV ($\sqrt{s_{NN}} = 6.3$ and 7.6 GeV) are shown in figure 1. The full lines

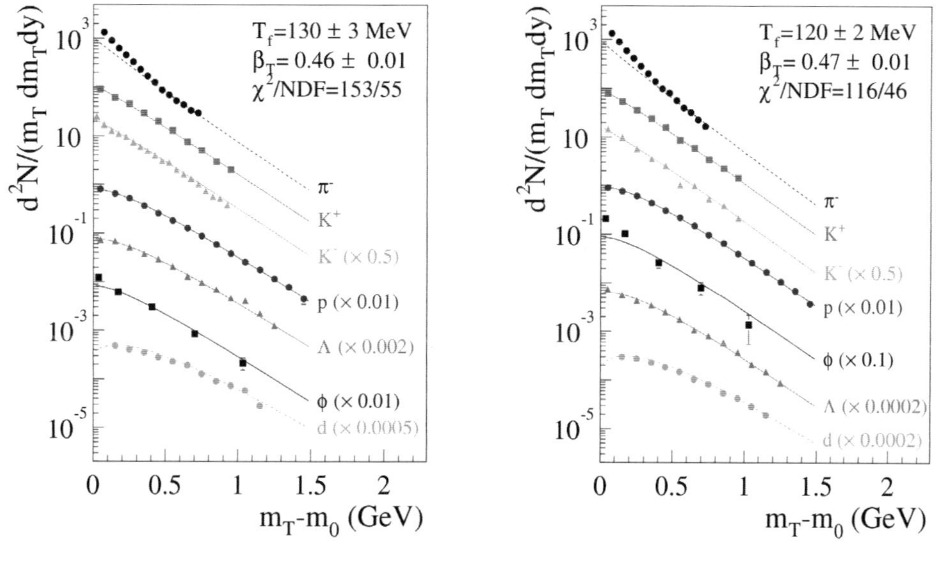

Figure 1. Transverse-mass spectra at midrapidity for various particle species in central Pb+Pb collisions at (left) 30 AGeV and (right) 20 AGeV beam energy. The full lines show the result of a simultaneous blast-wave fit [5] to the spectra. The pions were excluded from the fit.

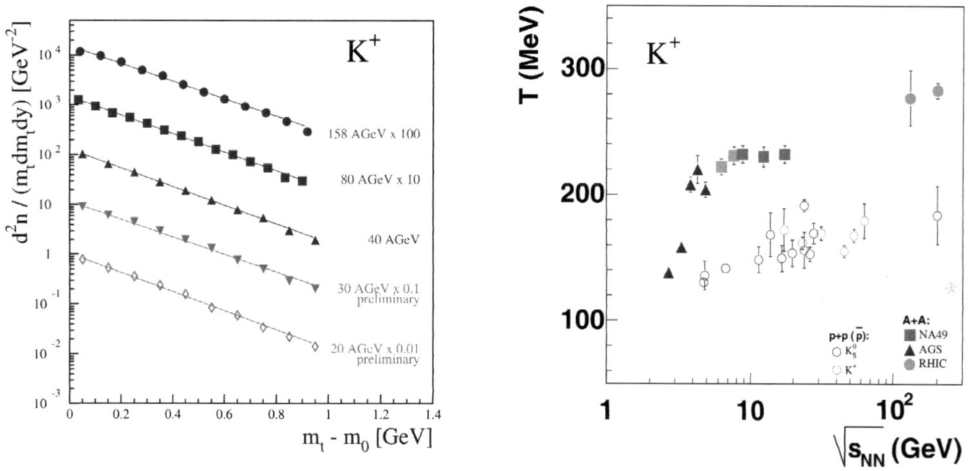

Figure 2. (Left) Transverse-mass spectra of K^+ in central Pb+Pb collisions at five different beam energies. The full lines show exponential fits to the spectra. (Right) Inverse slope parameter of K^+ in central Pb+Pb (Au+Au) as function of CM energy. The open points show results from pp collisions.

show the result of a simultaneous blast-wave fit with constant velocity [5] to the spectra. Due to the influence of feeddown to the their yield, pions were excluded from the fit. Although such a parametrisation provides only an approximate description of the hydrodynamical behaviour, the good overall agreement indicates a common kinetic freeze-out behaviour of the measured particles. This also holds for the higher SPS energies (40, 80 and 158 AGeV, not shown).

While generally, the transverse spectra are curved in logarithmic representation due to the effect of transverse flow, the kaons turn out to be almost perfectly exponential as demonstrated in figure 2 (left). Surprisingly, the kaon inverse slopes are nearly the same for all SPS energies

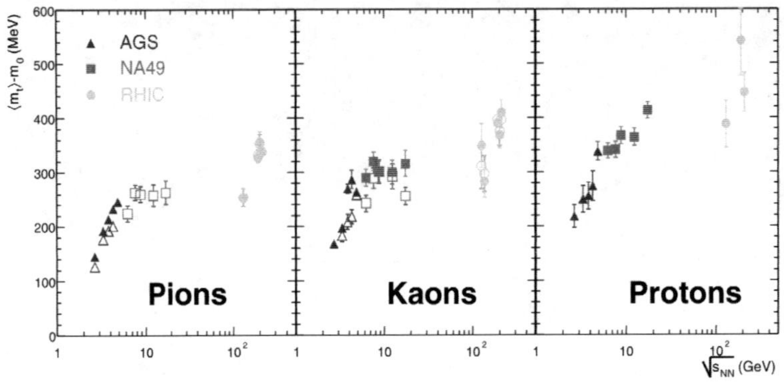

Figure 3. Mean transverse mass for pions, kaons and protons in central Pb+Pb (Au+Au) collisions as functions of CM energy

measured by NA49. The excitation function shown in figure 2 (right) exhibits a steep rise at AGS energies, followed by a plateau st SPS energies and a further rise toward RHIC. Since the apparent temperature is caused by the "true" temperature and the transverse flow, this behaviour indicates a soft region in the equation of state which can be attributed to the mixed phase of a first-order phase transition [6, 7].

For other particles than kaons, the inverse slope parameter is less meaningful due to the deviation of the transverse-mass spectra from pure exponential. We thus calculated the mean m_t instead. Its energy dependence is presented in figure 3 for pions, kaons and protons. In all cases, the plateau-like structure is again observed, strengthening the conclusion of non-continuous behaviour of both temperature and transverse flow.

3. Strangeness-to-pion ratio

The broad rapidity coverage of the NA49 detector gives access to integrated particle yields. Figure 4 shows the energy dependence of the yields of the main strangeness carriers K^+, K^- and Λ, normalised to the total pion yield. While K^+ and Λ exhibit a sharp maximum around 30 AGeV beam energy, the relative K^- yield evolves rather smoothly from AGS over SPS to RHIC energies. The full lines in figure 4 show the predictions of the hadron gas model (HGM), which smoothly parametrises the dependence of temperature and baryochemical potential on collision energy [8]. The comparison with the data shows that kaons are well described up to 30 AGeV but overestimated at higher SPS energies. In particular, the sharp peak in the K^+/π ratio cannot be accounted for by a smooth evolution of the system parameters as assumed by the model. For the Λ excitation function, the data at lower energies are less conclusive. While the overall description by the HGM is good, it predicts a maximum in the Λ/π ratio at top AGS energy, whereas NA49 observes a peak at 30 AGeV just as for the K^+.

Further information about the appropriateness of a purely hadronic description will arise from the measurement of the rare strange particles Ξ and Ω. At present, NA49 data are only available at 40 and 158 AGeV for these particles. Since the HGM predicts the maximum of the Ξ/π ratio to be above 40 AGeV beam energy, the Ξ measurement at lower energies could have discriminative power. Results of NA49 on Ξ production at low SPS energies are expected soon.

The measurement of the bulk strangeness carriers K and Λ allows the study of the total strangeness-to-pion ratio as a function of CM energy. Figure 5 presents the excitation function of the variable $E_S = (\Lambda + 2(K^+ + K^-))/\pi$, compared to the prediction of the statistical model of the early stage [9], which assumes a phase transition from hadronic matter to the quark-gluon

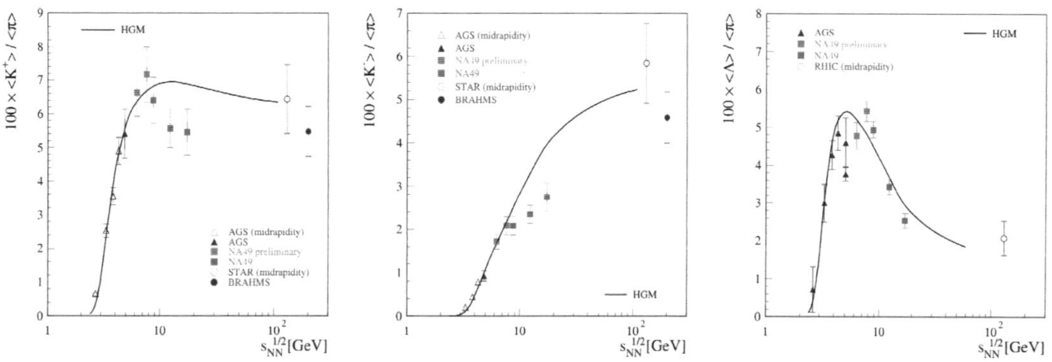

Figure 4. Inegrated yields of (left) K^+, (centre) K^- and (right) Λ in central Pb+Pb (Au+Au) collisions, normalised to the pion yields, as functions of CM energy. The full lines show the predictions of the hadron gas model [8]. NA49 data points correspond to 20, 30, 40, 80 and 158 AGeV, respectively.

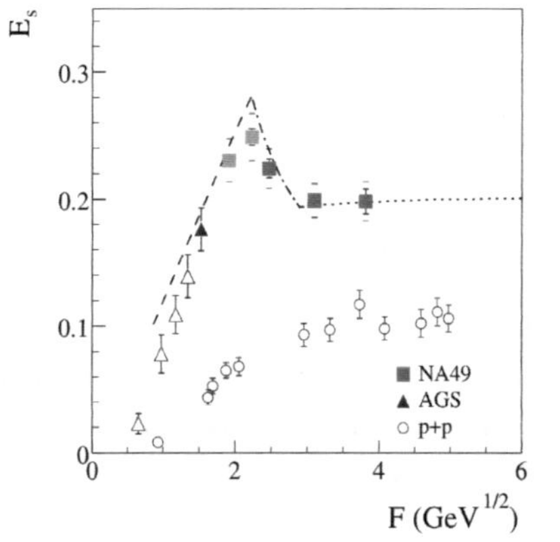

Figure 5. Total strangeness-to-pion ratio E_S in central Pb+Pb (Au+Au) collisions as function of the Fermi variable F. The lines shows the prediction of the statistical model of the early stage [9]. NA49 data points correspond to 20, 30, 40, 80 and 158 AGeV, respectively.

plasma to take place at 30 AGeV beam energy. The good agreement between model and data provides strong evidence for such a scenario.

4. Event-by-event fluctuations

Lattice QCD predicts a critical endpoint in the phase diagram of strongly interacting matter, separating a region of first-order phase transition at high baryochemical potential from a smooth crossover at higher collision energies [1]. If the system passes near this critical point, large fluctuations are expected. NA49 has studied the eventwise kaon to pion ratio at five different beam energies. Figure 6 shows the results of this study compared to the fluctuations obtained with the hadronic transport model UrQMD [10]. The dynamical fluctuations, calculated as the quadratic difference of the widths of the same-event and mixed-event K/π distributions, show a rise towards lower CM energies, which is not reproduced by UrQMD. In contrast, for the negative fluctuations seen in the proton-to-pion ratio, probably reflecting correlations due to resonance decays, perfect agreement between data and model is observed. A quantitative interpretation of the dynamical fluctuations in the K/π ratio is, however, still lacking.

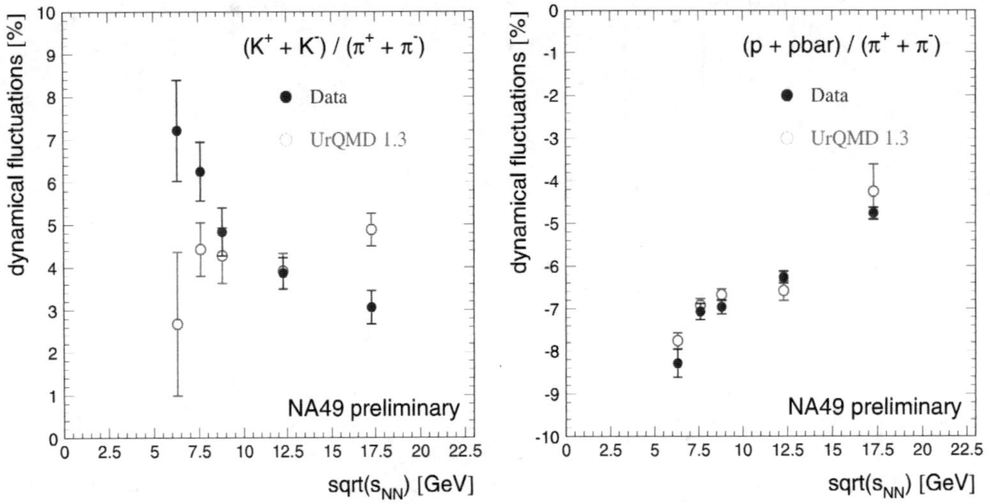

Figure 6. Dynamical event-by-event fluctuations of (left) the kaon-to-pion ratio and (right) the proton-to-pion ratio in central Pb+Pb collisions as functions of the CM energy. The NA49 data points correspond to 20, 30, 40, 80 and 158 AGeV, respectively.

Acknowledgments

This work was supported by the US Department of Energy Grant DE-FG03-97ER41020/A000, the Bundesministerium für Bildung und Forschung, Germany, the Polish State Committee for Scientific Research (2 P03B 130 23, SPB/CERN/P-03/Dz 446/2002-2004, 2 P03B 04123), the Hungarian Scientific Research Foundation (T032648, T032293, T043514), the Hungarian National Science Foundation, OTKA, (F034707), the Polish-German Foundation, and the Korea Research Foundation Grant (KRF-2003-070-C00015).

References

[1] Fodor Z and Katz S 2003 *Heavy Ion Phys.* **18** 41
Allton C R *et al.* 2002 *Phys. Rev.* D **66** 074507
[2] Koch P, Müller B and Rafelski J 1986 *Phys. Rep.* **142** 167
[3] Afanasiev S *et al.* (NA49 collaboration) 1999 *Nucl. Instrum. Methods* **A 430** 210
[4] Friese V *et al.* (NA49 collaboration) *J. Phys. G: Nucl. Phys.* **31**, to be published
[5] Schnedermann E, Sollfrank J and Heinz U 1993 *Phys. Rev.* C **48** 2462
[6] Gorenstein M, Gaździcki M and Bugaev K 2003 *Phys. Lett.* B **567** 175
[7] Hama Y *et al.* 2004 *Braz. J. Phys.* **34** 322
[8] Braun-Munzinger P *et al.* 2002 *Nucl. Phys.* **A 697** 902
[9] Gaździcki M and Gorenstein M 1999 *Acta Phys. Pol.* **B 30** 2705
[10] Bleicher M *et al.* 1999 *J. Phys. G: Nucl. Phys.* **25** 1859

Institute of Physics Publishing Journal of Physics: Conference Series **50** (2006) 349–352
doi:10.1088/1742-6596/50/1/046 5th International Conference on Physics and Astrophysics of Quark Gluon Plasma

Dimuon production in p-A and In-In collisions

J. Seixas[2,5] R. Arnaldi[9], K. Banicz[2,4], J. Castor[3], B. Chaurand[7], C. Cicalo[1], A. Colla[9], P. Cortese[9], S. Damjanovic[4], A. David[2,5], A. de Falco[1], A. Devaux[3], L. Ducroux[6], H. En'yo[8], A. Ferretti[9], M. Floris[1], P. Force[3], N. Guettet[2,3], A. Guichard[6], H. Gulkanian[10], J. Heuser[8], M. Keil[2,4], L. Kluberg[2,7], J. Lozano[5], C. Lourenço[2], F. Manso[3], A. Masoni[1], P. Martins[2,5], A. Neves[5], H. Ohnishi[8], C. Oppedisano[9], P. Parracho[2], P. Pillot[6], G. Puddu[1], E. Radermacher[2], P. Ramalhete[2,5], P. Rosinsky[2], E. Scomparin[9], S. Serci[1], R. Shahoyan[2,5], P. Sonderegger[5], H.J. Specht[4], R. Tieulent[6], G. Usai[1], R. Veenhof[2,5], H.K. Wöhri[2,5] (NA60 Collaboration)

[1] Univ. di Cagliari and INFN, Cagliari, Italy,
[2]CERN, Geneva, Switzerland,
[3]LPC, Univ. Blaise Pascal and CNRS-IN2P3, Clermont-Ferrand, France,
[4]Univ. Heidelberg, Heidelberg, Germany
[5]IST-CFTP, Lisbon, Portugal
[6]IPN-Lyon, Univ. Claude Bernard Lyon-I and CNRS-IN2P3, Lyon, France
[7]LLR, Ecole Polytechnique and CNRS-IN2P3, Palaiseau, France
[8]RIKEN, Wako, Saitama, Japan
[9]Univ. di Torino and INFN, Italy
[10]YerPhI, Yerevan, Armenia

E-mail: joao.seixas@cern.ch

Abstract. The NA60 experiment studies open charm and prompt dimuon production in proton-nucleus and nucleus-nucleus collisions at the CERN SPS. From June 2002 to November 2004 we collected data on dimuon production from proton-nucleus and Indium-Indium collisions. These data samples allow us to study low mass dimuon production (resonances and continuum), J/ψ suppression and open-charm production. We will also search for thermal dileptons, through the yield of "intermediate mass dimuons". In this paper we briefly overview some of the results obtained up to now, after shortly describing the detector.

Since twenty years many experiments studied proton-nucleus and nucleus-nucleus interations at high energies with the aim of finding signs of a transition from normal hadronic matter to a deconfined state composed of free quarks and gluons [1]. The existence of such a state, known as Quark-Gluon Plasma (QGP), is predicted by QCD, the theory of strong interactions [2]. However, high-energy central nucleus-nucleus collisions are very complex systems to study, and the hot and dense medium they produce is affected both by initial and final state effects. The early stages of the collision are expected to be dominated by hard QCD interactions among the deconfined partons. Hard and electromagnetic probes play a fundamental role in the understanding of the formed state since they are produced in the earliest stages of the interaction, while the soft hadronic physics is determined by the hadronization stage.

The suppression of J/ψ production was predicted as a signature of the deconfinement transition [3] and existing results clearly indicate that the produced medium undergoes a major change as the number of nucleons intervenient in the collisions is increased [4]. However, the variable which governs this behaviour is not known unambiguously and, thus, the exact mechanism responsible for the suppression is as yet not clear. Furthermore, a significant fraction ($\sim 30\,\%$) of the produced J/ψ's come from χ_c decays and, therefore, a proper understanding of the J/ψ suppression mechanism requires studying the production or suppression of the χ_c. Up to now, such experimental information is missing.

Building on previous studies by NA38 [5] and HELIOS-3 [6], NA50 has also shown [7] a significant enhancement of intermediate mass dimuons in heavy-ion collisions, a result which might be a manifestation of thermal dilepton production [8]. It is very important to perform a direct measurement in order to clarify this observation. D production should also be measured, to complement the J/ψ picture.

The deconfinement transition must be accompanied by the restoration of chiral symmetry. These two transitions could, in principle, occur for different values of temperature and/or baryon potential. However, lattice QCD results for zero baryon chemical potential have shown that both transitions happen at the same temperature [9] and that they are related to one another. CERES results [10] have shown that the low mass dilepton spectrum is strongly modified when going from p-Be collisions at 450 GeV to S-Au at 200 GeV or Pb-Au at 158 GeV. This observation has been attributed to a decrease of the ρ mass when the produced medium approaches chiral symmetry restoration [11] and to a broadening of the resonance due to in-medium interactions [12]. These results are very interesting and deserve to be studied in detail with improved experimental capabilities, in terms of statistics, mass resolution, etc.

NA60 was built to provide answers to the questions just mentioned. In order to fulfill these goals, NA60 complemented the muon spectrometer previously used by NA50 with a new target area, comprising a silicon beam tracker and silicon tracking vertex telescope, placed inside a 2.5 T dipole field. The beam tracker, operated at 130 K for increased radiation tolerance, is composed of two stations of back-to-back micro-strip sensors with 24 strips of 50 μm pitch, giving a 20 μm resolution on the transverse coordinates of the interaction point, at the target. During the 2003 Indium run the silicon tracking telescope was made of pixel detectors based on the ALICE1LHCb readout chips. The 2002 proton data sample was collected using micro-strip detectors, while in 2004 we complemented the previously used detectors with two new planes of ATLAS pixels, operated at 40 MHz, and some new strip planes.

The improvement achieved by the inclusion of the vertex telescope can be seen in Fig. 1, where the dimuon invariant mass spectrum for the p-A run in 2002 is represented before and after matching the muon spectrometer tracks with the corresponding vertex telescope tracks. During this running period, a 400 GeV proton beam (at intensities around 1–$3{\cdot}10^8$ protons/burst) interacted on Be, In and Pb targets, yielding 600 000 dimuons. After full reconstruction, vertex selection, phase space cuts and matching, 15 000 oposite-sign dimuons were kept. The mass resolution at the ϕ peak is ~ 30 MeV. The z-vertex resolution is ~ 600–900 μm, depending on the target position, which allows a clear separation of the individual targets, as shown in Fig. 1-right. The use of the different target materials simultaneously present in the target system allows the extraction of the nuclear dependence of the particle production cross-sections, using a fit of the form $\sigma_{pA} = \sigma_0 A^{\alpha}$. For the ω and ϕ mesons, in particular, we obtain $\alpha^{\omega} = 0.82 \pm 0.01$ and $\alpha^{\phi} = 0.91 \pm 0.02$ (statistical errors only), for the rapidity range $3.3 < y_{lab} < 4.2$ [13].

In the 2003 In-In run we collected $\sim 2.3 \times 10^8$ dimuon trigers with a beam intensity of $\sim 5{\times}10^7$ ions per 5-second burst. The resolution on the determination of the z coordinate of the interaction vertex is ~ 200 μm. This allows a precise selection of events having the primary vertex in one of the seven targets. After matching and combinatorial background subtraction we get the signal dimuon spectrum shown in Fig. 2 where, apart from the ϕ and ω peaks, the $\eta \to \mu\mu$ channel is also visible, for the first time in heavy ion collisions. The mass resolution at the ϕ

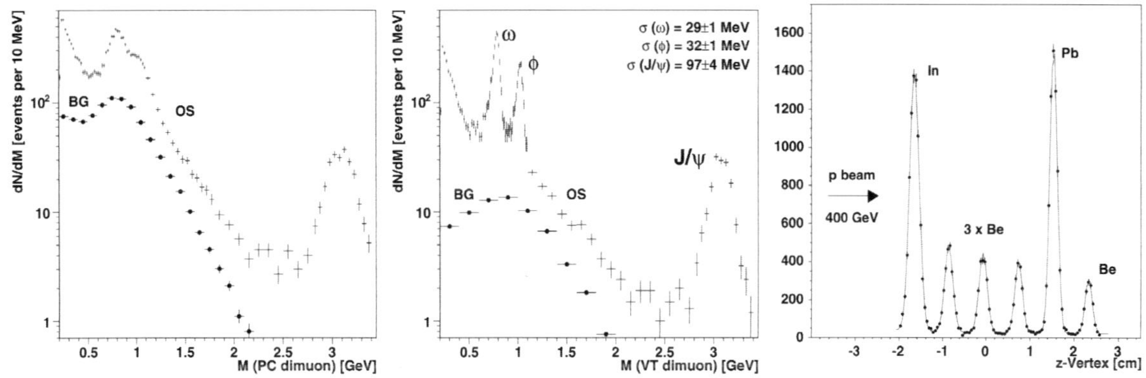

Figure 1. Opposite-sign (OS) and combinatorial background (BG) dimuon mass distributions before (left) and after (centre) muon track matching, integrating events from all the targets seen on the z-vertex distribution (right).

is 23 MeV. The ϕ transverse momentum distribution was studied by selecting signal dimuons in a narrow window around the ϕ mass. The background under the ϕ peak was evaluated and subtracted using two side windows symmetrically located around the peak. The probability that the detector accepts a ϕ meson with a certain p_T value depends on the corresponding dimuon rapidity. Therefore, we calculated the ϕ acceptance and pixel efficiency maps in y versus p_T cells, and made a 2-dimensional correction of the measured data. The analysis was done in four centrality bins [14]. The corrected 2-dimensional distributions were then projected on the p_T axis and fitted to the expression $1/p_T\, dN/dp_T \sim \exp(-m_T/T)$, in order to determine the inverse slope parameter T. The result is shown in Fig. 2-right where the measured T is plotted as a function of the number of participants. There is a clear increase of the inverse slope parameter between peripheral and central Indium-Indium collisions, from ~ 218 to ~ 255 MeV. Integrating over all collision centralities we obtain $T = 252 \pm 3$ MeV. In order to check the stability of the results, we have changed the position and width of the side bins used to subtract the background. The results are quite insensitive to such changes. Also the acceptance correction plays a minor

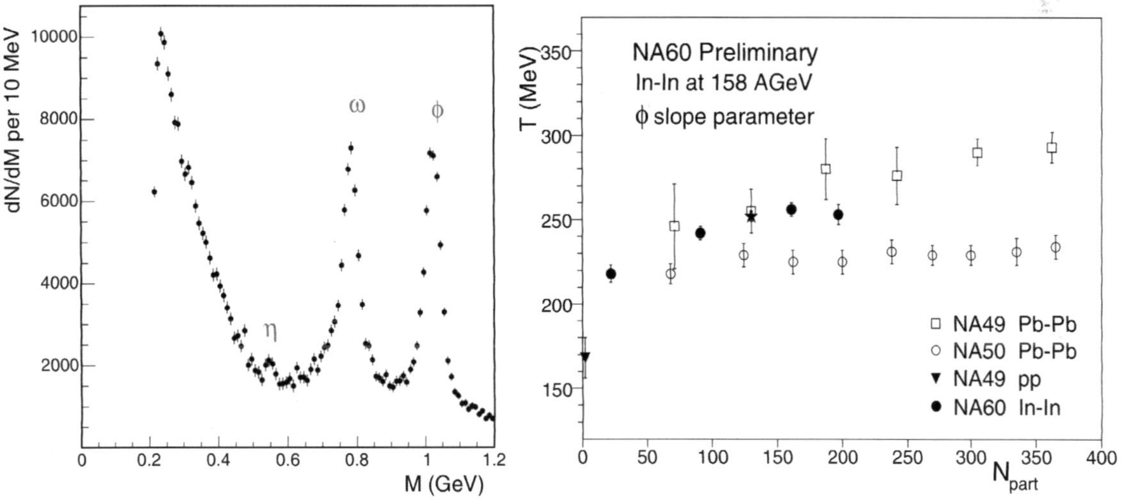

Figure 2. Left: Dimuon mass spectrum after matching and background subtraction. Right: ϕ inverse slope parameter, T, as a function of the number of participants.

role since, at the ϕ mass, it does not change by more than a factor of 2 over the full p_T range.

Using the information from the vertex telescope we are able to separate prompt and displaced muon pairs [15]. The dimuon mass distributions for the displaced and prompt selections are shown in Fig. 3-left. Their ratio, Fig. 3-right, shows clear minima in the regions dominated by the ω, ϕ and J/ψ resonances. The displaced sample dominates in the mass region where open charm is expected to contribute the most.

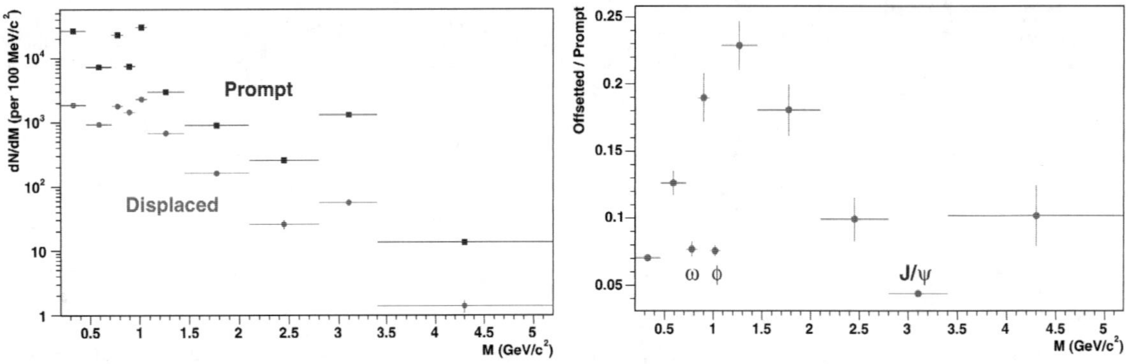

Figure 3. Mass spectra corresponding to "displaced" and "prompt" selections (left) and their ratio (right).

The results presented in this paper will be improved and extended in the near future, thanks to the much higher statistics collected by NA60 in 2004, with 400 GeV protons incident on an expanded target system. Our efforts are also concentrated on the determination and subtraction of the fake matches component, so that we can also address the physics of the dilepton continuum.

This work was partially supported by the Fundação para a Ciência e Tecnologia, Portugal.

References

[1] C. Lourenço, Lect. Notes Phys. **583**, 1 (2002)
[2] F. Karsch, Lect. Notes Phys. **583**, 583 (2002)
[3] T. Matsui, H. Satz, *Phys. Lett.* **B**178 (1986) 416.
[4] B. Alessandro *et al.* (NA50 Coll.), Eur. Phys. J. **C39** (2005) 335; M. Baldit *et al.* (NA50 Coll.), Phys. Lett. **B140** (1997) 337.
[5] C. Lourenço *et al.* (NA38 Coll.), Nucl. Phys. **A566** (1994) 77.
[6] A.L.S. Angelis *et al.* (HELIOS-3 Coll.), Eur. Phys. J. **C13** (2000) 433.
[7] M.C. Abreu *et al.* (NA50 Coll.), Eur. Phys. J. **C14** (2000) 443.
[8] R. Rapp, E.V. Shuryak, *Phys. Lett.* **B**473 (2000) 13.
[9] S. Digal, E. Laermann and H. Satz, *Nucl. Phys* **A**702 (2002) 159.
[10] G. Agakichiev *et al.* (CERES Coll.), Phys. Rev. Lett. **75** (1995) 1272; Phys. Lett. **B422** (1998) 405; B. Lenkeit *et al.* (CERES Coll.), Nucl. Phys. **A661** (1999) 23c.
[11] G.E. Brown and M. Rho *Phys. Repts.* 363 (2002) 85.
[12] R. Rapp and J. Wambach *Adv. Nucl. Phys.* **25**(2000)1.
[13] H.K. Wöhri *et al.* (NA60 Coll.), Proc. of the "Int. Conf. on Hard and Electromagnetic Probes of High Energy Nuclear Collisions (Hard Probes 2004)", Nov. 2004, Ericeira, Portugal, Eur. Phys. J. **C**, in print.
[14] G. Usai *et al.* (NA60 Coll.), Proc. of the "Int. Conf. on Hard and Electromagnetic Probes of High Energy Nuclear Collisions (Hard Probes 2004)", Nov. 2004, Ericeira, Portugal, Eur. Phys. J. **C**, in print.
[15] R. Shahoyan *et al.* (NA60 Coll.), Proc. of the "Int. Conf. on Hard and Electromagnetic Probes of High Energy Nuclear Collisions (Hard Probes 2004)", Nov. 2004, Ericeira, Portugal, Eur. Phys. J. **C**, in print.

Institute of Physics Publishing
doi:10.1088/1742-6596/50/1/047
Journal of Physics: Conference Series **50** (2006) 353–356
5th International Conference on Physics and Astrophysics of Quark Gluon Plasma

NA50 final results on charmonia suppression

C Quintans[6] on behalf of the NA50 Collaboration:
B Alessandro[10], C Alexa[3], R Arnaldi[10], M Atayan[12], S Beolè[10], V Boldea[3], P Bordalo[6,13], G Borges[6], C Castanier[2], J Castor[2], B Chaurand[9], B Cheynis[11], E Chiavassa[10], C Cicalò[4], M P Comets[8], S Constantinescu[3], P Cortese[1], A De Falco[4], N De Marco[10], G Dellacasa[1], A Devaux[2], S Dita[3], J Fargeix[2], P Force[2], M Gallio[10], C Gerschel[8], P Giubellino[10], M B Golubeva[7], A A Grigoryan[12], S Grigoryan[12], F F Guber[7], A Guichard[11], H Gulkanyan[12], M Idzik[10,14], D Jouan[8], T L Karavicheva[7], L Kluberg[9,5], A B Kurepin[7], Y Le Bornec[8], C Lourenço[5], M Mac Cormick[8], A Marzari-Chiesa[10], M Masera[10], A Masoni[4], M Monteno[10], A Musso[10], P Petiau[9], A Piccotti[10], J R Pizzi[11], F Prino[10], G Puddu[4], C Quintans[6], L Ramello[1], S Ramos[6,13], L Riccati[10], H Santos[6], P Saturnini[2], E Scomparin[10], S Serci[4], R Shahoyan[6,15], F Sigaudo[10], M Sitta[1], P Sonderegger[5,13], X Tarrago[8], N S Topilskaya[7], G L Usai[4], E Vercellin[10], L Villatte[8], N Willis[8] and T Wu[8]

[1] Università del Piemonte Orientale, Alessandria and INFN-Torino, Italy
[2] LPC, Univ. Blaise Pascal and CNRS-IN2P3, Aubière, France
[3] IFA, Bucharest, Romania
[4] Università di Cagliari/INFN, Cagliari, Italy
[5] CERN, Geneva, Switzerland
[6] LIP, Lisbon, Portugal
[7] INR Moscow, Russia
[8] IPN, Univ. de Paris-Sud and CNRS-IN2P3, Orsay, France
[9] Laboratoire Leprince-Ringuet, Ecole Polytechnique and CNRS-IN2P3, Palaiseau, France
[10] Università di Torino/INFN, Torino, Italy
[11] IPN Univ. Claude Bernard Lyon-I and CNRS-IN2P3, Villeurbanne, France
[12] YerPhI, Yerevan, Armenia

Abstract. The most recent and final results from the NA50 experiment on charmonia production in Pb-Pb interactions at 158 GeV/c per nucleon are presented. A strong J/ψ suppression is observed, which increases with the centrality of the collisions.

The J/ψ production is found to be anomalously suppressed starting at mid-centralities, when compared to a reference obtained from proton-induced collisions. The most recent developments on the determination of this normal absorption curve are explained. It is also shown that for sulphur-induced reactions there is full agreement with the extrapolated normal p-A behaviour.

The suppression of $\psi\prime$ production in heavy ion collisions (Pb-Pb and S-U) is also presented. It is seen to increase with the centrality of the collision and to be significantly stronger than the one measured in proton-induced reactions.

[13] also at IST, Universidade Técnica de Lisboa, Lisbon, Portugal
[14] also at AGH University of Science and Technology, Faculty of Physics and Applied Computer Science, Krakow, Poland
[15] on leave of absence of YerPhI, Yerevan, Armenia

1. Introduction

The prediction that the J/ψ vector-meson is suppressed in a quark gluon plasma medium is investigated by the NA50 experiment at the CERN/SPS. Ultra-relativistic Pb-Pb collisions are used to form a very dense medium emitting signals from its formation, cooling and hadronization successive phases, which could reveal the hadronic or quarkonic nature of the initially formed matter.

The NA50 experiment is devoted to the detection of muon pairs. It allows to study specific experimental signatures like charmonium suppression, but also charm, strangeness and thermal dimuon production. The experimental apparatus consists of beam detectors, a multiplicity detector (MD), an electromagnetic (EC) and a very forward hadronic (ZDC) calorimeters, followed by a dimuon spectrometer. The latter includes a carbon core absorber and an air-core toroidal magnet in between two sets of multi-wire proportional chambers and trigger hodoscopes. For the data taking periods of 1998 and 2000, the MD was used as a vertex detector, particularly useful to unambiguously locate the interaction vertex of the most peripheral Pb-Pb collisions. For the 2000 data, the last and best period (from the point of view of data quality), the target region up to the EC central hole was put under vacuum, in order to avoid spurious lead-air interactions. The strong suppression of the J/ψ production in central lead-lead collisions, observed for the first time in the 1995 NA50 data [1], has now been confirmed in all the subsequent sets of improved quality data.

2. J/ψ suppression in Pb-Pb

In order to study J/ψ suppression as a function of centrality, NA50 has 3 independent centrality estimators: the neutral transverse energy measured by the EC, the multiplicity of charged particles from the MD, and the energy of the spectator nucleons given by the ZDC.

A reference process is also needed to study the centrality dependent charmonia suppression. Drell-Yan dimuons are used as a reference, since this is a well-known process that scales with the number of nucleon-nucleon collisions, from p-p up to Pb-Pb, as has been experimentally shown [2]. The number of Drell-Yan events is obtained from the same dimuon mass spectra as the J/ψ events, with the same selection criteria, and thus most of the systematical errors cancel out in their ratio. The drawback of using Drell-Yan is the relatively poor statistics of the collected event samples.

3. The normal J/ψ absorption

J/ψ suppression in Pb-Pb collisions increases with the centrality. While this is partly expected from nuclear absorption mechanisms, as can be inferred from the study of proton-nucleus collisions (the so-called normal nuclear absorption), there is an anomalous extra amount of suppression in Pb-Pb, setting in from mid-centralities on. The normal absorption cross-section, σ^{ψ}_{abs}, of the J/ψ vector-meson has been previously calculated from the high statistics NA50 p-A data at 450 GeV/c, together with NA51 p-p and p-d data and NA38 S-U data at 200 GeV/nucleon. Although these data sets had different beam energies, it was shown that their independently extracted σ^{ψ}_{abs} are compatible within errors, thus allowing for a simultaneous fit of the J/ψ/DY ratios. Using the Glauber formalism, we obtain $\sigma^{\psi}_{abs} = 4.18 \pm 0.35$ mb, and a rescaling factor to go from 450 to 200 GeV. The fit parameters are used to compute the normal nuclear absorption curve, where we include the appropriate energy rescaling from 200 to 158 GeV/nucleon [3].

The normal absorption curve can be obtained in an alternative way, using exclusively proton-nucleus data [4]. In this new procedure, the high statistics NA50 J/ψ/DY p-A data at 450 and 400 GeV/c and the NA51 p-p and p-d at 450 GeV/c are fitted simultaneously using again the Glauber model which leads to $\sigma^{\psi}_{abs} = 4.18 \pm 0.35$ mb (non-coincidently the same as in the previous method). The normalization of the absorption curve is obtained by further considering together with these data sets the low statistics NA38 and NA3 p-A data at 200 GeV/nucleon, and by redoing the Glauber fit of the $B_{\mu\mu}\sigma(\mathrm{J}/\psi)$ as a function of A, with fixed σ^{ψ}_{abs} (see figure 1). The significantly smaller J/ψ rescaling factor to go from 200 to 158 GeV is obtained from the Schuler parameterization (while for the Drell-Yan it is obtained theoretically).

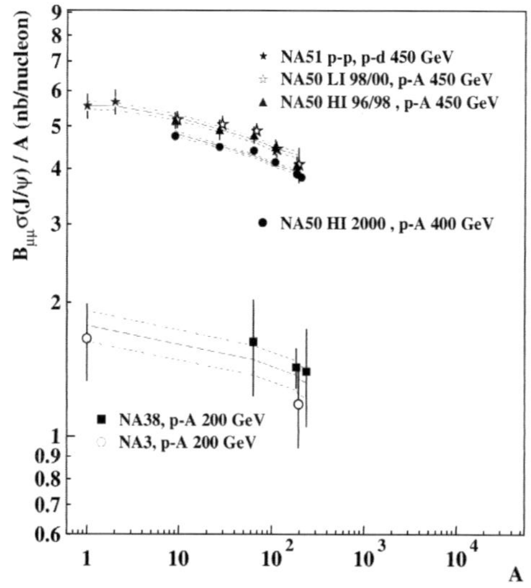

Figure 1. J/ψ absolute cross-sections in p-A collisions, at 450, 400 and 200 GeV/c. Fit imposing the slope extracted from p-A data at 450/400 GeV/c only.

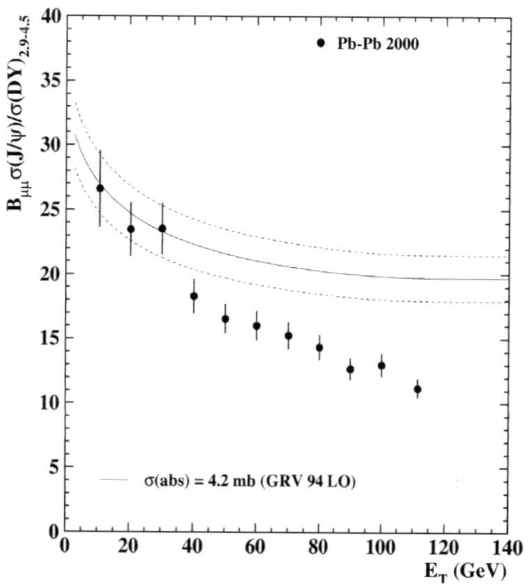

Figure 2. Jψ/DY ratio as a function of E_T, in Pb-Pb collisions. Normal absorption curve extracted from p-A data only.

4. Studies as a function of centrality

Figure 2 shows J/ψ/DY as a function of centrality in Pb-Pb collisions, from the analysis of the 2000 data set, together with the normal absorption curve obtained from proton data only. The normal nuclear absorption curves from each of the methods explained above exactly coincide, the second method being affected by a larger systematical error band (9% as compared to 4% in the most central region), thus showing that J/ψ suppresssion in S-U collisions is fully compatible with the normal behaviour observed in p-A collisions.

Three independent analyses were performed using the 2000 data set, as a function of the neutral transverse energy, as a function of the energy of the spectator nucleons and as a function of the charged particles multiplicity. All the analyses show a similar J/ψ/DY suppression, that increases with the centrality of the collisions. For the most central collisions, this suppression is much stronger than what one would expect from an extrapolation of the normal nuclear absorption behaviour.

The analyses of 1998 and 2000 data sets are in very good agreement, allowing for the averaging of J/ψ/DY results in each centrality region, thus increasing the statistical precision of the result.

5. $\psi\prime$ suppression

The $\psi\prime$ vector-meson production is more difficult to analyse than the J/ψ, since it has a small dimuon cross-section, and several physics processes superimpose in its mass region. The estimation of the Drell-Yan contribution under the $\psi\prime$ peak suffers from a 7% uncertainty, depending on the parton distribution functions (PDF) set chosen. Besides, one expects for the $\psi\prime$ a stronger suppression effect as compared to the J/ψ, since it is a less bound charmonium state. In order to minimize the influence of the PDF set chosen in the $\psi\prime$ normalization, the high mass Drell-Yan (4.2 to 7.0 GeV/c^2) is used as reference.

A method of analysis similar to the one used for J/ψ is applied, showing that the $\psi\prime$ is already suppressed in peripheral Pb-Pb collisions, and this effect increases with centrality. Although the statistical error bars are large, there is a very good agreement between the 1998 and the 2000 data analyses. The results were averaged to increase statistical accuracy. The $\psi\prime$ absorption is also shown to be much stronger for Pb-Pb and S-U collisions than for proton induced collisions.

6. Conclusions

A detailed study of J/ψ suppression in Pb-Pb collisions is performed using as reference the Drell-Yan process. While in peripheral Pb-Pb collisions J/ψ production agrees with expectations inferred from p-A interactions, for mid-central collisions data exhibit a clear departure from this normal behaviour which, furthermore, increases with centrality. The $\psi\prime$ is also shown to be increasingly suppressed with centrality, with peripheral Pb-Pb collisions producing less resonant events than one would expect from a simple extrapolation of the p-A absorption behaviour.

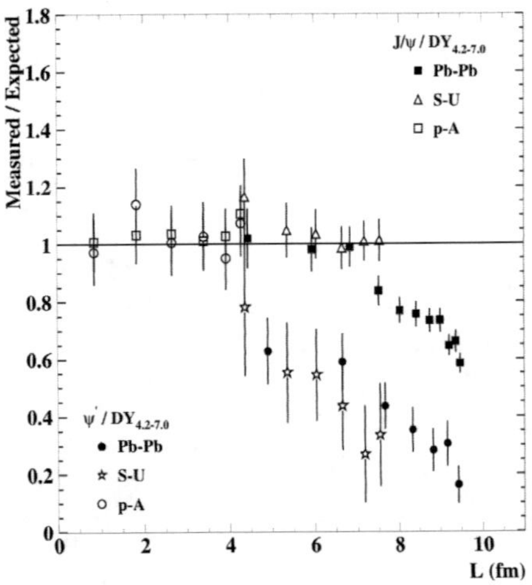

Figure 3. Measured values of J/ψ/DY and $\psi\prime$/DY divided by the expected values as deduced from p-A measurements.

By dividing the measured charmonia/DY ratios by the corresponding expected values as deduced from p-A measurements, one can directly compare the J/ψ and $\psi\prime$ suppressions as a function of centrality. Figure 3 shows these measured/expected ratios for proton, sulphur and lead induced collisions, showing that in nucleus-nucleus collisions the $\psi\prime$ departs from the normal behaviour for smaller L values than the J/ψ vector-meson.

Acknowledgments

This work is partially supported by the Fundação para a Ciência e Tecnologia, Portugal.

References

[1] Abreu M C *et al* 1997 *Phys. Lett.* B **410** 327
[2] Abreu M C *et al* 1997 *Phys. Lett.* B **410** 337
[3] Alessandro B *et al* 2004 *Eur. Phys. J.* C **33** 31
[4] Borges G *et al* 2004 *Proc. Int. Conf. Hard and Electromagnetic Probes of High Energy Nuclear Collisions 2004 (Ericeira, Portugal).* Accepted for publication in *Eur. Phys. J.* C
[5] Alessandro B *et al* 2005 *Eur. Phys. J.* C **39** 335
[6] Santos H *et al* 2004 *J. Phys.* G **30** S1175

Institute of Physics Publishing Journal of Physics: Conference Series **50** (2006) 357–360
doi:10.1088/1742-6596/50/1/048 5th International Conference on Physics and Astrophysics of Quark Gluon Plasma

The CBM experiment at FAIR

P. Senger[1] **for the CBM Collaboration**

[1] Gesellschaft für Schwerionenforschung, Planckstr.1, 64291 Darmstadt

E-mail: jpcs@iop.org

Abstract. The study of high-energy nucleus-nucleus collisions will be one of the major activities at the future accelerator complex FAIR in Darmstadt. The goal of the research program is the exploration of the QCD phase diagram in the region of high baryon densities.

1. The future Facility for Antiproton and Ion Research (FAIR)

The future Facility for Antiproton and Ion Research (FAIR) in Darmstadt will provide unique research opportunities in the fields of nuclear, hadron, atomic and plasma physics [1]. The accelerators will deliver primary beams (protons up to 90 GeV, Uranium up to 35 AGeV, nuclei with Z/A = 0.5 up to 45 AGeV) and secondary beams (rare isotopes and antiprotons) with high intensity and quality. The facility comprises a double-ring synchrotron, rings for accumulation, cooling and storage of primary and secondary beams, and dedicated detector arrangements. The research program includes the study of nuclei far from stability, hadron physics with antiproton beams, the study of compressed nuclear matter, the investigation of plasmas induced by ion and laser beams and atomic physics.

2. Exploring the QCD phase diagram with heavy-ion collisions

The exploration of the phase diagram of strongly interacting matter is a major field of modern high-energy physics. Of particular interest is the transition from hadrons to partonic degrees of freedom which is expected to occur at high temperatures or high baryon densities. These phases play an important role in the early universe and in the core of neutron stars [2]. The discovery of this phase transition will shed light on fundamental but still puzzling aspects of Quantum Chromo Dynamics (QCD): confinement and chiral symmetry breaking. In particular, at high baryon densities one expects new phases of strongly interacting matter [3]. The scientific progress in this exciting field, QCD at high baryon densities, is driven by new experimental data.

The aim of the planned Compressed Baryonic Matter (CBM) experiment at FAIR is to explore the QCD phase diagram at high net baryon densities and moderate temperatures in nucleus-nucleus collisions. This approach is complementary to the studies of matter at high temperatures and low net baryon densities performed at RHIC and LHC. Our current knowledge on the QCD phasediagram is illustrated in figure 1. The data points correspond to chemical freeze-out and result from a statistical analysis of particle ratios measured in Pb+Pb and Au+Au collisions at SIS, AGS, SPS and RHIC. The phase boundary between quark-gluon matter and hadronic and the location of the critical endpoint as shown in figure 1 is predicted by recent lattice QCD calculations which indicate that for values of μ_B larger than about 400 MeV the phase transition is first order, whereas for μ_B smaller than 400 MeV there is a smooth cross over from the hadronic to the partonic phase (dotted line).

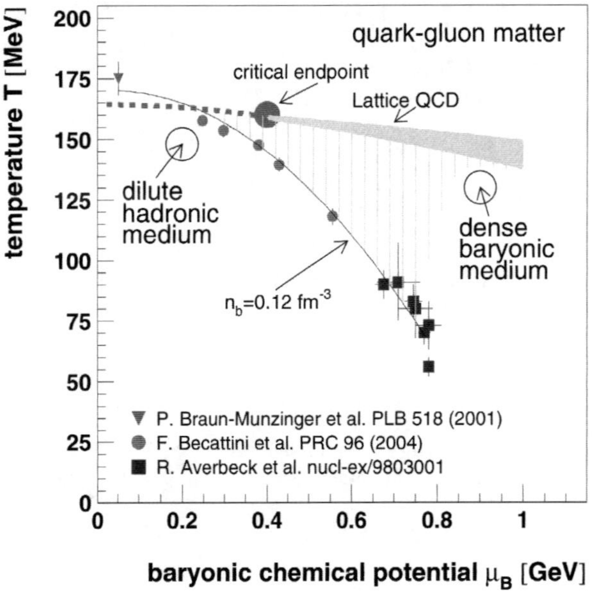

Figure 1. The phase diagram of strongly interacting matter plotted as a function of temperature and baryon chemical potential. Full symbols: freeze-out points obtained with a statistical model analysis from particle ratios measured in heavy collisions [4, 5, 6]. The critical endpoint at $\mu_B \approx 400$ MeV is predicted by lattice QCD calculations [7, 8].

3. The experimental observables

The CBM research program includes the search for the deconfinement phase transition at high baryon densities, the study of chiral symmetry restoration in superdense baryonic matter, the search for the critical endpoint, and the study of the nuclear equation of state at high densities. The major challenge is to find diagnostic probes which are connected to chiral symmetry restoration and to the deconfinement phase transition. The observation of in-medium modifications of hadrons might be a signature for the onset of chiral symmetry restoration [9]. The in-medium spectral function of short-lived vector mesons can be measured directly via their decay into dilepton pairs. Since leptons are essentially unaffected by the passage through the high-density matter, they provide, as a penetrating probe, almost undistorted information on the conditions in the interior of the collision zone.

The anomalous suppression of charmonium due to screening effects in the Quark Gluon Plasma (QGP) was predicted to be an experimental signal of the QGP [10]. Particles containing heavy quarks like charm are produced in the early stage of the collision. At FAIR, open and hidden charm production will be studied at beam energies close to the kinematical threshold, and the production mechanisms of D and J/ψ mesons will be sensitive to the conditions inside the early fireball.

One of the early predictions for a QGP signal was the increased production of strangeness in the deconfined phase resulting in an enhanced yield of strange particles after hadronization [11]. This effect was expected to be even more pronounced for multistrange hyperons. Recently, data on the excitation function of strangeness production measured by NA49 have revived the discussion on the role of strangeness as a signature for a deconfinement phase transition [12].

Many signals for the the quark-gluon-plasma (QGP) have been proposed and are still under discussion, although the hope for finding a "smoking gun" has not yet become true. The discovery of the critical endpoint of the deconfinement phase transition would be such a direct indication for the existence of a new phase. Theoretical investigations suggest that particle density fluctuations occur in the vicinity of the critical endpoint, which might be observed experimentally as nonstatistical event-by-event fluctuations of observables [13].

4. The experimental setup

The study of those very different observables calls for a modular setup and different running conditions. The measurement of $J\psi$ mesons via their electron-positron decay requires (i) high reaction rates of up to 10^7 events/second to compensate for the low production cross sections at threshold, (ii) an online event selection to reduce the data flow down to an archiving rate of 25 kHz, and (iii) a detector system for electron identification and pion suppression (electron/pion ratio of 10^{-4} required). The detection of open charm (D mesons) via hadronic decays requires the measurement of displaced vertices with an accuracy of about 50 μm along the beam axis to suppress the combinatorial background of prompt pions and kaons. The measurement of low-mass vector mesons requires the rejection of close pairs (i.e. the detection of soft electrons) to reduce the combinatorial background caused by electrons/positrons from π^0 Dalitz decays and gamma conversions. The study of observables event-by-event requires particle identification and a large acceptance which does not vary with beam energy.

In particular the charm measurements require unprecedented detector performances concerning rate capabilities and radiation hardness. The detector signals have to be processed by a high-speed data acquisition and an online event selection system. A sketch of the CBM detector is shown in figure 2.

The measurement of particle tracks and the determination of primary and secondary vertices will be performed with a Silicon Tracking System (STS. The current STS layout consists of minimum 7 layers (2 pixel and 5 strip) and is placed inside a magnetic dipole field which provides the bending power required for momentum determination with an accuracy of $\Delta p/p = 1$ %. The STS has to fulfill the following requirements: material budget below 0.3% radiation length per layer to reduce multiple scattering, hit resolution of about 10 μm (for the pixel sensors) to achieve a vertex resolution of about 50 μm along the beam axis, radiation hardness up to a dose of 50 MRad corresponding to the dose accumulated in ten years of running, and read-out times of about 25 ns to accommodate reaction rates of 10 MHz.

The RICH detector is designed to provide identification of electrons and suppression of pions in the momentum range of electrons from low-mass vector-meson decays. The actual layout of the RICH detector consists of a radiator (3 gases under study: N_2, 40% He + 60 %CH_4, and 50% N_2 + 50 %CH_4, length 2.2 m), a mirror (3 mm Be covered with 0.5 mm glass, radius 4.5 m), and a photon detector composed of about 100.000 photomultiplier (PM) channels (6 mm diameter, quantum efficiency 20 %). The glass window of the PMs is covered with wave-length shifter (WLS) films in order to increase the absorption of Cherenkov photons.

Three Transition Radiation Detector stations will serve for particle tracking and for the identification of high energy electrons and positrons ($\gamma > 2000$) which are used to reconstruct J/ψ mesons. According to simulations which are based on the experience obtained with the development of the TRD for ALICE and of the TRT for ATLAS, pion suppression factors of up to 200 (for momenta above 2 GeV/c) at an electron efficiency of better than 90% can be achieved.

The major technical challenge is to develop highly granular and fast gaseous detectors which can stand the high-rate environment of CBM in particular for the inner part of the detector planes covering forward emission angles. For example, at small forward angles and at a distance of 4 m from the target, we expect particle rates of more than 100 kHz/cm^2 for 10 MHz minimum bias Au+Au collisions at 25 AGeV. In a central collision, particle densities of about 0.05/cm^2 are reached. In order to keep the occupancy below 5% the size of a single cell should be

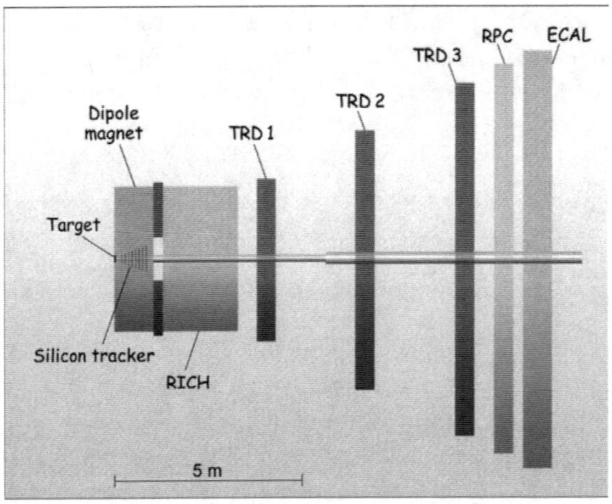

Figure 2. Sketch of the planned Compressed Baryonic Matter (CBM) experiment. The setup consists of a superconducting dipole magnet with a Silicon tracker System inside, a Rich Imaging Cherenkov detector (RICH), the three Transition Radiation Detectors (TRD), the Time-Of Flight (TOF) wall which is a Resistive Plate Chamber (RPC) and the electromagnetic Calorimeter (ECAL). The total length of the setup is approximately 12 m.

about 1 cm^2. Various prototypes of fast Multi Wire Proportional Chambers (MWPC) and Gas Electron Multipliers (GEM) have been tested with proton and pion beams up to intensities of 100 kHz/cm^2 and no major deterioration of the performance has been observed.

An array of Resistive Plate Chambers will be used for hadron identification via TOF measurements. The TOF wall is located about 10 m downstream of the target and covers an active area of about 120 m^2. The required time resolution is about 80 ps. For 10 MHz minimum bias Au+Au collisions the innermost part of the detector has to work at rates up to 20 kHz/cm^2. At small deflection angles the pad size is about 5 cm^2 corresponding to an occupancy of below 5% for central Au+Au collisions at 25 AGeV. With a small-size prototype a time resolution of about 90 ps has been achieved at a rate of 25 kHz/cm^2.

The electromagnetic calorimeter will be used to measure direct photons, neutral mesons decaying into photons, electrons and muons. Simulations and R&D have been started based on the shashlik type of detector modules as used in HERA-B, PHENIX and LHCb. Particular emphasis is put on a good energy resolution and a high pion suppression factor.

5. References

[1] http://www.gsi.de/GSI-Future/cdr/
[2] F. Weber, J. Phys. G: Nucl. Part. Phys. 27 (2001) 465
[3] F. Wilczek, Physics Today 53 (2000) 22 and hep-ph/0003183
[4] F. Becattini et al., Phys.Rev. C69 (2004) 024905
[5] J. Cleymans and K. Redlich, Phys. Rev. Lett. 81 (1998) 5284
[6] P. Braun-Munzinger, Nucl. Phys. A 681 (2001) 119c
[7] Z. Fodor and S.D. Katz, hep-lat/0402006, and JHEP 0404 (2004) 050
[8] S. Ejiri et al., hep-lat/0312006
[9] R. Rapp and J. Wambach, Nucl. Phys. A 661 (1999) 33c
[10] T. Matsui and H. Satz, Phys. Lett. B 178 (1986) 416
[11] J. Rafelski and B. Müller, Phy. Rev. Lett. 48 (1982) 1066
[12] M. Gazdzicki et al., J.Phys. G: Nucl. Part. Phys. 30 (2004) 701c
[13] M. Stephanov, K. Rajagopal, E. Shuryak, Phy. Rev. D 60 (1999) 114028
[14] http://www-linux.gsi.de/~hoehne/report/cbmtsr_public.pdf

Physics of the Muon Spectrometer of the ALICE Experiment [1]

Ginés MARTINEZ [2]
for the ALICE Collaboration
SUBATECH (EMN-UN-IN2P3), 4 rue Alfred Kastler BP20722, 44307 Nantes Cedex 3

Abstract.
The main goal of the ALICE Muon spectrometer experiment is the measurement of heavy quark production in p+p, p+A and A+A collisions at LHC energies, via the muonic channel. Physics motivations and expected performances have been presented in this talk.

1. Physics motivations

Heavy ion collisions (HIC) at relativistic energies are a privileged tool for creating very hot and dense matter in a laboratory. In particular, lattice chromo-dynamics (lQCD) predicts a cross-over toward a new state of matter called Quark Gluon Plasma (QGP) at a temperature $\sim$ 170 MeV for vanishing chemical potential μ_B [1, 2, 3]. Heavy ion collisions allow to experimentally study the properties of this new state of matter. This experimental program started in the mid 80's with fixed target heavy ion experiments at the AGS and SPS [4] and continued with the physics program developed at the RHIC collider (BNL) [5]. Heavy ion collisions at the future Large Hadron Collider (LHC) at CERN will open new experimental insights in the study of hadronic matter at high temperature. The ALICE experiment will be the only experiment at the LHC devoted to heavy ion physics, whereas the ATLAS and the CMS experiments plan to develop a heavy-ion program [6, 7] in parallel with their main physics goals.

The LHC collider will provide lead (argon) high luminosity beams $<\mathcal{L}>= 5\cdot10^{26}\mathrm{cm}^{-2}\mathrm{s}^{-1}$ ($<\mathcal{L}>= 5\cdot10^{28}\mathrm{cm}^{-2}\mathrm{s}^{-1}$) at $\sqrt{s}$ =5.5A TeV ($\sqrt{s}$ = 6.3A TeV). In addition, the LHC will deliver proton beams $<\mathcal{L}>= 3\cdot10^{30}\mathrm{cm}^{-2}\mathrm{s}^{-1}$ at $\sqrt{s}$ =14 TeV and d+Pb beams $<\mathcal{L}>= 8\cdot10^{28}\mathrm{cm}^{-2}\mathrm{s}^{-1}$ at $\sqrt{s}$ ~6.2A TeV, providing a solid baseline for the study of medium effects in HIC [8] [3]. At such ultra-relativistic energies new phenomena emerge, improving the experimental conditions for studying the hadronic matter in HIC:

- **Initial conditions.** The initial conditions will be under control by the gluon saturation scenario. At these energies the initial nucleus-nucleus interaction can be viewed as *weak* interactions of a huge number of *small* x gluons which are freed in the beginning of the collision leading to the formation of a big gluonic ball [10, 11]. Up to 8000 gluons in the

[1] Talk presented in the ICPAQGP Conference, February 8-12, 2005, Salt Lake City, Kolkata, India. Web page of the conference : http://www.veccal.ernet.in/~icpaqgp/
[2] martinez@in2p3.fr
[3] It should be noted that d+Pb and Ar+Ar beams will require an upgrade of the LHC collider. Reported luminosities are very preliminary [9].

early stage of the collisions are predicted [12]. Such a system will rapidly evolve toward a equilibrium [13]. Most of these processes (as secondary interaction of minijets) leading to thermalization will be governed by hard processes ($\alpha_s < 1$).

- **Equilibrated matter.** After equilibration of the initial gluonic ball, a hot and long-lived hadronic matter will be formed (hotter and longer-lived than at lower energy HIC). The increase of the beam energy will favor the creation of vanishing baryonic potential hadronic matter with a temperature around 0.5-1 GeV, well above the critical temperature predicted by lQCD.
- **Observables.** The experience acquired during the last 20 years of Heavy-Ion Physics in the relativistic regime has shown the necessity to measure most of the probes of the reaction dynamics, from hard processes and QGP formation observables, until the freeze-out of the expanding hadron gas. A coherent explanation of the full set of observables will be the only way to study the properties of the ephemeral QGP. Final states probes like particle multiplicities, hadron p_T distributions, particle ratios, strangeness production, azimuthal asymmetries, etc ... will shed light on the condition of the phase transition and the dynamical evolution of the expanding hadron gas. Penetrating probes, as real and virtual photon production, charmonium production, and light vector meson properties will provide informations about the QGP formation and properties. Moreover, the study of the QGP at LHC energies will be enriched by exploiting new probes which can be efficiently studied in this energy regime:
  (i) Hard processes leading to very fast leading partons will provide information about the QGP through the interaction of jets with the surrounding partonic matter [14]. Huge parton energy losses (around 1-3 GeV/fm) in the QGP will modify the hadronization process of these very energetic jets, inducing a jet-quenching and a suppression of high p_T hadron production. At the LHC, jet production cross-section in the very high p_T range (p_T >50 GeV) will increase by more of 4 orders of magnitude with respect to RHIC beam energies;
  (ii) The Debye screening of bottomonium states in QGP will be studied. Suppression of the Upsilon family production will be exploited due to its sensitivity to the density of color charges in the medium [15];
  (iii) Open charm and open beauty production will be exploited as new probes of the strong interacting system of partons [17, 16]. Open charm (beauty) production cross-section will increase by a factor 10 (100) with respect to RHIC energies;
  (iv) The production of massive electroweak bosons (W^{+-} and Z) will open the possibility to check the validity of the glauber scaling in HIC. These bosons do not interact with the surrounding medium and they are produced in hard parton collisions;
  (v) Finally, the huge particle multiplicity at the LHC will make possible to measure a large number of observables on an event-by-event basis, increasing the sensitivity to non-statistical fluctuations predicted to occur in a phase transition scenario.

2. Heavy Quark Production at the LHC

Heavy quarks (charm and beauty quarks) in HIC are produced in the first stages of the collisions and then they coexist with the surrounding medium due to their long life-time. Production rates, transverse momentum (p_T) and rapidity (y) distributions, quarkonia production rates, etc ... will allow for probing the properties of the medium, as the QGP.

2.1. Initial hard production

Charm quarks will be copiously produced in HIC at LHC energies: more than 100 $c\bar{c}$ pairs per central Pb+Pb collision and around 5 $b\bar{b}$ pairs per collision are expected [19]. Production of heavy quarks will be dominated by prompt parton-parton scattering. Open heavy flavor production will be dominated by gluon fusion. Next leading-order (NLO) diagrams noticeably contribute to the heavy quark production cross-section [18] although the LO one particle differential cross-section shapes are not appreciably modified by the NLO contribution. It has been observed that pQCD calculations underestimate the measured differential cross-sections in $p\bar{p}$ collisions at Tevatron energies, although they are compatible within uncertainties [20]. All those experimental data concerns high p_T production (larger than 5 GeV/c). At LHC energies, heavy quark production will explore a x and transfered momentum Q^2 domain which has not been well studied. For instance in rapidities $2.5 < |y| < 4$ at LHC charm production will be sensible to $x \sim 10^{-4}$ and $Q^2 \sim 10$ GeV^2. In the case of HIC, one should expect gluon shadowing which will modify the heavy quark production based on glauber scaling. It is expected a heavy quark yield suppression of about 20% (10%) for charm (beauty) production in p+Pb collisions at LHC energies [21] due to gluon shadowing.

About 1% of the heavy quark pairs will end in the formation of a colorless bound state: quarkonium. The process of direct formation of quarkonia from the initially produced $Q\bar{Q}$ pair is a non-perturbative process. Models like Color Evaporation Model (CEM) or non-relativistic QCD [19] have been successfully tested in pp and $p\bar{p}$ collisions. In HIC direct production of quarkonia will take place inside the impinging nuclei, leading to an initial nuclear absorption of quarkonia (normal suppression in HIC). The nuclear absorption increases with the enhancement of the relative rapidity between the quarkonia and the impinging nuclei. For this reason, nuclear absorption becomes higher at LHC energies, a factor 2 higher than at SPS energies. The nuclear absorption exhibits a relatively flat distribution as a function of the HIC centrality for central and semi-central collisions and becomes maximal for central Pb+Pb collisions at the LHC, where the survival probability of the charmonium is $\sim$ 25% (50% for bottomonium) [19].

2.2. *Pre-equilibrium production*

Pre-equilibrium stage at LHC energies will be dominated by secondary mini-jet interactions which could noticeably contribute to the total production cross-section of charm quark pairs [22]. One could expect up to 200 $c\bar{c}$ in central Pb+Pb collisions taking into account the production of heavy quarks during the thermalization phase.

2.3. *Heavy quarks in hot and dense medium*

Heavy quarks will be then embedded in a matter mainly formed by gluons and light quarks (u, d, s). It is a open question how heavy quarks will behave in such a medium. Will heavy quarks thermalize? Will heavy quarks develop collective motion? Some models assume that heavy quarks behave like *Brownian* particles ($m_Q > m_q$) [23] and their transverse momentum (p_T) and rapidity distributions, collective motion ... will probe the properties of the QGP. Although pQCD heavy quark interaction cross-sections with the medium seems too small for reaching full thermalization of heavy quarks, non-perturbative effects in QGP could noticeably increase the heavy-quark cross-section, leading to their full thermalization [24]. Other models do not address the problem of thermalization and directly assume statistical coalescence of heavy quarks [31, 26, 28, 29].

The study of the heavy quark bound states will allow for probing the medium via the Debye screening [3, 27], the gluon dissociation of quarkonia [28], kinetic recombination of heavy quarks [29] and/or their statistical hadronization [31, 26]. It is expected that charmonium production in HIC at LHC will be dominated by recombination of $c\bar{c}$ pairs in the QGP or at the hadronization. Due to the huge production rate of charmed quarks at LHC, these models predict a spectacular enhancement of the charmonium yield in central Pb+Pb collisions, more than a factor 10. Based on these models, charmonium yield will exhibit centrality dependence proportional to $< N_{c\bar{c}} >^2$ [26, 28, 29], which will be a very clear signature of QGP. In the case of charmonium, a non-negligible production could occur after hadronization of the QGP due to $D\bar{D}$ annihilation [30, 31]. For the bottomonium production at the LHC, the recombination mechanism will play a less important role due to the lower bottom pair production rate. Recombination mechanism would compete with the suppression mechanism due to Debye screening and/or gluon dissociation. Charm measurements at RHIC would help to predict what should occur with beauty at LHC since production rates per central HIC are similar.

2.4. Secondary production

There will be a production of J/ψ due to charmonium decay modes of the B mesons. At LHC energies the relative contribution of J/ψ from B-meson decay should be about 25% [17]. Upsilon family production will be increased by radiative decay of χ_b quarkonia. About 60% of the total Upsilon family production will originate from these resonances.

3. Production of electroweak bosons

Electroweak bosons ($W^{+,-}$, Z) will be abundantly produced at LHC and this will allow for several precision measurements, for instance high precision measurement of the W mass [33]. However, proton luminosities in ALICE experimental region will be much lower ($<\mathcal{L}>= 3\cdot 10^{30}\text{cm}^{-2}\text{s}^{-1}$) than in other LHC experiments, due to ALICE detector performances. At leading-order, W bosons are produced in quark - antiquark collisions: $u\bar{d}\rightarrow W^+$, $c\bar{s}\rightarrow W^+$ etc ... The LO production cross-section of W^{+-} obtained from Pythia [34] amounts to ~17 nb in p+p collisions at 14 TeV [35], leading to a production of half a millions W bosons decaying in the muonic channel in one year of data taken in the ALICE interaction point. At these energies, contributions to W production from higher order diagrams should be about 10% [36]. In proton+proton collisions, there will be more u than d valence quarks, leading to a enhancement of W^+ with respect W^- which is more pronounced at high rapidities where the W-boson production is dominated by valence-quarks. At mid-rapidity the production of W^+ and W^- is almost symmetric.

In HIC, W bosons production will not be affected by the medium effects since W bosons will leave the hot and dense bulk of matter without any secondary interaction. In this respect, W bosons production will be a very powerful tool for checking the validity of the glauber scaling hypothesis at these energies. Since shadowing could be different in quarks and gluons, measurements of W production in d+Pb (or p+Pb) will be necessary. W-boson production could provide a powerful normalization of very high p_T (p_T ~30-40 GeV/c) beauty and charm production where a suppression could be expected due to jet quenching of heavy quarks. The asymmetry between W^+ and W^- in HIC will be given by the isospin of the nuclei and the elementary asymmetry in p+p, n+n, p+n and n+p collisions [35].

4. Muon Spectrometer

In the framework of the ALICE physics program [6], the goal of the Muon spectrometer of ALICE [37] is the study of open heavy flavor production and quarkonia production (J/ψ, ψ' and $\Upsilon(1S)$, $\Upsilon(2S)$ and $\Upsilon(3S)$) via the muonic channel. For HIC the dependence with the collision centrality and with the reaction plane will also be studied. The study of electroweak bosons is also foreseen and preliminary studies are under progress [35].

The main experimental requirement is to measure the quarkonia production in central Pb+Pb collisions at LHC energies, down to very low p_T, since low p_T quarkonia will be sensitive to medium effects like heavy-quark potential screening. Since muons are passively identified by the absorber technique, a Lorentz boost is needed to be able to measure quarkonia at low p_T. On the other hand, the Muon spectrometer has to be as close as possible to the physics of the QGP which occurs in the mid-rapidity region. As a compromise, the muon spectrometer allows for measuring muons and quarkonia production in an intermediate rapidity range $-4.0 < y < -2.5$.

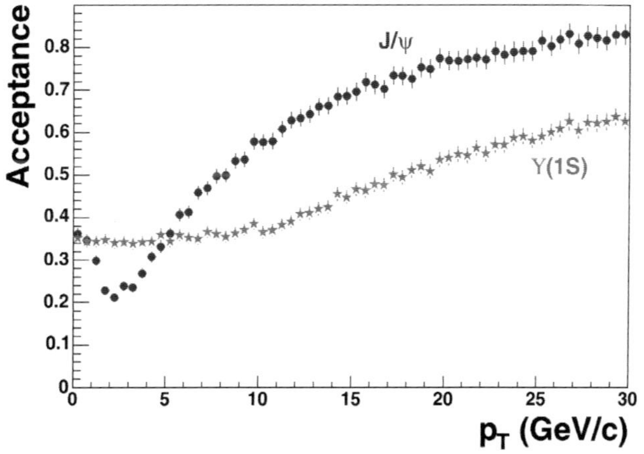

Figure 1. Acceptance of the Muon spectrometer as a function of the transverse momentum for J/ψ and $\Upsilon(1S)$ in the rapidity range $-4.0 < y < -2.5$, via their muon pair decay and with a muon low p_T threshold equal to 1 GeV/c.

The acceptance plot in this rapidity range for J/ψ and $\Upsilon(1S)$ mesons decaying into muon pairs is presented in Fig. 1. The muon spectrometer will be a unique apparatus at the LHC to measure charmonia production down to $p_T = 0$ and will cover a rapidity range which completes the one measured by the ALICE central barrel, and by the CMS and ATLAS experiments. The other important experimental requirement for the muon spectrometer is to be able to disentangle the different resonances of the Υ family. In particular the separation between the resonance $\Upsilon(2S)$ and $\Upsilon(3S)$ (ΔM $\sim$ 300 MeV/c^2) imposes to the spectrometer an invariant mass resolution about 100 MeV/c^2 in the Υ mass region.

In order to reach these requirements, the muon spectrometer is located downstream (side C of point 2 at LHC) of the ALICE detector covering the angular range $171.^\circ < \theta < 178.^\circ$, (-4.0< y <-2.5) consisting of 3 absorbers, a muon magnet, a trigger system and a tracking system. A more detailed description of the Muon spectrometer is given in these proceedings [37].

5. Physics program in the Muon spectrometer of ALICE

The Muon physics program is focused on the measurement of heavy flavor production in p+p, p+A like and A+A collisions at LHC energies. Many studies have been undertaken [40, 42, 43] mainly in the framework of the Technical Design Report [38] and the Physics Performance Report of ALICE [39] and other studies are still in progress.

5.1. Quarkonia measurements

Invariant mass analysis of muon pairs will provide a direct measurement of the quarkonia production: J/ψ, ψ', $\Upsilon(1S)$, $\Upsilon(2S)$, $\Upsilon(3S)$. Quarkonia measurement in central Pb+Pb collisions will be very challenging due to the huge low-energy background [38, 44] and the large muon combinatorial background (see in Tab. 1 the quarkonia signal to background ratio and the significance from reference [40]). In the case of J/ψ, a ratio signal to background (S/B) in the order of $\sim$ 0.15 is expected in most central (0-10%) Pb+Pb collisions at 5.5A TeV (see Fig. 2). During the first Pb+Pb run with nominal luminosity at the LHC [39] (this corresponds to a period of 10^6 s and an average luminosity of $5 \cdot 10^{26}$ cm^{-2} s^{-1}) we expect to measure about $\sim 6 \cdot 10^5$ of J/ψ and $\sim 6500\ \Upsilon(1S)$ [4]. Those numbers would fluctuate depending on the physics of the heavy quark production and quarkonia production in heavy ion collisions: from total suppression due to color screening, to enhancement due to recombination. Measurement of the ratio of $\Upsilon(2S)/\Upsilon(1S)$ will provide a very powerful experimental observable to constraint the different models on quarkonia suppression in the QGP [41, 43]. The measurement of J/ψ elliptic flow will also be possible and the study of very high p_T J/ψ will represent a very promising hard probe. The increase of luminosity for the Ar+Ar will allow for cumulating about a factor 5 more quarkonia statistics than in Pb+Pb LHC run

	b (fm)	0-3	3-6	6-9	9-12	12-16
J/ψ	S/B	0.167	0.214	0.425	1.237	6.243
	$S/\sqrt{S+B}$	111.3	180.4	213.8	193.4	94.95
ψ'	S/B	0.009	0.011	0.021	0.063	0.273
	$S/\sqrt{S+B}$	4.185	6.902	8.604	9.641	7.171
Υ	S/B	2.084	2.732	4.31	7.977	12.01
	$S/\sqrt{S+B}$	27.39	41.71	40.03	27.16	10.42
Υ'	S/B	0.807	1.043	1.661	2.871	4.319
	$S/\sqrt{S+B}$	11.68	18.26	18.48	13.02	5.077
Υ''	S/B	0.566	0.722	1.18	1.936	3.024
	$S/\sqrt{S+B}$	7.951	12.55	13	9.274	3.73

Table 1. Preliminary quarkonia signal over background (S/B) and significance ($S/\sqrt{S+B}$) for quarkonium resonances measured versus centrality in the ALICE forward muon spectrometer [40]. The input cross-sections are taken from [19]. Shadowing is taken into account. Any other suppression or enhancement effects are not included. The numbers correspond to one month of Pb-Pb data taking and are extracted with a 2σ mass cut.

In the case of proton-proton run at 14 TeV [39] (this corresponds to a period of 10^7 s and an average luminosity of $3 \cdot 10^{30}$ cm^{-2} s^{-1}) , statistics per LHC run will be $\sim 2.5 \cdot 10^6$ J/ψ, and $\sim 27000\ \Upsilon(1S)$. During the first 10 years of LHC running it is planned to deliver deuterium lead collisions at $\sqrt{s}$=6.2A TeV and the cumulated statistic per d+Pb LHC run will be $\sim 1.5 \cdot 10^6$ J/ψ's and $\sim 15000\ \Upsilon(1S)$'s [5]

5.2. Open heavy flavor measurements

The measurement of open heavy flavors will also be a priority of our physics program. Being very interesting in its own, it will allow for a normalization of the quarkonia production rates. Different alternative analysis will be applied, which exploit the muon production from open charm and beauty mesons via their semi-leptonic decay.

[4] Glauber scaling and shadowing have been taken into account for this estimation

[5] We have not taken into account in this estimation the fact that the center-of-mass frame in d+Pb collisions will present a relative rapidity of ± 0.12 (depending on the deuterium or lead beam direction: d+Pb or Pb+d) with respect the laboratory frame due to the asymmetry in the deuterium and lead beam energies. This will slightly change the acceptance of the Muon spectrometer and could slightly increase or decrease the reported statistics. Since we are interested to measure shadowing for small x values inside the lead nucleus, the lead beam will come from the back of the spectrometer, and the spectrometer will cover the rapidity range $-3.88 < y < -2.28$. This effect would be more pronounced in p+Pb collisions.

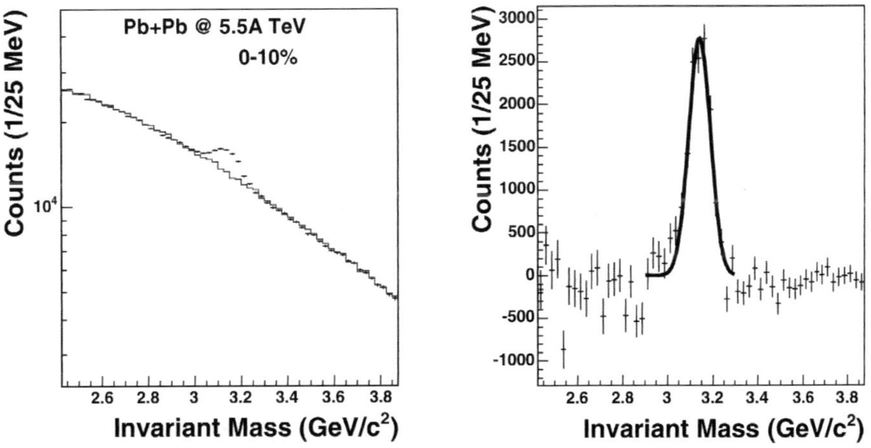

Figure 2. Invariant mass distribution of unlike-sign muon pairs in the Muon spectrometer for Pb+Pb collisions at LHC energies in the most central collisions (0-10%).

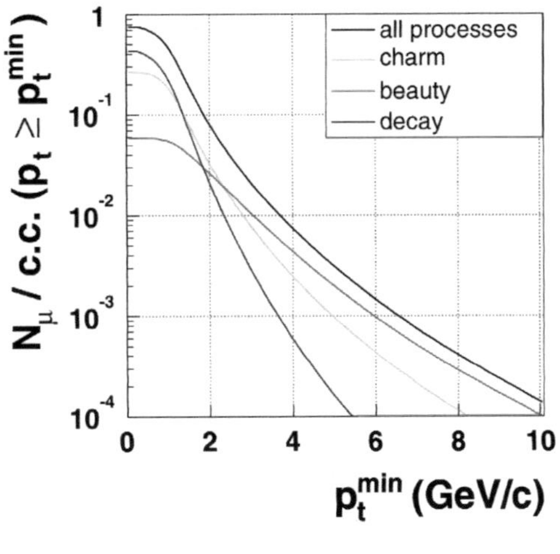

Figure 3. Integrated single muon p_T distribution in most central Pb+Pb collisions at 5.5A TeV [42].

Single muon measurements. The measurement of muon p_T distributions will provide the first measurement of heavy quark production at high rapidities at LHC. For p_T larger than 5-8 GeV/c muon production will be dominated by the semi-muonic decay of D and B mesons. The expected integrated single muon contribution in the most central Pb+Pb collisions at 5.5A TeV is presented in Fig. 3. For very high p_T a noticeable contribution to the single muon p_T distribution will be originated from the muonic decay of the W-bosons. This contribution will be maximum in the p_T range between 30 and 40 GeV/c. In Fig. 4 the differential cross-section for the muonic channel in the Muon spectrometer acceptance is represented. During a p+p run at 14 TeV it is expected to detect around 50000 muons coming from the decay of the W boson.

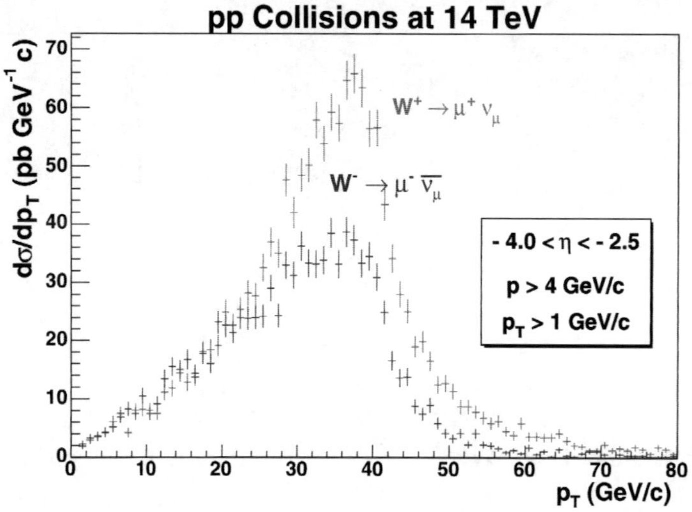

Figure 4. Single muon p_T distribution coming from the W muon decay in the Muon spectrometer acceptance for p+p collisions at 14 TeV [35].

Muon pairs measurements. Like and unlike sign muon pairs originating from the same hard scattering or same heavy quark, will present a residual correlation in the low (1-3 GeV/c^2) and high (4-8 GeV/c^2) invariant mass (M_{inv}) regions. Correlated unlike-sign muon pairs in high M_{inv} region will be mainly produced by semi-muonic decays of $D - \bar{D}$ and $B - \bar{B}$ mesons from the same hard scattering:

$\mu^+ + ...$

D° (or D^+, D^*, D^{*+}, D_s^+) + ...

Hadron collision $\Longrightarrow c + \bar{c} + ...$ Residual $\mu^+\mu^-$ correlation $D - \bar{D}$

$\bar{D}^\circ$ (or D^-, $\bar{D}^*$, D^{*-} D_s^-) + ...

$\mu^- + ...$

$\mu^- + ...$

$\bar{B}^\circ$ (or B^-, $\bar{B}^*$, $\bar{B}_s^\circ$, ...) + ...

Hadron collision $\Longrightarrow b + \bar{b} + ...$ Residual $\mu^+\mu^-$ correlation $B - \bar{B}$

B° (or B^+, B^*, B_s°, ...) + ...

$\mu^+ + ...$

The low M_{inv} region will be populated by $B-D$ or $\bar{B}-\bar{D}$ semi-muonic decays from the same bottom quark fragmentation. Like-sign muon pairs will be dominated by $B-\bar{D}$ and, more exotic mechanism, like $B^0-\bar{B}^0$ oscillations. The study of such correlations will allow for measuring the beauty production cross-section [42] and to study the relative contribution of NLO diagrams to the total beauty production cross-section. In HIC, the $\mu^+\mu^-$ residual correlations could be modified by final state effects. In this respect, $\mu^+\mu^-$ residual correlations could become a probe of the heavy quark re-scattering in the QGP medium.

Correlation with the central barrel. In the same way, residual correlation between unlike sign muons and electrons (or Kaons) can be exploited to get similar physics informations. Since electrons and kaons are detected in the ALICE central barrel, such a correlation will be sensible to heavy quark production in an intermediate rapidity range $-1.0 < y < -2.5$. Additionally, unlike-sign $e-\mu$ pairs cannot not be generated by any resonance, nor thermal or direct production [17].

Multi-correlation of muons. Residual correlation of 3 (or more) muons will only be sensible to beauty production. In particular, 3 muon correlation will probe the beauty production decaying in the charmonium channel which will noticeably contribute to the J/ψ total production cross-section. Unfortunately, the combinatorial background of 3-muon correlations in Pb+Pb central collisions will be huge and this technique could only be efficiently applied in p+p, d+Pb and (perhaps) in Ar+Ar collisions [9].

6. Conclusions and Perspectives

In summary, the muon spectrometer supports an ambitious physics program in the ALICE experiment, focused on heavy quark physics for studying QGP at the LHC collider. The muon spectrometer has entered in its construction phase and will be ready in 2007 at the interaction point 2 of the LHC for the first data taking.

7. Acknowledgments

I would like to thank my colleagues of ALICE collaboration for fruitful discussions. In particular, I would like to thank Philippe Crochet, Christian Finck, Rachid Guernane, Smbat Grigoryan and Andreas Morsch for helping me in the preparation of this talk. I would like to thank Laurent Aphecetche, Zaida Conesa del Valle, Philippe Crochet and Christian Finck for carefully reading this manuscript.

8. References

[1] F. Karsch *et al.*, hep-ph/0103314 (2001).
[2] S. Ejiri *et al.*, hep-lat/0408046 (2004).
[3] R. Gavai and E. Laermann in these proceedings.
[4] H. Satz, Nucl. Phys. A715 (2004) 3c.
[5] T.K. Hemmick, J. Phys.G: Nucl. Part. Phys. 30 (2004) S659.
[6] Y. Schutz in these proceedings.
[7] H. Takai, J. Phys.G: Nucl. Part. Phys. 30 (2004) S1105 and LHCC 2004-009/I-013 (2004).
[8] ALICE collaboration, ALICE Physics Performace Report Vol I, J. Phys. G30 (2004) 1517.
[9] A. Morsch, Private Communication (2005).
[10] L. Mc. Lerran, hep-ph/0104285 (2001).
[11] E. Iancu and R. Venugopalan, hep-ph/0303204 (2003).
[12] A. Krasnitz and R. Venugopalan, Phys. Rev. Lett. 86 (2001) 1717.
[13] R. Baier, A.H. Mueller, D. Shiff and D.T. Son, hep-ph/0009237 (2000).
[14] Urs. Wiedemann, Nucl. Phys. B588 (2000) 303; X.-N. Wang, Phys. Rev. C58 (1998) 2321; X.-N. Wang and Z. Huang, Phys. Rev. C55 (1997) 3047.
[15] T. Matsui and H. Satz,Phys. Lett B178 (1986) 416.

[16] S.M.H. Wong, Phys. Rev. C58 (1998) 2358; E. Shuryak, Phys. Rev. C55 (1997) 961; Z. Lin, R. Vogt, and X.-N. Wang, Phys. ReV. C57 (1998) 899.
[17] Ph. Crochet, nucl-ex/0503008 (2005).
[18] P. Nason *et al.*, Nucl. Phys. B327 (1989) 49.
[19] M. Bedjidian *et al.*, CERN Yellow Report hep-ph/0311048 (2003).
[20] CDF Collaboration, Phys. Rev. D65 (2002) 052005; CDF Collaboration, Phys. Rev. Lett. 91 (2003) 241804; D0 Collaboration, Phys. Lett. B487 (2000) 264.
[21] J. Raufeisen, hep-ph/0403195 and J. Phys. G30 (2004) S1159.
[22] B. Muller and X.N. Wang, Phys. Rev. Lett. 68 (1992) 2437.
[23] P.B. Gossiaux, hep-ph/0411324 (2004).
[24] H. van Hees and R. Rapp, nucl-ex/0412015 and Phys. Rev. C71 (2005) 034907.
[25] P. Braun-Munzinger and J. Stachel, Phys. Lett. B490 (2000) 196.
[26] A. Andronic *et al.*, nucl-ex/0303036 and Phys. Lett. B571 (2003) 36.
[27] S. Datta *et al.*, J. Phys.G: Nucl. Part. Phys. 30 (2004) S1347.
[28] L. Grandchamp *et al.*, Phys. Rev. Lett. 92 (2004) 212301.
[29] R.L. Thews *et al.*, Phys. Rev. C63 (2001) 054905.
[30] C.M. Ko *et al.*, Phys. Lett. B444 (1998) 237.
[31] P. Braun-Munzinger and K. Redlich, Eur. Phys. J. C16 (2000) 519.
[32] B.W. Zhang *et al.*, Phys. Rev. Lett. 93 (2004) 072301.
[33] ATLAS Collaboration, ATLAS Detector and Physics Performance Technical Design Report, http://atlas.web.cern.ch/Atlas/GROUPS/PHYSICS/TDR/access.html (1999).
[34] PYTHIA http://www.thep.lu.se/~torbjorn/Pythia.html
[35] Z. Conesa del Valle, in preparation (2005).
[36] S. Frixione and M.L. Mangano CERN-PH-TH/2004-081 and hep-ph/0405130 (2004).
[37] Ch. Finck in these proceedings.
[38] ALICE collaboration, CERN/LHCC 99-22 (1999).
[39] ALICE Collaboration, ALICE Physics Performance Report Vol.2, in preparation (2004).
[40] S. Grigoryan Private Communication and ALICE Internal note, in preparation (2004).
[41] J.F. Gunion and R. Vogt, Nucl. Phys. B492 (1997) 301.
[42] R. Guernane *et al.*, ALICE-INT-2004, in preparation (2004).
[43] E. Dumonteil and Ph. Crochet, ALICE-INT-2005-002 (2005).
[44] A. Zintchenko *et al.*, ALICE-INT-2003-006 (2003).

Institute of Physics Publishing Journal of Physics: Conference Series **50** (2006) 371–376
doi:10.1088/1742-6596/50/1/050 5th International Conference on Physics and Astrophysics of Quark Gluon Plasma

The CBM Experiment @ FAIR - New challenges for Front-End Electronics, Data Acquisition and Trigger Systems

Walter F.J. Müller[1] **for the CBM collaboration**

[1] GSI, Darmstadt

E-mail: w.f.j.mueller@gsi.de

Abstract. The 'Compressed Baryonic Matter' (CBM) experiment at the new 'Facility for Antiproton and Ion Research' (FAIR) in Darmstadt is designed to study the properties of highly compressed baryonic matter produced in nucleus-nucleus collisions in the 10 to 45 A GeV energy range. One of the key observables is hidden (J/ψ) and open (D^0, $D^{\pm}$) charm production. To achieve an adequate sensitivity extremely high interaction rates of up to 10^7 events/second are required, resulting in major technological challenges for the detectors, front-end electronics and data processing. The front-end electronics will be self-triggered, autonomously detect particle hits, and output hit parameter together with a precise absolute time-stamp. Several layers of feature extraction and event selection will reduce the primary data flow of about 1 TByte/sec to a level of 1 GByte/sec. This new architecture avoids many limitations of conventional DAQ/Trigger systems and is for example essential for open charm detection, which requires the reconstruction of displaced vertices, in a high-rate heavy ion environment.

Introduction

The exploration of the phase diagram of strongly interacting matter is the leading theme in heavy-ion physics. Of particular interest is the transition from hadronic to partonic degrees of freedom which is expected to occur at high temperatures and/or high baryon densities. The experiments at SPS, RICH, and LHC explore the regime of low baryon density and high temperature, and thus probe the conditions important in the development of the early universe. The planned 'Compressed Baryonic Matter' (CBM) experiment at the new 'Facility for Antiproton and Ion Research' (FAIR) in Darmstadt focuses at the regime of high baryon density and modest temperatures, realized in nature for example in neutron stars. It is designed to study the properties of highly compressed baryonic matter produced in nucleus-nucleus collisions in the 10 to 45 A GeV energy range. The key observables are

- Low-mass vector mesons. The in-medium spectral function of short-lived vector mesons will be measured directly via their decay into dileption pairs.
- Charm production. The effective masses of D mesons are expected to be modified in dense matter, which leads to a change of the relative abundance of charmonium and D mesons. The anomalous suppression of charmonium due to the screening effects of a partonic medium allows to probe the onset of QGP formation. It is thus important to measure the production of open charm (D^0, $D^{\pm}$) as well as hidden charm (J/ψ).

- Strangeness production.
- Event-by-event fluctuations.

A more detailed account of the scientific program and the experimental setup of CBM is given in a separate contribution to these proceedings [1], a full description is given in the recent 'CBM Technical Status Report' [2].

Data Acquisition and Event Selection

Many of the signatures pursued with the CBM experiment are based on rare processes. To achieve an adequate sensitivity, the detector systems are designed to operate at interaction rates of up to 10 MHz for A-A collisions and up to several 100 MHz for p-p and p-A collisions. It is the task of the data acquisition and event selection system to identify the candidate events for the physics signals under study and send them to archival storage. The most challenging aspect is here the measurement of open and hidden charm production in heavy ion collisions down to very low cross sections. The D mesons will be identified via the displaced vertices of their hadronic decays, the decision for selecting candidate events thus requires tracking, primary vertex reconstruction, and secondary vertex finding in the STS. In addition, the system has to be configurable to handle a wide range of physics signals, ranging from D and J/ψ in A-A collisions over low-mass dileptons to Υ in p-p and p-A collisions.

The conventional system design with triggered front-end electronics allows to keep the event information for a limited time, usually a few μs, in the front-end electronics while a fast first level trigger decision is determined from a subset of the data. Upon a positive trigger decision, the data acquisition system transports the selected event to higher level trigger processing or archival storage. A system with such a fixed trigger latency constraint is not well matched to the complex algorithms needed for a D trigger, especially in the case of heavy ion interactions, where the multiplicities and thus the numerical effort needed for a decision varies strongly from event to event.

The concept adopted for CBM will use self-triggered front-end electronics, where each particle hit is autonomously detected and the measured hit parameters are stored with precise timestamps in large buffer pools. The event building, done by evaluating the time correlation of hits, and the selection of interesting events is then performed by processing resources accessing these buffers via a high speed network fabric. The large size of the buffer pool ensures that the essential performance factor is the total computational throughput rather than decision latency. Since we avoid dedicated trigger data-paths, all detectors can contribute to event selection decisions at all levels, yielding the required flexibility to cope with different operation modes.

In this approach we have no physical trigger signal which prompts a data acquisition system to read a selected event and transport it to further processing or storage. The role of the data acquisition system is to transport data from the front-end to processing resources and finally to archival storage. The event selection is done in several layers of processing resources, reminiscent of the trigger level hierarchy in conventional systems.

One consequence of using self-triggered front-end electronics is a much higher data flow coming from the front-ends on the detector. For CBM a data rate of about 1 TByte/sec is expected. However, communication cost is currently improving faster over time than processing cost, an observation sometimes termed Gilder's law, making such a concept not only feasible but also cost effective.

Overall Architecture

The communication and processing needed between the front-end electronics, generating digitized detector information, and the archival storage, where the complete context of selected candidate events is recorded, can be structured and organized in several ways. The current baseline solution for CBM is guided by two principles: processing is done after event building and it is done in a structured processor farm. It is well adapted to the type of processing needed in the CBM experiment and leads to a straightforward and modular architecture.

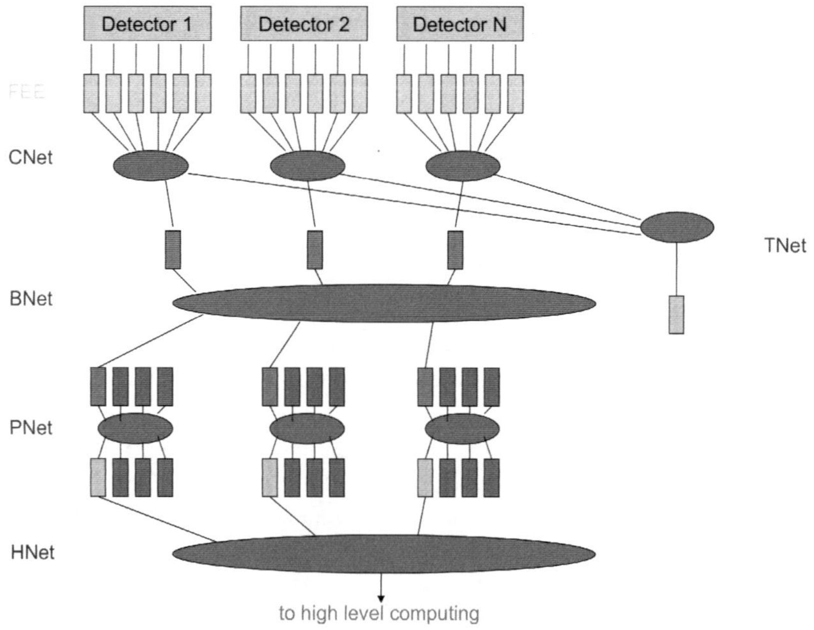

Figure 1. CBM overall data processing architecture

A logical data flow diagram is shown in Fig. 1, indicating the data sources and processing elements as boxes and every form of interconnection networks as ovals. The main components are:

- **Front-end electronics (FEE):** The front-end detects autonomously every particle hit and sends the hit parameters together with a precise timestamp and channel address information over the concentrator network (CNet) to a buffer pool.
  A rough estimate for the data volume generated by a detector channel can be deduced from typical CBM operation and detector parameters: 10 MHz interaction rate, 10 % occupancy for central collisions, a ratio of 1/4 for minimum bias to central multiplicity, and a typical cluster size of 3 fired electronics channels per physical particle hit gives a channel count rate of about 750 kHz. Assuming 8 byte per hit yields a data flow of about 6 MB/sec and channel. For a typical FEE unit with 16 channels this results in a data rate of 100 MB/sec which can be transported over a single GBit serial link.
- **Clock and time distribution (TNet):** The timestamps of each hit are used to associate hits with events but also in drift and flight time measurements. Thus not only a time scale, in practice a frequency, but also information about the absolute time has to be communicated to all front-end units. The most stringent requirements come from the detectors involved in time-of-flight measurements, where the contribution from the clock jitter should be below 25 ps sigma.
  The most straightforward approach is to distribute a common clock frequency and to provide a mechanism for broadcasting information with clock cycle precise latency to

all units. The minimal required functionality is a global clock reset at the begin of the measurement, or alternatively, distribution of tick marks every second as provided by the planned campus-wide frequency and time normal. The TNet is thus a dedicated broadcast network, connecting a central controller logically with all front-end units. The last hop to the front-end units may be implemented with the part of the CNet infrastructure, as indicated by the connection of TNet to CNet in Fig. 1.

- **Concentrator Network (CNet):** The role of the concentrator network is to collect the data from the individual front-end units and aggregate the traffic on a set of high speed links which connect the detector with the area where the data buffers and the data processing is located. A rough estimate for the total data rate is 1 TB/sec which could be finally transported off the detector with about 1000 links with 10 Gbps each.
  The simplest implementation of the CNet is a collection of independent concentrator trees, one for each high speed link. However, a better load balancing or an appropriate degree of failure tolerance is likely to call for a more connected topology.
  In addition to the hit data transport from front-end to buffering, other communication tasks like control traffic or time distribution can be handled by the CNet infrastructure. Such an integrated approach is especially useful in conjunction with low-cost optical links.
- **Active Buffers:** The next stage in the data flow is a large buffer pool. The units are indicated as magenta boxes in Fig. 1. They are dubbed 'active buffers' because the data is not only stored but potentially also reformatted and reorganized. They are also hand-over points between different types of networks, thus logically separating them and allowing to use different technologies in CNet, BNet, and PNet.
- **Build Network (BNet):** The data arrives from the detector in about 10^3 parallel streams, each reflecting a small section of a detector sub-system. For the event selection processing, the information of an event has to be assembled in a farm node. This data reorganization is performed by the build network and the active buffers.
  In a conventional system, a trigger already defines the context of an event, so all further data processing and transport can be organized in terms of events starting at the FEE. In our case, the FEE sends a stream of time-stamped hits, and it is one of the tasks of the data processing, to first identify at what times interactions occur, and in a second step, to associate the hits with those events.
  This *event tagging* processing can be done before or after data traverses the BNet. In the first case, event tagging is handled in the active buffers, and the entities being assembled in the BNet transfers are indeed events. In the second case, only the timestamp information is available, and it is thus natural to assemble all the data of a time interval.
  A strict event-by-event approach would lead at the nominal Au+Au interaction rate of 10 MHz to a message rate of 10^{10} messages per second with an average message size of 100 Bytes. However, because the transfer latency is uncritical, it is possible to choose a bigger dispatching unit, either an interval of events, or in the simplest case, a time interval containing a significant number of events. This aggregation reduces the message rate, increases message size, and because event size fluctuations average out, also yields a more uniform message size distribution. A reasonable choice is an event interval of about 100 events or equivalently a time interval in the order of 10 μs.
  The simplest solution is a time interval based build logic with a shaping and traffic scheduling setup similar to the one developed for the LHCb first level trigger [3]. More involved solutions, which support building event intervals and the suppression of event incoherent backgrounds, are currently being investigated.
  Fig. 1 indicates that source as well as destination of a BNet transfer is an active buffer. They implement the protocol used on the BNet and are responsible for organization of the data flow, in particular for traffic shaping and appropriate scheduling of transfers. Because

the actual traffic seen by the BNet can be controlled to a large degree and adapted to a given networking technology, it is assumed that the BNet can be based on a commercial off-the-shelf (COTS) technology. Plausible candidates are Ethernet, Infiniband, or the emerging Advanced Switching Interconnect (ASI).
Fig. 1 only indicates the logical data flow, not a concrete network topology. It is possible to merge the functionality of the two active buffer layers, one interfacing CNet to Bnet and one interfacing BNet to PNet, into a single entity, leading to a system with half as many BNet physical ports and bidirectional traffic on all BNet links. Also, it is possible to factorize the network in several ways, which allows to build the BNet with a set of medium-sized switches and thus to exploit the usually significantly better price/port ratio of smaller switches.

- **Processing Resources:** The first level of event selection processing has to handle the full event rate, and depending on how many detector sub-systems are involved in the decision, a substantial fraction of the total data volume. A very rough estimate shows, that processing a data flow on the scale of a TByte/sec is likely to require a computational bandwidth on the scale of 10^{15} operations/sec. With todays technology, the most promising approach is a hybrid system using a combination of hardware processors, implemented with programmable logic components like FPGA's, and software processors, implemented with commodity PC's. The kernels of algorithms which allow highly parallel execution are done in hardware processors, the rest in software processors. The aim is to execute most of the operations on hardware processors, which offer the best price/performance ratio for computational bandwidth, but to keep most of the code volume on software processors, which offer much easier program development and maintenance.
Since programmable logic devices are essentially arrays of simple structures, it is expected that they scale well and thus density and speed will improve as the underlying silicon technology evolves. The development for software processors over the relevant time scale till the design freeze of the CBM data processing is likely to be more complex. The evolution of single processor speed has reached apparent limits of complexity and power dissipation, calling for changes in concepts and architectures. This trend is already apparent in recent developments like the compute ASIC for IBM's Blue Gene system or the Sony-Toshiba-IBM Cell processor. It is also expected, that conventional fixed instruction set processor and programmable logic concepts will be coupled, resulting new forms of configurable computing, like processors with dynamically adaptable instruction sets. The current R&D efforts are based on physically separated hardware and software processors because these are the mainstream products available today and the near future. The overall architecture is, however, easily adaptable to more integrated forms of configurable computing.
- **Processing Network (PNet):** The processing resources are grouped in farm nodes. Each farm node is organized around a local processing network which provides the communication between the associated processing resources and active buffers, which act as central data repository and as gateway to the BNet. The PNet is thus structured into many local networks.
The number of hardware and software processors aggregated to one farm node is determined by the amount of resources needed to efficiently handle all the algorithms needed for an event selection decision. Each hardware processor will have a dedicated configuration to execute a particular algorithm, a total about a dozen different configurations may be needed. A rough estimate gives, that an about equal number of software processors is needed for a balanced load of the whole system.
It is expected that the resources of a farm are concentrated in a crate or are at least in close proximity. The PNet can therefore use technologies designed for short distance chip-to-chip and board-to-board interconnects, plausible candidates are from todays perspective PCIexpress or ASI. As stated already for the BNet, Fig. 1 only indicates the logical data

flow, not a concrete network topology. The PNet can be a homogeneous, single technology, switch based star network as suggested by the figure, but many other topologies are possible.

- **High-level Network (HNet):** The task of the event selection processing described up to now is to perform a first reduction, similar to the 'level 1 trigger' in conventional systems. A further reduction will be needed to reduce the data volume to a level suitable for archival storage. The HNet provides the connection to this high-level computing.

Summary

The measurement of open charm is the key factor shaping the architecture of front-end electronics, data acquisition, and event processing in CBM, leading to an architecture with self-triggered front ends without a conventional trigger.

The key R&D areas for CBM are the development of self-triggered front-end electronics with adequate output bandwidth, of a data processing system which allows to combine the advantages of hardware processors build from programmable logic and software processors, and last but not least, of highly efficient feature extraction and event selection algorithms adapted to such a processing environment.

Acknowledgements

This work is supported by the EU Integrated Infrastructure Initiative Hadron-Physics Project under Contract number RII3-CT-2004-506078.

References

[1] P. Senger, contribution to these proceedings

[2] Compressed Baryonic Matter Experiment - Technical Status Report. January 2005. http://www.gsi.de/documents/DOC-2005-Feb-447-1.pdf

[3] D. Atanasov, I. Kisel V. Lindenstruth, H. Müller, D. Altmann, A. G. Elias, F. V. dos Santos: A scalable 1 MHz Trigger Farm Prototype with Event-Coherent DMA input, Proceedings of the 13th Real Time Conference 2003, Valencia Spain

Institute of Physics Publishing
Journal of Physics: Conference Series **50** (2006) 377–380
doi:10.1088/1742-6596/50/1/051
5th International Conference on Physics and Astrophysics of Quark Gluon Plasma

Design, construction and front end electronics of second station of Forward Muon Spectrometer in ALICE.

Pratap Bhattacharya [1]
Saha Institute of Nuclear Physics, 1/AF, Bidhannagar, Kolkata -700 064, India.
E-mail: pratap.bhattacharya@saha.ac.in

Abstract. Forward muon spectrometer in ALICE in the LHC experiments of CERN will look for signals of quark gluon plasma (QGP) formation in Pb-Pb collisions at 5.6 TeV/nucleon energy. This spectrometer is designed specifically to look for decays of heavy quarkoniums through the correlated muon pairs. The design and construction details with emphasis on the R & D contributions from the nodal agency, Saha Institute of Nuclear Physics (SINP) will be presented.

1. Introduction

The tracking stations in the Forward Muon Spectrometer (FMS) of A Large Ion Collider Experiment (ALICE) [1] is intended to look for heavy quark resonance suppression, assumed to be one of the major probes of quark gluon plasma expected to be formed in relativistic heavy ion collisions. FMS will study the flight path of muon pairs from the decay of heavy quarkia (e.g., $J/\psi,\Upsilon$) produced in pp, pA and AA collisions through a dipole magnetic field and will measure the complete spectrum of heavy quark vector mesons as well as the invariant mass continuum in the $\mu^+\mu^-$ channel. Details of the FMS are described by Finck [2].

To achieve a good invariant mass resolutions of about 100 MeV to separate the upsilon resonances, stringent requirements of muon tracking (spatial resolution of around 100 μm, low density ($< 2\%$ of radiation length) and high granularity and full geometrical coverage were met by optimized design of cathode pad chambers (CPC) to be used as muon tracking stations. There are five stations in the FMS of ALICE. SINP is responsible for design, fabrication and installation of the second station.

2. Design

Design of the 1st and 2nd tracking stations followed a common strategy - a quadrant structure for the CPC's. The station 2 lay out is shown in Fig. 1, the eight quadrants make up the two CPC detectors, Station 3, Station 4 as described by Finck [2]. The detector is placed at a distance 6860 mm from the vertex. The inner radius of the detector is 231 mm while the outer radius is 1190 mm. The thickness of the chamber is 75 mm including the readout electronics. However, the total thickness of the station is about 300 mm. The parameters

[1] By whom the talk was delivered

critical for achieving the stringent requirements were tested with appropriate prototypes for spatial resolution, mechanical stability and temperature, electronic noise etc. Several prototypes for on-beam tests with relativistic protons and pions and also for mechanical properties were reported [3, 4, 5].

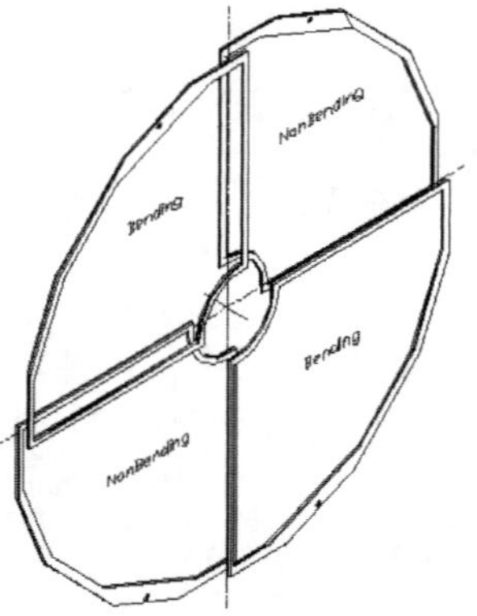

Figure 1. Lay out of quadrants of detectors 3-4 in station 2 of dimuon spectrometer

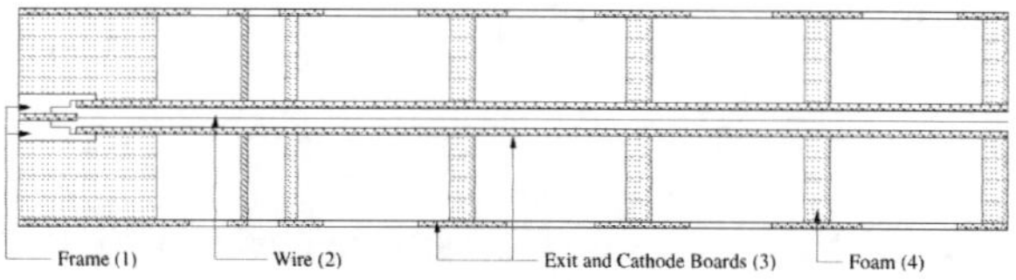

Figure 2. Mechanical structure of the cathode pad chamber.

3. Construction

The construction of the chamber 0 followed the freezing of designs as per the Technical Design Report [6]. The design of the mechanical structure of the CPC is shown in Fig. 2. The two cathodes, made of 0.4mm FR4 glass epoxy boards, are stabilized by making mechanically rigid sandwich structure with 0.4 mm FR 4 exit boards and 30 mm of rohacell material (density 51kg/m^3), with holes to pass the signal carrying flexible PCB's. The elastic properties were simulated and studied with mechanical prototypes [3]. Because of the large size, almost all components have to be segmented. The plastic frames are made in six parts by machining in CNC milling machines and joined together on a bi-surface Aluminium master tooling plate to ensure uniformity of the spacer between anode and cathode planes and identity of the quadrants. The cathode PCB's are segmented into eight fragments, the edges being cut with a 20 μm accuracy with a special cutting tool with microscopic and CCD camera controlled cutter. The cathode segments were glued together after alignment under a CCD camera. The electronic signals are carried by flexible kapton PCB's and those are soldered to the back of the cathodes. Each soldering joint have been visually checked under CCD camera and electrically monitored. The rohacell was glued to the cathode PCB in controlled climate of dust (10000/m^3 temperature (24°C) and humidity, on a granite reference surface with surface planarity better than 5 μm.

The mechanical and electronic exit boards are glued to the rohacell and plastic frame. The PCB's to solder the anode wires are glued with care to make those flush with the plastic frame. The anode wires are soldered and glued to the anode wire pads from an aluminium transfer frame. The construction followed stringent quality control at every stage and a production data base is maintained.

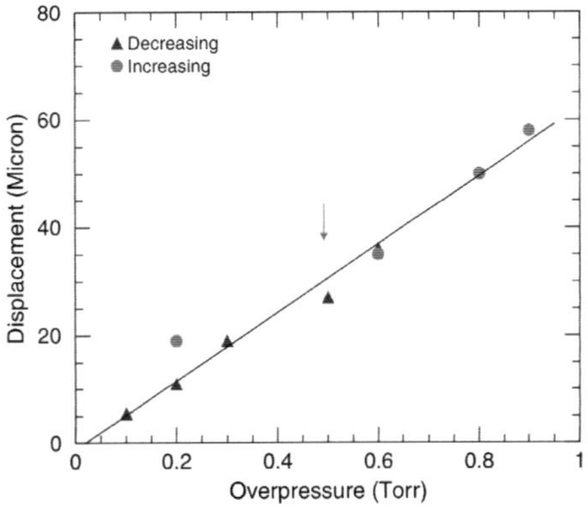

Figure 3. Elastic deformation of the detector against over-pressure

4. Performance

The full sized chamber 0 has been extensively tested in-house and on line with relativistic proton and pion beams. The elastic deformation (shown in Fig. 3) and the surface planarity (shown in Fig. 4) were measured and found to be consistent with a 15% variation in the gain of the detector (shown in Fig. 5). The electronic noise was found to be below 1000 e^- $r.m.s$. A position resolution of 50 μm is obtained with the detector.

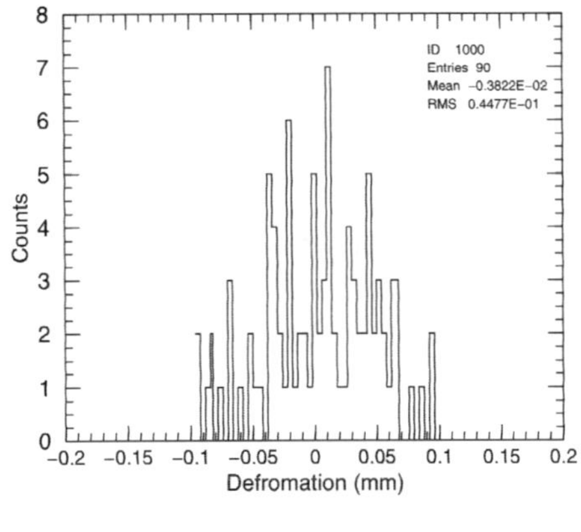

Figure 4. Surface planarity of the cathode surface.

The front end electronic (FEE) chip to amplify the charge signals on cathode pads and multiplex 16 channel read out is also fabricated in 1.2 μm CMOS technology at the Indian foundry at Chandigarh. The pre-production chip, MANAS 1.2 -3 (Fig. 6) showed a gain dispersion of about 3.6%. Improved process parameters in the production are expected to

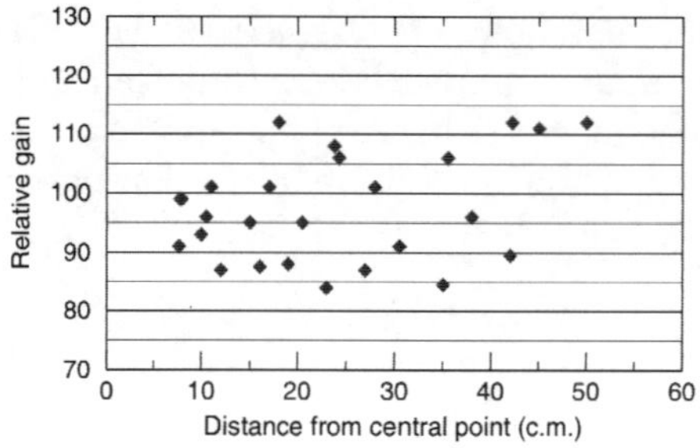

Figure 5. Gain variation in the detector as a function of distance from the central fixation point.

reduce the gain dispersion to below 2% level. Results from small sample batch of 5 production level chips were presented and indicated the expected reduction in gain spread.

Figure 6. Manas 1.2-3 plastic packaged chip

5. Conclusion

Chamber 0 of the 2nd muon tracking station fully satisfies the design parameters. After some minor design and process changes to ensure stability and long life of detectors, the production of eight quadrants can proceed in full swing to meet the assembly deadline in early 2006. Front end chips are under production and sample tests showed full compliance with the required 2% gain spread.

References

[1] ALICE Technical Proposal: CERN/LHCC/95-71 (1995)
[2] : proceedings of this conference
[3] P. Bhattacharya *et al.* ALICE-INT-2003-003
[4] P. Bhattacharya *et al.* ALICE-INT-2003-046
[5] M.P. Comets *et al.* ALICE-INT-2003-035
[6] ALICE Collaboration: Technical Design Report of the Dimuon Forward Spectrometer,CERN/LHCC 96-32 and CERN/LHCC 99-22, Addendum to the Technical Design Report of the Dimuon Forward Spectrometer,CERN/LHCC 2000-46

The dimuon High Level Trigger

Z. Z. Vilakazi (for the ALICE Collaboration)

Department of Physics and Centre for High Performance Computing, University of Cape Town, Private Bag Rondebosch, Cape Town, 7700, Republic of South Africa

E-mail: vilakazi@hep.phy.uct.ac.za

Abstract. The CERN-ALICE experiment will produce global data sizes in excess of 75 Mbytes/event, which at event rates of 200-1000 Hz will yield global data streams of 15 Gbyte/s. Online processing – in the form of pattern recognition – is necessary to filter interesting (sub-) events and/or compress data correctly by modelling techniques. A layer of trigger – selecting topologically distinct physics signals (J/ψ) and γ – is currently being developed and benchmarked under stringent conditions. An emphasis is placed on the implementation of this high level trigger (HLT) for muon reconstruction efficiency studies. Investigations thus have alluded to the HLT being able to reduce background by a factor of 4 for low transverse momentum muons and a factor of 100 for high Pt single muons. Furthermore the invocation of the HLT will help refine the resolution of the Pt cut efficiency.

1. Introduction

The CERN-ALICE experiment [1] will study physics associated with heavy-ion collision processes at centre-of-mass energies of about 5.5 TeV per nucleon pair and 14 TeV for a pp collision process. The ALICE detector will yield global event sizes (for Pb-Pb central events) typically in the range of 75 Mbytes/event [2]. The maximum allowed DAQ bandwidth to permanent storage is set at 1.2 GB/s, thus meaning that the ALICE central detectors can only run at 20 Hz.

Of particular importance is the implication this has on the dimuon spectrometer [3] – a sub-detector located at the backward part of the ALICE detector and covers an angular region of $171^\circ < \theta < 178^\circ$, corresponding to the (pseudo) rapidity range $-2.5 \geq \eta \geq -4.0$. The dimuon spectrometer is tasked with the responsibility of measuring the colour screening effects using heavy-quark resonances J/ψ andΥ. These resonances will be studied through their decay, as a function of their centrality, into dimuons. However, the task of identifying muon candidates is made difficult by the effect of large particle multiplicities and pile-up in the spectrometer, thus requiring the use of an online event filter – the high-level trigger – whose basic functionalities are implemented in a form resembling that of a massive parallel computing system. The dimuon High level trigger (dHLT) – will, upon its realisation – have to perform fast, online pattern recognition in real time [4] and will thus, in the process help in the selection of topologically distinct signals corresponding to good physics events. As a result, the dHLT will help reduce data produced by the detectors without any loss of good physics events. This, it is envisaged will result also in a great reduction in taping costs.

2. Architecture and data flow

Studies of the HLT architecture and data flow processes for the global ALICE event readout are outlined in detail by the HLT-TPC group [5], where it is shown how data get transferred via optical fibres, into the ReadOut Receiver card (RORC) of the DAQ system (D-RORC). These data streams are also replicated into the HLT-RORC. The HLT-RORC has an embedded FPGA co-processor which does all data intensive tasks for local pattern recognition. Also embedded in the architecture of the HLT-RORC is an external memory to be used as a storage of look-up tables and these can be used to perform a variety applications.

Likewise, the data flow inside the HLT and its boundaries is driven by the physics requirements and operating scenarios. Of importance are the following considerations to be taken into account when implementing/optimising the high level trigger: (i) Fast data rates coming from the front end electronics, and (ii) data rates that DAQ can handle. In particular, we are interested in maximal rates from the spectrometer which must be handled. These numbers therefore affect the following design choices for the dHLT:

- Algorithm design: Fast and effective algorithms must be implemented to cope with high data rates and remove a high percentage of backgrounds.
- The number of processors (CPUs) is required to scale with required data rates and algorithm speed.
- Required buffer sizes and network capacity will have to scale proportionally with incoming data rates.

The raw data from the dimuon cathode readout is then sent out into the HLT computing cluster. These events are first validated by the fast (1.2 μs) level 0 (L0) trigger [6], whose main function is to perform – as described in detail below – a "loose" cut in transverse momentum. The next level of triggering in the data-flow chain involves the pattern recognition algorithms performed either on the FPGA or on PC's. In accordance with the dHLT design [7], data from Resistive Plate Chambers (RPC) [3] come in the form of a selection of trigger events from a transverse momentum Pt-cut on a single muon. However, the limited segmentation of the RPC and the muon filter [8] do not allow for a sharp Pt cut. Thus for this purpose, the correlation of information/hits from the RPC and the tracking information will be used in order to validate the HLT trigger decision. To this end, the main dHLT algorithm consists of three successive steps; namely, clusterisation or fast hit reconstruction on the tracking planes, application of the fast online track reconstruction algorithm and improved transverse computation subsequent to the validation and/or rejection of a muon event.

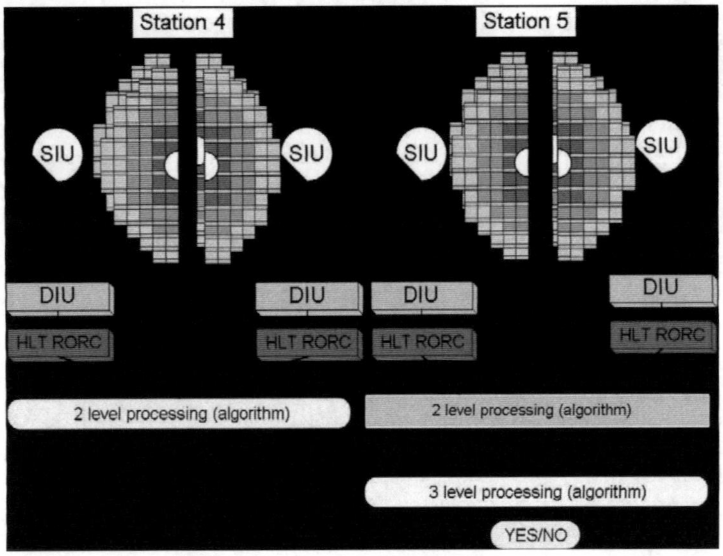

Figure 1: Layout of the dimuon HLT architecture

Data from the detectors is shipped out via the source interface unit (SIU) across an optical fibre network into the HLT processing farm as shown in Figure 1. The first stage of HLT processing – cluster finding – is performed directly on the front end processors (FEP) which are connected to terminators (DIU) of the optical detector data link (DDL) transmitting data at 100 MB/s. After the execution of the two-dimensional cluster finder, the FEPs will ship data – in the form of reconstructed space points – one level down the logical trigger hierarchy, where online HLT track reconstruction takes place. Trigger information will be made available by means of data shipped by the trigger stations. Thus the online tracker will rely on the information given by the combination of available trigger and clustered space points. The most important feature of the online tracker is the refinement of the algorithm, which is naturally evolving over time and will be adjusted accordingly. Furthermore, the number of compute nodes at the tracking stages depends on the tracking performance and/or the type of algorithm used and can thus also be adjusted dynamically. The application of the dHLT has to be carried out under (semi-) real time conditions and thus it is essential that information about processing time (latency) both for the cluster finder and the tracker be ascertained. Using a 2 GHz processor, total dHLT latency (both for the cluster finder and the tracker) of 20 ms was obtained. Of course, when cognizance of Moore's law is taken, then this CPU latency will drop – at least by a factor of three when data taking commences at the start of the LHC physics run.

3. HLT Physics performance

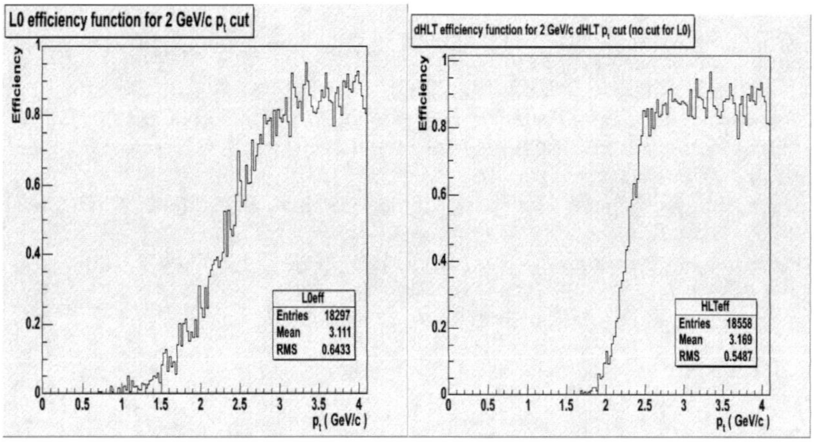

Figure 2. Efficiency distribution function of single muons as a function of transverse momentum.

The results of the HLT transverse momentum cut efficiency are shown in Figure 2. What is clearly elucidated in the figure is how efficient the HLT (right panel) is in obtaining a sharper and a more refined Pt cut than the hardware L0 trigger (left panel). These results show that the use of the HLT greatly improves the momentum cut resolution of the spectrometer, meaning therefore that the dHLT will, indeed, introduce a higher and more accurately selectivity on incoming raw data streams. Furthermore, the High Level Trigger tracking abilities can also be used for online reconstruction of the invariant mass spectrum of the heavy quarkonia resonances. Earlier investigations into the feasibility of the HLT [6] have alluded to the fact that the HLT will also decrease the data acquisition rate by a factor of four and 100 for Pt-cuts of 1 GeV and 2 GeV, respectively. These results, representing an extreme case of a central Pb-Pb collision, are shown in Table 1.

	L0	Dimuon HLT
Low pt cut	*2000 Hz*	*500 Hz*
High pt cut	550 Hz	Few Hz

Table 1. Acquisition rates for the both the L0 trigger and the HLT for the low and high Pt single muon cuts.

4. Conclusions

The dimuon HLT is an online buffering filter with a primary task of rejecting background coming mainly from low Pt contamination. To this end, first studies of the HLT have shown that background rejection of a factor of 4 and 100 can be achieved when HLT Pt-cuts of 1 GeV and 2 GeV, respectively, are invoked. Further, the HLT can also improve the resolution of the Pt cut efficiency and thus making it easier for online mass reconstruction of the J/ψ andΥ resonances – currently being planned for investigation at the dimuon spectrometer of the ALICE detector. The HLT is currently undergoing stringent benchmark tests in preparation for implementation at the LHC.

References

[1] ALICE Collaboration, *Technical Proposal*, CERN/LHCC/95-71 (1995)
[2] ALICE Collaboration, CERN/LHCC/2003-062 (2003)
[3] The Dimuon Forward Spectrometer, *Technical Design Report*, CERN/LHCC/99-22 (1999)
[4] C. Loizides et.al., Online Pattern Recognition for the ALICE High level Trigger, Proceedings ACAT02, Moscow, June 2002
[5] T. Alt. *et.al.*, Proceedings of Quark Matter 04, J.Phys. G: Nucl. Part. Phys. 30 S1097-S1100 (2004)
[6] O. Roig and P. Dupieux, *Alice Internal Note*, ALICE-INT-1998-08 (1998)
[7] F. Manso, *Alice Internal Note*, ALICE-INT-2002-04 (2002)
[8] A. Baldisseri, Proceedings of Quark Matter 02, Nucl. Phys. A715 839c (2002)

Institute of Physics Publishing
doi:10.1088/1742-6596/50/1/053
Journal of Physics: Conference Series **50** (2006) 385–388
5th International Conference on Physics and Astrophysics of Quark Gluon Plasma

Jet Physics in Heavy Ion Collisions with Compact Muon Solenoid detector at the LHC

Igor Lokhtin (for the CMS Collaboration)

Skobeltsyn Institute of Nuclear Physics, Moscow State University, Moscow 119992, Russia

E-mail: Igor.Lokhtin@cern.ch

Abstract. The status of CMS jet simulations and physics analysis in heavy ion collisions is presented. Jet reconstruction and high-p_T particle tracking in the high multiplicity environment of heavy ion collisions at the LHC using the CMS calorimetry and tracking system are described. The Monte Carlo tools used to simulate jet quenching are discussed.

1. Introduction

One of the important tools for studying properties of the quark-gluon plasma (QGP) in ultrarelativistic heavy ion collisions is QCD jet production. Medium-induced energy loss of energetic partons, "jet quenching", should be very different in cold nuclear matter and in the QGP, resulting in many observable phenomena [1]. Recent RHIC data on high-p_T particle production are in agreement with the jet quenching hypothesis [2]. However direct event-by-event reconstruction of jets is not available in the RHIC experiments. At the LHC, a new regime of heavy ion physics will be reached at $\sqrt{s_{\rm NN}} = 5.5$ TeV where hard and semi-hard particle production can dominate over the underlying soft events. The initial gluon densities in Pb + Pb reactions at the LHC are expected to be significantly higher than at RHIC, implying stronger partonic energy loss, observable in various new channels [3].

2. CMS detector

The Compact Muon Solenoid (CMS) is a general purpose detector designed primarily to search for the Higgs boson in proton-proton collisions at the LHC [4]. The detector is optimized for accurate measurements of the characteristics of high-energy leptons, photons and hadronic jets in a large acceptance, providing unique capabilities for "hard probes" in both pp and AA collisions [5]. A detailed description of the detector elements can be found in the corresponding Technical Design Reports [6, 7, 8, 9]. The central element of CMS is a 13 m long, 6 m diameter, high-field (4 T) solenoid with an internal radius of ≈ 3 m. The tracker and muon chambers cover the pseudorapidity region $|\eta| < 2.4$ while the electromagnetic (ECAL) and hadron (HCAL) calorimeters reach $\eta = \pm 3$ and $\eta = \pm 5.2$ respectively. A pair of quartz-fibre very-forward (HF) calorimeters, located at ± 11 m from the interaction point, cover the region $3 < |\eta| < 5.2$. In addition, the quartz-fibre calorimeter CASTOR covers the region $5.3 < |\eta| < 6.4$. The high precision tracker is composed of silicon pixel and strip counters and allows track momenta to be determined with a resolution better than 2% for tracks with p_T between 0.5 GeV/c and a few tens of GeV/c.

3. Jet production in heavy ion collisions at CMS

The following probes for studying jet quenching with CMS has been proposed:

- high-p_T jet pair production [10];
- jets tagged by a leading charged hadron or neutral pion [11];
- B-jets tagged by a leading muon [12];
- jets produced opposite a gauge boson in γ+jet [13, 14] and γ^*/Z^0+jet [15, 16, 17] final states;
- high mass dimuons from semileptonic B and D decays [18, 19] and secondary J/ψ [19];
- inclusive high-p_T particle spectra [20];
- energy flow measurements [21, 22]

Channel	Time=1.2×10^6 s, $\sigma_{AA} = A^2\sigma_{pp}$
jet+jet, $E_T^{\rm jet} > 100$ GeV	4×10^6
jet tagged by $h^\pm/\pi^0$, $E_T^{\rm jet} > 100$ GeV, $z > 0.5$	2×10^5
B-jet tagged by μ, $E_T^{\rm jet} > 50$ GeV, $z > 0.3$	2×10^4

Table 1. Expected rates for some jet channels in a one month Pb + Pb run ($z = E_T^{\rm leader}/E_T^{\rm jet}$).

The dependence of these probes on event centrality (determined using very forward CMS calorimetry [23]) and their azimuthal distributions (the event plane can be reconstructed using energy flow in CMS endcaps [24]) will carry information about the properties of the QGP at the LHC. The event rates for some channels, including hard jets, in a one month Pb + Pb run (assuming two weeks of data taking and not including reconstruction and trigger efficiencies) in the CMS acceptance estimated with PYTHIA 6.2 [25] are presented in table 1.

4. Jet reconstruction with calorimetry

The main difficulty of QCD jet recognition in heavy ion collisions arises from the "false" jet background – transverse energy fluctuations coming from the high event multiplicity. In the CMS heavy ion programme, the sliding window-type jet finding algorithm has been developed [5, 3] to search for "jet-like" clusters above the average energy and to subtract the η-dependent background from the underlying event. Figure 1 shows the linear correlation between reconstructed (full GEANT-based simulation) and generated (PYTHIA 6.2) transverse energies of jets with cone radius $R = 0.5$ in Pb + Pb ($dN^\pm(y = 0)/dy = 5000$) and pp events. The purity of jet reconstruction, defined as the number of events with a true QCD jet divided by the number of events with reconstructed jets, becomes ~ 1 at 100 GeV (figure 2). Note also that the fine angular resolution, $\Delta\varphi \sim \Delta\eta \sim 0.3$ at $E_T^{\rm MCjet} = 100$ GeV, less than the azimuthal size of a calorimeter tower. Further development of the jet reconstruction algorithm using tracker information for the jet energy correction is under way.

5. Charged particle tracking in jets

Track finding in heavy ion collisions is difficult due to the large number of tracks in an event. In addition to the primary tracks, the CMS tracker is occupied by secondaries produced by interactions with the detector material. The CMS track reconstruction algorithm, originally developed for pp collisions, is based on Kalman Filtering and includes seed generation, track propagation, trajectory updating and smoothing. In order to reduce the combinatorial background during track seeding in heavy ion collisions, the modified track finder also includes primary vertex finding and restriction of the vertex region [26]. The requirement for tracks to leave the tracker trough the outermost layer leads to a minimum transverse momentum cutoff of $p_T > 1$ GeV/c for the track to be considered reconstructable. Given this constraint, the track

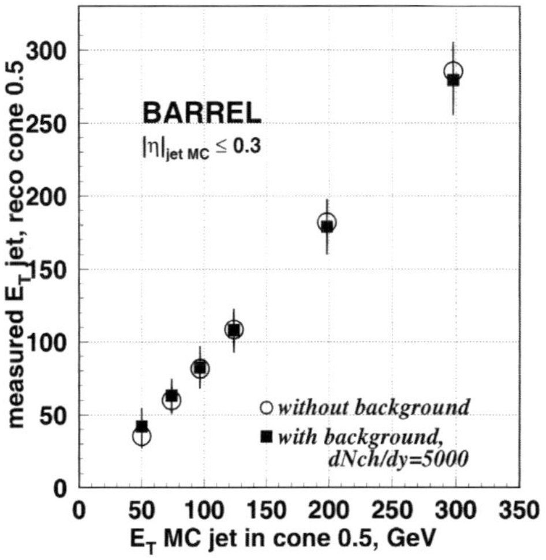

Figure 1. Correlation between reconstructed and generated jet transverse energies.

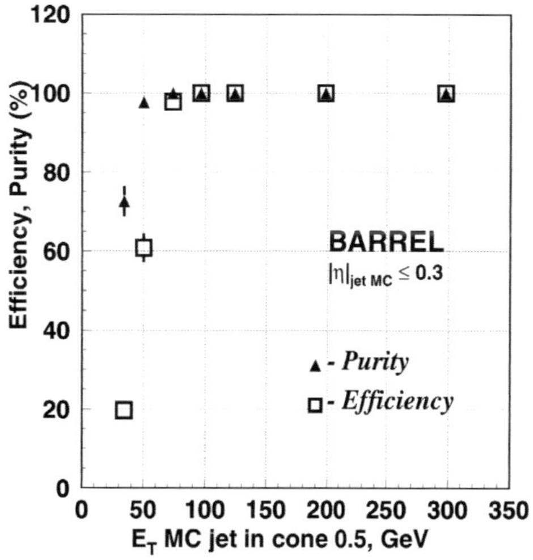

Figure 2. Purity and efficiency of jet reconstruction versus generated jet energy.

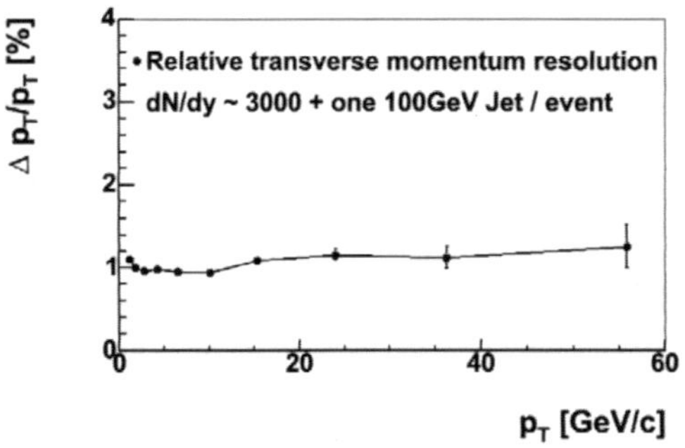

Figure 3. Relative transverse momentum resolution of tracks inside a 100 GeV jet.

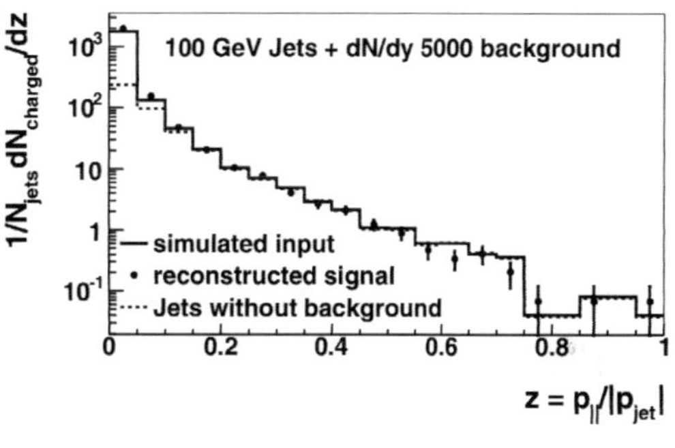

Figure 4. Distribution over momentum fraction z along the 100 GeV jet axis.

finder gives about $\approx 80\%$ reconstruction efficiency and low fake rate even at high track densities of $dN^{\pm}(y=0)/dy = 3000-5000$. The momentum resolution is less than 2% for $p_T < 100$ GeV/c (figure 3). As an example, the reconstructed jet fragmentation function for $E_T^{\rm jet} = 100$ GeV using CMS tracker is shown in Figure figure 4.

6. Monte-Carlo tools to simulate jet quenching

In most available event generators for LHC energies such important medium-induced effects such as jet quenching and elliptic flow are not included or are implemented poorly. Thus, in order to test the sensitivity of LHC observables to QGP formation, and to study the corresponding experimental capabilities of real detectors, the creation of adequate fast Monte-Carlo tools is necessary. One such tool developed recently to simulate rescattering and medium-induced partonic energy loss, is the fast event generator PYQUEN [27], implemented to modify a standard PYTHIA [25] jet event. The code HYDJET merges a fast generator of flow effects [28] with PYQUEN by simulating full heavy ion events as a superposition of soft, hydro-type state and hard multi-jets. These tools are very useful for detailed testing of the CMS capability to observe various probes of jet quenching and preparations for the heavy ion physics part of CMS Technical Design Report.

7. Conclusions

With its large acceptance, nearly hermetic fine granularity hadronic and electromagnetic calorimetry, and good muon and tracking systems, CMS is an excellent device for the study of medium-induced energy loss by light and heavy quarks ("jet quenching") at the LHC. Adequate jet reconstruction and high-p_T particle tracking with CMS in the high multiplicity environment are possible. Significant progress in development of Monte-Carlo tools to simulate jet quenching in CMS has been achieved.

Acknowledgments

The author wishes to express his gratitude to the members of CMS Collaboration, especially to Daniel Denegri, Christof Roland, Sergey Petrushanko, Ludmila Sarycheva, Alexander Snigirev, Constantin Teplov, Irina Vardanyan, Ramona Vogt and Boleslaw Wyslouch, for support and useful discussion. The author thanks the organizers of ICPAQGP 2005 for the warm welcome and hospitality. I also gratefully acknowledge partial support from Russian Foundation for Basic Research (grant N 04-02-16333).

References

[1] Baier R, Schiff D and Zakharov B G 2000 *Annual Rev. Nucl. Part. Sci.* **50** 37
[2] Wang X-N 2004 *Phys. Lett.* B **579** 299
[3] Accardi A.*et al* 2003 *Preprint* hep-ph/0310274
[4] *CMS Collaboration Technical Proposal* 1994 CERN/LHCC 94-38
[5] Baur G *et al Heavy Ion Physics Programme in CMS* 2000 CMS Note 2000/060
[6] *CMS HCAL Technical Design Report* 1997 CERN/LHCC 97-31
[7] *CMS ECAL Technical Design Report* 1997 CERN/LHCC 97-32
[8] *CMS MUON Technical Design Report* 1997 CERN/LHCC 97-33
[9] *CMS Tracker Technical Design Report* 1998 CERN/LHCC 98-6
[10] Lokhtin I P and Snigirev A M 2000 *Eur. J. Phys.* C **16** 527
[11] Lokhtin I P and Snigirev A M 2003 *Phys. Lett.* B **567** 39
[12] Lokhtin I P, Sarycheva L I, Snigirev A M and Teplov K Yu 2004 *Eur. J. Phys.* C **37** 465
[13] Wang X-N, Huang Z and Sarcevic I 1996 *Phys. Rev. Lett.* **231** 77
[14] Kodolova O L *et al* 1999 CMS Note 1999/063
[15] Kartvelishvili V, Kvatadze R and Shanidze R 1995 *Phys. Lett.* B **356** 589
[16] Srivastava D K, Gale C and Awes T C *Phys. Rev.* C 2003 **67** 054904
[17] Lokhtin I P, Sherstnev A V and Snigirev A M 2003 *Phys. Lett.* B **599** 260
[18] Lin Z and Vogt R 1999 *Nucl. Phys.* B **544** 339
[19] Lokhtin I P and Snigirev A M 2001 *J. Phys. G: Nucl. Phys.* **27** 2365
[20] Gyulassy M and Wang X-N 1992 *Phys. Lett.* B **68** 1480
[21] Lokhtin I P, Shmatov S V and Zarubin P I 2002 *Preprint* hep-ph/0212055
[22] Lokhtin I P, Snigirev A M and Sarycheva L I 2004 *Eur. J. Phys.* C **36** 375
[23] Damgov I *et al* 2001 CMS Note 2001/055
[24] Lokhtin I P, Petrushanko S V, Sarycheva L I and Snigirev A M 2003 CMS Note 2003/019
[25] Sjostrand T 2001 *Comp. Phys. Com.* **135** 238
[26] Roland C 2003 CMS Rapid Note 2003/003
[27] Lokhtin I P and Snigirev A M 2004 *Preprint* hep-ph/0406038
[28] Lokhtin I P and Snigirev A M 2003 *Preprint* hep-ph/0312204

Institute of Physics Publishing
doi:10.1088/1742-6596/50/1/054

Journal of Physics: Conference Series **50** (2006) 389–392
5th International Conference on Physics and Astrophysics of Quark Gluon Plasma

Probing dense baryonic matter with dileptons

Joachim Stroth, for the HADES and CBM collaborations

Institut für Kernphysik, Johann Wolfgang Goethe-Universität Frankfurt,
Max von Laue Strasse 1, 60438 Frankfurt, Germany

E-mail: j.stroth@gsi.de

Abstract. The role of dilepton spectroscopy in the exploration of exotic phases of nuclear matter is discussed. Special emphasis is put on the region of moderate beam energies where comparatively long-lived states of compressed matter are created. The opportunities at the future FAIR facility are outlined.

1. Introduction

At zero temperature and in absence of strong gravitational forces nuclear matter saturates at the nuclear ground state density of 0.16 fm^{-3}. Transitions to different phases of nuclear matter can be induced in collision experiments with heavy ions in the nuclear overlap region while the matter is heated up and compressed. During the liquid to gas phase transition, the relevant microscopic degrees of freedom are nucleons [1]. The transition temperature amounts to about 6-8 MeV [2] and consequently the mean thermal energy of a constituent baryon is well below the threshold for inelastic processes (i.e. the pion mass). The transition from a hadron gas to a deconfined phase of quarks and gluons occurs at a temperature of $\simeq$ 170 MeV where thermal excitation of pionic modes is copious and the confining potential substantially screened. It is characterized by a dramatic change of the microscopic degrees of freedom from hadrons to quarks and gluons [3]. This phase transition region is accessible with ultra-relativistic beam energies and currently investigated at the RHIC in the region of small net-baryon density. A phase transition to a deconfined color super-conducting phase is conjectured for the region of high net-baryon density and very low temperatures [4]. The exploration of the phase boundary and the search for an onset of confinement towards this region of the phase diagram is the goal of the experiments described here.

It has been argued that the hadron gas phase formed in heavy ion collisions can be subdivided into a region where copious inelastic collisions are rapidly changing its chemical composition and a region where the chemical composition is changed only by decay processes of the short-lived resonances [5]. The composition of the hadron gas strongly depends on temperature and density: at low temperature (or beam energy) it is dominated by the primordial nucleons, at high temperature the meson density exceeds the baryon density by factors and the baryon to anti-baryon ratio approaches unity [6]. There is sound theoretical understanding as well as experimental indications that the properties of the constituent hadrons are substantially modified in this phase [7]. The relation of these medium-modifications to a possible onset of deconfinement and a partial restoration of the chiral symmetry is a central question of modern nuclear physics. An experimental challenge is the investigation of the microscopic properties of such a resonance gas. As a matter of fact, the direct observation of resonances, or more general of short-lived hadronic states, formed in the hadron gas can only be achieved by a complete detection of their decay products. However, final state interactions of the decay products is of concern in case they interact strongly with the surrounding medium. Hadrons, which can decay into a purely leptonic final state are therefore favorable probes for hot and dense nuclear matter. Vector mesons are in particular suited as they can exclusively decay into two charged leptons. In the following section the importance of medium modifications in dielectron spectroscopy is discussed.

2. The HADES spectrometer

The HADES spectrometer, which is operational at GSI since 2002, is optimized to detect electron pairs from vector meson decay in heavy ion reactions at energies up to 2 AGeV [8]. To cope with the extremely small production probability of such pairs – i.e. of order 10^{-6} per central heavy ion collision for the lightest vector mesons ρ or ω – the detector operates at trigger rates above 10 kHz and achieves high pair acceptance of around 40%. Moreover, the spectrometer uses a second level trigger system to reject before mass storage those events, where no electron candidate was found by the on-line image processing units. To prepare such a decision, information from the ring imaging Cherenkov detector, pre-shower and time-of-flight detectors are evaluated and the respective hit information is matched between the detectors. The experimental program of HADES was started investigating $C + C$ collisions at 1 AGeV and 2 AGeV. In two production runs 650 and 220 million events where taken, respectively. A $Ca+Ca$ run at 2 AGeV is planned for this year. In parallel to the heavy ion programme, the HADES collaboration pursues an experimental programme focusing on electron pair production in elementary reaction using d, p and π beams. These measurements will provide further information to better understand the role of meson baryon coupling in the electron pair emission out of nuclear fireballs.

A first preliminary result of the HADES collaboration was presented on the INPC 2004, The the electron pair yield measured in $C + C$ collisions at 2 AGeV [9] is shown in the right panel of Fig. 1. Since the outer tracking system was not fully in place at the time of data taking, the analysis was carried out using position information behind the magnetic field from hits in the time-of-flight and pre-shower detectors only. In this way, a phi symmetric response of the spectrometer was achieved at the price of a reduced momentum resolution of $\approx 8\%$ at 500 MeV/c and an increased contribution from misidentified tracks. Electron identification was accomplished by matching identified rings in the RICH with track segments reconstructed in the drift chambers before the magnetic field. To further improve the reconstruction purity, these candidates where matched with hits in the detector systems behind the field region and by applying momentum dependent velocity conditions. The remaining contribution of misidentified tracks for this analysis amounts to about 2%. In Fig.1 the experimental data is shown together with a result obtained by analyzing the output of a UrQMD calculation, after it was propagated in the experimental setup using GEANT and corrections to simulate the detector response had been applied. Each of the two spectra is normalized to the number of reactions. For the UrQMD events, the experimental trigger condition, where a minimum of 4 active time-of-flight detector cells were required, was applied also. The systematic error due to the normalization procedure is indicated by the shaded arrow band. An additional systematic error is quoted for the the low-mass region and is the result of an imperfect modeling of the experimental detector response in the simulation. In the UrQMD simulation, only electron pairs from π^0 and η decay after "freeze-out" are treated. The results suggests that there is little room for substantial pair enhancement in the mass region between π Dalitz and vector meson decay, however, more substantial conclusions can only be drawn once the final result is available.

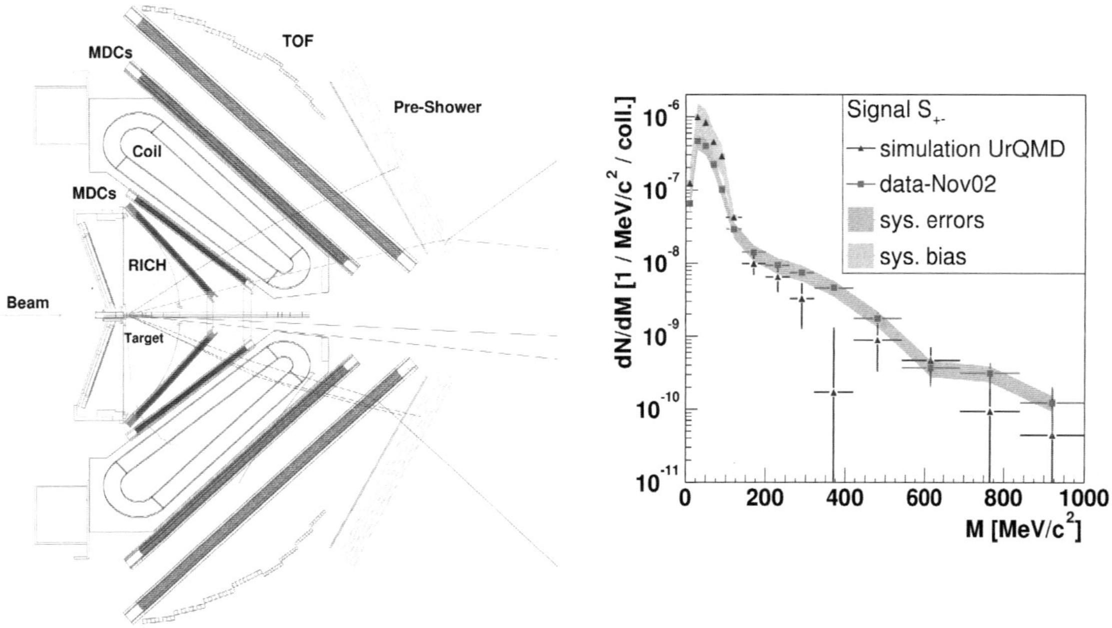

Figure 1. (**left panel**) Lateral cross section through the HADES spectrometer. Momentum measurement for charged particles is achieved by tracking the particles in front of and behind a toroidal field generated by six super-conducting coils arranged around the beam axis. In total 4 planes of low-mass multiwire drift chambers (MDC) allow to determine the particle trajectory before entering and after leaving the high-field region with a maximum field strength of 1.4 Tesla. The invariant mass of electron pairs can thus be reconstructed with a resolution of around 1% in the low mass vector meson region. For on- and off-line electron identification a ring imaging cherenkov detector with a solid CsJ photo cathode (RICH) is used. It is supplemented by a electromagnetic pre shower detector in the forward hemisphere to further discriminate against fast pions. General particle identification is accomplished by a time-of-flight system (TOF). The tracks shown stem from an omega decay in the target region as simulated in the GEANT framework. (**right panel**) Experimental (circles) and simulated (triangles) invariant mass distribution of electron pairs from $C + C$ collisions at 2 AGeV. The colored bands indicate systematical uncertainties, the error bars of the data points refer to statistical uncertainties.

3. Dilepton spectroscopy at FAIR

In the comming years HADES will fully exploit the possibilities of dielectron measurements in heavy ion collisions up to 2 AGeV. With the upcoming new facility FAIR, a new possibility to systematically study the microscopic properties of dense baryonic matter up to beam energies of 35 AGeV will open. Measurements will be possible in an energy range, which was not used for such studies before and a full excitation function for electron pair production up to the lowest SPS energies will be within reach. According to the current FAIR project plans the 100 Tm will be operational a few years before the 300 Tm ring will deliver heavy ion beams up to energies of 35 AGeV. Monte Carlo studies have shown, that the HADES spectrometer is able to run also at energies of 8 AGeV and with medium heavy nuclei.The then increasing boost in the laboratory frame of particles moves the accepted phase space to the backward rapidity region and thus the apparent particle multiplicity in the spectrometer at tolerable values.

For the higher beam energies, measurements will be conducted with the new CBM spectrometer [10]. This multi-purpose detector, which is described elsewhere in this volume,

Figure 2. Layout of the experimental setups forseen for dense matter physics at the future accelerator laboratory FAIR. Behind the HADES spectrometer the future CBM experiment is seen. It combines a compact tracking sation in a strong magnetic dipole field with Cherenkov, transition radiation, resistive plate detectors and a calorimeter (see contribution to this volume).

is designed to accomplish this task using a completely different approach than HADES. The concept is to perform tracking of all charged particles emitted from the reaction zone directly behind the target in a compact magnetic field using silicon pixel/micro-strip detectors. Detectors needed for particle identification will follow behind. For electron identification a RICH detector and several planes of transition radiation detectors are foreseen. At these higher energies, efficient π rejection is a crucial issue. To achieve this the active (track by track) recognition of background events from π^0 decay is mandatory. Since the electron pairs from such decays are opened up in the magnetic field before reaching the PID detectors, these tracks will have to be rejected by means of their topology. Hence, efficient tracking of all charged tracks with high resolution has to be achieved. The optimal design of a silicon tracking system for this task is currently a matter of detailed simulation studies. With the high rate capability and a momentum resolution of $\simeq 1\%$ at 1 GeV/c momentum, systematic investigation will be possible. The option to run CBM as a muon spectrometer is currently also investigated.

4. Summary

An experimental campaign to systematically investigate the microscopic properties of hadrons in dense nuclear matter with penetrating probes has been started at GSI in 2002 with the going-online of the HADES spectrometer. With the future FAIR facility such experiments can be conducted up to beam energies of 35 AGeV. The planned CBM experiment will address electron pairs, then extending the measurements up to the J/Ψ mesons, as one essential observable but also concentrate on open charm production and other observables accessible through hadronic final states.

References

[1] S. Das Gupta, A. Z. Mekjian and M. B. Tsang, arXiv:nucl-th/0009033.
[2] W. Trautmann, Nucl. Phys. A **752** (2005) 407.
[3] E. Laermann and O. Philipsen, Ann. Rev. Nucl. Part. Sci. **53** (2003) 163.
[4] M. G. Alford, K. Rajagopal and F. Wilczek, Nucl. Phys. B **537** (1999) 443.
[5] P. Braun-Munzinger, K. Redlich and J. Stachel, arXiv:nucl-th/0304013.
[6] R. Stock, J. Phys. G **30** (2004) S633.
[7] R. Rapp and J. Wambach, Adv. Nucl. Phys. **25** (2000) 1.
[8] C. Garabatos [HADES Collaboration], Nucl. Phys. Proc. Suppl. **61B** (1998) 607.
[9] T. Eberl *et al.* [HADES Collaboration], Nucl. Phys. A in press (2005).
[10] C. Hohne [CBM Collaboration], Nucl. Phys. A **749** (2005) 141.

Institute of Physics Publishing
doi:10.1088/1742-6596/50/1/055
Journal of Physics: Conference Series **50** (2006) 393–396
5th International Conference on Physics and Astrophysics of Quark Gluon Plasma

Dielectron physics with ALICE Transition Radiation Detector

Prashant Shukla (for the ALICE TRD Collaboration)
Physikalisches Institut der Universität Heidelberg, Philosophenweg 12, 69120 Heidelberg, Germany
E-mail: shukla@physi.uni-heidelberg.de

Abstract. The Transition Radiation Detector (TRD) of ALICE will serve to identify and track electrons in the central region. In this contribution, after a brief introduction of TRD, we review the detector performance concerning electron/pion identification as well as J/ψ and Υ detection.

1. Introduction

The ALICE experiment will study the collision process of lead nuclei at a center of mass energy per nucleon pair of 5.5 TeV at the CERN Large Hadron Collider (LHC). These collisions will create very high energy density well above the estimated threshold of forming Quark Gluon Plasma (QGP). One of the most interesting probes of QGP are heavy vector resonances $J/\Psi, \Psi^{'}, \Upsilon, \Upsilon^{'}, \Upsilon''$ through their leptonic decays. The complexity of heavy quark physics at LHC energies calls for comprehensive measurements of heavy quark and quarkonia production in pp, pA and AA collisions. The Transition Radiation Detector (TRD) [1, 2] of ALICE is placed in the Central Barrel. Via the dielectron channel, the production of light and heavy vector mesons as well as the continuum will be measured. The single electron channel will measure the semi-leptonic decays of hadrons with open charm (D) and open beauty (B) using the displaced vertex information provided by the ITS. The correlated production of $D\bar{D}$ and $B\bar{B}$ can be studied via coincidences of electrons in the central barrel and muons in the forward muon arm. The task of TRD is to identify electrons from the dominant background of pions. In addition, it will serve as a trigger for electrons with momenta higher than 2 GeV/c.

2. Transition Radiation Detector

The TRD is placed between Time Projection Chamber(TPC) and Time of Flight (TOF) and covers a pseudorapidity range $|\eta| \leq 0.9$. The TRD is segmented into 18 sectors in azimuth, each subdivided into 5 sections along the beam axis and six layers of chambers in the radial direction. This adds up to 540 detector modules. The largest chamber dimension is 120 x 159 cm^2. A schematic view of a detector module and its principle is shown in Fig.1. Each module comprises a radiator (4.8 cm, fiber/foam) followed by an ionization and drift gap (3 cm) and a multiwire proportional counter (0.7 cm) filled with Xe/CO_2(85:15). When an electron passes through the radiator it generates Transition Radiation (TR). The electron while traversing the gas produces ionization as do pions. Both the ionization electrons and those from the absorption of the TR photons drift towards the amplification region and produce avalanches around anode wires, which induces signal on the cathode pads. The pad signals are sampled in 20 time bins.

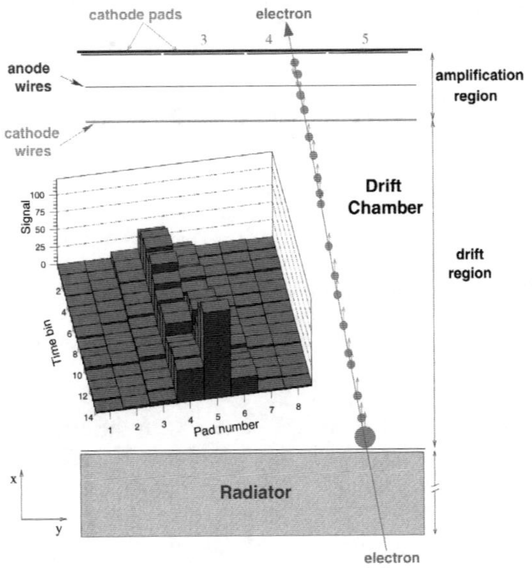

Figure 1. TRD module and its principle.

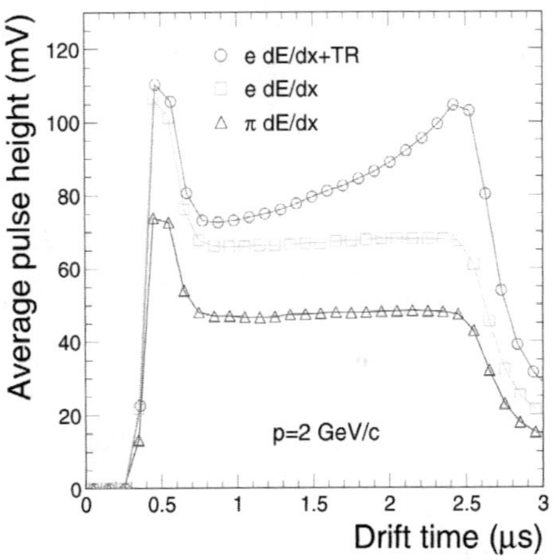

Figure 2. A pulse height spectrum from TRD module.

Figure 2 shows average pulse height distributions measured in the TRD chamber as a function of drift time. The peak at early drift times is generated by the charge deposited in the amplification region. While the pion spectrum is rather flat at larger drift times, the electron spectrum is peaked at late times since larger energy is deposited towards longer drift times corresponding to TR photons absorbed predominantly at the entrance of the chamber.

An online trigger is required to tag events, with high p_t electrons. The 20 time samples of the ADC are fed into a digital chip for zero suppression, gain calibration and tail cancellation of the signal. The data are subsequently processed by the local tracking unit (LTU) on the same chip, which calculates the hit position in each time bin and builds local tracklets. An electron likelihood factor based on integrated charge is associated to each tracklet. The information from the different LTUs is subsequently fed into the global tracking unit (GTU). The GTU associates tracklets to tracks and fits the track parameters and applies a cut in the transverse momentum. The track electron likelihood function is derived from the tracklet likelihood information. The performance of this online trigger was extensively studied in simulations [2].

3. Electron/pion identification

The pion efficiency is calculated for different electron efficiencies by the likelihood method using measured data. The required pion rejection factor for quarkonia measurement should be 100 at 90 % electron efficiency. Figure 3 shows the measured and simulated pion efficiency as a function of number of layers. The performance of six layers fulfills the requirement on pion rejection.

Figure 4 shows the pion efficiency as a function of momentum. The filled triangles (L_Q) are obtained by likelihood on measured integrated charge. The hollow triangles are simulations where the foam/fiber radiator is modelled by a stack of variable number of foils. Filled circles are obtained by bidimensional likelihood (L_{QX}), for which the distribution of the time bin of the maximum amplitude is used together with the integrated charge [3]. An improvement of the

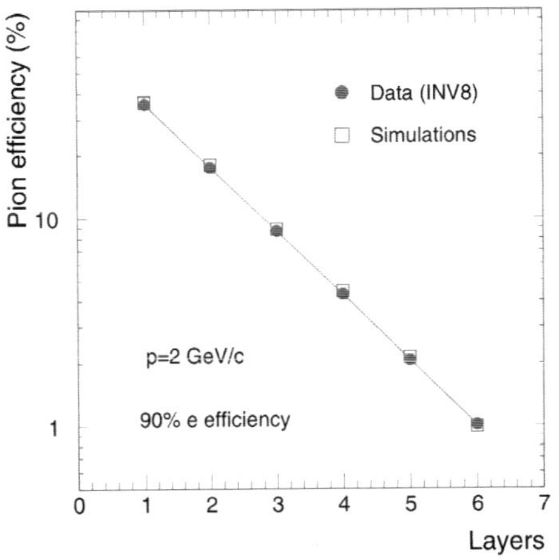

Figure 3. Pion efficiency as a function of number of layers.

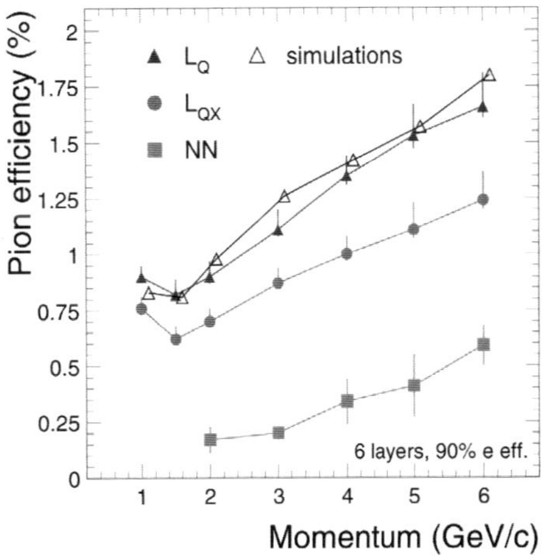

Figure 4. Pion efficiency as a function of momentum.

pion rejection by a factor up to 1.3 is achieved with this method. Solid squares are obtained with a neural network (NN) approach which further improves the pion rejection factor up to 4.5 times. The results presented above are for the ideal case of isolated tracks. However, for real events in ALICE a degradation of the performance is expected as a function of multiplicity [2].

4. Quarkonia signal $J/\Psi, \Upsilon$

The position resolution from one TRD chamber is about 400 μm in ϕ-direction. The combined ITS-TPC-TRD momentum resolution dp_t/p_t at $p_t = 1$ GeV/c and 5 GeV/c has been shown to be 1% and 1.8 %, respectively [4]. Figure 5 shows a dielectron invariant mass simulation with AliRoot [5]. Charged pions are generated with a parameterized HIJING model for a rapidity density dN/dy = 2000. The signals in the dielectron channel amount to 0.03 J/Ψ and 4×10^{-4} Υ. The combinatorial background is due to the semielectronic deacys from 115 $c\bar{c}$ and 5 $b\bar{b}$ included per event. The solid line is derived by TPC information only, which contains substantial amount of misidentified pions. The dashed line includes TRD which reduces the combinatorial background by more than an order of magnitude.

The signal to background ratio, S/B as function of a single track p_t-cut is shown in Fig. 6. The background from semileptonic decays of D and B mesons as well as misidentified pions are concentrated at lower transverse momentum p_t. The S/B can therefore be optimized by a judicious choice of such a cut. Here, the two curves TPC2 and TPC3 represent two different pion efficiencies based on different assumed TPC pion rejection. The combined TRD-TPC mass resolution for J/Ψ and Υ is about $\sigma^M_{J/\Psi} \sim 27$ MeV/c^2 and $\sigma^M_{\Upsilon} \sim 75$ MeV/c^2, respectively. These values indicate the Υ-family to be well resolved.

5. Conclusions

The ALICE TRD is well suited to study quarkonia production in the high multiplicity environment of Pb+Pb collisions at LHC. Test measurements have shown that the required electron/pion separation can be achieved. At the trigger level, rare heavy mesons like Υ can be selected. In conjunction with other detectors in the central barrel, the signal from the Υ family

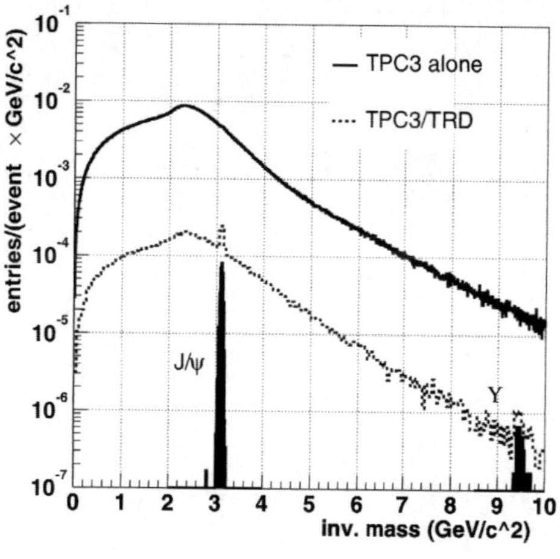

Figure 5. Dielectron mass spectrum.

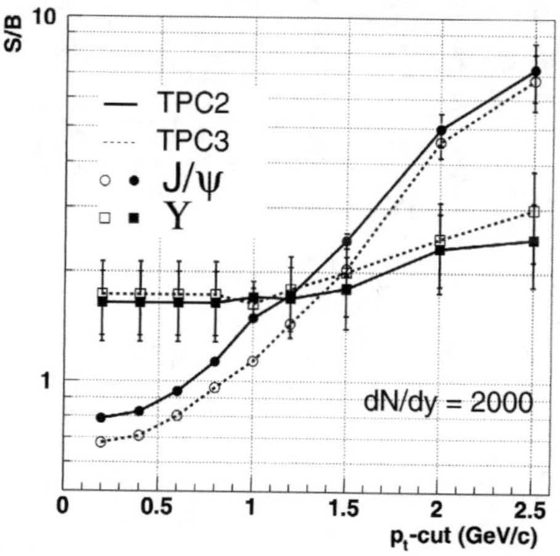

Figure 6. The signal to background ratio.

can be observed with a mass resolution of 1 %.

ALICE TRD Collaboration

C. Adler[1], A. Andronic[2], V. Angelov[3], H. Appelshäuser[4], S. Bablok[11], C. Baumann[5], R. Bailhache[2], C. Blume[4], P. Braun-Munzinger[2], D. Bucher[5], O. Busch[2], V. Cătănescu[1,6], V. Chepurnov[7], S. Chernenko[7], P. Christakoglou[9], E.S. Conner[11], H. Daues[2], J. de Cuveland[3], D. Emschermann[1], O. Fateev[7], Y. Foka[2], S. Freuen[1], C. Garabatos[2], R. Glasow[5], D. Gottschalk[3], H. Gottschlag[5], J.F. Grosse-Oetringhaus[5], T. Gunji[8], M. Gutfleisch[3], H. Hamagaki[8], G. Hartung[10], J. Hehner[2], N. Heine[5], N. Herrmann[1], H. Höbbel[3], M. Hoppe[5], U. Kebschull[3], R. Keidel[11], E. Kislov[7], V. Kiworra[3], C. Klein-Bösing[5], E. Kofler[11], T. Krawutschke[10], V. Lindenstruth[3], C. Lippmann[2], T. Mahmoud[1], P. Malzacher[2], A. Marin[2], J. Mercado[1], D. Miskowiec[2], Y. Morino[8], K. Oyama[1], Yu. Panebratsev[7], V. Petracek[1], A. Petridis[9], M. Petrovici[6], S. Radomski[2], A. Radu[2,6], C. Reichling[3], K. Reygers[5], I. Rusanov[1], S. Saito[8], A. Sandoval[2], R. Santo[5], R. Schicker[1], R. Schneider[3], K. Schwarz[2], P. Shukla[1], R.S. Simon[2], L. Smykov[7], H.K. Soltveit[1], W. Sommer[4], J. Stachel[1], H. Stelzer[2], M.R. Stockmeier[1], G. Stoicea[6], H. Tilsner[3], G. Tsiledakis[2], M. Vassiliou[9], W. Verhoeven[5], B. Vulpescu[1], J.P. Wessels[5], A. Wilk[5], B. Windelband[1], Yu. Zanevsky[7], V. Yurevich[7].

[1]PI Universität Heidelberg, Germany, [2]GSI Darmstadt, Germany, [3]KIP Heidelberg, Germany, [4]IKF Universität Frankfurt am Main, Germany, [5]IKP Universität Münster, Germany, [6]NIPNE Bucharest, Romania, [7]JINR Dubna, Russia, [8]University of Tokyo, Japan, [9]University of Athens, Greece, [10]Fachhochschule Köln, Germany, [11]Fachhochschule Worms, Germany.

References

[1] ALICE Collaboration 1999 *A Transition Radiation Detector, Technical Proposal* Addendum 2, CERN/LHCC/1999-13
[2] ALICE Collaboration 2001 *Transition Radiation Detector, Technical Design Report* CERN/LHCC/2001-021
[3] Andronic A et al. 2004 *Nucl. Instr. Meth.* A **522** 40 (*Preprint* physics/0402131)
[4] Safarik K 2005 *Nucl. Phys. A* **749** 229
[5] Mahmoud T 2004 *Doctoral thesis: Development of readout chamber of the ALICE Transition Detector and Evaluation of its Physics Performance* (University of Heidelberg)

Institute of Physics Publishing Journal of Physics: Conference Series **50** (2006) 397–401
doi:10.1088/1742-6596/50/1/056 5th International Conference on Physics and Astrophysics of Quark Gluon Plasma

The Muon Spectrometer of the ALICE

Christian Finck for ALICE Muon Spectrometer collaboration
SUBATECH (EMN-UN-IN2P3), 4 rue Alfred Kastler BP20722, 44307 Nantes Cedex 3
E-mail: Christian.Finck@subatech.in2p3.fr

Abstract. The Muon Spectrometer of the ALICE experiment is a unique tool to study of the heavy quarks in pp, pA and AA collisions at LHC energies, via the muonic channel. The spectrometer consists of different absorbers, a dipole, five tracking stations and two trigger stations.

1. Introduction

The future Large Hadron Collider (LHC) at CERN will provide beam energies of 14TeV for p-p and 5.5TeV for Pb-Pb collisions. The ALICE experiment will be the only experiment at LHC dedicated to the heavy ion physics. The goal of the Muon spectrometer of ALICE is to study the open heavy flavor production and the quarkonia production (J/ψ, ψ' and $\Upsilon(1S)$, $\Upsilon(2S)$ and $\Upsilon(3S)$) via the muonic channel.

The muon spectrometer will be able to measure the quarkonia production in central Pb-Pb collisions at LHC energies, down to very low p_T, since low p_T quarkonia will be sensitive to medium effects like heavy-quark potential screening and/or statistical recombination. Since muons are passively identified by the absorber technique, a Lorentz boost is needed to be able to measure quarkonia at low p_T. The muon spectrometer will be a unique apparatus at LHC to measure charmonia production at $p_T \sim 0$ with a large rapidity range $-4.0 < \eta < -2.5$ [1].

The muon spectrometer consists of absorbers, a muon magnet, a muon filter (iron wall), a trigger system and a tracking system. The layout of the muon spectrometer is presented on Fig. 1 [2].

2. Absorbers

Absorbers reduce the initial flux of primary hadrons from nucleus-nucleus collisions by a factor 100, and they reduce the low energy particles flow produced in secondary interactions (mainly low energy electrons). The front absorber is the most critical component and it has been designed for minimizing the invariant mass resolution deterioration of the spectrometer due to straggling and multi-scattering. This imposes an upper limit of the amount of material ($\lambda_I \sim 10$), and requires that components with low Z are located close to the interaction point (IP), whereas high Z components are placed close to the spectrometer. Moreover, muons from hadronic weak decay are optimally suppressed by placing the front absorber as close as possible to the interaction point. The distance from the IP is, however, limited to 90 cm since physics performances of the ALICE central barrel should not be deteriorated by the presence of any absorber. The front absorber is the main contributor to the invariant mass resolution of the spectrometer in

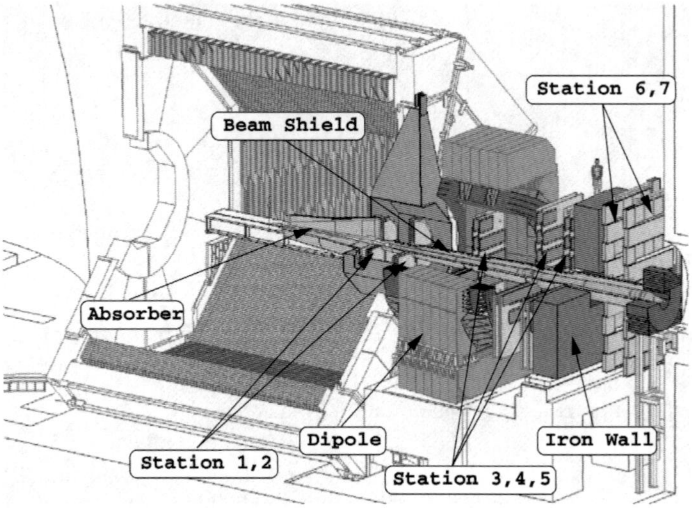

Figure 1. Layout of the muon spectrometer. The tracking system consists of stations 1 to 5 and trigger system of stations 6 and 7.

the Υ region, with a quadratic contribution equal to about 80 MeV/c^2. The absorber (shield) around the beam pipe is crucial to reduce the low energy background in the tracking and trigger chambers due to secondary interactions along the pipe. Finally, an 120 cm thick iron wall is located between the tracking stations (stations 1-5 in fig. 1) and the trigger stations (stations 6 and 7 in fig. 1) allows for reducing the low energy particle background in the trigger chambers. At present, the absorbers are under construction at CERN.

3. Muon Dipole Magnet

Muon momenta are determined by muon tracking in a magnetic field generated by a warm dipole of 820 tons, the largest warm dipole in the world with a nominal field of 0.7 T and a field integral along beam axis $\int |B| dz \sim 3$ Tm. The yoke was provided by the Dubna collaboration whereas the two coils were manufactured by SigmaPhi company at Vannes (France). The magnetic field is directed in the horizontal plane perpendicular to the beam direction (x axis) defining a bending plane (zy plane) and a non bending plane (xz plane). The muon magnet has been assembled and has been tested in the cavern (point 2). It's presently dismounted and is being mounted in its nominal position. The procedure will last until end of June 2005. During summer the field's mapping will be performed.

4. Trigger system

Muon detection based on absorption technique allows for a very efficient triggering on high p_T muons. This is crucial in order to take advantage of the full luminosity of the heavy ion beams at LHC, taking into account that muon acquisition system is limited to an event rate of 1kHz. The trigger system (stations 6 and 7 in fig.1) consists of 4 planes of 18 Resistive Plate Chambers (RPC) each, located between 16 m and 17 m from the IP (just behind the iron wall) and operating in the streamer mode with a gas mixture of Ar, CH_2F_4, C_4H_{10} and SF_6 [3] [1]. Signals in individual strips of the RPC are treated by a dual threshold discriminator (ADULT) [4]. The time resolution of the 2 mm gap RPC equipped with the ADULT chip is in the order of 1 ns. Information from the 21,000 strips are locally processed by the so-called local cards [5]

[1] For the pp runs, an avalanche operating mode is being investigated to reduce the aging effect of RPC's

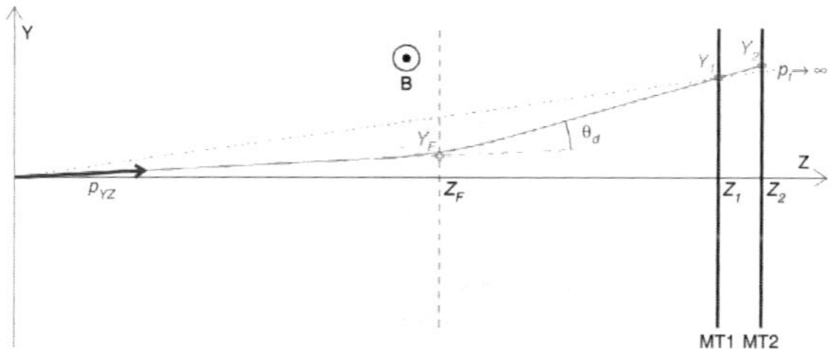

Figure 2. Tranverse momentum determination with trigger stations.

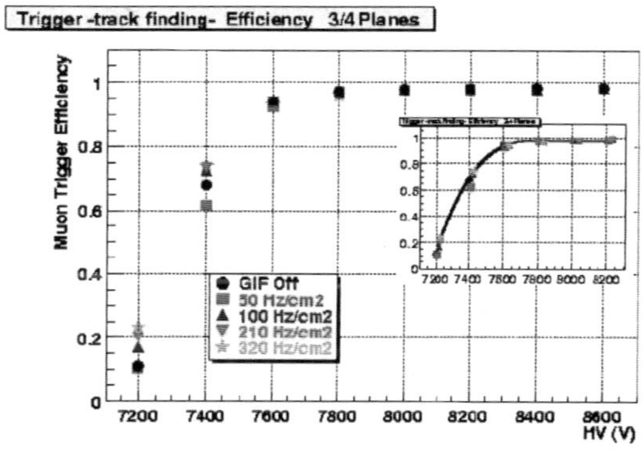

Figure 3. Reconstruction track efficiency of equipped RPC at GIF for different background rates.

(234 cards), determining roughly the transverse momentum of the muon track (see fig 2). The 234 local cards are dispatched in 16 crates, each equipped by a controller (called regional card). The global cards collect the full information from regional boards and determine the trigger condition of the event in less than 700 ns. The data from the regional crates and from the global card are steered to the readout card (Dimuon Arm Readout Controller). Different trigger types are possible: low p_T pairs (above 1 GeV/c for J/ψ studies), high p_T pairs (above 2 GeV/c for Υ studies), like sign muon pairs, single muons, etc... The muon trigger delivers information to the central trigger processor for the generation of the ALICE level 0 trigger. A full test of the RPC equipped with the front-end electronics and with the local boards has been performed at GIF in 2002 [6]. Reconstruction track efficiency was found to be higher than 98 % (see Fig. 3 [7]).

5. Tracking system

Muon transverse momenta in the bending plane ($|p_{zy}|$) are determined by tracking muons along the magnetic field. A momentum resolution of about 1% is needed to achieve the required resolution in the Υ invariant mass region ($\sim$ 100 MeV/c^2). The tracking system is made of 5 stations with two detection planes consisting of 5 mm gas gap drift multi-wire proportional chambers with segmented cathode plane (cathode pad chambers, CPC). The thickness of each chamber is below 3% radiation length. The first two stations are placed in front of the muon magnet at a distance of $\sim$5.4 m and $\sim$6.8 m respectively from the IP. They consist of four detection planes made of four CPC's each with a quadrant design. Stations 3, 4 and 5 are placed at a distance of $\sim$9.7 m (inside the muon magnet), $\sim$12.6 m and $\sim$14.2 m from the IP. A modular design was chosen for these stations, consisting of rectangular CPC called *slat* (36, 52 and 52 slats for stations 3, 4 and 5 respectively). Different pad densities are present in the CPC depending on the station and pseudo-rapidity positions, ranging from 4×6 mm^2 for pads closest to the beam-pipe in station 1, to 5×100 mm^2 for stations 3, 4 and 5 in the bending plane. An initial direction of the muon is determined from the track parameters of the muon and from the position of the IP, taking into account the multi-scattering in the absorber. In order to keep the invariant mass resolution in the Υ region below 100 MeV/c^2, a spatial resolution of 1 mm is required in the non-bending plane (x direction) and a resolution below 100 μm is required for the bending plane (y direction). Several campaigns of testbeam were performed at the PS and the SPS facilities (CERN), the spatial resolution in the bending plane of the tracking chamber is better than the required value (see Fig. 4), with an efficiency of 98 % [8].

The muon tracking system consists of 10^6 channels. The front-end card, so-called MANU

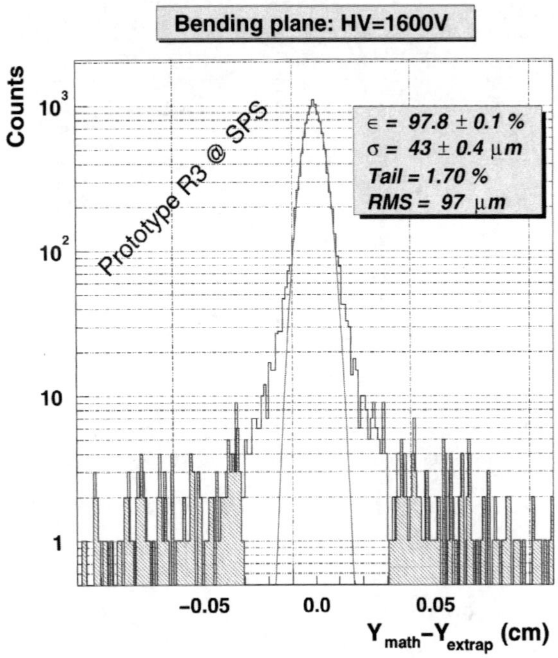

Figure 4. Spatial resolution in the bending plane of the CPC under muon beam at 100 GeV/c in SPS. The tail is the proportion of event outside the window ± 300 μm.

board (64 channels) [9] consist of 4 MANAS chips (pre-amplifier and shaper chip [10]), several ADC's and the controller chip (Muon Arm Readout Controller) [9]. The latter takes in charge the zero suppression and the communication at the DSP level. The noise level of the full electronics chain is around one thousand electrons. The data flow is gathered with the help of 100 front-end CROCUS board placed in 20 crates. Each crate has its concentrator board CROCUS [9] which communicates with the Alice - DAQ [11].

6. Conclusions and Perspectives

In summary, the muon spectrometer has entered in its construction phase. The passive elements are presently under installation (absorber and dipole). Most of the active elements are under production (RPC, CPC, front-end chips, controller boards) or under final design and/or test (readout boards). And the muon spectrometer should be ready in 2007 at the interaction point 2 of LHC for the first data taking.

[1] G. Martinez for ALICE - Muon spectrometer collaboration, in this conference.
[2] ALICE collaboration, CERN/LHCC 99-22 (1999).
[3] R. Arnaldi et al. for the ALICE collaboration, N.I.M. A 451 (2000) 462.
[4] R. Arnaldi et al. for the ALICE collaboration, N.I.M. A 457 (2001) 117.
[5] G. Blanchard, Ph. Crochet and P. Dupieux, PRR local trigger electronics, ALICE-EN-010 (2003)
[6] P. Dupieux and B. Forestier for the ALICE collaboration, IEEE-TNS 51 n3 (2004) 375.
[7] ALICE collaboration, CERN/LHCC 96-32 (1996).
[8] K. Boudjemline et al. for ALICE - Muon station 3, 4 and 5 collaboration, ALICE-INT 02-23 (2002); M.P. Comets et al. ALICE-INT 03-35 (2003); N. Willis et al., ALICE-INT 03-30 (2003).
[9] P. Courtat, B. Espagnon et al., ALICE-INT 04-026 (2004).
[10] P. Bhattacharya for ALICE - Muon spectrometer collaboration, in this conference.

Institute of Physics Publishing
doi:10.1088/1742-6596/50/1/057

Journal of Physics: Conference Series **50** (2006) 402–405
5th International Conference on Physics and Astrophysics of Quark Gluon Plasma

Statistical production of pentaquarks in high-energy nuclear collisions

Jørgen Randrup
Lawrence Berkeley National Laboratory, Berkeley, California 94720, USA
E-mail: Randrup@LBL.gov

Abstract. The production yield of exotic pentaquarks in high-energy nuclear collisions is estimated on the basis of the standard statistical description of the hadronic freeze-out.

1. Introduction

An increasing number of investigators are reporting evidence for the production of pentaquarks (for recent reviews of the experimental status, see Refs. [1, 2]). The term pentaquark refers to a novel type of baryon whose simplest Fock components contain four quarks and one antiquark, thus making their combinations of baryon number, electric charge, and strangeness fall outside the familiar baryon classfication scheme and hence rendering them exotic.

Because of their unique combination of attributes, the pentaquarks would be particularly useful as diagnostic tools in the exploration of the hadronization dynamics in high-energy nuclear collisions. It is therefore of interest to search for them in the data and, for such an undertaking, it would be useful to have a rough idea of their production rate.

Consequently, we present theoretical estimates of pentaquark production in high-energy nuclear collisions. For this purpose, we adopt a statistical framework which has proven to account rather well for the hadronic production yields (for a review, see Ref. [3]). On this basis, we seek to estimate the production rates of various pentaquarks of current interest.

2. Statistical model

The statistical model treats an ideal gas of hadronic species $\{i\}$ within a volume V. The partition function then factorizes, $\mathcal{Z} = \Pi_i \mathcal{Z}_i$ where a hadron of specie i has mass m_i, spin degeneracy g_i, baryon number B_i, charge Q_i, and strangeness S_i. Furthermore, the environment is characterized by the freeze-out temperature T and the three chemical potentials $\{\mu_B, \mu_Q, \mu_S\}$ which enter through the corresponding fugacity, $\lambda_i = \exp(-(B_i\mu_B + Q_i\mu_Q + S_i\mu_S)/T)$. The values used are illustrated in Fig. 1. The partition function for the specie i is then given by

$$\ln \mathcal{Z}_i(T, V, \{\mu\}) = \pm g_i \frac{VT^3}{2\pi^2} \sum_{n=1}^{\infty} \frac{(\pm\lambda_i)^n}{n^4} \left(\frac{nm_i}{T}\right)^2 K_2\left(\frac{nm_i}{T}\right) , \tag{1}$$

where we have exhibited the regular combination $x^2K(x) \to 2$ for $x \to 0$. The upper sign holds for bosons (*i.e.* mesons), while the lower sign holds for fermions (*i.e.* baryons) and thus yields

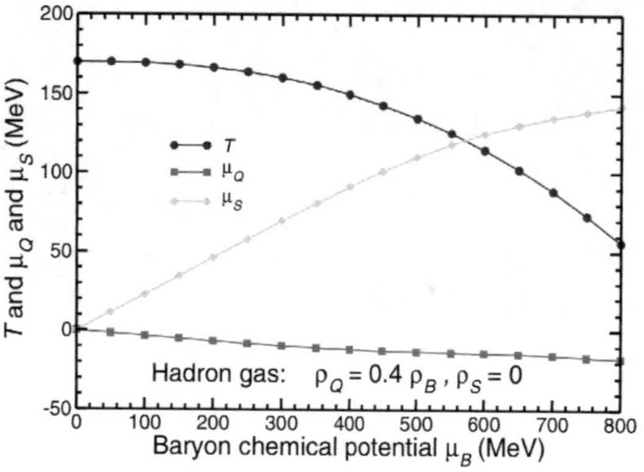

Figure 1. The adopted dependence of the temperature T and the chemical potentials for charge and strangeness, μ_Q and μ_S, on the specified baryon chemical potential μ_B. The temperature represents a fit to the freeze-out temperatures extracted from the various experiments [3], while the chemical potentials have been determined by demanding that $\langle Q\rangle$=0.4$\langle B\rangle$, as is appropriate for Au+Au, and $\langle S\rangle$=0 (which yields an increasing μ_S that compensates for the hyperons).

an alternating series; the classical result is obtained by retaining only the first term, $n = 1$. The corresponding spatial density of a given specie, $n_i = N_i/V$, then readily follows,

$$n_i(T,\{\mu\}) \;=\; \pm g_i \frac{T^3}{2\pi^2} \sum_{n=1}^{\infty} \frac{(\pm\lambda_i)^n}{n^3} \left(\frac{nm_i}{T}\right)^2 K_2\!\left(\frac{nm_i}{T}\right) \;\approx\; g_i \left(\frac{m_i T}{2\pi}\right)^{\frac{3}{2}} \lambda_i \,\mathrm{e}^{-m_i/T} \;, \tag{2}$$

where the last expression emerges in the large-mass limit, $m_i \gg T$. Figure 2 shows that quantum statistics is important only for pions, while the large-mass limit tends to be inadequate.

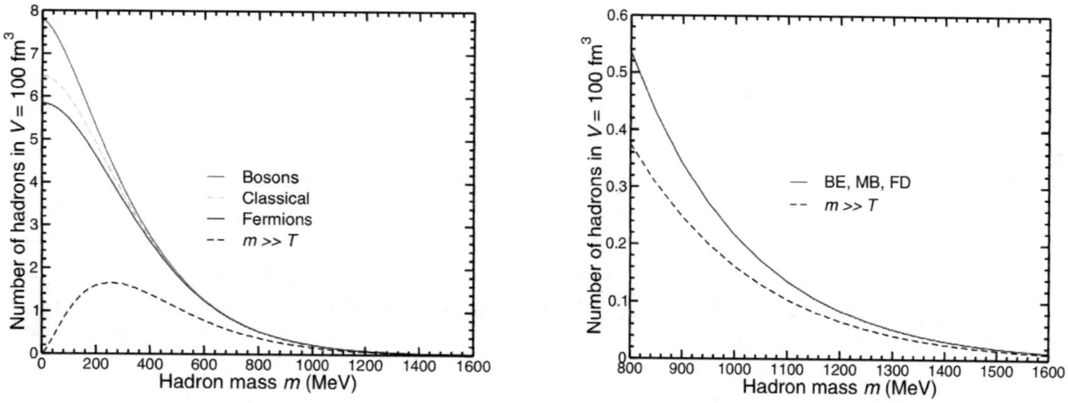

Figure 2. The average number of hadrons in a volume of V=100 fm^3 as a function of the hadron mass m at a temperature of T=170 MeV and vanishing chemical potentials. The quantum-statistical values are generally close to the classical result, except for pions, while the large-mass approximation is not quantitatively useful, being still about 20% too low for Θ^+.

3. Results

Most pentaquark research has focussed on the $\Theta^+(1530)$ for which quantitative predictions were made by Diakonov *et al.* [4]. Contrary to the familiar hyperons which have negative strangeness, Θ^+ has *positive* strangeness, S=+1, as well as positive electric charge, Q=+1. It follows that its leading Fock component is $|uudd\bar{s}\rangle$. Production of Θ^+ in high-energy nuclear collisions was first discussed in Ref. [5] where statistical estimates suggested that there might be about one Θ^+ per unit rapidity at central rapidity in a head-on Au+Au collision at the top RHIC energy (though only certain decay branches would be readily identifiable in the data analysis). More

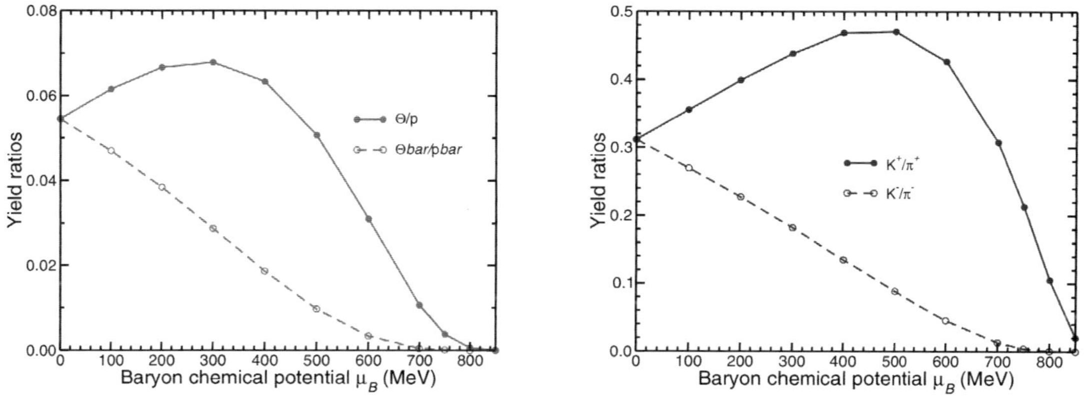

Figure 3. *Left:* The abundance of Θ^+ relative to the proton (solid) as well as the corresponding ratio for the antiparticles, $\bar{\Theta}$ to $\bar{p}$ (dashed), as functions of the baryon chemical potential μ_B. *Right:* The similar plots for K^+ relative to π^+ (solid) and K^- relative to π^- (dashed).

refined studies based on microscopic transport treatment were subsequently made by Liu and Ko [6] and the statistical studies were broadened by Letessier *et al.* [7].

Figure 3 shows the statistical yields of Θ^+ relative to that of the proton, as well as the ratio between the K^+ and π^+ yields. The behavior of these ratios is easy to understand: as μ_B is increased the value of μ_S also goes up, see Fig. 1, thus introducing a favorable bias for positive strangeness, a trend that continues until the ultimate decrease of the temperature disfavors the larger masses. Since no special strangeness suppression has been included, the $K^+ : \pi^+$ ratio exceeds the data. Therefore, in order to reduce the sensitivity of the results to the somewhat uncertain chemical conditions, we prefer to use double yield ratios for which the various attributes counterbalance. The result is then approximately independent of the fugacities and behave as $\sim \exp(-\Delta m/T)$, where Δm is the mismatch in masses. For the double ratio $\Theta^+ : p$ to $K^+ : \pi^+$ we have $\Delta m = 250$ MeV which thus dominates the dependence on μ_B, as borne out in Fig. 4. Indeed, the double ratio is approximately constant, until the value of μ_B has become so large that the temperature has dropped significantly (see Fig. 1).

The NA49 Collaboration recently reported the observation of a somewhat heavier exotic baryon, $\Sigma^{--}(1862)$, in proton-proton collisions at the CERN SPS [8]. It has both charge and strangeness equal to minus two which thus implies a minimal composition given by $|ddss\bar{u}\rangle$. Figure 4 includes the statistical result for the double ratio $\Sigma^{--} : \Lambda$ to $K^- : \pi^+$ for which $\Delta m = 392$ MeV. While exhibiting a behavior that is qualitatively similar to the double ratio for Θ^+, the larger value of Δm causes a reduction in the result by a factor of about two.

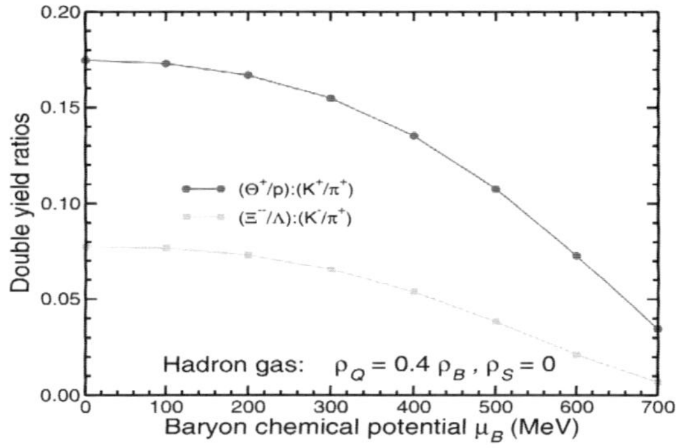

Figure 4. Double yield ratios for which all the fugacities approximately counterbalance: The ratio between the two yield ratios $\Theta^+ : p$ and $K^+ : \pi^+$ shown in Fig. 3 (upper curve), as well as $\Sigma^{--} : \Lambda$ divided by $K^- : \pi^+$ (lower curve), as functions of the baryon chemical potential μ_B. Their form is approximately given by $\sim \exp(-\Delta m/T(\mu_B))$.

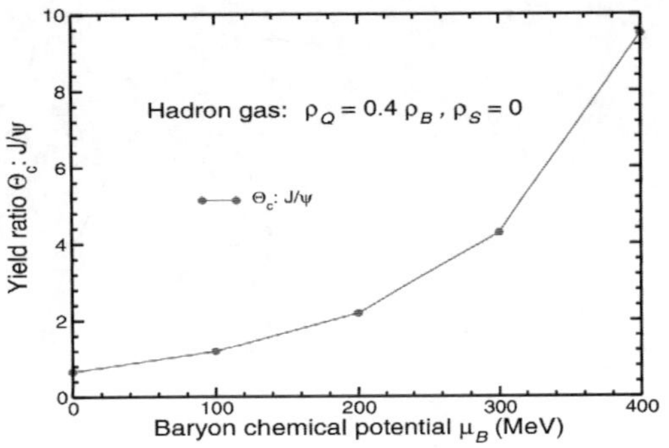

Figure 5. The yield of Θ_c relative to the yield of J/ψ as a function of μ_B.

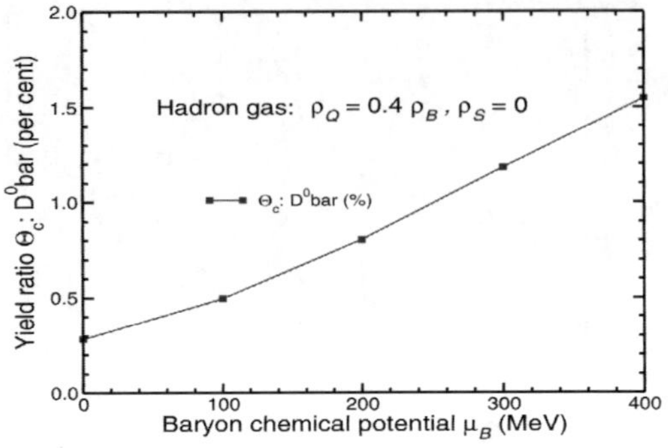

Figure 6. The yield of Θ_c relative to the yield of $\bar{D}^0$ (in percent) as a function of μ_B.

The H1 Collaboration at HERA has reported evidence for an anti-charmed pentaquark, $\Theta_c(3099)$ with C=−1, B=+1 and Q=S=0, implying a minimal composition of $|uudd\bar{c}\rangle$ [9]. Therefore, the Θ_c abundance in a baryon-free environment is determined primarily by its mass and it should thus be similar to that of the J/ψ, which is readily observable at both SPS and RHIC. Furthermore, in environments with $\mu_B > 0$ the Θ_c should be more abundant than the J/ψ, as shown in Fig. 5. The Θ_c yield may also be compared with that of the $\bar{D}^0$ which has the same charge, strangeness, and charm, as done in Fig. 6. At $\mu_B = 0$ its relative abundance is $\approx 0.3\%$, in accordance with the large-mass formula, and the ratio increases with μ_B.

4. Concluding remarks

Pentaquarks are of particular interest for high-energy heavy-ion physics, because they carry exotic combinations of attributes and may therefore be utilized as probes to provide novel information about the hadronization dynamics. Elementary statistical estimates of their production yield suggest that these hadrons should be detectable in currently studied high-energy nuclear collisions. Furthermore, high-energy nuclear collisions may provide additional or novel evidence regarding pentaquarks. Thus, the observation of pentaquarks that have already been reported would add independent evidence for their existence. More importantly, nuclear collisions may permit the production of pentaquark species, such as the $\bar{\Theta}$, that cannot be reached by the available elementary reactions, thus contributing novel data.

A pentaquark search that would establish an upper limit significantly below the statistical expectation would also be valuable: *Either* the pentaquark production is significantly suppressed in nuclear collisions which would constrain the dynamical mechanisms. *Or* their absence would constitute independent evidence *against* their existence, which would clearly be important. In conclusion, whatever the case may turn out to be, it seems worthwhile to undertake pentaquark searches in high-energy nuclear collisions.

References

[1] Hicks K 2004, `hep-ex/0412048`; Hicks K 2005, `hep-ex/0501018`
[2] Carman D S 2005, *Eur. Phys. J.* A **24S1** 15
[3] Braun-Munzinger P, Redlich K, Stachel J 2003, *Quark Gluon Plasma 3* (World Scientific); `nucl-th/0304013`
[4] Diakonov D, Petrov V, Polyakov M 1997, *Z. Phys.* A **359** 305
[5] Randrup J 2003, *Phys. Rev.* C **68** 31903
[6] Liu W and Ko C M 2003, *Phys. Rev.* C **68** 45203
[7] Letessier J *et al.* 2003, *Phys. Rev.* C **68** 61901
[8] Alt C *et al.* (NA49 Collaboration) 2004, *Phys. Rev. Lett.* **92** 42003
[9] Aktas A *et al.* (H1 Collaboration) 2004, *Phys. Lett.* B **588** 17

Institute of Physics Publishing Journal of Physics: Conference Series **50** (2006) 406–409
doi:10.1088/1742-6596/50/1/058 5th International Conference on Physics and Astrophysics of Quark Gluon Plasma

The Horn and the Thermal Model.

J. Cleymans[a], H. Oeschler[b], K. Redlich[c], S. Wheaton[a]

[a] UCT-CERN Research Centre and Department of Physics, University of Cape Town, Rondebosch 7701, Cape Town, South Africa
[b] Institut für Kernphysik, Technische Universität Darmstadt, D-64289 Darmstadt, Germany,
[c] Institute of Theoretical Physics, University of Wrocław, Pl-45204 Wrocław, Poland

Abstract. The recently discovered sharp peak in the K^+/π^+ ratio in relativistic heavy-ion collisions is discussed in the framework of the thermal model. In this model a rapid change is expected as the hadronic gas undergoes a transition from a baryon-dominated to a meson-dominated gas. The transition occurs at a temperature T = 140 MeV and baryon chemical potential μ_B = 410 MeV corresponding to an incident energy of $\sqrt{s_{NN}}$ = 8.2 GeV.

The thermal model has been extremely successful in bringing order to a very large number of experimental results on particle yields in relativistic heavy-ion collisions. The results for the temperature and baryon chemical potential have been found to be consistent with having $E/N = 1$ GeV from the lowest beam energies up to the highest ones. This is illustrated in Fig. 1 which combines results from SIS up to RHIC.
The NA49 Collaboration has recently performed a series of measurements of Pb-Pb collisions at 20, 30, 40, 80 and 158 AGeV beam energies. When these results are combined with measurements at lower beam energies from the AGS and SIS they reveal an unusually sharp variation with beam energy in the $\Lambda/\langle\pi\rangle$, with $\langle\pi\rangle \equiv 3/2(\pi^+ + \pi^-)$, and in the K^+/π^+ ratios. Such a strong variation with energy does not occur in pp collisions and therefore indicates a major difference in heavy-ion collisions. This transition has been referred to in Ref. [1] as the "horn". A strong variation with energy of the $\Lambda/\langle\pi\rangle$ ratio has been predicted on the basis of arguments put forward in [2]. In [3] another, less spectacular, possibility for the origin of the sharp maximum, namely as being due to the transition from a baryon-dominated to a meson-dominated hadronic gas has been suggested; the distinction being based on whether the entropy of the hadronic gas is dominated by baryons or by mesons. For this purpose various quantities along the freeze-out curve [4] as a function of $\sqrt{s_{NN}}$ have been studied in [3].

In the thermal model a steep rise at low energies and a subsequent flattening off leading to a mild maximum in the K^+/π^+ ratio, was predicted many years ago [6, 5, 8]. The sharpness of the observed peak therefore comes as a surprise. On the other hand, a sharp peak in the $\Lambda/\langle\pi\rangle$ ratio was predicted by the thermal model [8] and is in good agreement with the data. While the thermal model cannot explain the sharpness of the peak in the K^+/π^+ ratio, there are nevertheless several phenomena, giving rise to the rapid change, which warrant a closer look at the model. To get a better estimate of the thermal parameters we have analyzed the entropy density as a function of beam energy following the freeze-out curve given in [4] (see Fig. 2). There is a clear transition from a meson to a baryon-dominated hadronic gas at $\sqrt{s_{NN}}$ = 8.2 GeV. Above this value the entropy is carried mainly by mesonic degrees of freedom. It is remarkable that the entropy density divided by T^3 is constant over the entire freeze-out

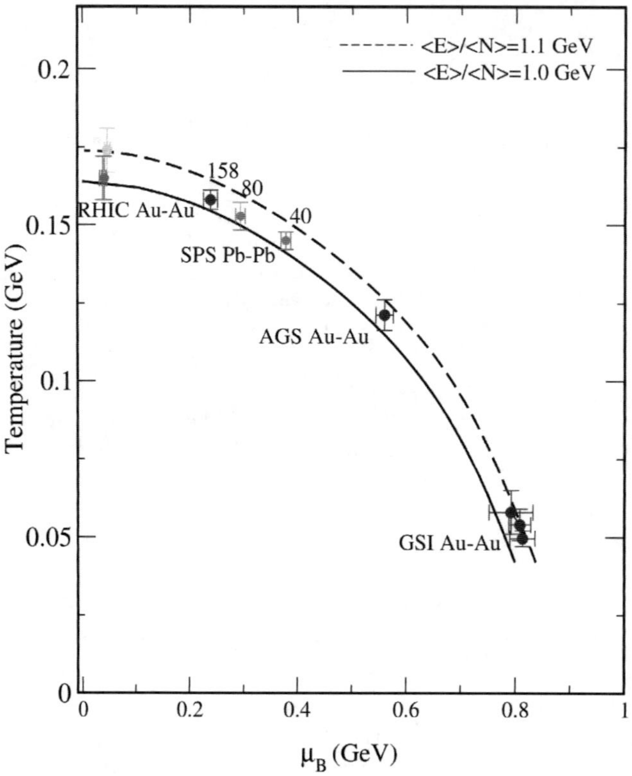

Figure 1. The chemical freeze-out points together with curves at fixed values of E/N.

curve, except for the low-energy, SIS, energy region. The line denoting the transition from a baryon-dominated to a meson-dominated hadron gas is shown in Fig. 2. This line crosses the freeze-out curve at a temperature of T = 140 MeV, when the baryon chemical potential equals μ_B = 410 MeV. The corresponding invariant energy is $\sqrt{s_{NN}}$ = 8.2 GeV. The strong decrease in the net baryon density is due to the fact that low energies are characterized by a very low multiplicity of mesons and, correspondingly, a very large baryon-to-meson ratio. As a consequence, the baryon chemical potential is also very large. As the beam energy is increased, meson production increases and the baryon chemical potential decreases. The number of strange baryons produced in heavy-ion collisions at different collision energies will follow the net baryon density since a large baryon chemical potential will also enhance the number of hyperons. As is well-known [9, 10], the thermal model description leads to a mild maximum in the K^+/π^+ ratio which does not reproduce the so-called "horn" observed by the NA49 collaboration [1]. The observed deviations at the highest SPS energy have been interpreted as a lack of full chemical equilibrium in the strangeness sector, leading to a strangeness suppression factor, γ_s, deviating from its equilibrium value by about thirty percent. Detailed fits using the thermal model in the region of the "horn" show rapid variations in γ_s [10] which do not lend themselves to any interpretation. There is no corresponding peak in the K^-/π^- ratio because the production of K^- is not tied to that of baryons. As the relative number of baryons decreases with increasing energy, there is no corresponding decrease in the number of K^- as is the case with K^+ as these

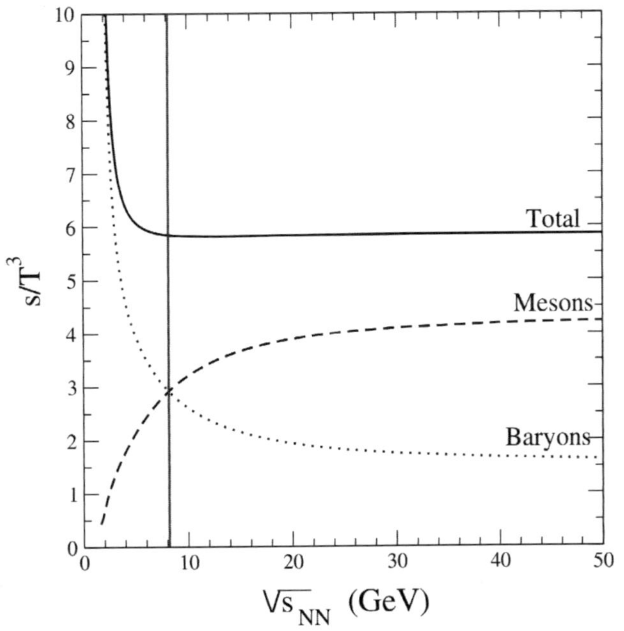

Figure 2. The entropy density normalised to T^3 as a function of the beam energy as calculated in the thermal model using THERMUS [7].

must be balanced by strange baryons.

It is worth noting that the maxima in the ratios for multi-strange baryons occur at ever higher beam energies. The higher the strangeness content of the baryon, the higher in energy is the maximum. This behavior is due to a combination of the facts that strangeness has to be balanced, the baryon chemical potential decreases rapidly and the multi-strange baryons have successively higher thresholds. The dependence of the baryon chemical potential on the beam energy is shown in Fig. 3. It is to be expected that if these maxima do not all occur at the same temperature, i.e. at the same beam energy, then the case for a phase transition is not very strong.

In conclusion, while the thermal model cannot explain the sharpness of the peak in the K^+/π^+ ratio, its position corresponds precisely to a transition from a baryon-dominated to a meson-dominated hadronic gas. This transition occurs at a temperature $T = 140$ MeV, a baryon chemical potential $\mu_B = 410$ MeV and an energy $\sqrt{s_{NN}} = 8.2$ GeV. In the thermal model this transition leads to a sharp peak in the $\Lambda/\langle\pi\rangle$ ratio, and to moderate peaks in the K^+/π^+, Ξ^-/π^+ and Ω^-/π^+ ratios. Furthermore, these peaks are at different energies in the thermal model. The thermal model predicts that the maxima in the $\Lambda/\langle\pi\rangle$, Ξ^-/π^+ and Ω^-/π^+ occur at increasing beam energies.

If the change in properties of the above excitation functions were associated with a genuine

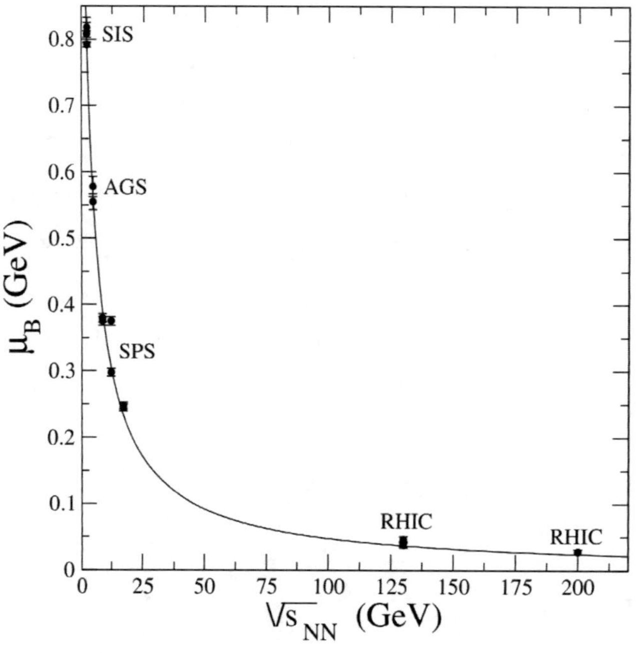

Figure 3. The energy dependence of the baryon chemical potential at freeze-out.

deconfinement phase transition one would expect these changes to occur at the same beam energy. It is clear that more data are needed to clarify the precise nature of the sharp variation observed by the NA49 collaboration.

References

[1] M. Gaździcki, (NA49 Collaboration), J. Phys. G: Nucl. Part. Phys. **30** (2004) S701.
[2] M. Gaździcki and M.I. Gorenstein, Acta Phys. Polonica B **30** (1999) 2705.
[3] J. Cleymans, H. Oeschler, K. Redlich, S. Wheaton, hep-ph/0411187, Phys. Lett. B (to be published).
[4] J. Cleymans and K. Redlich, Phys. Rev. Lett. **81** (1998) 5284.
[5] F. Becattini, J. Cleymans, A. Keränen, E. Suhonen, K. Redlich, Phys. Rev. **C64** (2001) 024901.
[6] J. Cleymans, H. Oeschler and K. Redlich, Phys. Rev. **C59** (1999) 1663.
[7] S. Wheaton and J. Cleymans, THERMUS - A Thermal Model Package for ROOT, hep-ph/0407174.
[8] P. Braun-Munzinger, J. Cleymans, H. Oeschler, K. Redlich, Nucl. Phys. **A697** (2002) 902.
[9] P. Braun-Munzinger, K. Redlich and J. Stachel, nucl-th/0304013 in *Quark-Gluon Plasma 3*, (eds.) R. Hwa and X.N. Wang.
[10] F. Becattini, M. Gaździcki, A. Keränen, J. Manninen, R. Stock, Phys. Rev. **C69** (2004) 024905.

Institute of Physics Publishing
doi:10.1088/1742-6596/50/1/059
Journal of Physics: Conference Series **50** (2006) 410–413
5th International Conference on Physics and Astrophysics of Quark Gluon Plasma

Excitation Function of the Longitudinal Expansion in Central Nuclear Collisions

Marcus Bleicher

Institut für Theoretische Physik, J. W. Goethe Universität, 60054 Frankfurt am Main, Germany

Abstract. Longitudinal hadron spectra from Proton-Proton (pp) and nucleus-nucleus (AA) collisions from $E_{lab} = 2$ AGeV to $\sqrt{s} = 200$ AGeV are investigated. The widths of the rapidity spectra for various particle species increases monotonously with energy. The present calculation indicates no sign of a step like behaviour as excepted from the Kaon transverse mass systematics. For Pions, the transport simulation is consistent with a Landau type scaling of the rapidity widths, both in central AA reactions and in pp collisions. However, other hadron species do not follow the Landau scaling. The present model predicts a decreasing rapidity width with particle mass for newly produced particles, not supporting a Landau type flow interpretation.

Based on recent lattice QCD (lQCD) calculations it has been speculated that partonic degrees of freedom might already lead to visible effects at ~ 5 $A{\cdot}$GeV [1]. Especially the hardening of the measured transverse mass (m_t) spectra in central Au+Au collisions relative to pp interactions [2, 3] around AGS energies obtained great interest and was studied in detail [4]. This increase of the inverse slope parameter T is commonly attributed to strong collective flow, which is absent in the respective pp or pA collisions. It has also been proposed [5] to interpret the high and approximately constant $K^{\pm}$ slopes above ~ 30 AGeV – the 'step' – as an indication of the phase transition.
This interpretation seems supported by microscopic transport simulations

- indicating the increasing importance of sub-hadronic degrees of freedom above AGS energies [6] and
- from the comparison of the thermodynamic parameters T and μ_B extracted from the transport models in the central overlap region [7] with the experimental systematics on chemical freeze-out configurations [8, 9, 10] in the T, μ_B plane.

Let us now explore whether a similar 'step' is also present in the excitation function of longitudinal observables. For the present study we employ the UrQMD model (v2.2) [11, 12]. It takes into account the formation and multiple rescattering of hadrons and dynamically describes the generation of pressure in the hadronic expansion phase. It involves also interactions of (di-)quarks, however gluonic degrees of freedom are not treated explicitly, but are implicitly present in strings. This simplified treatment is generally accepted to describe Proton-Proton and Proton-nucleus interactions.

It became popular to interpret relativistic heavy ion reactions with Landau's hydrodynamical model [13, 14, 15, 16, 17, 18] (for recent applications of this model to relativistic nucleus-nucleus interactions see [19, 20, 21, 22]). Therefore we will use this simple hydrodynamical picture as a baseline for the model and data comparison. The main physics assumptions of Landau's

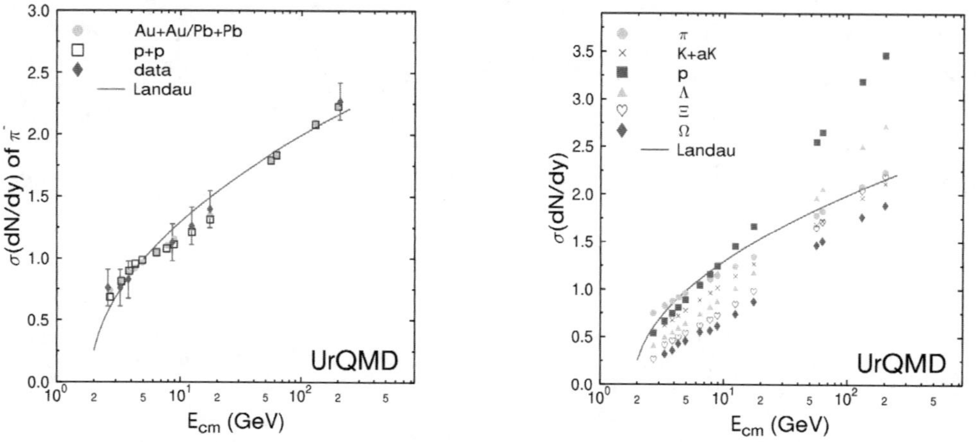

Figure 1. Left: The root mean square of the rapidity distribution of negatively charged Pions in central Au+Au/Pb+Pb and Proton+Proton reactions as a function of the center of mass energy. UrQMD calculations for Au+Au/Pb+Pb are denoted by full circles, the pp results are shown by open squares. The prediction from Landau's model is given by the line (Eq. 1). Data [22] are depicted by full diamonds. Right: The root mean square of the rapidity distribution of Pions, Kaons, Protons, Lambdas, Cascades and Omega hadrons in central Au+Au/Pb+Pb reactions as a function of center of mass energy. UrQMD calculations for Au+Au/Pb+Pb are denoted by symbols. The prediction from Landau's model is given by the line.

picture are: The collision of two Lorentz-contracted hadrons or nuclei leads to full thermalization in a volume of size $Vm_p/\sqrt{s}$. This justifies the use of thermodynamics and establishes the system size and energy dependence. A simple equation of state $p = \epsilon/3$ is assumed. Chemical potentials are usually assumed to vanish. The main results derived from these assumptions are: A universal formula for the produced entropy, determined mainly by the initial Lorentz contraction and Gaussian rapidity distributions, at least for newly produced particles. The results can be summarised in the energy dependent rapidity density [17]:

$$\frac{dN}{dy} = \frac{Ks^{1/4}}{\sqrt{2\pi L}} \exp(-\frac{y^2}{2L}) \quad \text{with} \quad L = \sigma_y^2 = \ln(\sqrt{s}/2m_p). \tag{1}$$

As depicted in Fig. 1 (left) the UrQMD predictions (full circles) for the rapidity widths of negatively charged Pions in Au+Au (Pb+Pb) reactions are in line with the experimental data [22] (full diamonds) and Landau's hydrodynamical model (full line). A rather surprising observation is that the calculated rapidity widths of π^- in pp interactions (open squares) are identical to the AA results.

The rhs. of Fig. 1 shows the rapidity widths of different particle species as a function of energy. Here the calculation (for hadrons other than Pions) differs considerably from the Landau model: (I) Hadrons containing (initial) up or down quarks show a strong increase of the rapidity widths with energy (leading particle effect). (II) Hadrons without initial up or downs quarks show a decreasing rapidity width with increasing mass at fixed energy.

The second feature is shown in detail in Fig. 2 for central Au+Au reactions at $\sqrt{s}$ = 200 AGeV. Here the root mean square of hadrons without initial quarks is given as a function of particle mass. Beginning from Pions with rapidity width of 2.2 units up to anti-Xi baryons with a rapidity width of only 1.65 units. Within 10% deviations, this results in Bjorken-plateau of $y_{cm} \pm 0.75$.

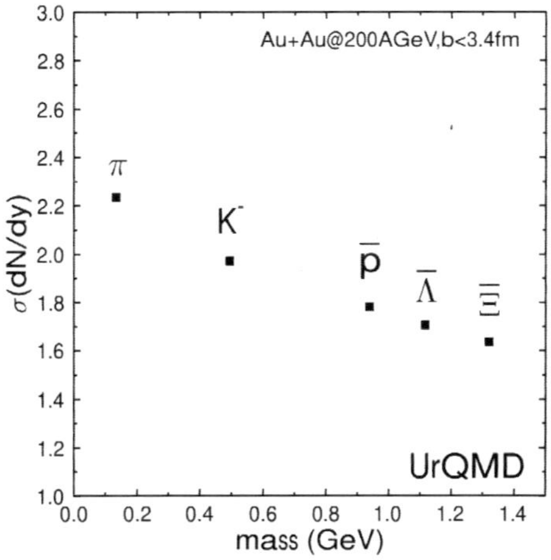

Figure 2. Root mean square of the rapidity distribution as a function of particle mass in central Au+Au reactions at $\sqrt{s} = 200$ AGeV.

In conclusion, we have explored the excitation functions of the rapidity widths of Pions, Kaons, Protons, Lambdas and Xis in pp and/or AA collisions. The rapidity spectra of Pions for all investigated systems and energies can be described well by Gaussians. The energy dependence of the width of the Pion rapidity distribution follows the prediction of Landau's hydrodynamical model, both in nucleus-nucleus and Proton-Proton interactions. In nucleus-nucleus reactions, the width of all other investigated hadrons deviates from the Landau picture. For newly produced hadrons, the present calculation shows a strong mass dependence of the rapidity width, allowing for a Bjorken-plateau only in a narrow window ($y_{cm} \pm 0.75$ at the highest RHIC energy) around midrapidity. We do not observe any irregularities (e.g. steps or peaks) in the rapidity width of all investigated hadrons.

Acknowledgements
This work was supported by GSI, DFG and BMBF. This work used computational resources provided by the Center for Scientific Computing at Frankfurt (CSC).

References
[1] E. L. Bratkovskaya *et al.*, Phys. Rev. Lett. **92**, 032302 (2004)
[2] V. Friese *et al.*, NA49 Collaboration, J. Phys. G **30**, 119 (2004).
[3] M. I. Gorenstein, M. Gaździcki, and K. Bugaev, Phys. Lett. B **567**, 175 (2003).
[4] E. L. Bratkovskaya *et al.*, Phys. Rev. C **69**, 054907 (2004)
[5] M. Gazdzicki and M. I. Gorenstein, Acta Phys. Polon. B **30**, 2705 (1999).
[6] H. Weber, C. Ernst, M. Bleicher *et al.*, Phys. Lett. B **442**, 443 (1998).
[7] L. V. Bravina *et al.*, Phys. Rev. C **60**, 024904 (1999), Nucl. Phys. A **698**, 383 (2002).
[8] P. Braun-Munzinger and J. Stachel, Nucl. Phys. A **606**, 320 (1996).
[9] P. Braun-Munzinger and J. Stachel, Nucl. Phys. A **638**, 3 (1998).
[10] J. Cleymans and K. Redlich, Phys. Rev. C **60**, 054908 (1999).
[11] S.A. Bass *et al.*, Prog. Part. Nucl. Phys. **42**, 255 (1998).
[12] M. Bleicher *et al.*, J. Phys. G **25**, 1859 (1999).
[13] E. Fermi, Prog. Theor. Phys. **5**, 570 (1950).
[14] L. D. Landau, Izv. Akad. Nauk Ser. Fiz. **17**, 51 (1953).
[15] S. Z. Belenkij and L. D. Landau, Usp. Fiz. Nauk **56**, 309 (1955).
[16] P. Carruthers and M. Doung-van, Phys. Rev. D **8**, 859 (1973).
[17] P. Carruthers, Annals N.Y.Acad.Sci. 229, 91 (1974).

[18] P. A. Carruthers, LA-UR-81-2221
[19] J. Stachel and P. Braun-Munzinger, Phys. Lett. B **216**, 1 (1989).
[20] P. Steinberg, arXiv:nucl-ex/0405022.
[21] M. Murray, arXiv:nucl-ex/0404007.
[22] G. Roland, Quark Matter 2004.

Institute of Physics Publishing Journal of Physics: Conference Series **50** (2006) 414–417
doi:10.1088/1742-6596/50/1/060 5th International Conference on Physics and Astrophysics of Quark Gluon Plasma

Effect of curvature on confinement-deconfinement phase transition.

Ashok Goyal[1,3] **and Deepak Chandra**[2,3]

[1] Department of Physics and Astrophysics, University of Delhi, Delhi-110007, India.
[2] S. G. T. B. Khalsa College, University of Delhi, Delhi-110007, India.
[3] Inter University Center for Astronomy and Astrophysics, Ganeshkhind, Pune-411007, India.

E-mail: agoyal@iucaa.ernet.in deepak@iucaa.ernet.in

Abstract. First order confinement-deconfinement phase transition is studied through hadronic bubble nucleation in an expanding quark-gluon plasma in the context of heavy ion collisions. For this study we consider interacting quark and hadron gas and incorporate the effects of curvature energy. We find that the interactions have the effect of hastening the phase transition whereas the curvature energy has mixed behaviour. Lowering surface tension has the effect of increasing super cooling and slowing down the process of hadronisation in contrast to the case of early Universe. Higher values of bag pressure tend to speed up transition. An interesting feature is the beginning of hadronisation process as soon as the QGP is formed.

1. Introduction

The heavy ion collision experiments are aimed at studying the behavior of quantum chromodynamics (QCD) at high energies [1]. A hot baryon free plasma of quarks and gluons (QGP) is expected to be created whereas at lesser energies, it shows large stopping of the nuclei and hint at the baryon rich matter. The phase transition from QGP to hadron resonance gas (HRG) has been extensively studied for the early Universe where the time scale is of the order of $10^{-6} - 10^{-7}$ s [2] [3] [4]. What emerges is a QGP fireball of $\sim 150 fm^3$ created at the initial temperature of $\sim 300-350$ MeV, which is about twice the expected critical temperature T_c. The effect of curvature energy term has been studied earlier by Mardor and Svetitsky [6] and it has been pointed out that the curvature energy term $8\pi\gamma r$, in addition to the surface energy $4\pi r^2\sigma$ (where σ and γ are the surface and curvature energy densities respectively) plays an important role [6] [7] [8] [9]. A first order thermodynamic phase transition would proceed through the nucleation of hadronic bubbles. Csernai and Kapusta [5] have computed a nucleation rate and applied it to a first order phase transition in a set of rate equations. We examine here the dynamics of the phase transition by treating the QGP as a gas of massless u,d quarks, massive s quarks and massless gluons. QCD interactions are treated perturbatively to order g^3 and the long range confinement effects are parameterized by the bag pressure B. For the hadron phase we use the known masses of the low lying 33 baryons and 45 mesons whose masses and degeneracy factors are taken from the Particle Data Group Summary [10]. We take baryons and antibaryons to have the same size as protons, given by $V_p = m_p/4B$.

2. Bubble nucleation

The critical radius for bubble nucleation is obtained by extremizing the thermodynamic work expanded to create a bubble, i.e.

$$W \equiv -\frac{4\pi}{3} r^3 (P_h - P_q) + 4\pi r^2 \sigma - 8\pi(\gamma_q - \gamma_h) r, \tag{1}$$

where P_h and P_q are the pressures in the hadronic and quark phase respectively, and $\gamma = (\gamma_q - \gamma_h)$ is the curvature coefficient. Following Mardor and Svetitsky [6] the curvature energy in the MIT model for massless quarks is given by

$$E_c = \frac{gr}{3\pi} \int_0^\infty dkk\{1 + exp(k - \mu_B)/T\}^{-1}. \tag{2}$$

We get $\gamma \simeq \gamma_q \equiv \frac{T^2}{16}$, assuming $\gamma_h << \gamma_q$ and $\mu_B << T$. The critical radii for extremums (a maxima) and (a minima) below T_c are given as

$$r_{c+} = \frac{\sigma}{\Delta P}(1 + \sqrt{1 - \beta}) \tag{3}$$

and

$$r_{c-} = \frac{\sigma}{\Delta P}(1 - \sqrt{1 - \beta}) \tag{4}$$

respectively, where $\Delta P = P_h - P_q$ and $\beta = \frac{2\Delta P}{\sigma^2}\gamma$. Since $\beta > 1$, a real solution exists only if $\beta \leq 1$. The equilibrium (minima) bubbles do not grow. There is a extremum (a minima) in the free energy even above the critical temperature,

$$r_{c+>} = \frac{\sigma}{\Delta P}(-1 + \sqrt{1 + \beta}). \tag{5}$$

Significantly, the restriction of $\beta \leq 1$ is not present for these bubbles. This ensures that the phase transition actually begins well above the critical temperature having interesting consequences.

The bubble nucleation rate at temperature T is given by $I = I_0 e^{-W_c/T}$ where I_0 is the prefactor having dimensions of T^4. The prefactor used traditionally in the early Universe studies is given by $I_0 = (W_c/2\pi T)^{3/2} T^4$. Csernai and Kapusta have calculated this in a course grain effective field theoretic approximation to QCD and give

$$I_0 = \frac{16}{3\pi}\left(\frac{\sigma}{3T}\right)^{3/2} \frac{\sigma \eta_q r_c}{\xi_q^4 (\Delta w)^2}, \tag{6}$$

where $\eta_q = 14.4T^3$ is the shear viscosity in the plasma phase, ξ_q is a correlation length in the plasma phase, and Δw is the difference in the enthalpy densities of the two phases.

Taking contribution of all the critical bubbles, hadron fraction $h(t)$ is now given by

$$h(t) = \int_{t_c}^{t} dt' I_+(T(t'))\{1 - h(t')\} V(t', t) + \int_{t_c}^{t} dt' I_-(T(t'))\{1 - h(t')\} \frac{4\pi}{3} r_{c-}^3 (T(t'))$$

$$+ \int_{t_0}^{t_c} dt' I_{+>}(T(t'))\{1 - h(t')\} \frac{4\pi}{3} r_{c+>}^3 (T(t')). \tag{7}$$

The dynamical equation using the longitudinal scaling hydrodynamics of Bjorken is given by

$$\frac{de}{dt} = -\frac{w}{t} \tag{8}$$

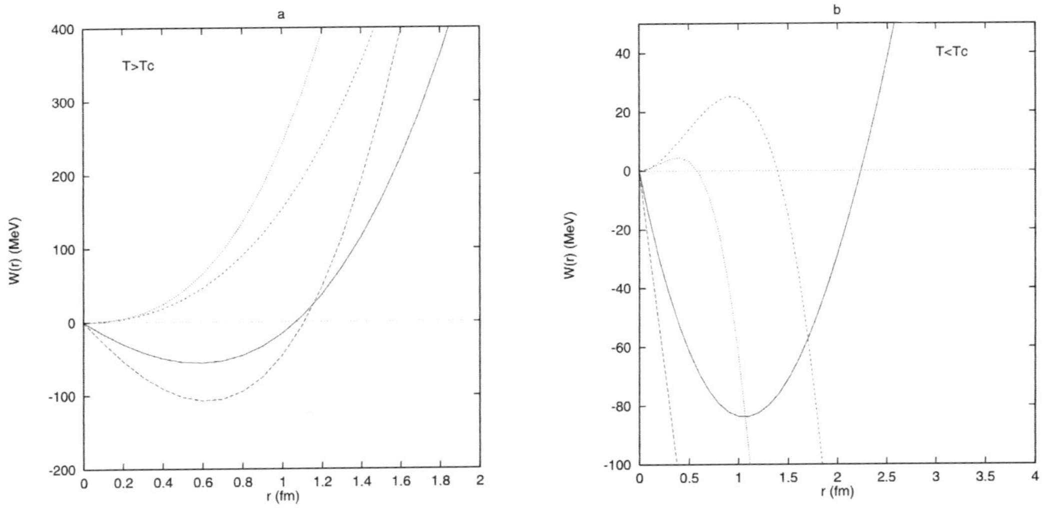

Figure 1. Free energy W in units of MeV as a function of bubble radius r in fm. In fig.1a Solid (with γ) and dashed (without γ) curves are for $B^{1/4} = 235MeV$ and $\sigma \sim 7MeVfm^{-2}$. Long dashed (with γ) and dotted (without γ) curves are for $B^{1/4} = 300MeV$ and $\sigma \sim 7MeVfm^{-2}$. curves in 1a are at a temperature of $1.01T_c$. curves in 1b are labelled as in 1a except that they are at a temperature of $0.99T_c$.

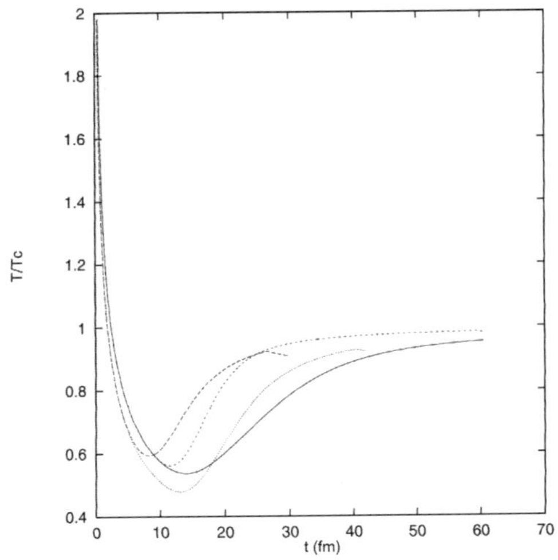

Figure 2. Temperature T/T_c as a function of time t in fm. Solid, dashed, dotted and long dashed curves are labelled as in fig. 1a.

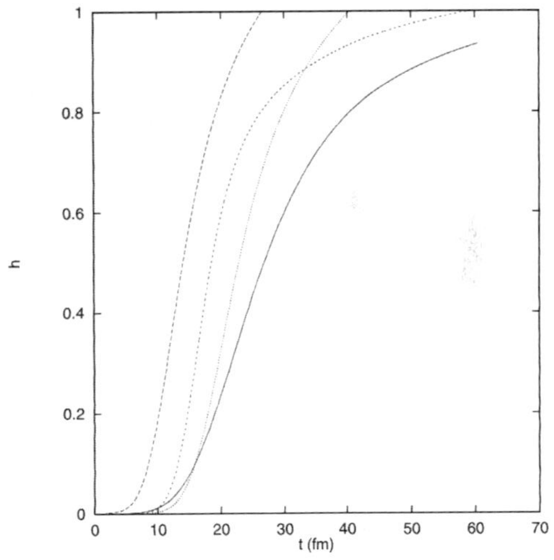

Figure 3. The hadron fraction h as a function of time t in fm. Curves are labelled as in fig. 1a.

where $e(T)$ and $w(T)$ are the energy and enthalpy densities respectively given by $e(T) = h(t)e_h(t) + (1 - h(t))e_q(t)$, likewise for enthalpy density. The growth velocity of bubbles is taken to be $v(T) = v_0(1 - T/T_c)^{3/2}$ where $v_0 = 3c$ for $T > (2/3)T_c$. We have assumed that for $T < (2/3)T_c$, $v = c/\sqrt{3}$. The volume of an expanding bubble at time t nucleated at time t' is given by

$$V(t',t) = \frac{4\pi}{3}\{r_{c+}(T(t')) + \int_{t'}^{t} dt'' v(T(t''))\}^3. \quad (9)$$

We use the parameters $\xi_q = 0.7fm$ and $\eta_q = 14.4T^3$.

3. Results and discussion

We find that the inclusion of the interactions in both the phases lowers the critical temperature, so higher values of the bag pressure are required for having reasonable values of T_c. In fig.1a it is clearly seen that equilibrium sized hadron bubbles appear even above the critical temperature. In fig.1b we find that below the critical temperature the curvature effect produces a minima as well as a maxima in the free energy. In fig.2 the degree of supercooling depends on the parameters and can be a significant fraction of T_c. The important feature to see here is that the prefactor I_0 is the deciding factor for supercooling and the duration of the transition rather than the exponent in contrast to early Universe studies. The inclusion of γ term in this case actually reduces the time to complete the phase transition. This can be contrasted with the case corresponding to the other set of parameters in fig.1b. Comparing with the no curvature case, it is seen that increasing B speeds up the phase transition whearas decreasing it slows it down if the γ term is included. We also see that the decrease in the surface tension increases the supercooling of the plasma by significant amount (can be close to 50 percent). However, increasing B reduces supercooling when γ is included. Finally, fig.3 shows the hadron fraction h as a function of time t. Figure shows that for low bag pressures, the effect of γ is to delay the phase transition (solid curve) but for larger values of B we see that the curvature correction actually speeds up the process of hadronization (long dashed curve).

Acknowledgements

This work was partially supported under the SERC Scheme of The Department of Science and Technology (DST), India. One of us (A.G.) is thankful to U.G.C. for partial financial support.

References

[1] J. D. Bjorken, *Phys. Rev.* D**27**, 140 (1983)

[2] C.J. Hogan, *Phys. Lett.* B**133**, 172 (1983);
K. Kajantie and M. Kurki-Suonio, *Phys. Rev.* D**34**, 1719 (1996);
G.M. Fuller,G.J. Mathews and C.R. Alcock, *Phys. Rev.* D**37**, 1380 (1988).

[3] D. Chandra and A. Goyal, *Phys. Rev.* D**62**, 063505 (2000).

[4] E. Witten, *Phys. Rev.* D**30**, 272 (1984); A. Applegate and C.J. Hogan, *Phys. Rev.* D**31**, 3037 (1985);
E. Farhi and R.L. Jaffe, *Phys. Rev.* D**30**, 2379 (1984).

[5] L.P. Csernai and J.I. Kapusta, *Phys. Rev. Lett.* **69**, 737 (1992); *Phys. Rev.* D**46**, 1379 (1992) and the references cited therein.

[6] I. Mardor and B. Svetitsky, *Phys. Rev.* D**44**, 878 (1991).

[7] J. Madsen, *Phys. Rev. Lett.* **70**, 391 (1993).

[8] J. E. Horvath, *Phys. Rev.* D**49**, 5590 (1994).

[9] A. Goyal and D. Chandra, *Astron. Astrophys.* **330**, 10 (1998).

[10] Particle Data Group, *Phys. Rev.* D**50**, (1994)

Consequences of Quark-Hadron Phase Transition in Dense Stars

Abhijit Bhattacharyya[a], Sanjay K. Ghosh[b], and Sibaji Raha[b]

[a] Department of Physics, Scottish Church College, 1 & 3, Urquhart Square, Kolkata - 700 006, INDIA

[b] Department of Physics, Bose Institute, 93/1, A.P.C. Road, Kolkata - 700 009, INDIA

E-mail: bhattacharyyaabhijit_10@yahoo.co.uk

Abstract. The deconfinement phase transition in a rotating compact star and its consequences on neutrino emission have been studied. It has been found that rotating twin star solutions are obtained upto $\Omega \approx 4000s^{-1}$. For faster rotating stars no twin solution is obtained. The neutrino emission shows a strong directionality.

1. Introduction

Compact stars are thought to be the best possible laboratory to study the strongly interacting matter at super nuclear densities. The matter density near the core of a compact star can be $8-10$ times that of normal nuclear matter. Different exotic phase transitions may take place in the strongly interacting matter [1] at such high densities. In this work we look at some of the consequences of deconfinement phase transition inside a rotating compact star.

Recently some of us have used the NLZ model for the hadronic sector and MIT Bag model for the quark sector to look at the quark-hadron phase transition in static compact stars [2]. In that work it was found that there was a solution for the third family of stars known as twin stars [3]. Here we employ the same EOS for a rapidly rotating star. The basic motivation is to study the fate of the twin stars in a rotating model and also to study the neutrino emission from theses stars as a possible signature of the phase transition. The organisation of the paper is as follows. In section 2, we discuss the properties of the rotating stars. In section 3, we study the process of neutrino emission associated with a phase transition to the quark phase. Finally, in section 4, we present our conclusion.

2. Equation of State and Star Properties

In this work we use a variant of non-linear Walecka model for the hadronic sector and the MIT bag model ($B^{\frac{1}{4}} = 180MeV$) for the quark sector. A mixed phase has been formulated using Gibbs construction. The details of the models and the parameter sets may be obtained in ref. [2, 4]. In figure 1 we plot the Equation of State (EOS) obtained from the above mentioned model. From figure 1 one can see that the phase transition starts at around $\epsilon_c = 4 \times 10^{14} gm/cm^3$ and ends around $\epsilon_c = 1 \times 10^{15} gm/cm^3$.

The EOS obtained above is used to solve the Einstein's equations for the rotating star. The metric for a stationary rotating star can be written as [5]

$$ds^2 = -e^{\gamma+\rho}dt^2 + e^{2\alpha}\left(dr^2 + r^2 d\theta^2\right) + e^{\gamma-\rho}r^2 sin^2\theta\,(d\phi - \omega dt)^2 \tag{1}$$

where α, γ, ρ and ω are the gravitational potentials which depend on r and θ only. The Einstein's equations for the three potentials γ, ρ and ω have been solved by Komatsu *et.al.* [6] using Green's function technique. The fourth potential α has been determined from other potentials. All the physical quantities can be calculated from these potentials [5]. Solution of the potentials, and hence the calculation of physical quantities, is numerically quite an involved process. In this work we have used the **'rns'** code [5, 6] for this purpose.

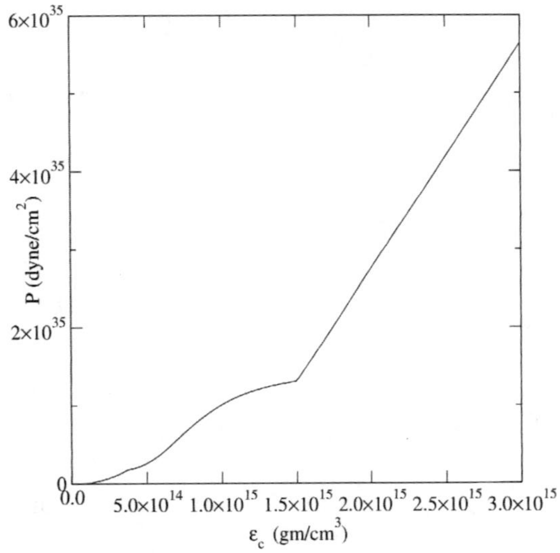

Figure 1. Equation of state of the model considered here.

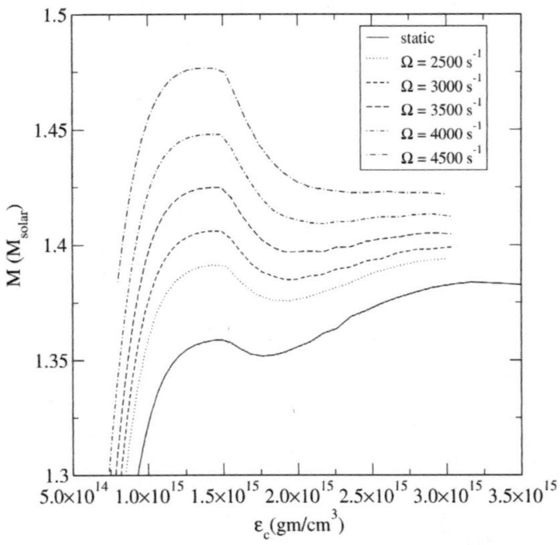

Figure 2. Mass-ϵ_c plots for different Ω.

In figure 2, we have plotted the mass - ϵ_c curves for different angular frequency Ω. Normally, the mass - ϵ_c curve for neutron stars shows a maximum beyond which no stable configuration is supposed to exist. Harrison *et al.* [7] showed that for a smooth equation of state no stable stellar configuration exists for central densities above that corresponding to the maximum mass limit. Recently Glendenning *et al.* [8] have found that if there is a first order phase transition inside the neutron star, this inference may not be valid. As a result a stable configuration for a third family of stars, with higher central densities, may arise. In the present work also, we have found the existence of stars which belong to the third family, known as twin stars. These results may be seen in figure 2. However, as we increase Ω, the third family solution becomes less probable, as can be seen from the plots, and it vanishes just after $\Omega = 4000s^{-1}$. This is natural as with increasing angular velocity, the phase boundary moves outward, matter gets redistributed to larger equatorial radii and the twin star solution vanishes.

The appearance of the twin star solution has other interesting consequences as well. In figure 3 we have plotted the moment of inertia (I) as a function of Ω for different rest masses, *i.e.*, for both normal and supermassive sequences. For the normal sequence, the moment of inertia increases with the angular velocity monotonically. However, for a supermassive sequence, there are two different branches; for one branch of the curve the moment of inertia increases with Ω, whereas, for the other, the moment of inertia decreases with increase in Ω. This anomalous behaviour of moment of inertia is known as backbending and can be attributed to the phase transition from incompressible nuclear matter to highly compressible quark matter [9]. As pointed out by several authors, this feature could be a possible signature of the quark-hadron phase transition. Here we would like to emphasize that even for supermassive sequence, the backbending is only observed for the cases where density in the core is such that it facilitates a large quark phase region inside the star. For example, though the range of rest mass $1.35M_\odot < M_0 < 1.48M_\odot$ lies in the supermassive domain, no backbending has been found for masses within this range. Higher mass ($1.6M_\odot$ and $1.7M_\odot$) sequences correspond to larger I and Ω. Due to the redistribution of matter, as mentioned earlier, the $dI/d\Omega > 0$ part, corresponding to the quark core (as seen for masses $1.48M_\odot - 1.5M_\odot$), does not appear.

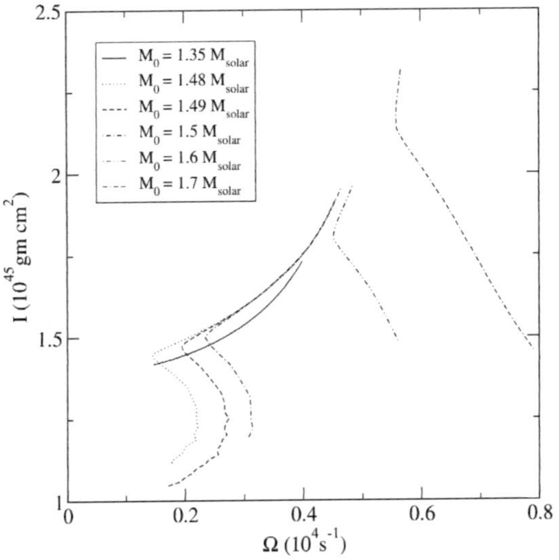

Figure 3. I as a function of Ω for different rest masses.

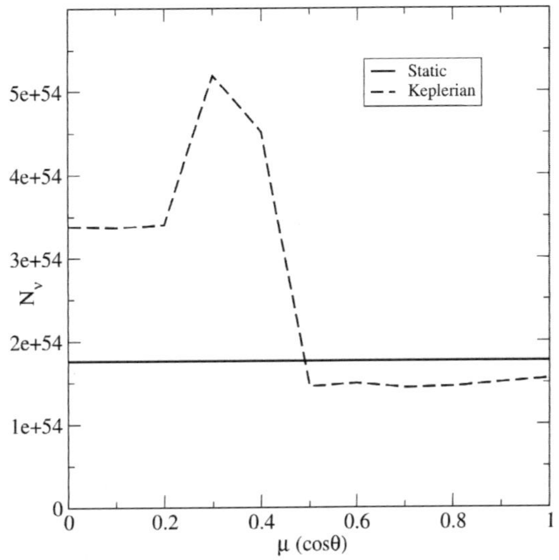

Figure 4. Neutrino emissivity as a function of μ.

3. Neutrino Emission

In this section we look at the neutrino emission from a rotating compact star due to the phase transition *only*. The neutrino emission from the equilibrated matter is much smaller in magnitude and has been neglected [10].

The reactions that we consider here are

$$\begin{aligned} u + d &\leftrightarrow u + s; \\ d(s) \rightarrow u + e^- + \bar{\nu}_e \quad &; \quad u + e^- \rightarrow d(s) + \nu_e; \\ d(s)e^+ \rightarrow u + e^+ + \bar{\nu}_e \quad &; \quad u \rightarrow d(s) + e^+ + \nu_e; \end{aligned} \tag{2}$$

We solve the rate equations for these reactions self consistently. The rate of change of the u quark density is given by

$$\begin{aligned} \frac{dn_u(t)}{dt} &= R_{d\rightarrow u}(e^-) + R_{s\rightarrow u}(e^-) - R_{u\rightarrow d}(e^-) - R_{u\rightarrow s}(e^-) \\ &+ R_{d\rightarrow u}(e^+) + R_{s\rightarrow u}(e^+) - R_{u\rightarrow d}(e^+) - R_{u\rightarrow s}(e^+) \end{aligned} \tag{3}$$

Similarly, one can obtain the other rate equations. Solving the above equations one can obtain the neutrino emission rate as a function of density. The emission rate thus obtained can be folded with the baryon density profile of the star as a function of θ to get the number of neutrinos emitted as a function of θ. In figure 4, we have plotted the number of neutrinos emitted as a function μ ($= cos\theta$, $\mu = 0$ corresponds to the equator and $\mu = 1$ corresponds to the pole). From the figure one can see that as one moves from the static limit to the keplerian limit the uniformity of the neutrino emission is lost. The neutrino emission is dominated over a small angle. This happens because of the redistribution of matter due to rotation. In this particular case the neutrino emission is constrained mostly within the range 60^0 and 80^0.

4. Conclusion

To summarise, we have studied the effect of deconfinement transition in a rotating compact star. The third family of star *i.e.* the twin star solution is expected to occur due to the substantial change in the sound speed across the phase boundary. In the present model, the twin star solution is obtained for the static as well as for the rotating case. However, with the increase in angular velocity Ω, the mass as well as ϵ_c of the star increases, which implies that the twin star solutions for higher central densities vanishes beyond a certain Ω. At this point, most of the star, in the present study is found to be made up of quark matter. So the faster rotating stars have larger domain of mixed phase and do not exhibit the third family of stars.

The emission of neutrinos are also modified due to the rotation. For a compact star rotating fast enough the emission of neutrinos are constrained within a very small angle. One should also study other models and look into the angular distribution of the emitted neutrinos. Work in these directions are in progress.

Acknowledgements AB would like to thank University Grants Commission for partial support through the grant PSW-083/03-04.

[1] Alam J, Raha S and Sinha B 1996 *Phys. Rep.* **273** 243
[2] Mishustin I N, Hanauske M, Bhattacharyya A, Satarov L M, Stoecker H and Greiner W 2003 *Phys. Lett.* B **552** 1
[3] Schertler K, Greiner C, Schaffner-Bielich J and Thoma M 2000 *Nucl. Phy.* A **677** 463
[4] Bhattacharyya A, Ghosh S K, Hanauske M and Raha S, to appear in *Phys. Rev. C*
[5] Cook G B, Shapiro S L and Teukolsky S A 1994 *Ap. Jr.* **424** 823
[6] Komatsu H, Eriguchi Y and Hachisu I 1989 *Mon. Not. R. Astr. Soc.* **237** 355
[7] Harrison B K, Thorne K S, Wakano M and Wheeler J A 1965 Gravitation Theory and Gravitational Collapse, (University of Chicago Press, Chicago)
[8] Glendenning N K and Kettner C, 2000 *Astron. Astrophys.* **353** L9
[9] Glendenning N K, Pei S and Weber F 1997 *Phys. Rev. Lett.* **79** 1603
[10] Ghosh S K, Phatak S C and Sahu P K 1995 *Nucl. Phys.* A **596** 670

Institute of Physics Publishing Journal of Physics: Conference Series **50** (2006) 422–425
doi:10.1088/1742-6596/50/1/062 5th International Conference on Physics and Astrophysics of Quark Gluon Plasma

Whitening of the Quark-Gluon Plasma

Stanislaw Mrówczyński

Sołtan Institute for Nuclear Studies
ul. Hoża 69, PL - 00-681 Warsaw, Poland
and Institute of Physics, Świętokrzyska Academy
ul. Świętokrzyska 15, PL - 25-406 Kielce, Poland

E-mail: mrow@fuw.edu.pl

Abstract.
Parton-parton collisions do not neutralize local color charges in the plasma but they only redistribute them among momentum modes. Color diffusion and conductivity are discussed as processes responsible for the plasma whitening. The conductivity and diffusion coefficients are computed and the time evolution of the color density is studied. The conductivity is shown to be more effective than the diffusion, it whitens the plasma even before the momentum degrees of freedom are thermalized.

1. Introduction

This talk is based on two works done in collaboration with Cristina Manuel [1, 2]. In the first paper [1], we have analyzed the local equilibrium of the quark-gluon plasma which is defined as a state of maximal local entropy. Using the kinetic equations with the collision terms of the Waldmann-Snider form, we have proved that such a state is generically *colorful*, *i.e.* the color four-current is non-vanishing. Thus, the collisions, which are responsible for equilibration of the parton momenta, do *not* neutralize the local color density. Since the color current is (covariantly) conserved in every collision process, the inter-parton collisions redistribute the color charges among various momentum modes but they do not change a local macroscopic color charge. Consequently, if the color charges are not homogeneously distributed in the process of the plasma production due to, say, statistical fluctuations, the inter-parton collisions will not neutralize the system. On the other hand, the global equilibrium of the quark-gluon plasma is locally colorless because of the maximum entropy principle. We assume here that the system does not curry a global color charge and that it does not experience an external chromodynamic field. Once the inter-parton collisions are not responsible for the neutralization of the local charges, one has to invoke other collective mechanism to whiten the plasma. This is the subject of the second article [2].

Local charges are neutralized due to the currents that flow in the system. We consider the diffusive currents generated by the charge density gradient (Fick's law) and the ohmic currents caused by the chromoelectric field (Ohm's law) which is, in turn, induced by the charge density. The color conductivity of the quark-gluon plasma has been studied for long time [3, 4, 5, 6, 7, 8, 9, 10], but only recently the problem has been well understood [11, 12, 13, 14, 15, 16, 17, 18]. As far as we know, the color diffusion was only briefly discussed by Heiselberg [10]. In all these papers, the plasma near the colorless global equilibrium was studied. We are, however, interested in the plasma that is locally colorful. Thus, in Sec. 2 we derive the diffusion and conductivity coefficients in such a plasma, and then, in Sec. 3 the temporal evolution of the color charge density is considered.

2. Color Diffusion and conductivity coefficients

Using the transport theory, we derive here the diffusive and ohmic currents in a plasma that is locally colorful. The transport equations of quarks, antiquarks and gluons read

$$(D^0 + \mathbf{v}\cdot\mathbf{D})Q(\mathbf{p},x) - \frac{g}{2}\{\mathbf{E}+\mathbf{v}\times\mathbf{B}, \nabla_p Q(\mathbf{p},x)\} = C[Q,\bar{Q},G]\,, \tag{1}$$

$$(D^0 + \mathbf{v}\cdot\mathbf{D})\bar{Q}(\mathbf{p},x) + \frac{g}{2}\{\mathbf{E}+\mathbf{v}\times\mathbf{B}, \nabla_p \bar{Q}(\mathbf{p},x)\} = \bar{C}[Q,\bar{Q},G]\,, \tag{2}$$

$$(D^0 + \mathbf{v}\cdot\mathbf{D})G(\mathbf{p},x) - \frac{g}{2}\{\mathbf{E}+\mathbf{v}\times\mathbf{B}, \nabla_p G(\mathbf{p},x)\} = C_g[Q,\bar{Q},G]\,. \tag{3}$$

The (anti-)quark on-mass-shell distribution functions $Q(\mathbf{p},x)$ and $\bar{Q}(\mathbf{p},x)$, which are $N_c \times N_c$ hermitean matrices, belong to the fundamental representation of the SU(N_c) group, while the gluon distribution function $G(\mathbf{p},x)$, which is a $(N_c^2-1)\times(N_c^2-1)$ matrix, belongs to the adjoint representation. The covariant derivative $D^\mu \equiv \partial^\mu - ig[A^\mu(x), \cdots]$, the chromoelectric and chromomagnetic fields, $\mathbf{E}$ and $\mathbf{B}$, which enter the transport equations also belong to either the fundamental or adjoint representation, correspondingly. To simplify the notation, we use the same symbols D^0, $\mathbf{D}$, $\mathbf{E}$ and $\mathbf{B}$ to denote a given quantity in the fundamental or adjoint representation. $x \equiv (t,\mathbf{x})$ denotes the four-position while $\mathbf{p}$ is the three-momentum. Because the partons are assumed to be massless, the velocity $\mathbf{v}$ equals $\mathbf{p}/|\mathbf{p}|$. The collision terms C, $\bar{C}$ and C_g will be discussed later on.

We are interested in a state close to the colorful local equilibrium. When the effects of quantum statistics are neglected, the (on-mass-shell) local equilibrium distribution functions are [1]

$$Q^{\rm eq}(\mathbf{p},x) = \exp\Big[-\beta(x)\big(u_\nu(x)p^\nu - \mu_b(x) - \widetilde{\mu}(x)\big)\Big]\,, \tag{4}$$

$$\bar{Q}^{\rm eq}(\mathbf{p},x) = \exp\Big[-\beta(x)\big(u_\nu(x)p^\nu + \mu_b(x) + \widetilde{\mu}(x)\big)\Big]\,, \tag{5}$$

$$G^{\rm eq}(\mathbf{p},x) = \exp\Big[-\beta(x)\big(u_\nu(x)p^\nu - \widetilde{\mu}_g(x)\big)\Big]\,, \tag{6}$$

where $p^\mu = (E_p, \mathbf{p})$, and $E_p = |\mathbf{p}|$ for massless quarks and antiquarks, and for gluons; $\beta = 1/T$, u^μ, μ_b denote, respectively, the inverse temperature, hydrodynamic velocity and baryon chemical potential; the colored chemical potentials of quarks ($\widetilde{\mu}$) and of gluons ($\widetilde{\mu}_g$) obey the relationship

$$\widetilde{\mu}_g(x) = 2T^a {\rm Tr}[\tau^a \widetilde{\mu}(x)] = \mu_a(x) T^a\,, \tag{7}$$

where τ^a, T^a with $a = 1, \ldots, N_c^2-1$ are the SU(N_c) group generators in the fundamental and adjoint representations, normalized as ${\rm Tr}[\tau^a\tau^b] = \frac{1}{2}\delta^{ab}$ and ${\rm Tr}[T^aT^b] = N_c\delta^{ab}$. In the local equilibrium state there is a non-vanishing color charge density and current. However, the current vanishes in the local rest frame because the distribution functions are locally isotropic.

We further consider small deviations from local equilibrium, and we write down the quark distribution function as $Q(\mathbf{p},x) = Q^{\rm eq}(\mathbf{p},x) + \delta Q(\mathbf{p},x)$. Assuming that

$$|Q^{\rm eq}| \gg |\delta Q|\,, \qquad |D^\mu Q^{\rm eq}| \gg |D^\mu \delta Q|\,, \qquad |\nabla_p Q^{\rm eq}| \gg |\nabla_p \delta Q|\,, \tag{8}$$

and taking into account the local isotropy of the equilibrium state, the transport equation (1) can be approximated as

$$\left(D^0 + \mathbf{v} \cdot \mathbf{D}\right) Q^{\text{eq}} + \frac{g}{2}\{\mathbf{E}, \nabla_p Q^{\text{eq}}\} = L[\delta Q] \ , \tag{9}$$

where $L[\delta Q]$ is the collision term linearized around the local equilibrium distribution function. Analogous equations hold for antiquarks and gluons. Let us recall here that the collision terms evaluated for the local equilibrium distribution functions (2) vanish [1].

To get the transport coefficients of color diffusion and conductivity we assume that the linearized collision terms L, $\bar{L}$ and L_g satisfy the relationship

$$\int \frac{d^3p}{(2\pi)^3} \, \mathbf{v} \, L[\delta Q] = -\gamma \int \frac{d^3p}{(2\pi)^3} \, \mathbf{v} \, \delta Q \ , \tag{10}$$

where $1/\gamma$ is the characteristic relaxation time which, for simplicity, is assumed to be the same for quarks, antiquarks and gluons. The relation is obeyed [15] by the Waldmann-Snider collision term linearized around global (colorless) equilibrium. The relationship (10) is also trivially satisfied by the collision term in the relaxation time approximation, which, unfortunately, violates the covariant current conservation. However, the collision term can be easily improved. In analogy to the so-called BGK collision term [19], one constructs the expression [2] which obeys both the covariant current conservation and the relation (10).

Using Eq. (10), the color current (in the adjoint representation), which is generated by deviations from equilibrium, gets the form

$$\mathbf{j}^a(x) = -d \, \mathbf{D}^{ab} \rho^b(x) + \sigma^{ab} \mathbf{E}^b(x) \ , \tag{11}$$

where $\mathbf{D}^{ab} = \delta^{ab}\nabla + g f^{acb}\mathbf{A}^c$, the diffusion constant equals $d = 1/3\gamma$ and the color conductivity tensor reads

$$\sigma^{ab} = \frac{g^2}{3\gamma} \int \frac{d^3p}{(2\pi)^3} \frac{1}{E_p} \left(\mathrm{Tr}[\{\tau^a, \tau^b\}(Q^{\text{eq}} + \bar{Q}^{\text{eq}})] + \mathrm{Tr}[\{T^a, T^b\} G^{\text{eq}}]\right) . \tag{12}$$

When the equilibrium is colorless, the conductivity σ is proportional to the unit matrix in color space.

3. Temporal evolution of the color density

The color current generated by both the gradient of color density and the chromoelectric field is given by Eq. (11). The covariant current conservation $D^0\rho + \mathbf{D}\mathbf{j} = 0$ combined with the Gauss law $\mathbf{D}\mathbf{E}(x) = \rho(x)$, provides the equation

$$[D^0 - d\,\mathbf{D}^2 + \sigma]\rho(x) = 0 \ , \tag{13}$$

where the term $[\mathbf{D}, \sigma]\,\mathbf{E}$ has been neglected. It can be shown [2] that $[\mathbf{D}, \sigma]\,\mathbf{E} \ll \sigma\mathbf{D}\mathbf{E}$ in the small coupling limit. Since the commutator $[D^0, \sigma]$ can be also shown to be small, Eq. (13) is solved by

$$\rho(x) = e^{-\sigma t}\, n(x) \tag{14}$$

with n satisfying the diffusion equation

$$[D^0 - d\,\mathbf{D}^2] n(x) = 0 \ . \tag{15}$$

$\rho(x)$ and $n(x)$ obey the initial condition $\rho(0, \mathbf{x}) = n(0, \mathbf{x}) = \rho_0(\mathbf{x})$. One shows [2] that the conductivity dominates over the diffusion for the long wavelength charge density modes with the wave vector $\mathbf{k}$ obeying $\mathbf{k}^2 < \sigma/d$.

4. Discussion

The question arises what is the characteristic time of plasma whitening. The only reliable estimates of the time scales of interest have been found for the perturbative quark-gluon plasma close to the global (colorless) equilibrium. The characteristic time scale of the color conductivity and of the parton-parton collisions at momentum transfer of order g^2T are [8, 15]

$$\frac{1}{\sigma} \sim \frac{\ln(1/g)}{T}, \quad t_{\rm soft} \sim \frac{1}{g^2 \ln(1/g)\, T}. \tag{16}$$

As seen, $1/\sigma \ll t_{\rm soft}$. Since the evolution of color degrees of freedom is dominated by the soft collisions, the relaxation time $1/\gamma$ is of order $t_{\rm soft}$. Our derivation of the color conductivity and the further analysis apply for $t \geq 1/\gamma$ as $1/\gamma$ is the minimal time scale which enters the linearized transport equation (9) with the collision term (10). Taking into account the above estimate of color conductivity, the charge density is found to vanish at $t \geq 1/\gamma$. Because the characteristic time scale of color dissipation cannot be shorter than $t_{\rm soft}$, the whitening of the quark-gluon plasma occurs at $t \sim t_{\rm soft}$. For shorter times the color density is expected to oscillate.

To equilibrate the system's momentum degrees of freedom, there are needed parton-parton collisions at the momentum transfer of order T. Such collisions occur at the time scale [15]

$$t_{\rm hard} \sim \frac{1}{g^4 \ln(1/g)\, T}, \tag{17}$$

which in the perturbative limit is much longer than $t_{\rm soft}$. Thus, we conclude that the plasma becomes white first and then the momentum degrees of freedom are thermalized.

[1] Manuel C and Mrówczyński St 2003 *Phys. Rev.* **D68** 094010
[2] Manuel C and Mrówczyński St 2004 *Phys. Rev.* **D70** 094019
[3] Heinz U 1986 *Annals Phys.* **168** 148
[4] Czyz W and Florkowski W 1986 *Acta Phys. Polon.* **B17** 819
[5] Dyrek A and Florkowski F 1987 *Phys. Rev.* **D36** 2172
[6] Mrówczyński St 1988 *Acta Phys. Polon.* **B19** 91
[7] Mrówczyński St 1990 *Adv. Ser. Direct. High Energy Phys.* **6** 185
[8] Selikhov AV and Gyulassy M 1993 *Phys. Lett.* **B316** 373
[9] Selikhov AV and Gyulassy M 1994 *Phys. Rev.* **C49** 1726
[10] Heiselberg H 1994 *Phys. Rev. Lett.* **72** 3013
[11] Bödeker D 1998 *Phys. Lett.* **426** 351
[12] Bödeker D 1999 *Nucl. Phys.* **B559** 502
[13] Litim DF and Manuel C 1999 *Phys. Rev. Lett.* **82** 4981
[14] Litim DF and Manuel C 1999 *Nucl. Phys.* **B562** 237
[15] Arnold P, Son DT and Yaffe LG 1999 *Phys. Rev.* **D59** 105020
[16] Blaizot JP and Iancu E 1999 *Nucl. Phys.* **B557** 183
[17] Arnold P and Yaffe LG 2000 *Phys. Rev.* **D62** 125014
[18] Martinez Resco JM and Valle Basagoiti MA 2001 *Phys. Rev.* **D63** 056008
[19] Bhatnagar PL, Gross EP and Krook M 1954 *Phys. Rev.* **94** 511

Institute of Physics Publishing Journal of Physics: Conference Series **50** (2006) 426–429
doi:10.1088/1742-6596/50/1/063 5th International Conference on Physics and Astrophysics of Quark Gluon Plasma

A transport coefficient: the electrical conductivity

Sourendu Gupta

Department of Theoretical Physics, Tata Institute of Fundamental Research, Homi Bhabha Road, Mumbai 400005, India.

E-mail: sgupta@tifr.res.in

Abstract. I describe the lattice determination of the electrical conductivity of the quark gluon plasma [6]. Since this is the first extraction of a transport coefficient with a degree of control over errors, I next use this to make estimates of other transport related quantities using simple kinetic theory formulæ. The resulting estimates are applied to fluctuations, ultra-soft photon spectra and the viscosity. Dimming of ultra-soft photons is exponential in the mean free path, and hence is a very sensitive probe of transport.

1. Introduction

Transport coefficients have recently been computed to leading logarithmic order in weak coupling perturbation theory [1, 2]. However, at temeperatures of interest to heavy-ion collisions experiments, the gauge coupling $g = \mathcal{O}(1)$ and this computation is ineffective. At the other extreme of $g \to \infty$, the AdS/CFT correspondence has been exploited to bound the shear viscosity $\eta/S \geq 1/4\pi$ [3], where S is the extropy density. This result is for a highly supersymmetric version of QCD and needs to be corrected in order to connect with experiments.

There is clearly a case for a lattice computation, if one can have control over the process of analytic continuation from Euclidean to Minkowski space-time. Direct computations of the shear viscosity were tried [4, 5], but were inconclusive because statistical problems beset even the Euclidean computations of the correlator of the energy-momentum tensor, thus making it hard to check the analytic continuation.

In this work [6] I shift the focus slightly and compute the electrical conductivity, σ, of the quark-gluon plasma. This involves a lattice computation of the correlator of the electromagnetic current. Since this is no harder than the extraction of the ρ-meson mass, statistical errors in the Euclidean computation are under control, and effort can be concentrated on the harder job of analytical continuation. We are able to demonstrate control over this process, and thereby extract σ.

This is interesting in itself for heavy-ion collisions because several properties of soft photons in media can be inferred once σ is known. Since this is the first statistically reliable extraction of a transport coefficient from lattice QCD, one can use this to estimate many different quantities which depend on transport, and are important for heavy-ion physics. With σ in hand, and using two well-specified approximations, we are able to predict the value of η, and compare the lattice results with others.

The lattice computations described in Section 2 are from [6]. The material in Section 3 is partly new.

2. The lattice computation

All lattice computations of transport coefficients are performed in the context of linear response theory. The response, A, of a system to a time-dependent force F is given, in the frequency domain by $A(\omega) = \chi(\omega)F(\omega)$. The response function $\chi(\omega)$ can then be written in terms of the retarded correlation function of A, and the transport coefficient can be extracted using Kubo formulæ. For σ, the Kubo formula is

$$\sigma(T) = \frac{1}{6}\frac{\partial}{\partial\omega}\rho_{EM}{}^{i}_{i}(\omega, \mathbf{0}, T)\bigg|_{\omega=0}, \tag{1}$$

where the sum is over spatial polarisations of the imaginary part of the retarded photon propagator, *i.e.*, the spectral density, $\rho^{\mu\nu}_{EM}$, of the electromagnetic current correlator. The usual QED Ward identity gives $\rho^{00}(\omega, \mathbf{0}, T) = 2\pi\chi_Q\omega\delta(\omega) = 0$, where χ_Q is the usual charge susceptibility.

A lattice computation proceeds from the spectral representation for Euclidean current correlators—

$$G_{EM}(t,T) = \int_0^\infty \frac{d\omega}{2\pi} K(\omega, t, T)\rho_{EM}(\omega, T), \tag{2}$$

where K is the free propagator, G_{EM} is the product of the zero (spatial) momentum vector correlator summed over all spatial polarisations, G_V, and the EM vertex factor $C_{EM} = 4\pi\alpha\sum_f e_f^2$, where e_f is the charge of a quark of flavour f. For later use we note that $C_{EM} \approx 1/20$ for two flavours.

The main problem in this determination is that ρ_{EM} is needed at an infinite number of points and G_{EM} is known only at N_t different values. The solution is to constrain the function ρ_{EM} through an informed guess, and use a Bayesian method to extract it.

We note a second complication. A free field theory computation shows that $\rho_{EM}(\omega, T)$ increases as ω^3 at large ω. This means that a naive application of the Kubo formulæ would violate causality [7]. We try to solve this problem by using $\Delta G_{EM}(t,T)$ which is the difference of the measured and free-field theory propagators to extract $\Delta\rho_{EM}$.

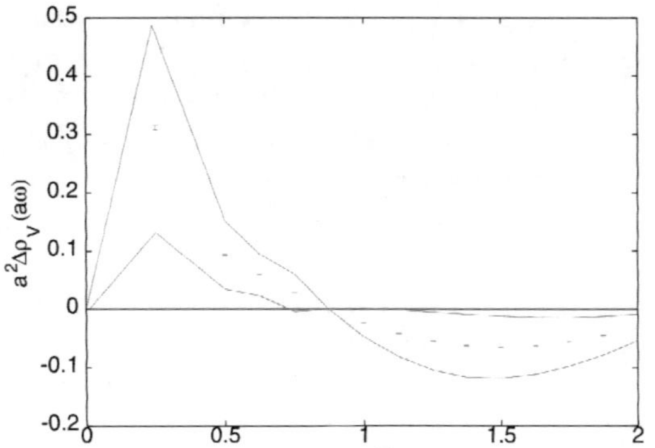

Figure 1. The spectral density $\Delta\rho_{EM}$ for $T = 2T_c$ on a $12 \times 26^2 \times 48$ lattice with quark mass $0.03T_c$. See the text for a discussion of statistical and systematic errors.

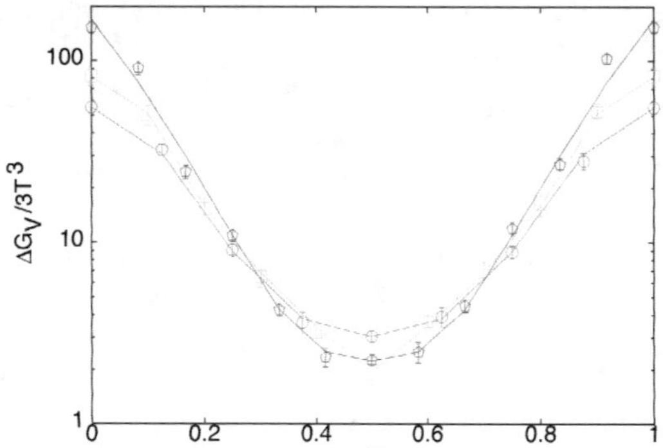

Figure 2. The propagator ΔG_{EM} obtained through the fit of $\Delta\rho_{EM}$ compared to the data for $T = 2T_c$ and $N_t = 8$ (circles), 10 (squares) and 12 (pentagons) with quark mass $0.03T_c$.

Using a variety of Bayesian methods [6] to extract $\Delta G_{EM}(t,T)$ we found a peak at $\omega/(2\pi T) \approx 0.15$ [6] and a smooth drop to zero at $\omega = 0$ (see Figure 1). The checks which have been performed on this extraction are—

(i) The statistical errors are negligible, as seen in Figures 1 and 2.

(ii) The position of the peak seen in Figure 1 is not strongly dependent on the lattice spacing, as checked by using $N_t = 8$, 10 and 12.

(iii) Variations in the results are investigated due to algorithmic changes in the Bayesian prior and the truncation and discreitzation of eq. (2). The width of the band of systematic errors shown in Figure 1 is thrice the width due to such variation.

It would be useful to have similiar tests of robustness on all uses of Bayesian methods including the maximum entropy method (MEM).

The goodness of fit to the Euclidean data can be judged from Figure 2. We draw attention to the small error bars on this data when compared to the errors on typical measurements of Euclidean correlators of energy momentum tensor.

Having established the existence of the peak, further information is extracted using a parametrization of the peak in $\Delta\rho_{EM}$ as the ratio of two polynomials, with that in the numerator being of lower order than the denonimator, so that the integrals used in the Kramers-Krönig relation converge. The denominator has to be at least a quartic polynomial in order to smoothly connect to the weak-coupling picture of a pinch singularity giving rise to the peak along real ω [8]. (We note in this connection that earlier attempts to extract transport quantities have worked with the full ρ_{EM} rather than with $\Delta\rho_{EM}$. This error can strongly suppress estimates of the transport coefficients. Since ρ_{EM} increases with ω, this biases the fitted peak towards larger ω, thus decreasing the slope at the origin for any fit form with one peak.) Using these fits our estimate of the electrical conductivity is—

$$\frac{\sigma(T)}{T} = C_{EM} \times \begin{cases} 7.5 \pm 0.8 & (T = 1.5T_c) \\ 7.7 \pm 0.6 & (T = 2T_c) \\ 7.0 \pm 0.4 & (T = 3T_c) \end{cases} \tag{3}$$

3. Appplications

3.1. Viscosity

In kinetic theory one can write the Drude formula for the electrical conductivity—

$$\sigma = \frac{C_{EM} S_q \tau_q}{m}, \tag{4}$$

where S_q is the entropy density in quarks, m their screening mass and τ_q their mean free time. Using the lattice result for σ one gets $\tau_q = 0.30 \pm 0.03$ fm. While charge is transported through quarks, momentum is transported mainly through gluons. Due to colour factors the mean free time for gluons is about half the mean free time for quarks. Then using a kinetic theory formula for the viscosity along with such an estimate of τ_g one gets

$$\frac{\eta}{S} \approx 0.21 \pm 0.02 > \frac{1}{4\pi}. \tag{5}$$

Only lattice errors are shown in this estimate. Errors on the kinetic theory are harder to extract. The lattice + kinetic theory result is 2.5 times the AdS/CFT limit.

3.2. Photon dimming

Non-zero electrical conductivity leads to screening of photons with energy below the peak of the spectral function found earlier, *i.e.*, for $\omega < 300$ MeV. This is a direct consequence of Maxwell's equations. This range of screened energies lies far below the excess which may have been seen by the CERN NA49 experiment. It seems possible that direct photons at such low-energy could be distinguished from photons coming from decays of pions at RHIC-II. One method for using an observation of photon dimming to extract the conductivity has been suggested [9].

The mean free path, ℓ, of these ultra-soft photons is given by the formula

$$\ell = \frac{\tau_q}{C_{EM}} \approx 6 \text{ fm}. \quad (6)$$

Since the fireball at RHIC is about 7 fm in size, one could expect dimming by a factor of $\exp(-7/6)$, *i.e.*, by a factor of 3.2. There is a self-consistency argument here involving only the data. If the viscosity is small, then mean free times of quarks and gluons are small, and the mean free path of photons can only be about 20 ($= 1/C_{EM}$) times larger. However, dimming is exponential in the mean free path. For example, the mean free path as predicted in AdS/CFT would lead to dimming by a factor of 18.8. Thus, the use of photon dimming to extract σ is an important cross check of values of η extracted by other means.

3.3. Fluctuations

The study of event-to-event fluctuations in limited rapidity bins is expected to give information on the phase diagram of QCD as well as make contact between experiments and basic lattice QCD computations. However, fluctuations of conserved quantities in a limited region of space are expected to even out by transport processes such as diffusion. The computation of σ is relevant not only to the diffusion of charge, but also of baryon number and isospin.

The diffusion constant can be obtained using the kinetic theory formula [10]

$$D = \tau_q c_s^2 \approx 0.1 \text{ fm}, \quad (7)$$

where we have taken the square of the speed of sound $c_s^2 = 0.30 \pm 0.01$ at $T = 2T_c$, as shown by recent lattice computations [11]. If the plasma is highly dissipative, then D must be small. When D decreases, the fluctuation are more localized at the same time. Small viscosity implies smaller mean free path, and hence longer persistence of the fluctuation signal. In the approach of [10] this would mean that the rapidity width of a fluctuation is $\Delta Y \leq 2D/\tau_0 \approx 0.4$ for fluctuations created at time $\tau_0 = 0.5$ fm.

3.4. A brave new world

When weak coupling was the only tool to investigate transport properties, it was felt that the QGP would be weakly dissipative. Now with other tools at our disposal, the question seems more open, and preliminary evidence is that dissipation is strong. The first lattice results on transport [6] pin down many aspects of transport, as outlined in this section, and lead to the hope that multiple experimental constraints on transport are feasible.

References

[1] Arnold P, Moore G D and Yaffe L G 2000 *J. H. E. P.* **0011** 001
[2] Arnold P, Moore G D and Yaffe L G 2001 *J. H. E. P.* **0111** 057
[3] Kovtun P, Son D T and Starinets A O 2003 *J. H. E. P.* **0310** 064
[4] Karsch F and Wyld H W 1987 *Phys. Rev.* D **35** 2518
[5] Nakamura A and Sakai S 2005 *Phys. Rev. Lett.* **94** 072305
[6] Gupta S 2004 *Phys. Lett.* B **597** 57
[7] Hilgevoord J 1960 *Dispersion Relations and Causal Description* (Amsterdam: North-Holland)
[8] Aarts G and Resco J M M (2002) *J. H. E. P.* **0204** 053
[9] Gupta S 2004 Here's looking at you, fireball *Preprint* hep-ph/0411355
[10] Aziz M A and Gavin S 2004 *Phys. Rev.* C **70** 034905
[11] Gavai R V, Gupta S and Mukherjee S 2004 The speed of sound and specific heat in the QCD plasma: hydrodynamics, fluctuations and conformal symmetry *Preprint* hep-lat/0412036

Institute of Physics Publishing Journal of Physics: Conference Series **50** (2006) 430–433
doi:10.1088/1742-6596/50/1/064 5th International Conference on Physics and Astrophysics of Quark Gluon Plasma

Chiral dynamics and operator relations at non-zero chemical potential

Sourendu Gupta and Rajarshi Ray[1]

Department of Theoretical Physics, Tata Institute of Fundamental Research, Homi Bhabha Road, Mumbai 400005, India.

E-mail: sgupta@tifr.res.in,rajarshi@theory.tifr.res.in

Abstract. We discuss Taylor expansions of operator expectation values in QCD with respect to chemical potentials of quarks. Maxwell's relations between coefficients and Ward identities between series are used to relate the operators which give the Taylor coefficients of the series for the chiral condensate, the pseudoscalar susceptibility and the mass dependence of quark number susceptibilities. Through such relations the physics of chiral dynamics are explored. The renormalized expectation values of the chiral condensate and its Taylor coefficients are extracted from simulation.

1. Introduction

Direct lattice simulations at non-zero chemical potential are possible only along certain subspaces of the full parameter space, such as $\mathrm{Re}\,\mu_f = 0$ or $\mu_u = -\mu_d$. However, Taylor expansions around $\mu_f = 0$ have proved to be useful in studying the physics for general μ_f. Here we discuss the two flavour case $f \in \{u, d\}$ and with degenerate quark masses $m_u = m_d = m$. We shall freely go from the parameter space expressed in terms of μ_u and μ_d to the isoscalar $\mu_0 = (\mu_u + \mu_d)/2$ and the isovector $\mu_3 = (\mu_u - \mu_d)/2$ chemical potentials. These are simply related to the baryon and electric charge chemical potentials. Since we perform Taylor expansions around the point $\mu_f = 0$, all expectation values are computable by the standard methods of lattice gauge theory.

Various operators related to chiral dynamics in QCD are connected among themselves via chiral Ward identities and Maxwell relations, such that there are only few independent Taylor coefficients. For us the three independent parameters of interest in the chiral sector are the single linear response coefficient of the chiral condensate to μ, and two quadratic response coefficients (QRCs). The linear coefficient vanishes by symmetry in an expansion around $\mu = 0$. Thus, the two chiral QRCs encapsulate the physics arising from the influence of baryon dynamics due to $\mu \neq 0$ on chiral fluctuations.

2. Taylor Expansion

2.1. *Chiral condensate*

Here we consider only the isoscalar condensate $C_S(m, T, \mu_u, \mu_d) = \frac{1}{2}\left[\langle \overline{u}u\rangle + \langle \overline{d}d\rangle\right]$, which can also be written as, $C_S(m, T, \mu_u, \mu_d) = \frac{1}{2V_4}\left(\frac{\partial \log Z}{\partial m_u} + \frac{\partial \log Z}{\partial m_d}\right)\Big|_{m_u=m_d=m}$ where V_4 is the 4-volume

[1] By whom the talk was delivered

of the lattice. In our present computations the volumes are taken large enough, but with non-zero quark masses. Since we work with staggered quarks, the mass renormalization is multiplicative, and so are the renormalization of the condensate and its Taylor coefficients. We are interested only on the variation of the condensate with μ at constant T, so we simply work with either of the ratios

$$\frac{C_S(m_R,T,\mu_u,\mu_d;a)}{C_S(m_R,0,0,0;a)} \text{ (Z scheme)} \quad \text{or} \quad \frac{C_S(m_R,T,\mu_u,\mu_d;a)}{C_S(m_R,T,0,0;a)} \text{ (T scheme)}. \tag{1}$$

The computations have to be performed at fixed renormalized mass m_R. One expects the T scheme to be undefined for $T > T_c$ when $m_R = 0$, since the chiral condensate then vanishes. We perform a formal expansion here which will later be renormalized in either of these schemes.

$$\begin{aligned} C_S(m,T,\mu_0) &= C_S^0 + \left(C_S^{20} + C_S^{11}\right)\frac{\mu_0^2}{2} + \cdots \\ C_S(m,T,\mu_3) &= C_S^0 + \left(C_S^{20} - C_S^{11}\right)\frac{\mu_3^2}{2} + \cdots \end{aligned} \tag{2}$$

where C_S^{20} and C_S^{11} are respectively the diagonal and off-diagonal coefficients.

2.2. Chiral Ward identities

Chiral Ward identities are consequences of certain operator equalities which follow from chiral symmetries. The prototypical chiral Ward identity is $C_S(T,\mu) = m\chi_\pi(T,\mu)$, where χ_π is the pseudo-scalar susceptibility. Taylor expansions of both sides of this identity can then be equated term by term. A second chiral Ward identity [1] relates the isovector scalar susceptibility to the mass derivative of the condensate: $\partial C_S/\partial m = -\chi_\epsilon$.

We show later that in the continuum limit of the high temperature phase the second derivatives of the condensate with respect to the chemical potential vanish. As a result, the the pion susceptibilities are insensitive to chemical potential. Also, in this phase, symmetry arguments show that $\chi_\pi = \chi_\epsilon$ [2]. As a result, the scalar susceptibility is also independent of the isoscalar chemical potential, at least to quadratic order. Below T_c this chain of logic does not hold. The scalar susceptibility is interesting because of speculation about the massless modes at the critical end point [3]. We will discuss it at greater length elsewhere.

2.3. Maxwell relations

A Maxwell relation is the equality of two distinct physical interpretations of a mixed derivative obtained by interchanging the order of the derivatives. From the Taylor expansion of the chiral condensate in eq. (2) we can find Maxwell relations with the change of quark number susceptibilities (QNS) with the quark mass. The leading order relation $\frac{\partial C_S}{\partial \mu} = \frac{\partial n}{\partial m}$ was first noted in [4]. It is trivially true at $\mu = 0$ since the first derivative on the left vanishes, as does n for all quark masses. The second derivatives give two non-trivial Maxwell relations and consequent relations using the chiral Ward identities discussed earlier,

$$\begin{aligned} C_S^{20} &= \frac{\partial^2 C_S}{\partial \mu_u^2} = m\frac{\partial^2 \chi_\pi}{\partial \mu_u^2} = \frac{\partial \chi_{uu}}{\partial m}, \\ C_S^{11} &= \frac{\partial^2 C_S}{\partial \mu_u \partial \mu_d} = m\frac{\partial^2 \chi_\pi}{\partial \mu_u \partial \mu_d} = \frac{\partial \chi_{ud}}{\partial m}. \end{aligned} \tag{3}$$

Here, as a byproduct of our computation of the renormalized chiral condensate, we shall give the continuum limit of the derivative of the susceptibility. Also, the relative rates of strange and light quark production in heavy-ion collisions i.e. the Wroblewski parameter on the lattice [5] is

$$\lambda_s = \frac{\chi_{ss}}{\chi_{uu}}, \tag{4}$$

with obvious extensions to the production rates of heavier quarks. Since it is hard to perform lattice computations at realistic values of light quark masses due to constraints of computer time, one can lighten the computational burden by using a Taylor series for the mass dependence of the susceptibilities utilizing the above Maxwell relation. Then one can compute λ_s at some reasonably light quark mass, corresponding to, say, the pion mass being two to three times heavier than in the real world, and extrapolate to the physical quark mass values using the Maxwell relation. Extrapolating the results of our measurements at heavier quark masses to the realistic values we find the change in $\lambda_s < 1\%$.

3. Results

3.1. Continuum limit in quenched QCD

3.1.1. $T > T_c$ We used stored gauge configurations from the study in [6] for our measurements. These were obtained on $N_t = 4, 8, 10, 12$ and 14 lattices for temperatures $T = 1.5T_c, 2T_c$ and $3T_c$ respectively.

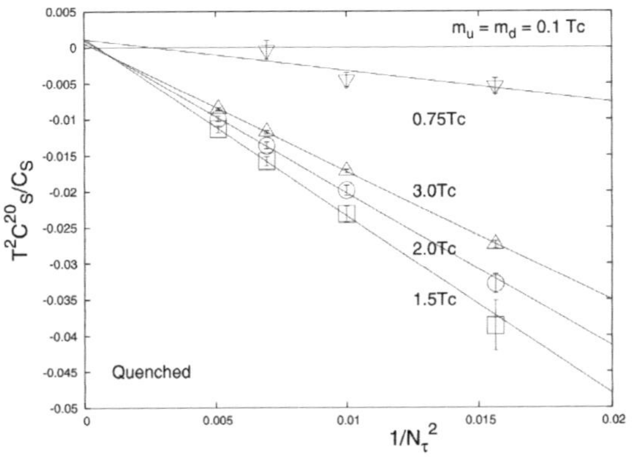

Figure 1. The diagonal QRC in the T scheme, at fixed $T/T_c = 1.5$ holding the renormalized quark mass fixed by fixing the bare mass to be $m/T_c = 0.1$. The lines show the continuum extrapolation.

The diagonal QRC in the T scheme is shown in Figure 1. It is clear that there is significant μ-dependence on coarser lattices. However, this Taylor coefficient vanishes at the 99% confidence limit on extrapolation to the continuum. The identities in eq. (3) then imply that χ_{uu} is insensitive to changes in the quark mass and χ_π is insensitive to μ in this range of temperatures. Since χ_{uu} agrees with a perturbative evaluation for $T \geq 1.5T_c$, its insensitivity to m can be understood from the fact that the effective infrared cutoff is given by the Matsubara frequency πT and not by m when $m/T_c < \pi T/T_c$.

3.1.2. $T < T_c$ We also made a series of simulations at fixed $T/T_c = 0.75$ for $N_t = 6$, 8, 10 and 12 in quenched QCD with the Wilson action. We found that the continuum limit of both $\langle\mathcal{O}_2\rangle$ and $\langle\mathcal{O}_{11}\rangle$ were consistent with zero within reasonably small errors. However, note that at coarser lattices the values were increasing with increasing temperature, but below T_c the behaviour is reversed. We then do a dynamical simulation near T_c to investigate this further.

3.2. Dynamical staggered quarks

We give the results of a series of computations with two flavours of dynamical quarks as the temperature varies between $0.75T_c$ and $2T_c$. The simulations were performed using the R-algorithm with two flavours of quarks with bare mass of $0.1T_c$. Three lattice sizes were used in most of the simulations, namely 4×8^3, 4×12^3 and 4×16^3. Lattice computations with dynamical quarks are still too costly for the continuum limit to be taken easily. However, in order to have some idea of the continuum limit we have also performed two further simulations with 6×12^3 and 8×16^3 lattices at $T = 0.9T_c$. In addition, we have results at T_c on 6×12^3 lattices.

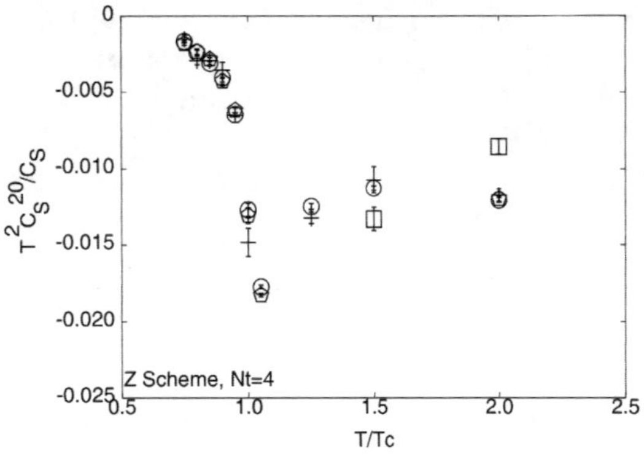

Figure 2. The diagonal QRC as a function of the temperature for $a = 1/4T$. Pluses denote results obtained on 4×8^3 lattices, pentagons on 4×12^3 and circles on 4×16^3. The boxes represent the quenched QCD value.

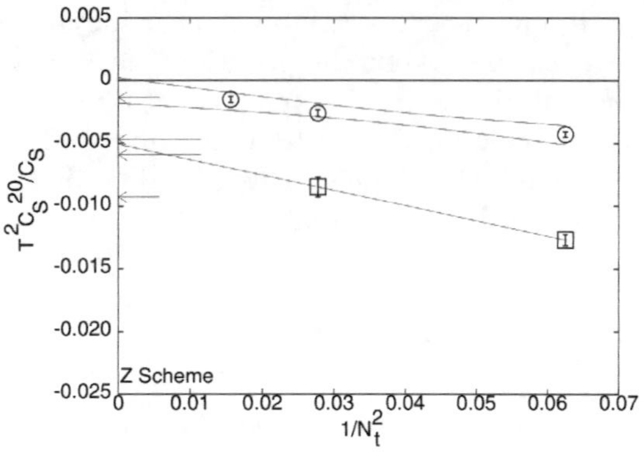

Figure 3. The continuum extrapolation of the diagonal QRC. The circles denote data for $T = 0.9T_c$, and the 3-σ band for the continuum extrapolation is shown. The boxes represent data at T_c, and the 1-σ (nearer arrows) and 3-σ (further arrows) limits of the extrapolation are shown.

Consistent with our previous observations, C_S^1 was seen to vanish with small errors across the full range of temperatures considered. The diagonal QRC was negative and differed significantly from zero, as shown in Figure 2. Eq. (3) then implies the supression of pion fluctuations and increase in pion mass as one approches the critical end point. Note a small shift of the peak from T_c. This may be an indicator of a crossover transition. For comparison, results from the quenched theory at the same bare quark mass are also shown. The T-dependence is seen to be a little different; this could reflect either the difference in the value of m_π/m_ρ or be a quenching artifact. The off-diagonal QRC was consistent with zero within small errors above T_c. Near and below T_c larger fluctuations are seen. Although still consistent with zero at the 3-σ level, the averages increase by several orders of magnitude compared to $T > T_c$, and seem to be comparable in magnitude to the diagonal QRC. Reduction of the noise in measurement is required to get a clearer signal. Any non-zero value of C_S^{11} would imply an assymetry of the phase diagram in the μ_0 and μ_3 directions (see eq. 2)

The results of our preliminary investigation of the continuum limit with dynamical quarks are shown in Figure 3. At $0.9T_c$ the 3-σ error band on the extrapolation of the diagonal QRC is consistent with zero. At T_c the continuum extrapolation perforce had to be performed with data from only two lattice spacings. With this caveat, our data currently points to a non-zero continuum extrapolation of renormalized C_S^{20} at T_c.

Acknowledgments

For more details of this work please refer to [7]

References

[1] Patel A 1984 *Phys. Lett.*B **141** 244
[2] Gupta S 1999 *Phys. Rev.*D **64** 094505
[3] Hatta Y and Ikeda T 2003 *Phys. Rev.*D **67** 014028
[4] Kogut J B *et al.*1983 *Nucl. Phys.*B **225** 93
[5] Gavai R V and Gupta S 2002 *Phys. Rev.*D **65** 094515
[6] Gavai R V and Gupta S 2003 *Phys. Rev.*D **67** 034501
[7] Gupta S and Ray R 2004 *Phys. Rev.*D **70** 114015

Institute of Physics Publishing Journal of Physics: Conference Series **50** (2006) 434–437
doi:10.1088/1742-6596/50/1/065 5th International Conference on Physics and Astrophysics of Quark Gluon Plasma

Strings with a confining core in a Quark-Gluon Plasma

Biswanath Layek, Ananta P. Mishra and Ajit M. Srivastava
Institute of Physics, Sachivalaya Marg, Bhubaneswar 751005, India

Presented by Biswanath Layek

E-mail: `layek@iopb.res.in`

Abstract. We consider the intersection of N different interfaces interpolating between different Z_N vacua of an SU(N) gauge theory using the Polyakov loop order parameter. Topological arguments show that at such a string-like junction, the order parameter should vanish, implying that the core of this string is in the confining phase. Using the effective potential for the Polyakov loop proposed by Pisarski for QCD, we use numerical minimization technique and estimate the energy per unit length of the core of this string to be about 2.7 GeV/fm at a temperature about twice the critical temperature. For the parameters used, the interface tension is obtained to be about 7 GeV/fm^2. Lattice simulation of pure gauge theories should be able to investigate properties of these strings. With the interpretation that quark contributions lead to explicit breaking of this Z_N symmetry, such *QGP* strings may play important role in the evolution of the quark-gluon plasma phase and in the dynamics of quark-hadron transition.

1. The Polyakov loop model

For pure SU(N) gauge theory, an order parameter for the confinement-deconfinement phase transition is the Polyakov loop $l(x)$ which is defined as [1],

$$l(x) = \frac{1}{N} tr\Big(Pexp\Big(ig\int_0^\beta A_0(x,\tau)d\tau\Big)\Big). \tag{1}$$

For temperatures above T_c, non-zero value of the expectation value of $l(x)$ ($\equiv l_0$ (say)) breaks the $Z(N)$ symmetry spontaneously and gives rise to N degenerate vacua. For the case of QCD with $N = 3$, effective Lagrangian density is written as [2],

$$L = \frac{N}{g^2}|\partial_\mu l|^2 T^2 - V(l). \tag{2}$$

$$V(l) = \Big(-\frac{b_2}{2}|l|^2 - \frac{b_3}{6}(l^3 + (l^*)^3) + \frac{1}{4}(|l|^2)^2\Big) b_4 T^4. \tag{3}$$

Values of various parameters [3, 4] in Eq.(2) and Eq.(3) are fixed by making correspondence to lattice results. The potential in Eq.(3) gives rise to $Z(3)$ degenerate vacua for $T > T_c$. The $Z(3)$ interface solution will correspond to a planar solution (say, in x-y plane) where l starts at one of the minimum of $V(l)$ at $z = -\infty$ and ends up at the other minimum of $V(l)$ at $z = +\infty$.

Consider now the junction of all three different $Z(3)$ interfaces, say with the line like junction being along the z axis. The interfaces then will be perpendicular to the x-y plane. It is immediately clear that as one considers a closed loop in the physical space encircling the z axis, then one encircles $l = 0$ point in the complex l plane. It is then obvious from the continuity of l that l must vanish along the z axis. These are topological strings (We call it QGP string to distinguish from the QCD string.). Potential $V(l)$ in Eq.(3) is shown below. An important thing

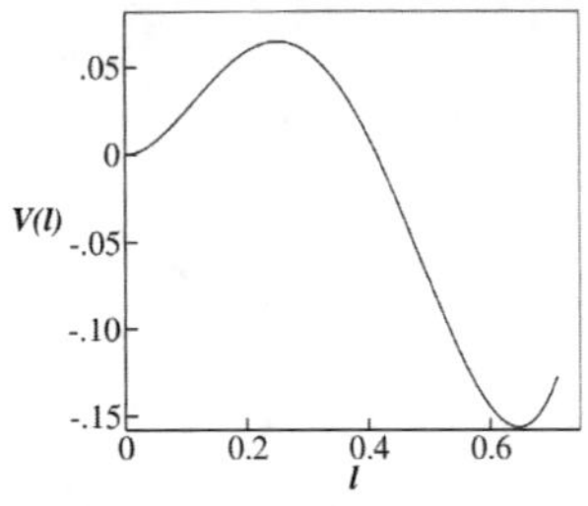

Figure 1. shows the plot of $V(l)$ in $\theta = 0$ direction (with $l = |l|e^{i\theta}$) for $T \simeq 185MeV$ ($T_c = 182MeV$).

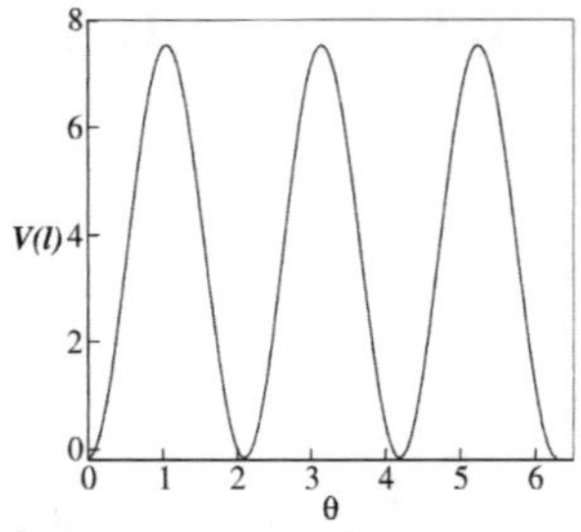

Figure 2. The $Z(3)$ structure of the vacuum can be seen in in the plot of the potential $V(l)$ as a function of θ for fixed $|l| = l_0$.

to note from Fig.1,2, is that the barrier height between different $Z(3)$ vacua is much higher than the barrier height between the metastable vacuum at $l = 0$ and the true vacuum. This situation is in complete contrast to the standard axionic models where the central bump is higher than the barrier between different $Z(N)$ vacua. The solution for kink solitons are well known in such axionic models. Such a solution cannot be found here, as the height of barrier between the $Z(3)$ vacua is much higher than the central bump. Fig.3 shows the surface plot of the potential $V(l)$ in Eq.(3) in the complex l plane for $T = 400MeV$. From Fig.3, it may give the impression that the $Z(3)$ interface will have $l = 0$ in the middle of the interface. However, as we discuss below, the actual domain wall cannot have $l = 0$ inside it.

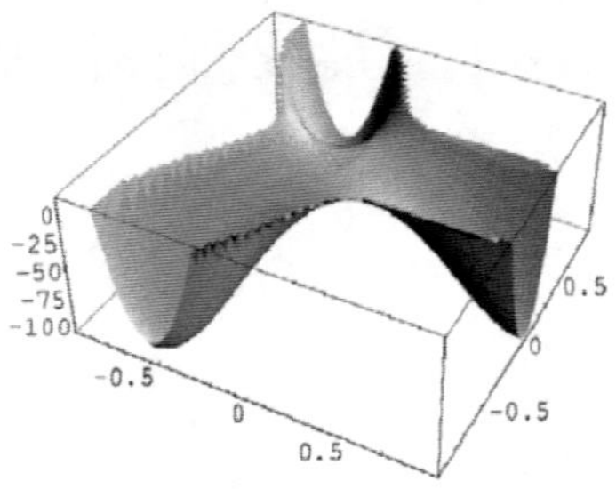

Figure 3. Surface plot of the potential $V(l)$ (in units of T_c^4) in Eq.(3) in the complex l plane for $T = 400MeV$. Plot range in the V axis is restricted to make the $Z(3)$ barriers and the central bump distinctly visible.

2. Numerical techniques and the $Z(3)$ interface profile

For a planar wall in the x-y plane we can write down the field equation as,

$$\frac{d^2 l_i}{dz^2} = \frac{g^2}{NT^2} \frac{\partial V(l_1, l_2)}{\partial l_i}, \qquad (l = l_1 + i l_2). \tag{4}$$

Using generalization of standard techniques [5], we interpret z as time, and l_1 and l_2 as the two dimensional position space for a particle which is moving under the influence of potential

$= -V(l_1, l_2)(\frac{g^2}{NT^2})$. Domain wall solution will then correspond to the particle trajectory which for $z \to -\infty$ approaches one of the minima of $V(l)$, while for $z \to +\infty$ it approaches a different minima of $V(l)$. Fig.4 shows the plot of the inverted potential, i.e. $-V(l_1, l_2)$. Minima of V now become maxima of $-V$. As we mentioned above, the domain wall solution will correspond to the particle trajectory starting at the top of one of the hills, say, at P, and ending at the top of another hill, say at Q. With this picture it becomes immediately obvious that the domain wall solution cannot go through $l = 0$. The particle starting at P, and rolling down to $l = 0$ cannot turn back to end up at Q. It will rather go to the other side and roll away downward, as shown by the dashed curve in Fig.4. To end up at Q, the trajectory must loop back before reaching $l = 0$, as shown by the solid curve in Fig.4. Thus, l must remain non-zero inside the domain wall. However, in contrast to the real scalar field case, it is very difficult to numerically

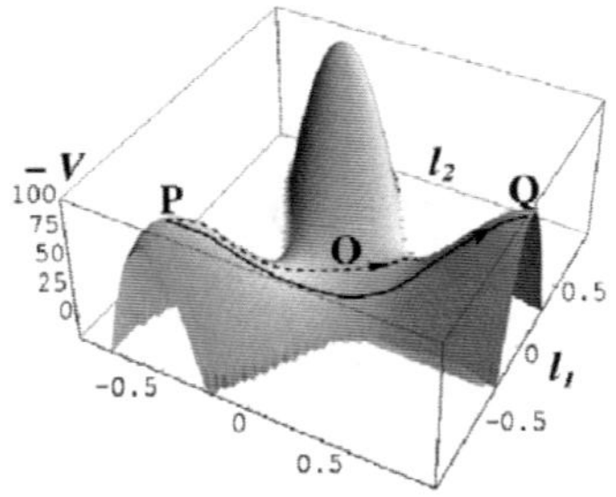

Figure 4. Plot of the inverted potential, i.e. $-V(l_1, l_2)$. Plot range is restricted for negative values to show the shape of the potential clearly.

solve Eq.(4) to get this "generalized" bounce solution. In the absence of such a technique for complex $l(x)$, we resort to numerical minimization [3] of the energy to determine the appropriate solutions. For time independent case, the energy density from Eq.(2),(3) is,

$$\mathcal{E} = \frac{NT^2}{g^2} | \nabla l |^2 + V(l). \tag{5}$$

For the energy minimization (see ref.[3] for detailed discussion), we have used a code as was used in ref.[6]. To determine the domain wall solution we only need to consider the profile of l in one dimension (along z). We fix the values of l at the two boundaries of the one dimensional lattice as $l = l_{01}$ and $l = l_{02}$ where l_{01} and l_{02} are the values of l corresponding to the two distinct minima of $V(l)$. For the intermediate lattice points we use an interpolating configuration to get the final wall solution (Fig.5). The surface tension of the wall in this case is found out to be about 7 GeV/fm^2. Note, l remains non-zero in the profile of the wall, as we argued above.

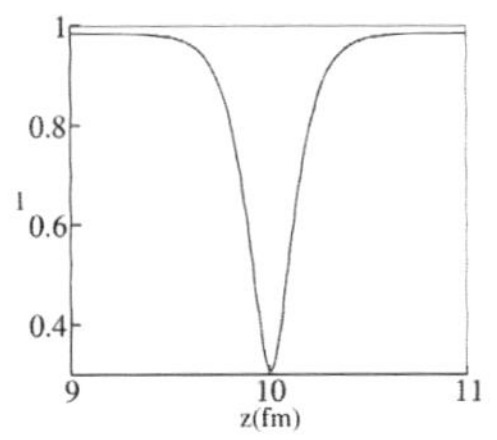

Figure 5. The profile for the domain wall solution (centered near $z = 10$ fm) for $T = 400MeV$. Note that l remains non-zero inside the wall, with the lowest value of l being about 0.3. Wall thickness (where $l \simeq 0.9$) is seen to be about 0.5 fm.

3. Junctions of $Z(3)$ interfaces, the string profile
To determine the profile of the QGP string, we use the numerical minimization with a two dimensional lattice. The string is taken to be perpendicular to the lattice. To get the string

configuration (see, ref. [3] for detailed discussion), we fix the center of the string [6] at the middle of an elementary lattice square. We then fix l at the four lattice points forming this particular lattice square. Everywhere else l is fluctuated, and the energy is minimized to get lowest energy string profile. Fig.6 shows the surface plot of $-l$ showing clearly the string, connected to three interfaces. To separate out the string core energy contribution from the contribution of the energy of the interfaces, we use following parametrization of $E(r)$,

$$E(r) = E_0(r), \quad 0 < r < r_0 \tag{6}$$

$$E(r) = E_0(r_0) + 3\sigma(r - r_0), \quad r > r_0 \tag{7}$$

Here, $E_0(r)$ denotes the core energy contribution which should be the dominant contribution up to some distance r_0. Beyond r_0, the linear contribution of interfaces becomes significant. By plotting $E(r)$ vs. r, we can get σ as well as the core energy $E_0(r_0)$. Fig.7 shows this plot. We have fitted the large r part of the plot with a straight line (dashed line). Its slope is found to be about 23 GeV/fm^2, giving the value of $\sigma \simeq 7.7$ GeV/fm^2 in reasonably good agreement with our numerical estimate for the wall tension given in the previous section. The core energy $E_0(r_0)$ is found to be about 2.7 GeV/fm.

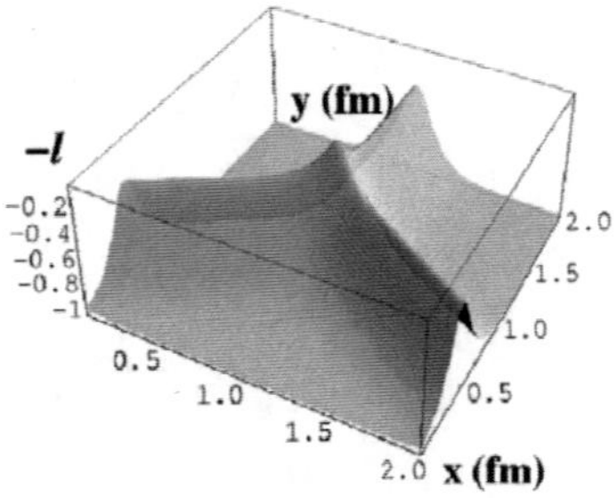

Figure 6. Surface plot of $-l$ for a small portion of the two dimensional lattice.

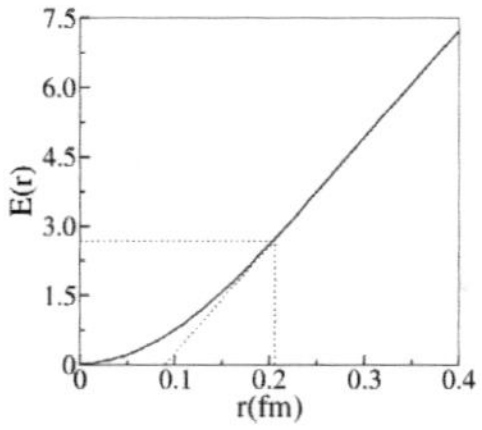

Figure 7. Plot of $E(r)$ in GeV/fm vs. r.

4. Conclusions

We have discussed special configurations of junctions of $Z(N)$ interfaces for an SU(N) gauge theory and have shown that these correspond to topological strings which have confining phase in the core. The energy of the string is found out to be about 2.7 GeV/fm at $T \simeq 2T_c$. Lattice simulation of pure gauge theories should be able to investigate properties of these strings. With the interpretation that quark contributions lead to explicit breaking of this Z_N symmetry [2, 4], such QGP strings may play important role in the evolution of the quark-gluon plasma phase and in the dynamics of quark-hadron transition.

Acknowledgments

I would like to thank the organizers for a very stimulating and enjoyable conference.

[1] McLerran L D and Svetitsky B 1981 *Phys. Rev.* D **24** 450
[2] Pisarski R D 2000 *Phys. Rev.* D **62** 111501
[3] Layek B, Mishra A P and Srivastava A M 2005 *Preprint* hep-ph/0502250
[4] Dumitru A and Pisarksi R D 2001 *Phys. Lett.* B **504** 282
[5] Coleman S 1985 *Aspects of symmetry* Cambridge University Press
[6] Srivastava A M 1993 *Phys. Rev.* D **47** 1324

Institute of Physics Publishing Journal of Physics: Conference Series **50** (2006) 438–441
doi:10.1088/1742-6596/50/1/066 5th International Conference on Physics and Astrophysics of Quark Gluon Plasma

Dynamical screening in a quark gluon plasma

Munshi G. Mustafa† [3] , Purnendu Chakraborty† , and Markus H. Thoma‡

† Theory Division, Saha Institute of Nuclear Physics, 1/AF Bidhannagar, Kolkata, India.
‡Centre for Interdisciplinary Plasma Science, Max-Planck-Institut für extraterrestrische Physik, P.O. Box 1312, 85741 Garching, Germany

Abstract. We calculate the screening potential of a fast parton moving through a quark-gluon plasma in the framework of semi-classical transport theory. We found an anisotropic potential showing a minimum in the direction of the parton velocity. Possible consequences of this potential on binary states in a quark-gluon plasma are discussed.

Screening of charges in a plasma is an important collective effect in plasma physics. In a classical isotropic and homogeneous plasma the screening potential of a point-like test charge Q at rest is modified from Coulomb potential into a Yukawa potential [1] as

$$\phi(r) = \frac{Q}{r}\exp(-m_D r) \tag{1}$$

with the Debye mass (inverse screening length) m_D ($\hbar = c = k_B = 1$). In the quark-gluon plasma the Debye mass of a chromoelectric charge follows from the static limit of the longitudinal polarization tensor which in the high temperature limit and one loop order is given by [2, 3, 4, 5],

$$\Pi_{00}(\omega = 0, k) = -m_D^2 = -g^2T^2\left(1 + \frac{n_f}{6}\right), \tag{2}$$

where g is the strong coupling constant and n_f the number of light quark flavors in the QGP with $m_q \ll T$.

The modification of the confinement potential below the critical temperature into a Yukawa potential above the critical temperature might have important consequences for the discovery of the QGP in relativistic heavy-ion collisions. Bound states of heavy quarks, in particular the J/ψ meson, which are produced in the initial hard scattering processes of the collision, will be dissociated in the QGP due to screening of the quark potential and break-up by energetic gluons [6]. Hence the suppression of J/ψ mesons have been proposed as one of the most promising signatures for the QGP formation [7]. On the other hand, the formation of colored bound states, *e.g.*, qq, $\bar{q}\bar{q}$, gg, of partons at rest has also been claimed [8] above the critical temperature ($2T_c$ - $3T_c$) by analyzing lattice data, indicating that the plasma behaves as a strongly coupled quark-gluon plasma.

Earlier in most calculations of the screening potential in the QGP, the test charge was assumed to be at rest. However, quarks and gluons coming from initial hard processes receive a transverse momentum which causes them to propagate through the QGP [9]. In addition, hydrodynamical

[3] mustafa@theory.saha.ernet.in

models predict a radial outward flow in the fireball [10]. Hence, it is of great interest to estimate the screening potential of a parton moving [11] relatively to the QGP. Chu and Matsui [12] have used the Vlasov equation to investigate dynamic Debye screening for a heavy quark-antiquark pair traversing a quark-gluon plasma. They found that the screening potential becomes strongly anisotropic.

The screening potential of a moving charge Q with velocity v follows from the linearized Vlasov and Poisson equations as [13, 14]

$$\phi(\vec{r},t;\vec{v}) = \frac{Q}{2\pi^2}\int d^3k \frac{\exp\left[-i\vec{k}\cdot(\vec{r}-\vec{v}t)\right]}{k^2 \mathrm{Re}[\epsilon_l(\omega=\vec{k}\cdot\vec{v},k)]}. \tag{3}$$

The dielectric function following from the semi-classical Vlasov equation describing a collisionless plasma is related to the high-temperature limit of the polarization tensor. For example, the longitudinal dielectric function following from the Vlasov equation is given by [3, 4, 5]

$$\epsilon_l(\omega,k) = 1 - \frac{\Pi_{00}(\omega,k)}{k^2} = 1 + \frac{m_D^2}{k^2}\left(1 - \frac{\omega}{2k}\ln\frac{\omega+k}{\omega-k}\right), \tag{4}$$

where the only non-classical inputs are Fermi and Bose distributions instead of the Boltzmann distribution. The gluon self-energies derived within the hard thermal loop approximation [15] have been shown to be gauge invariant and the dielectric functions obtained from these are therefore also gauge invariant.

For small velocities, $v \to 0$, i.e. $\omega \ll k$, we obtain

$$\epsilon_l(\omega \ll k) = 1 + \frac{m_D^2}{k^2}, \tag{5}$$

from which again the (shifted) Yukawa potential results [14]

$$\phi(\vec{r},t;\vec{v}) = \frac{Q}{|\vec{r}-\vec{v}t|}\exp(-m_D|\vec{r}-\vec{v}t|). \tag{6}$$

In the general case, for parton velocities v between 0 and 1, we have to solve (3) together with (4) numerically. Since the potential is not isotropic anymore due to the velocity vector $\vec{v}$, we will restrict ourselves only to two cases, $\vec{r}$ parallel to $\vec{v}$ and $\vec{r}$ perpendicular to $\vec{v}$, i.e., for illustration we consider [11] the screening potential only in the direction of the moving parton or perpendicular to it.

In Fig.1 the screening potential ϕ/Q in $\vec{v}$-direction is shown as a function of $r' = r - vt$, where $r = |\vec{r}|$, between 0 and 6 fm for various velocities. For illustration we have chosen a strong fine structure constant $\alpha_s = g^2/(4\pi) = 0.3$, a temperature $T = 0.25$ GeV, and the number of quark flavors $n_f = 2$. The shifted potentials depend only on v and not on t as it should be the case in a homogeneous and isotropic plasma. For $r' < 1$ fm one observes that the fall-off of the potential is stronger than for a parton at rest. The reason for this behavior is the fact that there is a stronger screening in the direction of the moving parton due to an enhancement of the particle density in the rest frame of the moving parton.

In addition, a minimum in the screening potential at $r' > 1$ fm shows up. For example, for $v = 0.8$ this minimum is at about 1.5 fm with a depth of about 8 MeV. A minimum in the screening potential is also known from non-relativistic, complex plasmas, where an attractive potential even between equal charges can be found if the finite extension of the charges is considered [16]. A similar screening potential was found for a color charge at rest in Ref.[17], where a polarization tensor beyond the high-temperature limit was used. However, this approach has its limitation as a gauge dependent and incomplete (within the order of the coupling constant) approximation for the polarization tensor was used [15]. Obviously, a minimum in the interparticle potential in a relativistic or non-relativistic plasma is a general feature if one goes beyond the Debye-Hückel approximation by either taking quantum effects, finite velocities, or finite sizes of the particles into account.

We also note that Chu and Matsui [12] did not report the existence of a minimum in the potential of a quark traversing the QGP. However, in their Fig.1(d) a negative value of the potential of a fast quark ($v = 0.9$) in the direction of the quark velocity is shown. Since the potential has to tend to zero for large distances, this implies a minimum in the screening potential. This minimum was not found because the potential was plotted only for the limited range $0 < r < 1/m_d \simeq 0.35$ fm for our choice of the parameters.

The minimum could give rise to bound states, e.g., of diquarks, if thermal fluctuations do not destroy them. The two-body potential, associated with the dipole fields created by two test charges Q_1 and Q_2 at $\vec{r}_1$ and $\vec{r}_2$ with velocities $\vec{v}_1$ and $\vec{v}_2$, can be written as

$$\begin{aligned}\Phi(\vec{r}_1 - \vec{r}_2, \vec{v}_1 - \vec{v}_2, t) &= \frac{Q_1 Q_2}{4\pi^2} \int d^3k \left\{ \frac{e^{[i\vec{k}\cdot(-(\vec{r}_1-\vec{r}_2)-(\vec{v}_1-\vec{v}_2)t)]}}{k^2 \mathrm{Re}[\epsilon_l(\omega = \vec{k}\cdot\vec{v}_1, k)]} \right. \\ &\left. + \frac{e^{[i\vec{k}\cdot((\vec{r}_1-\vec{r}_2)-(\vec{v}_1-\vec{v}_2)t)]}}{k^2 \mathrm{Re}[\epsilon_l(\omega = \vec{k}\cdot\vec{v}_2, k)]} \right\}. \end{aligned} \tag{7}$$

For comoving quarks, $\vec{v}_1 = \vec{v}_2$, this two-body potential reduces to the one-body potential (3), showing the attraction between the quarks which could give rise to a bound state. Colored bound states, e.g., diquarks, of partons at rest have also been claimed by analyzing lattice data [8]. For a comoving diquark system, where the interaction between two color charges $Q_1 Q_2 > 0$, the two body potential shows an attractive minimum and may therefore indicate a possible appearance of a diquark bound state as compared to partons at rest where the two body potential is only repulsive. On the other hand, for a quark-antiquark system, where $Q_1 Q_2 < 0$, the two-body potential is inverted, showing a maximum. This may lead to short living mesonic resonances and an enhancement of the attraction between quarks and antiquarks of mesonic states moving through the QGP.

The results for $\vec{r}$ perpendicular to $\vec{v}$ are shown in Fig.2, where the potential is shown as a function of $|\vec{r} - \vec{v}t| = \sqrt{r^2 + v^2t^2}$ between 0.1 and 1 fm. Here we consider only the case $t = 0$ since at $t > 0$ and $v > 0$ there is no singularity in the potential due to $\sqrt{r^2 + v^2t^2} > 0$ for all r. Hence the potential is cut-off artificially at small distances if plotted as a function of $\sqrt{r^2 + v^2t^2}$. In contrast to the parallel case (Fig.1) the fall-off of the potential at larger values of v is less steep, i.e. the screening is reduced as it is expected since the formation of the screening cloud is suppressed at large velocities. Also no minimum in the potential is found.

Summarizing, we have calculated the screening potential of a fast color charge moving through the QGP from semi-classical transport theory corresponding to the high-temperature limit. As in Ref.[12] the potential is found to be strongly anisotropic. The screening is reduced in the perpendicular direction of the moving parton but increased in the direction of the moving parton, which may lead to a modification of the J/ψ suppression. In addition, we found a new feature of the screening potential of a fast parton in a QGP: a minimum in the potential shows up which could give rise to bound states of, for example, diquarks if not destroyed by thermal fluctuations. For a quark-antiquark pair this minimum turns into a maximum which could cause short living mesonic resonances. Combining the effect of reduced screening in perpendicular direction and the presence of a maximum in parallel direction we expect a stronger binding of moving J/ψ mesons with respect to the QGP than of J/ψ mesons at rest. The consequences, e.g, for the J/ψ yield should be investigated in more detail using hydrodynamical models or event generators for

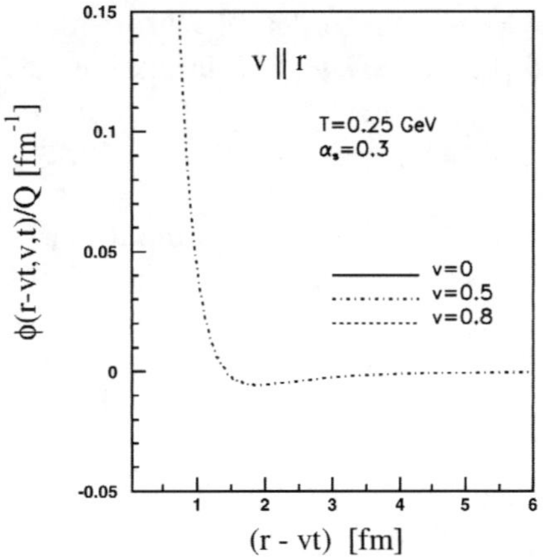

Figure 1. Screening potential parallel to the velocity of the moving parton in a QGP as a function of $r - vt$ (0 to 6 fm) for $v = 0$, 0.5, and 0.8.

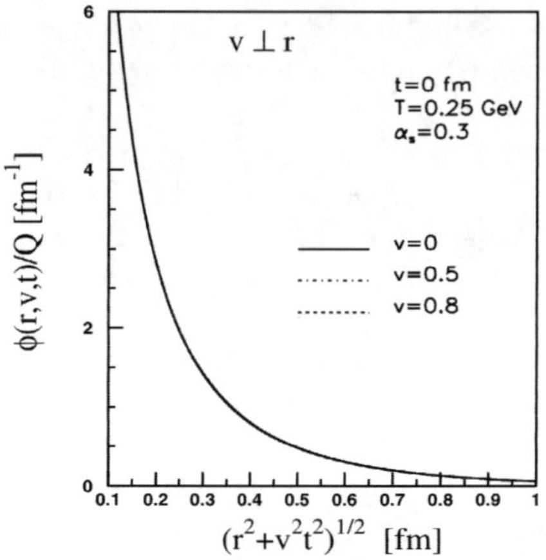

Figure 2. Screening potential perpendicular to the velocity of the moving parton in a QGP as a function of $|\vec{r}-\vec{v}t|$ (0.1 to 1 fm) for $v = 0$, 0.5, and 0.8.

the space-time evolution of the fireball. Finally, let us note that our results also apply to other ultrarelativistic plasmas such as an electron-positron plasma in Supernova explosions. In this case one simply has to replace the Debye mass m_D by eT, where e is the electron charge.

References

[1] Pitaevskii L P and Lifshitz E M, *Physical Kinetics* (Pergamon Press, Oxford, 1981).
[2] Kajantie K and Kapusta J I, Ann. Phys. (N.Y.) **160**, 477 (1985).
[3] Silin V P, Sov. Phys. JETP **11**, 1136 (1960).
[4] Klimov V V, Sov. Phys. JETP **55**, 199 (1982).
[5] Weldon H A, Phys. Rev. D **26**, 1394 (1982).
[6] Patra B K and Srivastava D K, Phys. Lett. B **505**, 113 (2001).
[7] Matsui T and Satz H, Phys. Lett. B **178**, 416 (1986).
[8] Shuryak E V and Zahed I, Phys. Rev Dbf 70, 054507 (2004).
[9] Gyulassy M, Plümer M, Thoma M H, and Wang X N, Nucl. Phys. **A538**, 37c (1992).
[10] Huovinen P, Kolb P F, Heinz U W, Ruuskanen P V and Voloshin S A, Phys. Lett. B **503**, 58 (2001).
[11] Mustafa M G, Thoma M H and Chakraborty P, Phys. Rev. C **71** , 017901 (2005).
[12] Chu M C and Matsui T, Phys. Rev. D **39**, 1892 (1989).
[13] Ichimaru S, *Basic Principles of Plasma Physics* (Benjamin, Reading, 1973).
[14] Spatschek K H, *Theoretische Plasmaphysik* (Teubner, Stuttgart, 1990).
[15] Braaten E and Pisarski R, Nucl. Phys. B337, 569 (1990).
[16] Tsytovich V N, JETP Lett. **78**, 1283 (2003).
[17] Gale C and Kapusta J, Phys. Lett. B **198**, 89 (1987).

Jet Characterization at RHIC Energy

P.K. Sahu and S.C. Phatak
Institute of Physics, Sachivalaya Marg
Bhubaneswar - 751 005, India
E-mail: pradip@iopb.res.in

Abstract. Jet quenching is considered to be one of the signatures of the formation of quark gluon plasma. In order to investigate the jet quenching, it is necessary to know the properties of jets produced in relativistic heavy ion collisions. In this calculation, we propose that determination of flow parameters may be used to identify and characterize the jets.

1. Introduction

The quenching of jets in relativistic heavy ion collisions has been proposed to be one of the signatures of the formation of quark gluon plasma [1]-[3]. In order to understand the jet quenching, it is necessary to know the properties of jets. In nucleus-nucleus collisions the experimental identification of jets and determination of their properties is quite difficult task because during the collision process, a large number of hadrons are produced and the particles belonging to a jet, constitute a very small fraction of these. Recently, a method [4] of extracting the properties of the jets present in a heavy ion collision event has been proposed. In this method, a transverse momentum cut has been used to reduce the number of background particles which are used for the analysis. This is based on the azimuthal flow parameters of the particles produced in a single event. This treatment is simple and is able to identify the events having single jet and estimate the number of particles in the jet as well as the jet opening angle reasonably well for different transverse momentum cuts. Basically, the method is successful when the number of jet particles are 5-10% of total number of particles in an event. Using transverse energy cut ensures that low transverse momentum particles, which are abundant and do not belong to jet are removed. At the same time, one has to ensure that the number of background particles are not too small since the fluctuation effects would dominate the flow coefficients and the determination of the properties of the jet will not be accurate.

In the following, the method([4]) is briefly discussed in Sec II. It is applied to the synthetic data as well as simulated data[5] in Section III and summary is presented in Section IV.

2. The Method

The flow coefficients for particles distributed in azimuthal angle ϕ are given by

$$v_m = \Big\langle \cos(m(\phi - \phi_E)) \Big\rangle, \tag{1}$$

where the angular brackets stand for averaging over all the particles. The flow coefficients v_m's are measured with respect to the reaction plane, defined by azimuthal angle ϕ_E. Since the reaction plane is not readily defined in an experiment, it is convenient to compute the flow coefficients using the pairwise correlations of all the particles[4]. It turns out that the computation of pairwise correlated quantities not only removes the uncertainty in the knowledge of the reaction plane but also helps in reducing the statistical fluctuations. In this case the square of the flow coefficients v_m^2 are given by

$$v_m^2 = \Big\langle \cos(m(\phi_i - \phi_j)) \Big\rangle_{i,j} = \frac{\sum_{i,j} \cos(m(\phi_i - \phi_j))}{N^2}. \quad (2)$$

It has been shown in ref([4]) that if the data consists of the number of background particles (N_b) distributed uniformly over the azimuthal angle and the number of jet particles (N_j) confined within a small azimuthal angle $\Delta\phi$, the theoretical estimate of the flow coefficients is

$$v_m^2 = \frac{N_j^2}{N^2}\Big[1 - m^2\sigma^2\Big] = \frac{N_j^2}{N^2}\Big[1 - \frac{\Delta\phi^2 m^2}{12}\Big]. \quad (3)$$

Here σ is the variance of the azimuthal distribution of the jet particles and thelast part in equation above results if the jet particles are uniformly distributed within $\Delta\phi$.

One can compute the flow coefficients from the data using eq(2) and plot v_m^2 *vs* m^2. A linear fit to the square of the flow coefficients would then yield the number of jet particles and the jet opening angle $\Delta\phi$. One should note that:

(i) Even in the absence of jet particles, the flow coefficients are nonzero. However, in this case, the number of jet particles computed from the intercept are small and consistent with zero.

(ii) If N_j is very small in comparison with the total number of particles, the flow coefficients would be small. In that case, the fluctuations dominate and the result is close to the case when there are no jet particles.

(iii) The expression in eq(3) is approximate and valid when $\Delta\phi^2 m^2 << 12$. Thus, not many values of m can be included in the fitting procedure.

(iv) If the data has more than one jet, one does not get a simple expression for flow coefficients. In such a case, the procedure outlined above is not expected to work. However, it has been shown[4] that if the two jets are back-to-back, the flow coefficients for even m are much larger than those for odd m.

3. The Calculations and Discussions

The background particles in an event are constructed by taking the output from an event generator. For this purpose we have used the generator[5] which uses dynamical hadron string cascade model. For our analysis, we have considered 130 GeV collision of gold nuclei. To these 'background' particles, the jet particles are added by hand. We choose different numbers of jet particles and assume that their transverse momenta are larger than 2 GeV. The direction of the jet as well as the azimuthal angle of the jet particles are chosen randomly. We, however assume that the probability of the azimuthal angle of the jet particles is constant over the opening angle $\Delta\phi$. In the following we shall consider the analysis when the event has only a jet (no background) and both jet and background particles are present.

3.1. Jet Only Events

Let us now consider the case where there are no background particles. This is an ideal case where we expect to obtain the best results from our analysis. The results display for the extracted values of number of jet particles (N_o) and the opening angle ($\Delta\phi_o$) and compare these with the corresponding input values (N_g and $\Delta\phi_g$ respectively) in Figs. 1 and 2. The calculations are repeated a large number of times with different sets of background particles as well as jet particles so as to estimate the possible error in the extracted quantities. These results seem to indicate that it may be possible to determine the number of jet particles reasonably well. As for opening angles, we find that the determination may not be very much reliable if the opening angle exceeds 0.75.

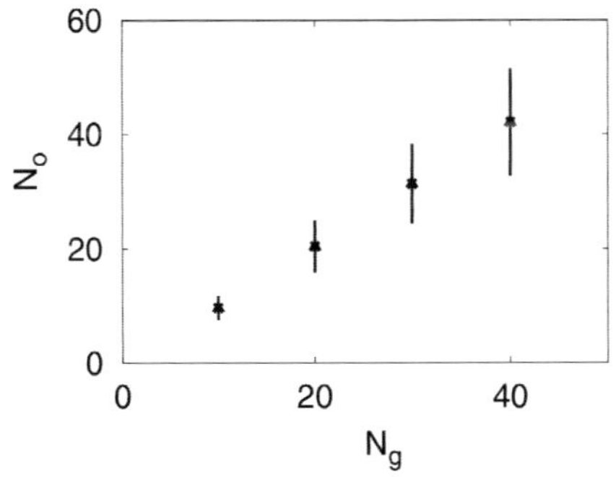

Figure 1. Extracted number of jet particles (N_o) *vs* the corresponding input values (N_g).

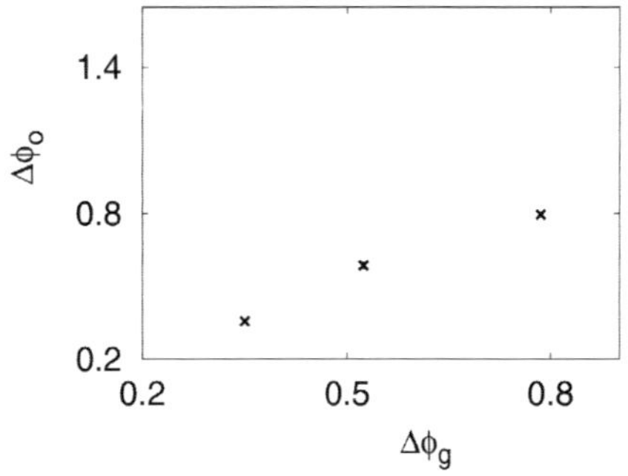

Figure 2. Extracted opening angle ($\Delta\phi_o$) *vs* the corresponding input values ($\Delta\phi_g$).

3.2. Events With Jets and Background

We now consider the analysis of events in which a jet is produced in the presence of 'background' particles. For this analysis, a large number of events (central collisions) have been considered. The number of particles produced in a central event are large and this may make the determination of jet difficult. It is clear that, for our method to work, the number of background particles should not be very large, so that the jet is not completely swamped by the background particles. One way of controlling the number of background particles is to remove the particles having transverse momenta cutoff larger than 1.5 and 2 GeV, these choices are given in ref[4]. We have therefore analyzed the events by including the background particles having transverse momenta larger than 1.5 and 2 GeV.

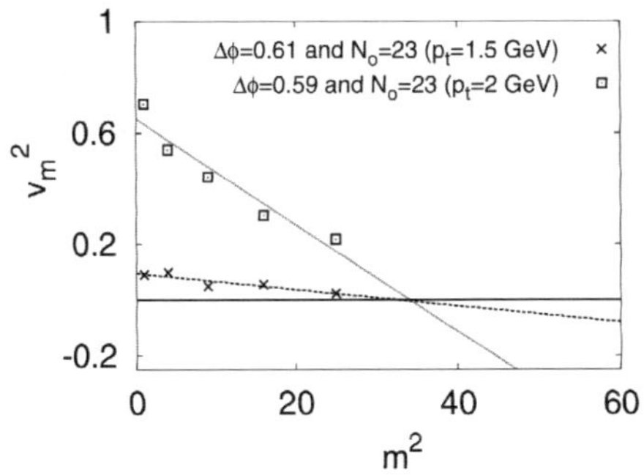

Figure 3. For single jet case, v_m^2 *vs* m^2 for the central collisions (N=74 and 28 for p_t=1.5 and 2 GeV, respectively).

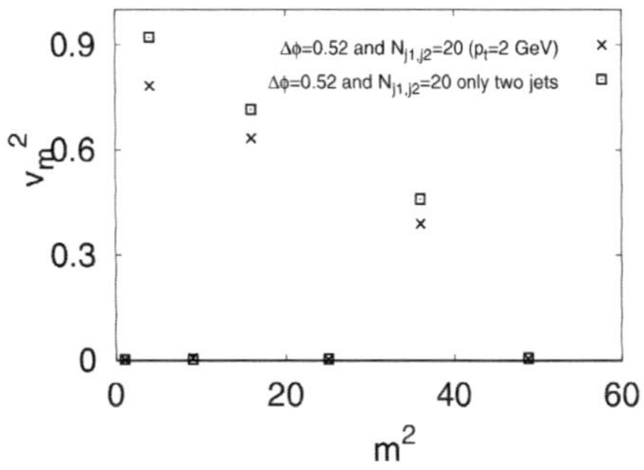

Figure 4. For two jets case, v_m^2 *vs* m^2 for the central collisions (background particles with two jets and only two jets without background), if the two jets are back-to-back.

Fig. 3 displays a plot of the square of the flow coefficients v_m *vs* m^2 for the two transverse momentum cuts. Here the opening angle is $\pi/6$ and the number of jet particles is 20 for single jet case. We find that in this figure the extracted values of number of jet particles as well as the opening angle are within 25 % of the input values. The plots in Fig. 3 are for the analysis of one particular event. We have repeated the analysis for a large number of such events and these are displayed in Figs. 5 and 6.

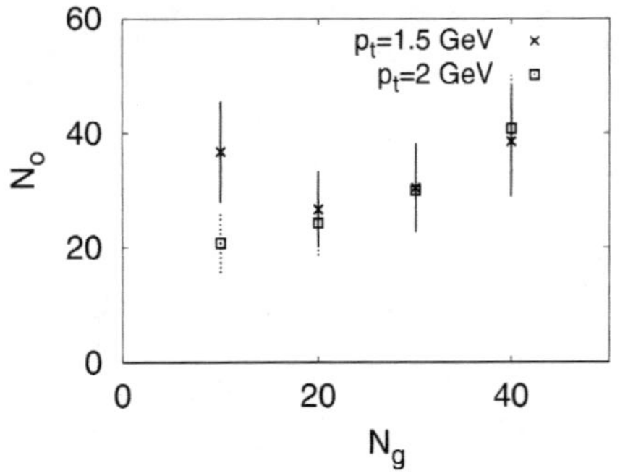

Figure 5. The extracted numbers of jet particles *vs* the corresponding input values are shown for $p_t = 1.5$ GeV and $p_t = 2$ GeV, respectively.

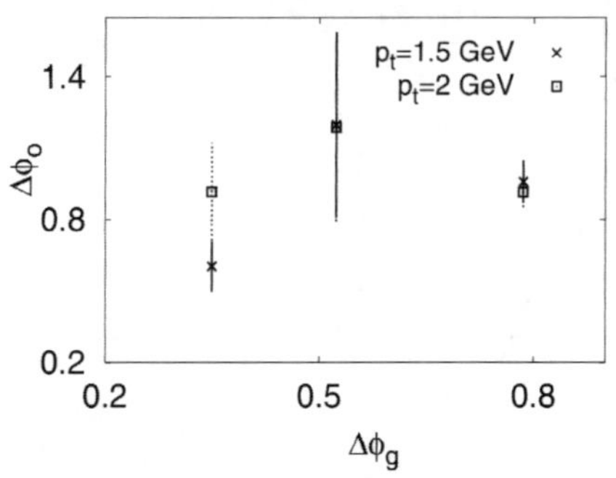

Figure 6. The extracted opening angles *vs* the corresponding input values are displayed for $p_t = 1.5$ GeV and $p_t = 2$ GeV, respectively.

The plots in Figs. 5 and 6 show that there is a linear correlation between the particle numbers extracted and the corresponding input numbers when the number of jet particles are larger than 10. Also, the correlation is better for 2 GeV cutoff. The extracted values of the number of jet particles for different input opening angles have similar behavior. Thus, it seems that the extraction of the number of jet particles is quite robust but that of the jet opening angle from the flow analysis is less reliable.

Fig. 4 displays a plot of v_m *vs* m^2 for two jet case, where the opening angle is $\pi/6$, the number of jet particles in each jet is 20 and the angle between the two jets is π. It clearly shows in this figure that the flow coefficients for odd m's are generally smaller than those of even m's as discussed in ref.[4].

4. Summary

By using flow analysis method we have shown that when the data has a jet with the number of jet particles in excess of 10, the higher flow coefficients are large enough to be measured. In particular, we show that transverse momentum cuts between 1.5 and 2 GeV can be used to enhance the effect of jet particles on the flow coefficients. Further, by analyzing the flow coefficients of different orders, we show that it is possible to extract the information of the number of particles in the jet and jet opening angle.

We have also considered the case when two jets are present. For such a case, we have argued that an important information to be determined in this case is the angle between the two jets. We have shown that it may be possible to estimate this angle between the two jets. This method is useful to apply the RHIC data for the identification of jets.

References

[1] Wang X N and Gyulassy M, 1992 *Phys. Rev. Lett.* **68** 1480

[2] Gyulassy M, 2000 *Nucl. Phys.*B **571** 197

[3] Muller B, 2003 *Phys. Rev.* C **67** 061901

[4] Phatak S C and Sahu P K, 2004 *Phys. Rev.* C **69** 024901

[5] Nara Y, Otuka N, Ohnishi A, Niita K, and Chiba S, (2000) *Phys. Rev.* C **61** 024901

Pions emerging from an Arbitrarily Disoriented Chiral Condensate.

B.Bambah[1]and
K.V.S. Shiv Chaitanya
[1] School of Physics
University of Hyderabad, Hyderabad-500 046,India
E-mail: bbsp@uohyd.ernet.in

C.Mukku[2]
[2] Department of Mathematics Panjab University, Chandigarh-160014,India
[2] Department of Computer Science & Applications
Panjab University, Chandigarh-160014,India
E-mail: c_mukku@yahoo.com

Relativistic heavy ion collisions either with fixed target or with head on collisions are providing results which suggest that the chiral phase transition does take place with temperatures crossing 170Mev. The comparatively large sizes of the heavy ions involved in the collisions along with their energy/nucleon, provides a large interaction region for the quarks and gluons. This region is where the Quark Gluon Plasma (QGP) forms. The picture we therefore have is that of a collision region where a QGP is formed in a highly non-equilibrium situation with chiral symmetry being restored. This plasma then expands and cools, causing the restored chiral symmetry to be broken once again. In a recent paper [1], we have given a quantum field theoretical model providing multiplicities of pions which would be observed if the QGP undergoes a rapid quench in its expansion or if the expansion is adiabatic. It was also shown how the rapid quench scenario, which is appropriate for the formation of the disoriented chiral condensate (DCC) has a enhancement in the multiplicities of the observed pions. Thus the enhancement signatures in the pion multiplicities are signals for the formation of the DCC also. In our analysis, we had assumed the expansion of the QGP to be isotropic for simplicity. However, recent results from RHIC suggest that there may be anisotropy built in from the nature of the collisions. Not all collisions will be head on with zero impact parameter. There will be instances where there will be non-zero impact parameter causing the impact region to be elliptical and the corresponding QGP to have an elliptical flow. From the nature of our quantum field theoretical model, borrowing from the studies of scalar fields on expanding universes, such as the Friedmann-Robertson-Walker universe, it is easy to extend the study to include anisotropy such as that caused by an elliptic flow. We shall report these results elsewhere, in a separate communication [2]. Here, we shall be examining another aspect of the restoration of chiral symmetry. The particular form of our model allows for many metastable states through which the DCC can form. In particular, we find that mixing between various erstwhile pion states in the collision region can produce different signals in the multiplicities of the final state pions.

We start with an $O(4)$ quartet of scalar fields in order that we can construct the dynamics of quenched pions in the formation of a disoriented chiral condensate. The background field is the classical disoriented vacuum and the Hamiltonian of the quantum fluctuations around the background field is derived from the $O(4)$ sigma model keeping one-loop quantum corrections (quadratic order in the fluctuations)[1]. The resulting dynamical Hamiltonian for the pion and sigma fields fields in terms of the neutral , charged pion and sigma creation and annihilation operators ($a, a^\dagger, c, c^\dagger, b\ b^\dagger$, and $d, d^\dagger$) is[1]:

$$\begin{aligned}
H &= \int \frac{d^3\underline{k}}{(2\pi)^3}\frac{1}{2}\{\frac{\omega_\pi}{a^3}(a_k^\dagger a_k + a_k a_k^\dagger) + \frac{\omega_\pi}{2a^3}(\frac{\Omega_\pi^2}{\omega_\pi^2} - 1)(a_k^\dagger a_k + a_k a_k^\dagger + a_{-k}a_k + a_{-k}^\dagger a_k^\dagger) \qquad (1)\\
&+ \frac{\omega_\Sigma}{a^3}(d_k^\dagger d_k + d_k d_k^\dagger) + \frac{\omega_\Sigma}{2a^3}(\frac{\Omega_\Sigma^2}{\omega_\Sigma^2} - 1)(d_k^\dagger d_k + d_k d_k^\dagger + d_{-k}d_k + d_{-k}^\dagger d_k^\dagger)\}\\
&+ \frac{\omega_\pi}{a^3}(b_k^\dagger b_k + c_k c_k^\dagger) + \frac{\omega_\pi}{2a^3}(\frac{\Omega_{\pi\pm}^2}{\omega_\pi^2} - 1)(b_k^\dagger b_k + c_k c_k^\dagger + b_{-k}c_k + c_{-k}^\dagger b_k^\dagger)\\
&+ \frac{\lambda a^3 v^2 cos^2(\rho) sin^2(\theta)}{4\omega_\pi}(b_k b_{-k} + b_k c_k^\dagger + c_k^\dagger b_k + c_k c_{-k} + c_k^\dagger c_{-k}^\dagger + c_k b_k^\dagger + b_k^\dagger c_k + b_k^\dagger b_{-k}^\dagger))\\
&+ \frac{\lambda a^3 v^2 cos(\rho) sin(\rho) sin^2(\theta)}{2\omega_\pi}\left(b_k a_{-k} + b_k a_k^\dagger + c_k^\dagger a_k + c_k a_{-k} + c_k^\dagger a_{-k}^\dagger + c_k a_k^\dagger + b_k^\dagger a_k + b_k^\dagger a_{-k}^\dagger\right)\\
&+ \frac{\lambda a^3 v^2 sin(\rho) sin(\theta) cos(\theta)}{\sqrt{\omega_\pi \omega_\Sigma}}\left(d_k a_{-k} + d_k a_k^\dagger + d_k^\dagger a_k + d_k^\dagger a_{-k}^\dagger\right)\\
&+ \frac{\lambda a^3 v^2 cos(\rho) sin(\theta) cos(\theta)}{\sqrt{\omega_\pi \omega_\Sigma}}\left(b_k d_{-k} + b_k d_k^\dagger + c_k^\dagger d_k + c_k d_{-k} + c_k^\dagger d_{-k}^\dagger + c_k d_k^\dagger + b_k^\dagger d_k + b_k^\dagger d_{-k}^\dagger\right)\}
\end{aligned} \qquad (2)$$

where

$$\frac{\omega_\pi^2(k)}{a^6} \equiv \frac{\omega_{\pi_0}^2(k)}{a^6} = \frac{\omega_{\pi_\pm}^2(k)}{a^6} = (m_\pi^2 + \frac{k^2}{a^2}) : \frac{\omega_\Sigma^2(k)}{a^6} = (m_\Sigma^2 + \frac{k^2}{a^2}) \qquad (3)$$

and

$$\begin{aligned}
\frac{\Omega_{\pi_0}^2 - \omega_{\pi_0}^2}{a^6} &= \lambda[(< \Phi^2 > - v^2) + 2v_3^2]\\
\frac{\Omega_{\pi\pm}^2 - \omega_{\pi\pm}^2}{a^6} &= \lambda[(< \Phi^2 > - v^2) + 2v_+ v_-]\\
\frac{\Omega_\Sigma^2 - \omega_\Sigma^2}{a^6} &= \lambda[(< \Phi^2 > - v^2) + 2\sigma^2]
\end{aligned} \qquad (4)$$

The background field in the disoriented phase is given by :

$$\begin{pmatrix} v_+ \\ v_- \\ v_3 = v \\ \sigma \end{pmatrix} \equiv < \Phi >, \qquad (5)$$

In keeping with the disorientation of the vacuum on a three sphere we parametrize the background field through three angles:

$$< \Phi > = \begin{pmatrix} vCos(\rho)Sin(\theta)Sin(\alpha) \\ vCos(\rho)Sin(\theta)Cos(\alpha) \\ vSin(\rho)Sin(\theta) \\ vCos(\theta) \end{pmatrix}. \qquad (6)$$

The dynamical evolution of a system governed by this Hamiltonian is considered from a non equilibrium state state of restored symmetry $< \Phi >= 0$ at high temperatures to the equilibrium state of broken symmetry $< \Phi >= v$ due to the subsequent expansion and cooling of the QGP. The time evolution can be modeled through a time dependence of the parameter $< \Phi > (t)$, which depends on the way in which the system relaxes to equilibrium. In addition there is a time dependence of the expansion factor $a(t)$ which allows for the rate of the expansion of the plasma. The following scenarios are possible:

A. A sudden quench from a state of restored symmetry to a state of broken symmetry in a rapidly cooling expanding plasma, the configuration of the field lags behind the expansion of the plasma.If τ is the time the state spends in the symmetry restored phase, a quench can be modeled by assuming that for 0 the vacuum expectation value $< \Phi^2 >= 0$ and for $t > \frac{\tau}{2}$ the vacuum relaxes to its value $< \Phi^2 >=< v^2 >$. In such a case the dynamical evolution of $< \Phi >$ could be modeled by $< \Phi > (t) = v\Theta[(t - \tau)]$ In addition in a realistic scenario the expansion coefficient $a(t)$ must also be a function of time and has been modeled appropriately in ref [1]. For a sudden quench scenario the Hamiltonian given in equation 1 diagonalizes to

$$H = \int \frac{d^3k}{(2\pi)^3} \frac{1}{2a^3} \{\Omega_\pi(k,t)\{(A_k^\dagger A_k + \frac{1}{2}) + (C_k^\dagger C_k + B_k^\dagger B_k + 1)\} + \Omega_\Sigma(k,t)(D_k^\dagger D_k + \frac{1}{2})\}. \quad (7)$$

by a unitary matrix given by a squeezed transformation

$$U(r,t) = e^{\int \frac{d^3k}{(2\pi)^3} r(k,t)\{(a_k^\dagger a_{-k}^\dagger - a_k a_{-k}) + (d_k^\dagger d_{-k}^\dagger - d_k d_{-k}) + (c_k b_{-k} + b_k c_{-k}) - (c_k^\dagger b_{-k}^\dagger + b_k^\dagger c_{-k}^\dagger)\}} \quad (8)$$

where r_k is the squeezing parameter related to the physical variables $\Omega_\pi(k,t)$ and $\omega_\pi(k)$ through

$$Tanh(2r_k) = \frac{(\frac{\Omega_k(t)}{\omega_\pi})^2 - 1}{(\frac{\Omega_k(t)}{\omega_\pi})^2 + 1} \quad (9)$$

The quench consists of a rapid change of frequency of the time dependent harmonic oscillator from $\Omega_\pi(k,t))$ to $\omega_\pi(k)$ though the time evolution of $< \Phi > (t)$ and $a(t)$. It can be shown that for a sudden frequency jump, such as that of a quench, there is substantial squeezing, resulting in an enhancement of low momentum modes signalling a DCC. The details of the evolution of the Hamiltonian of a system undergoing a sudden quench (enhanced squeezing) also show up in the difference in multiplicity distributions of the charged and neutral pions illustrated in Figure 1.

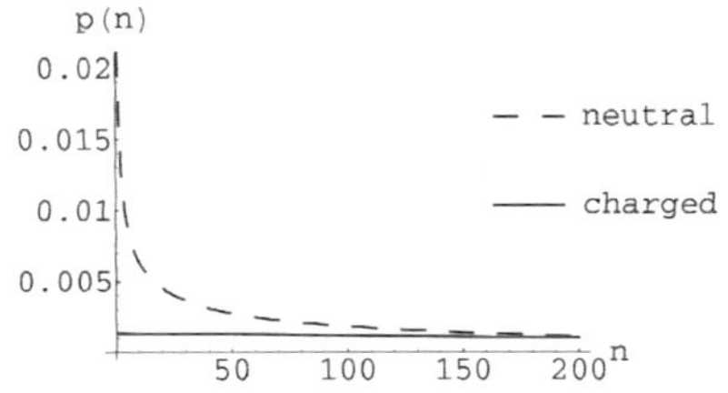

Figure 1. Shows the variation of P_{n0} (solid line)and P_{nc}(dashed line) with n for the **quenched** limit ($r_0 = 4$)

B. Another non-equilibrium situation that can arise is one where the system can go through a metastable disordered vacuum given by eqn[6] and then relax by quantum fluctuations to an equilibrium configuration. Here θ and ρ measure the degree of disorientation of the condensate in isospin space (for charge conservation $\alpha = \frac{\pi}{4}$) . The disorientation can be in the neutral sector $\rho = \frac{\pi}{2}$ in which case the angle θ mixes the neutral pion and sigma field resulting in oscillations and enhancement of neutral pions over the charged pions, we will call this case 1 . The disorientation can be in the charged sector $\theta = \frac{\pi}{2}$, where the angle ρ mixes the charged pions and the neutral pions resulting in charged oscillations and enhancement of charged pions over the neutral pions we will call this case 2. As an illustration of these effects we examine case 1 briefly. All the details of case 1 and 2 will be given in an expanded later communication [2]. For case 1 the examination of the Hamiltonian (obtained by putting $\rho = \frac{\pi}{2}$) shows a non-zero mixing term coming from the π_0-Σ sector. The misalignment of the vacuum through an angle θ induces a mixing of the two fields. The mixed fields are

$$\begin{pmatrix} A_{\theta(k)} \\ D_{\theta(k)} \end{pmatrix} = \begin{pmatrix} Cos(\theta) & Sin(\theta) \\ -Sin(\theta) & Cos(\theta) \end{pmatrix} \begin{pmatrix} a_k \\ d_k \end{pmatrix} \tag{10}$$

The diagonalization procedure for this case involves two squeezing transformations of the mixed fields

$$\begin{aligned} F_k &= \mu A_{\theta(k)} + \nu A^{\dagger}_{\theta(-k)} \\ G_k &= \rho D_{\theta(k)} + \sigma D^{\dagger}_{\theta(-k)}. \end{aligned} \tag{11}$$

The diagonalized form is

$$H_{\pi/2} = \int \frac{d^3\underline{k}}{(2\pi)^3} \{ \frac{\Omega_{\pi\pm}}{a^3} (C^{\dagger}_k C_k + B^{\dagger}_k B_k + 1) + \frac{\sqrt{\Omega_\pi \Omega_\Sigma}}{4a^3} \left((F^{\dagger}_k F_k + \frac{1}{2}) + (G^{\dagger}_k G_k + \frac{1}{2}) \right) \} \tag{12}$$

During time evolution again we have a time dependent frequency for the state of mixed neutral pions and sigma which changes from $\frac{\sqrt{\Omega_\pi(k,t)\Omega_\Sigma(k,t)}}{4a^3}$ to $\frac{\sqrt{\omega_\pi(k,t)\omega_\Sigma(k,t)}}{4a^3}$ this time through the time evolution of $\theta(t)$ from the disoriented value to 0 and $a(t)$. There is again an enhancement of neutral pions as a result of the DCC formation and furthermore a mixing in the neutral sector . In addition there are oscillations in the number of neutral pions as the disorientation θ changes with time shown in Figure 2.

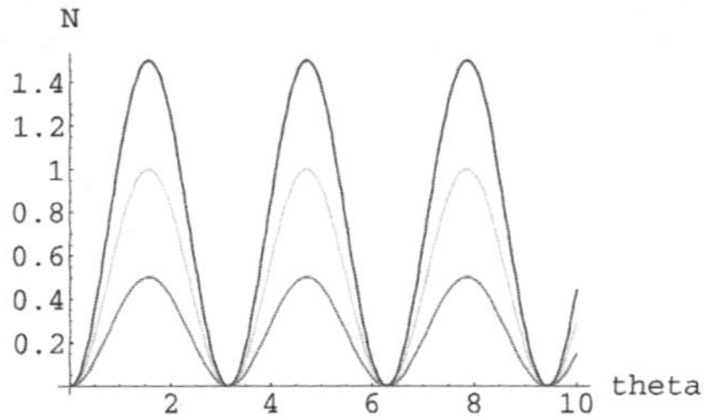

Figure 2. Shows the oscillations in the number of neutral pions as a function of the disorientation θ for the value of the squeezing parameters "$red = \nu = .5, green = \nu = 1, blue = \nu = 1.5$

A case of particular interest , in view of latest preliminary experimental results is case 2 i.e. $\theta = \frac{\pi}{2}$ because there is a mixing of charged and neutral fields due to the disorientation. This gives rise to interesting charge-neutral fluctuations which can be measured in a DCC detection experiment[3]. A full discussion of the detailed theory is given in [2].

To conclude we have modeled the evolution of the disoriented chiral condensate through both a sudden quench and with a transition through a metastable state with arbitrary disorientation and have shown that the total multiplicity distributions of charged and neutral pions functions are dramatic characteristic signals for the DCC and are related directly to the way in which the DCC forms. These are unambiguous, therefore they must be examined thoroughly in searches for the DCC [4]. We are encouraged by the first preliminary data of references [3] and [4], from the analysis of the WA98 experiment at the CERN SPS in which some events show an excess of photons (neutral pion excess) within the overlap region of charged and photon multiplicity detectors, the preliminary results given in ref [4] are shown in figure 3. In view of the results

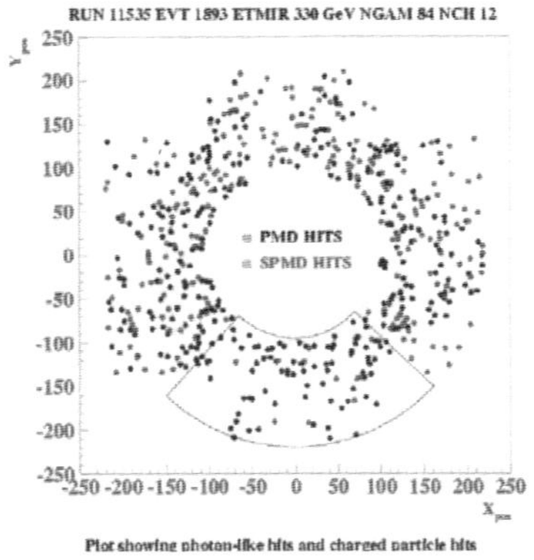

Figure 3. Shows the photon hits and charged particle hits in WA98 from ref. 4

presented in this section, we hope that the future will bring more exotic events with charge excess and neutral and charged multiplicity oscillations.

References

1. B. Bambah and C. Mukku, Phys. Rev. D. **70** (3) 0340001 (2004).
   B.A. Bambah and C.Mukku, Annals of Physics 314, 34,(2004)
2. B. A. Bambah , C. Mukku and K.V.S Shiv Chaitanya *Dynamics of the Dcc with Arbitrary Disorientation in Isospin Space*(In preparation).
3. T. Nayak ,Pramana 57 ,285,(2001).
4. M.M. Aggarwal et al., e-print archive:nucl-ex/0012004.
   Madan M Aggarwal, Pramana **60**, No.5, 987 , (2003)

Institute of Physics Publishing Journal of Physics: Conference Series **50** (2006) 451–454
doi:10.1088/1742-6596/50/1/069 5th International Conference on Physics and Astrophysics of Quark Gluon Plasma

Diquark interactions in quark-gluon plasma and their role in diquark stars

R S Kaushal, D Parashar, A K Sisodiya and V S Bhasin
Department of Physics and Astrophysics, University of Delhi, Delhi -110007, India.

Abstract. Within the framework of a polynomial field theory the diquark interaction energies in quark-gluon plasma are explicitly calculated. In particular, two- and three-diquark interaction energies are computed using ϕ^4 – theory and ϕ^6 – terms in the effective Lagrangian and their role in the stability of diquark stars is pointed out.

1. Introduction

It is well known that a composite structure of two quarks, called 'diquark' has played an important role in the studies of quark structure of hadrons [1]. The diquark system has also played an important role in recent years [2, 3, 4] in the studies of quark-gluon plasma (QGP). While the context in which the diquarks are used in the studies of elementary particle physics is slightly different, the role of diquarks is investigated by Donogue and Sateesh (DS) mainly within the framework ϕ^4-terms in the effective Lagrangian density

$$L_{eff} = \frac{1}{2}\left(\partial_\mu \phi^\dagger \partial^\mu \phi - m_{dq}^2 \phi^2\right) - \lambda(\phi^\dagger \phi)^2 - \beta(\phi^\dagger \phi)^3. \tag{1}$$

In this description the role of a particular type of diquarks, namely, spin-zero, color antitriplet diquarks has been investigated. On the other, hand this field-theoretic description of a system of diquarks, however, requires the involvement of all types of diquarks. In other words, it becomes rather essential to account for the plurality of color-spin states of two- and three-diquark states which, in fact, corresponds to ϕ^4- and ϕ^6-terms in (1), respectively.

In this paper, we highlight the relative importance of these states by way of investigating the interaction energies between the quarks. For this purpose, we work within the framework of a constituent quark model and make use of generalized Pauli principle.

2. Single diquark and double-diquark interaction energies

There exit four possible states of a pair of quarks (i.e., diquark) with color and spin quantum numbers $\{C, S\} = \{\bar{3},\ 0\}, \{\bar{3},\ 1\}, \{6,\ 0\}$ and $\{6,\ 1\}$. Here, assuming that only (u, d) flavor (I = 1/2) are present in QGP and taking the spatial configuration as symmetric, we write the interaction Hamiltonian in the spirit of constituent quark model as

$$H_{\text{int}} = -A\sum_{i,j} (\vec{\lambda}_i.\vec{\lambda}_j)(\vec{\sigma}_i.\vec{\sigma}_j), \tag{2}$$

where the indices i, j are used to label the interacting quarks and $\vec{\sigma}$ and $\vec{\lambda}$ are the usual Pauli spin and Gell-Mann color SU(3) matrices. In (2), the strength parameter A is determined to be 75/4 using the known values of N and Δ masses. Using (2), we have computed the interaction energy corresponding to different color-spin states of single diquark and the results so obtained are just half of the corresponding values given by DS [3]. The reason for this is that we have avoided the double counting of the terms while carrying out operation with H_{int}.

For the computation of interaction energies between two-diquark composites the following irreducible representations in both the color and the spin-space of the dq-dq system are considered:

$$|\bar{3}>_{dq_1} \otimes |\bar{3}>_{dq_2} \equiv |3_a>_{dq_1dq_2} \oplus |6_{as}>_{dq_1dq_2}, \tag{3a}$$

$$|6>_{dq_1} \otimes 16>_{dq_2} \equiv |15_s>_{dq_1dq_2} \oplus |15_{sa}>_{dq_1dq_2} \oplus |6_s>_{dq_1dq_2}, \tag{3b}$$

$$|\bar{3}>_{dq_1} \otimes |16>_{dq_2} \equiv |3_s>_{dq_1dq_2} \oplus |15_{as}>_{dq_1dq_2}, \tag{3c}$$

(For color states);

$$|1, S=0>_{dq_1} \otimes |1, S=0>_{dq_2} \equiv |1_a, S=0>_{dq_1dq_2}, \tag{3d}$$

$$|3, S=1>_{dq_1} \otimes |3, S=1>_{dq_2} \equiv |5, S=2>_{dq_1dq_2} \oplus |3_s, S=1>_{dq_1dq_2} \oplus |1_s, S=0>_{dq_1dq_2}, \tag{3e}$$

$$|1, S=0>_{dq_1} \otimes |3, S=1>_{dq_2} \equiv |3_s, S=1>_{dq_1dq_2}. \tag{3f}$$

(For spin states).

Here the subscripts dq_i (i = 1, 2) and dq_1dq_2 denote the single and double diquark states respectively. Note that such a labelling of color-spin states of two diquarks is done in order to account for their distinguishability. We further consider the following possible quark configurations in the two-diquark system:

Case I: Each diquark of the two-diquark composite consists of the same flavor of quarks (either u or d), viz., $< u_1u_2u_3u_4 \mid H_{int} \mid u_1u_2u_3u_4 >$ or $< d_1d_2d_3d_4 \mid H_{int} \mid d_1d_2d_3d_4 >$.

Case II: One diquark of the two-diquark composite contains the same flavor of quarks (uu or dd) and the other diquark contains the different flavor of quarks (ud); thus the possible configurations turn out to be $< u_1u_2u_3d_4 \mid H_{int} \mid u_1u_2u_3d_4 >$, $< u_1u_2d_3u_4 \mid H_{int} \mid u_1u_2d_3u_4 >$, $< u_1d_2u_3u_4 \mid H_{int} \mid u_1d_2u_3u_4 >$, $< d_1u_2u_3u_4 \mid H_{int} \mid d_1u_2u_3u_4 >$, $< d_1d_2d_3u_4 \mid H_{int} \mid d_1d_2d_3u_4 >$, $< d_1d_2u_3d_4 \mid H_{int} \mid d_1d_2u_3d_4 >$, $< d_1u_2d_3d_4 \mid H_{int} \mid d_1u_2d_3d_4 >$ or $< u_1d_2d_3d_4 \mid H_{int} \mid u_1d_2d_3d_4 >$.

Case III: One diquark of the two-diquark composite contains both the quarks of the same-flavor (say, uu) while the other one contains quarks of different flavor (say, dd) or vice-versa. The possible configurations are $< u_1u_2d_3d_4 \mid H_{int} \mid u_1u_2d_3d_4 >$ or $< d_1d_2u_3u_4 \mid H_{int} \mid d_1d_2u_3u_4 >$.

Case IV: Each diquark of the two-diquark composite is a hybrid diquark containing two quarks of different flavors. The possible configurations are $< u_1d_2u_3d_4 \mid H_{int} \mid u_1d_2u_3d_4 >$, $< d_1u_2d_3u_4 \mid H_{int} \mid d_1u_2d_3u_4 >$, $< d_1u_2u_3d_4 \mid H_{int} \mid d_1u_2u_3d_4 >$ or $< u_1d_2d_3u_4 \mid H_{int} \mid u_1d_2d_3u_4 >$.

Out of several possible color-spin states (eq. (3a) - (3f)), the interaction energies corresponding to minimum energy state and those of the DS analogue state in cases I-IV are given in table I. To give a feeling for the contribution of the ϕ^6-term in (1) we computed the interaction energy only for one three-diquark color-spin state, namely {3, 0}. This state is derived from (3a) and (3d). The three-diquark interaction energy corresponding to ϕ^6-term in (1) turns out to be (-65A/16) in this state.

3. Equation of state (EOS) for diquark matter

For this purpose, we compute the parameter λ in (1) by comparing the results of section 2 with those obtained in the bag model (cf. Ref. (3)). Thus, we have, $\lambda = (\Delta E)_{dq_1dq_2} E^2 R^3 / 0.34$, where $(\Delta E)_{dq_1dq_2}$ is the two-diquark interaction energy. The values of λ corresponding to DS analogue and minimum interaction energy state of cases I-IV are given in table 1.

Next we compute the total energy of the diquark system from [2, 3]

$$E_{dq} = \int d^3k(k^2+m_q^2)^{1/2} n(k) + (3\lambda/2V)\left[\int d^3k\{n(k)/(k^2+m_{dq}^2)^{1/2}\}\right]^2, \quad (4)$$

where for n(k) a Gaussian distribution (with width parameter a) namely, $n(k) = \{N/2(2\pi a^2 m_{dq}^2)^{3/2}\}\exp\left[-k^2/2a^2 m_{dq}^2\right]$ is used. The expression for E_{dq} in terms of quark mass m_q and quark number density $\rho(=N/V)$ turns out to be

$$E_{dq} = (1.49 N m_q/4\pi)\left[2\sqrt{2\pi}a^{-3}I_1 + (3\lambda/(1.49)^3)a^{-6}(\rho/m_q^3)I_2^2\right], \quad (5)$$

where $I_1 = \int_0^\infty dk k^2(k^2+1)^{1/2}\exp(-k^2/2a^2)$ and $I_2 = \int_0^\infty dk k^2(k^2+1)^{-1/2}\exp(-k^2/2a^2)$. In order to derive the EOS, we calculate the pressure for the diquark gas using $p = (-dE_{dq}/dV)N$, and the resulting expression is given by

$$p = -\sqrt{2/\pi}a^{-3}I_1\rho_m - (3\lambda/\pi m_q^3)a^{-6}I_2^2\rho_m^2, \quad (6)$$

where $\rho_m(= N m_{dq}/2V)$ is the diquark matter density. In fact the Gaussian width parameter *a* is fixed such that the energy of the diquark gas is minimum. For this purpose, one should have $(dE_{dq}/da) = 0$ and $(d^2E_{dq}/da^2) > 0$. The first condition yields

$$\rho/m_q^3 = \frac{(3.4)\sqrt{2\pi}a^3(3a^2I_1 - I_3)}{(3\lambda)I_2(I_4 - 3a^2I_2)}. \quad (7)$$

Corresponding to different values of λ of table 1 the pressure *p*, as a function of matter density ρ_m is plotted in figure 1 for the diquark gas.

Table 1. Values of parameter λ corresponding to the DS analogue and the minimum interaction energy states.

	DS analogue state		Minimum interaction energy state	
	$(\Delta E)\ dq_1dq_2$	λ	$(\Delta E)\ dq_1dq_2$	λ
Case I	15A/3	11.58	8A/3	6.18
Case II	5A/2	5.79	2A/3	1.54
Case III	-(A/8)	-0.29	4A/3	3.08
Case IV	3A	6.95	2A/3	1.54

4. Discussion of results

With a view to exploring the role of diquarks in the quark-gluon plasma via a new phase of diquark-gluon plasma we have computed the diquark interaction energies within the framework of a polynomial field theory. For this purpose, in addition to the ϕ^4-term which describes the two diquark interactions and is in accordance with earlier works [2, 3], an estimation of ϕ^6-term (corresponding to the three diquark interactions) is also carried out. We have worked in the spirit of constituent quark model and used the generalized Pauli principle. Corresponding to various allowed color-spin state of the diquark, the diquark interaction energies are computed [5]. The equations of state thus derived for various values of λ, while conform to the trend of earlier works, however, are found somewhat more steep in general for the allowed values of the parameter a (cf. eq. (7)), viz., 0< a < 0.074. For smaller values of a both pressure and density, in fact, attain more or less constant values. However some interesting features in the EOS are expected in the range 0.06 < a < 0.074. Hopefully, these features in the EOS, when used in the Tolman- Oppenheimer- Volkoff equations will give rise reasonable values of mass and radius for a stable configuration of a diquark star. Such studies are in progress.

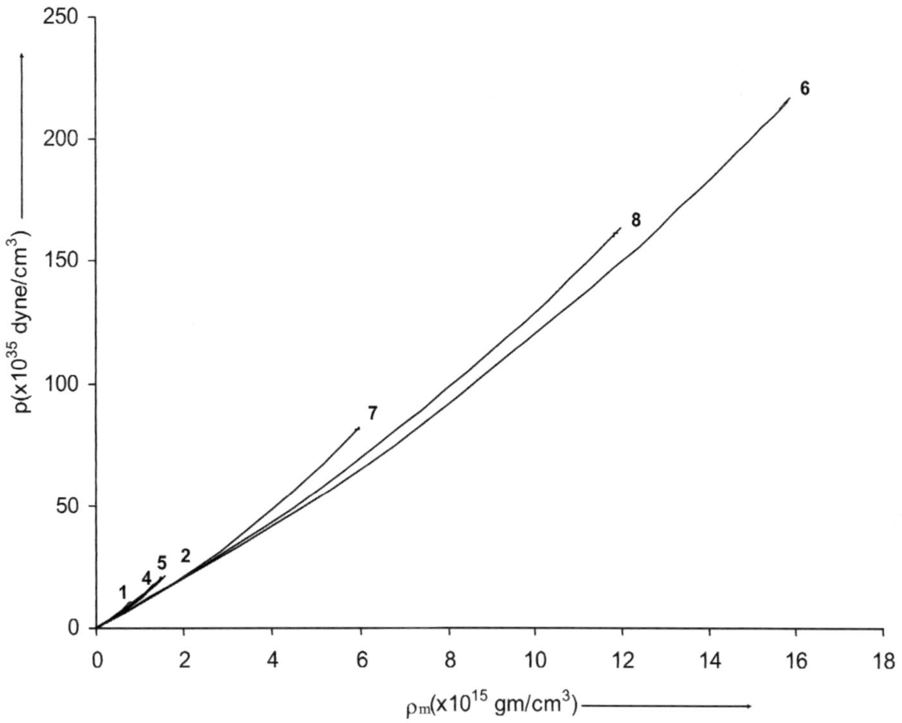

Figure 1. The equation of state for diquark matter. Curves 1, 2 and 4 correspond to the DS analogue state of the two-diquark system in cases I, II, and IV respectively, whereas the curves 5-8, correspond the configurations of minimum energy in all the four cases.

5. References

[1] See, for example, Anselmino M *et al.* 1993 *Rev. Mod. Phys.* **65** 1199 and the references therein

[2] Karn S K, Kaushal R S and Mathur Y K 2000 *Euro. Phys. J.* C **14** 487; also see, 1996 *Z. Phys.* C **72** 297

[3] Donoghue J F and Sateesh 1988 *Phys. Rev.* D **38** 360; Kaster D and Traschen J 1991 *Phys. Rev.* D **44** 3791

[4] Ekelin S and Fredriksson S 1986 *Phys. Rev. Lett.* **56** 2428; Horvath J E *et al.* 1992 *Phys. Rev.* D **46** 4754; Horvath J E *et al.* 2002 Quark-diquark matter equation of state and compact star structure *Preprint* astro-ph / 0203069 and the references therein

[5] Sisodiya A K, Bhasin V S and Kaushal R S On diquark clustering in quark-gluon plasma to be published

Institute of Physics Publishing Journal of Physics: Conference Series **50** (2006) 455–471
doi:10.1088/1742-6596/50/1/070 5th International Conference on Physics and Astrophysics of Quark Gluon Plasma

Exploring the Quark Gluon Plasma with Bikash Sinha

L. McLerran

RIKEN BNL Research Center, Brookhaven National Laboratory,
Upton, NY 11973, USA
Physics Department, Brookhaven National Laboratory, Upton, NY 11973, USA

Abstract. This paper presents a personal account of the scientific and professional adventures of Bikash Sinha on the occasion of the celebration of his 60'th birth anniversary held at Kolkata in February, 2005.

1. Introduction

It is a great honor for me to begin this meeting celebrating the accomplishments of Bikash Sinha.

I have heard rumors that Bikash may be a little nervous about what we are about to say. Perhaps he is feeling like a 60 year old virgin on her wedding night. She has a vague idea of what will come, but not specific knowledge.

I guess part of this nervousness arises from the great pleasure that all of his friends have had in putting together these presentations. It must make Bikash nervous when people are laughing and standing around their computer screens, and when you come into the room the laughing stops and the screens go blank. But don't worry Bikash, there is nothing you can do.

I must say, that all conspiracies require inside information. I thank Yogendra Viyogi and Jan-e Alam for helping to provide this, and particularly Bikash's wife, Aparna, for providing beautiful family photos. Also my very good friend Dinesh Srivastava is responsible for all of the many insights he has provided me into the works of Bikash, and more generally to the impact Bikash has had on science in India.

This talk is about the growth of a young determined boy into a research scientist and one of the great movers of Indian nuclear science. You can see this transition in the photos below:

In the photo 1a, you see a child whose posture and facial expression shows his determination, a child for whom much is expected, and a child who will deliver those expectations.

In the photo 1 b, Bikash is an adolescent posing next to one of the great men of Indian science, S. N. Bose. Is it any accident that Bikash became interested in the properties of a matter at finite temperature, given the great discovery of Bose about statistics of bosons and the applications of this discovery for finite temperature systems?

The next photo 2a of Bikash and a cricket racket I like very much, as it shows Bikash simply enjoying what life and youth have given him.

The last photo 2b represents public recognition for his service and accomplishments in science and administration. He is receiving the Padma Shri award from the president of India. For those non-Indians who do not understand this award, it is a really big deal. It is given for extraordinary accomplishment in the fields of the arts, sciences and human endeavor. (A few days ago, I read

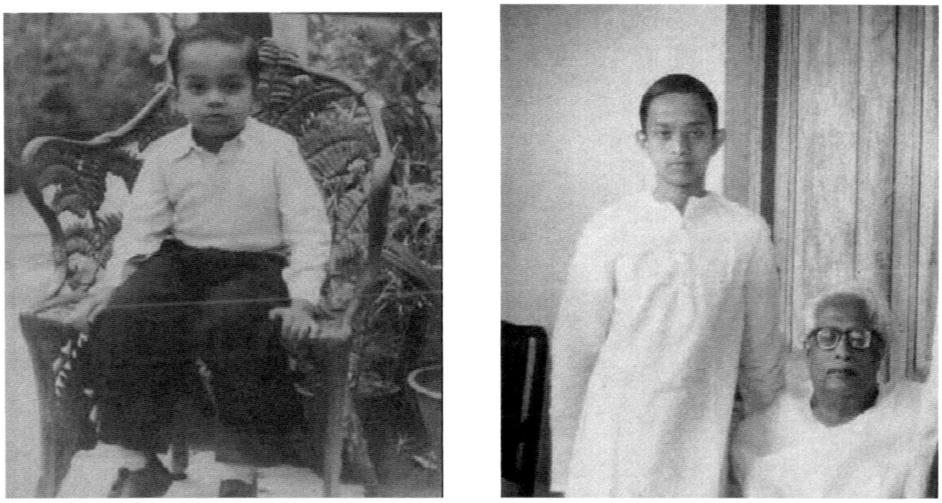

a b

Figure 1. (a) Bikash as a very young boy (b) Bikash as a teenager with the famous physicist S. N. Bose.

letters to the editor in one of the local newspapers where people were protesting giving the award to a movie star because the award was reserved for serious accomplishment.)

The point of this talk is to describe Bikash's journey in science, his accomplishments, what he has built, and what he is building.

2. Education at University of Calcutta

Bikash received his early science education at the University of Calcutta. The physics department there is shown in the picture 3. He received a Bachelor of Science with Honors in Physics in 1964.

The University of Calcutta was the first modern university in India, founded in 1857. Among the faculty associated with the University of Calcutta were three Nobel prize winners: Rabindrath Tagore, the poet philosopher and nationalist; Sir. C. V. Raman, famous for his discovery of Raman scattering; and Amartya Sen the mathematical economist famous for his work on welfare economics. In addition, both M. Saha, famous for his work on stellar dynamics, and S. N. Bose, the discoverer of Bose statistics were faculty members. Sir C. V. Raman and M. Saha are shown in 4a-b.

The university of Calcutta began the first medical school in Asia. It had the first science department in India. It had the first women's colleague in India. It provided the initial impetus for both the Saha and Bose Institutes.

3. Cambridge University and the University of London

Bikash went off to Cambridge University where he received his BA in 1967 and an MA in 1968 in Natural Sciences (with a physics Tripos). I think this is where he refined his famous and distinguished English accent. He went to London University and received a PhD in 1970. He became a Senior Research Fellow from 1970-1976 at King's College at the University of London.

Bikash's research was on the optical potential.[1]-[3] In particular, he recognized the importance of 2 body interactions including saturation effects which limits the nucleons from

a

b

Figure 2. (a) Bikash as young man enjoying life (b) Bikash receiving the Padma Shri award.

Figure 3. a Th physics department at the University of Calcutta.

getting too close to one another. This idea of saturation has important consequences in another area now: the limiting of the growth of gluons of fixed size in a high energy hadronic wavefunction.

During 1970-1976, Bikash worked hard and was very productive, publishing 9 Physics Letters articles, 4 Nuclear Physics Articles, and 10 in"other journals" such as Physical Review Letters, Physical Review C, and Physics Reports. (I am a senior editor for Nuclear Physics A, so I am personally predisposed articles of high quality written for Elsevier journals.)

Perhaps most notable during his time in London, he began his long collaboration with Dinesh Srivastava.[4] In 1973 they wrote their first paper together: The Energy Dependence of the

a b

Figure 4. (a) Sir C. V. Raman (b) Prof. M. Saha

Optical Potential.

My impression about Bikash and Dinesh can be summarised in imagery of two famous heros of the western movies of my youth: Lone Ranger and Tonto – analogous heroes must occur in Bollywood classics. The Lone Ranger is the central character, distinguished by the fact that in all situations, he is the hero. He is the origin of the phrase "Who was that masked man?" (I have never understood why he needed the mask.) His faithful companion is Tonto, an Indian brave (in politically correct modern English, a Native American warrior). Tonto would almost always work behind the scenes to insure that the Lone Ranger ended up the hero of each episode. Many people suspected Tonto of having great intelligence.

Of course, there are always times of stress where the Lone Ranger and Tonto are surrounded by Hostile Indians (perhaps from somewhere out west.) In this case, Tonto seems to want to make a run for it. As President Nixon's head of the Justice Department said to staff members of the president who were under indictment for misbehavior during the Watergate scandal: "When the going gets tough, the tough get going." I do not think that the double meaning was intended.

4. Back to Bombay

In 1976, Dr. R. Rammana invited Bikash to join the Nuclear Physics Division of the Bhabha Atomic Research Center. It was here that his interest developed on the Quark Gluon Plasma.

He began his program at BARC studying hot spots in nuclear collisions and density dependent delta function interactions. A photo of BARC is shown in Fig. 5.

Figure 5. The Homi Bhabha Atomic Research Center

In 1983 he wrote his first paper on the QGP.[5] The title and abstract of the paper are:

Universal Signals of the QGP

Abstract:

It is shown that the ratio of production rates of photon to muon pairs and pions to muon pairs from a QGP are independent of the space-time evolution of the plasma fireball and thus are universal signals of the quark-gluon plasma

The idea behind this work is that pions reflect the entropy produced in the collisions, and this is conserved in late stages of the collision. Energetic photons and dileptons are made early in the collisions and do not rescatter, so they reflect the initial conditions.

This is a great paper. It is seminal. It remains modern even more than 20 years after its publication. It has provided the basis of much of the later work by the Calcutta group.[6]-[13]

5. "Oh, Calcutta"

In 1984, Bikash moved to Calcutta to become head of Research Facilities and Computer at the Variable Energy Cyclotron Center in Calcutta. As anyone knows, establishing a group and taking a leadership role involves taking responsibility for young people. This reminds me of the famous American movie starring the present governor of California, Arnold Schwartzenegger, called Kindergarten Cop. It is the story of an undercover policeman who has to do secret work as a teacher in a kindergarten.

At Calcutta, Bikash began to make his mark: He established a research group on Quark Gluon Plasma. He arranged the first Indian school on the Quark Gluon Plasma in India in 1986. He took a leadership role in developing the talents of "young brilliant scientists". (The quotes are present because these are Bikash's own words.). He initiated participation in the WA 80-95 experiments at CERN in the search for direct photons, a search which is now a component of every QGP experiment.[14] In 1988 he organized the first of the International Conference on the Physics and Astrophysics of the Quark Gluon Plasma at the Tata Institute in Bombay.

In 1987, he began his long collaboration with Shibaji Raha.[6] – which could be illustrated as a scene from the movie King Kong vs Godzilla. These guys did very good physics, but I think like all good scientists, they had to work a little to communicate their ideas to one another. They had no problem communicating their ideas to others.

6. The ICPAQGP Series

The International Conference on the Physics and Astrophysics began with a very exciting meeting at the Tata Institute in Mumbai in 1988. Excellent people came to this meeting. Leon van Hove's regal presence and inimitable style were memorable. Angela Olinto and Charles Alcock were much excited about strange quark matter, and the possible formation of strange stars. This had an impact on Bikash and Shibaji's work, and has led to the experiment which will soon be underway in Darjeeling. From this beginning, Charles tried to interpret these strange stars as the origin of dark matter, and he began a now famous experiment to use gravitational lensing to search for such objects.

Ed Shuryak, Mitya Diakonov and Genya Zinoviev all arrived from Russia. They had many adventures, including when they went to the Taj Mahal after the meeting, they took a cab from Delhi, and said they had just come from Bombay. The taxi driver asked them what they did. They answered they were physicists. Then the driver said,"Oh, you must have been at the ICPAQGP." I learned from this story to never underestimate the influence of Bikash.

Before the meeting, there was a school in Jaipur at the Alsisas Havelli. My wife and I were there together and fell in love with the old style of architecture at this classic building. There were peacocks flying around in the garden, and monkeys in the trees. We got to know and appreciate Bikash as a result of this school. We got to taste the delicate concoctions of the famous Polo Bar at the Rambaugh Palace. (During this trip, my wife Alice, rode her first camel, and developed a love affair with this disgusting beast which has lasted for the past almost 20 years. I can tell you many stories about Alice and camels.)

The picture, Fig. 6 below was on the poster of this meeting, and I guess has been the icon for this meeting ever since. When I first saw the photo of the statue, I did not look carefully.

Figure 6. Creating the Universe in Calcutta.

Dinesh, at the meeting in Bombay, told me to look at it carefully, then smiled in a Dinesh sort of way and said "See, they are creating the universe". I had this poster on display in my office in Fermilab for several months after the meeting. These offices were made of glass so passers-by could see the poster. It was eventually stolen, I suspect by someone who found it to be too "spicy". Too bad, I really loved that poster.

Of course there were a number of such meetings afterwards: Calcutta in 1993, Jaipur in 1997 and 2001 and the meeting beginning tomorrow in Calcutta. All of the meetings have had excellent science, and the organization was always spectacular. Every meeting has a set of stories attached to it, and an exposure to a different aspect of Indian culture and science.

One story is in Jaipur in 1997, I was with my very good friend Jean-Paul Blaizot, who did not know Bikash well at the time. He was curious about Bikash because he knew I had such a high level of respect for what he was doing. One night at the first Jaipur meeting, we went to an organizational meeting for the next conference in the series, and Bikash was sitting surrounded by a circle of his Indian colleagues. He was quietly listening, and at the end of the meeting went back over his understanding of how to proceed, and then did a check off of everyone's assignments. He was firm and assertive in the Bikash style, which all of his colleagues understand. It was quite clear that Bikash was bringing order from chaos. Jean-Paul understood.

7. QGP and Electromagnetic Probes

Let me begin by quoting the famous translation of Ovid's The Metamorphoses by Garth and Dreyden:

"Before the sea, and this terrestrial ball,
And Heaven's high canopy that covers all,
One was the face of Nature, if a face,
Rather a crude and indigested mass,
A lifeless lump, un-fashioned and un-framed,
Of jarring seeds, and justly Chaos named,
No sun was lighted up, the world to view,
No moon did yet her blunted horns renew,
Nor yet was earth suspended in the sky,
Nor poised, did on her foundations lie,
Nor sea about the shores their arms had thrown,
But earth and air and water were on,
This air was devoid of light and earth unstable,
And waters dark abyss un-navigable"

In the figure below, you see some of the experimental results which were stimulated by Bikash's work on the Quark Gluon Plasma, Fig 7. In the plot of the WA80-95 data, there is a depletion of photons relative to what had been expected from hadron gas models. This was predicted by Bikash and Dinesh.[5]-[10] In CERES, there is an excess of dileptons which has been attributed to the QGP. In the results from PHENIX, one has just measured the hard photons and they are in agreement with computations from perturbative QCD.

The work of Bikash and colleagues in India has not been restricted to theory. Calcutta has built essential elements of the photon detectors for WA80-95, STAR and ALICE. Without their efforts, results such as these shown in the figures would not exist.

In the Fig. 8 below, I show the STAR Photon Multiplicity Detector 8. The first results from the PMD will be shown in the meeting over the next week.

8. Bikash Sinha: Cosmology and the QGP

One of the most beautiful and poetic conceptions of cosmology comes from the Chandogya Upanishad. This was written 3000 years ago. To me it presents in metaphor the deepest problems of modern cosmology:

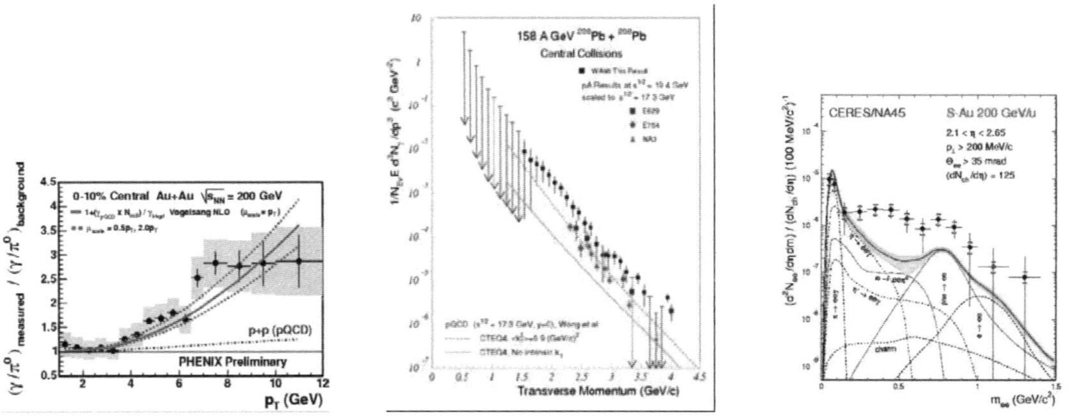

(a) (b) (c)

Figure 7. a Hard photons measured at RHIC b Soft photons measured at CERN c Soft dileptons measured at CERN .

Figure 8. The Photon Multiplicity Detector in STAR .

In the beginning, the world was just being. Some people would no doubt say, this world was just non-being, and from non-being was produced. But how could that be so? How could being be produced from non-being? On the contrary, the world was being alone. One being without a second.

Being thought to itself: "May I be many, may I procreate". It produced fire. Fire thought to itself "May I be many, may I procreate". Fire produced water. Therefore when a person perspires, it is from fire that water is produced. Water thought to itself: "May I be many, may I procreate". Water produced food. And when it rains, there is abundant food, for it is from water from which eating is produced.

Being thought to itself: "Having entered into these three divinities by means of this living self, let me develop names and forms"

Figure 9. The goddess Lakshmi in a reproduction of an old painting.

A Dialog Between a Student and a Teacher.

Bring me a fig from that tree
It is here.

Break it.
It is broken.

What do you see now?
Very fine seeds.

Now break a seed.
It is broken.

What do you see?
Nothing at all.

In truth, that subtle essence which you do not perceive is from what this giant fig tree arises. Believe me, that which is subtle essence, this whole world has that essence for itself.

Fig. 9 shows a beautiful painting of Lakshmi. She, among other things, is the goddess of light.

9. Bikash Sinha: The Astrophysics Connection
In every ICPAQGP, there has a been a major amount of time devoted to the discussion of the astrophysics of the Quark Gluon Plasma. (See the Fig. 10) One of the reasons for this is that in

Figure 10. We begin discussing the big bang.

the early universe, the Quark Gluon Plasma was the matter at times less than about 10^{-5} *sec*, as shown in Fig. 11.

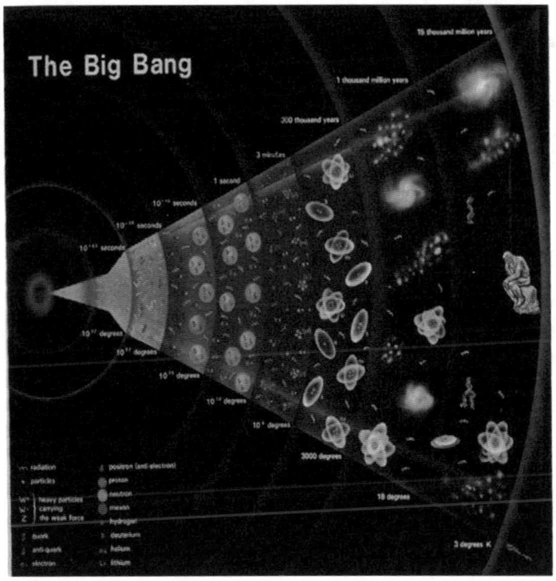

Figure 11. A pictorial history of the big bang.

At the time of the first ICPAQGP, there was considerable speculation that a first order phase transition to the QGP might generate large scale density fluctuations. This is illustrated in Fig.

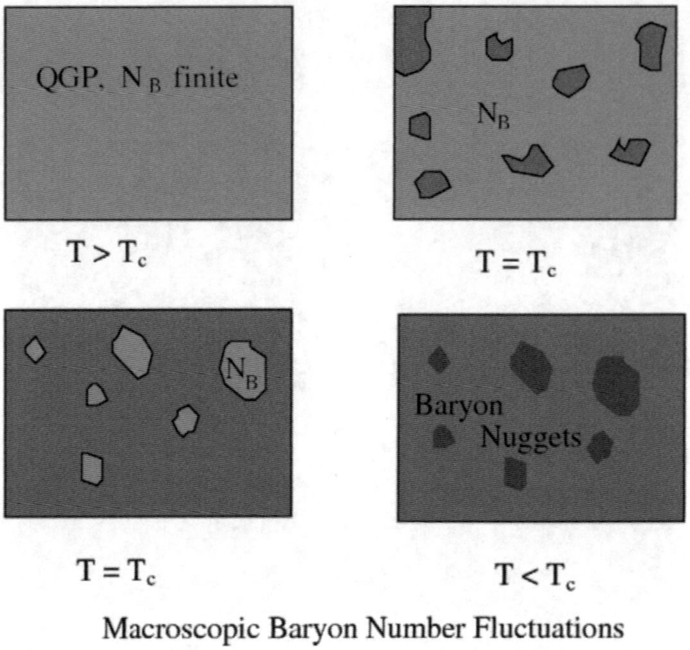

Figure 12. A hypothetical QGP phase transition in the early universe.

12. For temperatures $T > T_c$, the phase transition temperature, the system is homogeneous. A little below T_C, nucleation centers for, the system reheats to T_c, and a mixed phase of QGP and hadronic phase is made. Initially, the regions of hadronic gas are very small, but as the system expands, more and more of the volume becomes hadronic gas and less QGP. The size scale of these droplets can grow to be quite large, of the order of the size of the universe at $t \sim 10^{-5}$ *sec*

This can result in density fluctuations which survive long after the transition. To understand this, recall that in the deconfined phase, the mass of a quark is of order T_c, so the probability to have a baryon number fluctuation is $e^{-M_q/T} \sim 1$. On the other hand, in the hadron gas, the baryon has of order N_c quarks, and the mass is of order $M_P \sim N_c T$ so that the probability for a baryon number fluctuations is of order e^{-N_c}, which is very small. Most of the baryon number gets trapped in the region of the QGP. After the QGP has evaporated, the baryons must diffuse to uniformity, but diffusion can take a very long time.

Early estimates of this time suggested that one might have baryon number inhomogeneities which could last until nucleosynthesis, and the various element abundances might be affected. I think that we now believe that the QGP phase transitions is either not a first order phase transition, or is so weak, that this does not occur.

Bikash was interested in this scenario from a different point of view.[16]-[19] He wanted to use the big bang to make strangelets. If strange quark matter is the stable ground state of nuclear matter, one can make droplets out of such matter. The decay time for ordinary matter into the droplets is long enough so that in the absence of some seed of very high energy density QGP, matter is stable. Such strangelets have roughly equal numbers of up down and strange quark, whose charge adds up to zero, as shown in Fig. 13. This means that as a function of baryon number, the Coulomb self energy is never very large. The Coulomb self energy is what makes large nuclei unstable. Here, when the size of a droplet is large enough that the Coulomb self energy is substantial, the droplet is larger than the Bohr orbit of an electron, and the system becomes neutralized by the electrons. One can therefore make droplets of strangelets, which are

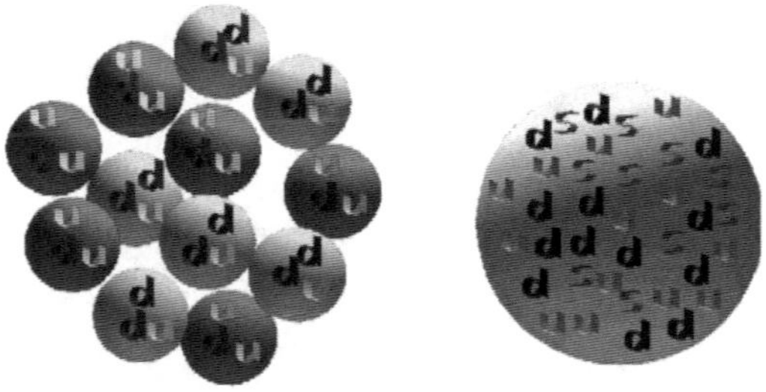

Figure 13. Nuclear matter and strange quark matter.

at nuclear matter density, of any large size one can imagine. For example, one can make a house or a planet out of strange quark matter.

It was hypothesized that in the big bang one might make such strange quark droplets as shown in Fig. 14. This is probably not so likely in light of what we now know about the QGP

Figure 14. The primeval QGP making strangelets.

phase transition. Nevertheless, they might be made in neutron star collisions, or the formation of black holes, or even possibly very rarely in heavy ion collisions, as illustrated below 15.

There have been experiments to search for small strangelets, in heavy ion collisions with negative results. There may be some evidence from Centauro events in cosmic ray collisions.

Bikash was involved in trying to see if one could make and detect strangelets.

He was excited by the work that his friend Charles Alcock did to look for strange stars using gravitational lensing. Together with Sibaji Raha, he pushed for an Indian cosmic ray experiment to look for strangelets. This experiment is near Darjeeling, and will begin operation soon.[20]

10. Bikash Sinha: Nurturing the Young

Why do good people like Bikash get involved in the world of physics politics and administration? Surely not because they enjoy it! Two reasons drive Bikash. The first is that if one wants to understand nature, one must build structure and organize people to carry out research. The

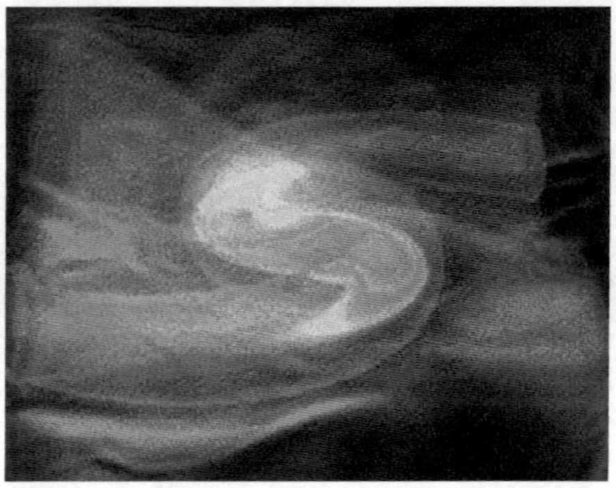

Figure 15. The a pictorial representation of the evolution of the collision of neutron stars.

a

b

Figure 16. a Bikash protecting the young. b Bikash cutting red tape.

a

b

Figure 17. a Bikash enjoying the cultural side of life. b Bikash greeting the Nobel prize winner Abdus Salam to India.

second is, that if a scientific field is to flourish, it must have bright young people entering the field.

These young people must be protected from the maelstrom and chaos of funding fluctuations, and positions must be made for them, Fig. 16a.

This necessarily involves competition from other with different agendas. One must make sure that when they are in their most productive scientific career, that they are not overloaded with bureaucratic chores, Fig. 16b.

One must also expose young people to culture. The young people in science come from many back grounds. Physicists share a common love of music, art, and literature. Bikash has always encouraged this appreciation and understanding, Fig. 17a.

At the first ICAQGP, I remember a wonderful get together in Mumbai where one of Bikash's friends entertained us with sitar music. Finally, one must broaden the physics horizons of the young. This is sometime done by bringing young colleagues into contact with the distinguished senior members of our community, Fig. 17b.

The major nuclear physics collaborators of Bikash, many of whom WERE young people (some still are) are:

V. R. W. Edwards, D. Srivastava, F. Duggan, R. J. Griffiths, S. Moszowski, S. Raha, A. K. Chaudhuri, D. N. Basu, B. Datta, S. Chakrabarty, J. Alam, P. Sarkar, S. Chattopadhyay, M. Mustafa, B. Dutta Roy, B. Patra, S. Banerjee, S. K. Ghosh, B. Mohanty, and A. Rahaman.

and of course

the members of WA80-05, STAR, and ALICE.

Bikash has been an author of over 150 peer reviewed scientific publications.

11. Bikash Sinha: The Larger World of Science

Bikash has written 33 articles of general interest which have been published in the media. Some of the titles I found most interesting were:

- "The Changing Scenario of Nuclear Physics", Science Today (1979)
- "Nuclear Power in India", Weekly DESH (1983)
- "The Craziness Necessary for Research Exists in Calcutta More Than Anywhere Else"
- "Why Are We Wasting Our Talent". The Telegraph (1991)
- "The Soviets Do Not Mind Shedding Tears in Public", The Telegraph (1991)
- "Sales Talk and Vodka Among the Test Tubes", The Telegraph (1993)
- "Electrified by a Nuclear Vision", The Telegraph (1993)
- "Onuclearo is not a Nightmare", Business Economic" (2000)

I love some of the titles of these articles. There is nice irony of "Why Are We Wasting Our Talent?" There is gentle teasing of Bikash's colleague when he talks about the crazy people from Calcutta. I really like "Onuclearo", but in spite of my best efforts was not able to figure out what it was. I guess it really is a nightmare.

Bikash also was central in establishing the Radiation Medicine Center in Kolkata as part of VECC.

He also has done some recent work about abnormal helium abundance in thermal hot springs in Bakreswar and Tantloi, and their correlation with geological activity. I come originally from the Pacific Northwest where there is a lot of geological activity, so I read these articles with much interest. (Perhaps the bottom line is that if your colleagues start talking with high squeaky voices, run like hell?)[21]

12. Bikash Sinha: Major Honors
Bikash has won so many honors for his work, that if he wore all the medals he would probably embarrass a Russian General. These include

- S. N. Bose Centenary Award 1994
- Fellow of Indian Academy of Sciences, Bangalore
- Fellow of Indian National Academy of Sciences, New Delhi
- Fellow of Indian Academy for Sciences, Allahabad
- DAE Raja Ramanna Prize 2001
- Pandya Endowment Memorial Lecture Award 2001
- Rais Ahmed Memorial Lecture Award 2001
- Fellow of 3'd World Academy of Sciences 2002
- Padma Shri Award

I include in the photos Fig 18, pictures of Bikash at celebrations of the Padma Shri award. The photo in Fig. 2b, shows him actually receiving the award. This is a really big deal!

a b

Figure 18. (a) Bikash is a bit nervous?. (b) Party time!.

13. Bikash Sinha: Major Accomplishments
One of the major accomplishments of Bikash's career is nearing completion now here in Calcutta. It is the building of the superconducting cyclotron. I show picture of the magnet coil, and the wrapping of wires around the magnet 19.

Since it is superconducting, one needs a really big refrigerator, also shown in Fig. 20. There is also a photo of visiting dignitaries associated with the cyclotron: Shri Satyabrata Mookherjee: Honorable Minister of State, Statistics and Program Implementation, Planning, Atomic Energy, Space, Commerce and Industry; Dr. Anil Kakodkar, Chairman AEC and Secretary, Dept. of Energy.

I understand that the cool down has started, and perhaps the magnet will have been energized.

a b

Figure 19. (a) The superconducting magnet. (b) Wrapping wire around the magnet.

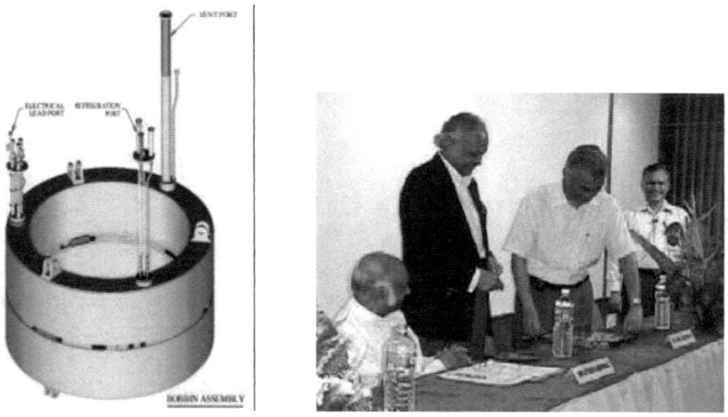

a b

Figure 20. a The refrigerator for the magnet. b VIPs

14. Bikash Sinha: Major Responsibilities

We all know Bikash Sinha is a very busy man. He is

- Director: Saha Institute
- Director: VECC
- Vice chancellor, West Bengal University of Technology
- Scientific Advisory Committee to Cabinet, Govt. of India, 1997-present

I am really surprised by the amount of responsibility Bikash has taken, and the substantial and important accomplishments he has achieved. A great man leaves more than the sum of his accomplishments. There is an intangible essence which exists in the minds and spirits of his colleagues who love and respect him.

15. Acknowledgments

I thank Yogendra Viyogi, Jan-e Alam, Dinesh Srivastava and Aparna Sinha for their help in putting this talk together. This manuscript has been authorized by Brookhaven Science

Associates, LLC under Contract No. DE-AC02-98CH1-0886 with the U.S. Department of Energy. The United States Government retains, and the publisher, by accepting the article for publication, acknowledges, a world-wide license to publish or reproduce the published form of this manuscript, or allow others to do so, for the United States Government purposes.

[1] "Optical Model for Proton Scattering from 40 Ca", Bikash Sinha and G.L. Thomas, *Phys. Rev. Lett.* **26**, 325 (1971).

[2] "Nucleus-Nucleus Optical Potential using a Density Dependent Two-Body Interaction", Bikash Sinha, *Phys. Rev. Letts.* **33**, 600 (1974).

[3] "The Optical Potential and Nuclear Structure", Bikash Sinha, *Phys. Reports* **20**, 1 (1975).

[4] "Energy Dependence of the Optical Potential", Bikash Sinha, D.K. Srivastava and N.K. Ganguly, *Phys. Letts.* **B43**, 113 (1973).

[5] "Universal Signals of Quark-Gluon Plasma", Bikash Sinha, *Phys. Lett.* **B128**, 91 (1983).

[6] "Quark-Gluon Plasma Diagnostics : Measuring π^0/γ Ratio with Dileptons", S. Raha and Bikash Sinha, *Phys. Rev. Lett.* **58**, 101 (1987).

[7] "Dileptons and Photons in High Energy Heavy Ion Reactions: A Review". S. Raha and Bikash Sinha, *Int. J. Mod. Phys.* **A6**, 517 (1991).

[8] "Single Photons from S + Au Collisions at the CERN Super Proton Synchrotron and the Quark-Hadron Phase Transition", D. K. Srivastava and Bikash Sinha, *Phys. Rev. Lett.* **73**, 2421 (1994).

[9] "Electromagnetic Probes of Quark Gluon Plasma", J. Alam, Bikash Sinha and S. Raha, *Phys. Rept.* **273**, 243 (1996).

[10] "Excess Production of Low Mass Lepton Paris in S + Au Collisions at the CERN Super Proton Synchrotron and the Quark - Hadron Phase Transition", D.K. Srivastava, Bikash Sinha and C. Gale, *Phys. Rev.* **C53**, 567 (1996).

[11] "Central Pb + Pb Collisions at 158 AGeV/c Studied by pi - pi Interferometry, WA98 Collaboration" M. M. Aggarwal, Bikash Sinha, et al, *Eur. Phys. J.* **C16**, 445 (2000).

[12] "A Second Look at Single Photon Production in S + Au Collisions at 200 AGeV and Implications for Quark Hadron Phase Transition", D. K. Srivastava and Bikash Sinha, *Eur. Phys. J.* **C 12**, 109 (2000).

[13] "Photon Interferometry and Size of the Hot Zone in Relativistic Heavy Ion Collisions", Jan-e Alam, Bedangadas Mohanty, Pradip Roy, Sourav Sarkar and Bikash Sinha, *Phys. Rev.* **C67**, 054902, (2003)

[14] "A Preshower Photon Multiplicity Detector for the WA98 Experiment", M.M. Aggarwal, Bikash Sinha et al, *Nucl. Instrum. Meth.* **A424**, 395 (1999).

[15] "The Star Photon Multiplicty Detector", M.M. Aggarwal, Bikash Sinha et. al., *Nucl. Instrum. Meth.* **A499**, 751 (2003).

[16] "Survivability of Cosmological Quark Nuggets in the Chromoelectric Flux Tube Fission Model of Baryon Evaporation", P. Bhattacharjee, J. Alam, Bikash Sinha and S. Raha, *Phys Rev.* **D48**, 4630 (1993).

[17] "Quark Nuggets as Baryonic Dark Matter", J.Alam, S.Raha and Bikash Sinha, *Astrophysical Journal* **513**, 572 (1999).

[18] "Relics of the Cosmological QCD Phase Transition", A. Bhattacharyya, J. Alam, S. Sarkar, P. Roy, Bikash Sinha, S. Raha, and P. Bhattacharjee, *Phys. Rev.* **D61**, 083509 (2000).

[19] "Massive Compact Halo Objects from the Relics of the Cosmic Quark-Hadron Transition", S. Banerjee, A. Bhattacharyya, S. K. Ghosh, S. Raha, Bikash Sinha, and Hiroshi Toki, *Mon. Not. R. Astron. Soc.* **340**, 284 (2003).

[20] "Strange Quark Matter in Cosmic Ray Flux and Exotic Events", S. Banerjee, S. Ghosh, A. Mazumdar, S. Raha, D. Syam, *Astrophys. Space. Sci.* **274**, 655 (2000).

[21] "Large-scale Helium Escape from Earth Surface around Bakreswar - Tantloi Geothermal Area in Birbhum District, West Bengal and Dumka District, Jharkhand, India," Debasis Ghose, D.P. Chowdhury and Bikash Sinha, *Current Science*, **82**, 993 (2002).

AUTHOR INDEX

AUTHOR INDEX

AUTHOR INDEX

Institute of Physics
Dirac House, Temple Back
Bristol BS1 6BE UK